MW00844461

Adult Neurogenesis

COLD SPRING HARBOR MONOGRAPH SERIES

Adult Neurogenesis

EDITED BY

Fred H. Gage
*Salk Institute for
Biological Studies*

Gerd Kempermann
*Center for Regenerative
Therapies Dresden*

Hongjun Song
Johns Hopkins School of Medicine

COLD SPRING HARBOR LABORATORY PRESS
Cold Spring Harbor, New York • www.cshlpress.com

Adult Neurogenesis

Monograph 52

Publisher	John Inglis
Acquisition Editor	Alex Gann
Development Director	Jan Argentine
Project Coordinators	Mary Cozza and Joan Ebert
Production Editor	Kaaren Hegquist
Desktop Editor	Lauren Heller
Production Manager	Denise Weiss
Marketing Manager	Ingrid Benirschke
Sales Manager	Elizabeth Powers
Cover Designer	Ed Atkeson

Front cover artwork: Newborn neurons in the adult mouse hippocampus labeled by retrovirus-mediated expression of fluorescent proteins (Green = GFP; Red= NeuN). Image courtesy of Chunmei Zhao, Salk Institute for Biological Studies, La Jolla, California.

Library of Congress Cataloging-in-Publication Data

Adult neurogenesis / edited by Fred H. Gage, Gerd Kempermann, Hongjun Song.
    p. ; cm. -- (Cold Spring Harbor monograph series ; 52)
  Includes bibliographical references and index.
  ISBN 978-0-87969-784-6 (hardcover : alk. paper)
  1. Developmental neurobiology. I. Gage, F. (Fred), 1950- II. Kempermann, Gerd. III.
Song, Hongjun. IV. Cold Spring Harbor Laboratory. V. Series.

  [DNLM: 1. Neurons--physiology. 2. Adult. 3. Brain--physiology. 4. Cell Proliferation.
5. Regeneration--physiology. WL 102.5 A244 2008]

  QP363.5.A35 2008
  612.8--dc22

                        2007028689

10 9 8 7 6 5 4 3 2 1

All World Wide Web addresses are accurate to the best of our knowledge at the time of printing.

All Cold Spring Harbor Laboratory Press publications may be ordered directly from Cold Spring Harbor Laboratory Press, 500 Sunnyside Blvd., Woodbury, New York 11797-2924. Phone: 1-800-843-4388 in Continental U.S. and Canada. All other locations: (516) 422-4100. FAX: (516) 422-4097. E-mail: cshpress@cshl.edu. For a complete catalog of all Cold Spring Harbor Laboratory Press publications, visit our World Wide Web Site http://www.cshpress.com/.

We dedicate this monograph to the memory of Peter Eriksson, a good friend and colleague who passed away unexpectedly. He generously contributed to our knowledge of Human Adult Neurogenesis.

Contents

FUNCTIONAL SIGNIFICANCE

NEUROLOGICAL DISEASES

COMPARATIVE NEUROGENESIS

Preface

THE TERM "ADULT NEUROGENESIS" IS USED TO DESCRIBE the observation that, in the adult mammalian brain, new neurons are born from stem cells residing in discrete locations and these new neurons migrate, differentiate, and mature into newly integrated, functioning cells. By virtue of this definition, adult neurogenesis is a process, not an event, and as such, can be dissected and examined in evermore discrete components. In general, researchers seek a complete understanding of not only the details of these separate components but also the purpose and function of this process as a whole. Once the tools became available to monitor and measure adult neurogenesis, the interest in this process grew enormously, not the least because the birth and integration of new neurons in the adult brain constitute the most extreme cases of neuroplasticity in the adult brain. While the phenomenon is interesting enough to investigate and understand in the normal, healthy brain, the fact that this process is also disrupted in many disease states adds substantially to the numbers of those studying adult neurogenesis. As a result, a new way of looking at brain therapy has emerged that incorporates the potential of generating new neurons in the context of aging and disease into the search for a strategy for "self-repair."

The idea for this book originated from a meeting on adult neurogenesis in the adult brain held at the Banbury Conference Center at Cold Spring Harbor Laboratory in February 2006. In the secluded and intimate setting of this event, the organizers sought to assemble an overview of the field as it stood at the time. The likely impermanence of this contribution did not deter us because it seemed necessary to bring together a number of leading researchers to make an attempt to define our growing field. The great success of the conference made it clear that the conclusions from the meeting should be disseminated to a wider audience in the form of a book. This decision also allowed us to expand the range of topics beyond those covered in the meeting and recruit more colleagues who had made important contributions to the field.

The 30 chapters in this volume provide an incomplete yet valuable overview of the field of adult neurogenesis research. Wherever possible, we teamed up authors on related topics who have either not yet worked together or did so long ago. Our aim was to help integrate the field by mapping its current scope and its diverging ideas, and we hope our selections do not reflect too much of our personal opinions and biases.

We would like to thank the staff at Cold Spring Harbor Laboratory Press for their advice and diligence, particularly John Inglis, Alex Gann, Denise Weiss, Kaaren Hegquist, Lauren Heller, Mary Cozza, and Joan Ebert. We would also like to thank the authors of all the chapters in the monograph for their thoughtful and scholarly presentations of often controversial and still emerging concepts surrounding this new field of adult neurogenesis.

FRED H. GAGE
GERD KEMPERMANN
HONGJUN SONG

1

Adult Neurogenesis: A Prologue

Fred H. Gage
Laboratory of Genetics LOG-G
Salk Institute for Biological Studies
La Jolla, California 92037

Hongjun Song
Institute for Cell Engineering
Department of Neurology
Johns Hopkins University School of Medicine
Baltimore, Maryland 21205

Gerd Kempermann
Genomics of Regeneration
Center for Regenerative Therapies Dresden
01307 Dresden, Germany

NEW IDEAS PASS THROUGH A SERIES OF STAGES from initial rejection to skepticism, to reluctant acceptance (without true belief in its importance), to a final casual acknowledgment of the obvious. It is fair to say that the acceptance of the idea that new neurons are generated in the adult brain of all mammals has been a slow process, and along the way, the idea has been met with skepticism and resistance. It is still not yet casually accepted as obvious. Rather, adult neurogenesis remains in the stage of reluctant acceptance, without a clear understanding of its importance, but the search for its function is in full gear.

Joseph Altman's original observations in the 1960s were met with significant reservation, as were attempted confirmations by a handful of investigators in the next 20 years. Somehow, Fernando Nottebohm and Steve Goldman's observation of neurogenesis in the brains of adult canaries was received more positively but—because it took place in birds—was not considered as much of a threat to the prevailing belief (often even termed "dogma") that there are no new neurons in the adult mammalian brain.

Why this resistance to the capacity of the adult brain to generate new neurons? It was well accepted that other systems, like blood, liver, and skin, could generate new cells, so why not the brain? The most straightforward explanation is that the brain is not just any organ. At a philosophical and metaphysical level, the brain is thought to be the place where the very essence of the individual resides. If neurons were being added to this structure, the reasoning might have gone, the consequence would be an instability of this "essence." Who we are would change with time and development. Furthermore, one of the structures where neurogenesis persists is the hippocampus, a structure involved in learning and memory: Wouldn't new neurons that are being added to the structure destabilize memories or their recall? In the 1960–1980s, much of conceptual modeling of brain function used computers as an analogy, with their complex wiring, feedforward/feedback mechanisms, and oscillations. The idea of structural plasticity provided more of a problem to these types of models rather than a resolution. On top of all these forces working against the acceptance of neurogenesis, there was the more practical problem that neurons are big and complex with thousands of connections. Furthermore, neurons are "postmitotic" (a judgment that still holds). How could they divide to make a new neuron? Where could the new cells come from if not from a dividing neuron? The idea of stem cells in the brain, to say nothing of their proven existence throughout adulthood, was not even on the horizon. But the problem of dividing neurons was solved between 1992 and 1995, when it was discovered that the subventricular zone of the lateral ventricles and the subgranular zone of the hippocampal dentate gyrus contained self-renewing stem cells from which the newly born neurons were generated. Therefore, although the computer:brain analogy is still dominant (albeit problematic), the complexity of incorporating neurogenesis into the model is now less problematic than originally considered, because neurogenesis only occurs in two regions of the brain, and the hippocampus (dentate gyrus) is thought to be involved more in forming new memories than in storing them. Therefore, updating the "software" by adding new neurons can fit into the computer model, for those who are so inclined.

The field has grown dramatically during the last 15 years. The growth has been further amplified by evidence of negative and positive regulation of adult neurogenesis through experience and activity, and more and more species, including humans, were added to the list of mammals showing adult neurogenesis, with fundamentally similar anatomical features. More recently, links to pathology have been made for several diseases, and the cellular and molecular mechanisms underlying the control of adult neurogenesis are being revealed. Intriguingly, although these mechanisms are similar to those seen in neuronal development during the embryonic and fetal periods, "adult neurogenesis" is also quite distinct from earlier development. It is precisely the interaction with the environment and activity, be it in the physiological or pathological context, that sets adult neurogenesis apart from intrauterine and early postnatal brain development.

Today, researchers entering the field are faced with an overwhelming number of publications. At the time of this writing, a simple search in PubMed using the key word "adult neurogenesis" generated more than 2000 entries, and the list of factors and issues that are brought into connection with "new neurons in the adult brain" is ever increasing. As so often is the case in emerging fields, there are times when increasing knowledge is accompanied by less rather than more insight. Apparent contradictions remain and speculative beliefs stand beside solidly grounded facts. We have much work to do.

As a function of reading through the chapters in this volume, four guiding principles seem to emerge with regard to adult neurogenesis:

- Adult neurogenesis is neuronal development under the conditions of the adult brain. It is therefore a process, not an event. And it is an exceptional process that by and large appears to be limited to only two "canonical" neurogenic regions: the hippocampus and the olfactory bulb.

- Adult neurogenesis appears to persist throughout life, but it does not produce great numbers of neurons after early adulthood. Adult neurogenesis is not a mass phenomenon but appears to make a qualitative rather than quantitative contribution.

- Although the exact function that adult neurogenesis has in normal behavior is not yet clear, it is now rather obvious that it has at least one role and might have several. Given that adult neurogenesis is the one dynamic process in the adult brain that affects the brain's structure in the most comprehensive way (because entire new neurons are formed), understanding the functions of adult neurogenesis will likely reveal insights into how brain plasticity is linked to function in general.

- Although there is good evidence that adult neurogenesis is stimulated by all kinds of pathologies, there is almost no indication that this neurogenesis is regenerative or even restorative. Rather, the new neurons appear to make a contribution to normal brain function, presumably in the sense of structural plasticity.

The first section of this volume begins with Methods. Seven chapters lay out the methodological foundations of research on adult neurogenesis and establish the crucial link to neuronal stem cell biology. The different methods that are used to identify newborn neurons are presented and discussed and their benefits and limitations are highlighted. Because the most commonly used labeling technique is based on the lasting incorporation of thymidine analog bromodeoxyuridine (BrdU), the merits and pitfalls of this method receive particular attention. Nowakowski and Hayes (Chapter 2) cover the issues related to cell cycle kinetics, whereas Rakic et al. (Chapter 4) focus on the problem of false-positive labeling, particularly in cases of pathology. Peterson and Kuhn (Chapter 3) discuss the problems that rank around this technique, particularly with respect to microscopy. Beyond BrdU, novel genetic tools based on reporter genes have been developed. Some of these are summarized by Rakic et al. (Chapter 4). Enikolopov and Overstreet-Wadiche (Chapter 5) discuss the possibilities of using transgenic reporter gene mice in full detail. Zhao (Chapter 6) introduces the growing use of viral vectors, be it for labeling purposes or for the introduction of genetic manipulations. One additional methodological chapter by Ray (Chapter 8) deals with adult-generated neuronal precursor cells in vitro. We acknowledge that neuronal precursor cell biology includes many aspects that are relevant to adult neurogenesis which cannot be covered here at an appropriate length. We decided to consider "neurogenesis" here to stand primarily for the in vivo process. Nevertheless, adult neurogenesis cannot be understood without reference to precursor cell biology. In some sense, adult neurogenesis is an example of applied precursor cell biology. And for more biotechnologically inclined researchers, adult neurogenesis might even serve as a good model system from which to learn how to "make" neurons from neuronal precursor cells.

Three chapters in the second section (Basic Processes) cover the phenomenology of adult neurogenesis: in the dentate gyrus of the hippocampus by Kempermann et al. (Chapter 9), in the subventricular zone and olfactory bulb by Lim et al. (Chapter 10), and in the stem cell niches by Ma et al. (Chapter 11).

In the third section (Molecular and Physiological Mechanisms), four separate chapters cover a variety of themes: proliferation, migration, and differentiation by Lie and Götz (Chapter 12), proneurogenic genes by

Pleasure and Buchen (Chapter 13), the balance of trophic support and cell death by Kuhn (Chapter 14), and maturation and integration by Bischofberger and Schinder (Chapter 15).

Regulation, the fourth section, covers the regulation of adult neurogenesis by epigenetic factors (Hsieh and Schneider, Chapter 16), activity dependency in aging (Kempermann, Chapter 17), stress (Lucassen et al. Chapter 18), and neurotransmitters (Jang et al. Chapter 19).

Three chapters in the fifth section (Functional Significance) attempt to summarize the current and growing knowledge of the functional significance of adult neurogenesis: Lledo (Chapter 20) focuses on the olfactory system and Abrous and Wojtowicz (Chapter 21) focus on hippocampal learning. Aimone and Wiskott (Chapter 22) add to this section by reviewing the current computational models being developed to represent adult neurogenesis.

Neurological Diseases, the sixth section, addresses the possible role of adult neurogenesis in disease. Sahay et al. (Chapter 23) focus on depression and anxiety, whereas Brundin et al. (Chapter 24) review the current evidence of an involvement of neurogenesis in neurodegenerative diseases. Jessberger and Parent (Chapter 25) review neurogenesis and epilepsy, and Lindvall and Kokaia (Chapter 26) cover the effects of stroke on adult neurogenesis.

Finally, Comparative Neurogenesis provides several reviews of the comparative nature of neurogenesis: Zupanc reviews neurogenesis in teleost fish (Chapter 27), and Goldman focuses on birds (Chapter 28). Eriksson et al. review the existing information on neurogenesis in the adult human brain (Chapter 29), and Lipp et al. contrast the differences in adult neurogenesis seen in a variety of feral and domestic species (Chapter 30).

With these 30 chapters, we have attempted to take a snapshot of the current field of adult neurogenesis, realizing that some areas are not covered completely or really at all. For example, adult neurogenesis in insects and invertebrates is not included. Furthermore, since this is such a fast-moving field and new data are being published even as we rush this volume to print, we are already inevitably behind in our reviews. As with some of the other monographs in the Cold Spring Harbor Laboratory series, this volume on adult neurogenesis will likely be updated in the coming years, and perhaps additional meetings at the Banbury Center and elsewhere will catalyze a greater synthesis in the future.

2

Numerology of Neurogenesis: Characterizing the Cell Cycle of Neurostem Cells

Richard S. Nowakowski and Nancy L. Hayes
Department of Neuroscience and Cell Biology
UMDNJ-Robert Wood Johnson Medical School
Piscataway, New Jersey 08854

DURING DEVELOPMENT, NEUROGENESIS IS A MULTISTEP PROCESS that includes cell proliferation, cell cycle exit, a choice between survival and death, cell migration, cell differentiation, and cell-fate decisions, including neuron versus glia and neuronal cell class decisions (for review, see Nowakowski et al. 2002). The same multiple steps and associated decisions occur during adult neurogenesis but with several significant differences, the most important being that (1) there are fewer proliferating cells during adult neurogenesis and (2) the selection of neuronal cell classes produced is limited. With respect to the ultimate outcome—the production of functional neurons—each step in this multistep process is, in effect, a possible site of regulation. The complexity of these regulatory steps is described in the other chapters of this book. In this chapter, we deal with the early steps in the process of neurogenesis, i.e., cell proliferation and cell cycle exit. We discuss how the number of cells produced during neurogenesis is regulated by the proliferative capacity of a population of dividing cells. The proliferative capacity, in turn, is determined by the length of the cell cycle, the number of proliferating cells, and the proportion of daughter cells that exit versus reenter the cell cycle. In addition, we review some of the methods for measuring these properties and discuss some of the pitfalls that are commonly encountered.

CELL CYCLE CHARACTERIZATION

Neurogenesis is driven by cell proliferation, and the core process of cell proliferation is the cell cycle. Conceptually, the cell cycle is simple (Fig. 1A). It begins when a "new cell" is created by a mitotic division. At the beginning, the cell resides in the G_1 phase in a normal diploid state (2N). The cell then enters the DNA synthetic or S phase of the cell cycle where it doubles its genome to become tetraploid (4N). After a brief pause (G_2), the cell then enters mitosis (M phase) where it divides to produce two daughter cells, each in the 2N state. The daughter cells then reenter G_1. About half-way through G_1, each daughter cell passes through a restriction checkpoint at which each of the two daughter cells decides independently whether or not it will reenter the cell cycle to repeat the process or exit the cell cycle and stop dividing. We refer to cells that are proliferating as P cells and those that exit (or quit) the cell cycle as Q cells. Since there are two daughter cells and two choices (reenter vs. exit), there are four possible combinations of decisions. Two of these combinations have equivalent outcomes, and thus there are three distinct types of cell divisions (Fig. 1B):

1. *Both daughter cells reenter the cell cycle.* This is "symmetric nonterminal" cell division producing exponential growth of the population. If continued, there will be 2^N cells for each founder cell after N cell cycles.
2. *Neither daughter cell reenters the cell cycle.* This is "symmetric terminal" cell division and terminates growth. The two daughter cells

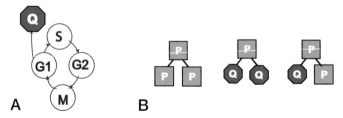

Figure 1. (A) A diagram of the four phases of the cell cycle, G_1, S, G_2, and M. During G_1, there is a restriction checkpoint at which the cells in G_1 make a decision as to whether they will exit the cell cycle to become a Q cell or reenter S to synthesize DNA and divide again. (B) After each division, the fates of the two daughter cells define three types of cell divisions: Symmetrical nonterminal producing two cells that both reenter the cell cycle, symmetrical terminal producing two cells that both exit the cell cycle, and asymmetrical producing one cell that reenters and one cell that exits. (Q) Cell that quits the cell cycle; (P) cell that is proliferating, i.e., in one of the four phases of the cell cycle.

persist as "postproliferative" or "postmitotic" cells and the one proliferating cell "disappears," terminating the proliferative lineage.

3. *One daughter cell reenters the cell cycle and the other exits the cell cycle.* This is "asymmetric" cell division, producing what is called "steady-state" growth. The proliferating population remains at a constant size, and the output is one new cell per proliferating cell, i.e., at the end of each cell cycle, one postproliferative cell and one proliferating cell reenter the cell cycle. If continued, for each founder cell after N cell cycles, there will be N postproliferative cells and one proliferating cell.

In different tissues and at different times, the time that it takes for a cell to complete the cell cycle varies. In addition, the length of time that a cell spends in each of the phases of the cell cycle also varies. For example, during the development of the CNS, the cell cycle is known to lengthen considerably. The best data are from the neocortex (for review, see Nowakowski et al. 2002) and the retina (Alexiades and Cepko 1996). In both of these tissues, the length of the cell cycle more than doubles during the period of time that neurons are being produced. During neocortical development, at the onset of neuron production, there are almost three cell cycles per day, whereas at the end of neuron production, there is only a little over one cell cycle per day (Takahashi et al. 1995). From a biological perspective, this is an important variable because the number of cell cycles that elapses during a specific time is directly related to the output of the proliferative population (Nowakowski et al. 2002).

CELL CYCLE AND CNS DEVELOPMENT

In the context of the development of the brain, the regulations of the length of the cell cycle and of the exit/reentry decisions work together to produce the appropriate specific complement neurons and neuron classes that comprise any given region of the CNS. To understand how this works, consider the need to produce a population of 250,000 neurons, i.e., about the number of neurons in the dentate gyrus (DG) of a typical adult mouse (Wimer and Wimer 1985, 1989; Wimer et al. 1988).

Two extreme and clearly distinct scenarios accomplish this: (1) If a single cell were to divide in a symmetrical nonterminal fashion 17 times followed by 1 symmetrical terminal division, 2^{18} (or ~260,000) cells would be produced. This is the fastest possible route by which to produce so many cells from a single founder cell. (2) If a single cell were to undergo *only* asymmetric cell divisions, it would produce only one postproliferative cell per cell cycle, and thus, 250,000 cell cycles would be required to produce the entire population. This is the slowest possible way to produce this population.

A third scenario, which is considerably more realistic, is an intermediate one, i.e., a mixture of both types of symmetric and asymmetric cell divisions, and is similar to what is believed to happen in the developing neocortex (Nowakowski et al. 2002). At the onset, there would be a period of symmetric cell divisions to produce a "founder" population. At this time, some of these cells would begin to produce daughter cells that exit the cell cycle. Thus, at least some of the cells would need to divide either asymmetrically or in a symmetrical terminal fashion. A retroviral study shows that both occur (Cai et al. 1996). For the next several cell cycles, the proportion of daughter cells that exit the cell cycle (the Q cells) remains small, i.e., less than 0.5 (or 50%). This means that the proportion of daughter cells that reenter the cell cycle, i.e., the P cells, would be greater than 0.5. As long as these conditions are met, more P cells would be present at the end of each cell cycle than there had been at the beginning, and thus, the proliferative population would continue to expand even as neurons are being produced. At some point, however, more daughter cells will exit the cell cycle than reenter, i.e., $Q > P$. When this point is reached, the proliferative population begins to disappear, eventually completely exhausting itself. In the developing neocortex (for review, see Nowakowski et al. 2002), the path from $Q = 0 \rightarrow Q = 1$ takes 11 cell cycles. The point of steady state, i.e., $Q = P = 0.5$, occurs at about cell cycle 7. The result is an amplification of approximately 125-fold. This more realistic scenario introduces two major regulatory points into the process: (1) the transition occurring at the end of the production of the founder cell population and (2) the Q/P changes during the period of neuron production. Manipulating these two regulatory points can, in principle, lead to the widely variant neuron numbers of the neocortex of different species (Nowakowski et al. 2002). The developing mouse DG has a much longer period of neurogenesis (Angevine and Sidman 1961), during which a similar scenario albeit involving more cell cycles and a different $Q = 0 \rightarrow Q = 1$ pathway is presumably occurring.

CELL CYCLE AND ADULT NEUROGENESIS IN THE DENTATE GYRUS

Given the developmental scenarios described above, how should we think about adult neurogenesis in the DG? First, there are two significant differences to consider: (1) The proliferative populations responsible for adult neurogenesis DG persist for a large proportion of adult life, whereas the proliferative populations responsible for developmental neurogenesis are transient and most disappear as the animal matures. (2) Both the number of proliferating cells and the number of neurons produced by the adult proliferative populations are both small compared to those numbers in developing proliferative populations.

Single Population Model: Steady-state Growth

The persistence of the adult proliferative populations over a long period can mean that either the population is "self-renewing" *or* it is being constantly renewed by an outside source or by cells that were previously not proliferating. The easiest way to think about self-renewal is to consider a population that is in "steady-state" growth. There are two distinct ways to achieve steady-state growth. The simplest such population, as discussed in scenario 3 above, consists entirely of cells that divide asymmetrically, i.e., with each pass through the cell cycle, one daughter cell reenters the S phase and remains a proliferative (or P) cell, and the other daughter cell exits the cell cycle to become a postmitotic (or Q) cell (Fig. 2A). If the proliferative population of the adult DG was of this type, then each proliferative cell in the DG would have exactly the same behavior, at least with respect to the fates of the two daughters that are produced at each cell cycle. In Figure 2B, an alternative form of steady-state growth is presented. Here, the fates of the two daughter cells are not correlated (i.e., during early G_1, each daughter cell interacts independently

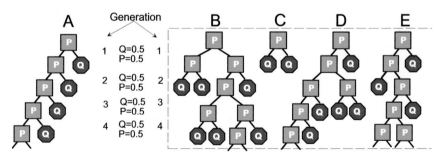

Figure 2. Two populations undergoing steady-state growth. At every cell cycle, both have a constant number of proliferative cells (P) and both produce the same number of postproliferative cells (Q). (*A*) All cells divide asymmetrically; (*B*) population is a mixture (1:2:1) of all three possible types of cell divisions. Note that despite the lineage differences, both scenarios have exactly the same number of proliferating cells and produce at each cell cycle. Retroviral methods, however, can distinguish them (Cai et al. 2002).

with the environment to make its decision whether or not to exit the cell cycle), and all three types of cell divisions (asymmetric, symmetric terminal, and symmetric nonterminal) occur. However, because the mixture of cell division is 1:2:1, both the size of the proliferative population and the output are constant over the life span of the proliferative population. This type of steady-state growth is known to occur in the crypts of the small intestine (Potten and Loeffler 1990) and seems likely to exist in the adult DG as well.

As a first approximation, steady-state growth with its intrinsic self-renewal property accounts well for what is known about adult neurogenesis, i.e., a long-lasting proliferating population produces postproliferative cells over a long period of time. There are at least two reasons, however, to consider that this does not tell the whole story. First, the 50–60 cell cycle "Hayflick limit" (Hayflick and Moorhead 1961; Hayflick 1965, 2000) could make the rate of self-renewal too low to be self-sustaining. If this is the case, then an alternative way of thinking about adult neurogenesis is needed (discussed in the next section). Second, we also know that the adult neurogenetic populations decline with age (Kronenberg et al. 2006). This would not happen if there were true steady-steady growth. The easiest way to account for this is to imagine that occasionally one of the proliferating cells dies and thus, the population is reduced. Alternatively, if the balance of Q and P were shifted even slightly toward Q, then the proliferating population would decline.

Two-population Model: Steady-state Population + Replenishing Population

If there is a significant amount of cell death in the proliferating population (for any reason), then some modification of the single population model must be introduced to compensate for the cell loss and to keep the size of the proliferative population constant (i.e., to maintain net steady state). The simplest modification is to postulate two discrete populations of proliferative cells in the adult DG: a "progenitor" population that produces Q cells and in which cell death occurs and a "replenishing" stem cell population in which there is no cell death. The proliferative behavior of the progenitor population is similar to that of the single-population model (either alternative shown in Figure 2 and described in text above); the difference is that there is some cell death among the proliferating cells (shown as stars in Fig. 3). To compensate for this cell death a second, smaller, population of stem cells continually replenishes the

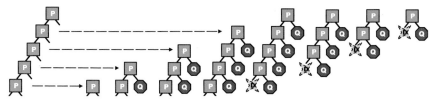

Figure 3. If there is cell death in the proliferating population, then there must be a proportionate number of proliferating cells that divide symmetrically in order to keep the size of the proliferative population constant. In this schematic diagram of the two-population model, a proportion of 20% cell death is shown in a pure asymmetrically dividing progenitor population (i.e., 20% of the proliferating cells of the progenitor population die per cell cycle), and thus the replenishing population would need to comprise 20% of the total proliferative population (i.e., would need to divide symmetrically) in order to maintain the size of the proliferating population at a constant level. Note that without replenishment, the progenitor population would become extinct.

progenitor population by symmetric nonterminal cell division. The replenishment is necessary to keep the proliferating population at a constant size (see Fig. 3). Furthermore, to balance out the cell death without expanding the population, the proportion of proliferating cells that comprises the replenishing stem cell population must be equal to the proportion of the total proliferative population that dies per cell cycle. Finally, a change in the ability of the replenishing population could contribute to the changes during aging.

The example shown in Figure 3 uses a high proportion (20% per cell cycle) of cell death for convenience in illustrating the concept. If the rate of cell death were smaller, e.g., 5% per cell cycle, then the size of the replenishing stem cell population would be commensurately smaller, e.g., 5% of the total population. In any event, each cell of the progenitor population has a limited life span. In contrast, the life span of cells of the replenishing stem cell population must be unbounded in order to be able to persist to replenish the progenitor population for the lifetime of the animal. Thus, the replenishing stem cell population differs from the progenitor population in four essential ways: (1) It is much smaller than the progenitor population, (2) its proliferative behavior is restricted to symmetric nonterminal cell division, (3) its life span is unlimited, and (4) no cell death occurs within it. Importantly, for the simplest form of this model to work (i.e., for steady state to be maintained in the proliferative population), three constraints apply: (1) The size of the replenishing stem population must be equal to the number of proliferating cells that die in the steady-state population at each cell cycle, (2) the rate of cell death (i.e., amount of cell death per cell cycle) must be constant, and (3) the length of the cell cycle (Tc) must be the same for both populations. If any one of these constraints were to be changed, then one of the other constraints must be changed in order to bring the system back into balance. For example, if the Tc for the replenishing stem cell population were to be much longer than the Tc for the progenitor population, then the number of replenishing stem cells would have to be commensurately larger for the right number of replacement cells to be produced at the right time. This leads to the interesting and speculative prediction that if the replenishing stem cell population had a very long cell cycle, it would need to be very large in order to be able to provide sufficient replenishment of the progenitor population.

Measuring the Cell Cycle

The number of cells produced by a proliferating population is derived from three properties that together define the proliferative capacity of that population: (1) the length of the cell cycle, (2) the number of proliferating cells in the population, and (3) the proportion of daughter cells that reenter versus exit the proliferating population. During development, all three of these properties change dynamically (Nowakowski et al. 2002), and the output of the proliferative population changes accordingly.

As the core process of neurogenesis, the cell cycle and its associated parameters dictate the output capacity of the proliferative population. If the proliferative population approximates steady state, then there will be one output cell per cell cycle per cell in the proliferative population. In other words, a great deal can be understood if we know the length of the cell cycle and the number of proliferating cells. Typically, "neurogenesis" is measured with the S-phase marker bromodeoxyuridine (BrdU). A single injection of BrdU will label all of the cells in S phase at the time of the injection. The number of labeled cells detected a short time (e.g., 30 minutes; see Fig. 5D–F and associated text below) after the injection will therefore represent the number of cells in S phase. Now, the number of S-phase cells is proportional to the total number of proliferating cells. The important question is what is the proportionality? If the proliferating cells are randomly distributed in the cell cycle, the number of cells in each phase will be proportional to the length of that phase, i.e., $N_S = N_{Tot} * Ts/Tc$. In the developing CNS, two methods are traditionally used to measure the length of the cell cycle: the cumulative labeling method and the percent labeled mitoses method (Sidman 1970; Nowakowski et al. 2002).

The cumulative labeling method is easy to understand (Fig. 4A). A single exposure to BrdU is given to label all of the cells in S phase; then repeated exposures to BrdU are given to label the additional cells entering S phase until all of the cells are labeled and a plateau is reached. The time to reach the plateau is Tc-Ts, the y intercept is Ts/Tc, and the slope is 1/Tc (Nowakowski et al. 1989). With two equations and two unknowns, both Ts and Tc can be derived. If the proliferating cells are positionally intermixed with nonproliferating cells, then an additional factor—the ratio of the proliferating cells to the whole population or growth fraction (GF)—must be added to the equation for $N_{LI} = GF * N_{Tot} * (Ts + T_{survival}/Tc)$. Since GF is the plateau, the y intercept becomes GF $*$ Ts/Tc.

The percent labeled mitosis (PLM) method is less frequently used and is a bit more cumbersome. For this method, a single injection of an S-phase label is given, and changes in the number of labeled cells in the M phase are used as a window into the dynamics of the cell cycle. By progressively increasing the length of the survival time and counting the percentage of the metaphase cells that are labeled, one can "watch" the wave of labeled cells move through M phase (Cai et al. 1997). The parameters of the cell cycle can be derived from a graph of the mitotic labeling index (MLI) against time (Fig. 4B). The time to the appearance of the first labeled metaphase approximates G_2, and the time from the first labeled metaphase until *all* of the metaphase figures are labeled is the length of M. The length of the S phase is the time from the first drop of the labeled mitosis curve from 100% minus the length of G_2. Finally, the length of the whole cell cycle is measured by waiting until the wave begins its second pass through the M phase, i.e., Tc is equal to the time between the first and second rises of the MLI curve. The complexity of the PLM method and the large number of animals required make it difficult to use, but one component of it is essential for studying cell proliferation. One should always include an experiment in which the survival time is long enough to show that labeled mitoses result from the incorporation of the S-phase marker. This is a positive demonstration that the same cells that are synthesizing DNA also are dividing. This is essential because there are situations in which DNA synthesis occurs but cells are not dividing (e.g., see Goto et al. 2002; Kuan et al. 2004; Burns et al. 2007) and hence, no cell proliferation (or neurogenesis) could possibly be occurring. In other words, it must be confirmed that BrdU incorporation is due to the S phase of the cell cycle, i.e., replication of the genome in

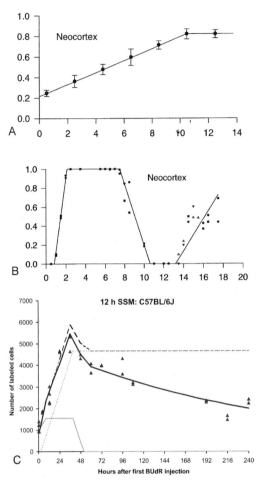

Figure 4. Graphs of data from three methods for measuring the length of the cell cycle. (*A*) The cumulative labeling method involves making repeated methods of an S-phase label (usually BrdU or [³H]thymidine) taking samples at intervals from 0.5 hour through a time that is longer than Tc-Ts, at which time, the entire proliferative population becomes labeled. (Reprinted, with permission, from Cai et al. 1997.) (*B*) The percent labeled mitosis (PLM) method uses a single injection of an S-phase marker and the metaphase figures in M as a window into the progression of the labeled population through the cell cycle. The first labeled metaphase figure appears at the end of G_2 as the labeled cells that were at the end of S enter M. The time for the percentage of labeled metaphase figures to reach 100% is the length of the M phase, and the time until the percentage begins to drop from 100% is G_2 + M + S. The total length of the whole cell cycle is the time from the first rise above 0% to the second rise above 0%. (Reprinted, with permission, from Cai et al. 1997.) (*C*) The Saturate and Survive Method (SSM) was devised for studying adult neurogenesis in the dentate gyrus (DG) where the proliferating and nonproliferating cells

anticipation of cell division. Once this is done, additional experiments are needed to trace the life history of the cells that are produced to determine that the detected cell proliferation indeed leads to neurogenesis, i.e., to ascertain that the appropriate "milestones" associated with neuron maturation occur (Kempermann et al. 2004).

During adult neurogenesis, however, the proliferative populations are small and generally mixed with other nonproliferating cells and, in particular, the cells that have just left the cell cycle. This is because the postproliferative cells do not migrate away from the proliferating cells. This situation means that a plateau in the cumulative labeling methods would never be reached; indeed, a labeling index is difficult to calculate meaningfully, particularly in the DG. To overcome this limitation for the adult DG, we developed the "Saturate and Survive Method" (SSM) (Fig. 4C) (Hayes and Nowakowski 2002). For this method, multiple BrdU injections are given over a period of time equal to T_c-T_s to label the entire proliferating population, and numbers of labeled cells are evaluated with increasing survival times. After the final injection, the number of labeled cells continues to increase as the labeled cells divide. Since the population is in approximate steady state, with each cell division, on average, half of the daughter cells exits the cell cycle and the other half continues to divide, resulting in a linear increase in the number of labeled cells. After three to four cell divisions, the cells that continue to divide dilute the amount of incorporated BrdU below the level of detection so that the number of labeled cells detected decreases. In contrast, the cells that have already exited the cell cycle retain sufficient label for detection (dashed line in Fig. 4C); however, the decline in the number of labeled cells continues for several days, indicating that cell death/survival decisions affect the postproliferative population. From the shape of the SSM curve, it can be estimated that in the DG of the C57BL/6J mouse, T_c is about 14 hours, T_s is

---

are intermixed. For this method, the proliferative population is first labeled to "saturation," and then the number of labeled cells is monitored over a period of time. The number of labeled cells initially increases as the labeled cells divide. At each cell cycle, some of the labeled cells exit the cell cycle and retain their label. After three to four cell divisions, the label in the daughter cells becomes diluted below the level of detection, and the labeled cells that are still proliferating "disappear" from the point of view of the assay. At about the same time, the effects of cell death become evident and the proportion of cells that survive can also be assessed with the SSM methods. (Reprinted, with permission, from Hayes and Nowakowski 2002 © Society for Neuroscience.)

about 7 hours, and the total proliferating cell population is approximately 1800 cells (Hayes and Nowakowski 2002). The half-life of the postproliferative cells is about 8 days with a plateau of surviving cells persisting at approximately 15% of the size of the proliferative population. Since these postproliferative cells are the surviving output of three to four cell cycles, this means that about 5% of the cells produced at each cell cycle survive to contribute to the adult population. Interestingly, in another strain of mouse, BALB/cByJ, the SSM gives almost identical results except that the total number of proliferating cells is only about 800 or about half of that found in C57BL/6J (Hayes and Nowakowski 2002).

What About the Subventricular Zone and the Rostral Migratory Stream?

A second widely recognized site of adult neurogenesis is the subventricular zone (SVZ) and associated rostral migratory stream (RMS) that produce cells destined for the olfactory bulb (OB) (Doetsch et al. 1999; Gage 2000). It is known that the cell cycle of the proliferating cells of the SVZ and RMS lengthens as the cells move closer to the OB (Smith and Luskin 1998). However, the large size of the combined SVZ and RMS and the diverse cell types generated make it difficult to understand the lineage relationships among the cells that occupy this large proliferative population. Thus, at this point, although the principles that are illustrated in Figures 2 and 3 certainly must apply, it is likely that the cell cycle and cell cycle exit are controlled quite differently in this complex structure and genetic analysis confirms this (Zheng et al. 2005). The recent discovery that the SVZ has a mosaic structure such that different regions produce different cell types (Merkle et al. 2007) may provide the insight and tools to address these issues.

Practical Considerations: Some Dos and Don'ts

The SSM curve (Fig. 4C) shows the fate over time of a group of labeled cells that represent all of the proliferating cells present during one cycle. This method is not suitable for routine quantification of cell proliferation occurring during neurogenesis as it is both time- and animal-intensive. What are the possible shortcuts and what are their advantages and drawbacks?

   The most direct experiment is simply a single injection of an S-phase marker (e.g., BrdU or [^{3}H]thymidine) with a short survival time between the injection and sacrifice. In both adult and embryonic mice, the label is incorporated rapidly and labeled cells are detected within 10–15 minutes after the injection. Thus, a short survival time of 30 minutes provides ample time for incorporation (Fig. 5D), and, most importantly, the result is the labeling of all of the cells in S phase at the time of the injection. This is a biologically meaningful measure. A major and important caveat regarding this simple experiment is that it is known that the ratio of Ts/Tc can and does change during development. We know that during neocortical development, Ts is relatively constant, but Tc lengthens, mostly from a lengthening of TG1 (Takahashi et al. 1995; Nowakowski et al. 2002). In addition, both Ts and Tc lengthen in the retina (Alexiades and Cepko 1996). The effects of changes in Ts and/or Tc are shown in Figure 5, A–C. For a proliferating population of constant size, either a lengthening of Ts or a shortening of Tc will increase the number of labeled cells from a pulse exposure, whereas either a shortening of Ts or a lengthening of Tc will decrease the number of labeled cells. For example, if Ts is 50% of Tc, then a single injection of BrdU will label half of the cells (Fig. 5A). If Ts is changed, but Tc remains the same, then the

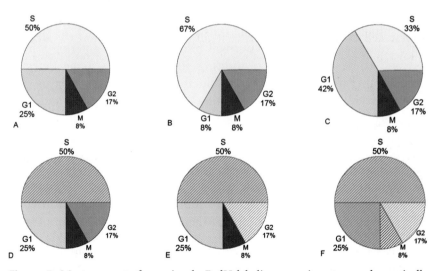

Figure 5. Measurements from simple BrdU-labeling experiments are dramatically influenced by changes in the length of the phases of the cell cycle even if the cell cycle length itself is not changed (A–C). The experimental parameters can also affect the number of labeled cells, again even if the underlying biology is unchanged (D–F). (A–C) Effects of changes in the length of the S phase; (D–F) effects of changes in the length of survival after a single injection (D–E) or multiple injections of an S-phase marker (for details see the text).

proportion of S-phase cells will be either increased (Fig. 5B) or decreased (Fig. 5C). Note that such changes in parameters of the cell cycle would not change the number of proliferating cells or the proliferative capacity of the population, but they would change the outcome of an experiment, i.e., the number of labeled cells. In other words, without knowledge of the cell cycle parameters, an increase or decrease in the number of labeled cells could be misinterpreted (e.g., Fig. 5, B vs. C). There are no available data on whether or not the relative lengths of Ts or Tc change during adult neurogenesis or as a result of pathology.

A second major caveat is that the number of labeled cells is also affected by the experimental parameters. The properties of the S-phase labels, i.e., the time required for them to be incorporated to levels of detection and the time that they are available for incorporation (Hayes and Nowakowski 2000), make the dynamics complex; however, the single experimental parameter that most influences the number of cells labeled by a single injection of an S-phase marker is the interval between the injection and sacrifice. As mentioned above, 10–15 minutes survival is sufficient time for the incorporation of an S-phase label to detection (Miller and Nowakowski 1988; Takahashi et al. 1992; Hayes and Nowakowski 2000). This means that a survival time of 0.5 hour is more than sufficient to label all of the cells in S phase (Fig. 5D). As the cell cycle continues, a longer survival will result in (1) cells entering S from G_1, and some of them incorporating sufficient label to be detected because the tracer remains available for several hours (Hayes and Nowakowski 2000), and (2) cells exiting S to enter G_2 (Fig. 5E) and, with time, progressing to M where they will divide, further increasing the number of labeled cells. In short, as the survival time is lengthened, the number of labeled cells will increase (Fig. 5E–F). To be biologically meaningful, the survival time should be long enough to provide time for incorporation, but significantly shorter than G_2, which in mouse embryos and in the mouse adult DG G_2 is about 2 hours (Fig. 5E) (Takahashi et al. 1995; Hayes and Nowakowski 2002). After 1–4 hours, the amount of label available for incorporation is reduced (Nowakowski and Rakic 1974; Hayes and Nowakowski 2000) so that the cells entering S from G_1 no longer become detectably labeled. Moreover, cells divide as they pass through M, some exiting the cell cycle, and others reentering S. Together, this means that long survival times (e.g., 12 or 24 hours) result in a mixture of types of labeled cells that is difficult if not impossible to unravel. The biologically relevant alternative is cumulative labeling, i.e., additional injections of the S-phase marker so that the cells entering S phase will also become labeled. If both the time between injections is less than Ts and the survival time is at least Tc-Ts, then *all* of the proliferating cells will become labeled (Fig. 5F). The cumulative labeling method (see above) is the most common cell cycle length measuring method (Nowakowski et al. 1989) and is also a convenient way to estimate the total size of the proliferative population using the initial part of the SSM curve (Hayes and Nowakowski 2002).

Putting It All Together

Although it may be succinctly designated by the word "neurogenesis," the production of new neurons is a complex and dynamic multistep process at the core of which is the cell cycle, itself a dynamic and changing process. For that reason, no single experiment—e.g., in one tissue, at one age, etc.—is sufficient to understand how neurogenesis contributes to the CNS. A second important question to keep in mind is to what extent does the behavior of an individual proliferating cell reflect the behavior of the proliferating population. This may be different, for example, during development versus in the adult. Fortunately, neurogenesis is a process that can be quantified and described mathematically and that allows one to compare data between conditions and, most importantly, to generate hypotheses that can be tested experimentally. The ultimate question, of course, is how many new, functional neurons are produced? The answer depends first on the proliferative capacity of the proliferative population: How many new cells can it produce per cell cycle? This capacity is determined by the size of the proliferative population (N_P) and the dynamics of the cell cycle. The initial size of the proliferating population is set up during development and then is modified at each cell cycle by the proportion of cells that reenter (P) versus exit (Q) the cell cycle, the number of proliferating cells that die (or otherwise disappear from the active proliferating population), and the number of dormant cells that might "awaken" and begin to proliferate. The proportion of cells exiting the population at each cell cycle (Q) determines the output (O) for the population (Nowakowski et al. 2002). Thus, the total output of new non-proliferating cells per cell cycle is simply $O_T = N_P * Q$. Since not all of the cells produced will survive (S) and not all surviving cells of the lineage will become neurons (L_N), the net long-term neuron production—neurogenesis (NG)—per cell cycle becomes $NG_T = N_P * Q * L_N * S$. The final note is that a clear distinction must be made between time and the number of cell cycles. Because the cell cycle is the core of neurogenesis, results per cell cycle are meaningful, whereas results per unit time can be distorted by the length of the cell cycle (Tc).

ACKNOWLEDGMENTS

This work has been supported by grants to R.S.N. from the National Institutes of Health (NIH) (NS049445 and EY015647), the NJ Commission on Spinal Cord Research, and the NJ Commission on Science and Technology and a grant to N.L.H. from the NIH (MH63957).

REFERENCES

Alexiades M.R. and Cepko C. 1996. Quantitative analysis of proliferation and cell cycle length during development of the rat retina. *Dev. Dyn.* **205:** 293–307.

Angevine J.B.J. and Sidman R.L. 1961. Autoradiographic study of cell migration during histogenesis of cerebral cortex in the mouse. *Nature* **192:** 766–768.

Burns K.A., Ayoub A.E., Breunig J.J., Adhami F., Weng W.L., Colbert M.C., Rakic P., and Kuan C.Y. 2007. Nestin-CreER mice reveal DNA synthesis by nonapoptotic neurons following cerebral ischemia-hypoxia. *Cereb. Cortex* (in press).

Cai L., Hayes N.L., and Nowakowski R.S. 1996. Clone size during neocortical development: Agreement of experiment and prediction. *Soc. Neurosci. Abstr.* **22:** 524.

———. 1997. Local homogeneity of cell cycle length in developing mouse cortex. *J. Neurosci.* **17:** 2079–2087.

Cai L., Hayes N.L., Takahashi T., Caviness V.S., Jr., and Nowakowski R.S. 2002. Size distribution of retrovirally marked lineages matches prediction from population measurements of cell cycle behavior. *J. Neurosci. Res.* **69:** 731–744.

Doetsch F., Caille I., Lim D.A., Garcia-Verdugo J.M., and Alvarez-Buylla A. 1999. Subventricular zone astrocytes are neural stem cells in the adult mammalian brain. *Cell* **97:** 703–716.

Gage F.H. 2000. Mammalian neural stem cells. *Science* **287:** 1433–1438.

Goto T., Takahashi T., Miyama S., Nowakowski R.S., Bhide P.G., and Caviness V.S., Jr. 2002. Developmental regulation of the effects of fibroblast growth factor-2 and 1-octanol on neuronogenesis: Implications for a hypothesis relating to mitogen-antimitogen opposition. *J. Neurosci. Res.* **69:** 714–722.

Hayes N.L. and Nowakowski R.S. 2000. Exploiting the dynamics of S-phase tracers in developing brain: Interkinetic nuclear migration for cells entering versus leaving the S-phase. *Dev. Neurosci.* **22:** 44–55.

———. 2002. Dynamics of cell proliferation in the adult dentate gyrus of two inbred strains of mice. *Brain Res. Dev. Brain Res.* **134:** 77–85.

Hayflick L. 1965. The limited in vitro lifetime of human diploid cell strains. *Exp. Cell Res.* **36:** 614–636.

———. 2000. The illusion of cell immortality. *Br. J. Cancer* **83:** 841–846.

Hayflick L. and Moorhead P.S. 1961. The serial cultivation of human diploid cell strains. *Exp. Cell Res.* **25:** 585–621.

Kempermann G., Jessberger S., Steiner B., and Kronenberg G. 2004. Milestones of neuronal development in the adult hippocampus. *Trends Neurosci.* **27:** 447–452.

Kronenberg G., Bick-Sander A., Bunk E., Wolf C., Ehninger D., and Kempermann G. 2006. Physical exercise prevents age-related decline in precursor cell activity in the mouse dentate gyrus. *Neurobiol. Aging* **27:** 1505–1513.

Kuan C.Y., Schloemer A.J., Lu A., Burns K.A., Weng W.L., Williams M.T., Strauss K.I., Vorhees C.V., Flavell R.A., Davis R.J., Sharp F.R., and Rakic P. 2004. Hypoxia-ischemia induces DNA synthesis without cell proliferation in dying neurons in adult rodent brain. *J. Neurosci.* **24:** 10763–10772.

Merkle F.T., Mirzadeh Z., and Alvarez-Buylla A. 2007. Mosaic organization of neural stem cells in the adult brain. *Science* **317:** 381–384.

Miller M.W. and Nowakowski R.S. 1988. Use of bromodeoxyuridine-immunohistochemistry to examine the proliferation, migration and time of origin of cells in the central nervous system. *Brain Res.* **457:** 44–52.

Nowakowski R.S. and Rakic P. 1974. Clearance rate of exogenous 3H-thymidine from the plasma of pregnant rhesus monkeys. *Cell Tissue Kinet.* **7:** 189–194.

Nowakowski R.S., Lewin S.B., and Miller M.W. 1989. Bromodeoxyuridine immunohistochemical determination of the lengths of the cell cycle and the DNA-synthetic phase for an anatomically defined population. *J. Neurocytol.* **18:** 311–318.

Nowakowski R.S., Caviness V.S., Jr., Takahashi T., and Hayes N.L. 2002. Population dynamics during cell proliferation and neuronogenesis in the developing murine neocortex. *Results Prob. Cell Differ.* **39:** 1–25.

Potten C.S. and Loeffler M. 1990. Stem cells: Attributes, cycles, spirals, pitfalls and uncertainties. Lessons for and from the crypt. *Development* **110:** 1001–1020.

Sidman R.L. 1970. Autoradiographic methods and principles for study of the nervous system with thymidine-H3. In *Contemporary research methods in neuroanatomy* (ed. W.J.H. Nauta and S.O.E. Ebbesson), pp. 252–274. Springer, New York.

Smith C.M. and Luskin M.B. 1998. Cell cycle length of olfactory bulb neuronal progenitors in the rostral migratory stream. *Dev. Dyn.* **213:** 220–227.

Takahashi T., Nowakowski R.S., and Caviness V.S., Jr. 1992. BUdR as an S-phase marker for quantitative studies of cytokinetic behaviour in the murine cerebral ventricular zone. *J. Neurocytol.* **21:** 185–197.

———. 1995. The cell cycle of the pseudostratified ventricular epithelium of the embryonic murine cerebral wall. *J. Neurosci.* **15:** 6046–6057.

Wimer R.E. and Wimer C.C. 1985. Three sex dimorphisms in the granule cell layer of the hippocampus in house mice. *Brain Res.* **328:** 105–109.

———. 1989. On the sources of strain and sex differences in granule cell number in the dentate area of house mice. *Brain Res. Dev. Brain Res.* **48:** 167–176.

Wimer R.E., Wimer C.C., and Alameddine L. 1988. On the development of strain and sex differences in granule cell number in the area dentata of house mice. *Brain Res.* **470:** 191–197.

Zheng W., Hayes N.L., and Nowakowski R.S. 2005. The genetics of adult neurogenesis: Strain differences in the adult mouse rostral migratory stream. *Soc. Neurosci. Abstr. 2005* Program No. 141.12. (CD-ROM).

3

Detection and Phenotypic Characterization of Adult Neurogenesis

H. Georg Kuhn
Institute for Neuroscience and Physiology
Göteborg University
S-40530 Göteborg, Sweden

Daniel A. Peterson
Center for Stem Cell and Regenerative Medicine
Rosalind Franklin University of Medicine and Science
North Chicago, Illinois 60064

ADVANCES IN OUR UNDERSTANDING OF THE EXTENT and regulation of adult neurogenesis have been dependent on continued improvements in the detection and quantification of critical events in neurogenesis. To date, no specific and exclusive stem cell marker has been described that would allow for prospective studies of neurogenesis. As a result, detection of neurogenic events has depended on a combination of labeling approaches that document the two critical events in neurogenesis: the generation of new cells and their subsequent progression through lineage commitment to a mature neuron.

Detection of neurogenesis in vivo requires the ability to image at a cellular resolution. Although advances in noninvasive imaging approaches, such as magnetic resonance imaging (MRI), show promise for longitudinal studies of neurogenesis, the lack of suitable resolution to characterize individual cells limits the information that can be obtained. In vivo microscopy, using deeply penetrating UV illumination with mulitphoton microscopy or by the recently available endoscopic confocal microscopy, may provide new opportunities for longitudinal studies of neurogenesis in the living animal with single-cell resolution. These latter microscopy approaches are particularly compelling when coupled with transgenic mice expressing phenotype-specific fluorescent reporter genes. However, at present, the predominant approach for studies of neurogenesis relies on traditional histological methods of fixation, production of tissue sections, staining, and microscopic analysis.

This chapter discusses methodological considerations for in vivo detection of neurogenesis in the adult brain according to our current state of knowledge. First, detection of newly generated cells is evaluated and the strengths of using exogenous or endogenous markers of the cell cycle are discussed. Next, the individual phenotype markers that contribute to resolving stages of neuronal lineage commitment and their value in combinatorial staining are addressed. The accurate analysis of the cell phenotype is subject to a number of potential artifacts. These potential pitfalls are elaborated along with suggestions for accurate detection and reliable quantification of cell number. Neurogenesis is modulated by many environmental variables, and accurate detection combined with a precise readout of changes in cell number is critical for reliable data in studies of neurogenesis.

CELL CYCLE PROGRESSION

Thymidine Analogs as Exogenous Markers of DNA Replication

Stem cells are defined by their ability to self-renew and the capacity of their progeny to adopt multiple lineages. In addition to self-renewal, expansion of progenitor populations by repeated entry into the cell cycle is a fundamental feature of neurogenesis. As a result, analysis of cell cycle activity is a critical component of neurogenesis studies. The cell cycle (Fig. 1A) consists of a sequence of distinct stages. DNA replication during the S phase of the cell cycle has been utilized as an opportunity to birth-date proliferating cells by supplying a thymidine analog that can be incorporated into the cell (Fig. 1B). The thymidine analog is retained in the postmitotic cell, permitting a cohort of dividing cells, labeled at a known time, to be subsequently detected.

The ability to label a cohort of dividing cells has provided a useful tool to assess the existence of adult neurogenesis and changes in neurogenesis under different conditions. The original approach to detect incorporation of [3H]thymidine using autoradiography had two serious drawbacks. The first was that it is difficult, using bright-field microscopy, to detect the

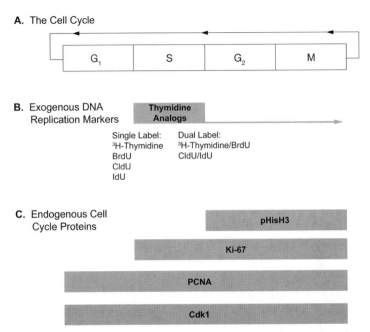

A. The Cell Cycle

| G₁ | S | G₂ | M |

B. Exogenous DNA
Replication Markers

Thymidine Analogs

Single Label: Dual Label:
³H-Thymidine   ³H-Thymidine/BrdU
BrdU   CldU/IdU
CldU
IdU

C. Endogenous Cell
Cycle Proteins

pHisH3

Ki-67

PCNA

Cdk1

Figure 1. Detection of neurogenic proliferation. (*A*) The mitotic cell cycle consists of the regulated transition of a cell through the sequential phases of G_1, S, G_2, and M from which the cell can remain postmitotic or reenter cell cycle again. (*B*) DNA replication occurs during S phase (synthesis), and this essential step of the cell cycle can be detected by the availability of thymidine analogs that are incorporated during DNA replication and can be subsequently detected by immunocytochemistry. Thymidine analogs allow birth-dating as they define the time of incorporation, but they do not discriminate subsequent phases of cell cycle (illustrated by the tail extending on into G_2 and M). DNA replication can be probed again to determine if a labeled cell has again reentered cell cycle pairing with either [³H]thymidine and bromodeoxyuridine (BrdU) or chlorodeoxyuridine (CldU) and iododeoxyuridine (IdU). (*C*) Endogenous cell cycle proteins are expressed at different stages of cell cycle progression, and their expression offers a tool to examine regulation of cell cycle progression. Some of these, such as Cdk1 and proliferating cell nuclear antigen (PCNA), are broadly expressed throughout the cell cycle and thus provide low temporal resolution about specific cell cycle progression. Ki-67 closely corresponds to thymidine analog detection as it is expressed both during S phase and through the remainder of the cell cycle. More specific temporal resolution is offered by phosphohistone H3 (pHisH3) which is only expressed following S phase.

deposition of silver grains in combination with immunoperoxidase labeling for phenotypic markers due to their overlap in the cell obscuring visualization. A second methodological concern is that the thin sections needed for an autoradiographic study limit the use of stereological quantitation.

Both of these drawbacks have been overcome through the use of the thymidine analog bromodeoxyuridine (BrdU). Like [^{3}H]thymidine, BrdU is incorporated during DNA replication, but its presence in the nucleus can be detected by immunohistochemistry, permitting single labeling or multiple labeling with phenotypic markers detected by bright-field or fluorescence microscopy. As thicker sections can be used, detection of BrdU by immunohistochemistry is suitable for design-based stereological quantification. Thymidine analogs are used in most study designs of in vivo neurogenesis. Table 1 presents a summary of study design considerations.

Although BrdU administration provides the opportunity to birth-date a cohort of proliferating cells and to study their fate at a later date, this approach is of limited temporal resolution with respect to both uptake and detection. There is no evidence that cells in the adult dentate gyrus (DG) or subventricular zone are synchronized with respect to their proliferative state. Thus, administration of BrdU will label a distribution of cells in S phase where any individual cell may completely or only partial overlap the availability of BrdU in tissue. In addition, once incorporated, the presence of BrdU does not discriminate between the cells' subsequent replicative states. As a result, additional cell cycle activity in a BrdU-labeled cell would not be detected. Early studies in the field of adult neurogenesis did not fully account for these considerations and tended to adopt a BrdU administration paradigm which ensured that an adequate population would be labeled to facilitate analysis. In recent years, it has become evident that such broad study designs, with BrdU given for days or weeks, severely limited the ability to discriminate between proliferation and survival of newly generated cells. Therefore, current studies typically have shorter administration periods and choose times to take the tissue for histology that permit discrimination of the different phases of neurogenesis. Nevertheless, when proliferation is slow or cells are few, such as in aging studies, it may be necessary to administer BrdU for a longer time to obtain an adequate cell sample size for study.

The retention of BrdU prevents detection of progression into the G_2 phase of cell cycle or the subsequent reentry of that cell into cell cycle (Fig. 1B). To achieve this degree of temporal discrimination, the examination of endogenous cell cycle proteins is recommended as discussed below. Commonly, a combination of BrdU and endogenous cell cycle markers have been used to calculate cell cycle kinetics. However, the question of cell cycle reentry can be more efficiently addressed by the use of sequential administration of halogenated thymidine analogs. Replacement of the bromo group with iodo or chloro groups results in thymidine analogs that behave the same as BrdU, but whose incorporation can

Table 1. Suggested parameters for in vivo neurogenesis studies using thymidine analogs to detect newly generated cells

	Thymidine analog injection frequency	Postinjection examination interval	Cells identified	Key considerations
Proliferation	• 50 mg/kg BrdU or equimolar equivalent of IdU or CldU • 1–3 injections at 2-hr intervals	2 hr after last injection	Cohort of proliferating cells	• Validate against false-positive artifacts as described in text • Conditions of low proliferation (such as aging) may require more injections to establish a detectable cohort
Cell cycle reentry	• 1–3 injections of IdU at 2-hr intervals, then: • 1–3 injections of CldU at 2-hr intervals	• Variable interval between analogs based on study question • 2 hr after last injection	Subpopulation of rapidly dividing cells	• Administration of analogs must be equimolar to allow comparison between each cohort of proliferating cells • Order of analog administration is unimportant and could be reversed to CldU first and then IdU • Interval between delivery provides an index of cycle length
Fate specification	• 50 mg/kg BrdU or equimolar equivalent of IdU or CldU • 1–3 injections at 2-hr intervals	~3–10 days after last injection	Recently generated cells coexpressing early lineage markers	• Validate that detection is in appropriate cell compartments for confirming true coexpression as described in text • Lineage commitment may produce a transient overlap of phenotypic markers

(continued)

Table 1. (*continued*)

	Thymidine analog injection frequency	Postinjection examination interval	Cells identified	Key considerations
Neuronal maturation	• 50 mg/kg BrdU or equimolar equivalent of IdU or CldU • 1–3 injections at 2-hr intervals	2–4 weeks after last injection	Newly generated cells differentiating into phenotypically mature neurons	• Cells observed may express various differentiation markers requiring multiple labeling approaches • Cells observed reflect the net result of continued proliferation following labeling and cell loss
Survival	• 50 mg/kg BrdU or equimolar equivalent of IdU or CldU • 1–3 injections at 2-hr intervals	Varies depending on study question	• Intervals of <10 days address short- and long-term survival • Intervals >30 days assess integration	• Study could combine IdU/CldU delivery to provide baseline data for comparison to proliferation within the same animal under the same conditions • Assessment of integration should be combined with a functional readout
Rare/Valuable subjects	• 1–3 injections of IdU at 2-hr intervals, then: • 1–3 injections of CldU at 2-hr intervals	• 2–4 weeks after IdU injection (first cohort) • 2 hr after CldU injection (second cohort)	• Differentiation from cells in the first cohort • Proliferation from cells in the second cohort	• Permits efficient use of rare or valuable subjects for measuring both proliferation and differentiation • Permits use of same animal for increased statistical power

be discriminated immunohistochemically (Vega and Peterson 2005). Thus, by timing the delivery of each analog administered, it is possible to address questions of cell cycle kinetics, particularly the frequency to reenter the cell cycle in a given cell population (see also Fig. 3) (Thomas et al. 2007). However, it is critical for the administration of the halogenated thymidine analogs to be given at an equimolar ratio for valid comparisons to be made (Vega and Peterson 2005).

Possible Artifacts to Thymidine Analog Incorporation

Thymidine analog antibodies have been generated from a variety of species as monoclonal and polyclonal antibodies. Binding of the antibody to labeled DNA requires denaturation of the DNA to remove histones, reduce tertiary DNA structures, or even generate single-stranded DNA. Denaturation is usually performed by exposing cells or tissue to hydrochloric acid, heat, enzymatic digestion, or a combination, depending on the specific antibody and tissue (Kass et al. 2000). Some suppliers mention in their antibody specifications a possible cross-reactivity with methylated DNA that could lead to unspecific labeling of all nuclei and therefore recommend specific pretreatment.

Several conditions can create false-positive signals, producing sporadic reports of neurogenesis in novel regions or in response to injury that demonstrate the need for careful controls in tissue preparation (Grassi Zucconi and Giuditta 2002; Rakic 2002; Kuan et al. 2004). The use of alternative fixatives should be compared to results from paraformaldehyde perfusion-fixed tissue, and the use of antigen retrieval treatments should be compared against endogenous cell cycle marker staining to assess the risk of false-positive detection. The specificity of primary antibodies to BrdU-labeled DNA should be determined for each tissue through the use of a short BrdU pulse-chase interval for labeling, with the results compared to coexpression with endogenous cell cycle markers for confirmation. Similar results should be obtained in tissue perfused after a 2-hour BrdU pulse-chase interval (Fig. 1). Through these tests, widespread unspecific labeling could be unmasked.

Incorporation of thymidine analogs also can occur during DNA repair, leading to concerns that some observations of cell proliferation could be due to this modest incorporation, rather than transition through S phase. Clearly, complete replication of the genome during cell division will incorporate thymidine analogs to a much higher extent than limited DNA repair (Cooper-Kuhn and Kuhn 2002). Nevertheless, it was suggested that immunohistology could also detect incorporation from DNA repair (Kuan et al. 2004). Whether this occurs at detectable levels depends largely on the extent of DNA damage and BrdU dose/availability. Cells that undergo extensive DNA repair can reside in lesioned tissue, for example, after radiation or ischemia. Damaged cells can also up-regulate cell cycle markers and even undergo aberrant cell cycle entry, making the distinction between apoptosis and cell cycle entry even more difficult (Busser et al. 1998; Yang et al. 2001). Although a more recent study suggests that mistaken apoptosis may not be a great concern in the adult brain (Bauer and Patterson 2005), investigators should still use caution in interpretation. The use of a lower dose of thymidine analog (50 mg/kg for BrdU or equivalent) is recommended to reduce the probability of detecting lower levels of non-S-phase thymidine analog incorporation, particularly as these lower doses are equally effective at detecting proliferation (Burns and Kuan 2005).

Endogenous Cell Cycle Proteins

Whereas thymidine analog incorporation provides a temporal signature for DNA replication, detection of endogenous cell cycle proteins identifies the population of cells engaged in the process of cell cycle at that time. As illustrated in Figure 1, a variety of cell cycle proteins have been used to detect separate phases of cell cycle progression (for comprehensive reviews of endogenous proteins in adult neurogenesis, see Eisch and Mandyam 2004, 2007). The pattern of protein expression provides a valuable tool for examining how changes to the microenvironmental niche might specifically regulate cell proliferation.

Labeling for endogenous cell cycle proteins has an advantage in that only currently cycling cells are detected, whereas with thymidine analog delivery, the event of DNA replication obscures subsequent discrimination of cell cycle activity (unless temporal separation of chlorodeoxyuridine [CldU] and iododeoxyuridine [IdU] probes are used as discussed above). Another practical advantage of endogenous protein detection is in studies where it is not feasible to deliver thymidine analogs, such as in use of available human postmortem tissue. However, caution must be used in equating thymidine analog studies with the use of endogenous protein detection due to the fact that the cell cycle proteins reveal a wider population of cells activated for cell cycle, not all of which may necessarily proceed through DNA replication if held up at an upstream checkpoint.

Lineage Tracking by Retroviral Vectors

Retroviral vectors specifically infect dividing cells, and this property has been used with great success for in vivo delivery of transgenes to proliferating cells in neurogenic niches. Use of fluorescent reporter genes, such as green fluorescent protein (GFP), has revealed the subsequent morphological development and distribution of newly generated cells and has proven to be a useful lineage marker for studies of subsequent lineage commitment. The filling of cells with GFP has enabled specific studies of functional connections of newly generated cells and allowed their identification in slice preparations for studying their electrophysiological maturation and integration (van Praag et al. 2002; Laplagne et al. 2006). Despite the usefulness of retrovirus-mediated delivery of a fluorescent reporter gene for lineage marker studies, caution should be used as the transfer of fluorescent protein to a postmitotic neuron has been reported following infection of proliferating microglial cells (Ackerman et al. 2006).

LINEAGE COMMITMENT

In early studies of adult neurogenesis, detecting coexpression of BrdU with a marker of mature neuronal phenotype was the primary goal for establishing whether or not neurogenesis was occurring. As the field has matured, it has become increasingly important to define more than the final maturation as an endpoint and to address the characterization and regulation of the process. Studies of adult neurogenesis have borrowed from the collection of markers expressed in the developing nervous system and applied these to the postnatal brain. Just as in development, newly generated cells in the adult brain progress through a series of lineage commitment stages prior to the adoption of mature phenotype markers and evidence of function (Kempermann et al. 2004). The markers that are presently available are not strictly sequential in their expression but have varying amounts of overlap as lineage commitment proceeds (Fig. 2). It has also become evident that lineage commitment is indicated prior to terminal exit from cell cycle (Fig. 2A). Thus, the early division of adult neurogenesis into distinct proliferation, differentiation, and mature phases has become blurred and is now thought of as a continuum. Although the expression of recognized lineage commitment markers provides a useful construct for our understanding of the process, it is likely that their expression patterns and duration reflect molecular regulatory events that we are only beginning to understand. As such, markers of lineage commitment provide important signposts for organizing data about the regulation of neurogenesis. Useful new tools have recently been reported using transgenic mice where fluorescent reporter genes are under the control of promoters for relevant neurogenic progression, such as nestin and doublecortin (DCX) (Couillard-Despres et al. 2005; Encinas et al. 2006) (see also Fig. 3).

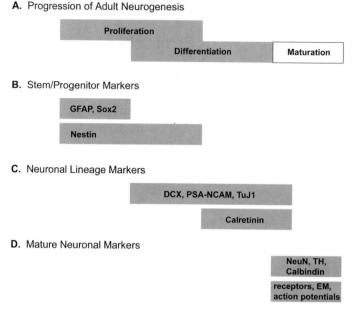

Figure 2. Detection of progression through lineage commitment. (*A*) Neurogenesis occurs by progression through sequential phases of *proliferation* to expand the population of neuronal progenitor cells, *differentiation* as these cells acquire lineage commitment and migrate to their final location, and *maturation* as the newly generated cells become structurally and functionally integrated. Initially thought to proceed in a strictly sequential fashion, it is now understood that these phases may overlap. (*B*) A schematic representation of proliferation identifies markers of stem/progenitor cells that are proliferative and whose expression may be confined to early period of proliferation (glial fibrillary acidic protein [GFAP] and Sox2) or extend further into the overlapping period with differentiation (nestin). (*C*) With increasing lineage commitment, newly generated cells begin to express characteristic markers of neuronal commitment, some of which begin expression while the cell has not yet terminally exited cell cycle. (*D*) Newly generated cells that survive this process ultimately express phenotypic markers and adopt functional characteristics of mature neurons.

Early Markers of Neuronal Stem Cells

The study of early events in stem cell biology, such as self-renewal and lineage commitment, requires the ability to detect these stem cells with suitable markers. However, for the purpose of specifically detecting neurogenesis, these markers are of limited value, since they can only provide an indication of whether a region is in principle able to generate new neurons. Furthermore, the multipotent character of stem cells makes it unclear to predict to what extent neuronal differentiation will be achieved. Currently, there is no single neuronal stem cell marker that is definitive for neuronal stem cells. The most widely used marker, nestin, is indicative of early neuroectodermal commitment, but it can also be detected in vascular structures. Therefore, the best practices for identifying early markers of neuronal stem cells rely on a combination of otherwise more ubiquitous markers. For example, the combination of glial fibrillary acidic protein (GFAP) and nestin appears to label multipotent cells in neurogenic regions. These cells represent astrocyte-like cells derived from a subset of astrocytes in the subgranular and subventricular zones (Alvarez-Buylla et al. 2002). These cells share many characteristics with radial glia cells, which act as neuronal progenitors during embryonic development (Malatesta et al. 2000, 2003). Neuronal stem cells can also be identified by other markers such as Sox-2 and Pax-6, but as these markers are also expressed widely outside the neurogenic regions, they are not suitable for single use but must be combined with other predictors of neurogenic activity (Stoykova and Gruss 1994; Komitova and Eriksson 2004; Nacher et al. 2005; Baer et al. 2006).

Early Markers of Neuronal Lineage Commitment

Detecting cells that have committed to neuronal development has become a very important indicator of ongoing adult neurogenesis. Markers expressed specifically during this phase could be good predictors of ongoing neurogenic activity, provided they fulfill specific criteria. Ideally, neuronal lineage commitment markers need to be transiently expressed only in the neuronally committed cell population, without expression in mature neurons or other brain cell types and without reexpression in neurons under conditions of neuronal injury or regeneration. When sufficiently validated, such endogenous markers could replace BrdU labeling for detecting neurogenesis, especially under conditions that preclude BrdU delivery, such as human tissue. Several such markers have been described in recent years. DCX, a microtubule-associated protein, is currently most widely used as it has been demonstrated to be transiently expressed in neuronal progenitor cells and immature neurons (Brown et al. 2003; Rao and Shetty 2004; Couillard-Despres et al. 2005). However, reports indicate ectopic expression in neocortical neurons and low-level expression in NG2-positive cells (presumptive glial progenitors), which caution against the use of DCX to determine neurogenesis in brain regions not confirmed by other methods (Nacher et al. 2001). The polysialylated form of the neural cell adhesion molecule (PSA-NCAM) and TOAD-64 (also known as TUC-4 and CRMP-4) are other examples.

Similar to DCX, transient expression patterns allow labeling of neuronal progenitor cells; however, reports on expression of PSA-NCAM in glial cells and in plasticity of mature neurons make these markers less specific for detection of neurogenesis (Nacher et al. 2000; Seki 2002; Nguyen et al. 2003; Bonfanti 2006). Another neuronal cytoskeletal protein, βIII-tubulin (also known as Tuj1), is expressed relatively early after neuronal commitment. However, βIII-tubulin is also reported to be expressed in a large number of mature neurons, and its transient expression character is thus unclear. When used in combination with birth-dating by thymidine labeling, most of these early neuronal lineage commitment markers can be reliably used to define early neuronal maturation in studies of adult neurogenesis.

Markers of Maturing Neurons

Once progenitor cells reach maturity, they begin expressing neuronal markers such as neurofilaments, NeuN, calcium-binding proteins, and neurotransmitters. NeuN and neurofilaments can be considered as pan-neuronal markers; however, NeuN is absent from a few neuronal populations. Fortunately, only one of these cell types, olfactory periglomerular neurons, are generated in adulthood (Mullen et al. 1992; Winner et al. 2002). Calcium-binding proteins, neurotransmitter enzymes, and neuro-transmitters are produced by maturing neurons when the cells become electrophysiologically active. These molecules are specific to subclasses of neurons, and together with BrdU birth-dating, they allow quantification of new neurons. The use of retroviral labeling has also provided a means of confirming the ultrastructural phenotype of new neurons by virtue of cell filling, enabling the identification of dendritic processes bearing mature synapses. Similarly, this technique has been used to validate the electrophysiological maturation of new neurons (van Praag et al. 2002; Laplagne et al. 2006).

DETECTION OF MARKER COEXPRESSION

Multiple Labeling Approaches

Detection of proliferating cells using exogenous thymidine analog administration or endogenous cell cycle proteins can provide useful information about quantitative changes in proliferative activity when detected using single-label immunoperoxidase staining. To extract the most information about specific subpopulations that may be represented within the detected population, it is necessary to utilize multiple staining approaches.

Multiple labeling can be accomplished using immunohistochemical staining where the binding of each primary antibody is revealed using a different chromagen. The approach is popular, as the use of immunoperoxidase staining is widespread and protocols can be adapted to include additional chromagens. Imaging can also be performed on a simple bright-field microscope, and thus the material is photostable over time and no additional, expensive fluorescence equipment is required. A number of serious drawbacks with using bright-field microscopy limit its usefulness for studies of multiple labeling. Bright-field microscopy projects focused light through the specimen using diascopic illumination. As a result, cells above and below the focal plane cast shadows that obscure the clear resolution of cells in the focal plane. Given the excessive collapse of tissue section height that occurs in the preparation process, usually in excess of half of the original tissue thickness, cells are packed tightly in the z axis, compounding the problem of resolution and creating difficulties for subsequent demonstrations of three-dimensional colocalization and performing quantitative stereology (for a full discussion, see Peterson 2004). Discrimination of the chromagen colors may not be possible under these circumstances or if the chromagens coexist in the same cell compartment. Tissue is commonly overstained to achieve high contrast, adding to the difficulty in discriminating overlapping labels.

One common approach to circumvent this problem is to use Nissl staining (also known as cresyl violet or thionin staining) to bind cytoplasmic and nuclear RNA, producing a pattern of labeling in all cells. Experience with the morphological distribution of RNA permits identification of the cell type. Although there is a long heritage of using such chromatic labeling to assign cell phenotype, this approach is not favored for demonstrating cell phenotype in studies of cell lineage now that precise markers are available. For instance, BrdU detection in a cell that appears by Nissl morphology to be a mature neuron is quite weak evidence compared to the use of a specific marker of cell phenotype, such as NeuN, calbindin, and tyrosine hydroxylase. The use of Nissl staining is of no value in the case of cells with early lineage commitment that may not have a remarkable morphology. Finally, Nissl staining is of no use in cases where cell density is high, such as in the granule cell layers of the DG and olfactory bulb (OB).

Multiple immunofluorescence labeling has become the approach of choice to enable clear discrimination of labels. This methodology offers

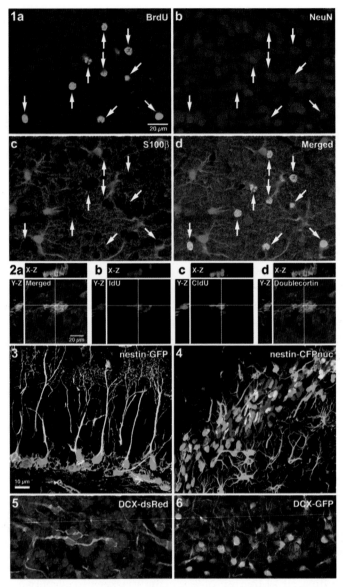

Figure 3. Detection of neurogenic markers. Detection of neurogenesis is commonly achieved by immunohistochemical detection of newly generated cells labeled with the thymidine analog BrdU (*1a*) assessed with their coexpression of mature neuronal markers, such as NeuN (*1b*), or mature glial markers, such as S100β for mature astrocytes (*1c*). Coexpression can be determined from the amount of signal overlap (*1d*). Full determination of colocalization requires discrimination of signal colocalization in three dimensions (*2a*), which can be illustrated by orthogonal projections of a point in space (shown as white crosshairs) with the projected side views in the *X-Z*

an advantage for detecting coexpression due to the ability to separately excite fluorophores and discriminate their emission based on spectral properties. Depending on the properties of the antibodies or dyes available and on suitable instrumentation, investigators can image three to four colors for analysis. In addition to using multiple immunofluorescence to extract additional information about neurogenesis from each section, it is possible to combine this labeling with endogenous expression of fluorescent reporter genes in transgenic mice (Fig. 3).

Determination of Coexpression in Three Dimensions

Fluorescence microscopy, particularly confocal microscopy, also facilitates the collection of expression data in three-dimensional stacks of images to provide evidence for coexpression in the same or distinct cell compartments. The importance of using exacting standards for identifying coexpression is well-illustrated by early reports in the field of adult neurogenesis that mistakenly report its occurrence in the cerebral cortex, when actually two closely apposed cells gave the appearance of the BrdU label in mature cortical neurons (Fig. 4). Largely as a result of these missteps, evidence for colocalization now requires well-documented three-dimensional colocalization (Kornack and Rakic 2001; Bhardwaj et al. 2006).

Assessment and quantification of marker coexpression should utilize three-dimensional data sets for the reasons discussed above. It is also good practice to document the three-dimensional data set when presenting

---

plane and the *Y-Z* plane allowing for interactive observation of signal colocalization. Merged images can then be shown separately for each signal, such as is illustrated here for the dual labeling with thymidine analogs IdU (*2b*) and CldU (*2c*) shown colocalized with doublecortin (*2d*). (Panels *2a–d*: Modified, with permission, from Vega and Peterson 2005.) Transgenic mice with fluorescent reporter genes can also be used to detect adult neurogenesis. Examples include filling of nestin expressing cells with GFP (*3*), colabled with GFAP (*red*) in the dentate gyrus (DG) or nuclear-localized cyan fluorescent protein (CFP) (*green*) in nestin-expressing cells in the rostral migratory stream (*4*) colabeled with BrdU (*red*) and GFAP (*blue*). (Panels *3* and *4* courtesy of Drs. J. Encinas and G. Enikolopov.) Another example of transgenic reporter mice is colabeling of endogenous DCX protein by fluorescent proteins expressed under the human DCX promoter. Note the fibrillary staining of endogenous microtubule-associated DCX colabels extensively with the diffuse cytoplasmic signal of DsRed (*5*) and eGFP (*6*). (Images taken from adult olfactory bulb tissue and provided courtesy of Drs. L. Aigner and S. Couillard-Despres.)

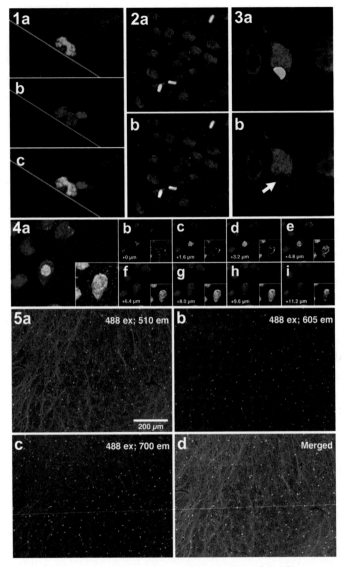

Figure 4. Potential detection artifacts. (*1*) Double labeling of dividing progenitor cells in the subgranular zone of the rat dentate gyrus (DG) with BrdU (*green*) and Ki-67 (*red*). BrdU was injected 2 hours before sacrifice. The staining indicates partial overlap of cells marked for BrdU (S phase) and for Ki-67 (cell cycle). (*2*) False double labeling of BrdU (*green*) and NeuN (*red*). The yellow signal (*2a*) can be attributed to autofluorescence of erythrocytes, since the same signal is detected in the blue channel yielding a white signal in *b*. (*3* and *4*) False double labeling of BrdU (*green*) and NeuN (*red*) can also be attributed to BrdU-positive satellite cells that are in close proximity to mature neurons. (*3a* and *b*) Side view of a satellite cell (BrdU in *green*) that

representative examples (Table 2). The most straightforward way to achieve this without the need for specialized software is to represent successive focal planes of a three-dimensional set as a series of panels in a figure (see Fig. 4). When such a sequence is presented in combination with a through-projection of the entire data set as a single image, the investigator is able to appreciate the overall coexpression of markers in space while having confidence that the spatial distribution of the signal validates the interpretation of coexpression. A popular alternative to this presentation is the generation of orthogonal projections at a defined coordinate within the three-dimensional data set (see Fig. 3). Orthogonal projections are calculated X-Z and Y-Z views (e.g., essentially side views, looking at the image set from "end-on") electronically generated to pass through a particular cell to illustrate colocalization at that spatial reference point. Software to generate orthogonal projections is a feature in the operating software for most confocal microscopes.

Artifacts in Fluorescence Detection

Despite the power of fluorescence detection of neurogenesis, there are two detection artifacts that have the potential to lead to misidentification and misinterpretation of neurogenesis. One of these artifacts is the result of emission from one fluorophore or dye contributing signal to another detection channel as a result of its extended emission spectral profile extending into the emission profile of another signal. This phenomenon

---

is very closely apposed to a NeuN-positive neuron (*red*). Note that the cytoplasm of the neuron appears to be indented by the attached satellite cell soma (arrow in *3b*). If the satellite cell and neuron are oriented along the *z* axis, regular fluorescence microscopy will determine colabeling (see *4a*), whereas *z*-stack analysis of confocal images would determine that the BrdU-positive nucleus (in focus in *e*) does not correspond to the neuronal nucleus in focus (*g*). Autofluorescence is a particularly difficult artifact in aged rodent tissue and primate tissue. Unstained sections mounted and imaged directly show autofluorescence in the olfactory bulb (OB) demonstrated by excitation illumination at 488 nm resulting in nonspecific emission of autofluorescing particles at 510 nm (*green, 5a*), 605 nm (*red, 5b*), and 700 nm (*blue, 5c*). Merging all channels of detection shows that the resulting combination of emission detection does not always produce the additive color of white as signal intensity differs between detection channels. The low-power view shown in panel *5a* would appear similar following staining for BrdU, and the false-positive autofluorescence would have to be carefully discriminated from true BrdU-positive signal to avoid overcounting.

Table 2. Best practices for detecting label coexpression

- Use fluorescence labeling when possible for precise discrimination of signal.
- Use appropriate high-numerical-aperture objective lenses for the best axial resolution (i.e., the narrowest depth of field).
- Use appropriate filters for specific fluorescence signal detection.
- Use entire dynamic range of signal detection to avoid oversaturating the image and obscuring coexpression.
- Collect a three-dimensional image stack to fully identify signal distribution within cell compartments.
- Validate that image does not contain signal bleed-through by collecting each channel with separate excitation.
- Validate that apparent colocalization is not autofluorescence by examining if low-wavelength excitation produces emission at longer wavelengths.

is commonly called bleed-through and can result in a false-positive interpretation of coexpression. Bleed-through is easily controlled for by choosing fluorophores or dyes with narrow emission spectra and always using sequential collection of each signal, rather than relying on simultaneous collection of all signals. Good microscopy practice dictates that the operator should validate the specificity of each signal being detected and checking in other detection channels for the possibility of bleed-through (Table 2). Sequential image collection may take longer, but the absence of artifacts justifies the extra effort.

Autofluorescence, especially in aging and primate tissue, is another serious artifact that can produce errors in interpretation. The natural accumulation of unprocessed cellular debris produces a broad emission spectrum when excited by a wide range of excitation wavelengths, with lower excitation wavelengths producing stronger emissions. The clumped nature of the autofluorescent emission can appear similar to nonsaturating levels of thymidine analog incorporation. As a result, the danger of misinterpreting autofluorescence as a true signal is greatest when the expected signal distribution is also punctate and primarily in the nucleus or perikaryon, such as with BrdU staining (Fig. 4). Various approaches have been proposed to quench endogenous autofluorescence prior to tissue staining; however, we have found none of these to be adequate to remove the detection problems. The best practice for imaging aging or primate tissue or tissue following experimental injury, where autofluorescence may be encountered, is to evaluate the extent of emission seen in all channels following excitation with a low wavelength (Table 2). Even with sequential collection, it is possible to acquire autofluorescence. Images should be examined for precise spatial coexpression of particles with relatively equal intensity and regard these as possible autofluorescence.

QUANTIFICATION OF NEUROGENESIS

To test hypotheses about changes in neurogenesis, it is necessary to generate reliable numbers that are suitable for statistical evaluation. The determination of cell number in tissue sections was long undertaken in a relatively naive fashion where it was assumed that all cells present could be readily observed and discriminated in a single focal plane. Even with thin sections under conditions of high staining contrast, this approach does not account for sectioning or detection artifacts. When changes in cell number are a major readout of the study results, methodological errors have the potential to seriously misrepresent the study outcome. As a result, the neuroscience community in recent years has increasingly moved to using quantitative approaches described as design-based stereology to reduce the contribution of errors and artifacts in the generation of numbers while at the same time systematically sampling through the entire tissue to obtain a reliable estimate of the total population of cells (Peterson 1999, 2004; Schmitz and Hof 2005). The production of estimates of all the labeled cells in a structure is a much more biologically relevant parameter than estimates based on a density (area, volume, or per section) that is subject to alteration between experimental conditions. The need to produce three-dimensional stacks of confocal images to accurately assess colocalization (discussed above) only requires careful standardization in sampling parameters to be ideally suited for using stereology (Table 3). It is important to note here that applying the disector counting principle to confocal image stacks is not, in itself, a proper implementation of stereology without appropriate tissue sampling and should not be described as such in reporting results. The image stacks themselves must be collected in a true systematic random fashion to be properly called stereology.

There are some challenges to implementing stereology in studies of adult neurogenesis, primarily due to the low abundance of newly generated cells in some neurogenic regions (particularly the hippocampal DG) and in regions being assessed for response of neural stem cells to injury or other experimental perturbation. As cells in such regions are relatively infrequent and often found in clusters, there has been considerable discussion about the use of stereology under these circumstances. One concern is the uncomfortably high level of within-sample variance produced

Table 3. Best practices for quantifying neurogenesis

- Use appropriate high-numerical-aperture oil immersion objective lenses for the best axial resolution (i.e., the narrowest depth of field).
- Validate that staining exists throughout the mounted section thickness so that there is no "dead zone" in the middle devoid of staining where no cells will be counted even though they may exist there.
- Use design-based stereological sampling to estimate total cell number independent from differences in region or volume between experimental and control conditions.
- Determine the thickness of mounted (not cut) tissue for use in calculating results.
- Use careful systematic sampling of the tissue when collecting image stacks to ensure compliance with stereological formulae.
- Confocal image stacks are ready-made for stereological counting, but use guard zones and disector probe counting rules to avoid artifacts and methodological biases.

by the distribution of these cells. In fact, this statistic merely reflects the realities of cell distribution.

Historically, within-sample variance has been used as a guide to assess whether sampling frequency has been adequate. This works well when there is an abundance of cells that are relatively homogeneous in distribution. As a result, within-sample variance statistics are of reduced value in judging suitability of sampling in studies of neurogenesis where cells are in low abundance. Investigators may need to oversample sections under these circumstances. A useful guide to the suitability of sampling can be obtained from determining the between-sample variance of the population estimates for all subjects within an experimental group. Some of this variance calculation will be due to inter-animal variation, some to differences in experimental manipulation if the animals are not naive controls, and the remainder will be due to sampling variance.

Another concern about using stereology to quantify neurogenesis has arisen from the amount of work required to sample relatively rare cell populations. Modifications of existing stereology methods are being devised to further automate this sampling and increase throughput to reduce the amount of work required to perform the sampling while retaining the three-dimensional sampling (using the optical disector probe) needed to avoid methodological errors in counting cells. These higher-throughput methods are currently being developed (D.A. Peterson, pers. comm.).

SUMMARY

Studies of adult neurogenesis have greatly expanded in the last decade, largely as a result of improved tools for detecting and quantifying neurogenesis. Occasions where original interpretations about neurogenesis have had to be reconsidered in light of subsequent reports indicate how careful attention to the strengths and limitations of detection and quantification methods are critical to substantiating study outcomes. Properly used and evaluated, the tools presently available to the research community provide ample resources to advance our understanding of the regulation of adult neurogenesis.

ACKNOWLEDGMENTS

We thank Drs. Juan Encinas and Grigori Enikolopov for permission to use previously unpublished images of their nestin-GFP and nestin-CFPnuc mice and Drs. Ludwig Aigner and Sebastien Couillard-Despres for permission to use previously unpublished images of doublecortin-GFP and doublecortin-dsRed mice.

REFERENCES

Ackman J.B., Siddiqi F., Walikonis R.S., and LoTurco J.J. 2006. Fusion of microglia with pyramidal neurons after retroviral infection. *J. Neurosci.* **26:** 11413–11422.

Alvarez-Buylla A., Seri B., and Doetsch F. 2002. Identification of neural stem cells in the adult vertebrate brain. *Brain Res. Bull.* **57:** 751–758.

Baer K., Eriksson P.S., Faull R.L., Rees M.I., and Curtis M.A. 2006. Sox-2 is expressed by glial and progenitor cells and Pax-6 is expressed by neuroblasts in the human subventricular zone. *Exp. Neurol.* **204:** 828–831.

Bauer S. and Patterson P.H. 2005. The cell cycle-apoptosis connection revisited in the adult brain. *J. Cell Biol.* **171:** 641–650.

Bhardwaj R.D., Curtis M.A., Spalding K.L., Buchholz B.A., Fink D., Bjork-Eriksson T., Nordborg C., Gage F.H., Druid H., Eriksson P.S., and Frisen J. 2006. Neocortical neurogenesis in humans is restricted to development. *Proc. Natl. Acad. Sci.* **103:** 12564–12568.

Bonfanti L. 2006. PSA-NCAM in mammalian structural plasticity and neurogenesis. *Prog. Neurobiol.* **80:** 129–164.

Brown J.P., Couillard-Despres S., Cooper-Kuhn C.M., Winkler J., Aigner L., and Kuhn H.G. 2003. Transient expression of doublecortin during adult neurogenesis. *J. Comp. Neurol.* **467:** 1–10.

Burns K.A. and Kuan C.Y. 2005. Low doses of bromo- and iododeoxyuridine produce near-saturation labeling of adult proliferative populations in the dentate gyrus. *Eur. J. Neurosci.* **21:** 803–807.

Busser J., Geldmacher D.S., and Herrup K. 1998. Ectopic cell cycle proteins predict the sites of neuronal cell death in Alzheimer's disease brain. *J. Neurosci.* **18:** 2801–2807.

Cooper-Kuhn C.M. and Kuhn H.G. 2002. Is it all DNA repair? Methodological considerations for detecting neurogenesis in the adult brain. *Brain Res. Dev. Brain Res.* **134:** 13–21.

Couillard-Despres S., Winner B., Schaubeck S., Aigner R., Vroemen M., Weidner N., Bogdahn U., Winkler J., Kuhn H.G., and Aigner L. 2005. Doublecortin expression levels in adult brain reflect neurogenesis. *Eur. J. Neurosci.* **21:** 1–14.

Eisch A.J. and Mandyam C.D. 2004. Beyond BrdU: Basic and clinical implications for analysis of endogenous cell cycle proteins. In *Focus on stem cell research* (ed. E.V. Greer), pp. 111–142. Nova Science Publishers, Hauppauge, New York.

———. 2007. Adult neurogenesis: Can analysis of cell cycle proteins move us "Beyond BrdU"? *Curr. Pharm. Biotechnol.* **8:** 147–165.

Encinas J.M., Vaahtokari A., and Enikolopov G. 2006. Fluoxetine targets early progenitor cells in the adult brain. *Proc. Natl. Acad. Sci.* **103:** 8233–8238.

Grassi Zucconi G. and Giuditta A. 2002. Is it only neurogenesis? *Rev. Neurosci.* **13:** 375–382.

Kass L., Varayoud J., Ortega H., Munoz de Toro M., and Luque E.H. 2000. Detection of bromodeoxyuridine in formalin-fixed tissue. DNA denaturation following microwave or enzymatic digestion pretreatment is required. *Eur. J. Histochem.* **44:** 185–191.

Kempermann G., Jessberger S., Steiner B., and Kronenberg G. 2004. Milestones of neuronal development in the adult hippocampus. *Trends Neurosci.* **27:** 447–452.

Komitova M. and Eriksson P.S. 2004. Sox-2 is expressed by neural progenitors and astroglia in the adult rat brain. *Neurosci. Lett.* **369:** 24–27.

Kornack D.R. and Rakic P. 2001. Cell proliferation without neurogenesis in adult primate neocortex. *Science* **294:** 2127–2130.

Kuan C.Y., Schloemer A.J., Lu A., Burns K.A., Weng W.L., Williams M.T., Strauss K.I., Vorhees C.V., Flavell R.A., Davis R.J., Sharp F.R., and Rakic P. 2004. Hypoxia-ischemia induces DNA synthesis without cell proliferation in dying neurons in adult rodent brain. *J. Neurosci.* **24:** 10763–10772.

Laplagne D.A., Esposito M.S., Piatti V.C., Morgenstern N.A., Zhao C., van Praag H., Gage F.H., and Schinder A.F. 2006. Functional convergence of neurons generated in the developing and adult hippocampus. *PLoS Biol.* **4:** e409.

Malatesta P., Hartfuss E., and Gotz M. 2000. Isolation of radial glial cells by fluorescent-activated cell sorting reveals a neuronal lineage. *Development* **127:** 5253–5263.

Malatesta P., Hack M.A., Hartfuss E., Kettenmann H., Klinkert W., Kirchhoff F., and Gotz M. 2003. Neuronal or glial progeny: Regional differences in radial glia fate. *Neuron* **37:** 751–764.

Mullen R.J., Buck C.R., and Smith A.M. 1992. NeuN, a neuronal specific nuclear protein in vertebrates. *Development* **116:** 201–211.

Nacher J., Crespo C., and McEwen B.S. 2001. Doublecortin expression in the adult rat telencephalon. *Eur. J. Neurosci.* **14:** 629–644.

Nacher J., Rosell D.R., and McEwen B.S. 2000. Widespread expression of rat collapsin response-mediated protein 4 in the telencephalon and other areas of the adult rat central nervous system. *J. Comp. Neurol.* **424:** 628–639.

Nacher J., Varea E., Blasco-Ibanez J.M., Castillo-Gomez E., Crespo C., Martinez-Guijarro F.J., and McEwen B.S. 2005. Expression of the transcription factor Pax 6 in the adult rat dentate gyrus. *J. Neurosci. Res.* **81:** 753–761.

Nguyen L., Rigo J.M., Malgrange B., Moonen G., and Belachew S. 2003. Untangling the functional potential of PSA-NCAM-expressing cells in CNS development and brain repair strategies. *Curr. Med. Chem.* **10:** 2185–2196.

Peterson D.A. 1999. Quantitative histology using confocal microscopy: Implementation of unbiased stereology procedures. *Methods* **18:** 493–507.

————. 2004. The use of fluorescent probes in cell counting procedures. In *Quantitative methods in neuroscience,* (ed. S. Evans et al.), pp. 85–114. Oxford University Press. United Kingdom.

Rakic P. 2002. Neurogenesis in adult primate neocortex: An evaluation of the evidence. *Nat. Rev. Neurosci.* **3:** 65–71.

Rao M.S. and Shetty A.K. 2004. Efficacy of doublecortin as a marker to analyse the absolute number and dendritic growth of newly generated neurons in the adult dentate gyrus. *Eur. J. Neurosci.* **19:** 234–246.

Schmitz C. and Hof P.R. 2005. Design-based stereology in neuroscience. *Neuroscience* **130:** 813–831.

Seki T. 2002. Expression patterns of immature neuronal markers PSA-NCAM, CRMP-4 and NeuroD in the hippocampus of young adult and aged rodents. *J. Neurosci. Res.* **70:** 327–334.

Stoykova A. and Gruss P. 1994. Roles of Pax-genes in developing and adult brain as suggested by expression patterns. *J. Neurosci.* **14:** 1395–1412.

Thomas R.M., Hotsenpiller G., and Peterson D.A. 2007. Acute psychosocial stress reduces cell survival in adult hippocampal neurogenesis but not initial proliferation. *J. Neurosci.* **27:** 2734–2743.

van Praag H., Schinder A.F., Christie B.R., Toni N., Palmer T.D., and Gage F.H. 2002. Functional neurogenesis in the adult hippocampus. *Nature* **415:** 1030–1034.

Vega C.J. and Peterson D.A. 2005. Stem cell proliferative history in tissue revealed by temporal halogenated thymidine analog discrimination. *Nat. Methods* **2:** 167–169.

Winner B., Cooper-Kuhn C.M., Aigner R., Winkler J., and Kuhn H.G. 2002. Long-term survival and cell death of newly generated neurons in the adult rat olfactory bulb. *Eur. J. Neurosci.* **16:** 1681–1689.

Yang Y., Geldmacher D.S., and Herrup K. 2001. DNA replication precedes neuronal cell death in Alzheimer's disease. *J. Neurosci.* **21:** 2661–2668.

4

Evolving Methods for the Labeling and Mutation of Postnatal Neuronal Precursor Cells: A Critical Review

Joshua J. Breunig and Pasko Rakic
Department of Neurobiology
Kavli Institute for Neuroscience
Yale University School of Medicine
New Haven, Connecticut 06520

Jeffrey D. Macklis
Departments of Neurosurgery and Neurology
Harvard Medical School, Massachusetts General Hospital
Department of Developmental and Regenerative Biology
Harvard Stem Cell Institute, Harvard University
Boston, Massachusetts 02114

AS RESEARCH ON POSTNATAL NEURONAL PROGENITOR, precursor, and stem cells progresses, methods of increasing sensitivity and complexity will be brought to bear in revealing how these cell types are maintained in the adult brain and how the brain adds neurons to mature circuits. Here, we review historical and current methods, such as bromodeoxyuridine (BrdU) labeling, and discuss several emerging genetic techniques, including viral vectors, small interfering RNAs (siRNAs), and inducible transgenic/knockout mice, that will be useful for the labeling and/or mutation of adult neuronal precursor cells (NPCs). As the complexity of these methods increases, so does the potential for misinterpretation of the results. The realization must be made that all methods have inherent disadvantages and confounds, preventing conclusive and definitive interpretations if used without cross-validation. We hope to give insight into how pitfalls might be avoided and provide a primer on additional methods that might be used in the pursuit of definitive results.

Adult Neurogenesis ©2008 Cold Spring Harbor Laboratory Press 978-087969-784-6

In the past decade, a newfound appreciation has developed for the regions displaying neurogenesis in the adult mammal (Gage 2000; Lledo et al. 2006). In two regions, the dentate gyrus (DG) of the hippocampus and the olfactory bulb (OB), neurons are continually added after birth (Lois and Alvarez-Buylla 1994; Kuhn et al. 1996). In the hippocampus, neurons are born in the subgranular zone (SGZ) from Gfap$^+$ precursor cells and migrate a short distance into the granule cell layer (GCL), where they integrate, sending an axon to CA3 and receiving input at their apical dendrite (Seri et al. 2001; van Praag et al. 2002). Cells destined for the OB are born in the subependymal zone of the lateral ventricle from Gfap$^+$ precursor cells (Doetsch et al. 1999). Newborn neurons migrate tangentially through the rostral migratory stream (RMS) and then radially in the OB, where they synaptically integrate (Lois and Alvarez-Buylla 1994; Belluzzi et al. 2003; Magavi et al. 2005). Furthermore, multipotent NPCs have been derived from almost every region examined (Gage 2000). (As the identity of "true" CNS stem cells is not yet known—especially in the postnatal brain—we use "precursor cells" as a generic term encompassing the entire lineage of neuronal stem and progenitor cells [Sohur et al. 2006].) The recent explosion is due in large part to the development of advanced techniques that allow for the precise labeling and manipulation of postnatal NPCs.

During the act of scientific experimentation, observation in and of itself is able to alter the phenomenon being observed. Examples of this include the observation of subatomic particles, where the observing light or radiation contains adequate energy to disrupt the natural path of matter or the measurement of electric circuits, where the very presence of electronics or voltmeters necessarily affects the current and voltage being measured. Thermometers absorb a finite amount of thermal energy to determine temperature. In these cases, this small observationally induced deviation from the natural state is referred to as the observer effect (often incorrectly confused with the Heisenberg uncertainty principle). In physics, simple calculations can often be used to minimize this phenomenon. However, in the study of postnatal precursor cells, the observer effect has the potential to become a considerable force. Every current technology for labeling and mutating NPCs carries with it the ability to fundamentally modify the natural state of the cell—frequently in unintended or unexpected ways. This includes methods ranging from thymidine analog labeling to emerging tamoxifen- and tetracycline-inducible genetic systems. If care is not taken, this could become a significant hindrance in the ability to interpret results from animal models of NPC behavior and apply them to the clinic. Here, we discuss ways to control for the observer effect and minimize its ability to alter experimental outcomes in the labeling and mutation of precursor cells. In addition, we highlight new techniques that might be used to allow for increased accuracy and sensitivity in this field.

CURRENT AND HISTORICAL METHODS FOR NEURONAL PRECURSOR IDENTIFICATION

Initially, precursor cells were examined ex post facto in fixed tissue. The development of the Golgi staining technique allowed for the examination of discrete populations of cells in brain slices. DNA stains allowed for the observation of mitotic figures. Cajal's (1913–1914) seminal work included the observation of tissue repair postlesion. However, these techniques, although powerful at the time—especially in Cajal's capable hands—were limited in that they provided only a snapshot and precluded easy determination of proliferative dynamics and an estimation of the total size of the postnatal population of dividing cells.

Then, Richard Sidman et al. (1959) began the practice of using tritiated thymidine ($[^3H]$thymidine) to study cell genesis in the developing nervous system, allowing for the observation of the origin, migration, and fate of newly born cells. $[^3H]$thymidine is incorporated into the DNA of cells during DNA synthesis, allowing for the relatively specific labeling of cells in S phase of the cell cycle. Incorporation of $[^3H]$thymidine is stoichiometric, meaning silver grains correlate with the percentage of DNA labeled. Thus, a cutoff was typically made by which only cells that had at least 50% of the maximal grain count were considered newly generated (Rakic 1973). It should be noted that this compound is potentially incorporated during DNA repair and during abortive mitosis characteristic of dying neurons. This later allowed for the initial observation of precursor cells in the postnatal brain in regions including the subependymal zone of the lateral ventricles and the DG of the hippocampus in rodents (Altman and Das 1965a,b) and primates (Nowakowski and Rakic 1981; Rakic and Nowakowski 1981). The radioactivity of this substance and the time-consuming nature of autoradiography, coupled with the inability to sample more than the upper few micrometers of a tissue section, were inherent limitations to this technique. In addition, $[^3H]$thymidine was found to up-regulate DNA-repair pathways (Sands et al. 1972). Nevertheless, this technique was invaluable for the study of the time of origin of neurons in a variety of species, including the rodent (Sidman et al. 1959; Angevine 1965) and nonhuman primate brain (Rakic 1973, 1985). In-depth postmortem surveys of these monkeys were made in many cases and no gross abnormalities were noticed in the more than 100 animals used (P. Rakic, pers. observation).

Decades later, the thymidine analog BrdU was used as an alternate method for labeling cells engaged in DNA synthesis (Miller and

Nowakowski 1988). Immunohistochemistry using a monoclonal antibody directed against this molecule gives this method several inherent advantages over [^{3}H]thymidine autoradiagraphy, including the ability to amplify the signal and detect labeled cells throughout thick tissue sections. In addition, fluorescently labeled secondary antibodies allow colocalization of BrdU with up to three cell-type markers (Magavi et al. 2000; Kornack and Rakic 2001; Magavi and Macklis 2002a,b). Precise colocalization remains a challenge in regions dense with cell bodies, especially if the marker does not closely overlap the nucleus, as is the case with the astrocytic marker glial fibrillary acidic protein (GFAP). This technique was pivotal in confirming the existence of neurogenesis in the human hippocampus and the lack of neurogenesis in the human neocortex (Eriksson et al. 1998). Recently, a C14-based method has been adapted for examining neurogenesis in the human brain as BrdU injections are understandably limited for ethical reasons (Spalding et al. 2005; Bhardwaj et al. 2006). However, compared with the stoichiometric nature of [^{3}H]thymidine autoradiography, immunohistochemical methods frequently use amplification, removing this ability (Rakic 2002a,b,c). In addition, the molecular structure of this analog is inherently different from the natural structure of thymidine (Stetson et al. 1988), which is believed to cause steric hindrance, thus affecting the natural conformation of the DNA molecule (Fig. 1A). Indeed, it is well known for its ability to sensitize cancer cells to radiation (Djordjevic and Szybalski 1960; Szybalski 1974). Therefore, at high enough doses, the abnormalities it causes in DNA transcription and protein translation might lead to mutation and cell toxicity, compromising the overall health of the organism. Furthermore, it appears that BrdU at currently recommended concentrations is selectively toxic to neurons in vitro (Caldwell et al. 2005). It has also been observed that BrdU can induce neuronal differentiation in vitro (Qu et al. 2004). In vivo BrdU induces abnormal proliferation (Goldsworthy 1992). BrdU is a mutagen, teratogen, and carcinogen and its toxicity will likely correlate with the percentage of cells incorporating this chemical and the percentage of thymidine nucleotides that it replaces within these cells (Bannigan and Langman 1979; Bannigan 1985; Bannigan et al. 1990; Nagao et al. 1998; Kolb et al. 1999; Kuwagata et al. 2004). Therefore, increasing dosages and frequency of injection will only exacerbate the cellular toxicity and adverse side effects, likely poisoning the very population under examination (Goldsworthy et al. 1992, 1993). It has been recommended that doses of 300 mg/kg be used (Cameron and McKay 2001), but other investigators have found that much lower doses (50 mg/kg) are adequate to reach near-saturation labeling (Burns and Kuan 2005). Considering the possible side effects on cell differentiation and toxicity, it seems prudent to err on the side of caution and limit injections to

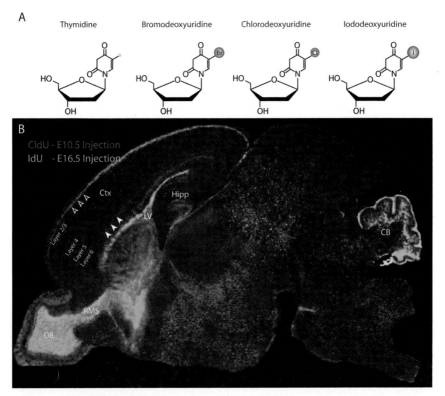

Figure 1. Thymidine analogs. (*A*) Structure of thymidine and the thymidine analogs bromodeoxyuridine (BrdU), chlorodeoxyuridine (CldU), and iododeoxyuridine (IdU). Halogens are not drawn to scale in comparison to the nucleotide chemical structure, but an attempt was made to portray the relative sizes of the halogens when compared with the methyl group for which they substitute. This is proposed to cause steric hindrance, affecting the natural structure of DNA and subsequent synthesis/repair or transcription and translation (see text). (*B*) Sagittal section of a P4 mouse brain injected at E10.5 with CldU and E16.5 with IdU. Note the lack of cross-reactivity between the two. For example, CldU labels a band of neurons in the deep layers of the cortex and IdU labels layer 2. Similarly, the Purkinje cell layer is labeled by the E10.5 injection and EGL/IGL/WM cells are labeled by the E16.5 injection. (OB) Olfactory bulb; (Ctx) cortex; (Hipp) hippocampus; (CB) cerebellum; (RMS) rostral migratory stream.

50–100 mg/kg per dose. Numerous smaller-dosage injections should be limited as well because they can lead to large cumulative doses that will have similar adverse effects. This was shown in the cerebellum by Mugnaini and colleagues (Sekerkova et al. 2004). In attempting to label cerebellar cell types, they noted significant defects caused by BrdU. This was only made obvious by the well-differentiated and laminated structure of this region. Purkinje cell bodies occupy a layer only one cell-body thick and have a well-characterized dendritic tree, allowing for easy determination of migration and differentiation abnormalities. The neurogenic regions and other regions of the forebrain do not have this characteristic because of the less precise placement and integration of newborn cells in the DG and OB. Thus, similar observations could be missed.

Recently, several groups have reported that chlorodeoxyuridine (CldU) and iododeoxyuridine (IdU) (both thymidine analogs with a structure similar to that of BrdU [Fig. 1A]) can be detected individually by different monoclonal antibodies to each in tissue labeled with both (Burns and Kuan 2005; Vega and Peterson 2005). This allows the ability to track two cohorts of temporally segregated (delineated by the separate injection times of each) S-phase cells over the course of time (Fig. 1B). The recommended precautions for BrdU should apply to these compounds as well.

An underappreciated phenomenon is the abortive cell cycle that some neurons enter following insult (Burns et al. 2007), during the disease process (Yang et al. 2001; Hoglinger et al. 2007), or before death (Kruman et al. 2004; Kuan et al. 2004). In certain cases, cycling neurons appear not to die (Burns et al. 2007) and can persist for months or even years (for review, see Herrup and Yang 2007; Yang et al. 2001; Hoglinger et al. 2007). As this cell cycle includes an S phase, in many cases, these neurons can incorporate BrdU or other thymidine analogs (Klein et al. 2002; Kuan et al. 2004; Burns et al. 2007; Hoglinger et al. 2007). In addition, they might express markers of immature neurons, such as βIII-tubulin (Kuan et al. 2004). A simple practice for excluding this possibility is to study the nature of BrdU incorporation from at least two times of investigation (Magavi et al. 2000; Sohur et al. 2006). For example, a short, 2–24-hour pulse-chase experiment would label only immature precursor cells or migrating neurons during the normal process of neurogenesis. Significant numbers of labeled nonmigratory neurons should only be labeled following relatively long survival times (>2 weeks). The newly developed CldU/IdU techniques will allow an added degree of ease and resolution in this process, as this method allows for the tracking of two temporally segregated populations of newborn cells in one animal (Fig. 1B). Additionally, preexisting neurons can be prelabeled by a variety of methods to avoid the risk of misinterpreting older cells as newborn (Magavi et al. 2000; Shin et al. 2000; Chen et al. 2004; Hoglinger et al. 2007). The stereotypical progression of neuronal differentiation from precursor cell to neuron allows for confirmation of neurogenesis when used in conjunction with BrdU and methods such as electrophysiology (van Praag et al. 2002) or retrograde labeling (Magavi et al. 2000; Chen et al. 2004; Hoglinger et al. 2007). However, many times, a single marker such as doublecortin (DCX), βIII-tubulin, or polysialic-acid-neural cell adhesion molecule (PSA-NCAM) is used in isolation as evidence for neurogenesis. If at all possible, these markers should be used together in combination with BrdU or other methods to confirm the specificity of these markers in the search for newborn cells (Magavi et al. 2000; Nacher et al. 2001). Furthermore, as these markers allow for the observation of cell morphology, a rigorous criterion should be used to identify the cells as migratory and not dying (Kuan et al. 2004), or up-regulating these markers due to some sort of plastic response (Nacher et al. 2001, 2002). Migratory cells should possess a small nucleus (in comparison with mature projection neurons) and a long leading process. In any case, expression of these markers in newborn cells should not be necessary for proof of neurogenesis, as it is known that in many cases, these cells do not survive or mature into functionally integrated neurons (Arvidsson et al. 2002; Parent et al. 2002).

The level of caution we suggest might seem to be extreme. However, this field has a rather large amount of unconfirmed or disputed findings owing to what seems to be primarily methodological problems. There have been reports of constitutive neurogenesis in the primate and rodent neocortex (Kaplan 1981; Gould et al. 1999, 2001), the amygdala (Bernier et al. 2002), area CA1 of the rodent (Rietze et al. 2000), the dorsal vagal complex of the brainstem (Bauer et al. 2005), the spinal cord (Yamamoto et al. 2001), and the substantia nigra (Zhao et al. 2003). To a large degree, these reports remain unconfirmed or have been directly challenged with negative findings, some showing the possible cause of the initial artifactual findings (Horner et al. 2000; Magavi et al. 2000; Ekdahl et al. 2001; Kornack and Rakic 2001; Lie et al. 2002; Koketsu et al. 2003; Frielingsdorf et al. 2004; Ackman et al. 2006; Hoglinger et al. 2007). A rigorous set of criteria in the interpretation and peer review of findings might prevent such disputes.

GENETIC METHODS FOR NPC LABELING AND MUTATION

Electroporation

One technique that has revolutionized the study of rodent embryonic neurogenesis in recent years is in utero electroporation (Fukuchi-Shimogori and Grove 2001; Tabata and Nakajima 2001; Molyneaux et al. 2005; M.R. Rasin et al. 2007). Plasmids containing a transgene (gain-of-function) or siRNA sequence (loss-of-function) and reporter are typically injected into the lateral ventricle of the embryo (Fig. 2A–C). Electrodes are then positioned on both sides of the embryo's head. An electrical field is created to electroporate the DNA through the membranes of the dividing ventricular zone (VZ) precursor cells. The promoter sequence in the plasmid is then activated, expressing the transgene/siRNA and the reporter gene, allowing for easy observation of cell-autonomous effects of genetic manipulation. Gross structures, such as the developing cortex, hippocampal primordium, or ganglionic eminences, can be effectively targeted. Although this method is ideal for studying early development, it has also been used to observe the results of transgenesis postnatally (Rasin et al. 2007). Embryos from separate mothers were electroporated with plasmids expressing one of the four isoforms of Numb. When compared with a control plasmid at P10, it was observed that radial glia were maintained in a cell-autonomous manner in the Numb electroporated animals (Fig. 2D–G). Retroviral transduction would have required viral production and packaging of four separate viruses, greatly increasing the time, workload, and safety precautions needed. Alternatively, deriving four separate transgenic mouse lines would increase the time and workload needed to observe this effect. However, the postnatal brain is

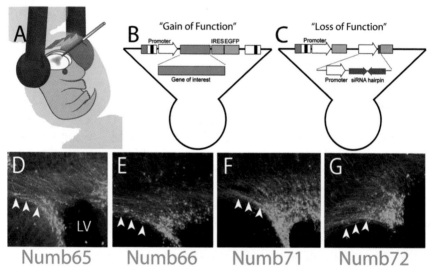

Figure 2. In utero electroporation. (*A*) During the procedure, the embryo is exposed, a solution containing the plasmid DNA is injected into the lateral ventricles, and a current is passed through the head, moving the negatively charged DNA with it and into the ventricular wall, permeating the cell membranes. (*B*) Typical plasmid design for gain-of-function experiments. The gene of interest is expressed by a strong promoter (EF1α, CAG, or CMV). An IRES sequence (see text) allows for expression of a reporter gene (EGFP depicted but may be substituted for red fluorescent protein (RFP), LacZ, or alkaline phosphatase). (*C*) Plasmid model to induce loss of function by siRNA. Alternatively, for a loss-of-function experiment, a dominant-negative protein can be expressed in the place of the gene of interest in *B*. (*D–G*) The four isoforms of Numb were electroporated into the cortical VZ at E16 and animals were sacrificed at P10. Note the maintenance of radial glia around the ventricles, which typically disappear perinatally (*white arrows*).

much less amenable to this method because of the damage caused by the voltage and thus, it loses its efficacy as the animal ages. This technique can be used in combination with inducible Cre technologies (discussed below). One could electroporate a CreER-expressing plasmid into floxed animals or electroporate a floxed transgene into CreER animals to facilitate the study of postnatal precursor cells derived from embryonic VZ precursors or other regions (Matsuda and Cepko 2007).

RNAi

RNA interference (RNAi) is a biological pathway that mediates posttranscriptional gene silencing. In short, double-stranded RNA molecules, known as siRNAs, mediate sequence-specific identification and degradation of targeted messenger RNA (mRNA) (Fire et al. 1998). This allows for knockdown of protein levels in targeted cells without the need for complex and time-consuming genetic engineering. siRNA technology has revolutionized biomedical research, and the seminal researchers in the field were recently awarded the 2006 Nobel Prize in Physiology or Medicine (Abbott 2006). This method, in combination with viral vectors, has led to several powerful insights into the regulation of integration of new neurons in the DG (Ge et al. 2006). Nevertheless, there is the demonstrated potential for off-target effects. One study showed that a short hairpin RNA (shRNA) against luciferase was sufficient to induce morphological, membrane property, and synaptic strength changes in neurons (Alvarez et al. 2006). As the rats used do not express luciferase, off-target effects are the only interpretation. So-called "scrambled" controls have little use in such situations. Rescue experiments in which cDNAs insensitive to the RNAi sequence are introduced into the RNAi-expressing cells are better alternatives. Cross-validation using other loss-of-function methods is an alternative solution (Alvarez et al. 2006).

Viral Vectors

A host of viral technologies have come into use in the study of postnatal neurogenesis because of their ability to genetically label/alter cells (Benraiss et al. 2001; van Praag et al. 2002; Consiglio et al. 2004; Magavi et al. 2005; Ackman et al. 2006). These include selected subtypes of the adeno-, lenti- and retrovirus families (Table 1). Each has advantages and disadvantages (Thomas et al. 2003). Adenovirus technology allows transduction of a range of cell types but often requires the use of shuttle vectors during the cloning process. In addition, the viral genome does not integrate into the genome and thus infection is transient. Lentivirus sim-

Table 1. Viral vectors for transducing neural stem cells

Vector	Genetic material	Integration of vector genome	Transduced cell types	Maximum insert size
Nonenveloped				
Adenovirus	dsDNA	none, episomal	broad range	~8 kb (helper virus permits larger sizes)
Enveloped				
Lentivirus	RNA	integrated, host genome	broad range	~8 kb
Retrovirus	RNA	integrated, host genome	dividing cells[a]	~7.5 kb

[a] A recent report by Loturco and colleagues (see text for discussion) indicates that retroviral infection can lead to spurious labeling of postmitotic cells due to fusion.

ilarly infects a range of cells but does not require shuttle vectors to clone in transgenes. Lentiviruses are reverse-transcribed into the genome, leading to permanent insertion into the host DNA (Thomas et al. 2003). Because of their ability to infect postmitotic cells, adeno- and lentiviruses are best used to overexpress secreted proteins or knockdown proteins in all cell types using siRNA or by expressing a dominant-negative protein (Thomas et al. 2003). Retroviruses specifically transduce dividing cells and are thus ideal for the study of proliferating cells (Lewis and Emerman 1994). Cells infected by any of these virus families can be labeled with a reporter gene such as enhanced green fluorescent protein (EGFP), β-galactosidase, or alkaline phosphatase. Transgenes can be overexpressed as well in combination with such a reporter gene by using an internal ribosome entry site (IRES) (a nucleotide sequence allowing for translation in the middle of mRNA, thus allowing for the expression of two distinct proteins from a single mRNA molecule) or second promoter. Alternatively, an siRNA sequence can be used to cell autonomously knock down a particular protein (Ge et al. 2006; Tashiro et al. 2006). The chief shortcoming of viral technologies is limited infectivity because of the minimal spread of virus in dense tissue, necessitating stereotaxic surgery to deliver the virus to the precise anatomical region. It should be noted that the random integration of lenti- and retroviruses into the genome can have unintended and adverse side effects. If the virus is inserted into the middle of a gene, it can alter or completely prevent the translation of the protein for which it codes. It has been observed that active loci are preferentially targeted by some viruses, increasing the potential for mutation of frequently transcribed genes (Schroder et al. 2002). Furthermore, as discussed in detail below, a recent report (Ackman et al. 2006) calls into question the reliability of retroviral transduction because of the observation that post-mitotic cells can be spuriously labeled by fused microglia.

Germ-line Transgenesis and Knockout Mice

Traditional knockout mouse technology allows for both the mutation of a gene of interest and the potential for labeling of all cells endogenously expressing this gene with a knocked-in marker gene. This powerful tool has led to many advancements in the understanding of embryonic neurogenesis and, in some cases, adult neurogenesis. However, this strategy is not ideal in many cases because of the fact that many genes that disrupt adult neuro-genesis are also critical for CNS development. Thus, the CNS is altered before birth and thus is not always well-suited for determining the normal function of the gene. Furthermore, the knockout is global and all cells are affected throughout development, so the organogenesis of other tissues can be perturbed. Often, heterozygous knockouts are used for studying adult neurogenesis, but gene dosage issues still apply and mutants can be signif-icantly altered from wild type because of developmental abnormalities.

Recombinase-based Systems

Cre/loxP and Flp recombinase systems allow for conditional mutagenesis, fate-mapping, and transgenesis (Branda and Dymecki 2004). Typically for studying NPCs, Cre recombinase is driven by a promoter active pre-dominantly in this population such as Nestin or Gfap. Although the tem-poral and spatial activation of Cre is regulated by this promoter, thus avoiding recombination before the onset of CNS development and dis-ruption of tissues outside of the nervous system, embryonic neurogenesis is often disrupted as is the case with traditional knockout technologies. Thus, homozygotes and even heterozygotes are often significantly differ-ent from wild-type littermates at birth. To date, no promoters have been found that show activity strictly in NPCs.

Inducible Genetic Technologies

Recently, inducible Cre recombinase systems—and the less often used Flp recombinase (Hunter et al. 2005; Matsuda and Cepko 2007)—for study-ing neurogenesis have been developed. In this system, an exogenous li-gand is introduced to initiate Cre recombination (Feil et al. 1996). The NPC lineage and active promoters are well-characterized. As with traditional Cre technologies, a promoter active in NPCs (e.g., the human Gfap promoter, which is active in embryonic and postnatal precursor cells, as well as mature astrocytes; Fig. 3A) is used to drive Cre-estrogen receptor (CreERT2) fusion protein expression in this population. However, Cre recombination is temporally controlled by ligand administration while still being spatially restricted to the population with the chosen promoter activity (Fig. 3A′). This allows a high degree of selectivity in targeting NPCs (Ganat et al. 2006). Cells can be inducibly mutated, a transgene can be expressed, or a reporter can be used to heritable fate map the cells (Fig. 3B–D). For example, huGfap promoter-driven CreERT2 was used to drive Cre, heritably labeling astrocytes and the putative NPCs in the subependymal zone and SGZ (Fig. 3E–F″,I). Alternate Nestin lines (Burns et al. 2007) have Cre expression limited to precursor and those cells expressing Nestin in that immediate lineage (ependymal cells and transit-amplifying cells). In both cases, all cells derived from these cells are heritably mutated or labeled with a reporter gene after recombination (Ganat et al. 2006; Burns et al. 2007). Two general types of CreER proteins are used, CreERT1 and CreERT2, differing primarily in their sensitivities to ligand (Indra et al. 1999; Leone et al. 2003). Quite a few mouse lines using this type of technology have been produced (Table 2).

There are inherent disadvantages to Cre technology. First, Cre toxicity is very apparent in proliferating cell populations such as NPCs. This can lead to increased cell death and other secondary side effects such as hydrocephalus caused by cell death of NPCs resulting from chromosomal abnormalities (Forni et al. 2006). Thus, tamoxifen-injected $Cre^+/wt/wt$ and $Cre^-/fl/fl$ or Cre^-/Tg controls must be used as well as noninduced $Cre^+/fl/fl$ or Cre^+/Tg mice. Second, it is challenging with current technologies to reliably label cells that are mutated. For example, it has been noted by Joyner that a high degree of recombination of a reporter allele was noted with little or no recombination of a floxed β-catenin allele (Joyner and Zervas 2006). Recombination of any paired *loxP* sites will be dependent on a host of factors, including Cre levels and the *loxP* integration site in the genome (Vooijs et al. 2001). Furthermore, it has become apparent to us as well as other investigators that because of this, reporter strains show varying degrees of efficiency. This is complicated by the fact that the Rosa26 promoter—frequently used for transgenesis—does not appear to be highly active in some cases in the astrocyte-like NPCs in the postnatal brain (F. Vaccarino, pers. comm.) or in neurons derived from them (Weber et al. 2001; Casanova et al. 2002; Shimshek et al. 2002), leading to a lack of reporter or transgene expression following tamoxifen administration. Thus, each reporter line must be recharacterized when crossed with a new Cre

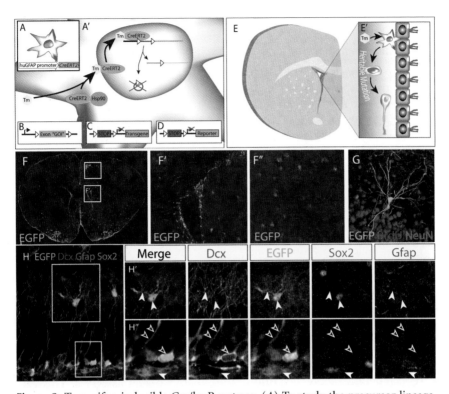

Figure 3. Tamoxifen-inducible Cre/loxP systems. (*A*) To study the precursor lineage, the human GFAP promoter was used to express a Cre-estrogen receptor T2 fusion protein in the precursor cell population. (*A'*) On tamoxifen administration, the fusion protein translocates the nucleus where it recombines paired *loxP* sites. (*B*) For generating a null allele, typically *loxP* sites flank the first exon or first several exons of the gene of interest. (*C*) For transgenic studies, a "stop" cassette is floxed. Recombination removes this transcriptional stop, allowing readthrough of the transgene or reporter (*D*). (*E*) Schematic of the postnatal SEZ, where on tamoxifen administration (*E'*), an astrocytic precursor cell gives rise to a transit-amplifying cell that divides to generate a new neuron. All are heritably mutated because of the recombination in the Gfap⁺ precursor cell. (*F*) Low magnification of the anterior forebrain showing EGFP reporter expression following postnatal administration of tamoxifen. (*F'*) Many reporter-positive cells line the SEZ germinal niche and are present in the white matter. (*F''*) Astrocytes express reporter as well. (*G*) A low level (<1% of labeled cells) of spurious recombination is often noted in such systems. Shown here is a fully ramified neuron present in the cortex 2 days after tamoxifen injection. (*H*) The SGZ displays robust reporter expression as well postnatal administration of tamoxifen. (*H'*) Example of an EGFP⁺/Sox2⁺/Gfap⁺ astrocyte. (*H''*) Dcx⁺/eGFP⁺ newborn neurons in the SGZ (*open arrowheads*) and an EGFP⁺/Gfap⁺/Sox2⁺ SGZ astroctye (*closed arrowhead*). (BrdU) Bromodeoxyuridine; (Dcx) doublecortin; (EGFP) enhanced green fluorescent protein; (Gfap) glial fibrillary acidic protein; (NeuN) neuronal nucleus; (Tm) tamoxifen.

Table 2. NSC-promoter tamoxifen-inducible and tetracycline-regulated mice

Name	Promoter	Type	Reference
Tamoxifen-inducible			
Ascl1-CreER™	BAC containing Ascl1 genomic	Cre-ERT1	Battiste et al. (2007)
GCE	huGFAP	Cre-ERT2	Ganat et al. (2006)
GFAP-CreERT2	huGFAP	Cre-ERT2	Hirrlinger et al. (2006)
GLAST::CreERT2	knockin to endogenous GLAST locus	Cre-ERT2	Mori et al. (2006)
Gli1-CreERT2	genomic fragment 5' of Gli1 (2.3 kb)	Cre-ERT2	Ahn and Joyner (2006)
Nestin-CreER	Nestin second intron/Hsp-68 minimal	Cre-ERT1	Burns et al. (2007)
Nestin-CreERT2	Xh5 plasmid (large fragment including Nestin promoter/introns/exons	Cre-ERT2	Battiste et al. (2007)
Nes-CreER(T2)	Nestin/Nestin second intron	Cre-ERT2	Imayoshi et al. (2006)
Nestin-Cre-ERT2	Nestin second intron/Hsp-68 minimal	Cre-ERT2	Carlen et al. (2006)
Nestin-creERtm	Nestin promoter and second intron	Cre-ERT1	Kuo et al. (2006)
Ngn2-CreER™	knockin to endogenous Ngn2 locus	Cre-ERT1	Raineteau et al. (2006)
Tetracycline-regulated			
Nestin-rtTA-M2	Nestin/Nestin second intron	rtTA-M2	Yu et al. (2005)
Nestin-tTA	Xh5 plasmid (see above)	tTA	Beech et al. (2005)

"driver" line. Finally, to mutate and mark a desired cell population or to create double knockouts, multiple pairs of *loxP* sites are often used. This raises the possibility that unpaired *loxP* sites can recombine (even on alternate chromosomes) which could lead to chromosomal instability and aberrant phenotypes (Van Deursen et al. 1995).

With inducible technologies, complications of strain, age, handling, etc., grow exponentially more dramatic. Replication of findings between groups will become increasingly more difficult. For example, it was previously mentioned that some Nestin-Cre and Nestin-CreER mouse lines cause perinatal hydrocephalus (Forni et al. 2006). Another group has recently used a similar Nestin-CreER line to ablate Numb in a Numblike-null background with the result being hydrocephalus (Kuo et al. 2006). Without the authors' proper presentation of tamoxifen-induced, Cre[+] controls, it would have been uncertain as to whether the hydrocephalus was due to Cre toxicity or Numb ablation. Furthermore, from litter to litter, there sometimes are phenotype discrepancies between identically handled animals with the same genotype, differing only by litter. In addition, differences in the strain and background of mice have led to an absence of reporter or transgene expression in certain cases (unpublished observations). Another factor to consider is leakiness of the system. Indeed, in both of the strains that we have handled (GCE and NestinCreER), we noted leakiness in the absence of tamoxifen and spurious labeling of fully elaborated, mature neurons in nonneurogenic regions by our reporter gene despite a survival time of less than 1 week after tamoxifen administration (Fig. 3H). Thus, observations of neurogenesis, especially in previously unreported regions, must be cross-validated.

Another problem unique to tamoxifen-inducible systems is the fact that tamoxifen is an estrogen antagonist. Estrogen is a known modifier of cell proliferation and NPC proliferation in particular (Martinez-Cerdeno et al. 2006). We did not observe any significant alterations in the proliferation of cells in the postnatal brain in our studies, but our delivery paradigms were designed to minimize acute effects. For example, 1 week was the shortest survival point and was chosen (1) to allow clearance of tamoxifen, permitting cells to return to normal, and (2) to allow recombination and subsequent transcription and translation in the case of our transgene or breakdown of the endogenous protein in the case of our inducible knockouts.

Tetracycline-regulated genetic modification is another method that has been used for conditional mutagenesis in the postnatal brain (Beech et al. 2004; Yu et al. 2005). In theory, this system allows for controlled temporal and spatial regulation of mutagenesis in a reversible manner. In practice, however, there is often notable leakiness and/or difficulties inducing gene expression because of the blood-brain barrier (Mansuy and Bujard 2000; Zhu et al. 2001). In addition, there are significant side effects attributed to long-term delivery of tetracycline or derivatives (Rifkin et al. 1994; Yrjanheikki et al. 1998). There are two primary subtypes of tetracycline-regulated mouse systems. The Tet-Off system (Fig. 4A) uses a tissue-specific promoter to drive expression of the tetracycline *trans*-activator (tTA) in the tissue of interest (Gossen and Bujard 1992; Furth et al. 1994). tTA binds tetracycline operator sequences (tetO) which allows expression of the gene of interest downstream from tetO. Administration of the synthetic derivative of tetracycline, deoxycycline (Dx), disallows the binding of tTA to the tetO sequence, preventing gene expression. Conversely, in the Tet-On system (Fig. 4B), a mutant tetracycline *trans*-activator (rtTA) is expressed instead of tTA and activates expression of the gene of interest only in the presence of Dx (Gossen et al. 1995). A bidirectional promoter (Fig. 4C) has been shown to allow simultaneous expression of a gene of interest and a reporter gene (Krestel et al. 2001).

OTHER TECHNIQUES FOR THE OBSERVATION OF NPC BEHAVIOR

Cell transplantation allows for the observation of the effects of host environment on NPCs (Snyder et al. 1997; Zigova and Newman 2002). Cells

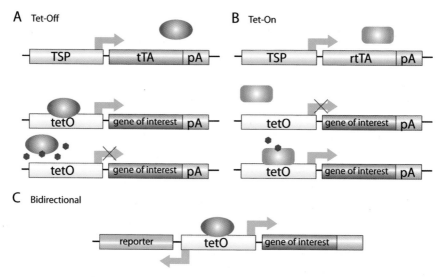

Figure 4. Tet-off, Tet-On, and bidirection tetracycline-regulated systems. (*A*) In the Tet-Off system, a tissue-specific promoter (TSP), allowing spatial regulation of expression, drives expression of the tetracycline transcriptional activator (tTA). This allows transcription of the gene of interest. Dox administration interferes with tTA binding to tetO, preventing gene expression. (*B*) The reverse is true in the rtTA system in which Dox administration allows for rtTA-tetO binding and subsequent gene expression. (*C*) Bidirection constructs allow for simultaneous expression of a gene of interest and reporter for marking the affected cells. (pA) poly(A) tail.

for transplantation can be isolated from numerous sources including the embryonic brain (Shin et al. 2000), the postnatal brain (Suhonen et al. 1996), embryonic precursor cells (Tabar et al. 2005), or immortalized cell lines (Englund et al. 2002). Prior to transplantation, cells can be transfected or transduced to express transgenes/reporters (Cao et al. 2002). Alternatively, cells can be isolated from transgenic or mutant animals. However, numerous passages or certain treatments (Bjorklund et al. 2002; Roy et al. 2006) can alter the genetic integrity of NPCs, causing cells to become tumorigenic. Another factor that must be considered is that fusion of NPCs in vitro has been reported (Chen et al. 2006; Jessberger et al. 2007). Furthermore, reports of transdifferentiation of nonneuronal cells remain controversial because of the observation of fusion under similar conditions, the dubious expression of neuronal markers, and/or spurious transfer of genetic/chemical label (Alvarez-Dolado et al. 2003; Lu et al. 2004; Neuhuber et al. 2004; Burns et al. 2006; Choi et al. 2006). The use of genetic labels and/or differing host/donor labels, i.e., transplantation of green fluorescent protein (GFP)-labeled donor cells into a ubiquitously cyan fluorescent protein (CFP)-expressing host, is becoming the gold standard for validating transplantation experiments (MacLaren et al. 2006).

CONCLUSIONS: FALSE-POSITIVE EXAMPLES AND PERSPECTIVES

A prime example of false positives from leading-edge techniques comes from a report by Loturco and colleagues noting that postmitotic neurons could be spuriously labeled by an EGFP-expressing retrovirus through a neuronal-microglial fusion (Ackman et al. 2006). Their intent was to label any postnatally generated neurons in the hope of determining whether they properly integrate. These authors elegantly showed that virus presence stimulated microglia to proliferate and fuse with existing pyramidal cells (Fig. 5A). No evidence of cortical neurogenesis was found despite injecting 134 rats. This was noted in vivo and in vitro. In vivo, fused microglia were only noted in the apical dendrites. However, in vitro, neurons with smaller extra nuclei were observed at the longest time point. In alternate experiments, it was shown that the fused microglia had likely proliferated as they incorporated BrdU. Thus, the specter is raised that the distal fusion event is a precursor to somal fusion, which has been the primary type of fusion noted by Alvarez-Dolado et al. (2003). It should be noted that it is possible that such a somal fusion could lead to a BrdU$^+$ nucleus in the cell body of a neuron, thus giving a false positive for neurogenesis. Of course, as shown by this group's example, false positives can be avoided by the multiple levels of analyses recommended above, including temporal analysis of progressive differentiation by developmental

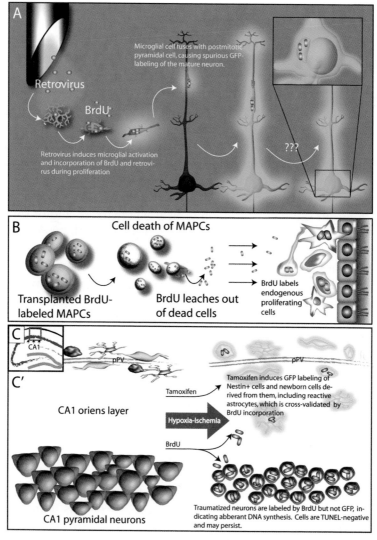

Figure 5. Examples of spurious labeling by retrovirus and BrdU. (*A*) Loturco et al. (2006) reported that injection of an EGFP retrovirus activated microglia (which could be labeled with BrdU, indicating proliferation), causing a neuronal-microglial fusion. This fusion causes EGFP labeling of the entire neuron. Somal fusion was seen in vitro and the presence of BrdU-labeled nuclei in fused neurons in vivo is thus a possibility. (*B*) In another report, Burns et al. (2006) reported that thymidine analogs could be transferred from prelabeled host cells to proliferating donor cells. (*C*) Boxed region is magnified in *C'*. (*C'*) Summary of results from Burns et al. (2007), showing that inducible labeling of newborn cells with EGFP cross-validates the finding that postmitotic neurons can incorporate BrdU in a nonproliferative manner. Following hypoxia-ischemia, newborn glia in the

markers, emergence of connectivity, and multiple complementary modes of rigorous testing for the nature of cell labeling.

Another methodological warning shows that BrdU might not be a reliable marker of newly generated cells for unexpected reasons. This was observed in experiments by Verfaillie and colleagues whereby cells were labeled with thymidine analogs and subjected to treatments that killed these cells. Then, the dead cells were transplanted. Nevertheless, numerous labeled neurons and glia were observed. Exhaustive and comprehensive experiments led to the interpretation that the thymidine analogs leached from dead transplanted donor cells into proliferating host cells (Fig. 5B) (Burns et al. 2006). This indicates that past transplant experiments that relied on BrdU labeling might need to be reexamined with more current techniques. Furthermore, large doses of BrdU coupled with the high turnover rate in the neurogenic regions could lead to a false readout of proliferative dynamics and overall neurogenesis as dying cells could leach BrdU into the local precursor cells (Burns et al. 2006). Thus, in a local precursor cell niche, dividing precursor cells could theoretically be labeled despite the fact that BrdU delivery was given in the distant past.

Cross-validation of results is paramount to definitive interpretation of the data. For example, we have recently shown that inducible genetic labeling of newborn cells can be a suitable means of cross-validating thymidine analog cell labeling or retroviral transduction of novel neurogenic regions. After ischemia-hypoxia challenge to the hippocampus (Fig. 5C), many BrdU-labeled hippocampal pyramidal neurons could be found, but the lack of EGFP reporter positivity in these neurons confirmed that this labeling was attributable to nonproliferative DNA synthesis (Fig. 5C′) (Burns et al. 2007). (Neighboring newborn glia were readily labeled with this method as were newborn subependymal [SEZ] cells.) This was further confirmed by a lack of migrating cells and the fact that the neurons were labeled with BrdU within 6 days of the insult—a period of time that is likely too short to allow for migration and integration of newborn pyramidal cells.

The purpose of this chapter was to provide a basic introduction to the advantages and caveats of a wide variety of techniques for the study

---

surrounding region incorporate BrdU and express EGFP after tamoxifen induction. However, CA1 pyramidal cells only incorporate BrdU and do not express EGFP after 6 days. (BrdU) Bromodeoxyuridine; (MAPCs) multipotent adult precursor cells; (pPV) posterior periventricle.

Table 3. Summary of methods: Advantages and warnings

Method	Nature	Means of delivery	Cell types targeted	Advantages	Caveats
Thymidine analogs					
Tritiated Thymidine	radioactive nucleotide	peripheral or local injection	cells synthesizing DNA	widespread labeling of proliferating cells; stoichiometric detection ratio	• taken up by cells undergoing abortive mitosis and by cells repairing DNA • causes DNA strand breaks • dilutes during every replication cycle
BrdU, CldU, Idu	halogenated thymidine analog	peripheral or local injection	cells synthesizing DNA	widespread labeling of proliferating cells; detection methods amplify signal; allows for phenotyping of labeled cells in thick tissue sections.	• taken up by cells undergoing abortive mitosis and by cells repairing DNA • causes DNA strand breaks, DNA transcription errors • highly toxic, mutagenic • dilutes over time • amplification in detection method obscures nature of DNA synthesis
Viral vectors					
Adenovirus	viral dsDNA	stereotaxic, focal injection	broad range	widespread infection of most tissues; pantrophic	• transient tranduction • subcloning of transgenes typically requires shuttle

Lentivirus	viral RNA	stereotaxic, focal injection	broad range	persistent genetic alteration of most tissues; pantrophic	• not specific for newborn cells • focal injection by its nature causes lesion • genomic integration disrupts host DNA at insertion site
Retrovirus	viral RNA	stereotaxic, focal injection	dividing cells*	persistent genetic alteration of dividing transduced cells; pantrophic	• not specific for newborn cells • focal injection by its nature causes lesion • genomic integration disrupts host DNA at insertion site • induced fusion reported • focal injection by its nature causes lesion
Genetic Knockin	engineered genomic modification	germ line	all	allows for observation of expression pattern of endogenous gene	• typically disrupts gene/promoter • specificity of phenotype determined by endogenous gene regulation

(continued)

Table 3. (*continued*)

Method	Nature	Means of delivery	Cell types targeted	Advantages	Caveats
					• can cause developmental abnormalities
Knockout	engineered genomic modification	germ line	all	allows for observation of null or haploinsufficent mutation	• not cell-type-specific • can cause developmental abnormalities
Transgenic	engineered genomic modification	germ line	all	allows for gain-of-function/cell labeling experiments in vivo	• insertion sites often random • not cell-type-specific (depending on promoter) • can cause developmental abnormalities
Cre/loxP	engineered genomic modification	germ line	typically cell-type-specific	powerful, conditional control of gene expression	• Cre toxicity • no postnatal precursor cell-specific promoters are known, thus causing developmental abnormalities • temporal control based on Cre-driver promoter

rtTa/tTa	engineered genomic modification	germ line	typically cell-type-specific	allows for conditional, neversible control of gene expression	• frequently leaky • difficult to manipulate in the postnatal brain • Lox-related side effects
Inducible Cre	engineered genomic modification	germ line	typically cell-type-specific	powerful, inducible control of gene expression	• Cre toxicity • tamoxifen-related side effects • recombination ratio suffers compared with traditional Cre driver lines
Electroporation	DNA electroporation	focal injection	typically VZ neural precursor cells	fast, powerful system for gain- and loss-of-function studies of neural precursor cells, eschewing the need for complicated genetic engineering of mice/crosses or viral production	• typically transfects a small region • electroporation method limited to in utero/perinatal period • can cause focal (injection site) abnormalities

(continued)

Table 3. (*continued*)

Method	Nature	Means of delivery	Cell types targeted	Advantages	Caveats
Miscellaneous					
Transplantation	chimeric animal-animal transplantation	focal (NPCs)/ systemic injection (BMSCs)	population initially isolated	allows for observation of cell-autonomous and noncell-autonomous effects of different host environments; often combined with genetic/viral technologies	• may provoke immune reaction • focal injection by its nature causes lesion • fusion a possibility • thymidine analogs can transfer from dying cells to proliferating host cells • genetic integrity can become compromised after multiple passages

of postnatal NPCs. Because of limited space, some important techniques might have been omitted or covered incompletely. It should also be noted that the possible combinations of techniques discussed herein could fill many pages. The hope is that this primer will allow authors and reviewers to meet high scientific criteria in the advancement of this fast-evolving field through the ability to adopt new techniques and cross-validate results in a rigorous and definitive manner.

REFERENCES

Abbott A. 2006. Youthful duo snags a swift Nobel for RNA control of genes. *Nature* **443:** 488.

Ackman J.B., Siddiqi F., Walikonis R.S., and LoTurco J.J. 2006. Fusion of microglia with pyramidal neurons after retroviral infection. *J. Neurosci.* **26:** 11413–11422.

Ahn S. and Joyner A.L. 2005. In vivo analysis of quiescent adult neural stem cells responding to sonic hedgehog. *Nature* **437:** 894–897.

Altman J. and Das G.D. 1965a. Post-natal origin of microneurones in the rat brain. *Nature* **207:** 953–956.

———. 1965b. Autoradiographic and histological evidence of postnatal hippocampal neurogenesis in rats. *J. Comp. Neurol.* **124:** 319–335.

Alvarez-Dolado M., Pardal R., Garcia-Verdugo J.M., Fike J.R., Lee H.O., Pfeffer K., Lois C., Morrison S.J., and Alvarez-Buylla A. 2003. Fusion of bone-marrow-derived cells with Purkinje neurons, cardiomyocytes and hepatocytes. *Nature* **425:** 968–973.

Alvarez V.A., Ridenour D.A., and Sabatini B.L. 2006. Retraction of synapses and dendritic spines induced by off-target effects of RNA interference. *J. Neurosci.* **26:** 7820–7825.

Angevine J.B., Jr. 1965. Time of neuron origin in the hippocampal region. An autoradiographic study in the mouse. *Exp. Neurol. Suppl.* **2:** 1–70.

Arvidsson A., Collin T., Kirik D., Kokaia Z., and Lindvall O. 2002. Neuronal replacement from endogenous precursors in the adult brain after stroke. *Nat. Med.* **8:** 963–970.

Bannigan J. and Langman J. 1979. The cellular effect of 5-bromodeoxyuridine on the mammalian embryo. *J. Embryol. Exp. Morphol.* **50:** 123–135.

Bannigan J.G. 1985. The effects of 5-bromodeoxyuridine on fusion of the cranial neural folds in the mouse embryo. *Teratology* **32:** 229–239.

Bannigan J.G., Cottell D.C., and Morris A. 1990. Study of the mechanisms of BUdR-induced cleft palate in the mouse. *Teratology* **42:** 79–89.

Battiste J., Helms A.W., Kim E.J., Savage T.K., Lagace D.C., Mandyam C.D., Eisch A.J., Miyoshi G., and Johnson J.E. 2007. Ascl1 defines sequentially generated lineage-restricted neuronal and oligodendrocyte precursor cells in the spinal cord. *Development* **134:** 285–293.

Bauer S., Hay M., Amilhon B., Jean A., and Moyse E. 2005. In vivo neurogenesis in the dorsal vagal complex of the adult rat brainstem. *Neuroscience* **130:** 75–90.

Beech R.D., Cleary M.A., Treloar H.B., Eisch A.J., Harrist A.V., et al. 2004. Nestin promoter/enhancer directs transgene expression to precursors of adult generated periglomerular neurons. *J. Comp. Neurol.* **475:** 128–141.

Belluzzi O., Benedusi M., Ackman J., and LoTurco J.J. 2003. Electrophysiological differentiation of new neurons in the olfactory bulb. *J. Neurosci.* **23:** 10411–10418.

Benraiss A., Chmielnicki E., Lerner K., Roh D., and Goldman S.A. 2001. Adenoviral brain-derived neurotrophic factor induces both neostriatal and olfactory neuronal recruitment from endogenous progenitor cells in the adult forebrain. *J. Neurosci.* **21:** 6718–6731.

Bernier P.J., Bedard A., Vinet J., Levesque M., and Parent A. 2002. Newly generated neurons in the amygdala and adjoining cortex of adult primates. *Proc. Natl. Acad. Sci.* **99:** 11464–11469.

Bhardwaj R.D., Curtis M.A., Spalding K.L., Buchholz B.A., Fink D., Bjork-Eriksson T., Nordborg C., Gage F.H., Druid H., Eriksson P.S., and Frisen J. 2006. Neocortical neurogenesis in humans is restricted to development. *Proc. Natl. Acad. Sci.* **103:** 12564–12568.

Bjorklund L.M., Sanchez-Pernaute R., Chung S., Andersson T., Chen I.Y., McNaught K.S., Brownell A.L., Jenkins B.G., Wahlested I.C., Kim K.S., and Isacson O. 2002. Embryonic stem cells develop into functional dopaminergic neurons after transplantation in a Parkinson rat model. *Proc. Natl. Acad. Sci.* **99:** 2344–2349.

Branda C.S. and Dymecki S.M. 2004. Talking about a revolution: The impact of site-specific recombinases on genetic analyses in mice. *Dev. Cell* **6:** 7–28.

Burns K.A. and Kuan C.Y. 2005. Low doses of bromo- and iododeoxyuridine produce near-saturation labeling of adult proliferative populations in the dentate gyrus. *Eur. J. Neurosci.* **21:** 803–807.

Burns K.A., Ayoub A.E., Breunig J.J., Adhami F., Weng W.L., Colbert M.C., Rakic N.R., and Kuan C.Y. 2007. Nestin-CreER mice reveal DNA synthesis by nonapoptotic neurons following cerebral ischemia-hypoxia. *Cereb. Cortex.* (in press).

Burns T.C., Ortiz-Gonzalez X.R., Gutierrez-Perez M., Keene C.D., Sharda R., Demorest Z.L., Jiang Y., Nelson-Holte M., Soriano M., and Nakagawa Y. et al. 2006. Thymidine analogs are transferred from prelabeled donor to host cells in the central nervous system after transplantation: A word of caution. *Stem Cells* **24:** 1121–1127.

Cajal R. 1913–1914. *Estudios sobre la degeneración y regeneración del sistema nervioso.* Moya, Madrid.

Caldwell M.A., He X., and Svendsen C.N. 2005. 5-Bromo-2′-deoxyuridine is selectively toxic to neuronal precursors in vitro. *Eur. J. Neurosci.* **22:** 2965–2970.

Cameron H.A. and McKay R.D. 2001. Adult neurogenesis produces a large pool of new granule cells in the dentate gyrus. *J. Comp. Neurol.* **435:** 406–417.

Cao Q.L., Onifer S.M., and Whittemore S.R. 2002. Labeling stem cells in vitro for identification of their differentiated phenotypes after grafting into the CNS. *Methods Mol. Biol.* **198:** 307–318.

Carlén M., Meletis K., Barnabé-Heider F., and Frisén J. 2006. Genetic visualization of neurogenesis. *Exp. Cell Res.* **312:** 2851–2859.

Casanova E., Fehsenfeld S., Lemberger T., Shimshek D.R., Sprengel R., and Manta Madiotis T. 2002. ER-based double iCre fusion protein allows partial recombination in forebrain. *Genesis* **34:** 208–214.

Chen J., Magavi S.S., and Macklis J.D. 2004. Neurogenesis of corticospinal motor neurons extending spinal projections in adult mice. *Proc. Natl. Acad. Sci.* **101:** 16357–16362.

Chen K.A., Laywell E.D., Marshall G., Walton N., Zheng T., and Steindler D.A. 2006. Fusion of neural stem cells in culture. *Exp. Neurol.* **198:** 129–135.

Choi C.B., Cho Y.K., Prakash K.V., Jee B.K., Han C.W., Park Y.K., Kim H.Y., Lee K.H., Chung N., and Rua H.K. 2006. Analysis of neuron-like differentiation of human bone marrow mesenchymal stem cells. *Biochem. Biophys. Res. Commun* **350:** 138–146.

Consiglio A., Gritti A., Dolcetta D., Follenzi A., Bordignon C., Gage F.H., Vescovi A.L., and Naldini L. 2004. Robust in vivo gene transfer into adult mammalian neural stem cells by lentiviral vectors. *Proc. Natl. Acad. Sci.* **101:** 14835–14840.

Djordjevic B. and Szybalski W. 1960. Genetics of human cell lines. III. Incorporation of 5-bromo- and 5-iododeoxyuridine into the deoxyribonucleic acid of human cells and its effect on radiation sensitivity. *J. Exp. Med.* **112:** 509–531.

Doetsch F., Caille I., Lim D.A., Garcia-Verdugo J.M., and Alvarez-Buylla A. 1999. Subventricular zone astrocytes are neural stem cells in the adult mammalian brain. *Cell* **97:** 703–716.

Ekdahl C.T., Mohapel P., Elmer E., and Lindvall O. 2001. Caspase inhibitors increase short-term survival of progenitor-cell progeny in the adult rat dentate gyrus following status epilepticus. *Eur. J. Neurosci.* **14:** 937–945.

Englund U., Bjorklund A., Wictorin K., Lindvall O., and Kokaia M. 2002. Grafted neural stem cells develop into functional pyramidal neurons and integrate into host cortical circuitry. *Proc. Natl. Acad. Sci.* **99:** 17089–17094.

Eriksson P.S., Perfilieva E., Bjork-Eriksson T., Alborn A.M., Nordborg C., Peterson D.A., and Gage F.H. 1998. Neurogenesis in the adult human hippocampus. *Nat. Med.* **4:** 1313–1317.

Feil R., Brocard J., Mascrez B., LeMeur M., Metzger D., and Chambon P. 1996. Ligand-activated site-specific recombination in mice. *Proc. Natl. Acad. Sci.* **93:** 10887–10890.

Fire A., Xu S., Montgomery M.K., Kostas S.A., Driver S.E., and Mello C.C. 1998. Potent and specific genetic interference by double-stranded RNA in *Caenorhabditis elegans. Nature* **391:** 806–811.

Forni P.E., Scuoppo C., Imayoshi I., Taulli R., Dastru W., Sala V., Betz U.A., Muzzi R., Martinuzzi D., Vercelli A.E., Kageyama R., and Ponzetto C. 2006. High levels of Cre expression in neuronal progenitors cause defects in brain development leading to microencephaly and hydrocephaly. *J. Neurosci.* **26:** 9593–9602.

Frielingsdorf H., Schwarz K., Brundin P., and Mohapel P. 2004. No evidence for new dopaminergic neurons in the adult mammalian substantia nigra. *Proc. Natl. Acad. Sci.* **101:** 10177–10182.

Fukuchi-Shimogori T. and Grove E.A. 2001. Neocortex patterning by the secreted signaling molecule FGF8. *Science* **294:** 1071–1074.

Furth P.A., St Onge L., Boger H., Gruss P., Gossen M., Kistner A., Bujand H., and Hennighausen L. 1994. Temporal control of gene expression in transgenic mice by a tetracycline-responsive promoter. *Proc. Natl. Acad. Sci.* **91:** 9302–9306.

Gage F.H. 2000. Mammalian neural stem cells. *Science* **287:** 1433–1438.

Ganat Y.M., Silbereis J., Cave C., Ngu H., Anderson G.M., Ohkubo Y., Ment L.R., and Vaccarino F.M. 2006. Early postnatal astroglial cells produce multilineage precursors and neural stem cells in vivo. *J. Neurosci.* **26:** 8609–8621.

Ge S., Goh E.L., Sailor K.A., Kitabatake Y., Ming G.L., and Song H. 2006. GABA regulates synaptic integration of newly generated neurons in the adult brain. *Nature* **439:** 589–593.

Goldsworthy T.L., Butterworth B.E., and Maronpot R.R. 1993. Concepts, labeling procedures, and design of cell proliferation studies relating to carcinogenesis. *Environ. Health Perspect.* (suppl. 5) **101:** 59–65.

Goldsworthy T.L., Dunn, C.S., and Popp, J.A. 1992. Dose effects of bromodeoxyuridine (BRUD) on rodent hepatocyte proliferation measurements. *Toxicologist* **12:** 265.

Gossen M. and Bujard H. 1992. Tight control of gene expression in mammalian cells by tetracycline-responsive promoters. *Proc. Natl. Acad. Sci.* **89:** 5547–5551.

Gossen M., Freundlieb S., Bender G., Muller G., Hillen W., and Bujard H. 1995. Transcriptional activation by tetracyclines in mammalian cells. *Science* **268**: 1766–1769.

Gould E., Reeves A.J., Graziano M.S., and Gross C.G. 1999. Neurogenesis in the neocortex of adult primates. *Science* **286**: 548–552.

Gould E., Vail N., Wagers M., and Gross C.G. 2001. Adult-generated hippocampal and neocortical neurons in macaques have a transient existence. *Proc. Natl. Acad. Sci.* **98**: 10910–10917.

Ganat Y.M., Silbereis J., Cave C., Ngu H., Anderson G.M., Ohkubo Y., Ment L.R., and Vaccarino F.M. 2006. Early postnatal astroglial cells produce multilineage precursors and neural stem cells in vivo. *J. Neurosci.* **26**: 8609–8621.

Herrup K. and Yang Y. 2007. Cell cycle regulation in the post-mitotic neuron: Oxymoron or new biology? *Nat. Neurosci. Rev.* **8**: 368–378.

Hirrlinger P.G., Scheller A., Braun C., Hirrlinger J., and Kirchhoff F. 2006. Temporal control of gene recombination in astrocytes by transgenic expression of the tamoxifen-inducible DNA recombinase variant CreERT2. *Glia* **54**: 11–20.

Hoglinger G.U., Breunig J.J., Depboylu C., Rouaux C., Michel P.P., Alvarez-Fischer D., Boutillier A.L., Degregori J., Oertel W.H., Rakic P., Hirsch E.C., and Hunot S. 2007. The pRb/E2F cell-cycle pathway mediates cell death in Parkinson's disease. *Proc. Natl. Acad. Sci.* **104**: 3585–3590.

Horner P.J., Power A.E., Kempermann G., Kuhn H.G., Palmer T.D., Winkler J., Thal L.J., and Gage F.H. 2000. Proliferation and differentiation of progenitor cells throughout the intact adult rat spinal cord. *J. Neurosci.* **20**: 2218–2228.

Hunter N.L., Awatramani R.B., Farley F.W., and Dymecki S.M. 2005. Ligand-activated Flpe for temporally regulated gene modifications. *Genesis* **41**: 99–109.

Imayoshi I., Ohtsuka T., Metzger D., Chambon P., and Kageyama R. 2006. Temporal regulation of Cre recombinase activity in neural stem cells. *Genesis* **44**: 233–238.

Indra A.K., Warot X., Brocard J., Bornert J.M., Xiao J.H., Chambon P., and Metzger D. 1999. Temporally-controlled site-specific mutagenesis in the basal layer of the epidermis: Comparison of the recombinase activity of the tamoxifen-inducible Cre-ER(T) and Cre-ER(T2) recombinases. *Nucleic Acids Res.* **27**: 4324–4327.

Jessberger S., Clemenson Jr. G.D., and Gage F.H. 2007. Spontaneous fusion and non-clonal growth of adult neural stem cells. *Stem Cells* **25**: 871–874.

Joyner A.L. and Zervas M. 2006. Genetic inducible fate mapping in mouse: Establishing genetic lineages and defining genetic neuroanatomy in the nervous system. *Dev. Dyn.* **235**: 2376–2385.

Kaplan M.S. 1981. Neurogenesis in the 3-month-old rat visual cortex. *J. Comp. Neurol.* **195**: 323–338.

Klein J.A., Longo-Guess C.M., Rossmann M.P., Seburn K.L., Hurd R.E., Frankel W.N., Bronson R.T., and Ackermann S.L. 2002. The harlequin mouse mutation downregulates apoptosis-inducing factor. *Nature* **419**: 367–374.

Koketsu D., Mikami A., Miyamoto Y., and Hisatsune T. 2003. Nonrenewal of neurons in the cerebral neocortex of adult macaque monkeys. *J. Neurosci.* **23**: 937–942.

Kolb B., Pedersen B., Ballermann M., Gibb R., and Whishaw I.Q. 1999. Embryonic and postnatal injections of bromodeoxyuridine produce age-dependent morphological and behavioral abnormalities. *J. Neurosci.* **19**: 2337–2346.

Kornack D.R. and Rakic P. 2001. Cell proliferation without neurogenesis in adult primate neocortex. *Science* **294**: 2127–2130.

Krestel H.E., Mayford M., Seeburg P.H., and Sprengel R. 2001. A GFP-equipped bidirectional expression module well suited for monitoring tetracycline-regulated gene expression in mouse. *Nucleic Acids Res.* **29:** E39.

Kruman, I.I., Wersto R.P., Cardozo-Pelaez F., Smilenov L., Chan S.L., Chrest F.J., Emokpae R., Gorospe M., and Mattson M.P. 2004. Cell cycle activation linked to neuronal cell death initiated by DNA damage. *Neuron* **41:** 549–561.

Kuan C.Y., Schloemer A.J., Lu A., Burns K.A., Weng W.L., Williams M.T., Strauss K.I., Vorhees C.V., Flavell R.A., Davis R.J., et al. 2004. Hypoxia-ischemia induces DNA synthesis without cell proliferation in dying neurons in adult rodent brain. *J. Neurosci.* **24:** 10763–10772.

Kuhn H.G., Dickinson-Anson H., and Gage F.H. 1996. Neurogenesis in the dentate gyrus of the adult rat: Age-related decrease of neuronal progenitor proliferation. *J. Neurosci.* **16:** 2027–2033.

Kuo C.T., Mirzadeh Z., Soriano-Navarro M., Rasin M., Wang D., Shen J., Sestan N., Garcia-Verdugo J., Alrarez-Buylla J., et al. 2006. Postnatal deletion of Numb/Numblike reveals repair and remodeling capacity in the subventricular neurogenic niche. *Cell* **127:** 1253–1264.

Kuwagata M., Muneoka K.T., Ogawa T., Takigawa M., and Nagao T. 2004. Locomotor hyperactivity following prenatal exposure to 5-bromo-2′-deoxyuridine: Neurochemical and behavioral evidence of dopaminergic and serotonergic alterations. *Toxicol. Lett.* **152:** 63–71.

Leone D.P., Genoud S., Atanasoski S., Grausenburger R., Berger P., Metzger D., Macklin W.B., Chambon P., and Suter U. 2003. Tamoxifen-inducible glia-specific Cre mice for somatic mutagenesis in oligodendrocytes and Schwann cells. *Mol. Cell Neurosci.* **22:** 430–440.

Lewis P.F. and Emerman M. 1994. Passage through mitosis is required for oncoretroviruses but not for the human immunodeficiency virus. *J. Virol.* **68:** 510–516.

Lie D.C., Dziewczapolski G., Willhoite A.R., Kaspar B.K., Shults C.W., and Gage F.H. 2002. The adult substantia nigra contains progenitor cells with neurogenic potential. *J. Neurosci.* **22:** 6639–6649.

Lledo P.M., Alonso M., and Grubb M.S. 2006. Adult neurogenesis and functional plasticity in neuronal circuits. *Nat. Rev. Neurosci* **7:** 179–193.

Lois C. and Alvarez-Buylla A. 1994. Long-distance neuronal migration in the adult mammalian brain. *Science* **264:** 1145–1148.

Lu P., Blesch A., and Tuszynski M.H. 2004. Induction of bone marrow stromal cells to neurons: Differentiation, transdifferentiation, or artifact? *J. Neurosci. Res.* **77:** 174–191.

MacLaren R.E., Pearson R.A., MacNeil A., Douglas R.H., Salt T.E., Akimoto M., Swaroop A., Sowden J.C., and Ali R.R. 2006. Retinal repair by transplantation of photoreceptor precursors. *Nature* **444:** 203–207.

Magavi S.S. and Macklis J.D. 2002a. Immunocytochemical analysis of neuronal differentiation. *Methods Mol. Biol.* **198:** 291–297.

———. 2002b. Identification of newborn cells by BrdU labeling and immunocytochemistry in vivo. *Methods Mol. Biol.* **198:** 283–290.

Magavi S.S., Leavitt B.R., and Macklis J.D. 2000. Induction of neurogenesis in the neocortex of adult mice. *Nature* **405:** 951–955.

Magavi S.S., Mitchell B.D., Szentirmai O., Carter B.S., and Macklis J.D. 2005. Adult-born and preexisting olfactory granule neurons undergo distinct experience-dependent modifications of their olfactory responses in vivo. *J. Neurosci.* **25:** 10729–10739.

Mansuy I.M. and Bujard H. 2000. Tetracycline-regulated gene expression in the brain. *Curr. Opin. Neurobiol.* **10:** 593–596.

Martinez-Cerdeno V., Noctor S.C., and Kriegstein A.R. 2006. Estradiol stimulates pro-genitor cell division in the ventricular and subventricular zones of the embryonic neo-cortex. *Eur. J. Neurosci.* **24:** 3475–3488.

Matsuda T. and Cepko C.L. 2007. Controlled expression of transgenes introduced by in vivo electroporation. *Proc. Natl. Acad. Sci.* **104:** 1027–1032.

Miller M.W. and Nowakowski R.S. 1988. Use of bromodeoxyuridine-immunohistochemistry to examine the proliferation, migration and time of origin of cells in the central ner-vous system. *Brain Res.* **457:** 44–52.

Molyneaux B.J., Arlotta P., Hirata T., Hibi M., and Macklis J.D. 2005. Fezl is required for the birth and specification of corticospinal motor neurons. *Neuron* **47:** 817–831.

Mori T., Tanaka K., Buffo A., Wurst W., Kühn R., and Götz M. 2006. Inducible gene dele-tion in astroglia and radial glia—a valuable tool for functional and lineage analysis. *Glia* **54:** 21–34.

Nacher J., Crespo C., and McEwen B.S. 2001. Doublecortin expression in the adult rat telencephalon. *Eur. J. Neurosci* **14:** 629–644.

Nacher J., Alonso-Llosa G., Rosell D., and McEwen B. 2002. PSA-NCAM expression in the piriform cortex of the adult rat. Modulation by NMDA receptor antagonist admin-istration. *Brain Res.* **927:** 111–121.

Nagao T., Kuwagata M., and Saito Y. 1998. Effects of prenatal exposure to 5-bromo-2′-deoxyuridine on the developing brain and reproductive function in male mouse offspring. *Reprod. Toxicol.* **12:** 477–487.

Neuhuber B., Gallo G., Howard L., Kostura L., Mackay A., and Fischer I. 2004. Reevaluation of in vitro differentiation protocols for bone marrow stromal cells: Disruption of actin cytoskeleton induces rapid morphological changes and mimics neuronal phenotype. *J. Neurosci. Res.* **77:** 192–204.

Nowakowski R.S. and Rakic P. 1981. The site of origin and route and rate of migration of neurons to the hippocampal region of the rhesus monkey. *J. Comp. Neurol.* **196:** 129–154.

Parent J.M., Vexler Z.S., Gong C., Derugin N., and Ferriero D.M. 2002. Rat forebrain neurogenesis and striatal neuron replacement after focal stroke. *Ann. Neurol.* **52:** 802–813.

Qu T.Y., Dong X.J., Sugaya I., Vaghani A., Pulido J., and Sugava K. 2004. Bromodeoxyuridine increases multipotency of human bone marrow-derived stem cells. *Restor. Neurol. Neurosci.* **22:** 459–468.

Raineteau O., Hugel S., Ozen I., Rietschin L., Sigrist M., Arber S., and Gähwiler B.H. 2006. Conditional labeling of newborn granule cells to visualize their integration into established circuits in hippocampal slice cultures. *Mol. Cell. Neurosci.* **32:** 344–355.

Rakic P. 1973. Kinetics of proliferation and latency between final cell division and onset of differentiation of cerebellar stellate and basket neurons. *J. Comp. Neurol.* **147:** 523–546.

———. 1985. Limits of neurogenesis in primates. *Science* **227:** 1054–1056.

———. 2002a. Adult neurogenesis in mammals: An identity crisis. *J. Neurosci.* **22:** 614–618.

———. 2002b. Neurogenesis in adult primates. *Prog. Brain Res.* **138:** 3–14.

———. 2002c. Neurogenesis in adult primate neocortex: An evaluation of the evidence. *Nat. Rev. Neurosci.* **3:** 65–71.

Rakic P. and Nowakowski R.S. 1981. The time of origin of neurons in the hippocampal region of the rhesus monkey. *J. Comp. Neurol.* **196:** 99–128.

Rasin M.R., Gazula V.R., Breunig J.J., Kwan K.Y., Johnson M.B., Liu-Chen S., Li H.S., Jan L.Y., Jan Y.N., Rakic P., and Sestan N. 2007. Numb and Numbl are required for maintenance of cadherin-based adhesion and polarity of neural progenitors. *Nat. Neurosci.* **10**: 819–827.

Rietze R., Poulin P., and Weiss S. 2000. Mitotically active cells that generate neurons and astrocytes are present in multiple regions of the adult mouse hippocampus. *J. Comp. Neurol.* **424**: 397–408.

Rifkin B.R., Vernillo A.T., Golub L.M., and Ramamurthy N.S. 1994. Modulation of bone resorption by tetracyclines. *Ann. N. Y. Acad. Sci.* **732**: 165–180.

Roy N.S., Cleren C., Singh S.K., Yang L., Beal M.F., and Goldman S.A. 2006. Functional engraftment of human ES cell-derived dopaminergic neurons enriched by coculture with telomerase-immortalized midbrain astrocytes. *Nat. Med.* **12**: 1259–1268.

Sands J.A., Snipes W., and Person S. 1972. Mutagenesis by tritium: Decays originating from growth and storage in tritiated water and from chemostatic growth in the presence of tritiated nucleic acid precursors. *Int. J. Radiat. Biol. Relat. Stud. Phys. Chem. Med.* **22**: 197–202.

Schroder A.R., Shinn P., Chen H., Berry C., Ecker J.R., and Bushman F. 2002. HIV-1 integration in the human genome favors active genes and local hotspots. *Cell* **110**: 521–529.

Sekerkova G., Ilijic E., and Mugnaini E. 2004. Bromodeoxyuridine administered during neurogenesis of the projection neurons causes cerebellar defects in rat. *J. Comp. Neurol.* **470**: 221–239.

Seri B., Garcia-Verdugo J.M., McEwen B.S., and Alvarez-Buylla A. 2001. Astrocytes give rise to new neurons in the adult mammalian hippocampus. *J. Neurosci.* **21**: 7153–7160.

Shimshek D.R., Kim J., Hubner M.R., Spergel D.J., Buchholz F., Casanova E., Stewart A.F., Seeburg P.H., and Sprengel R. 2002. Codon-improved Cre recombinase (iCre) expression in the mouse. *Genesis* **32**: 19–26.

Shin J.J., Fricker-Gates R.A., Perez F.A., Leavitt B.R., Zurakowski D., and Macklis J.D. 2000. Transplanted neuroblasts differentiate appropriately into projection neurons with correct neurotransmitter and receptor phenotype in neocortex undergoing targeted projection neuron degeneration. *J. Neurosci* **20**: 7404–7416.

Sidman R.L., Miale I.L., and Feder N. 1959. Cell proliferation and migration in the primitive ependymal zone: An autoradiographic study of histogenesis in the nervous system. *Exp. Neurol.* **1**: 322–333.

Snyder E.Y., Yoon C., Flax J.D., and Macklis J.D. 1997. Multipotent neural precursors can differentiate toward replacement of neurons undergoing targeted apoptotic degeneration in adult mouse neocortex. *Proc. Natl. Acad. Sci.* **94**: 11663–11668.

Sohur U.S., Emsley J.G., Mitchell B.D., and Macklis J.D. 2006. Adult neurogenesis and cellular brain repair with neural progenitors, precursors and stem cells. *Philos. Trans. R. Soc. Lond. B. Biol. Sci.* **361**: 1477–1497.

Spalding K.L., Bhardwaj R.D., Buchholz B.A., Druid H., and Frisen J. 2005. Retrospective birth dating of cells in humans. *Cell* **122**: 133–143.

Stetson P.L., Maybaum J., Wagner J.G., Averill D.R., Wollner I.S., Knol J.A., Johnson N.J., Yang Z.M., Preiskorn D., Smith P., et al. 1988. Tissue-specific pharmacodynamics of 5-bromo-2'-deoxyuridine incorporation into DNA in VX2 tumor-bearing rabbits. *Cancer Res.* **48**: 6900–6905.

Suhonen J.O., Peterson D.A., Ray J., and Gage F.H. 1996. Differentiation of adult hippocampus-derived progenitors into olfactory neurons in vivo. *Nature* **383**: 624–627.

Szybalski W. 1974. X-ray sensitization by halopyrimidines. *Cancer Chemother. Rep.* **58**: 539–557.

Tabar V., Panagiotakos G., Greenberg E.D., Chan B.K., Sadelain M., Gutin R.H., and Studer L. 2005. Migration and differentiation of neural precursors derived from human embryonic stem cells in the rat brain. *Nat. Biotechnol.* **23:** 601–606.

Tabata H. and Nakajima K. 2001. Efficient in utero gene transfer system to the developing mouse brain using electroporation: Visualization of neuronal migration in the developing cortex. *Neuroscience* **103:** 865–872.

Tashiro A., Sandler V.M., Toni N., Zhao C., and Gage F.H. 2006. NMDA-receptor-mediated, cell-specific integration of new neurons in adult dentate gyrus. *Nature* **442:** 929–933.

Thomas C.E., Ehrhardt A., and Kay M.A. 2003. Progress and problems with the use of viral vectors for gene therapy. *Nat. Rev. Genet.* **4:** 346–358.

Van Deursen J., Fornerod M., Van Rees B., and Grosveld G. 1995. Cre-mediated site-specific translocation between nonhomologous mouse chromosomes. *Proc. Natl. Acad. Sci.* **92:** 7376–7380.

van Praag H., Schinder A.F., Christie B.R., Toni N., Palmer T.D., and Gage F.H. 2002. Functional neurogenesis in the adult hippocampus. *Nature* **415:** 1030–1034.

Vega C.J. and Peterson D.A. 2005. Stem cell proliferative history in tissue revealed by temporal halogenated thymidine analog discrimination. *Nat. Methods* **2:** 167–169.

Vooijs M., Jonkers J., and Berns A. 2001. A highly efficient ligand-regulated Cre recombinase mouse line shows that LoxP recombination is position dependent. *EMBO Rep.* **2:** 292–297.

Weber P., Metzger D., and Chambon P. 2001. Temporally controlled targeted somatic mutagenesis in the mouse brain. *Eur. J. Neurosci.* **14:** 1777–1783.

Yamamoto S., Yamamoto N., Kitamura T., Nakamura K., and Nakafuku M. 2001. Proliferation of parenchymal neural progenitors in response to injury in the adult rat spinal cord. *Exp. Neurol.* **172:** 115–127.

Yang Y., Geldmacher D.S., and Herrup K. 2001. DNA replication precedes neuronal cell death in Alzheimer's disease. *J. Neurosci.* **21:** 2661–2668.

Yrjanheikki J., Keinanen R., Pellikka M., Hokfelt T., and Koistinaho J. 1998. Tetracyclines inhibit microglial activation and are neuroprotective in global brain ischemia. *Proc. Natl. Acad. Sci.* **95:** 15769–15774.

Yu T.S., Dandekar M., Monteggia L.M., Parada L.F., and Kernie S.G. 2005. Temporally regulated expression of Cre recombinase in neural stem cells. *Genesis* **41:** 147–153.

Zhao M., Momma S., Delfani K., Carlen M., Cassidy R.M., Johansson C.B., Brismar H., Shupliakov O., Frisen J., and Janson A.M. 2003. Evidence for neurogenesis in the adult mammalian substantia nigra. *Proc. Natl. Acad. Sci.* **100:** 7925–7930.

Zhu Z., Ma B., Homer R.J., Zheng T., and Elias J.A. 2001. Use of the tetracycline-controlled transcriptional silencer (tTS) to eliminate transgene leak in inducible over-expression transgenic mice. *J. Biol. Chem.* **276:** 25222–25229.

Zigova T. and Newman M.B. 2002. Transplantation into neonatal rat brain as a tool to study properties of stem cells. *Methods Mol. Biol.* **198:** 341–356.

5

The Use of Reporter Mouse Lines to Study Adult Neurogenesis

Grigori Enikolopov

Cold Spring Harbor Laboratory
Cold Spring Harbor, New York 11724

Linda Overstreet-Wadiche

Department of Neurobiology
University of Alabama at Birmingham
Birmingham, Alabama 35294

A LONG-STANDING PROBLEM IN THE FIELD of adult neurogenesis has been the need to identify newborn neurons and their precursors within a much larger population of preexisting mature neurons and glia. If these nascent cells could be identified, it would be possible to visualize and enumerate such cells in vivo, to access them for electrophysiological and molecular studies, to identify their connections in the neuronal networks, and to alter their activity and function. Several strategies have been developed to solve this problem of finding the proverbial needle in a haystack. Methods such as labeling with thymidine analogs, phenotypic analysis based on the expression of developmental markers, and retro- and lentiviral labeling have each had an important role in advancing our understanding of the proliferation and maturation of newborn neurons in the adult brain. As with all methods, these techniques have advantages and limits that demarcate their appropriate application. In this review, we focus on genetic approaches to studying adult mammalian neurogenesis, describing reporter lines of transgenic mice and summarizing recent advances that employ these emerging technologies.

The general strategy of these genetic approaches is to drive the expression of "live" markers such as green fluorescent protein (GFP) in a defined population of neurons, neuronal progenitors, or stem cells. Cytoplasmic expression of fluorescent proteins (FPs) allows the full morphology of labeled cells to be visualized, whereas nuclear expression of such proteins facilitates cell enumeration. FP expression also allows labeled cells to be identified and accessed in live animals and in acute brain slices, thus permitting electrophysiological analysis. Creation of new genetic tools for neurogenesis research has been propelled by rapid technical advancements in mouse genetics. This includes the proliferation of fluorescent genetically encoded markers (Giepmans et al. 2006), the use of regulated site-specific recombinases for inducible genetic fate mapping (Joyner and Zervas 2006), the advent of binary and ternary genetic systems for cell labeling (Miyoshi and Fishell 2006), generation of large collections of mouse lines carrying FPs driven by cell-specific promoters (Gong et al. 2003; Valenzuela et al. 2003), and techniques to accelerate the production of mutant mouse lines (Poueymirou et al. 2007).

Genetic approaches are critically dependent on the availability of promoter elements that direct transgene expression to defined cell populations in the nervous system. An important new source of reporter mouse lines comes from large-scale efforts to generate collections of transgenic lines expressing GFP in restricted cell populations in the adult brain. This includes The Gene Expression Nervous System Atlas (GENSAT) initiative to produce transgenic mice expressing GFP within bacterial artificial chromosome (BAC) vectors (Gong et al. 2003), generation of a genome-wide collection of knockout animals and embryonic stem (ES) cells in which actual genes are substituted by genes coding for FPs (the VelociGene collection; Valenzuela et al. 2003), and the Pleiades Promoter Project (http://www.pleiades.org/), which uses characterized promoter elements to drive FP expression in transgenic lines. Here, we focus on the currently used transgenic reporter mouse lines, most of which have been generated using well-characterized promoters and regulatory elements from the nestin, doublecortin (DCX), and pro-opiomelanocortin (POMC) genes. These lines allow selective identification of cells at specific developmental stages in neurogenic regions of the adult brain and have gained widespread use as tools for investigating adult neurogenesis.

NESTIN-BASED TRANSGENIC REPORTER LINES

Nestin (for neuroepithelial stem cells) is an intermediate filament protein whose expression marks stem and progenitor cells in the developing and adult CNS and PNS. Nestin was originally identified in the embryonic brain as a marker of neuroepithelial precursor cells (Lendahl et al. 1990). It continues to be expressed in undifferentiated cells of the developing nervous system and, in the adult brain, marks the regions of continuous neurogenesis. Nestin is also present in the embryonic myoblasts, tooth buds, and developing testis, among other embryonic tissues. In the adult organism, nestin can be detected in stem/progenitor-like cells in a range of tissues (retina, mammary gland, testis, pituitary, liver, pancreas, hair follicle), including cells of nonectodermal origin. Nestin expression is usually down-regulated upon differentiation, but it can be transiently reactivated upon injury or stress (e.g., in the reactive astrocytes and in the ependymal cells of the spinal cord).

The key regulatory elements of the nestin gene that direct its expression in the developing and adult nervous system are located in the second intron. These enhancer elements include binding sites for class III POU and group B1 Sox transcription factors and nuclear hormone receptors. A 257-bp fragment encompassing these binding sites directs expression of a transgene in the developing nervous system in a transient transgenic assay (Josephson et al. 1998; Yaworsky and Kappen 1999), and a 636-bp fragment is sufficient to drive transgene expression in the neurogenic regions of the adult CNS (Kawaguchi et al. 2001). Note that although the intronic sequences are evolutionarily highly conserved, reliable expression in the CNS of adult transgenic mice was only observed with the rat (but not human or mouse) enhancer elements (Johansson et al. 2002). Expression of the endogenous nestin gene is driven by its Sp1-containing promoter; however, in transgenic constructs, it can be replaced by heterologous promoters (e.g., from heat-shock protein 86 gene, *hsp68*, or herpesvirus thymidine kinase gene, *tk*) without apparent loss of specificity, provided the enhancer elements from the second intron are present (Zimmerman et al. 1994; Josephson et al. 1998; Kawaguchi et al. 2001).

Regulatory elements of the nestin gene have been used by several groups to generate reporter transgenic lines that express GFP or other fluorescent proteins in neural stem and progenitor cells of the adult nervous system (Yamaguchi et al. 2000; Kawaguchi et al. 2001; Mignone et al. 2004). The precise arrangements of the regulatory elements driving expression of the reporter transgene were different for each line, for example, 636 bp of the second intron enhancer of the rat nestin gene followed by the minimal hsp68 promoter (Kawaguchi et al. 2001), 1.8 kb of the second intron in combination with 2.5 kb of the rat nestin promoter (Yamaguchi et al. 2000), or 1.8 kb of the second intron and 5.8 kb of the nestin promoter (Mignone et al. 2004). However, the overall patterns of reporter expression in the zones of adult neurogenesis (subventricular zone [SVZ]; rostral migratory stream [RMS]; olfactory bulb [OB]; and subgranular zone [SGZ]) are remarkably similar for these lines.

Within the neurogenic regions of the brain, nestin-driven FP transgenes are expressed in stem and early progenitor cells (Fig. 1A). This conclusion is based on the following observations:

1. FPs are expressed in cells whose morphology and immunophenotype are consistent with them being stem or early progenitor cells, for example, radial glia-like cells in the SGZ that express nestin, glial fibrillary acidic protein (GFAP), vimentin, brain lipid-binding protein (BLBP), and Sox2; GFAP-expressing cells in the SVZ and RMS; or DCX-expressing cells in the RMS or dentate gyrus (DG).
2. FPs are not expressed in cells that have undergone differentiation (i.e., express neuronal, astrocytic, or oligodendrocytic markers).
3. FPs are expressed in most of the nestin-expressing cells of the SVZ, RMS, OB, and DG; moreover, in most cases, there is a correspondence between the strength of the GFP signal and nestin immunoreactivity. Note, however, that the overlap is sometimes not complete (e.g., some of the GFP-expressing cells in the OB are only weakly positive for nestin (Yamaguchi et al. 2000; Mignone et al. 2004). This may reflect the specificity of the enhancer, sensitivity of detecting GFP versus the endogenous protein, or the turnover of the transgene or nestin RNA or protein.
4. Transcriptional profiles of the populations of nestin-GFP cells with different GFP signal strengths are characteristic of cells with stem cell properties (G. Enikolopov, unpubl.).
5. Nestin-GFP cells isolated from the SVZ or the DG are capable of forming neurospheres that, when plated onto an appropriate substrate, generate differentiated neurons, astrocytes, and oligodendrocytes.
6. Nestin-GFP cells are strongly enriched in neurosphere-forming cells (up to 1400-fold); conversely, most of the neurosphere-forming cells of the adult brain are nestin-GFP-positive (Sawamoto et al. 2001; Mignone et al. 2004).
7. Isolated nestin-GFP cells or differentiated cells derived from nestin-GFP cells can form neurons in vivo after transplantation into the embryonic chick neuronal tube (G. Enikolopov et al., unpubl.) or striatum of adult rats with Parkinsonian lesions (Sawamoto et al. 2001).

Enhancer elements residing in the second intron of the nestin gene are thus necessary and sufficient to direct the transgene expression into

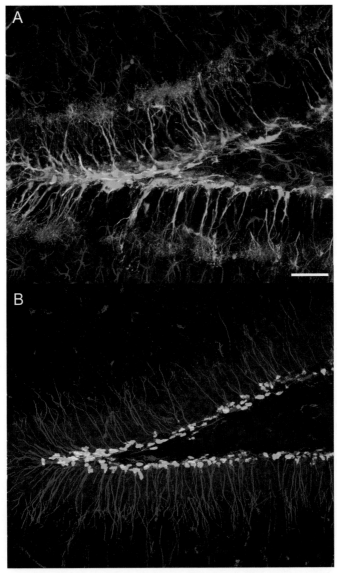

Figure 1. Neuronal stem and progenitor cells are visualized in nestin-GFP and nestin-CFPnuc mice. (*A*) Neuronal stem cells and progenitors (*green*) in the dentate gyrus (DG) of a nestin-GFP mouse. (*Red*) Immunostaining for GFAP. (*B*) Neuronal stem and progenitor cells (*green*) in the DG of a nestin-CFPnuc mouse. (*Red*) Immunostaining for nestin; (*blue*) DAPI. Bar, 100 μm in both *A* and *B*.

neuronal stem and progenitor cells of the adult nervous system, confirming their utility for studies on adult neurogenesis. Indeed, the nestin enhancer has been used to generate other lines of fluorescent reporter mice (Encinas et al. 2006; Tran et al. 2007) and mice carrying Cre recombinase (Tronche et al. 1999; Imayoshi et al. 2006; Burns et al. 2007) or tetracycline-responsive transcription activators (Mitsuhashi et al. 2001; Beech et al. 2004; Yu et al. 2005). Together, these are building an impressive collection of tools for the study of adult neurogenesis.

Interestingly, nestin-FP transgenic animals also report the presence of stem/progenitor-like cells in other (nonneural) tissues. They have been used to identify multipotent cells in the hair follicle (Li et al. 2003; Amoh et al. 2005; J.L. Mignone et al., in prep.), oval cells in the liver (Gleiberman et al. 2005), satellite cells in the skeletal muscle (Day et al. 2007), mural cells in the kidney medulla (Patschan et al. 2007), precursors to insulin-secreting β cells in the pancreas (Seaberg et al. 2004; Carriere et al. 2007), precursors to Leydig cells in the testis (Davidoff et al. 2004), and stem cells in the adult pituitary (A.S. Gleiberman et al., in prep.). Thus, expression of nestin may define an undifferentiated state for various types of stem/ progenitor cells in nonneural tissues, and the use of transgenic reporters may enable their identification, visualization, and isolation.

USING NESTIN-BASED REPORTER LINES TO QUANTIFY NEURONAL STEM AND PROGENITOR CELLS

Quantitative analysis of neurogenesis involves immunophenotyping followed by stereological enumeration of cells in each subclass. However, accurate analysis of neuronal precursors is often problematic: Complex morphology of stem and progenitor cells (consider, for example, that GFAP, nestin, and vimentin, often used to identify stem-like cells in the DG, are only present in the processes but not in the soma of these cells), their high density, the need to apply double and triple immunolabeling, and uncertainties in defining distinct boundaries between different subclasses of precursors present real challenges when precise counts are required. This problem is especially acute when analyzing young brain tissue, where the density of stem and progenitor cells is particularly high, or when the changes evoked by a neurogenic or antineurogenic stimulus are low (note that the majority of the stimuli which can induce adult neurogenesis often increase the number of newly generated cells only by 30–50%). Together, this reduces the precision and confidence in evaluating changes in particular subclasses of precursor cells in the brain. These challenges of counting immunophenotyped cells stand in contrast to the bromodeoxyuridine (BrdU) or thymidine labeling of cells: In these cases, the signal is restricted to the nucleus (and thus converted into a dot in the visual field), which greatly facilitates scoring of the signal and accurate enumeration of dividing and newborn cells.

For the quantitative analysis of adult neurogenesis, transgenic reporter lines in which the reporter FP is directed to the nucleus have been generated. In one such line, the reporter, a cyan fluorescent protein (CFP), is fused to the SV40 nuclear localization signal (Encinas et al. 2006); in another line, GFP is fused to histone H2B (G. Enikolopov, unpubl.). In both cases, the FP becomes translocated into the cell nucleus. Both lines use the same nestin regulatory elements as the nestin-GFP line (Mignone et al. 2004), and the reporter is expressed, as expected, in the developing nervous system and in the SVZ, RMS, OB, and SGZ of the adult brain. Importantly, however, the distribution of stem and progenitor cells in the neurogenic areas of reporter mice can be visualized as a dotted pattern corresponding to the nuclei of these cells (Fig. 2). This nuclear representation of stem/progenitor cells greatly reduces the complexity of their distribution pattern and permits their unambiguous enumeration, thus capturing the power of BrdU- or thymidine-based enumeration of nuclei. In pilot experiments, we carefully compared the structures of the SVZ and DG as revealed by immunochemistry for nestin and by expression of nestin-GFP, nestin-CFPnuc, or nestin-H2B-GFP. Whereas we were unable to generate accurate counts of nestin- or nestin-GFP-positive cells (particularly in the young brain), we were able to unambiguously count all of the labeled nuclei in the SVZ and DG of the nestin-CFPnuc and nestin-H2B-GFP mice (Encinas et al. 2006; J.M. Encinas and G. Enikolopov, unpubl.).

We have now successfully used the reporter mice with nuclear signal localization for quantitative analysis of adult and early postnatal neurogenesis. For instance, we were able to estimate the changes induced by the selective serotonin reuptake inhibitor (SSRI) antidepressant fluoxetine (Encinas et al. 2006), other antidepressant drugs and treatments, or radiation. Using these reporters, it is possible to reliably detect small (5–10%) statistically significant changes in subclasses of neuronal precursors using only 4–5 mice per group (G. Enikolopov et al., unpubl.).

For experiments that require both morphological and quantitative analysis of neurogenesis, use of two reporter lines, nestin-GFP and nestin-CFPnuc, is particularly helpful. In nestin-GFP mice, the fluorescent signal highlights all of the soma and the processes of stem and early progenitor cells (Fig. 1A), and these mice are very well suited for the studies of the morphology of neuronal precursors in the developing and adult brain. In contrast, in nestin-CFPnuc mice, the signal is localized in the

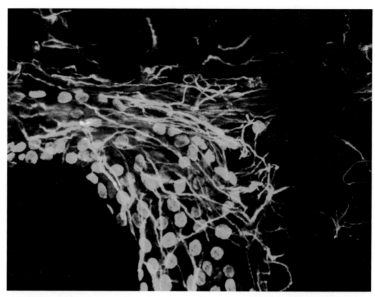

Figure 2. Neuronal stem and progenitor cells in the SVZ of nestin-CFPnuc mice. (*Blue*) Immunostaining for GFAP; (*red*) BrdU; (*green*) nestin-CFPnuc.

cell nucleus and the distribution of the stem and progenitor cells is visualized as a punctate pattern (Fig. 1B). Thus, these two reporter lines complement each other and allow visualization and counting of neural stem and progenitor cells. Importantly, crosses between these two lines allow simultaneous visualization of the soma and the nuclei of stem and progenitor cells; thus, we were able to follow the morphological changes in these cells while enumerating them (Fig. 3) (A.-S. Chiang et al., unpubl.).

In summary, the use of nestin-based transgenic reporter lines with cytoplasmic or nuclear localization of the fluorescent signal circumvents several obstacles in assessing changes in cell number during neurogenesis, for example, high cell density that hinders precise counts or uncertainty in attributing precursor cells to a particular class. It reduces the complex distribution pattern of precursor cells to a readily quantifiable punctate pattern of labeled nuclei. It allows unambiguous enumeration of cells in a particular precursor class and can be used to analyze changes induced by a wide range of stimuli in the developing or adult brain.

USING NESTIN-BASED TRANSGENIC LINES
TO ANALYZE NEURONAL MATURATION

Nestin-GFP mice have also been used to study the electrophysiological properties of neuronal precursors and potential signaling mechanisms

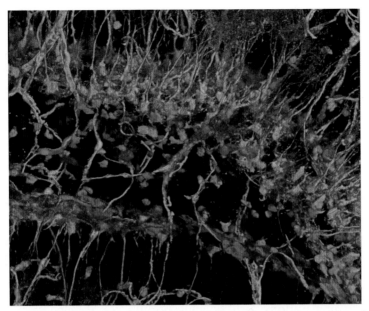

Figure 3. In the hybrid nestin-GFP/nestin-CFPnuc mice, both the nuclei and soma of stem and progenitor cells can be visualized simultaneously.

involved in the maturation of newborn granule cells. Early studies showed that labeled cells in acute hippocampal slices can be broadly categorized into two classes, based on intrinsic membrane properties (Filippov et al. 2003; Fukuda et al. 2003). Cells with radial processes that expressed GFAP have passive membrane properties and low input resistance typical of astrocytes (type 1 cells), whereas smaller nestin-expressing cells with short processes have more immature neuronal features, such as higher input resistance and voltage-gated currents (type II cells). A subpopulation of the latter cells also receives GABAergic synaptic currents (Wang et al. 2005). Depolarizing GABAergic input may be important to promote neuronal differentiation of nestin-expressing progenitors (Tozuka et al. 2005) and maturation of newborn neurons (Ge et al. 2006, 2007; Overstreet-Wadiche and Westbrook 2006; Overstreet-Wadiche et al. 2006a).

DOUBLECORTIN-BASED TRANSGENIC LINES

DCX is a widely used marker for phenotypic identification of neuronal precursors and immature neurons. Mutations in the DCX gene result in cortical malformation disorders, such as lissencephaly and subcortical laminar heterotopia ("double cortex" syndrome) (des Portes et al. 1998). DCX is a microtubule-associated protein that is primarily implicated in neuronal migration and neurite outgrowth (Francis et al. 1999; Gleeson et al. 1999). These functions are likely mediated by its ability to stabilize mictrotubules and regulate interactions between the actin and tubulin cytoskeleton (Tsukada et al. 2005). DCX is not expressed exclusively in newborn cells, as it can be reexpressed in certain populations of mature postmitotic neurons that are undergoing structural plasticity (Nacher et al. 2004). However, in the DG and olfactory system, it is selectively and reliably expressed in neuronal precursors and newborn neurons (Brown et al. 2003; Rao and Shetty 2004; Couillard-Despres et al. 2005).

Regulatory sequences upstream of the human DCX gene can drive expression of reporter genes in embryonic and adult neuronal precursors (Karl et al. 2005). Recently, Couillard-Despres et al. (2006) generated transgenic mice using the identified 3.5-kb fragment upstream of the DCX ATG start codon to drive expression of GFP or *Discoma* sp. Reef coral red fluorescent protein (DsRed2). The transgenic markers are expressed in a temporal pattern similar to the endogenous DCX protein, thus providing a convenient method of identifying newborn cells in living preparations such as acute brain slices. Labeled cells have a variety of morphologies (Fig. 4A), consistent with the range of developmental stages that are characterized by DCX expression. Electrophysiological recordings of DsRed-labeled cells in the hippocampus of young and aged mice reveal high input resistances that correlate with cell morphology, consistent with newborn adult-generated granule cells that express polysialic-acid–neural cell adhesion molecule (PSA-NCAM) (Schmidt-Hieber et al. 2004), POMC-GFP (Overstreet et al. 2004), and retroviral vectors (Esposito et al. 2005). Although the number of labeled cells is reduced during aging, high input resistance is retained in immature cells in aged mice (Couillard-Despres et al. 2006), consistent with the idea that newborn granule cells undergo a stereotyped progression of maturation throughout the life of an animal (Esposito et al. 2005; Laplagne et al. 2006; Overstreet-Wadiche et al. 2006a).

POMC-GFP REPORTER MICE

In principle, transgenic methods can allow selective labeling of very specific populations of adult-generated cells as they progress through stages of neuronal maturation. In some cases, transgenic mice generated for other purposes can be useful in this regard. The discovery that newborn granule cells are labeled in POMC-GFP transgenic mice allows selective identification of early postmitotic granule cells. POMC is a prohormone

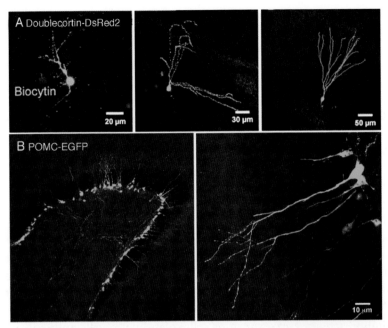

Figure 4. DCX and POMC reporters label newborn granule cells at various developmental stages. (*A*) Examples of individual Dcx-DsRed2-labeled granule cells that exhibit heterogeneous degrees of morphological development. Dendrite length and complexity are inversely correlated with input resistance and represent increasing stages of maturity (*left to right*). Dcx-DsRed2-expressing cells were filled with biocytin (shown in *green*) in hippocampal slices from 16–21-month-old mice. (Modified, with permission, from Couillard-Despres et al. 2006 [© Blackwell Publishing].) (*B*) A low-magnification image shows GFP expression in the dentate gyrus (DG) of POMC-GFP transgenic mice (*left*). (Modified, with permission, from Overstreet et al. 2004.) Most GFP-expressing granule cells have a homogeneous morphology with dendrites that branch within the inner molecular layer but do not extend to the outer molecular layer (*right*).

expressed in the pituitary and hypothalamus that regulates feeding and energy homeostasis. Transgenic mice with GFP expressed under the control of regulatory sequences in the POMC promoter have GFP colocalized with POMC peptides in those brain regions (Young et al. 1998). Independent lines of mice also have GFP expression in the DG that is not correlated with POMC mRNA or protein, but rather labels newborn granule cells (Overstreet et al. 2004). DG expression requires cryptic promoter sequences distinct from those required for expression in POMC-producing cells of the pituitary. Thus, POMC-GFP in the DG appears to be unrelated to expression of the POMC gene product. In a similar unexpected discovery, selective labeling of newborn granule cells occurs in the hippocampus of GAD67-GFP transgenic mice (C. Zhao et al., unpubl.). These unanticipated tools for studying adult neurogenesis suggest that there may be other alternative uses for the growing number of transgenic mice that are currently available. However, these results also highlight an important caveat of transgenic reporter mice: The expression pattern does not always precisely replicate the temporal and spatial pattern of the endogenous gene product. Thus, thorough characterization of a mouse line is necessary before it can be considered a reliable tool.

Characterization of hippocampal FP expression in the POMC-GFP mouse reveals that labeled cells have a generally homogeneous morphology, consisting of small cell bodies located near the SGZ and a primary dendrite that branches within the inner molecular layer (Fig. 4B) (Overstreet et al. 2004). GFP-labeled cells have membrane properties consistent with immature neurons, including a high input resistance, long time constant, and small action potentials. They express markers of immature granule cells, such as PSA-NCAM and DCX, but they lack GFAP and nestin immunoreactivity. GFP-expressing cells colabel with BrdU approximately 2 weeks after BrdU administration in adult mice, confirming that they are adult-generated newborn neurons. Furthermore, parallel to age- and exercise-induced changes in neuronal proliferation, the number of GFP-labeled cells declines with age and is enhanced by exercise and seizure activity (Overstreet-Wadiche et al. 2006b).

USING REPORTER LINES TO DEFINE STAGES OF ADULT NEUROGENESIS

The growing number of available mouse lines allows investigators to apply the most appropriate transgenic model to the question at hand. Understanding the stage-specific profile of transgenic labeling is critical for this choice. Figure 5 illustrates stages of granule cell maturation in the DG that can be identified with currently available mice (note that although for some lines the marker expression is thoroughly characterized with regard to specific stages of adult neurogenesis [e.g., for nestin-GFP, DCX-GFP, and POMC-GFP], we had to deduce the stages covered by marker expression in other lines).

Nestin-based transgenics reveal the earliest stages of the neuronal differentiation cascade in the DG. In all lines, FPs are expressed in the earliest progenitors, represented by the radial-like cells in the SGZ, and defined as quiescent neuronal progenitors (QNPs) in the scheme in

Figure 5. These cells have also been described as type-1 cells (Kronenberg et al. 2003; Kempermann et al. 2004), GFP-bright cells (Mignone et al. 2004), and rA cells (Seri et al. 2004). FPs are also expressed in progeny of these QNP cells, which still reside in the SGZ but lack the radial processes (amplifying neuronal progenitors [ANP] in Fig. 5; these cells have also been described as type-2 cells) (Kronenberg et al. 2003; Kempermann et al. 2004) and GFP-dim cells (Mignone et al. 2004). FP expression ceases as progenitors exit the cell cycle, and only in some transgenic lines can GFP be visualized, usually using immunocytochemistry, in postmitotic cells. A similar pattern of GFP expression is seen in transgenic mice carrying the Sox2-GFP transgene (Ellis et al. 2004; Brazel et al. 2005). Radial glia-like QNP cells are also marked in transgenic mice expressing GFP under the control of the human GFAP gene promoter (Nolte et al. 2001; Liu et al. 2006; Steiner et al. 2006). Very few progeny cells (ANPs) express GFP in these lines. BLBP-FP animals (Schmid et al. 2006) also have radial glia-like QNP cells labeled. FP expression in ANP cells is less clear. QNPs are also labeled in Hes-GFP mice (Ohtsuka et al. 2006), although GFP is also expressed in cells outside the SGZ and in the hilus in these mice.

DCX expression closely follows cessation of nestin expression; accordingly, there is a limited overlap between the transgene expression in DCX-GFP and nestin-GFP mice. Colabeling studies with BrdU and the cell cycle marker Ki-67 suggest that the majority (>70%) of DCX-expressing cells are postmitotic, whereas the remaining are precursors

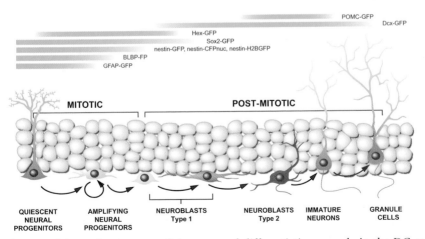

Figure 5. Schematic summary of the neuronal differentiation cascade in the DG as reflected in transgenic reporter mouse lines.

involved in the cell cycle (Plumpe et al. 2006). These authors also found that a large proportion of BrdU+ cells still expressed DCX even 30 days after a single BrdU injection (but see Brown et al. 2003). The prolonged window of expression (>1 month) presumably underlies the heterogeneity in dendrite morphologies and electrophysiological characteristics of DCX-GFP labeled cells (Fig. 4). The prolonged expression further suggests that DCX transgenics may be ideal for applications where visualization of newborn cells that encompass a variety of developmental stages is desired, such as in aging studies when reduced proliferation results in very few newborn neurons. DCX transgenics will also be useful to explore the synaptic properties of newborn granule cells that are sufficiently mature to receive glutamatergic input from the perforant path.

The temporal window of POMC-GFP expression is shorter than that seen in DCX transgenics. The uniform membrane properties and morphology of POMC-GFP-labeled cells suggest that they constitute a relatively brief developmental stage. At this stage, labeled granule cells have functional glutamate receptors, but they lack dendritic spines and synaptic input from the perforant path, likely because their short dendrites do not extend through the middle molecular layer (Overstreet et al. 2004). Thus, POMC-GFP mice are not useful for studying the properties of newly generated glutamatergic synapses under normal conditions. However, POMC-GFP-labeled cells are partially integrated into the neural network in the sense that they receive depolarizing GABAergic synaptic input (Overstreet Wadiche et al. 2005). In adults, peak colabeling between BrdU and GFP occurred at 12 days and declined before 30 days (Overstreet et al. 2004). The absence of BrdU colabeling at <3-day intervals and >30-day intervals indicates that labeled cells are not dividing precursors, nor mature granule cells. Thus, POMC-GFP-labeled cells comprise a subset of DCX-labeled cells that are early postmitotic neurons (Overstreet-Wadiche and Westbrook 2006). POMC-GFP mice are useful for studying newborn neurons at this particular developmental stage, a stage that appears to be highly sensitive to environmental conditions. For example, POMC-GFP-labeled cells in adults develop more slowly compared to those in neonates (Overstreet-Wadiche et al. 2006a). In addition, seizure activity can accelerate the maturation of POMC-GFP-labeled granule cells in adults (Overstreet-Wadiche et al. 2006b).

ADVANTAGES AND LIMITATIONS OF TRANSGENIC REPORTER SYSTEMS

Transgenic reporter lines with genetically encoded markers have several advantages when identifying newborn neurons and their precursors.

Similar to retroviral labeling, transgenic labeling allows full morphological analysis of labeled cells in fixed and in live preparations, with the additional advantage that quantitative analysis can be performed because expression levels of the marker are independent of injection parameters such as location and viral titer. Additionally, the transgenic reporter lines with nuclear FPs can be used for precise quantification of subtle changes in the size of particular populations of cells. Transgenic systems are also noninvasive; thus, there are no potential confounding factors resulting from inflammation and injury at the injection site. Furthermore, reporter mice can easily be crossed with other genetically modified mice to examine the role of specific genes or mutations on adult neurogenesis. It is important to consider, however, that transgenic systems generally use markers that are transiently expressed in individual cells during specific stages of their development. Because some developmentally regulated proteins might be reexpressed in mature neurons that may be undergoing structural or functional changes, specificity of the transgene expression in newborn cells must be confirmed by other approaches. Furthermore, since transgenic labeling identifies cells based on their ability to express a stage-specific marker, transgenic reporter lines provide a snapshot of a population of cells at a particular developmental stage. In contrast, retroviral or BrdU labeling permanently marks cells generated at a specific time ("birth date"), allowing individual cells to be tracked over time. The interpretations of the two types of labeling (genetic vs. retroviral or BrdU) are thus quite distinct and warrant careful consideration when choosing which approach to use in studying a continually regenerating and maturing population of newborn neurons and their precursors.

Newer transgenic approaches that exploit inducible and conditional expression systems, such as tamoxifen-dependent recombination (e.g., using Cre-ER, a fusion of Cre recombinase and the hormone-binding domain of the estrogen receptor) or tetracycline-dependent gene expression or recombination (e.g., using Tet-based activators), may combine the advantages of both labeling strategies to provide even more powerful methods to label cells and manipulate adult neurogenesis.

SUMMARY

Transgenic mouse lines expressing genetically encoded reporters have become an important tool for studying adult neurogenesis. Their use provides the ability to visualize neuronal stem and progenitor cells at different stages of neuronal differentiation, to examine their properties in vivo, to quantify and isolate these cells, and to manipulate their activity.

Available reporter lines cover all of the important stages of neuronal maturation and have already provided a wealth of important information on stem and progenitor cells and newborn neurons in the adult brain. Further progress will arrive with the increased repertoire of fluorescent markers, sophisticated binary and ternary systems for cell labeling, and emerging collections of systematically generated transgenic reporter lines.

ACKNOWLEDGMENTS

We thank Drs. Juan Manuel Encinas, Julian Banerji, and Natalia Peunova for discussion and comments, and Drs. Encinas and Ann-Shyn Chiang for the images.

REFERENCES

Amoh Y., Li L., Katsuoka K., Penman S., and Hoffman R.M. 2005. Multipotent nestin-positive, keratin-negative hair-follicle bulge stem cells can form neurons. *Proc. Natl. Acad. Sci.* **102:** 5530–5534.

Beech R.D., Cleary M.A., Treloar H.B., Eisch A.J., Harrist A.V., Zhong W., Greer C.A., Duman R.S., and Picciotto M.R. 2004. Nestin promoter/enhancer directs transgene expression to precursors of adult generated periglomerular neurons. *J. Comp. Neurol.* **475:** 128–141.

Brazel C.Y., Limke T.L., Osborne J.K., Miura T., Cai J., Pevny L., and Rao M.S. 2005. Sox2 expression defines a heterogeneous population of neurosphere-forming cells in the adult murine brain. *Aging Cell* **4:** 197–207.

Brown J.P., Couillard-Despres S., Cooper-Kuhn C.M., Winkler J., Aigner L., and Kuhn H.G. 2003. Transient expression of doublecortin during adult neurogenesis. *J. Comp. Neurol.* **467:** 1–10.

Burns K.A., Ayoub A.E., Breunig J.J., Adhami F., Weng W.L., Colbert M.C., Rakic P., and Kuan C.Y. 2007. Nestin-CreER mice reveal DNA synthesis by nonapoptotic neurons following cerebral ischemia-hypoxia. *Cereb. Cortex* (in press).

Carrière C., Seeley E.S., Goetze T., Longnecker D.S., and Korc M. 2007. The Nestin progenitor lineage is the compartment of origin for pancreatic intraepithelial neoplasia. *Proc. Natl. Acad. Sci.* **104:** 4437–4442.

Couillard-Despres S., Winner B., Schaubeck S., Aigner R., Vroemen M., Weidner N., Bogdahn U., Winkler J., Kuhn H.G., and Aigner L. 2005. Doublecortin expression levels in adult brain reflect neurogenesis. *Eur. J. Neurosci.* **21:** 1–14.

Couillard-Despres S., Winner B., Karl C., Lindemann G., Schmid P., Aigner R., Laemke J., Bogdahn U., Winkler J., Bischofberger J., and Aigner L. 2006. Targeted transgene expression in neuronal precursors: Watching young neurons in the old brain. *Eur. J. Neurosci.* **24:** 1535–1545.

Davidoff M.S., Middendorff R., Enikolopov G., Riethmacher D., Holstein A.F., and Muller D. 2004. Progenitor cells of the testosterone-producing Leydig cells revealed. *J. Cell Biol.* **167:** 935–944.

Day K., Shefer G., Richardson J.B., Enikolopov G., and Yablonka-Reuveni Z. 2007. Nestin-GFP reporter expression defines the quiescent state of skeletal muscle satellite cells. *Dev. Biol.* **304:** 246–259.

des Portes V., Pinard J.M., Billuart P., Vinet M.C., Koulakoff A., Carrie A., Gelot A., Dupuis E., Motte J., Berwald-Netter Y., et al. 1998. A novel CNS gene required for neuronal migration and involved in X-linked subcortical laminar heterotopia and lissencephaly syndrome. *Cell* **92:** 51–61.

Ellis P., Fagan B.M., Magness S.T., Hutton S., Taranova O., Hayashi S., McMahon A., Rao M., and Pevny L. 2004. SOX2, a persistent marker for multipotential neural stem cells derived from embryonic stem cells, the embryo or the adult. *Dev. Neurosci.* **26:** 148–165.

Encinas J.M., Vaahtokari A., and Enikolopov G. 2006. Fluoxetine targets early progenitor cells in the adult brain. *Proc. Natl. Acad. Sci.* **103:** 8233–8238.

Esposito M.S., Piatti V.C., Laplagne D.A., Morgenstern N.A., Ferrari C.C., Pitossi F.J., and Schinder A.F. 2005. Neuronal differentiation in the adult hippocampus recapitulates embryonic development. *J. Neurosci.* **25:** 10074–10086.

Filippov V., Kronenberg G., Pivneva T., Reuter K., Steiner B., Wang L.P., Yamaguchi M., Kettenmann H., and Kempermann G. 2003. Subpopulation of nestin-expressing progenitor cells in the adult murine hippocampus shows electrophysiological and morphological characteristics of astrocytes. *Mol. Cell. Neurosci.* **23:** 373–382.

Francis F., Koulakoff A., Boucher D., Chafey P., Schaar B., Vinet M.C., Friocourt G., McDonnell N., Reiner O., Kahn A., et al. 1999. Doublecortin is a developmentally regulated, microtubule-associated protein expressed in migrating and differentiating neurons. *Neuron* **23:** 247–256.

Fukuda S., Kato F., Tozuka Y., Yamaguchi M., Miyamoto Y., and Hisatsune T. 2003. Two distinct subpopulations of nestin-positive cells in adult mouse dentate gyrus. *J. Neurosci.* **23:** 9357–9366.

Ge S., Pradhan D.A., Ming G.L., and Song H. 2007. GABA sets the tempo for activity-dependent adult neurogenesis. *Trends Neurosci.* **30:** 1–8.

Ge S., Goh E.L., Sailor K.A., Kitabatake Y., Ming G.L., and Song H. 2006. GABA regulates synaptic integration of newly generated neurons in the adult brain. *Nature* **439:** 589–593.

Giepmans B.N., Adams S.R., Ellisman M.H., and Tsien R.Y. 2006. The fluorescent toolbox for assessing protein location and function. *Science* **312:** 217–224.

Gleeson J.G., Lin P.T., Flanagan L.A., and Walsh C.A. 1999. Doublecortin is a microtubule-associated protein and is expressed widely by migrating neurons. *Neuron* **23:** 257–271.

Gleiberman A.S., Encinas J.M., Mignone J.L., Michurina T., Rosenfeld M.G., and Enikolopov G. 2005. Expression of nestin-green fluorescent protein transgene marks oval cells in the adult liver. *Dev. Dyn.* **234:** 413–421.

Gong S., Zheng C., Doughty M.L., Losos K., Didkovsky N., Schambra U.B., Nowak N.J., Joyner A., Leblanc G., Hatten M.E., and Heintz N. 2003. A gene expression atlas of the central nervous system based on bacterial artificial chromosomes. *Nature* **425:** 917–925.

Imayoshi I., Ohtsuka T., Metzger D., Chambon P., and Kageyama R. 2006. Temporal regulation of Cre recombinase activity in neural stem cells. *Genesis* **44:** 233–238.

Johansson C.B., Lothian C., Molin M., Okano H., and Lendahl U. 2002. Nestin enhancer requirements for expression in normal and injured adult CNS. *J. Neurosci. Res.* **69:** 784–794.

Josephson R., Muller T., Pickel J., Okabe S., Reynolds K., Turner P.A., Zimmer A., and McKay R.D. 1998. POU transcription factors control expression of CNS stem cell-specific genes. *Development* **125:** 3087–3100.

Joyner A.L. and Zervas M. 2006. Genetic inducible fate mapping in mouse: Establishing genetic lineages and defining genetic neuroanatomy in the nervous system. *Dev. Dyn.*

235: 2376–2385.

Karl C., Couillard-Despres S., Prang P., Munding M., Kilb W., Brigadski T., Plotz S., Mages W., Luhmann H., Winkler J., et al. 2005. Neuronal precursor-specific activity of a human doublecortin regulatory sequence. *J. Neurochem.* **92:** 264–282.

Kawaguchi A., Miyata T., Sawamoto K., Takashita N., Murayama A., Akamatsu W., Ogawa M., Okabe M., Tano Y., Goldman S.A., and Okano H. 2001. Nestin-EGFP transgenic mice: Visualization of the self-renewal and multipotency of CNS stem cells. *Mol. Cell. Neurosci.* **17:** 259–273.

Kempermann G., Jessberger S., Steiner B., and Kronenberg G. 2004. Milestones of neuronal development in the adult hippocampus. *Trends Neurosci.* **27:** 447–452.

Kronenberg G., Reuter K., Steiner B., Brandt M.D., Jessberger S., Yamaguchi M., and Kempermann G. 2003. Subpopulations of proliferating cells of the adult hippocampus respond differently to physiologic neurogenic stimuli. *J. Comp. Neurol.* **467:** 455–463.

Laplagne D.A., Esposito M.S., Piatti V.C., Morgenstern N.A., Zhao C., van Praag H., Gage F.H., and Schinder A.F. 2006. Functional convergence of neurons generated in the developing and adult hippocampus. *PLoS Biol.* **4:** e409.

Lendahl U., Zimmerman L.B., and McKay R.D. 1990. CNS stem cells express a new class of intermediate filament protein. *Cell* **60:** 585–595.

Li L., Mignone J., Yang M., Matic M., Penman S., Enikolopov G., and Hoffman R.M. 2003. Nestin expression in hair follicle sheath progenitor cells. *Proc. Natl. Acad. Sci.* **100:** 9958–9961.

Liu X., Bolteus A.J., Balkin D.M., Henschel O., and Bordey A. 2006. GFAP-expressing cells in the postnatal subventricular zone display a unique glial phenotype intermediate between radial glia and astrocytes. *Glia* **54:** 394–410.

Mignone J.L., Kukekov V., Chiang A.S., Steindler D., and Enikolopov G. 2004. Neural stem and progenitor cells in nestin-GFP transgenic mice. *J. Comp. Neurol.* **469:** 311–324.

Mitsuhashi T., Aoki Y., Eksioglu Y.Z., Takahashi T., Bhide P.G., Reeves S.A., and Caviness V.S., Jr. 2001. Overexpression of p27Kip1 lengthens the G1 phase in a mouse model that targets inducible gene expression to central nervous system progenitor cells. *Proc. Natl. Acad. Sci.* **98:** 6435–6440.

Miyoshi G. and Fishell G. 2006. Directing neuron-specific transgene expression in the mouse CNS. *Curr. Opin. Neurobiol.* **16:** 577–584.

Nacher J., Pham K., Gil-Fernandez V., and McEwen B.S. 2004. Chronic restraint stress and chronic corticosterone treatment modulate differentially the expression of molecules related to structural plasticity in the adult rat piriform cortex. *Neuroscience* **126:** 503–509.

Nolte C., Matyash M., Pivneva T., Schipke C.G., Ohlemeyer C., Hanisch U.K., Kirchhoff F., and Kettenmann H. 2001. GFAP promoter-controlled EGFP-expressing transgenic mice: A tool to visualize astrocytes and astrogliosis in living brain tissue. *Glia* **33:** 72–86.

Ohtsuka T., Imayoshi I., Shimojo H., Nishi E., Kageyama R., and McConnell S.K. 2006. Visualization of embryonic neural stem cells using Hes promoters in transgenic mice. *Mol. Cell. Neurosci.* **31:** 109–122.

Overstreet L.S., Hentges S.T., Bumaschny V.F., de Souza F.S., Smart J.L., Santangelo A.M., Low M.J., Westbrook G.L., and Rubinstein M. 2004. A transgenic marker for newly born granule cells in dentate gyrus. *J. Neurosci.* **24:** 3251-3259.

Overstreet-Wadiche L.S. and Westbrook G.L. 2006. Functional maturation of adult-generated granule cells. *Hippocampus* **16:** 208–215.

Overstreet-Wadiche L.S., Bensen A.L., and Westbrook G.L. 2006a. Delayed development of adult-generated granule cells in dentate gyrus. *J. Neurosci.* **26:** 2326–2334.

Overstreet-Wadiche L.S., Bromberg D.A., Bensen A.L., and Westbrook G.L. 2005.

GABAergic signaling to newborn neurons in dentate gyrus. *J. Neurophysiol.* **94:** 4528–4532.

————. 2006b. Seizures accelerate functional integration of adult-generated granule cells. *J. Neurosci.* **26:** 4095–4103.

Patschan D., Michurina T., Shi H.K., Dolff S., Brodsky S.V., Vasilieva T., Cohen-Gould L., Winaver J., Chander P.N., Enikolopov G., and Goligorsky M.S. 2007. Normal distribution and medullary-to-cortical shift of Nestin-expressing cells in acute renal ischemia. *Kidney Int.* **71:** 744–754.

Plumpe T., Ehninger D., Steiner B., Klempin F., Jessberger S., Brandt M., Romer B., Rodriguez G.R., Kronenberg G., and Kempermann G. 2006. Variability of doublecortin-associated dendrite maturation in adult hippocampal neurogenesis is independent of the regulation of precursor cell proliferation. *BMC Neurosci.* **7:** 77.

Poueymirou W.T., Auerbach W., Frendewey D., Hickey J.F., Escaravage J.M., Esau L., Dore A.T., Stevens S., Adams N.C., Dominguez M.G., et al. 2007. F0 generation mice fully derived from gene-targeted embryonic stem cells allowing immediate phenotypic analyses. *Nat. Biotechnol.* **25:** 91–99.

Rao M.S. and Shetty A.K. 2004. Efficacy of doublecortin as a marker to analyse the absolute number and dendritic growth of newly generated neurons in the adult dentate gyrus. *Eur. J. Neurosci.* **19:** 234–246.

Sawamoto K., Nakao N., Kakishita K., Ogawa Y., Toyama Y., Yamamoto A., Yamaguchi M., Mori K., Goldman S.A., Itakura T., and Okano H. 2001. Generation of dopaminergic neurons in the adult brain from mesencephalic precursor cells labeled with a nestin-GFP transgene. *J. Neurosci.* **21:** 3895–3903.

Schmid R.S., Yokota Y., and Anton E.S. 2006. Generation and characterization of brain lipid-binding protein promoter-based transgenic mouse models for the study of radial glia. *Glia* **53:** 345–351.

Schmidt-Hieber C., Jonas P., and Bischofberger J. 2004. Enhanced synaptic plasticity in newly generated granule cells of the adult hippocampus. *Nature* **429:** 184–187.

Seaberg R.M., Smukler S.R., Kieffer T.J., Enikolopov G., Asghar Z., Wheeler M.B., Korbutt G., and van der Kooy D. 2004. Clonal identification of multipotent precursors from adult mouse pancreas that generate neural and pancreatic lineages. *Nat. Biotechnol.* **22:** 1115–1124.

Seri B., Garcia-Verdugo J.M., Collado-Morente L., McEwen B.S., and Alvarez-Buylla A. 2004. Cell types, lineage, and architecture of the germinal zone in the adult dentate gyrus. *J. Comp. Neurol.* **478:** 359–378.

Steiner B., Klempin F., Wang L., Kott M., Kettenmann H., and Kempermann G. 2006. Type-2 cells as link between glial and neuronal lineage in adult hippocampal neurogenesis. *Glia* **54:** 805–814.

Tozuka Y., Fukuda S., Namba T., Seki T., and Hisatsune T. 2005. GABAergic excitation promotes neuronal differentiation in adult hippocampal progenitor cells. *Neuron* **47:** 803–815.

Tran P.B., Banisadr G., Ren D., Chenn A., and Miller R.J. 2007. Chemokine receptor expression by neural progenitor cells in neurogenic regions of mouse brain. *J. Comp. Neurol.* **500:** 1007–1033.

Tronche F., Kellendonk C., Kretz O., Gass P., Anlag K., Orban P.C., Bock R., Klein R., and Schutz G. 1999. Disruption of the glucocorticoid receptor gene in the nervous system results in reduced anxiety. *Nat. Genet.* **23:** 99–103.

Tsukada M., Prokscha A., Ungewickell E., and Eichele G. 2005. Doublecortin association with actin filaments is regulated by neurabin II. *J. Biol. Chem.* **280:** 11361–11368.

Valenzuela D.M., Murphy A.J., Frendewey D., Gale N.W., Economides A.N., Auerbach W., Poueymirou W.T., Adams N.C., Rojas J., Yasenchak J., et al. 2003. High-throughput engineering of the mouse genome coupled with high-resolution expression analysis. *Nat. Biotechnol.* **21:** 652–659.

Wang L.P., Kempermann G., and Kettenmann H. 2005. A subpopulation of precursor cells in the mouse dentate gyrus receives synaptic GABAergic input. *Mol. Cell. Neurosci.* **29:** 181–189.

Yamaguchi M., Saito H., Suzuki M., and Mori K. 2000. Visualization of neurogenesis in the central nervous system using nestin promoter-GFP transgenic mice. *Neuroreport* **11:** 1991–1996.

Yaworsky P.J. and Kappen C. 1999. Heterogeneity of neural progenitor cells revealed by enhancers in the nestin gene. *Dev. Biol.* **205:** 309–321.

Young J.I., Otero V., Cerdan M.G., Falzone T.L., Chan E.C., Low M.J., and Rubinstein M. 1998. Authentic cell-specific and developmentally regulated expression of pro-opiomelanocortin genomic fragments in hypothalamic and hindbrain neurons of transgenic mice. *J. Neurosci.* **18:** 6631–6640.

Yu T.S., Dandekar M., Monteggia L.M., Parada L.F., and Kernie S.G. 2005. Temporally regulated expression of Cre recombinase in neural stem cells. *Genesis* **41:** 147–153.

Zimmerman L., Parr B., Lendahl U., Cunningham M., McKay R., Gavin B., Mann J., Vassileva G., and McMahon A. 1994. Independent regulatory elements in the nestin gene direct transgene expression to neural stem cells or muscle precursors. *Neuron* **12:** 11–24.

6

Retrovirus-mediated Cell Labeling

Chunmei Zhao

Laboratory of Genetics
Salk Institute for Biological Studies
La Jolla, California 92037

REPLICATION-INCOMPETENT RECOMBINANT RETROVIRUSES allow specific labeling of dividing cells and their progeny. Retrovirus-mediated cell labeling was first applied in the field of neurogenesis to understand the lineage relationship of different cell types and the migration pattern of clonally related cells. The combination of retrovirus vectors and live cell markers enables functional studies of newborn neurons in the adult mammalian brain. In addition, retrovirus vectors can be modified to manipulate the expression of a gene of interest, thus determining its role in the process of neurogenesis. This chapter summarizes our current understanding of neurogenesis based on studies using retrovirus-mediated cell labeling.

A BRIEF INTRODUCTION OF RETROVIRUSES

Retroviruses are (+)-stranded RNA viruses that are characterized by their ability to generate double-stranded DNA (dsDNA) from their RNA genome through reverse transcription. On entry into host cells through the specific interaction between viral surface glycoprotein and host-cell membrane receptor, the retroviral RNA genome is transcribed into a dsDNA in the cytoplasm of the host cell. The newly synthesized dsDNA enters the nucleus, where it integrates into the chromosomal DNA of the host cell and becomes a provirus. Although retroviruses package two copies of RNA genomes in their virions, it is believed that each virion makes a single provirus (Coffin et al. 1997).

Retroviruses have been classified in different ways. They are now divided into three subfamilies: orthoretrovirinae, spumaretrovirinae, and unclassified retroviridae (http://www.ncbi.nlm.nih.gov/Taxonomy/Browser/). There are six genera in the orthoretrovirinae subfamily, including alpharetrovirus, betaretrovirus, gammaretrovirus, deltaretrovirus, episilonretrovirus, and lentivirus, of which the first five were previously known as oncoretroviruses. Oncoretroviruses, such as Moloney murine leukemia virus (Mo-MLV), enter the nucleus during the prophase–prometaphase transition at the onset of mitosis, when the nuclear envelope breaks down. Therefore, oncoretroviruses cannot transduce quiescent cells. In comparison, lentiviruses transduce both mitotic and quiescent cells (Coffin et al. 1997). Because of this fundamental difference in viral-genome transduction, recombinant vectors of oncoretroviruses and lentiviruses have different application potentials. Retroviral vectors in this chapter only concern those derived from oncoretroviruses.

RECOMBINANT RETROVIRUSES

Wild-type retroviral genomes mostly contain three genes encoding Gag, Pol, and Env—the structural, enzymatic, and envelope proteins, respectively. Most recombinant retroviruses are made replication incompetent through the separation of the viral genes *gag, pol,* and *env* from the recombinant viral genome. Therefore, the modified viruses cannot make new virus particles after they infect and integrate into a host cell. For the preparation of infectious virus particles, one could introduce the *gag, pol,* and *env* genes through separate plasmids or use a cell line that stably expresses these genes. Recombinant retroviruses can be pseudotyped with envelope proteins of other viruses. For example, VSV-G, the envelope glycoprotein of the rhabdovirus vesicular stomatitis virus, has been used for the preparation of recombinant Mo-MLV because VSV-G-pseudotyped virus has a broader host range and is more stable when concentrated through ultraspeed centrifugation (Burns et al. 1993).

Because the integration of oncoretroviruses is dependent on the M phase of the cell cycle, recombinant retroviruses that carry the coding sequences of histochemical markers (for example, the bacterial protein β-galactosidase [β-gal]) have been used to label dividing progenitors during neurogenesis. Because the marker gene is integrated into the host-cell genome, there is no concern of the signal being diluted out if the cell undergoes multiple rounds of cell division. The ubiquitous localization of the marker proteins, such as alkaline phosphatase (AP), β-gal, and green fluorescent protein (GFP), allows the visualization of labeled cells. Furthermore, the expression of live fluorescent markers, such as GFP, enables the detection of labeled cells in live cultures or brain slices. Both the long terminal repeat (LTR) of the retrovirus and exogenous promoters have been used to drive the expression of these marker genes. Most of the retroviral vectors used in the field of neurogenesis are based on the backbone of Mo-MLV, with only a few exceptions (Table 1).

LINEAGE STUDIES USING RETROVIRUS-MEDIATED CELL LABELING

The uses of replication-incompetent recombinant retroviruses in studying neurogenesis were first reported in the late 1980s. The bacterial gene *lacZ* encoding β-gal was used to label dividing progenitors and their progeny to study the lineage relationships of different cell types during cortical neurogenesis and retinal development (Price 1987; Price et al. 1987; Turner and Cepko 1987; Gray et al. 1988; Luskin et al. 1988; Walsh and Cepko 1988). Labeled cells that were within a certain distance were considered as a clone derived from a single cell. This criterion has been challenged by other studies, which are discussed below. It was shown that different types of cells in the rodent retina may share the same progenitor and cells within a single clone have a radial arrangement (Turner and Cepko 1987). A similar phenomenon was observed in the chicken optic tectum (Gray et al. 1988). Studies on embryonic cortical neurogenesis in mice, on the other hand, confirmed previous views that neuronal and glial progenitors diverge quite early during development (Luskin et al. 1988). It has also been suggested that clonally related cells in the rodent embryonic cortex have a relative radial arrangement with certain degrees of spread (Price and Thurlow 1988; Walsh and Cepko 1988).

These early studies all used retroviruses at low multiplicity of infection (moi) so that clusters of labeled cells can be easily identified as clones. However, a problem associated with this assumption is that this analysis would exclude any cell that might have migrated away from a cluster. In fact, later studies did suggest that about 50% of the clonally related cells were quite dispersed in the rat cortex (Walsh and Cepko 1992). Here, they used a retrovirus library in which the modified viral genomes contain different DNA tags that can be distinguished by PCR. Cells that had the same DNA tag were frequently found in different areas of the cortex. This observation was confirmed later with a retrovirus library that expressed AP as a histochemical marker (Reid et al. 1995). There remains a possibility of two cells independently infected by two

Table 1. List of retroviral vectors discussed in this chapter

Name	Backbone	Replication competent?	Promoter	Additional features	References
BAG	Mo-MLV	no	LTR		Price et al. (1987); Turner and Cepko (1987); Price and Thurlow (1988); Walsh and Cepko (1988); Luskin (1993)
LZ1	Mo-MLV	no	SV40 early		Luskin et al. (1988)
Unnamed	RSV	no	LTR		Gray et al. (1988)
Unnamed	Mo-MLV	no	β-actin		Gray et al. (1988)
BAG library	Mo-MLV	no	LTR	genetic tags	Walsh and Cepko (1992); Reid et al. (1995)
CXL87	SNV	no	LTR	cytoplasmic LacZ	Kornack and Rakic (1995)
LZ12	Mo-MLV	no	SV40 early	nuclear LacZ	Kornack and Rakic (1995)
RCAS	ALV	yes	LTR	control of viral entry[a]	Doetsch et al. (1999), Seri et al. (2001, 2004)
NIT-GFP	Mo-MLV	no	TRE	regulatable[b]	Noctor et al. (2001, 2002); van Praag et al. (2002); Esposito et al. (2005); Jakubs et al. (2006)

Name	Virus		Promoter		Reference
DAP	Mo-MLV	no	SV40 early		Petreanu and Alvarez-Buylla (2002)
LZRS-eGFP-GAP43	Mo-MLV	no	CA	membrane GFP	Carleton et al. (2003)
gapEGFPm4	Mo-MLV	no	CA		Belluzzi et al. (2003)
CAG-GFP	Mo-MLV	no	CAG	WPRE	van Praag et al. (2005); Zhao et al. (2006); Laplagne et al. (2006)
Unnamed	Mo-MLV	no	EF1α	shRNA	Ge et al. (2006)
CAG-GFP/Cre	Mo-MLV	no	CAG	WPRE	Tashiro et al. (2006)

(Mo-MLV) Moloney murine leukemia virus; (RSV) Rous sarcoma virus; (SNV) avian spleen necrosis virus; (ALV) avian leukosis virus; (LTR) long terminal repeat; (SV40) simian virus 40; (TRE) tetracycline (tet) response element; (CA) cytomegalovirus (CMV) enhancer and β-actin promoter; (CAG) CMV promoter, chicken β-actin promoter, and synthetic intron; (EF1α) elongation factor 1α; (WPRE) woodchuck hepatitis virus posttranscriptional regulatory element; (shRNA) small hairpin RNA (also known as siRNA, small inhibitory RNA). The marker genes were not listed here because the same viral vector can be easily modified to express different genes. Only direct studies are cited here. The origin of these retroviral vectors can be found in the references cited in these studies.

[a] This virus was used in combination with a *trans*-genic mouse line GFAP-tva in which the receptor for ALV is expressed only in glial fibrillary acidic protein (GFAP)[+] cells.

[b] In this vector, the expression of GFP is initiated by binding of tet trans-activator (tTA) to the TRE, which can be inhibited by tet and tet derivatives such as doxycycline. The expression of tTA is driven by the LTR of Mo-MLV.

different virions with the same genetic tag, therefore creating a false dispersed clone. However, this is unlikely to account for the high incidence of widespread clones. These studies suggest that there are both radial and nonradial migration patterns for cortical neurogenesis. This view is in contrast to the "radial unit hypothesis" that considers radial arrangement as the dominant mode of cortical neuron migration (Rakic 1995). Lineage analysis of rhesus monkey cortical neurogenesis suggests that there might be two different modes of cell divisions, asymmetric and symmetric divisions, which gave rise to radial and horizontal arrangements of clonally related cells, respectively (Kornack and Rakic 1995). It has also been suggested that the tangential movement of dividing cortical progenitors within the proliferative zone might have caused some spread of clonally related cells (Fishell et al. 1993; Rakic 1995). The wide spread of clones in the neocortex could be explained by the fact that a subpopulation of the cortical interneurons reach the neocortex from the proliferative zone of the basal ganglia through tangential migration (Anderson et al. 1997). Furthermore, as we now know, new neurons generated in the postnatal subventricular zones (SVZs) migrate 5–8 mm tangentially before they integrate into the olfactory bulb (OB) (Doetsch and Alvarez-Buylla 1996).

A recent study by Noctor et al. (2001) showed that radial glia cells are responsible for the radial migration pattern of cortical neurogenesis. By using both conventional confocal microscopy and time-lapse imaging, Noctor et al. (2001, 2002) showed that radial glia cells gave rise to neurons that migrate along the long thin fiber of the parent radial glia and that the only mitotic cells labeled were of radial-glia morphology. This study, along with others, established that radial glia cells are neuronal progenitor cells during embryonic cortical neurogenesis. In addition, it has been suggested that after the completion of cortical neurogenesis and neuronal migration, radial glia cells transform into astrocytes (Voigt 1989; Rakic 2003). This is inconsistent with the notion that neuronal and glial progenitors diverge early during development. Indeed, some retrovirus-labeled clones contained both neurons and astrocytes (Walsh and Cepko 1992; Reid et al. 1995).

Retrovirus-mediated cell labeling has also been used in a lineage study of adult SVZ neurogenesis. It has been suggested that SVZ progenitors bear characteristics of astrocytes and express the glial fibrillary acidic protein (GFAP). To determine whether GFAP-expressing cells could give rise to neurons, Doetsch et al. (1999) used a transgenic mouse line that expresses the receptor for the avian leukosis virus (also a retrovirus) under the human GFAP promoter and injected the avian leukosis virus containing the reporter gene alkaline phosphatase into the SVZ. Their results show that new neurons can be derived from a population of dividing cells in which the human GFAP promoter is active. A replication-competent virus was used in this study, which might have increased the efficiency of cell labeling (Doetsch et al. 1999). Because neuronal progenitors in the adult SVZ have been suggested to be relatively quiescent (Morshead et al. 1994; Doetsch et al. 1999), the likelihood of these cells being transduced by retrovirus is presumably quite low. A replication-competent retrovirus whose entry into host cells depends on a specific promoter allows the continuous labeling of a specific population of cells. The same virus was used to target progenitors in the subgranular zone (SGZ) of the adult hippocampus, and labeled granule neurons were found 4 weeks after viral injection, suggesting that the human GFAP promoter is also active in at least a subpopulation of progenitors in the adult hippocampus (Seri et al. 2001).

In summary, three different ways have been developed for lineage analysis in the nervous system using retrovirus-mediated cell labeling: in vivo clonal analysis, time-lapse imaging, and promoter-dependent lineage tracing. Clonal analysis can be used to determine the relative time window when the progenitors of different cell types in the nervous system diverge, and this technique can be applied at any time during development. However, this approach is flawed by the fact that retroviruses are often silenced in host cells, and we know little about the epigenetic regulations of retroviruses in neural progenitors. Time-lapse imaging of labeled progenitors in cultured brain slices gives a direct view of progenitor proliferation, differentiation, migration, and morphogenesis. Yet, brain slice cultures might not fully recapitulate the condition in the intact brain. Promoter-dependent lineage tracing reveals whether a certain cell type, defined by the specific promoter, can serve as the precursor of another cell type. It is critical to use a promoter of high specificity.

INTEGRATION OF RETROVIRUS-LABELED NEWBORN NEURONS

Because retrovirus-mediated cell labeling results in permanent labeling of dividing cells and their progeny, this technique is frequently used for studying the morphological and physiological development of newborn neurons during postnatal and adult neurogenesis (Fig. 1). The first evidence of functional integration of newborn neurons in an adult rodent brain came from a study in the adult hippocampus using the retroviral vector NIT-GFP, which expresses the neomycin phosphotransferase (neo) and the tetracycline (tet) *trans*-activator protein (tTA) through a

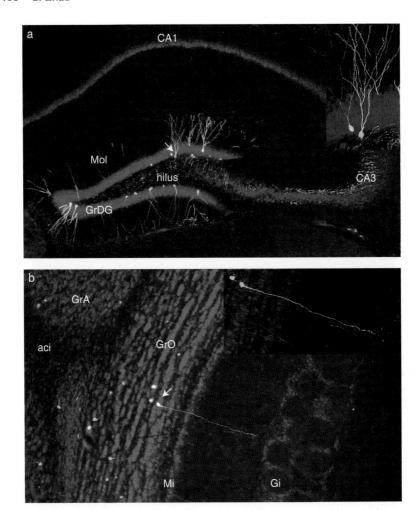

Figure 1. Retrovirus-mediated labeling of newborn neurons in the adult mouse brain. (*a*) Newborn neurons in adult hippocampus 56 days after viral injection. Note the growth of dendritic processes into the molecular layer and the projection of mossy fiber axons into field CA3 of hippocampus. (GrDG) Granule cell layer of the dentate gyrus (DG); (Mol) molecular layer of the DG. (*b*) Newborn granule neurons in adult olfactory bulb (OB) 15 days after viral injection. (aci) Anterior commissure, intrabulbar part; (Gl) glomerular layer of the OB; (GrA) granule cell layer of the accessory OB; (GrO) granule cell layer of the OB; (Mi) mitral cell layer of the OB. Insets are higher-magnification views of the cells indicated by arrows. (*Blue*) dapi; (*green*) GFP; (*red*) NeuN. Bars, 100 µm.

bicistronic cassette neo-ires-tTA, and GFP through the binding of tTA to the tet responsive element (TRE) (van Praag et al. 2002). At 4 weeks post-injection, a subpopulation of GFP$^+$ cells displayed morphological features typical of granule neurons, which can be evoked by perforant path stimulation. It was later shown that these newborn neurons follow a similar maturation process compared to granule neurons generated in the early postnatal hippocampus, being that they develop slow GABAergic responses, glutamatergic and fast GABAergic responses sequentially (Esposito et al. 2005).

Although the NIT-GFP vector provided invaluable advances in the field of neurogenesis (Noctor et al. 2001; van Praag et al. 2002; Esposito et al. 2005), the frequency of neuronal labeling by NIT-GFP is not consistent with that resulted from bromodeoxyuridine (BrdU) incorporation. For example, only 2.2% of GFP$^+$ cells expressed the panneuronal marker NeuN at 4 weeks after viral injection into the adult hippocampus, whereas 50–80% of BrdU-labeled cells were found to be positive for NeuN (Kempermann et al. 1997a; van Praag et al. 1999, 2002).

To more efficiently mark newborn neurons, we and other investigators have tried to use other ubiquitous promoters to drive the expression of GFP using the same Mo-MLV retroviral backbone (Ge et al. 2006; Zhao et al. 2006). We found that the compound promoter CAG, which contains the chicken actin promoter, minimum cytomegalovirus (CMV) enhancer, and a large synthetic intron, allowed efficient labeling of newborn neurons (Zhao et al. 2006). For example, more than 400 newborn neurons can be labeled by a single injection of CAG-GFP (10^5 colony forming unit) into the SGZ of 7–10-week mice. Furthermore, about 72% of GFP$^+$ cells are of neuronal phenotype, consistent with observations made with BrdU labeling. This vector allowed more systematic analyses of newborn neurons in the adult hippocampus. These neurons go through distinct stages of morphogenesis. They project mossy fiber axons into the distal area of CA3 within 16 days after birth, which is followed by a rapid growth of dendritic spines. Further structural modifications of newborn neurons can take place for several months and can be regulated by experience. In addition, newborn neurons in the adult appear to mature much slower compared to those in the early postnatal hippocampus (Zhao et al. 2006).

By labeling dentate granule neurons born in early embryonic/postnatal and adult brains with two different fluorescent proteins (GFP and red fluorescent protein [RFP]), Laplagne et al. (2006) were able to compare two distinct populations of neurons in the same mouse brain. Extensive electrophysiological studies showed that mature granule neurons exhibited similar GABAergic and glutamatergic afferent connectivities, whether they were born in embryonic, early postnatal, or adult brains.

The development of newborn neurons in the SVZ has also been studied in depth through retrovirus-mediated cell labeling. Luskin (1993) used the BAG β-gal-at-*gag* virus and showed that newly generated neurons in the postnatal rat SVZ took a restricted migrating route (which is later called the rostral migratory stream [RMS]) reaching the OB, where they become periglomerular and granule neurons. By 14 days after virus infection at P0, labeled periglomerular neurons in the rat OB displayed electrophysiological properties similar to unlabeled cells (Belluzzi et al. 2003). Newborn neurons in the OB develop functional GABA receptors before glutamate receptors (Belluzzi et al. 2003; Carleton et al. 2003). Detailed morphological analyses in the adult brain showed that new neurons originating from SVZ go through distinct stages of migration and morphogenesis. These neurons develop dendritic spines between 13 and 15 days after viral injection. Interestingly, there is a sharp decrease of newborn OB neurons 15 days after [^{3}H]thymidine labeling. The survival of newborn granule neurons after the formation of dendritic spines appears to depend on the input from olfactory activity (Petreanu and Alvarez-Buylla 2002).

Retrovirus-mediated cell labeling has enabled the examination of morphological development and functional integration of newborn neurons in vivo. The electrophysiological properties of new neurons have been characterized extensively at the input level. No study has attempted to address the functional output from these cells.

RETROVIRUS-MEDIATED GENE MANIPULATIONS

Retrovirus-mediated gene transfer not only allows the visualization of dividing cells and their progeny through the expression of molecular markers, but it can also be modified to alter the expression of a gene of interest so that the function of the gene can be studied in the context of progenitor proliferation, differentiation, and neuronal maturation. For example, one could overexpress different forms of a protein of interest, including wild-type, constitutively active, and dominant-negative forms. One could also knock down a gene of interest. Two recent studies have used different ways to down-regulate a gene of interest and studied the activity-dependent modulation of newborn neurons.

In the first study, Ge et al. (2006) used retrovirus-mediated expression of a short hairpin RNA (shRNA) that was designed to target the Na^+-K^+-$2Cl^-$ transporter NKCC1. Here, the vector expresses the shRNA under the human U6 promoter and GFP under the promoter of elongation factor 1α. In mature granule neurons, the neurotransmitter GABA elicits a hyperpolarization response through the flow of the negatively charged chloride ion (Cl^-) into the cells. In contrast, in immature granule neurons, GABA leads to the depolarization of the cells because these cells have a higher $[Cl^-]$ compared to the extracellular environment. The expression of NKCC1 is believed to be responsible for the high $[Cl^-]$ within immature cells. By down-regulating NKCC1 through shRNA, Ge et al. (2006) showed that newborn neurons with less NKCC1 indeed have lower $[Cl^-]$ compared to control cells and that these neurons hyperpolarized in response to tonic GABA activation. Interestingly, the targeted neurons do not mature normally, which is reflected in dramatic defects in both dendritic development and glutamatergic responses (Ge et al. 2006). These results suggest that the early GABAergic depolarization response has an important role in the development of newborn neurons born in the adult hippocampus.

In the second study, Tashiro et al. (2006) used a mouse line in which the DNA fragment encoding the transmembrane domain and the carboxy-terminal region of the N-methyl-D-aspartate (NMDA) receptor NR1 subunit is flanked by two loxP sites. A retrovirus expressing GFP/Cre fusion protein was delivered into the SGZ of adult hippocampus and induced the excision of the floxed NR1 fragment only in neuronal progenitors (Tashiro et al. 2006). By coinjection with the CAG-RFP virus, Tashiro et al. (2006) showed that deletion of NR1 in immature neurons increased cell death during the third week after virus injection, suggesting that functional NMDA receptor is important for the survival of newborn neurons. Furthermore, the survival rate of NR1 knockout neurons increased when an NMDA receptor antagonist, 3-(2-carboxypiperazin-4-yl) propyl-1-phosphonic acid (CPP), was administered to reduce the activity of wild-type cells, indicating a competitive mechanism of activity-dependent survival of new neurons. The timing of NMDA receptor-dependent survival coincides with dendritic spine growth and the onset of functional glutamatergic responses (Esposito et al. 2005; Ge et al. 2006; Tashiro et al. 2006; Zhao et al. 2006).

Both strategies allowed cell-autonomous analysis of individual neurons because only small populations of cells are altered and their surrounding environment is intact. These methods will no doubt be valuable for future mechanistic studies.

STUDYING NEUROGENESIS IN DISEASE MODELS

It is well known that the proliferation of neuronal progenitors and the survival of new neurons can be regulated by physiological and pathological conditions (Kempermann et al. 1997b; Gould et al. 1999; van Praag et al. 1999; Abrous et al. 2005; Ming and Song 2005). However, it is less clear how the integration of new neurons is affected by these conditions and whether the alteration of newborn neuron integration is involved in these processes.

Hippocampal neurogenesis decreases with aging, and this can be rescued to a certain extent by voluntary exercise. A pilot study showed that new dentate granule neurons in running old mice display morphological characteristics similar to those in young adult mice, suggesting that newborn neurons can also stably integrate in old hippocampus (van Praag et al. 2005).

Epilepsy is a pathological condition that has recently been associated with a dramatic increase in progenitor proliferation in the hippocampus (Parent et al. 1997; Jessberger and Kempermann 2003). Morphological abnormalities of seizure-induced newborn neurons have been reported, such as the growth of basal dendrites and ectopic migration of granule neurons (Parent et al. 1997; Shapiro and Ribak 2006). To study the function of seizure-induced neurogenesis, Jakubs et al. (2006) labeled new cells in the hippocampus with the retroviral vector NIT-GFP in a rat model of status epilepticus (SE). They showed that 4–6 weeks after viral labeling, seizure-induced new neurons displayed reduced excitatory input and increased inhibitory input compared to neurons generated in running rats (Jakubs et al. 2006). It remains to be determined whether this represents a mechanism of self-repair through the decreased excitability of newborn neurons; it is not known whether new neurons in epileptic brain also have lower excitability than those in a sedentary rat brain.

Using the retrovirus vector CAG-GFP, we examined the integration of newborn neurons in kainic acid (KA)-induced epilepsy in rats. A subpopulation of newborn neurons displayed growth of basal dendrites, consistent with earlier findings using immature neuronal markers. Furthermore, new neurons with aberrant dendrites seem to be stably integrated in the epileptic brain, as a similar percentage of labeled neurons were found to bear basal dendrites even 3 months after viral infection (Jessberger et al. 2007). Moreover, seizure activity appears to have differential effects on mature neurons, as we found that neurons born before seizure never developed basal dendrites. In contrast, we observed mossy fiber sprouting only in neurons at least 4 weeks old at the time of KA administration (Jessberger et al. 2007). This is consistent with the finding that mossy fiber sprouting was observed in GFP⁻ cells in POMC-GFP transgenic mice, in which GFP is expressed in immature neurons in the adult SGZ (Overstreet-Wadiche et al. 2006).

LENTIVIRUS-MEDIATED CELL LABELING

Because the integration of retrovirus depends on the mitotic activity of host cells, retrovirus-mediated cell labeling is biased toward cells with an active cell cycle, for example, the transit-amplifying cells in the SVZ. This is supported by the observation that very few cells were detected in the RMS 15 days after viral injection (Petreanu and Alvarez-Buylla 2002). In contrast to recombinant retroviruses, the newly transcribed DNA from a lentivirus could enter the host-cell nucleus at any time during the cell cycle. Therefore, lentiviral vectors can be used to target relatively quiescent cells. Using lentivirus-mediated expression of GFP in SVZ cells, Consiglio et al. (2004) showed that GFP-labeled cells could be found in the RMS even 6 months after viral injection, confirming that newborn olfactory neurons originate from the SVZ. However, this vector cannot distinguish neuronal progenitors from other cell types in the SVZ, because a ubiquitous promoter was used and all cells at the injection site expressed the GFP marker. Therefore, this approach cannot be adapted for studying hippocampal neurogenesis, because newborn neurons do not migrate far away from the site of proliferation. Instead, specific promoters of putative stem cell markers have been used to confer specificity to lentiviral vectors (H. Suh et al., unpubl.).

DISADVANTAGES OF RETROVIRUS-MEDIATED CELL LABELING

Retrovirus-mediated cell labeling in the adult rodent brain requires stereotaxic surgery to deliver the virus to the area of interest. The surgery itself can induce local inflammatory reactions. It has been reported that the combination of retrovirus and microglia proliferation could stimulate the fusion of labeled microglia with existing postmitotic neurons in the postnatal neocortex (Ackman et al. 2006). Therefore, one should take caution interpreting results obtained from retroviral labeling. Fortunately, it is relatively easy to determine whether newborn neurons identified through retroviral labeling are indeed derived from dividing progenitors in the adult brain. Because neurons go through extensive morphogenesis before maturation, a time-course study of retrovirus-labeled neurons provided unambiguous evidence that retrovirus-labeled cells were in fact derived from progenitor cells (Petreanu and Alvarez-Buylla 2002; Carleton et al. 2003; Esposito et al. 2005; Ge et al. 2006; Zhao et al. 2006).

Retroviruses are often silenced in host cells such as embryonic stem cells, embryonic carcinoma cells, and hematopoietic stem cells. This is complicated by the fact that two daughter cells from the same parent cell could undergo differential silencing and that expressing cells could also be silenced during differentiation (Ellis 2005). It has not been carefully investigated whether retroviral vectors are also silenced in neuronal stem cells or during their differentiation, but the low level of neuronal labeling of the NIT-GFP vector does suggest that the expression of retroviral vectors is not homogeneous in all cell types. Therefore, retrovirus and lentivirus vectors at the current stage are not optimal for lineage analysis in the nervous system, as one cannot be certain whether all progenies of a neuronal stem cell can be traced.

SUMMARY

Retrovirus-mediated gene expression can specifically label dividing cells and their progeny, making it an ideal tool for studying neurogenesis, especially in the adult brain, in which only a small population of cells are mitotic. However, one should bear in mind that we know little about epigenetic and promoter regulations of retroviruses in adult neuronal progenitors, especially when retroviral vectors are used in lineage and fate-mapping studies. Nevertheless, marker gene expression can be maintained in neurons, which allowed long-term systematic studies of newborn neuron integration. Furthermore, with the technical advancement in molecular biology and imaging, it is hoped that retrovirus-mediated cell labeling will help us understand the molecular mechanisms which regulate the self-renewal and differentiation of adult neuronal progenitors, as well as the maturation and integration of newborn neurons.

ACKNOWLEDGMENTS

The author thanks Rusty Gage, Ayumu Tashiro, and Xinwei Cao for comments.

REFERENCES

Abrous D.N., Koehl M., and Le Moal M. 2005. Adult neurogenesis: From precursors to network and physiology. *Physiol. Rev.* **85:** 523–569.
Ackman J.B., Siddiqi F., Walikonis R.S., and LoTurco J.J. 2006. Fusion of microglia with pyramidal neurons after retroviral infection. *J. Neurosci.* **26:** 11413–11422.

Anderson S.A., Eisenstat D.D., Shi L., and Rubenstein J.L. 1997. Interneuron migration from basal forebrain to neocortex: Dependence on Dlx genes. *Science* **278:**
474–476.

Belluzzi O., Benedusi M., Ackman J., and LoTurco J.J. 2003. Electrophysiological differentiation of new neurons in the olfactory bulb. *J. Neurosci.* **23:** 10411–10418.

Burns J.C., Friedmann T., Driever W., Burrascano M., and Yee J.K. 1993. Vesicular stomatitis virus G glycoprotein pseudotyped retroviral vectors: Concentration to very
high titer and efficient gene transfer into mammalian and nonmammalian cells. *Proc.
Natl. Acad. Sci.* **90:** 8033–8037.

Carleton A., Petreanu L.T., Lansford R., Alvarez-Buylla A., and Lledo P.M. 2003. Becoming
a new neuron in the adult olfactory bulb. *Nat. Neurosci.* **6:** 507–518.

Coffin J.M., Hughes S.H., and Varmus H.E., eds. 1997. *Retroviruses.* Cold Spring Harbor
Laboratory Press, Cold Spring Harbor, New York.

Consiglio A., Gritti A., Dolcetta D., Follenzi A., Bordignon C., Gage F.H., Vescovi A.L.,
and Naldini L. 2004. Robust in vivo gene transfer into adult mammalian neural stem
cells by lentiviral vectors. *Proc. Natl. Acad. Sci.* **101:** 14835–14840.

Doetsch F. and Alvarez-Buylla A. 1996. Network of tangential pathways for neuronal
migration in adult mammalian brain. *Proc. Natl. Acad. Sci.* **93:** 14895–14900.

Doetsch F., Caille I., Lim D.A., Garcia-Verdugo J.M., and Alvarez-Buylla A. 1999. Subventricular zone astrocytes are neural stem cells in the adult mammalian brain. *Cell* **97:**
703–716.

Ellis J. 2005. Silencing and variegation of gammaretrovirus and lentivirus vectors. *Hum.
Gene Ther.* **16:** 1241–1246.

Esposito M.S., Piatti V.C., Laplagne D.A., Morgenstern N.A., Ferrari C.C., Pitossi F.J., and
Schinder A.F. 2005. Neuronal differentiation in the adult hippocampus recapitulates
embryonic development. *J. Neurosci.* **25:** 10074–10086.

Fishell G., Mason C.A., and Hatten M.E. 1993. Dispersion of neural progenitors within
the germinal zones of the forebrain. *Nature* **362:** 636–638.

Ge S., Goh E.L., Sailor K.A., Kitabatake Y., Ming G.L., and Song H. 2006. GABA regulates
synaptic integration of newly generated neurons in the adult brain. *Nature* **439:**
589–593.

Gould E., Beylin A., Tanapat P., Reeves A., and Shors T.J. 1999. Learning enhances adult
neurogenesis in the hippocampal formation. *Nat. Neurosci.* **2:** 260–265.

Gray G.E., Glover J.C., Majors J., and Sanes J.R. 1988. Radial arrangement of clonally
related cells in the chicken optic tectum: Lineage analysis with a recombinant retrovirus. *Proc. Natl. Acad. Sci.* **85:** 7356–7360.

Jakubs K., Nanobashvili A., Bonde S., Ekdahl C.T., Kokaia Z., Kokaia M., and Lindvall O.
2006. Environment matters: Synaptic properties of neurons born in the epileptic adult
brain develop to reduce excitability. *Neuron* **52:** 1047–1059.

Jessberger S. and Kempermann G. 2003. Adult-born hippocampal neurons mature into
activity-dependent responsiveness. *Eur. J. Neurosci.* **18:** 2707–2712.

Jessberger S., Zhao C., Toni N., Clemenson G.D., Li Y., and Gage F.H. Seizure-associated,
aberrant neurogenesis in adult rats characterized with retrovirus-mediated cell labeling. *J. Neurosci.* (in press).

Kempermann G., Kuhn H.G., and Gage F.H. 1997a. Genetic influence on neurogenesis in
the dentate gyrus of adult mice. *Proc. Natl. Acad. Sci.* **94:** 10409–10414.

———. 1997b. More hippocampal neurons in adult mice living in an enriched environment. *Nature* **386:** 493–495.

Kornack D.R. and Rakic P. 1995. Radial and horizontal deployment of clonally related cells in the primate neocortex: Relationship to distinct mitotic lineages. *Neuron* 15: 311–321.

Laplagne D.A., Esposito M.S., Piatti V.C., Morgenstern N.A., Zhao C., van Praag H., Gage F.H., and Schinder A.F. 2006. Functional convergence of neurons generated in the developing and adult hippocampus. *PLoS Biol.* 4: e409.

Luskin M.B. 1993. Restricted proliferation and migration of postnatally generated neurons derived from the forebrain subventricular zone. *Neuron* 11: 173–189.

Luskin M.B., Pearlman A.L., and Sanes J.R. 1988. Cell lineage in the cerebral cortex of the mouse studied in vivo and in vitro with a recombinant retrovirus. *Neuron* 1: 635–647.

Ming G.L. and Song H. 2005. Adult neurogenesis in the mammalian central nervous system. *Annu. Rev. Neurosci.* 28: 223–250.

Morshead C.M., Reynolds B.A., Craig C.G., McBurney M.W., Staines W.A., Morassutti D., Weiss S., and van der Kooy D. 1994. Neural stem cells in the adult mammalian forebrain: A relatively quiescent subpopulation of subependymal cells. *Neuron* 13: 1071–1082.

Noctor S.C., Flint A.C., Weissman T.A., Dammerman R.S., and Kriegstein A.R. 2001. Neurons derived from radial glial cells establish radial units in neocortex. *Nature* 409: 714–720.

Noctor S.C., Flint A.C., Weissman T.A., Wong W.S., Clinton B.K., and Kriegstein A.R. 2002. Dividing precursor cells of the embryonic cortical ventricular zone have morphological and molecular characteristics of radial glia. *J. Neurosci.* 22: 3161–3173.

Overstreet-Wadiche L.S., Bromberg D.A., Bensen A.L., and Westbrook G.L. 2006. Seizures accelerate functional integration of adult-generated granule cells. *J. Neurosci.* 26: 4095–4103.

Parent J.M., Yu T.W., Leibowitz R.T., Geschwind D.H., Sloviter R.S., and Lowenstein D.H. 1997. Dentate granule cell neurogenesis is increased by seizures and contributes to aberrant network reorganization in the adult rat hippocampus. *J. Neurosci.* 17: 3727–3738.

Petreanu L. and Alvarez-Buylla A. 2002. Maturation and death of adult-born olfactory bulb granule neurons: Role of olfaction. *J. Neurosci.* 22: 6106–6113.

Price J. 1987. Retroviruses and the study of cell lineage. *Development* 101: 409–419.

Price J. and Thurlow L. 1988. Cell lineage in the rat cerebral cortex: A study using retroviral-mediated gene transfer. *Development* 104: 473–482.

Price J., Turner D., and Cepko C. 1987. Lineage analysis in the vertebrate nervous system by retrovirus-mediated gene transfer. *Proc. Natl. Acad. Sci.* 84: 156–160.

Rakic P. 1995. Radial versus tangential migration of neuronal clones in the developing cerebral cortex. *Proc. Natl. Acad. Sci.* 92: 11323–11327.

———. 2003. Developmental and evolutionary adaptations of cortical radial glia. *Cereb. Cortex* 13: 541–549.

Reid C.B., Liang I., and Walsh C. 1995. Systematic widespread clonal organization in cerebral cortex. *Neuron* 15: 299–310.

Seri B., Garcia-Verdugo J.M., McEwen B.S., and Alvarez-Buylla A. 2001. Astrocytes give rise to new neurons in the adult mammalian hippocampus. *J. Neurosci.* 21: 7153–7160.

Seri B., Garcia-Verdugo J.M., Collado-Morente L., McEwen B.S., and Alvarez-Buylla A. 2004. Cell types, lineage, and architecture of the germinal zone in the adult dentate gyrus. *J. Comp. Neurol.* 478: 359–378.

Shapiro L.A. and Ribak C.E. 2006. Newly born dentate granule neurons after pilocarpine-induced epilepsy have hilar basal dendrites with immature synapses. *Epilepsy Res.* 69: 53–66.

Tashiro A., Sandler V.M., Toni N., Zhao C., and Gage F.H. 2006. NMDA-receptor-mediated, cell-specific integration of new neurons in adult dentate gyrus. *Nature* **442:** 929–933.

Turner D.L. and Cepko C.L. 1987. A common progenitor for neurons and glia persists in rat retina late in development. *Nature* **328:** 131–136.

van Praag H., Kempermann G., and Gage F.H. 1999. Running increases cell proliferation and neurogenesis in the adult mouse dentate gyrus. *Nat. Neurosci.* **2:** 266–270.

van Praag H., Shubert T., Zhao C., and Gage F.H. 2005. Exercise enhances learning and hippocampal neurogenesis in aged mice. *J. Neurosci.* **25:** 8680–8685.

van Praag H., Schinder A.F., Christie B.R., Toni N., Palmer T.D., and Gage F.H. 2002. Functional neurogenesis in the adult hippocampus. *Nature* **415:** 1030–1034.

Voigt T. 1989. Development of glial cells in the cerebral wall of ferrets: Direct tracing of their transformation from radial glia into astrocytes. *J. Comp. Neurol.* **289:** 74–88.

Walsh C. and Cepko C.L. 1988. Clonally related cortical cells show several migration patterns. *Science* **241:** 1342–1345.

———. 1992. Widespread dispersion of neuronal clones across functional regions of the cerebral cortex. *Science* **255:** 434–440.

Zhao C., Teng E.M., Summers R.G., Jr., Ming G.L., and Gage F.H. 2006. Distinct morphological stages of dentate granule neuron maturation in the adult mouse hippocampus. *J. Neurosci.* **26:** 3–11.

7

Neurospheres

Ilyas Singec

Burnham Institute for Medical Research
Stem Cell and Regeneration Program
La Jolla, California 92037

Alfredo Quiñones-Hinojosa

Johns Hopkins University
Department of Neurosurgery
Cancer Research Building II
Baltimore, Maryland 21231

NEURAL STEM CELLS (NSCs) are ideal candidates to study fundamental cell biological and developmental questions under defined experimental conditions and to lay the foundation for cell-based therapies for neurological diseases. Somatic stem cells, such as NSCs, are primordial cells that generate all major cell types of a given organ during development, maintain homeostasis and integrity throughout life, and may initiate repair following injury. Stem cells have to make decisions as to whether they generate identical daughter cells by symmetric cell division or give rise to a more differentiated progeny by asymmetric cell division. It is conceivable to believe that NSCs and the regulation of their function (cell autonomous as well as non-cell autonomous) are complex and precisely controlled. In addition, since rapid phenotype changes are in the nature of stem cells, this adds another dimension of challenge to the work with NSCs. Therefore, the successful culture, controlled expansion, and ex vivo manipulation of bona fide stem cells are still ambitious tasks that necessitate continued improvement as our knowledge about stem cell biology increases. Better control and understanding of cell growth and the conditions that promote this growth, as well as differentiation of NSCs, are crucial elements before large-scale applications of stem cell therapy can be moved to the clinic.

Traditionally, because of the difficulties in growing neuronal cells, the NSC field has emerged from studies with immortalized cell lines. These cell lines were derived from spontaneously occurring tumors or were immortalized by genetic manipulation and introduction of oncogenes into neuronal cells (Frederiksen et al. 1988; Snyder et al. 1992; Gage et al. 1995; Villa et al. 2000). Technological progress, including the production of cell-type-specific monoclonal antibodies, availability of recombinant growth factors, and improved cell culture media for neuronal cells, paved the way for NSC growth without immortalization (Kohler and Milstein 1975; Lendahl et al. 1992; Ray et al. 1993; Kitchens et al. 1994). Yet, NSCs can be isolated from the developing and adult central nervous system (CNS) and perpetuated under serum-free conditions for extended periods of time in the presence of mitogenic factors such as fibroblast growth factor-2 (FGF-2) and/or epidermal growth factor (EGF).

NSCs have been cultured as monolayers on coated substrates (e.g., fibronectin and laminin) or as free-floating spheroid cell aggregates, the so-called neurospheres (Reynolds and Weiss 1992; Ray et al. 1993, Kitchens et al. 1994; Palmer et al. 1999; Song et al. 2002). Because of the relative ease, practical handling, and better survival of neuronal cells growing as neurospheres, this method quickly became popular. Similar sphere assays are now being used in cell cultures derived from other organ systems (e.g., mammospheres, cardiospheres, and pancreato-spheres) and in the cancer stem cell field (e.g., sarcospheres, which will be referred to as "cytospheres") (Fig. 1). Important biological conclusions have been drawn using the neurosphere assay. Despite some commonly used principles to grow neuro/cytospheres, considerable variation exists among the protocols published in the literature. This variability includes hormones and growth factor combinations, as well as their respective concentrations, cell density, duration in culture before quantification, medium volume, and surface area of the culture dish (Chaicana et al. 2006; Singec et al. 2004, 2006; Jessberger et al. 2007). In this chapter, we provide an overview of the current state of the neurosphere concept and its relationship to "stemness," highlight the potentials and pitfalls, and suggest criteria to uniformly calibrate this assay system.

THE PHILOSOPHY OF DEFINING AND THE ART OF CULTURING

Historically, the hierarchical concept of stem cells was formulated on the basis of work performed in the hematopoietic system in the early 1960s (Till and McCulloch 1961). A single hematopoietic stem cell (HSC) in the bone marrow is capable of giving rise to a cell clone through various

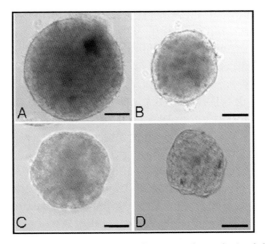

Figure 1. Phase-contrast images of typical neurosphere-derived human tissues. (A) Tumor neurosphere derived from a 9-year-old patient with a well-differentiated oligodendroglioma (WHO Grade II). (B) Tumor neurosphere derived from a 36-year-old patient with a well-differentiated ependymoma (WHO Grade II). (C) Tumor neurosphere derived from a 69-year-old patient with glioblastoma multiforme. (D) Neurosphere obtained from noncancer ependyma of a 1-year-old patient diagnosed with cortical dysplasia. Bars: (A–D) 100 μm.

progenitor cells that ultimately produces all blood cell lineages (erythroid, lymphoid, myeloid). In general, progenitor cells have a more restricted developmental potential but are more proliferative than quiescent bona fide stem cells. The functionality of the hierarchical concept in the hematopoietic field can be directly measured by serially reconstituting the blood system of lethally irradiated mice with single HSCs (Spangrude et al. 1988; Smith et al. 1991; Kiel et al. 2005). The elegance of these experiments, advanced methods to enrich and purify cells based on multiple specific surface markers with fluorescence-activated cell sorting (FACS), and convincing demonstrations of clonality in particular had a strong impact on other disciplines including the NSC field. Hence, a clonal NSC should be able to generate identical daughter cells (self-renewal) and differentiate into the major cell types of the brain: neurons, astrocytes, oligodendrocytes (multipotentiality or multilineage differentiation). As with most definitions, the theoretical definition of NSCs is clearly formulated, straightforward, and widely accepted. However, a practical approach and tools to study these key properties of NSC during experimentation remain controversial. There are different reasons and some inherent problems that are responsible for this situation, and they can be explained by comparing HSCs with NSCs. First, bona fide stem cells are *rare and quiescent* in the CNS and other organs, including the bone marrow. It is obvious that different tissues and their organ-specific stem cells have different cell-turnover rates to maintain homeostasis. For instance, the physiologic function of the blood system is dependent on an extremely high cell-turnover rate. In contrast, once the CNS is formed during development, continuous cell turnover is restricted to discrete brain regions such as the hippocampal dentate gyrus (DG) and olfactory bulb (OB). The population doubling time is faster and the capability of generating transit-amplifying progenitor cells is higher for HSCs than for NSCs. These differences seem to remain after isolation and culture of HSCs and NSCs. In addition, HSCs can be directly characterized by their function, reconstituting the blood system and rescuing lethally irradiated mice. No such functional assay has been developed for NSCs, although reports show that NSCs can widely populate the brain after grafting experiments (J.P. Lee et al. 2007). Second, bona fide HSCs can be isolated, purified, and enriched based on specific surface marker expression. Unfortunately, a sufficient characterization of NSCs including the identification of specific and sensitive cell surface markers that would allow for their prospective isolation has not yet been established. Third, purified and enriched populations of cells were used to establish clonal experiments and colony-forming assays in the HSC field. Similar clonal experiments and colony-forming assays are difficult to design if the starting cell population is heterogeneous and the stem cells relatively quiescent and prone to undergo apoptotic cell death or spontaneous differentiation upon stress during the initiation of the cell culture. The finding that neuronal cells can be grown as free-floating neurospheres and are composed of nestin-expressing cells implied to some researchers that neurospheres must represent clonal entities if single cells are plated into a cell culture dish and stimulated with growth factors (Reynolds and Weiss 1992; Reynolds et al. 1992). This conclusion was supported by experiments where true clonal experiments (one cell per well) were performed in order to demonstrate the existence of adult NSCs (Reynolds and Weiss 1996; Weiss et al. 1996). Again, without the tools to purify and enrich NSCs, clonal analysis is difficult to perform on a routine basis and not practical for every single experiment. The neurosphere assay became so popular because it seemed that the cardinal features of NSCs such as clonality, self-renewal, and multipotentiality could be studied conveniently and quickly without plating one cell per well. In addition, given the low percentage of bona fide stem cells in the initial cell suspension and the poor cell survival when plating a single cell per well, the use of the neurosphere assay with better cell survival rates ("community effect") appeared to be the method of choice. It seemed that some inherent problems to NSC cultures could be circumvented leading to experimental results in a relatively short period of time. If neurospheres were considered clonal structures, their presence as well as their size and number after a few passages would indicate self-renewal, proliferative activity, and "stemness" of neuronal cells grown under these conditions. Multipotentiality of NSCs was concluded when single free-floating neurospheres grown in suspension cultures were attached to coated culture dishes and induced to differentiate by removal of growth factors. Often, the application of serum, retinoic acid, or neurotrophic factors was used to promote neural differentiation of plated neurospheres ultimately generating neurons, astrocytes, and oligodendrocytes (Caldwell et al. 2001). Furthermore, for some scientists, it was an attractive idea to assume that the number of neurospheres formed in a culture dish reflects the in vivo number of NSCs (Morshead et al. 1998; Kippin et al. 2005). The capability of any given tissue to generate neurospheres was viewed as a key strategy to isolate and analyze the existence of NSCs. Besides these assumptions, considerable experimental variability was introduced into the "standard" neurosphere protocols used since 1992 (Reynolds and Weiss 1992; Chiasson et al. 1999; Represa et al. 2001; Seaberg and van der Kooy 2002; Kippin et al. 2005). However, it is always a good scientific practice to resist the tautology to believe that a method must be valuable, unambiguous, and correct because it has been used and prominently published so many times in the literature. We and other investigators have revisited the neurosphere concept, and recent progress has been made to understand and exploit the neurosphere assay in an appropriate and realistic way, with a critical eye and a rational mindset (Singec et al. 2004, 2006; Reynolds and Rietze 2005; Chaicana et al. 2006; Jessberger et al. 2007). This is particularly important because the conclusions drawn may be relevant for the treatment of human diseases including brain cancer.

OLD AND NEW CHALLENGES IN NSC CULTURE

Neuronal stem/progenitor cells can be isolated from the developing and adult CNS. Alternatively, NSCs can be derived from pluripotent embryonic stem cells and grown as neurospheres (Reubinoff et al. 2001; Singee et al. 2007). The old "no new neuron dogma" in neuroscience has been refuted by findings showing that discrete regions in the adult mammalian brain are capable of adult neurogenesis (Gage 2000; Ming and Song 2005). Given the strong interest in NSCs and their potential to model and treat neurological diseases, we are at a point where critical reevaluations of common cell culture tools are mandatory to further advance the field.

Mission: Fact or Artifact?

Basically, NSC cultures are important because of two reasons. First, NSCs studied in vitro under defined and controlled conditions are essential to complement in vivo experiments. Dissection of signaling pathways and mechanistic studies that are difficult to address in vivo can be approached in vitro when NSCs are directly accessible for experimentation (Song et al. 2002). The ex vivo manipulation of human NSCs with cell and molecular biological tools, as well as with pharmacological compounds, is expected to have a great impact on future clinical medicine. Second, reliable NSC cultures are necessary for a very practical reason, i.e., to multiply cell numbers over extended periods of time. Expanded and well-characterized NSCs could then be stored and used as a potential source for cell transplantation.

It is important to note that any cell culture condition is artificial when compared to the stem cell's highly intricate, tightly controlled, and privileged microenvironment in vivo, the stem cell niche (Palmer et al. 2000; Shen et al. 2006; Scadden 2006). The more we learn about the complexity of the stem cell niche, the more we become aware that not all cellular and molecular components of the stem cell niche can be modeled in vitro. However, with this caution in mind, a better handle on the control mechanisms of proliferation and differentiation of NSCs in vitro is of great importance. Dissociated single cells cultured on coated substrates and grown as two-dimensional monolayers is the most widely used cell culture technique. It is widely held that under this condition, all cells are reliably exposed to growth factors and other compounds in order to perpetuate a homogeneous cell population that would be the ideal scenario for biochemical studies. Monolayer cultures of neuronal stem/progenitor cells have been initiated from the developing and adult CNS, but considerable variability exists between the protocols used (Davis and Temple 1994; Johe et al. 1996; Palmer et al. 1999; Song et al. 2002). Among others, this variability includes the developmental stage and brain region to be dissociated, composition of the initial cell suspension, choice of the coating substrate, the use of FGF-2 and EGF alone or in combination, passage number, and the method to induce differentiation. Thus, monolayer NSC culture protocols significantly differ from each other, and results obtained under specific culture conditions may not be reproducible in similar but nonidentical experiments. The problem of using variable and nonuniformly calibrated protocols for NSC culture will also be discussed for the neurosphere assay (see below). Moreover, unlike immortalized cell lines, monolayer NSCs are more sensitive to repeated enzymatic treatment for cell passage, which can lead to apoptotic cell death or spontaneous differentiation.

Clonal experiments are the only way to determine the true developmental potential of a single cell (Anderson 2001). Dissociated cells plated at one cell per well or at "clonal" cell density on a coated substrate and continuous observation with time-lapse microscopy are currently the most reliable conditions for clonal analysis (Davis and Temple 1994; Parker et al. 2005; Shen et al. 2006; Singec et al. 2006). Experiments of this kind are challenging and time-consuming since bona fide stem cells are rare, quiescent, and difficult to grow. Therefore, the neurosphere assay became a popular shortcut for putative clonal experiments. Since there is no in vivo counterpart of neurospheres, the formation of free-floating proliferative cell clusters in uncoated culture dishes can be conceived as a cell culture artifact. The neurosphere assay is a simple, practical, and economic cell culture tool when used appropriately. In addition, cells plated at low cell densities survive better and display higher proliferative activity in neurosphere cultures as compared to monolayer cultures. Given these advantages, it was tempting to assume that single cells plated at "low or clonal cell densities" in suspension cultures would yield clonal neurospheres. The idea that clonal neurospheres can be grown in suspension cultures emerged from the original work by Reynolds and Weiss (1992) and has been used since then. However, attempts to experimentally determine if this assumption of clonality is true for neurospheres had been addressed many years later in coculture experiments (Represa et al. 2001; Imura et al. 2003; Morshead et al. 2003; Nunes et al. 2003). The shortcoming of these experiments is that unlabeled cells were cocultured with green fluorescent protein (GFP)-expressing cells. In consequence, the clonal growth and homogeneity of neurospheres cannot be detected with high enough sensitivity if, for instance, unlabeled cells are integrated into a GFP-positive sphere. Therefore, we performed experiments using NSCs from two different transgenic animals ubiquitously expressing LacZ (β-galactosidase) and enhanced GFP (EGFP), respectively (Singec et al. 2004, 2006). Surprisingly, cells plated at systematically varied densities (10^5 to 10^2 cells/cm^2 or 500–0.5 cells/μl) demonstrated that the majority of neurospheres were chimeric, hence composed of LacZ and EGFP-expressing cells (Fig. 2). Further analyses and direct

EGFP   β-gal   merge

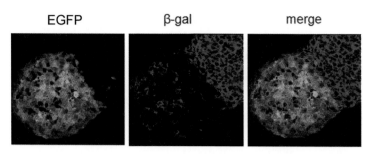

Figure 2. Confocal image showing ongoing neurosphere merging. Polyclonal neurospheres predominate in coculture experiments of genetically labeled cells expressing EGFP and β-gal, respectively.

observation with time-lapse video microscopy revealed that neurosphere cultures are highly dynamic. Despite their size and multicellular nature, neurospheres were highly motile structures constantly merging with each other and incorporating floating single cells. Time-lapse video microscopy showed that free-floating neurospheres are actively moving by beating cellular processes expressed on their surface (Fig. 3). These filopodia-like processes propel spheres through the culture medium. In addition, whole neurospheres plated on a coated substrate were also highly motile, migrating and exchanging cells with their environment. These results were confirmed in cultures derived from different species (mouse, rat, human) and a number of different brain regions (subventricular zone, striatum, midbrain, spinal cord) and ages (fetal, neonatal, adult). Merging is a common phenomenon of free-floating spheres (including mammospheres initiated from breast cancer tissue) or even embryoid bodies derived from human embryonic stem cells (I. Singec, unpubl. observation). In conclusion, these studies revealed that because of the dynamic nature of neurospheres, it is extremely difficult to maintain clonal boundaries in suspension cultures. Clonal analysis in conventional neurosphere cultures is highly risky and not compatible with the standards that are necessary for the stem cell field. Since polyclonal neurospheres are prevalent in this assay, additional and more rigorous experiments such as plating one cell per well or monolayer cultures with continuous observation should be performed for clonal studies. In a few studies, the semisolid substance methylcellulose was used to ensure putative clonal growth of neurospheres (Kukekov et al. 1997; Gritti et al. 1999). However, considering the highly motile nature of single cells and neurospheres on coated substrates, it remains to be demonstrated if methylcellulose or other semisolid matrices might in fact be helpful for clonal studies.

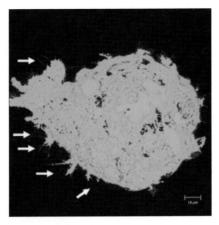

Figure 3. Confocal false-color image demonstrating cytoplasmic specializations that cells develop on the outer neurosphere surface. These cytoplasmic extensions (some indicated by white arrows) are motile and propel free-floating neurospheres through the culture medium (see also the time-lapse movies in Singec et al. [2006]).

As stated above, neurospheres are highly motile structures and prone to merge with each other. Interestingly, single cells in densely packed chimeric neurospheres seem not to fuse as shown by confocal microscopy and the absence of colocalization of transgenic markers at the single-cell level (Singec et al. 2006). Considering the occurrence of cell fusion in NSC cultures and in vivo after grafting into animals has recently become an important issue, since cells can exchange genetic material by fusion. Cell fusion may explain some of the published data previously interpreted as "transdifferentiation," namely, that somatic stem cells (e.g., NSCs) have a broad developmental potential and can contribute to unrelated cell lineages (for review, see Alvarez-Dolado 2007). Currently, it is not known if cell fusion is part of the stem cell's plasticity or an artifact (Alvarez-Dolado 2007; Jessberger et al. 2007).

Time: Short Term or Long Term?

Most laboratories using the neurosphere assay have presumed that repeated dissociation of spheres and cell passage can enrich for a proliferative population of clonally growing stem cells. The resultant formation of the so-called "secondary neurospheres" has been interpreted as an indicator of self-renewal. Moreover, the number of secondary spheres was then equated with the number of bona fide stem cells, and the size of these spheres was considered as a metric for proliferative activity. With our current knowledge about the true nature of neurospheres, namely, that neurospheres are polyclonal in suspension cultures, these assumptions appear to be misleading and inaccurate. Sphere formation is a common cell culture phenomenon when proliferative cells are plated into uncoated dishes. Importantly, sphere formation is not a metric and not a reliable and stringent method to study the key properties of NSCs (clonality, self-renewal, multipotentiality). After Singec et al. (2004) demonstrated the shortcomings of the neurosphere assay, Reynolds and Rietze (2005) suggested that proliferative progenitor cells can also form secondary and tertiary neurospheres indicating that self-renewal of NSCs should be studied over extended periods of time. These same authors (Reynolds and Rietze 2005) calculated that the frequency of bona fide stem cells is very low (which is not unexpected) and argued against the one-to-one relationship between sphere and stem cell numbers. Although mostly theoretical considerations are presented in this study, these cautionary notes are important and point to the right direction. One may still ask why it has taken so long (since 1992) to see the limitations and the variability of the neurosphere assays used. One answer might be that we are just beginning to understand the complexity of stem biology and we have probably overestimated our basic cell culture tools when using them on NSCs.

Neurosphere formation should be viewed as a dynamic process that reflects cell proliferation in combination with mixing with other spheres and cells. The neurosphere assay is a practical method to grow and expand proliferative cells for extended periods of time without exposing cells to stress during mechanical and enzymatic passage. For instance, Svendsen et al. (1998) demonstrated that using a tissue chopper to cut whole spheres into small pieces and replating them in the presence of FGF-2 and EGF is a powerful method to expand human neuronal cells. By using this simple technique, these authors were able to achieve an exponential 1.5-millionfold increase in precursor numbers over a period of less than 200 days. In contrast, only a 12-fold expansion in total cell number was possible with mechanical passage. Although neurosphere cells derived from fetal brain regions are considered as a relatively safe source for transplantation, in one report, abnormal growth of rodent neurospheres was suggested after passage number 10 (Morshead et al. 2002).

The neurosphere assay has been criticized because in contrast to monolayer cultures, cells growing in three dimensions might be more heterogeneous, prone to differentiate spontaneously, and less controllable with growth factors added to the medium. This shortcoming of the assay is particularly problematic when the size of neurospheres becomes too large, probably leaving the sphere core undersupplied with oxygen and nutrients. Ultrastructural analysis of EGF-expanded neurospheres derived from rat fetal striatum demonstrated different cell types in neurospheres, including some apoptotic and necrotic cells (Lobo et al. 2003).

It has been confirmed by different groups that neurospheres and their developmental potential change over time. For instance, neurospheres at low passage generate more neurons and fewer astrocytes, whereas higher-passage long-term spheres give rise to more astrocytes and fewer neurons (Carpenter et al. 1999; Wright et al. 2006; Anderson et al. 2007). Neurosphere cultures derived from the ventral midbrain initially produce tyrosine hydroxlase (TH)-expressing cells (presumably dopaminergic) but lose this potential rapidly after a few passages (Liste et al. 2004; Kim et al. 2007). Our knowledge about directed neuronal phenotype-specific differentiation of neurosphere cells is still poor. Clearly, the predominant transmitter phenotype of neuronal cells in short- and long-term cultures is γ-amino-n-butyric acid (GABA).

Some studies have addressed the question of cell senescence in neurosphere cultures. Cells grown as neurospheres can be expanded dramatically but eventually undergo senescence (Carpenter et al. 1999; Wright et al. 2006). It is noteworthy that leukemia inhibitory factor (LIF) can slow down the senescence process by increasing telomerase expression and maintaining high growth rates in long-term neurosphere cultures. The precise mechanism for this observation remains elusive. The selection for a specific progenitor population has been proposed, and since LIF is involved in glial differentiation (Koblar et al. 1998), this may explain why astrocyte numbers increase in neurosphere cultures over time at the expense of neuronal cells.

Space: Two-dimensional or three-dimensional?

In vivo cells are integrated into a complex three-dimensional environment directly interacting with neighboring cells and the ECM. As stated earlier, monolayer cultures are artificial and simplistic in that they represent a two-dimensional system with variable and limited interactions with their immediate microenvironment. We are learning that normal and cancerous cells in a three-dimensional context can behave very differently in terms of gene and receptor expression, cell adhesion molecules, cell signaling, proliferation, and contact inhibition as compared to a two-dimensional situation (Jacks and Weinberg 2002; Abbott 2003; G.Y. Lee et al. 2007). Some of these differences are also evident for NSCs cultured either as neurospheres or as monolayers. For instance, it has been shown that neurosphere cells exhibit a higher proliferation and survival rate than monolayer cells (Svendsen et al. 1998, Layer et al. 2002). Depending on the questions to be answered and the experimental design, neurosphere cultures may be more advantageous to use than monolayer cultures.

Formation of three-dimensional spheres is a long-noted cell culture phenomenon and has been used in different experimental paradigms much earlier than the first description of the neurosphere assay. For instance, dispersed cells of sponges and hydra were used to study the phenomenon of cell reaggregation and regeneration of complete organisms (Wilson et al. 1907; Gierer et al. 1972). These pioneering studies are considered to be early tissue-engineering experiments. Technically, the process of reaggregation can be facilitated by culturing cells in conical tubes within roller drums; other cell culture methods are described in more detail by Layer et al. (2002). Interestingly, phenotypically different cells isolated from specific tissues such as the avian embryonic retina can reaggregate in vitro in an organotypic fashion resembling the laminated in vivo organization of the retina (Layer et al. 2002). Thus, studying the organization principles of multicellular structures such as cytospheres may help to develop new ex vivo model systems based on stem cells.

CONCLUDING REMARKS

Sphere formation is a long-noted cell culture phenomenon and is currently undergoing a "renaissance" in many disciplines influenced by the popularity of the neurosphere assay. Cytospheres are now used across many different organ systems and in the cancer stem cell field. Therefore, it is important to be familiar with the biological and technical potentials and limitations of this useful cell culture tool. Many research groups continue to introduce experimental variability into the cytosphere assay, and the currently used protocols have not been standardized, thereby hampering the comparison of results obtained by different scientists. Caution needs to be taken if the cytosphere assay of any cell-plating density is being used to study the key properties of stem cells at the single cell level. Clonality, self-renewal, and multipotentiality of single stem cells cannot be studied reliably using sphere cultures because of the so far underestimated dynamics of suspension cultures, including the constant merging of free-floating spheres and the motility of single cells contributing to polyclonal sphere growth. Consequently, the number of spheres, their diameter, and formation after passage should not be used as a functional metric of single stem cells. Bulk cytosphere assays are helpful for long-term cell expansion, particularly for the multiplication of vulnerable human cells that are difficult to grow as monolayer cultures.

REFERENCES

Abbott A. 2003. Cell culture: Biology's new dimension. *Nature* **424**: 870–872.

Alvarez-Dolado M. 2007. Cell fusion: Biological perspectives and potential for regenerative medicine. *Front. Biosci.* **12**: 1–12.

Anderson D.J. 2001. Stem cells and pattern formation in the nervous system: The possible versus the actual. *Neuron* **30**: 19–35.

Anderson L., Burnstein R.M., He X., Luce R., Furlong R., Foltynie T., Sykacek P., Menon D.K., and Caldwell M.A. 2007. Gene expression changes in long term expanded human neural progenitor cells passaged by chopping lead to loss of neurogenic potential in vivo. *Exp. Neurol.* **204**: 512–524.

Caldwell M.A., He X., Wilkie N., Pollack S., Marshall G., Wafford K.A., and Svendsen C.N. 2001. Growth factors regulate the survival and fate of cells derived from human neurospheres. *Nat. Biotechnol.* **19**: 475–479.

Chaichana K., Zamora-Berridi G., Camara-Quintana J., and Quinones-Hinojosa A. 2006. Neurosphere assays: Growth factors and hormone differences in tumor and nontumor studies. *Stem Cells* **24**: 2851–2857.

Carpenter M.K., Cui X., Hu Z.Y., Jackson J., Sherman S., Seiger A., and Wahlberg L.U. 1999. In vitro expansion of a multipotent population of human neural progenitor cells. *Exp Neurol.* **158**: 265–278.

Chiasson B.J., Tropepe V., Morshead C.M., and van der Kooy D. 1999. Adult mammalian forebrain ependymal and subependymal cells demonstrate proliferative potential, but only subependymal cells have neural stem cell characteristics. *J. Neurosci.* **19**: 4462–4471.

Davis A.A. and Temple S. 1994. A self-renewing multipotential stem cell in embryonic rat cerebral cortex. *Nature* **372**: 263–266.

Frederiksen K., Jat P.S., Valtz N., Levy D., and McKay R. 1988. Immortalization of precursor cells from the mammalian CNS. *Neuron* **1**: 439–448.

Gage F.H. 2000. Mammalian neural stem cells. *Science* **287**: 1433–1438.

Gage F.H., Ray J., and Fisher L.J. 1995. Isolation, characterization, and use of stem cells from the CNS. *Annu. Rev. Neurosci.* **18**: 159–192.

Gierer A., Berking S., Bode H., David C.N., Flick K., Hansmann G., Schaller H., and Trenkner E. 1972. Regeneration of hydra from reaggregated cells. *Nat. New Biol.* **239**: 98–101.

Gritti A., Frolichsthal-Schoeller P., Galli R., Parati E.A., Cova L., Pagano S.F., Bjornson C.R., and Vescovi A.L. 1999. Epidermal and fibroblast growth factors behave as mitogenic regulators for a single multipotent stem cell-like population from the subventricular region of the adult mouse forebrain. *J. Neurosci.* **19**: 3287–3297.

Imura T., Kornblum H.I., and Sofroniew M.V. 2003. The predominant neural stem cell isolated from postnatal and adult forebrain but not early embryonic forebrain expresses GFAP. *J. Neurosci.* **23**: 2824–2832.

Jacks T. and Weinberg R.A. 2002. Taking the study of cancer cell survival to a new dimension. *Cell* **111**: 923–925.

Jessberger S., Clemenson G.D. Jr., and Gage F.H. 2007. Spontaneous fusion and nonclonal growth of adult neural stem cells. *Stem Cells* **25**: 871–874.

Johe K.K., Hazel T.G., Muller T., Dugich-Djordjevic M.M., and McKay R.D. 1996. Single factors direct the differentiation of stem cells from the fetal and adult central nervous system. *Genes Dev.* **10:** 3129–3140.

Kiel M.J., Yilmaz O.H., Iwashita T., Yilmaz O.H., Terhorst C., and Morrison S.J. 2005. SLAM family receptors distinguish hematopoietic stem and progenitor cells and reveal endothelial niches for stem cells. *Cell* **121:** 1109–1121.

Kim H.J., Sugimori M., Nakafuku M., and Svendsen C.N. 2007. Control of neurogenesis and tyrosine hydroxylase expression in neural progenitor cells through bHLH proteins and Nurr1. *Exp. Neurol.* **203:** 394–405.

Kippin T.E., Kapur S., and van der Kooy D. 2005. Dopamine specifically inhibits forebrain neural stem cell proliferation, suggesting a novel effect of antipsychotic drugs. *J. Neurosci.* **25:** 5815–5823.

Kitchens D.L., Snyder E.Y., and Gottlieb D.I. 1994. FGF and EGF are mitogens for immortalized neural progenitors. *J. Neurobiol.* **25:** 797–807.

Koblar S.A., Turnley A.M., Classon B.J., Reid K.L., Ware C.B., Cheema S.S., Murphy M., and Bartlett P.F. 1998. Neural precursor differentiation into astrocytes requires signaling through the leukemia inhibitory factor receptor. *Proc. Natl. Acad. Sci.* **95:** 3178–3181.

Kohler G. and Milstein C. 1975. Continuous cultures of fused cells secreting antibody of predefined specificity. *Nature* **256:** 495–497.

Kukekov V.G., Laywell E.D., Thomas L.B., and Steindler D.A. 1997. A nestin-negative precursor cell from the adult mouse brain gives rise to neurons and glia. *Glia* **21:** 399–407.

Layer P.G., Robitzki A., Rothermel A., and Willbold E. 2002. Of layers and spheres: the reaggregate approach in tissue engineering. *Trends Neurosci.* **25:** 131–134.

Lee G.Y., Kenny P.A., Lee E.H., and Bissell M.J. 2007. Three-dimensional culture models of normal and malignant breast epithelial cells. *Nat. Methods* **4:** 359–365.

Lee J.P., Jeyakumar M., Gonzalez R., Takahashi H., Lee P.J., Baek R.C., Clark D., Rose H., Fu G., Clarke J., McKercher S., Meerloo J., Muller F.J., Park K.I., Butters T.D., Dwek R.A., Schwartz P., Tong G., Wenger D., Lipton S.A., Seyfried T.N., Platt F.M., and Snyder E.Y. 2007. Stem cells act through multiple mechanisms to benefit mice with neurodegenerative metabolic disease. *Nat. Med.* **13:** 439–447.

Lendahl U., Zimmerman L.B., and McKay R.D. 1992. CNS stem cells express a new class of intermediate filament protein. *Cell* **60:** 585–595.

Liste I., Garcia-Garcia E., and Martinez-Serrano A. 2004. The generation of dopaminergic neurons by human neural stem cells is enhanced by Bcl-XL, both in vitro and in vivo. *J. Neurosci.* **24:** 10786–10795.

Lobo M.V., Alonso F.J., Rendondo C., Lopez-Toledano M.A., Caso E., Herranz A.S., Paino C.L., Reimers D., and Bazan E. 2003. Cellular characterization of epidermal growth factor-expanded free-floating neurospheres. *J. Histochem. Cytochem.* **51:** 89–103.

Ming G.L. and Song H. 2005. Adult neurogenesis in the mammalian central nervous system. *Annu. Rev. Neurosci.* **28:** 223–250.

Morshead C.M., Craig C.G., and van der Kooy D. 1998. In vivo clonal analyses reveal the properties of endogenous neural stem cell proliferation in the adult mammalian forebrain. *Development* **125:** 2251–2261.

Morshead C.M., Benveniste P., Iscove N.N., and van der Kooy D. 2002. Hematopoietic competence is a rare property of neural stem cells that may depend on genetic and epigenetic alterations. *Nat. Med.* **8:** 268–273.

Morshead C.M., Garcia A.D., Sofroniew M.V., and van der Kooy D. 2003. The ablation of glial fibrillary acidic protein-positive cells from the adult central nervous system results in the loss of forebrain neural stem cells but not retinal stem cells. *Eur. J. Neurosci.* **18:** 76–84.

Nunes M.C., Roy N.S., Keyoung H.M., Goodman R.R., McKhann G., II, Jiang L., Kang J., Nedergaard M., and Goldman S.A. 2003. Identification and isolation of multipotential neural progenitor cells from the subcortical white matter of the adult human brain. *Nat. Med.* **9:** 439–447.

Palmer T.D., Willhoite A.R., and Gage F.H. 2000. Vascular niche for adult hippocampal neurogenesis. *J. Comp. Neurol.* **425:** 479–494.

Palmer T.D., Markakis E.A., Willhoite A.R., Safar F., and Gage F.H. 1999. Fibroblast growth factor-2 activates a latent neurogenic program in neural stem cells from diverse regions of the adult CNS. *J. Neurosci.* **19:** 8487–8497.

Parker M.A., Anderson J.K., Corliss D.A., Abraria V.E., Sidman R.L., Park K.I., Teng Y.D., Cotanche D.A., and Snyder E.Y. 2005. Expression profile of an operationally-defined neural stem cell clone. *Exp. Neurol.* **194:** 320–332. Erratum in: *Exp. Neurol.* 2006. **201:** 275.

Ray J., Peterson D.A., Schinstine M., and Gage F.H. 1993. Proliferation, differentiation, and long-term culture of primary hippocampal neurons. *Proc. Natl. Acad. Sci.* **90:** 3602–3606.

Represa A., Shimazaki T., Simmonds M., and Weiss S. 2001. EGF-responsive neural stem cells are a transient population in the developing mouse spinal cord. *Eur. J. Neurosci.* **14:** 452–462.

Reubinoff B.E., Itsykson P., Turetsky T., Pera M.F., Reinhartz E., Itzik A., and Ben-Hur T. 2001. Neural progenitors from human embryonic stem cells. *Nat. Biotechnol.* **19:** 1134–1140.

Reynolds B.A. and Weiss S. 1992. Generation of neurons and astrocytes from isolated cells of the adult mammalian central nervous system. *Science* **255:** 1707–1710.

———. 1996. Clonal and population analyses demonstrate that an EGF-responsive mammalian embryonic CNS precursor is a stem cell. *Dev. Biol.* **175:** 1–13.

Reynolds B.A. and Rietze R.L. 2005. Neural stem cells and neurospheres—Reevaluating the relationship. *Nat. Methods* **2:** 333–336.

Reynolds B.A., Tetzlaff W., and Weiss S. 1992. A multipotent EGF-responsive striatal embryonic progenitor cell produces neurons and astrocytes. *J. Neurosci.* **12:** 4565–4574.

Scadden D.T. 2006. The stem-cell niche as an entity of action. *Nature* **441:** 1075–1079.

Seaberg R.M. and van der Kooy D. 2002. Adult rodent neurogenic regions: The ventricular subependyma contains neural stem cells, but the dentate gyrus contains restricted progenitors. *J. Neurosci.* **22:** 1784–1793.

Shen Q., Wang Y., Dimos J.T., Fasano C.A., Phoenix T.N., Lemischka I.R., Ivanova N.B., Stifani S., Morrisey E.E., and Temple S. 2006. The timing of cortical neurogenesis is encoded within lineages of individual progenitor cells. *Nat. Neurosci.* **9:** 743–751.

Singec I., Jandial R., Crain A., Nikkhah G., and Snyder E.Y. 2007. The leading edge of stem cell therapeutics. *Annu. Rev. Med.* **58:** 313–328.

Singec I., Meyer R.P., Volk B., Frotscher M., and Knoth R. 2004. Fusion of neurospheres and single cells in neural stem cell cultures. Program No. 825.4. Abstract Viewer/Itinerary Planner. Society for Neuroscience, Washington, D.C.

Singec I., Knoth R., Meyer R.P., Maciaczyk J., Volk B., Nikkhah G., Frotscher M., and Snyder E.Y. 2006. Defining the actual sensitivity and specificity of the neurosphere assay in stem cell biology. *Nat. Methods* **3:** 801–806.

Smith L.G., Weissman I.L., and Heimfeld S. 1991. Clonal analysis of hematopoietic stem-cell differentiation in vivo. *Proc. Natl. Acad. Sci.* **88:** 2788–2792.

Snyder E.Y., Deitcher D.L., Walsh C., Arnold-Aldea S., Hartwieg E.A., and Cepko C.L. 1992. Multipotent neural cell lines can engraft and participate in development of mouse cerebellum. *Cell* **68:** 33–51.

Song H., Stevens C.F., and Gage F.H. 2002. Astroglia induce neurogenesis from adult neural stem cells. *Nature* **417:** 39–44.

Spangrude G.J., Heimfeld S., and Weissman I.L. 1988. Purification and characterization of mouse hematopoietic stem cells. *Science* **241:** 58–62. Erratum in: *Science* 1989. **244:** 1030.

Svendsen C.N., ter Borg M.G., Armstrong R.J., Rosser A.E., Chandran S., Ostenfeld T., and Caldwell M.A. 1998. A new method for the rapid and long term growth of human neural precursor cells. *J. Neurosci. Methods* **85:** 141–152.

Till J.E. and McCulloch E.A. 1961. A direct measurement of the radiation sensitivity of normal mouse bone marrow cells. *Radiat. Res.* **14:** 1419–1430.

Villa A., Snyder E.Y., Vescovi A., and Martinez-Serrano A. 2000. Establishment and properties of a growth factor-dependent, perpetual neural stem cell line from the human CNS. *Exp. Neurol.* **161:** 67–84.

Weiss S., Reynolds B.A., Vescovi A.L., Morshead C., Craig C.G., and van der Kooy D. 1996. Is there a neural stem cell in the mammalian forebrain? *Trends Neurosci.* **19:** 387–393.

Wilson H.V. 1907. On some phenomena of coalescence and regeneration in sponges. *J. Exp. Zool.* **5:** 245–258.

Wright L.S., Prowse K.R., Wallace K., Linskens M.H., and Svendsen C.N. 2006. Human progenitor cells isolated from the developing cortex undergo decreased neurogenesis and eventual senescence following expansion in vitro. *Exp. Cell Res.* **312:** 2107–2120.

8

Monolayer Cultures of Neural Stem/Progenitor Cells

Jasodhara Ray
Laboratory of Genetics
Salk Institute for Biological Studies
La Jolla, California 92037

THE CENTRAL DOGMA IN NEUROSCIENCE "no new neurons after birth" existed for almost a century. Only in recent years was it believed that neurons are generated exclusively during the prenatal phase of development. The study of adult neurogenesis was started in earnest in 1990s, and it has now become clear that active neurogenesis, a process of generating functionally integrated neurons from undifferentiated multipotent stem or progenitor cells, continues in discrete regions of the adult CNS throughout the life of mammals, including humans.

During development, nerve cells in the mammalian CNS are generated by the proliferation of multipotent stem/progenitor cells that migrate, find their site of final destination, and ultimately terminally differentiate. Owing to their relative rarity and lack of specific phenotypic markers, putative stem cells have been characterized based on their functional criteria. The discovery of putative stem cells in a given tissue is usually contingent on the development of in vitro culture conditions enabling a rigorous characterization. According to these criteria, stem cells must demonstrate the ability to proliferate, self-renew over an extended period of time, and generate a large number of progeny (progenitor or precursor cells) that can differentiate into the primary cell types of the tissue from which it was generated (Gage 1998; Temple 2001a,b). The in vitro culture consists of both stem and progenitor cells, and the terms "stem, progenitor, and precursor cells" have been used interchangeably in the literature. In this chapter, I use the term "stem/ progenitor cells."

The molecular specification of neural stem/progenitor cells has been guided by extrinsic signals, intrinsic mechanisms, and regulated expression of transcription factors (Jessell 2000). It is now known that both regional specification of cells derived from distinct areas of the CNS (Hitoshi et al. 2002; Ostenfeld et al. 2002; Parmar et al. 2002) and species-specific differences do exist between cells involved in neurogenesis (Ray and Gage 2006). Since cultures provide a powerful tool to test hypotheses on in vivo properties of cells, there is intense interest in studying CNS stem/progenitor cells in culture models to understand the fundamental cellular, biochemical, and physiological bases of in situ regulation of cell growth, differentiation, and death. For decades, attempts have been made to isolate and culture CNS stem cells to understand the fundamental cellular, biochemical, and physiological properties of these cells in situ; intrinsic and extrinsic mechanisms that control various steps of neurogenesis, including proliferation, survival, fate choice, neuronal migration, integration, and synapse formation; and the factors involved in these processes. Although establishment of short-term cultures of proliferative neuroblasts and glioblasts or mature neurons and glia was successful (Walicke et al. 1986); Banker and Goslin 1989; Brewer 1997), the generation of long-term cultures of stem/progenitor cells that produce large quantities of cells was not successful until recently.

Neural tissues are composed of both neuronal and nonneuronal cells as well as connective and vascular tissues. When put in culture, cells lose their physiological connections, anchorages, and the humoral environments. Strategies to isolate and culture stem/progenitor cells from CNS involved (1) isolating intact cells from an intertwined network of thousands of adhesive contacts, (2) separating stem/progenitor cells from other brain cells and connective tissue debris, and (3) providing the appropriate environmental conditions, including specific nutrients, growth factors, pH (7.2–7.6), and osmolarity. Different in vitro culture conditions have been developed to provide exogenous factors that closely resemble the in vivo environments needed for the survival, proliferation, and differentiation of stem/progenitor cells. The optimum conditions necessary for the survival and growth of stem/progenitor cells vary from species to species (Reynolds et al. 1992; Ray et al. 1993; Gage et al. 1995a,b; Svendsen et al. 1997; Ostenfeld and Svendsen 2004; Flanagan et al. 2006; Ray and Gage 2006), and the parameters for each species should be explored before proceeding with a large-scale culture.

ANALYSIS OF CULTURE METHODOLOGY

Two methods commonly used to culture stem/progenitor cells are neurosphere and monolayer cultures (Gottlieb 2002). In neurosphere cultures, individual cells plated on nonadhesive substrate proliferate and generate suspended "free-floating" cultures of cells (Reynolds and Weiss 1992; Reynolds et al. 1992). In the adhesive cultures, progenitor cells grow as a monolayer on coated substrates such as polyornithine (PORN) and laminin (Ray et al. 1993; Gage et al. 1995b).

Since its first report, it was assumed that each neurosphere is generated from a stem cell or that there is a one-to-one correlation between neurospheres and stem cells (Reynolds et al. 1992). The ability to culture these cells as free-floating spheres led to the development of the neurosphere assay (NPA) that has been used like the colony-forming assay to determine whether or not a cell is stem-like (Reynolds and Rietze 2005; Jensen and Parmar 2006). Clonality is the heart of stem cell biology because it is the only way to determine the self-renewal capacity and the true potential of a given individual cell (Parker et al. 2005). There is now growing evidence that neural cells that are capable of proliferation and sphere formation are not all derived from stem cells, suggesting that just growth as a neurosphere does not fulfill the criteria of being neural stem cells (Seaberg and van der Kooy 2003; Parker et al. 2005; Jensen and Parmar 2006; Navarro-Galve and Martinez-Serrano 2006).

Despite their ease of use, several studies have reported that using neurospheres to study progenitors' properties has drawbacks (Jensen and Parmar 2006; Navarro-Galve and Martinez-Serrano 2006). One difficulty with the system comes from the inherent properties of the suspension culture; since it is not possible to look inside the tightly packed spheres, it is impossible to monitor the morphology and properties of individual cells during the culture period. Extensive cell death in tightly packed necrotic cores of neurospheres (Svendsen et al. 1997) and spatially restricted penetration of nutrients, oxygen, and effectors (e.g., drugs, reagents, and growth factors) may cause variability in the stem cells' ability to respond to environmental cues and thus compromise the accuracy of various studies, e.g., proliferation assay, effects of growth factors, and other reagents on cell survival, growth, and differentiation. Furthermore, neurospheres are heterogeneous in nature as each sphere contains cells at various stages of differentiation (Suslov et al. 2002), and this heterogeneity increases with sphere size. Heterogeneity also arises because the neurospheres are motile and merge with each other under culture conditions

(Jessberger et al. 2006; Singec et al. 2006). Both the composition of cell types and the properties of cells within each neurosphere are variable as cell density within the spheres is not constant; this alters the microenvironment which in turn may affect the proliferation capacity and positional cues to which the cells are exposed (Tropepe et al. 1999; Hack et al. 2004). Since neurosphere cultures tend to generate differentiated cells in their core, the interaction between stem/progenitor cells and differentiating cells may expose the stem cells to paracrine factors that promote differentiation (Reynolds and Rietze 2005). The heterogeneous nature and the problem associated with it is further elucidated by gene expression studies which showed that in a neural stem cell clone cultured as a neurosphere, the pattern shifted from "stem-like" to more "differentiated" (Parker et al. 2005; Navarro-Galve and Martinez-Serrano 2006). Thus, it is proposed that without functional validation, a neurosphere may be a poor model for predicting a stem cell's attributes because it consists of heterogeneous populations of cells and only a small proportion are truly "stem-like" (Reynolds and Rietze 2005).

Monolayer culture overcomes some of these problems by allowing the cells to remain more isolated and to be continuously nurtured by the factors in the medium, thereby maintaining a higher degree of homogeneity. The greater homogeneity does not imply that in monolayer cultures, all cells would be identical or all cells would be stem cells or progenitor cells. It should be noted that due to the asymmetric division of stem cells, long-term cultures derived from even single-cell clones will contain stem, lineage-restricted, and differentiated neurons and glia. However, the ease of monitoring the morphology of individual cells makes it easier to study the properties of a cell at a single-cell level. The lack of tight contacts between cells would minimize spontaneous differentiation. Stem/progenitor cells in monolayer cultures allow direct access for testing pharmacological agents and electrophysiological recordings. This chapter describes the principles involved and strategies used for culturing stem/progenitor cells from fetal and adult rodent brains as monolayer and discusses the rationale for using different reagents (medium, substrates, growth factors, etc.) for cell isolation and monolayer culturing followed by in vitro characterization of cells.

CULTURING STRATEGIES

Cultures of neural stem/progenitors from the adult CNS are largely established on the basis of their preferential growth over other cell types. The general strategy for culturing stem/progenitor cells involves the separation of cells from connective tissues by enzymatic degradation, partial purification from nonneuronal as well as lineage-restricted and differentiated neuronal cells, and plating on a substratum for attachment followed by culturing for the generation of large quantities of cells. Long-term culturing of primary stem/progenitor cells has been advanced by the development of serum-free medium and the findings that a number of mitogenic growth factors including epidermal growth factor (EGF) and basic fibroblast growth factor-2 (FGF-2) have proliferative effects on stem/progenitor cells (Gage et al. 1995a; Weiss et al. 1996a; McKay 1997).

RATIONALE FOR VARIOUS CULTURING APPROACHES

Isolation of Cells by Enzymatic Degradation of Tissue

The general scheme for culturing stem/progenitor cells from adult CNS regions involves enzymatic digestion of tissue, their partial purification from connective tissues, and other cells and plating (Fig. 1). Two common methods for tissue digestion are papain-protease-DNase I (PPD) (Gage et al. 1995b; Palmer et al. 1999) and trypsin-hyaluronidase-kynurenic acid digestion (Gritti et al. 1996; Weiss et al. 1996b). A number of other pro-teases alone or in combination have also been successfully used to isolate stem/progenitor cells from adult brain (Reynolds and Weiss 1992; Weiss et al. 1996b; Brewer 1997; Svendsen et al. 1997). Although almost equal numbers of cells can be isolated with different enzymes (trypsin/ proteinase K, protease XXIII, papain, collagenase, and dispase) from 6-week-old rats, the viability of cells isolated with papain alone was bet-ter (Brewer 1997). In our laboratory, we use the PPD digestion method (Gage et al. 1995b; Palmer et al. 1999; Ray and Gage 2006) for isolation of embryonic and adult stem/progenitor cells from different CNS regions and in all age groups. However, fetal tissue is soft and thus the enzymatic digestion step is not necessary to isolate cells (Ray et al. 1993).

Growth Medium and Supplements

Most methods to culture rodent or human stem/progenitor cells as monolayer use Dulbecco's modified Eagle's medium (DMEM), often combined (1:1) with Ham's F12 medium buffered for use at ambient CO_2. Cultures are grown in an atmosphere of 5% CO_2, 95% air. Neurobasal medium that has lower osmolarity than DMEM, lower con-centration of cysteine and glutamine, and no ferrous sulfate promotes survival of isolated cells in culture.

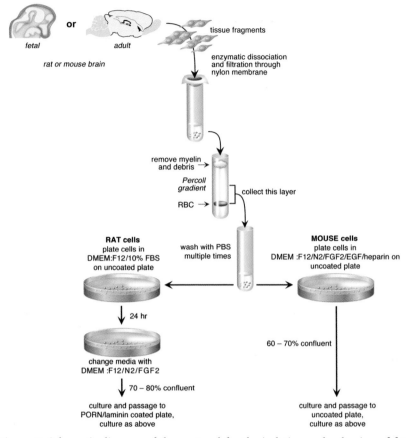

Figure 1. Schematic diagram of the protocol for the isolation and culturing of fetal and adult brain-derived stem/progenitor cells. (Modified, from permission, from Ray and Gage 2006.)

Serum-free medium are essential when tight control over the concentration of hormones, amino acids, and growth factors is needed. N2 and B27 supplements have been developed to culture neuronal cells under serum-free chemically defined conditions (Bottenstein and Sato 1979; Brewer et al. 1993). N2 supplement is composed of progesterone, sodium selenite, putrescine, glutamine, insulin, and transferrin (Bottenstein and Sato 1979; Ray et al. 1993). The B27 supplement contains a range of hormones, vitamins, anti-oxidants, essential fatty acids, and retinal acetate, in addition to the basic formulation of N2 (Brewer et al. 1993). These two supplements have been extensively used to culture fetal, embryonic, and adult CNS-derived stem/progenitor cells from various species including human.

The most commonly used base medium for culturing neural stem/progenitor cells from various species is made by adding an N2 supplement (available from suppliers as 100× stock solution) to DMEM: F12 medium containing 1 mM glutamine and an antibiotic-antimitotic (penicillin-streptomycin-fungizone). Following their initial plating, N2 medium increases both the survival and proliferation of stem/progenitor cells, whereas survival but not proliferation of neuronal cells is strongly increased in B27 medium (Svendsen et al. 1995). B27/neurobasal medium restricts glia growth, and the culture is enriched in both stem/progenitor and neuronal populations (Brewer et al. 1993). This medium is more commonly used to culture human cells (Walsh et al. 2005; Flanagan et al. 2006).

Growth Substrate for Cell Attachment

The type of substratum can greatly influence the morphology, metabolism, and ultrastructure of the cells. In addition, the composition of the substratum is important for the adhesion, survival, proliferation, and differentiation of stem/progenitor cells. Neuron-enriched cultures derived from both rodent and human CNS tissues have long been grown in serum-supplemented medium (Laerum et al. 1985). Serum contains a number of adhesion factors that aid in cell attachment and peptides that regulate growth and differentiation. However, the undefined components present in the serum brought unwanted variability and thus the use of serum became less desirable. It was subsequently noted that short exposure of cells to serum promoted the attachment of cells to the substratum without causing any deleterious effects on the cells. To initiate cultures, primary adult rat stem/progenitor cells were isolated and plated on uncoated plates in the presence of fetal bovine serum (FBS) for a short period of time (4–12 hours) and then cultured long term in serum-free medium (Gage et al. 1995b). Components in serum preferentially promote attachment of stem/progenitor cells over more differentiated neurons and glia or nonneuronal cells such as fibroblasts. Serum does not promote attachment of adult mouse stem/progenitor cells to the substrate (Ray and Gage 2006).

In the absence of serum, however, cells cannot attach to the plates, and it is necessary to provide an adhesive substrate. Culture plates were coated with polymer of basic amino acids such as polyornithin (PORN), poly-D-lysine (PDL), or poly-L-lysine (PLL) which are highly adhesive. It is essential to consider the type of research to be conducted in choosing the substrate. For example, PDL should not be used for electrophysiological studies as it alters cAMP synthesis and interferes with ion-channel functioning (Walsh et al. 2005). Treating the growth surface with extracellular matrix (ECM) molecules promotes cell adhesion, survival, differentiation, and neurite outgrowth. Among the ECM proteins, laminin and fibronectin have been used as substrate to culture rodent and human stem/progenitor cells (Ray et al. 1993; Vicario-Abejon et al. 2000). For long-term culturing, plates are coated with PORN followed by laminin. Adult stem/progenitor cells (secondary cultures) are passaged onto PORN/laminin (P/L)-coated plates (Gage et al. 1995b). Laminin enhances neurite elongation from neurons. Primary fetal rat stem/progenitor cells do not need serum for attachment, and both primary and secondary cells attach to P/L and form monolayer cultures (Ray et al. 1993). In contrast to rat cells, adult mouse stem/progenitor cells (secondary cells) attach to P/L, but they proliferate slowly on this substrate (Ray and Gage 2006). Fibronectin is more commonly used than laminin for culturing human cells (Walsh et al. 2005). However, laminin stimulated neurite outgrowth of human neurons to a greater extent than fibronectin (Flanagan et al. 2006). Note that the source of laminin is important: We have found that some batches of laminin induce cell differentiation instead of proliferation.

Growth Factors Used for Cell Proliferation

Stem/progenitor cells cultured from various adult rat and mouse CNS regions exhibit different properties, including their responsiveness to mitogenic growth factors and differentiation potentials (Temple 2001a,b; Ostenfeld et al. 2002). Two mitogenic growth factors, EGF and FGF-2, have been used to culture stem/progenitor cells from various species. At low concentrations (1 pg/ml to 5 ng/ml), FGF-2 has both survival and neurite elongation effects on rat neuronal progenitor cells, whereas at higher concentrations (10–100 ng/ml), it functions as a proliferative factor (Ray et al. 1993; Ray and Gage 1994). EGF, first used to culture stem cells from embryonic and adult mouse subependymal/forebrain regions as neurospheres (Reynolds et al. 1992; Reynolds and Weiss 1992), cannot generate monolayer cultures (Ray and Gage 2006). In contrast, FGF-2 alone was successful in establishing monolayer cultures of rat stem/progenitor cells from different embryonic and adult CNS regions (Ray et al. 1993; Ray and Gage 1994; Gage et al. 1995a,b; Palmer et al. 1995; Shihabuddin et al. 1997). FGF-2 can generate neurospheres from the lateral ventricle/forebrain of the adult mouse, but they survive only short term (Gritti et al. 1996). Direct comparison of survival and growth rates of embryonic rat or mouse striatum-derived neurospheres in EGF showed that although mouse cells can be expanded long term (~2 months), rat cells died between 3 and 4 weeks. EGF, FGF-2, or their combinations generate neurospheres from mouse and rat CNS, but FGF-2 has limited capability to expand them as neurospheres (Svendsen et al. 1997; Caldwell et al. 2004).

In some cases, FGF-2 and heparin are required for the proliferation of mouse and rat cells (Weiss et al. 1996b; Caldwell et al. 2004). Inclusion of heparin in the EGF + FGF-2 mixture caused mouse stem/progenitor cells to attach to the substratum, followed by their robust expansion as monolayer. In the presence of FGF-2 and heparin, mouse progenitors expressing FGF receptor-1 (FGFR-1) and extracellular heparan sulfate (HS) attach cells to each other to form monolayer cultures (Richard et al. 1995, 2000). To address whether the properties of rat and mouse cells differ from each other particularly in their growth, proliferation, and differentiation profiles, mouse and rat whole-brain-derived stem/progenitor cells were isolated and cultured as monolayer under different cultures conditions (Ray and Gage 2006). Stem/progenitor cells from both species isolated under the same conditions respond differently to exogenous factors, and their proliferative biochemical properties and differential potentials are intrinsically different. These results, for the first time, showed that cell properties vary between species and the characteristics of one type of cells cannot be extrapolated to other kinds of cells. Besides FGF-2 and EGF, other factors such as sonic hedgehog (Shh) and insulin-like growth factor-I (IGF-I) at high concentration have proliferative effects on rat stem/progenitor cells. However, these factors have not been used to establish long-term cultures (Aberg et al. 2003; Lai et al. 2003). Human stem/progenitor cells have been cultured as neurosphere (Svendsen and Smith 1999; Gottlieb 2002; Walsh et al. 2005) or monolayer cultures (Roy et al. 2000). For monolayer culturing, cells were plated on laminin and cultured with FGF-2 and a low level of FBS.

Plating Density

Defined media support neuronal cell cultures at high density (>100,000 cells/cm^2). To establish rapidly proliferating monolayer cultures in a short time, rat stem/progenitor cells are plated at high density (Gage et al. 1995a,b; Palmer et al. 1995; Shihabuddin et al. 1997). Survival of rat stem/progenitor cells grown as monolayer cultures is density-dependent, and when plated at <1000 cells/cm^2, they do not survive even in the presence of a high concentration of FGF-2. When the culture medium is supplemented with 25% condition medium (CM) collected from a high-density culture, both survival and growth improved. This observation led to the purification and characterization of an *N*-glycosylated form of cystatin C (CCg), which acts as an autocrine/paracrine cofactor(s) for the mitogenic activity of FGF-2 (Taupin et al. 2000). Mouse stem/progenitor cells also require high-density plating for initiation of growth in monolayer culture. These results may suggest that cell-cell contacts and autocrine/paracrine factors are important in the cell growth process. The morphology of fetal and adult rat and mouse hippocampal stem/progenitor cells in monolayer cultures is shown in Figure 2.

Establishment of Clonal Cultures

Two methods are generally used to generate clonal cultures: Genetic marking of cells and the limiting dilution method. It should be noted that although clonal cultures are derived from a single cell, due to asymmetric cell division, stem cells give rise to both stem cells and lineage-restricted and differentiated neurons and glia. Hence, upon long-term culturing, the clonal cultures become heterogeneous in nature.

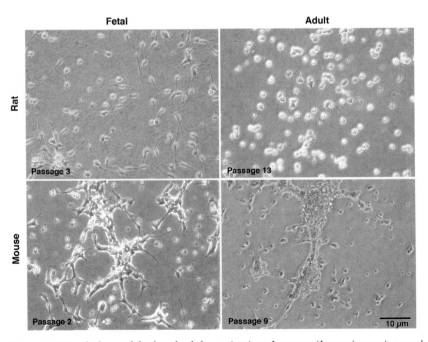

Figure 2. Morphology of fetal and adult rat (*top*) and mouse (*bottom*) stem/progenitor cells cultured as described in the text. Passage numbers are indicated in the panels.

Genetic marking of cells

A bulk population of stem/progenitor cells was infected with a retroviral vector expressing a reporter gene such as green fluorescent protein (GFP), *Escherichia coli* LacZ, or alkaline phosphatase genes. Retroviral vectors are derived from Moloney murine leukemia virus (Mo-MLV) and contain the gene of interest (transgene) and a selectable marker such as the neomycin-resistant gene (Somia and Verma 2000; Tashiro et al. 2006). Retroviral vectors can infect only dividing cells and integrate randomly in the cellular genome. As a result, all of the progeny of a single infected cell will inherit a unique and identifiable integration site (Cepko 1989), and the clonality of a population of cells can be determined by restriction enzyme digestion, followed by Southern blotting (Palmer et al. 1997, 1999). Cells are plated in the presence of the minimum amount of G418 needed to select for stable transfectants, and the medium is supplemented with 50% CM collected from a high-density stem cell culture.

Limiting Dilution Method

Clonal structures can be derived by the limiting dilution method (see Subprotocols).

IN VITRO DIFFERENTIATION OF STEM/PROGENITOR CELLS INTO NEURONS AND GLIA

A tight control of cell cycle progression is a prerequisite for the generation of appropriate cell fates during many developmental steps that transform proliferating undifferentiated tissues into fully differentiated and functional molecules. Although some of the molecular interactions and pathways have been elucidated in neural stem/progenitor cell differentiation, some important questions still remain unanswered. Little is known about the intrinsic factors that regulate stem cell maintenance, decide whether neurons or glia are generated, or control terminal differentiation. Recently, several patterning genes were found to control the cell proliferation rate of distinct CNS structures, and dual function molecules have been described that couple cell cycle and cell fate. Different reagents and protocols have been used for differentiation of neural stem/progenitor cells into neurons, astrocytes, or oligodendrocytes. The mode of action for some of these reagents is discussed below.

MECHANISMS OF ACTION OF VARIOUS DIFFERENTIATION FACTORS

Neuronal Differentiation

Retinoic Acid

Retinoic acid (RA) regulates the expression of several transcription factors that are important in neuronal determination, such as NeuroD and MASH1 (Johnson et al. 1992). Exposure to RA caused an up-regulation of NeuroD mRNA followed by an increase in p21 expression. These events cause a concurrent exit from the cell cycle and a decrease in the number of dividing cells. These changes are accompanied by a threefold increase in the number of cells differentiating into immature neurons (Kaplan et al. 1993; Takahashi et al. 1999). RA also sustains or up-regulates neurotrophin (NT) receptors trkA, trkB, trkC, and p75NGFR mRNA. However, the effect of RA on each receptor is cell-type-specific (Kaplan et al. 1993); the functional relevance of stimulating trk expression can be inferred by monitoring the expression of the immediate-early gene c-*fos* after NT treatment. The signaling cascade following c-*fos* induction is considered to be integral to the subsequent steps in differentiation (Sheng and Greenberg 1990).

Rat or mouse stem/progenitor cells treated with RA/FBS generate immature neurons and glia (mixed differentiation). For the generation of more mature neurons, cells were further treated with serum-supplemented medium containing brain-derived neurotrophic factor (BDNF). This treatment results in the generation of a large number of subtype neurons (Takahashi et al. 1999). Although RA is a strong differentiation factor for rat stem/progenitor cells, generating a large number of neurons and astrocytes, it differentiates mouse cells mostly to astrocytes (Ray and Gage 2006). However, few more matured neurons are generated in RA/FBS-treated mouse cells as indicated by cell morphology.

Forskolin

Forskolin enhances intracellular cAMP by activating adenyl cyclase. Signaling events can trigger cAMP-mediated signal transduction pathways, resulting in a differentiation response (Satoh et al. 1994; Joannides et al. 2004). Both neuronal and glial differentiations of rat and mouse stem/progenitor cells are influenced equally by forskolin. However, the patterns of neurite elongation of cells were different between the two cell types. TubβIII$^+$ cells generated in mouse cultures had one or more long neurites and many dendrites, whereas TubβIII$^+$ cells in rat cultures had short processes (Fig. 3) (Ray and Gage 2006).

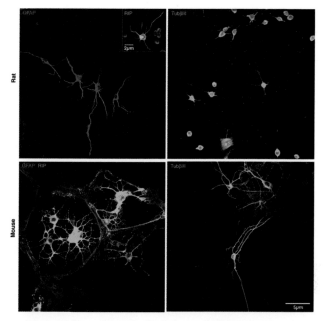

Figure. 3. (*Top*) Differentiation of adult rat hippocampal stem/progenitor cells with RA/forskolin/FBS. Cells are immunostained with antibodies against different markers. Multipotent rat stem cells differentiate into astrocytes (GFAP; *top left*), oligodendrocytes (RIP; *inset left*), and neurons (TubβIII; *top right*). (*Bottom*) Differentiation of adult mouse hippocampal stem/progenitor cells with forskolin/FBS. Cells are immunostained with antibodies against different markers. Cells differentiate into astrocytes (GFAP), oligodendrocytes (RIP; *bottom left*), and neurons (TubβIII; *bottom right*).

Valproic Acid

Valproic acid (VPA) is involved in neuronal differentiation of stem/progenitor cells. VPA acts on two phosphoprotein substrates of protein kinase C that have integral roles in CNS development and neuroplasticity in the adult brain. Chronic exposure of neuronal cells to therapeutic concentrations of VPA causes a significant inhibition of cell proliferation, an increase in growth-associated protein (GAP-43) expression, a downregulation of myristoylated alanine-rich C kinase substrate (MARCKS), and a highly significant increase in neurite outgrowth, all parameters suggestive of neuronal differentiation. Cell doubling time is increased, which is reflected in slower cell growth over time, and neurite outgrowth is increased in both length and number of neurites per cell (Watterson et al. 2002).

Insulin-like Growth Factors

IGF, IGF-I, or IGF-II has both proliferation and neuronal differentiation effects on adult rat stem/progenitor cells in a dose-dependent manner (Aberg et al. 2003). At a low dose (1 ng/ml), stem/progenitor cells differentiate into neurons. IGFs act via stimulation of the IGF-I receptor (IGFR-I), and the mitogen-activated protein kinase (MAPK) pathway has a major role in this process. A time-dependent effect has been described with an early stimulation of MAPK leading to proliferation, followed by a switch with late inhibition of MAPK which stimulates differentiation (Adi et al. 2002). IGFs or insulin activates the IGFR-I in stem/progenitor cells, which leads to the up-regulation of bone morphogenetic protein (BMP) antagonists such as Noggin, Smad6, and Smad7. These factors inhibit BMP signaling. The net effects are a decrease or block of astrocyte differentiation and an increase in neuronal and oligodendroglia differentiation (Hsieh et al. 2004). Further studies are needed to determine whether IGFs can directly promote oligodendroglia lineage commitment in a Noggin/Smad6/Smad7-independent manner.

Neuronal, Astrocytic, and Oligodendrocytic Differentiation

Platelet-derived Growth Factor

Platelet-derived growth factor (PDGF) induces both neuronal and astrocytic differentiation. PDGF causes neuronal differentiation by sustained activation of the Ras/ERK signal transduction pathway. It has also been shown to induce the production of oligodendrocytes and astrocytes. The production of neurons is diminished in response to PDGF, suggesting that PDGF drives multipotent adult neural stem/progenitors to produce oligodendrocytes, rather than neurons (Merkle and Alvarez-Buylla 2006). PDGF stimulates the proliferation of glia-restricted oligodendrocyte precursor cells (OPCs) that express a receptor for PDGF, PDGFR-α. Expression of transcription factors Nkx2.2 and Olig2 is required for oligodendrocyte production.

Leukemia-inhibitory Factor/Bone Morphogenetic Proteins

Leukemia-inhibitory factor/BMPs (LIF/BMPs) induce the differentiation of stem/progenitor cells into the astrocytic lineage. The cytokine LIF (leukemia inhibitory factor) acts through the gp130 signal-transducing subunit to activate the transcription factor signal transducer and activator of transcription 3 (STAT3). The gp130 cytokines also stimulate the prodifferentiation of the Ras-ERK pathway (Bonni et al. 1997). BMPs induce inhibitor of differentiation (Id) genes via the Smad pathway. Id proteins bind to ubiquitous basic helix-loop-helix (bHLH) factors. Id genes induced by BMP or other signals block entry into neural lineages, which is otherwise only partially prevented by LIF/STAT3 (Ying et al. 2003).

Ciliary Neurotrophic Factor

Binding of the ciliary neurotrophic factor (CNTF) stimulated the Janus kinase (JAK) signal transducer and activator of transcription (JAK-STAT) signaling pathway and selectively enhanced differentiation along a glial lineage (Bonni et al. 1997). CNTF inhibits differentiation of stem/progenitor cells along a neuronal lineage and promotes differentiation into astrocytes (Johe et al. 1996).

PROTOCOLS

Protocols for culturing, generation of clonal cultures, and characterization of stem/progenitor cells have been discussed recently (Ray and Gage 2006). A brief summary is given below.

Isolation and Propagation of Stem/Progenitor Cells

Isolation and culturing of rat and mouse neural stem/progenitor cells are outlined in Figure 1.

Adult Rat Stem/Progenitor Cell Culture

Dissect out brain regions and determine the weight of the tissue. Chop the tissue into small pieces, wash with Dulbecco's-PBS (phosphate-buffered saline) (D-PBS)/PSF (penicillin, streptomycin, fungizone), resuspend in PPD (0.01% papain, 0.1% neutral protease, and 0.01% DNase I in HBBS (HEPES-buffered balanced salt solution) supplemented with 12.4 mM MgSO$_4$) solution (10 ml/g tissue), and incubate in a 37°C water bath for 15–20 minutes. Dissociate tissue by repeated pipetting. When no tissue chunk is visible, mix with an equal volume of DMEM:F12/10% FCS (DMEM:F12/FCS), centrifuge, and resuspend pellet in the same medium. Mix cell suspension with an equal volume of Percoll solution (9 parts

Percoll + 1 part 10× D-PBS/PSF; 50% gradient). Fractionate the cell suspension by centrifugation in a Beckmann ultracentrifuge at 20,000g for 30 minutes at room temperature (Palmer et al. 1999). Remove the top layer containing myelin, debris, and differentiated cells. Collect the bottom layer, dilute with D-PBS/PSF, and centrifuge. Wash the pellet in D-PBS/PSF. Resuspend the cells in 1–2 ml of DMEM:F12/FBS medium and plate cells on uncoated tissue culture plates at high density. The next day, replace the medium with serum-free N2$^+$ medium (DMEM:F12 medium containing N2 supplement and 20 ng/ml FGF-2). Proliferating clusters of cells attached to the plates start to appear in about 3–5 days. Passage 60–70% confluent cultures onto P/L-coated plates and culture in N2$^+$ medium.

Fetal Rat Cultures

Dissect out fetal brain regions, suspend tissue in N2$^+$ medium, and triturate to make a single cell suspension. Plate cells onto P/L-coated plates and culture as described above for adult cells.

Fetal and Adult Mouse Stem/Progenitor Cultures

Isolate stem/progenitor cells from fetal or adult mouse brain regions by the same method described above for rat culture. Suspend isolated cells in N2 medium containing 20 ng/ml FGF-2, 20 ng/ml EGF, and 5 µg/ml heparin (N2^{+++}), plate onto uncoated tissue culture plates, and culture.

SUBPROTOCOLS

Preparation for Cultures and Differentiation of Rat and Mouse Cells

Coating of Tissue Culture Plates with PORN/Laminin

Add enough PORN (10 µg/ml in water) to cover plates and incubate overnight at room temperature. Wash plates, add laminin (5–10 µg/ml in PBS), and incubate overnight at 37°C. Store plates wrapped in Saran Wrap at −20°C until use.

Passaging, Freezing, and Reculturing of Cells

Trypsin is commonly used for detaching the cells from adhesive plates. However, be careful not to over-trypsinize since this can kill cells and reduce yield. Distribute trypsin (usually a mixture of trypsin and EDTA) evenly on top of cells and incubate to dislodge cells. Remove trypsin either by diluting the cell suspension or by treatment with a trypsin inhibitor. Plate cells as described above for culturing or freeze in appropriate medium containing 10% dimethylsulfoxide (DMSO) in liquid nitrogen. To reculture, thaw frozen cells quickly in a 37°C water bath, remove DMSO, plate, and culture.

Establishment of Clonal Cultures by the Limiting Dilution Method

Plate genetically marked stem/progenitor cells at a low density (1 cell/well; 96-well plate) in appropriate medium supplemented with 50% CM collected from high-density cultures and culture until clones appear. Expand clones and establish clonality by Southern blot analysis. For details, see Shihabuddin et al. (2000) and Scheel et al. (2005).

Differentiation of Stem/Progenitor Cells

Plate 100,000 cells/ml/well in P/L-coated 4-well chamber slides in $N2^+$ (rat) or $N2^{+++}$ (mouse) medium and culture for 24–72 hours. Remove medium and treat cells with differentiating agents as indicated below. The percentage of cells differentiating into a specific cell type varies with the method used.

Rat Cells

Mixed Differentiation

Incubate cells with 1 μM retinoic acid (RA) and 1% FBS for 4 days (Hsieh et al. 2004).

Neuronal Differentiation

Use one of the following conditions: (1) 1 μM RA, 5 μM forskolin in N2 medium containing 1% FBS for 4–6 days (Ray and Gage 2006). For generation of matured neurons, replace medium with 1% FBS-supplemented medium containing 20 ng/ml BDNF for 6 days (Takahashi et al. 1999); (2) 1 mM valproic acid in N2 medium for 4 days (Hsieh et al. 2004); or (3) 10 ng/ml PDGF AA, AB, or BB isoforms starting from 2 days prior to FGF-2 withdrawal and then for 6 days in the absence of FGF-2. Up to 80% of the cells express neuronal markers (Johe et al. 1996).

Astrocytic Differentiation

Use one of the following conditions: (1) 50 ng/ml LIF and 50 ng/ml BMP-2 in N2 medium for 6 days (Hsieh et al. 2004) or 5–10% FBS for 6 days; or (2) 10 ng/ml CNTF for 6 days (Johe et al. 1996).

Oligodendrocytic Differentiation

Use one of the following conditions: (1) 500 ng/ml IGF-I or IGF-II or 500 ng/ml insulin in insulin-free N2 medium for 6 days (Hsieh et al. 2004); or (2) 10 mM thyroid hormone T3 for 6 days (Johe et al. 1996).

Mouse Cells

Neuronal Differentiation

Use one of the following conditions: (1) 5 μM forskolin/1% FBS in N2 medium for 4–6 days (Ray and Gage 2006); or (2) 100 ng/ml BDNF in N2 medium containing 2% FBS for 7–8 days (Bull and Bartlett 2005).

Glial Differentiation

Use one of the following conditions: (1) 1 μM RA and 1% FBS in N2 medium for 4–6 days; or (2) 10% FBS in N2 medium for 4–6 days. Image analyses and quantification of cells with different phenotypes are performed as described previously (Lie et al. 2002; Ray and Gage 2006). The morphologies of adult rat and mouse hippocampus-derived stem/progenitor cells differentiated into neurons, astrocytes, and oligodendrocytes are shown in Figure 3.

SUMMARY

This chapter described the rationale for using different conditions (reagents) and methods for long-term culturing and differentiation of rat and mouse stem/progenitor cells as monolayer. Although stem/progenitor cells from rat and mouse brain tissue can be isolated by the same method, monolayer culturing of rat cells requires culturing conditions different from that of mouse cells. Stem/progenitor cells have been differentiated into neurons and glia by different reagents and their mode of action has been discussed. Isolated and cultured stem/progenitor cells provide an important source of cells for studying how intra- or extracellular stimuli influence the fate choice and the differentiation potentials. In addition, stem/progenitor cells can be grafted in animal models of neurodegenerative diseases with the ultimate aim of developing methods for cell replacement or gene therapy in humans.

ACKNOWLEDGMENTS

The author thanks Bobbi Miller and Ruth Oefner for their excellent technical assistance and Linda Kitabyashi for making the illustrations. The work was supported by grants from the National Institutes of Health (AGO 10435 and NIA 20938).

REFERENCES

Aberg M.A., Aberg N.D., Palmer T.D., Alborn A.M., Carlsson-Skwirut C., Bang P., Rosengren L.E., Olsson T., Gage F.H., and Eriksson P.S. 2003. IGF-I has a direct proliferative effect in adult hippocampal progenitor cells. *Mol. Cell. Neurosci.* **24:** 23–40.

Adi S., Bin-Abbas B., Wu N.Y., and Rosenthal S.M. 2002. Early stimulation and late inhibition of extracellular signal-regulated kinase 1/2 phosphorylation by IGF-I: A potential mechanism mediating the switch in IGF-I action on skeletal muscle cell differentiation. *Endocrinology* **143:** 511–516.

Banker G. and Goslin K. 1988. Developments in neural cell culture. *Nature* **336:** 185–186.

Bonni A., Sun Y., Nadal-Vicens M., Bhatt A., Frank D.A., Rozovsky I., Stahl N., Yancopoulos G.D., and Greenberg M.E. 1997. Regulation of gliogenesis in the central nervous system by the JAK-STAT signaling pathway. *Science* **278:** 477–483.

Bottenstein J.E. and Sato G.H. 1979. Growth of a rat neuroblastoma cell line in serum-free supplemented medium. *Proc. Natl. Acad. Sci.* **76:** 514–517.

Brewer G.J. 1997. Isolation and culture of adult rat hippocampal neurons. *J. Neurosci. Methods* **71:** 143–155.

Brewer G.J., Torricelli J.R., Evege E.K., and Price P.J. 1993. Optimized survival of hippocampal neurons in B27-supplemented Neurobasal, a new serum-free medium combination. *J. Neurosci. Res.* **35:** 567–576.

Bull N.D. and Bartlett P.F. 2005. The adult mouse hippocampal progenitor is neurogenic but not a stem cell. *J. Neurosci.* **25:** 10815–10821.

Caldwell M.A., Garcion E., terBorg M.G., He X., and Svendsen C.N. 2004. Heparin stabilizes FGF-2 and modulates striatal precursor cell behavior in response to EGF. *Exp. Neurol.* **188:** 408–420.

Cepko C.L. 1989. Immortalization of neural cells via retrovirus-mediated oncogene transduction. *Annu. Rev. Neurosci.* **12:** 47–65.

Flanagan L.A., Rebaza L.M., Derzic S., Schwartz P.H., and Monuki E.S. 2006. Regulation of human neural precursor cells by laminin and integrins. *J. Neurosci. Res.* **83:** 845–856.

Gage F.H. 1998. Stem cells of the central nervous system. *Curr. Opin. Neurobiol.* **8:** 671–676.

Gage F.H., Ray J., and Fisher L.J. 1995a. Isolation, characterization, and use of stem cells from the CNS. *Annu. Rev. Neurosci.* **18:** 159–192.

Gage F.H., Coates P.W., Palmer T.D., Kuhn H.G., Fisher L.J., Suhonen J.O., Peterson D.A., Suhr S.T., and Ray J. 1995b. Survival and differentiation of adult neuronal progenitor cells transplanted to the adult brain. *Proc. Natl. Acad. Sci.* **92:** 11879–11883.

Gottlieb D.I. 2002. Large-scale sources of neural stem cells. *Annu. Rev. Neurosci.* **25:** 381–407.

Gritti A., Parati E.A., Cova L., Frolichsthal P., Galli R., Wanke E., Faravelli L., Morassutti D.J., Roisen F., Nickel D.D., and Vescovi A.L. 1996. Multipotential stem cells from the adult mouse brain proliferate and self-renew in response to basic fibroblast growth factor. *J. Neurosci.* **16:** 1091–1100.

Hack M.A., Sugimori M., Lundberg C., Nakafuku M., and Gotz M. 2004. Regionalization and fate specification in neurospheres: The role of Olig2 and Pax6. *Mol. Cell. Neurosci.* **25:** 664–678.

Hitoshi S., Tropepe V., Ekker M., and van der Kooy D. 2002. Neural stem cell lineages are regionally specified, but not committed, within distinct compartments of the developing brain. *Development* **129:** 233–244.

Hsieh J., Aimone J.B., Kaspar B.K., Kuwabara T., Nakashima K., and Gage F.H. 2004. IGF-I instructs multipotent adult neural progenitor cells to become oligodendrocytes. *J. Cell Biol.* **164:** 111–122.

Jensen J.B. and Parmar M. 2006. Strengths and limitations of the neurosphere culture system. *Mol. Neurobiol.* **34:** 153–161.

Jessberger S., Clemenson Jr., G.D., and Gage F.H. 2006. Spontaneous fusion and nonclonal growth of adult neural stem cells. *Stem Cells* **25:** 871–874.

Jessell T.M. 2000. Neuronal specification in the spinal cord: Inductive signals and transcriptional codes. *Nat. Rev. Genet.* **1:** 20–29.

Joannides A., Gaughwin P., Schwiening C., Majed H., Sterling J., Compston A., and Chandran S. 2004. Efficient generation of neural precursors from adult human skin: Astrocytes promote neurogenesis from skin-derived stem cells. *Lancet* **364:** 172–178.

Johe K.K., Hazel T.G., Muller T., Dugich-Djordjevic M.M., and McKay R.D. 1996. Single factors direct the differentiation of stem cells from the fetal and adult central nervous system. *Genes Dev.* **10:** 3129–3140.

Johnson J.E., Zimmerman K., Saito T., and Anderson D.J. 1992. Induction and repression of mammalian achaete-scute homologue (MASH) gene expression during neuronal differentiation of P19 embryonal carcinoma cells. *Development* **114:** 75–87.

Kaplan D.R., Matsumoto K., Lucarelli E., and Thiele C.J. 1993. Induction of TrkB by retinoic acid mediates biologic responsiveness to BDNF and differentiation of human neuroblastoma cells. Eukaryotic Signal Transduction Group. *Neuron* **11:** 321–331.

Laerum O.D., Steinsvag S., and Bjerkvig R. 1985. Cell and tissue culture of the central nervous system: Recent developments and current applications. *Acta Neurol. Scand.* **72:** 529–549.

Lai K., Kaspar B.K., Gage F.H., and Schaffer D.V. 2003. Sonic hedgehog regulates adult neural progenitor proliferation in vitro and in vivo. *Nat. Neurosci.* **6:** 21–27.

Lie D.C., Dziewczapolski G., Willhoite A.R., Kaspar B.K., Shults C.W., and Gage F.H. 2002. The adult substantia nigra contains progenitor cells with neurogenic potential. *J. Neurosci.* **22:** 6639–6649.

McKay R. 1997. Stem cells in the central nervous system. *Science* **276:** 66–71.

Merkle F.T. and Alvarez-Buylla A. 2006. Neural stem cells in mammalian development. *Curr. Opin. Cell Biol.* **18:** 704–709.

Navarro-Galve B. and Martinez-Serrano A. 2006. "Is there any need to argue..." about the nature and genetic signature of in vitro neural stem cells? *Exp. Neurol.* **199**: 20–25.

Ostenfeld T. and Svendsen C.N. 2004. Requirement for neurogenesis to proceed through the division of neuronal progenitors following differentiation of epidermal growth factor and fibroblast growth factor-2-responsive human neural stem cells. *Stem Cells* **22**: 798–811.

Ostenfeld T., Joly E., Tai Y.T., Peters A., Caldwell M., Jauniaux E., and Svendsen C.N. 2002. Regional specification of rodent and human neurospheres. *Brain Res. Dev. Brain Res.* **134**: 43–55.

Palmer T.D., Ray J., and Gage F.H. 1995. FGF-2-responsive neuronal progenitors reside in proliferative and quiescent regions of the adult rodent brain. *Mol. Cell. Neurosci.* **6**: 474–486.

Palmer T.D., Takahashi J., and Gage F.H. 1997. The adult rat hippocampus contains primordial neural stem cells. *Mol. Cell. Neurosci.* **8**: 389–404.

Palmer T.D., Markakis E.A., Willhoite A.R., Safar F., and Gage F.H. 1999. Fibroblast growth factor-2 activates a latent neurogenic program in neural stem cells from diverse regions of the adult CNS. *J. Neurosci.* **19**: 8487–8497.

Parker M.A., Anderson J.K., Corliss D.A., Abraria V.E., Sidman R.L., Park K.I., Teng Y.D., Cotanche D.A., and Snyder E.Y. 2005. Expression profile of an operationally-defined neural stem cell clone. *Exp. Neurol.* **194**: 320–332.

Parmar M., Skogh C., Bjorklund A., and Campbell K. 2002. Regional specification of neurosphere cultures derived from subregions of the embryonic telencephalon. *Mol. Cell. Neurosci.* **21**: 645–656.

Ray J. and Gage F.H. 1994. Spinal cord neuroblasts proliferate in response to basic fibroblast growth factor. *J. Neurosci.* **14**: 3548–3564.

———. 1999. Neural stem cell isolation, characterization and transplantation. In *Modern techniques in neuroscience research* (ed. U. Windhorst and H. Johaansson), pp. 339–360. Springer-Verlag, New York.

———. 2006. Differential properties of adult rat and mouse brain-derived neural stem/progenitor cells. *Mol. Cell. Neurosci.* **31**: 560–573.

Ray J., Peterson D.A., Schinstine M., and Gage F.H. 1993. Proliferation, differentiation, and long-term culture of primary hippocampal neurons. *Proc. Natl. Acad. Sci.* **90**: 3602–3606.

Reynolds B.A. and Rietze R.L. 2005. Neural stem cells and neurospheres—Re-evaluating the relationship. *Nat. Methods* **2**: 333–336.

Reynolds B.A. and Weiss S. 1992. Generation of neurons and astrocytes from isolated cells of the adult mammalian central nervous system. *Science* **255**: 1707–1710.

Reynolds B.A., Tetzlaff W., and Weiss S. 1992. A multipotent EGF-responsive striatal embryonic progenitor cell produces neurons and astrocytes. *J. Neurosci.* **12**: 4565–4574.

Richard C., Liuzzo J.P., and Moscatelli D. 1995. Fibroblast growth factor-2 can mediate cell attachment by linking receptors and heparan sulfate proteoglycans on neighboring cells. *J. Biol. Chem.* **270**: 24188–24196.

Richard C., Roghani M., and Moscatelli D. 2000. Fibroblast growth factor (FGF)-2 mediates cell attachment through interactions with two FGF receptor-1 isoforms and extracellular matrix or cell-associated heparan sulfate proteoglycans. *Biochem. Biophys. Res. Commun.* **276**: 399–405.

Roy N.S., Wang S., Jiang L., Kang J., Benraiss A., Harrison-Restelli C., Fraser R.A., Couldwell W.T., Kawaguchi A., Okano H., Nedergaard M., and Goldman S.A. 2000. In vitro neurogenesis by progenitor cells isolated from the adult human hippocampus. *Nat. Med.* **6:** 271–277.

Satoh J., Tabira T., and Kim S.U. 1994. Rapidly proliferating glial cells isolated from adult mouse brain have a differentiative capacity in response to cyclic AMP. *Neurosci. Res.* **20:** 175–184.

Scheel J.R., Ray J., Gage F.H., and Barlow C. 2005. Quantitative analysis of gene expression in living adult neural stem cells by gene trapping. *Nat. Methods* **2:** 363–370.

Seaberg R.M. and van der Kooy D. 2003. Stem and progenitor cells: the premature desertion of rigorous definitions. *Trends Neurosci.* **26:** 125–131.

Sheng M. and Greenberg M.E. 1990. The regulation and function of c-*fos* and other immediate early genes in the nervous system. *Neuron* **4:** 477–485.

Shihabuddin L.S., Ray J., and Gage F.H. 1997. FGF-2 is sufficient to isolate progenitors found in the adult mammalian spinal cord. *Exp. Neurol.* **148:** 577–586.

Shihabuddin L.S., Horner P.J., Ray J., and Gage F.H. 2000. Adult spinal cord stem cells generate neurons after transplantation in the adult dentate gyrus. *J. Neurosci.* **20:** 8727–8735.

Singec I., Knoth R., Meyer R.P., Maciaczyk J., Volk B., Nikkhah G., Frotscher M., and Snyder E.Y. 2006. Defining the actual sensitivity and specificity of the neurosphere assay in stem cell biology. *Nat. Methods* **3:** 801–806.

Somia N. and Verma I.M. 2000. Gene therapy: Trials and tribulations. *Nat. Rev. Genet.* **1:** 91–99.

Suslov O.N., Kukekov V.G., Ignatova T.N., and Steindler D.A. 2002. Neural stem cell heterogeneity demonstrated by molecular phenotyping of clonal neurospheres. *Proc. Natl. Acad. Sci.* **99:** 14506–14511.

Svendsen C.N. and Smith A.G. 1999. New prospects for human stem-cell therapy in the nervous system. *Trends Neurosci.* **22:** 357–364.

Svendsen C.N., Fawcett J.W., Bentlage C., and Dunnett S.B. 1995. Increased survival of rat EGF-generated CNS precursor cells using B27 supplemented medium. *Exp. Brain. Res.* **102:** 407–414.

Svendsen C.N., Skepper J., Rosser A.E., ter Borg M.G., Tyres P., and Ryken T. 1997. Restricted growth potential of rat neural precursors as compared to mouse. *Brain Res. Dev. Brain Res.* **99:** 253–258.

Takahashi J., Palmer T.D., and Gage F.H. 1999. Retinoic acid and neurotrophins collaborate to regulate neurogenesis in adult-derived neural stem cell cultures. *J. Neurobiol.* **38:** 65–81.

Tashiro A., Zhao C., and Gage F.H. 2006. Retrovirus-mediated single-cell gene knockout technique in adult newborn neurons in vivo. *Nat. Protoc.* **1:** 3049–3055.

Taupin P., Ray J., Fischer W.H., Suhr S.T., Hakansson K., Grubb A., and Gage F.H. 2000. FGF-2-responsive neural stem cell proliferation requires CCg, a novel autocrine/paracrine cofactor. *Neuron* **28:** 385–397.

Temple S. 2001a. Stem cell plasticity—building the brain of our dreams. *Nat. Rev. Neurosci.* **2:** 513–520.

———. 2001b. The development of neural stem cells. *Nature* **414:** 112–117.

Tropepe V., Sibilia M., Ciruna B.G., Rossant J., Wagner E.F., and van der Kooy D. 1999. Distinct neural stem cells proliferate in response to EGF and FGF in the developing mouse telencephalon. *Dev. Biol.* **208:** 166–188.

Vicario-Abejon C., Collin C., Tsoulfas P., and McKay R.D. 2000. Hippocampal stem cells differentiate into excitatory and inhibitory neurons. *Eur. J. Neurosci.* **12:** 677–688.

Walicke P., Cowan W.M., Ueno N., Baird A., and Guillemin R. 1986. Fibroblast growth factor promotes survival of dissociated hippocampal neurons and enhances neurite extension. *Proc. Natl. Acad. Sci.* **83:** 3012–3016.

Walsh K., Megyesi J., and Hammond R. 2005. Human central nervous system tissue culture: A historical review and examination of recent advances. *Neurobiol. Dis.* **18:** 2–18.

Watterson J.M., Watson D.G., Meyer E.M., and Lenox R.H. 2002. A role for protein kinase C and its substrates in the action of valproic acid in the brain: Implications for neural plasticity. *Brain Res.* **934:** 69–80.

Weiss S., Reynolds B.A., Vescovi A.L., Morshead C., Craig C.G., and van der Kooy D. 1996a. Is there a neural stem cell in the mammalian forebrain? *Trends Neurosci.* **19:** 387–393.

Weiss S., Dunne C., Hewson J., Wohl C., Wheatley M., Peterson A.C., and Reynolds B.A. 1996b. Multipotent CNS stem cells are present in the adult mammalian spinal cord and ventricular neuroaxis. *J. Neurosci.* **16:** 7599–7609.

Ying Q.L., Nichols J., Chambers I., and Smith A. 2003. BMP induction of Id proteins suppresses differentiation and sustains embryonic stem cell self-renewal in collaboration with STAT3. *Cell* **115:** 281–292.

9

Neurogenesis in the Adult Hippocampus

Gerd Kempermann, Hongjun Song, and Fred H. Gage
Genomics of Regeneration
Center for Regenerative Therapies Dresden
01307 Dresden, Germany

As NOTED PREVIOUSLY IN THIS VOLUME, adult neurogenesis is a process, not an event. Adult neurogenesis comprises a series of sequential developmental events that are all necessary for the generation of new neurons under the conditions of the adult brain. In the original publications on adult neurogenesis, the precursor cell population from which neurogenesis originates was identified only by the detection of proliferative activity and the absence of mature neuronal markers (Altman and Das 1965; Kaplan and Hinds 1977; Cameron et al. 1993; Kuhn et al. 1996). The new neurons, in contrast, were identified by the presence of mature neuronal markers in cells that had been birthmarked with the thymidine-oder BrdU (bromodeoxyuridine) method (see Chapters 2 and 3) a couple of weeks earlier. The expression of the polysialilated neural cell adhesion molecule (PSA-NCAM) with neurogenesis has been noted early but could not be clearly linked to either proliferation or mature stage (Seki and Arai 1993a,b). PSA-NCAM expression was the first indication of the developmental events that take place, filling the gaps between the start and endpoint of development. Today, we have quite detailed knowledge about the course of neuronal development in the adult hippocampus, and although many detailed questions remain open, a clear overall picture has emerged (Kempermann et al. 2004; Abrous et al. 2005; Ming and Song 2005; Lledo et al. 2006). Although we coarsely talk about neurogenesis in the hippocampus, it should be noted that neurogenesis occurs only in the dentate gyrus (DG), not other regions; in an older nomenclature, the DG is not even part of the hippocampus proper. Although there are reasons to subsume the DG only to the hippocampal formation and not the hippocampus, we believe that from any functional perspective this distinction is awkward. The vote has long been made by the scientific audience: We talk about adult hippocampal neurogenesis, when we mean neurogenesis in the adult DG.

Adult hippocampal neurogenesis generates only one type of neuron: granule cells in the DG. To date, there is no conclusive evidence that other neuronal cell types could be generated under physiological conditions, although some as yet unconfirmed claims have been made (Rietze et al. 2000; Liu et al. 2003). Granule cells are the excitatory principal neurons of the DG. They receive input from the entorhinal cortex and send their axonal projection along the mossy fiber tract to area CA3, where they terminate in large synapse- and interneuron-rich structures, the so-called *boutons*. They provide excitatory input to the pyramidal cells of CA3 and neurons in the hilus regions. They fire very sparsely and their activity is modulated by a large number of interneurons in the DG and hilus area. The precursor cells, from which adult neurogenesis originates, reside in a narrow band of tissue between the granule cell layer and the hilus, the so-called subgranular zone (SGZ). The term was coined by the discoverer of adult hippocampal neuro-genesis, Josef Altman, in 1975 (Altman 1975). The original description of adult neurogenesis in the rodent brain was published by Josef Altman and his colleague Gopal Das in 1965 (Altman and Das 1965).

The SGZ contains a microenvironment that is permissive for neu-ronal development to occur. In analogy to other stem cell systems in the body, this microenvironment is called the neurogenic "niche." The niche comprises the precursor cells themselves, their immediate progeny and immature neurons, other glial cells and endothelia, very likely immune cells, microglia, and macrophages, and an extracellular matrix. The niche is surrounded by a common basal membrane (Mercier et al. 2002). Because of the prominent role that the vasculature appears to have in this context, the neurogenic niche has also been called the "vascular niche" (Palmer et al. 2000). The type-1 precursor cells, from which adult neuro-genesis originates, have end-feet on the vasculature in the SGZ (Filippov et al. 2003); vascular endothelial growth factor (VEGF) is a potent regu-lator of adult neurogenesis (Jin et al. 2002; Schanzer et al. 2004), and a complex relationship between endothelial cells and hippocampal precur-sor cells exists (Wurmser et al. 2004). In any case, the niche provides a unique milieu that allows neuronal development to occur. Many pieces of evidence point to the direction that it is the local astrocytes that have a key role in promoting neurogenesis. In vivo, the developing cells show a close spatial relationship with astrocytes (Shapiro et al. 2005; Plumpe et al. 2006). Ex vivo, astrocytes and astrocyte-derived factors are potent inducers of neurogenesis from hippocampal precursor cells (Song et al. 2002; Barkho et al. 2006).

The SGZ is also special in that it receives synaptic input from various other brain regions: dopaminergic fibers from the ventral tegmental area, serotonergic projections from the raphe nuclei, acetylcholinergic input from the septum, and GABAergic connections from local interneurons. Manipulations of all these neurotransmitter systems, e.g., by lesioning studies to the input structures or pharmacological intervention, have revealed a regulatory effect on adult neurogenesis, although the level of resolution is still too low to identify the specific contributions of the individual systems to the control of adult neurogenesis (Bengzon et al. 1997; Cooper-Kuhn et al. 2004; Dominguez-Escriba et al. 2006).

DISTINCT STEPS OF NEURONAL DEVELOPMENT

Adult neurogenesis can be divided into four phases: a precursor cell phase, an early survival phase, a postmitotic maturation phase, and a late survival phase. On the basis of cell morphology and a set of marker proteins, six distinct milestones can be identified, which to date still somewhat overemphasize the precursor cell stages of adult neurogenesis (Fig. 1) (Kempermann et al. 2004; Steiner et al. 2006). From a radial glia-like precursor cell, adult neurogenesis progresses over three identifiable progenitor stages associated with high proliferative activity to a postmitotic maturation phase and finally the existence of a new granule cell (Brandt et al. 2003; Filippov et al. 2003; Fukuda et al. 2003; Encinas et al. 2006; Steiner et al. 2006). Whereas on the precursor cell stage and early after cell cycle exit, large changes in cell numbers occur, the effects of development become more and more qualitative at later times.

The precursor cell phase serves the expansion of the pool of cells that might differentiate into neurons. The early survival phase marks the exit from the cell cycle: Most newborn cells are eliminated within days after they were born. The postmitotic maturation phase is associated with the establishment of functional connections, the growth of axon and dendrites, and synaptogenesis. The late survival phase represents a period of fine tuning. It has been estimated that the entire period of adult neurogenesis takes approximately 7 weeks. Characteristic electrophysiological patterns allow the assignment of functional states to the morphologically distinguishable steps of development.

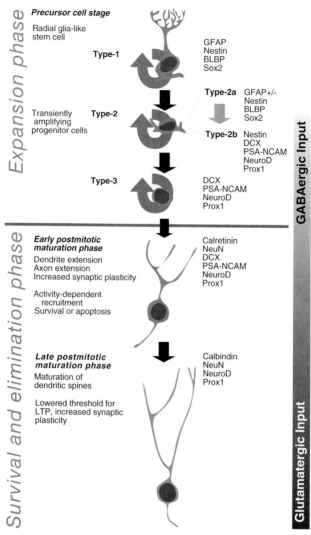

Figure 1. Developmental stages in the course of adult hippocampal neurogenesis. See text for details.

One central question in research on adult hippocampal neurogenesis is how similar to or distinct from embryonic and early postnatal neurogenesis is the DG? The DG develops in three distinctive waves of development, of which adult neurogenesis is the last (Altman and Bayer 1990a,b). The bulk of DG neurons is produced around P7. From a functional perspective, Laplagne et al. (2006) have argued that adult-generated neurons behave highly similar to those produced during the neonatal period, suggesting a homogeneous population. On the other hand, quality and quantity of extrinsic stimuli and memory contents that pass the DG will be dramatically different between postnatal and adult periods. In addition, the speed of maturation might differ (Overstreet-Wadiche et al. 2006a), although with respect to the influence of extrinsic stimuli (i.e., seizures) on differentiation speed, the data are not consistent (Jakubs et al. 2006; Overstreet-Wadiche et al. 2006b; Plumpe et al. 2006).

THE PRECURSOR CELL PHASE

A number of morphologically identifiable "types" of precursor cells are involved in the course of adult hippocampal neurogenesis. Such cell types do not actually constitute distinct populations of cells but rather reflect milestones of a developmental process.

Adult hippocampal neurogenesis originates from a population of precursor cells with glial properties. A subset of these shows morphological and antigenic characteristics of radial glia. Their cell body is found in the SGZ, and the process extends into the molecular layer. Not all radial elements show the same marker expression, and some markers for radial glia during embryonic development are absent. The astrocytic nature of hippocampal precursor cells has been demonstrated by Seri et al. (2001). These authors suppressed cell division by application of a cytostatic drug and found that the first cells that reappeared were proliferative astrocyte-like cells with radial morphology (Seri et al. 2001). The second line of evidence comes from experiments in which the receptor for an avian virus was expressed under the promoter of glial fibrillary acidic protein (GFAP) or nestin, so that astrocyte-like or nestin-expressing cells could be specifically infected by an otherwise inert virus. Transduced cells generated new neurons in vivo, demonstrating the developmental potential in vivo (Seri et al. 2001, 2004). The study was related to similar experiments in the subventricular zone (SVZ) olfactory bulb (OB) system (Doetsch et al. 1999a,b; Laywell et al. 2000).

Ex vivo, hippocampal precursor cells were first isolated by Ray et al. (1993) from the embryonic brain and by Palmer et al. (1995) from the adult rat brain. In culture, the precursor cells show signs of stemness, i.e., self-renewal and multipotency (Palmer et al. 1997). This claim has been disputed by other investigators (Seaberg and van der Kooy 2002; Bull and Bartlett 2005), but methodological and strain differences between the studies prevented closing the case. After careful microdissection of DG tissue and by the use of an enrichment procedure, it was found that the murine DG in fact contained "stem cells" in the stricter sense of the definition (Babu et al. 2007).

The radial glia-like cells, type-1 cells in our nomenclature, give rise to intermediate progenitor cells, type-2 cells, which show a high proliferative activity. A subset of these cells still express glial markers but lack the characteristic morphology of radial cells (type 2a). On the level of type-2 cells that together with type-1 cells express intermediate filament nestin, first indications of neuronal lineage choice appear. These markers comprise transcription factors NeuroD1 and Prox1. This cellular phenotype has been called type-2b cells (Steiner et al. 2006). Of these, Prox1 is specific to granule cell development. To date, a point-by-point comparison between adult neurogenesis and fetal and early postnatal neurogenesis in the DG is still lacking. Insight into the transcriptional control of the initiation of neuronal differentiation is scarce. From the available data, however, it is obvious that if a fate-choice decision is made at all, it must occur on the level of the type-2a cells. All later cells express NeuroD1 and Prox1, and there is no overlap between NeuroD1 and Prox1 and astrocytic markers at any time point. Tailless (Tlx) is a key candidate for a transcription factor involved in controlling the transition between glial and neuronal phenotype of the precursor cells (Shi et al. 2004). Chapter 13 covers the transcriptional control of hippocampal neurogenesis in greater detail.

On the level of type-2 cells, the developing cells also receive first synaptic input, which is GABAergic (Tozuka et al. 2005; Wang et al. 2005). They are first responsive to ambient GABA, and more and more respond to synaptic excitatory GABAergic input (Ge et al. 2006). Although type-1 cells can respond to extrinsic stimuli by increasing cell proliferation (Huttmann et al. 2003; Kunze et al. 2006), the burden of expansion lies on the type-2 cells. Type-2 cells respond to physiological stimuli such as voluntary wheel running (Kronenberg et al. 2003) or pharmacological stimulation via serotonin-dependent mechanisms (Encinas et al. 2006).

Among the neuronal lineage markers first appearing at the type-2b stage is doublecortin (DCX). DCX is expressed at the proliferative stage, even after nestin has been down-regulated (type-3 cells). Normally, type-3 cells show only little proliferative activity. Under pathological conditions, however, such as experimental seizures, type-3 cells show a disproportional increase in cell division (Jessberger et al. 2005). DCX expression extends from a proliferation stage, through cell cycle exit to a period of postmitotic maturation that lasts approximately 2–3 weeks (Brandt et al. 2003; Rao and Shetty 2004; Couillard-

Despres et al. 2005; Plumpe et al. 2006). DCX shows an almost complete overlap with PSA-NCAM, explaining the early observations mentioned above.

The precursor cell phase is characterized by the proliferative activity of the cells on the different stages of precursor cell development. The mechanisms underlying the proliferative activity itself are outlined in Chapter 12.

THE EARLY SURVIVAL PHASE

Very early after cell cycle exit, the new neurons express postmitotic markers such as NeuN and the transient marker calretinin (Brandt et al. 2003). Because type-3 cells are still proliferative, NeuN can be found in some cells as early as one day after the injection of the proliferation marker. In fact, the number of NeuN-positive new neurons is highest at very early timepoints and decreases dramatically within a few days. Thus, the majority of cells are eliminated well before they have made functional connections in the target area in CA3 or received correct dendritic input in the molecular layer of the DG. This elimination process is apoptotic (Biebl et al. 2000; Kuhn et al. 2005).

The initiation of dendrite development after the timepoint of cell cycle exit appears to be highly variable. The time course of dendritic development itself, in contrast, seems to follow a rather fixed temporal course. Within days after cell cycle exit, the new cells send their axon to target area CA3, where they form appropriate synapses. Accordingly, this phase is associated with the expression of collapsing response mediator protein (Crmp, also known as TOAD-64 or TUC-4), a molecule involved in axon path finding.

The main synaptic input to the new cells is GABAergic and remains excitatory. GABA switches to its inhibitory function only, when sufficient glutamatergic contact has been made and, presumably, when the cells have begun to develop their own glutamatergic neurotransmitter phenotype (Tozuka et al. 2005). GABA action itself drives neuronal maturation in these cells and steers the synaptic integration (Ge et al. 2006).

Quantitatively, most of the regulation occurs at this stage of neuronal development, not in the expansion phase as it is often assumed (Kempermann et al. 2006). The reason is that precursor cell proliferation generates a vast surplus of new neurons and that only a very small proportion survives for long periods of time (Kempermann et al. 2003). It seems that cells that have survived the first 2 weeks will be stably and persistently integrated into the network of the DG. After this timepoint, only very small changes in cell number occur. One other consequence of this observation is that adult neurogenesis lifelong contributes to growth of the DG and does not replace older cells (Crespo et al. 1986), although this growth has not been proven at later life stages, when the levels of adult neurogenesis are very low. Growth has been demonstrated in several studies for the first year in the life of a rodent (Altman and Das 1965; Bayer et al. 1982; Boss et al. 1985), but a modern stereological account is still lacking. Along similar lines of reasoning, it seems that stimuli that control the expansion phase tend to be rather nonspecific (e.g., the pan-synaptic activation in seizures, physical activity), whereas stimuli that are more specific to the hippocampus, in that they reflect hippocampus-dependent function, affect the survival phases. On a quantitative level, this has been demonstrated only for the early postmitotic period. Exposure to the complexity of an enriched environment or, at least in some studies, to learning stimuli of hippocampus-dependent learning tasks increases survival at this stage (Gould et al. 1999; Dobrossy et al. 2003). Presumably, similar effects are found at later stages as well but so far have not been measurable with the available methods. For more details on the mechanisms underlying cell cycle exit, migration, and early maturation, see Chapter 12.

POSTMITOTIC MATURATION PHASE

Serendipitously, it was found that the maturing cells up-regulate promoter activity of pro-opiomelanocortin (POMC), although the protein is not detectable in these cells. A transgenic mouse line expressing GFP (green fluorescent protein) under the POMC promoter has become a useful tool to study the electrophysiology of the immature neurons (Overstreet et al. 2004).

Details of spine development have largely been investigated by transducing a proliferating cell with a GFP-expressing retrovirus (Fig. 2) (Zhao et al. 2006). From these experiments, we know that axon elongation precedes spine formation on the dendrites. Axonal contact to CA3 is made about 10 days after labeling the proliferative cells, whereas the first spines appear almost 1 week later.

Functional maturation of the new neurons has now been characterized to a considerable degree (van Praag et al. 2002; Ambrogini et al. 2004; Schmidt-Hieber et al. 2004; Couillard-Despres et al. 2006). The cells progress from a state with high input resistance to the normal membrane properties of mature granule cells. For more details on the functional maturation of the new neurons, see Chapter 15.

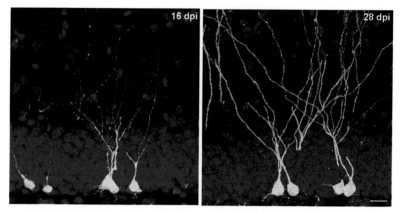

Figure 2. Dendrite development of newborn neurons in the dentate gyrus (DG). Proliferating precursor cells in the subgranular zone (SGZ) were labeled with a green fluorescent protein (GFP)-expressing retrovirus and analyzed at later timepoints. Here, GFP-positive new granule cells can be seen at about 2 weeks (16 days after injection, *left*) and 4 weeks (28 days after injection, *right*). During the early post-mitotic maturation phase, the cells develop the full morphology of hippocampal granule cells. It is noteworthy that the cells might show a slightly different pace of maturation. After 4 weeks, many cells have extended their dendritic tree far into the molecular layer. First, dendritic spines can be seen on the dendrites. For details of dendritic development in adult hippocampal neurogenesis, see Zhao et al. (2006). (Figure courtesy of Chunmei Zhao, Salk Institute.) Bar, 15 μm.

LATE MATURATION PHASE

We presently know least about the adaptive changes that occur late in neuronal development of adult neurogenesis. The period of calretinin expression lasts only about 3–4 weeks, roughly consistent with the temporal pattern of dendritic maturation. Presumably, after full structural integration into the existing network, the new cells switch their calcium-binding protein from calretinin to calbindin (Brandt et al. 2003). Still, it takes several more weeks until the new cells have become electrophysiologically indistinguishable from their older neighbors (van Praag et al. 2002; Ambrogini et al. 2004). Once glutamatergic synaptic connections have been made, the new neurons go through a phase of increased synaptic plasticity. The threshold to induce long-term potentiation (LTP) in the immature neurons is lower than that in mature granule cells (Wang et al. 2000; Schmidt-Hieber et al. 2004). This critical period lasts from about 1 to 1.5 months after the cells were generated (Ge et al. 2007). Some theories about the potential function of the new granule cells build upon this fact by arguing that the altered plastic properties help the DG to encode temporal information into memories to be stored (Aimone et al. 2006). Alternatively, the increased plasticity might serve the purpose to facilitate preferential integration of the new cells in order to achieve long-term changes in the network (Ramirez-Amaya et al. 2006; Wiskott et al. 2006; Kee et al. 2007). Possibly, both ideas are correct, and a specific transient function prepares the ground for an equally specific long-term function. Presumably, important regulatory events take place at this stage, but they will be effective more on a qualitative level than on a quantitative level.

CONTROL OF NEURONAL DEVELOPMENT

The inherent mechanisms that constitute the process of neuronal development must be distinguished from regulatory events that act upon these mechanisms. Transcriptional control of adult neurogenesis represents the backbone of neuronal development. Regulatory events do not change this backbone but modulate it. Transcriptional control thereby represents something like the home stretch of regulation. These mechanisms are described and discussed in Chapter 13. The following paragraph attempts to tie these distinct molecular mechanisms to the identifiable stages of development.

On the level of the precursor cells, basic helix-loop-helix factor Sox2 characterizes the stem-like cells with glial properties (D'Amour and Gage 2003; Steiner et al. 2006). Overlap between Sox2 and early neuronal markers is minimal. However, Sox2 is also found in $S100\beta$-positive astrocytes without precursor cell function.

As mentioned earlier, the transition between glial and neuronal phenotype might be controlled by Tlx (Shi et al. 2004). The earliest known neuronal factor is NeuroD1, which is recognized by a binding motif in the promoter region of the Dcx gene (Steiner et al. 2006). Parallel to NeuroD1 is found Prox1. Prox1 is highly specific to granule cells, but its function is not yet known (Pleasure et al. 2000).

This set of transcription factors is different from those of the subventricular system, where Pax6, Dlx, and Olig2 have prominent roles. Expression of Pax6 has been noted in the DG as well (Nacher et al. 2005), but its function is not yet clear. Olig2 is expressed in the DG but in cells outside the lineage that leads to granule cell development. It is thus assumed that new oligodendrocytes in the adult DG, which are very rare anyway (Kempermann et al. 2003; Steiner et al. 2004), originate from a distinct pool of precursor cells that are characterized by their expression of proteoglycan NG2.

REGULATION OF ADULT HIPPOCAMPAL NEUROGENESIS

Although there is no consistent use of the terminology, "control" and "regulation" of a biological process are not identical. Regulation means those processes that act upon the basic mechanisms that control neurogenesis. Regulation thus encompasses processes on many conceptual levels, from behavioral to molecular. Quantitatively, regulation of adult hippocampal neurogenesis mostly occurs on the level of survival of the newborn cells. Among 30 inbred strains of mice, the genetically determined level of survival explained 85% of the variance found in net neurogenesis, whereas cell proliferation explained only 19% (Kempermann et al. 2006). On the other hand, numerous studies reported examples of factors that influence cell proliferation. The current hypothesis is that this broad sensitivity of precursor cell proliferation is nonspecific, whereas survival-promoting effects depend on specifically hippocampal functional stimuli. Chapters 16 through 19 extend on this idea.

SUMMARY

Adult hippocampal neurogenesis is a multistep process that originates from a sequence of proliferative precursor cells and leads to the existence of a new granule cell in the DG. An expansion phase on the level of the precursor cells, during which proliferation is regulated by many nonspecific stimuli, gives way to a postmitotic maturation phase, during which only a subset of the newly generated cells survive. On the precursor cell level, the cascade originates in radial glia-like type-1 cells, presumably the highest ranking stem cell in this system. It gives rise to highly proliferative type-2 cells, which can be divided into a more glia-like (type-2a) and a neuronally determined phase (type-2b). Finally, a proliferative late precursor cell, type-3, exists that marks the exit from the cell cycle. The selective postmitotic survival is dependent on specific hippocampus-dependent stimuli and accounts for the greatest part of the neurogenic regulation. Morphological maturation finds its most visible expression in the extension of the dendrites and the emergence of dendritic spines. GABAergic input, first ambient, later synaptic, promotes neuronal maturation until regular glutamatergic input from the entorhinal cortex sets in. In a brief postmitotic interval, during which the new cells express the calcium-buffering protein calretinin, the new neurons also extend their axon to area CA3. This phase of early synaptic integration is also characterized by increased synaptic plasticity, presumably facilitating the survival-promoting effects of functional integration. At present, little is known about the details of neuronal maturation, but it seems that after a period of approximately 7 weeks, the new neurons become indistinguishable from their older neighbors. A number of transcription factors have been identified that can be linked to particular stages of neuronal development in the adult hippocampus, for example, granule-cell-specific factor Prox1, which is expressed very early on the level of type-2 progenitor cells and remains expressed in mature granule cells.

REFERENCES

Abrous D.N., Koehl M., and Le Moal M. 2005. Adult neurogenesis: From precursors to network and physiology. *Physiol. Rev.* **85:** 523–569.

Aimone J.B., Wiles J., and Gage F.H. 2006. Potential role for adult neurogenesis in the encoding of time in new memories. *Nat. Neurosci.* **9:** 723–727.

Altman J. 1975. Postnatal development of the hippocampal dentate gyrus under normal and experimental conditions. In *The hippocampus* (ed. R.L. Isaacson and K.H. Pribram), pp. 95–122. Plenum Press, New York.

Altman J. and Bayer S.A. 1990a. Migration and distribution of two populations of hippocampal progenitors during the perinatal and postnatal periods. *J. Comp. Neurol.* **301:** 365–381.

————. 1990b. Mosaic organization of the hippocampal neuroepithelium and the multiple germinal sources of dentate granule cells. *J. Comp. Neurol.* **301:** 325–342.

Altman J. and Das G.D. 1965. Autoradiographic and histologic evidence of postnatal neurogenesis in rats. *J. Comp. Neurol.* **124:** 319–335.

Ambrogini P., Lattanzi D., Ciuffoli S., Agostini D., Bertini L., Stocchi V., Santi S., and Cuppini R. 2004. Morpho-functional characterization of neuronal cells at different stages of maturation in granule cell layer of adult rat dentate gyrus. *Brain Res.* **1017:** 21–31.

Babu H., Cheung G., Kettenmann H., Palmer T.D., and Kempermann G. 2007. Enriched monolayer precursor cell cultures from micro-dissected adult mouse dentate gyrus yield functional granule cell-like neurons. *PLoS ONE* **2:** e388.

Barkho B.Z., Song H., Aimone J.B., Smrt R.D., Kuwabara T., Nakashima K., Gage F.H., and Zhao X. 2006. Identification of astrocyte-expressed factors that modulate neural stem/progenitor cell differentiation. *Stem Cells Dev.* **15:** 407–421.

Bayer S.A., Yackel J.W., and Puri P.S. 1982. Neurons in the rat dentate gyrus granular layer substantially increase during juvenile and adult life. *Science* **216:** 890–892.

Bengzon J., Kokaia Z., Elmér E., Nanobashvili A., Kokaia M., and Lindvall O. 1997. Apoptosis and proliferation of dentate gyrus neurons after single and intermittent limbic seizures. *Proc. Natl. Acad. Sci.* **94:** 10432–10437.

Biebl M., Cooper C.M., Winkler J., and Kuhn H.G. 2000. Analysis of neurogenesis and programmed cell death reveals a self-renewing capacity in the adult rat brain. *Neurosci. Lett.* **291:** 17–20.

Boss B.D., Peterson G.M., and Cowan W.M. 1985. On the number of neurons in the dentate gyrus of the rat. *Brain Res.* **338:** 144–150.

Brandt M.D., Jessberger S., Steiner B., Kronenberg G., Reuter K., Bick-Sander A., von der Behrens W., and Kempermann G. 2003. Transient calretinin expression defines early postmitotic step of neuronal differentiation in adult hippocampal neurogenesis of mice. *Mol. Cell. Neurosci.* **24:** 603–613.

Bull N.D. and Bartlett P.F. 2005. The adult mouse hippocampal progenitor is neurogenic but not a stem cell. *J. Neurosci.* **25:** 10815–10821.

Cameron H.A., Woolley C.S., McEwen B.S., and Gould E. 1993. Differentiation of newly born neurons and glia in the dentate gyrus of the adult rat. *Neuroscience* **56:** 337–344.

Cooper-Kuhn C.M., Winkler J., and Kuhn H.G. 2004. Decreased neurogenesis after cholinergic forebrain lesion in the adult rat. *J. Neurosci. Res.* **77:** 155–165.

Couillard-Despres S., Winner B., Schaubeck S., Aigner R., Vroemen M., Weidner N., Bogdahn U., Winkler J., Kuhn H.G., and Aigner L. 2005. Doublecortin expression levels in adult brain reflect neurogenesis. *Eur. J. Neurosci.* **21:** 1–14.

Couillard-Despres S., Winner B., Karl C., Lindemann G., Schmid P., Aigner R., Laemke J., Bogdahn U., Winkler J., Bischofberger J., and Aigner L. 2006. Targeted transgene expression in neuronal precursors: Watching young neurons in the old brain. *Eur. J. Neurosci.* **24:** 1535–1545.

Crespo D., Stanfield B.B., and Cowan W.M. 1986. Evidence that late-generated granule cells do not simply replace earlier formed neurons in the rat dentate gyrus. *Exp. Brain Res.* **62:** 541–548.

D'Amour K.A. and Gage F.H. 2003. Genetic and functional differences between multipotent neural and pluripotent embryonic stem cells. *Proc. Natl. Acad. Sci.* (Suppl 1) **100:** 11866–11872.

Dobrossy M.D., Drapeau E., Aurousseau C., Le Moal M., Piazza P.V., and Abrous D.N. 2003. Differential effects of learning on neurogenesis: Learning increases or decreases the number of newly born cells depending on their birth date. *Mol. Psychiatry* **8:** 974–982.

Doetsch F., Garcia-Verdugo J.M., and Alvarez-Buylla A. 1999a. Regeneration of a germinal layer in the adult mammalian brain. *Proc. Natl. Acad. Sci.* **96:** 11619–11624.

Doetsch F., Caille I., Lim D.A., Garcia-Verdugo J.M., and Alvarez-Buylla A. 1999b. Subventricular zone astrocytes are neural stem cells in the adult mammalian brain. *Cell* **97:** 703–716.

Dominguez-Escriba L., Hernandez-Rabaza V., Soriano-Navarro M., Barcia J.A., Romero F.J., Garcia-Verdugo J.M., and Canales J.J. 2006. Chronic cocaine exposure impairs progenitor proliferation but spares survival and maturation of neural precursors in adult rat dentate gyrus. *Eur. J. Neurosci.* **24:** 586–594.

Encinas J.M., Vaahtokari A., and Enikolopov G. 2006. Fluoxetine targets early progenitor cells in the adult brain. *Proc. Natl. Acad. Sci.* **103:** 8233–8238.

Filippov V., Kronenberg G., Pivneva T., Reuter K., Steiner B., Wang L.P., Yamaguchi M., Kettenmann H., and Kempermann G. 2003. Subpopulation of nestin-expressing progenitor cells in the adult murine hippocampus shows electrophysiological and morphological characteristics of astrocytes. *Mol. Cell. Neurosci.* **23:** 373–382.

Fukuda S., Kato F., Tozuka Y., Yamaguchi M., Miyamoto Y., and Hisatsune T. 2003. Two distinct subpopulations of nestin-positive cells in adult mouse dentate gyrus. *J. Neurosci.* **23:** 9357–9366.

Ge S., Yang C.H., Hsu K.S., Ming G.L., and Song H. 2007. A critical period for enhanced synaptic plasticity in newly generated neurons of the adult brain. *Neuron* **54:** 559–566.

Ge S., Goh E.L., Sailor K.A., Kitabatake Y., Ming G.L., and Song H. 2006. GABA regulates synaptic integration of newly generated neurons in the adult brain. *Nature* **439:** 589–593.

Gould E., Beylin A., Tanapat P., Reeves A., and Shors T.J. 1999. Learning enhances adult neurogenesis in the hippocampal formation. *Nat. Neurosci.* **2:** 260–265.

Huttmann K., Sadgrove M., Wallraff A., Hinterkeuser S., Kirchhoff F., Steinhauser C., and Gray W.P. 2003. Seizures preferentially stimulate proliferation of radial glia-like astrocytes in the adult dentate gyrus: Functional and immunocytochemical analysis. *Eur. J. Neurosci.* **18:** 2769–2778.

Jakubs K., Nanobashvili A., Bonde S., Ekdahl C.T., Kokaia Z., Kokaia M., and Lindvall O. 2006. Environment matters: Synaptic properties of neurons born in epileptic brain develop to reduce excitability. *Neuron* **52:** 1047–1059.

Jessberger S., Romer B., Babu H., and Kempermann G. 2005. Seizures induce proliferation and dispersion of doublecortin-positive hippocampal progenitor cells. *Exp. Neurol.* **196:** 342–351.

Jin K., Zhu Y., Sun Y., Mao X.O., Xie L., and Greenberg D.A. 2002. Vascular endothelial growth factor (VEGF) stimulates neurogenesis in vitro and in vivo. *Proc. Natl. Acad. Sci.* **99:** 11946–11950.

Kaplan M.S. and Hinds J.W. 1977. Neurogenesis in the adult rat: Electron microscopic analysis of light radioautographs. *Science* **197:** 1092–1094.

Kee N., Teixeira C.M., Wang A.H., and Frankland P.W. 2007. Preferential incorporation of adult-generated granule cells into spatial memory networks in the dentate gyrus. *Nat. Neurosci.* **10:** 355–362.

Kempermann G., Jessberger S., Steiner B., and Kronenberg G. 2004. Milestones of neuronal development in the adult hippocampus. *Trends Neurosci.* **27:** 447–452.

Kempermann G., Chesler E.J., Lu L., Williams R.W., and Gage F.H. 2006. Natural variation and genetic covariance in adult hippocampal neurogenesis. *Proc. Natl. Acad. Sci.* **103:** 780–785.

Kempermann G., Gast D., Kronenberg G., Yamaguchi M., and Gage F.H. 2003. Early determination and long-term persistence of adult-generated new neurons in the hippocampus of mice. *Development* **130:** 391–399.

Kronenberg G., Reuter K., Steiner B., Brandt M.D., Jessberger S., Yamaguchi M., and Kempermann G. 2003. Subpopulations of proliferating cells of the adult hippocampus respond differently to physiologic neurogenic stimuli. *J. Comp. Neurol.* **467:** 455–463.

Kuhn H.G., Dickinson-Anson H., and Gage F.H. 1996. Neurogenesis in the dentate gyrus of the adult rat: Age-related decrease of neuronal progenitor proliferation. *J. Neurosci.* **16:** 2027–2033.

Kuhn H.G., Biebl M., Wilhelm D., Li M., Friedlander R.M., and Winkler J. 2005. Increased generation of granule cells in adult Bcl-2-overexpressing mice: A role for cell death during continued hippocampal neurogenesis. *Eur. J. Neurosci.* **22:** 1907–1915.

Kunze A., Grass S., Witte O.W., Yamaguchi M., Kempermann G., and Redecker C. 2006. Proliferative response of distinct hippocampal progenitor cell populations after cortical infarcts in the adult brain. *Neurobiol. Dis.* **21:** 324–332.

Laplagne D.A., Esposito M.S., Piatti V.C., Morgenstern N.A., Zhao C., van Praag H., Gage F.H., and Schinder A.F. 2006. Functional convergence of neurons generated in the developing and adult hippocampus. *PLoS Biol* **4:** e409.

Laywell E.D., Rakic P., Kukekov V.G., Holland E.C., and Steindler D.A. 2000. Identification of a multipotent astrocytic stem cell in the immature and adult mouse brain. *Proc. Natl. Acad. Sci.* **97:** 13883–13888.

Liu S., Wang J., Zhu D., Fu Y., Lukowiak K., and Lu Y. 2003. Generation of functional inhibitory neurons in the adult rat hippocampus. *J. Neurosci.* **23:** 732–736.

Lledo P.M., Alonso M., and Grubb M.S. 2006. Adult neurogenesis and functional plasticity in neuronal circuits. *Nat. Rev. Neurosci.* **7:** 179–193.

Mercier F., Kitasako J.T., and Hatton G.I. 2002. Anatomy of the brain neurogenic zones revisited: Fractones and the fibroblast/macrophage network. *J. Comp. Neurol.* **451:** 170–188.

Ming G.L. and Song H. 2005. Adult neurogenesis in the mammalian central nervous system. *Annu. Rev. Neurosci.* **28:** 223–250.

Nacher J., Varea E., Blasco-Ibanez J.M., Castillo-Gomez E., Crespo C., Martinez-Guijarro F.J., and McEwen B.S. 2005. Expression of the transcription factor Pax 6 in the adult rat dentate gyrus. *J. Neurosci. Res.* **81:** 753–761.

Overstreet L.S., Hentges S.T., Bumaschny V.F., de Souza F.S., Smart J.L., Santangelo A.M., Low M.J., Westbrook G.L., and Rubinstein M. 2004. A transgenic marker for newly born granule cells in dentate gyrus. *J. Neurosci.* **24:** 3251–3259.

Overstreet-Wadiche L.S., Bensen A.L., and Westbrook G.L. 2006a. Delayed development of adult-generated granule cells in dentate gyrus. *J. Neurosci.* **26:** 2326–2334.

Overstreet-Wadiche L.S., Bromberg D.A., Bensen A.L., and Westbrook G.L. 2006b. Seizures accelerate functional integration of adult-generated granule cells. *J. Neurosci.* **26:** 4095–4103.

Palmer T.D., Ray J., and Gage F.H. 1995. FGF-2-responsive neuronal progenitors reside in proliferative and quiescent regions of the adult rodent brain. *Mol. Cell. Neurosci.* **6:** 474–486.

Palmer T.D., Takahashi J., and Gage F.H. 1997. The adult rat hippocampus contains premordial neural stem cells. *Mol. Cell. Neurosci.* **8:** 389–404.

Palmer T.D., Willhoite A.R., and Gage F.H. 2000. Vascular niche for adult hippocampal neurogenesis. *J. Comp. Neurol.* **425:** 479–494.

Pleasure S.J., Collins A.E., and Lowenstein D.H. 2000. Unique expression patterns of cell fate molecules delineate sequential stages of dentate gyrus development. *J. Neurosci.* **20:** 6095–6105.

Plumpe T., Ehninger D., Steiner B., Klempin F., Jessberger S., Brandt M., Romer B., Rodriguez G.R., Kronenberg G., and Kempermann G. 2006. Variability of doublecortin-associated dendrite maturation in adult hippocampal neurogenesis is independent of the regulation of precursor cell proliferation. *BMC Neurosci.* **7:** 77.

Ramirez-Amaya V., Marrone D.F., Gage F.H., Worley P.F., and Barnes C.A. 2006. Integration of new neurons into functional neural networks. **26:** 12237–12241.

Rao M.S. and Shetty A.K. 2004. Efficacy of doublecortin as a marker to analyse the absolute number and dendritic growth of newly generated neurons in the adult dentate gyrus. *Eur. J. Neurosci.* **19:** 234–246.

Ray J., Peterson D.A., Schinstine M., and Gage F.H. 1993. Proliferation, differentiation, and long-term culture of primary hippocampal neurons. *Proc. Natl. Acad. Sci.* **90:** 3602–3606.

Rietze R., Poulin P., and Weiss S. 2000. Mitotically active cells that generate neurons and astrocytes are present in multiple regions of the adult mouse hippocampus. *J. Comp. Neurol.* **424:** 397–408.

Schanzer A., Wachs F.P., Wilhelm D., Acker T., Cooper-Kuhn C., Beck H., Winkler J., Aigner L., Plate K.H., and Kuhn H.G. 2004. Direct stimulation of adult neural stem cells in vitro and neurogenesis in vivo by vascular endothelial growth factor. *Brain Pathol.* **14:** 237–248.

Schmidt-Hieber C., Jonas P., and Bischofberger J. 2004. Enhanced synaptic plasticity in newly generated granule cells of the adult hippocampus. *Nature* **429:** 184–187.

Seaberg R.M. and van der Kooy D. 2002. Adult rodent neurogenic regions: The ventricular subependyma contains neural stem cells, but the dentate gyrus contains restricted progenitors. *J. Neurosci.* **22:** 1784–1793.

Seki T. and Arai Y. 1993a. Distribution and possible roles of the highly polysialylated neural cell adhesion molecule (NCAM-H) in the developing and adult central nervous system. *Neurosci. Res.* **17:** 265–290.

————. 1993b. Highly polysialylated neural cell adhesion molecule (NCAM-H) is expressed by newly generated granule cells in the dentate gyrus of the adult rat. *J. Neurosci.* **13:** 2351–2358.

Seri B., Garcia-Verdugo J.M., McEwen B.S., and Alvarez-Buylla A. 2001. Astrocytes give rise to new neurons in the adult mammalian hippocampus. *J. Neurosci.* **21:** 7153–7160.

Seri B., Garcia-Verdugo J.M., Collado-Morente L., McEwen B.S., and Alvarez-Buylla A. 2004. Cell types, lineage, and architecture of the germinal zone in the adult dentate gyrus. *J. Comp. Neurol.* **478:** 359.

Shapiro L.A., Korn M.J., Shan Z., and Ribak C.E. 2005. GFAP-expressing radial glia-like cell bodies are involved in a one-to-one relationship with doublecortin-immunolabeled newborn neurons in the adult dentate gyrus. *Brain Res.* **1040:** 81–91.

Shi Y., Chichung Lie D., Taupin P., Nakashima K., Ray J., Yu R.T., Gage F.H., and Evans R.M. 2004. Expression and function of orphan nuclear receptor TLX in adult neural stem cells. *Nature* **427:** 78–83.

Song H., Stevens C.F., and Gage F.H. 2002. Astroglia induce neurogenesis from adult neural stem cells. *Nature* **417:** 39–44.

Steiner B., Klempin F., Wang L., Kott M., Kettenmann H., and Kempermann G. 2006. Type-2 cells as link between glial and neuronal lineage in adult hippocampal neurogenesis. *Glia* **54:** 805–814.

Steiner B., Kronenberg G., Jessberger S., Brandt M.D., Reuter K., and Kempermann G. 2004. Differential regulation of gliogenesis in the context of adult hippocampal neurogenesis in mice. *Glia* **46:** 41–52.

Tozuka Y., Fukuda S., Namba T., Seki T., and Hisatsune T. 2005. GABAergic excitation promotes neuronal differentiation in adult hippocampal progenitor cells. *Neuron* **47:** 803–815.

van Praag H., Schinder A.F., Christie B.R., Toni N., Palmer T.D., and Gage F.H. 2002. Functional neurogenesis in the adult hippocampus. *Nature* **415:** 1030–1034.

Wang L.P., Kempermann G., and Kettenmann H. 2005. A subpopulation of precursor cells in the mouse dentate gyrus receives synaptic GABAergic input. *Mol. Cell. Neurosci.* **29:** 181–189.

Wang S., Scott B.W., and Wojtowicz J.M. 2000. Heterogeneous properties of dentate granule neurons in the adult rat. *J. Neurobiol.* **42:** 248–257.

Wiskott L., Rasch M.J., and Kempermann G. 2006. A functional hypothesis for adult hippocampal neurogenesis: Avoidance of catastrophic interference in the dentate gyrus. *Hippocampus* **16:** 329–343.

Wurmser A.E., Nakashima K., Summers R.G., Toni N., D'Amour K.A., Lie D.C., and Gage F.H. 2004. Cell fusion-independent differentiation of neural stem cells to the endothelial lineage. *Nature* **430:** 350–356.

Zhao C., Teng E.M., Summers R.G., Jr., Ming G.L., and Gage F.H. 2006. Distinct morphological stages of dentate granule neuron maturation in the adult mouse hippocampus. *J. Neurosci.* **26:** 3–11.

10

Adult Subventricular Zone and Olfactory Bulb Neurogenesis

Daniel A. Lim,[1] Yin-Cheng Huang,[2] and Arturo Alvarez-Buylla[1]
[1]Department of Neurological Surgery
University of California
San Francisco, California 94143
[2]Chang-Gung Memorial Hospital
Fu-hsin St. Kweishan
Taoyuan, Taiwan 333

IN THE ADULT MAMMALIAN BRAIN, NEW NEURONS are added to the olfactory bulb (OB) throughout life. In rodents, the adult germinal region for OB neurogenesis is the subventricular zone (SVZ), a layer of cells found along the walls of the brain lateral ventricles (for review, see Alvarez-Buylla and Garcia-Verdugo 2002). Neuroblasts born in the SVZ migrate a relatively long distance into the OB where they then disperse radially and differentiate into interneurons. Most of these new OB neurons integrate into functional circuits (Belluzzi et al. 2003; Carleton et al. 2003), and about half survive long-term (Petreanu and Alvarez-Buylla 2002). SVZ cell proliferation is lifelong (Kuhn et al. 1996; Goldman et al. 1997; Molofsky et al. 2006), with thousands of new neurons generated daily for the mouse OB (Lois and Alvarez-Buylla 1994). The adult SVZ is also the birthplace of oligodendrocytes in both normal and diseased brain (Nait-Oumesmar et al. 1999; Picard-Riera et al. 2002; Menn et al. 2006; Parent et al. 2006). This profound level of continuous neurogenesis and concomitant oligodendrogliogenesis argues for the existence of a self-renewing multipotent precursor cell—or, neural stem cell (NSC)—within the SVZ. The SVZ-OB system is an attractive model in which to study neurogenesis and neuronal replacement as it includes the basic processes of NSC maintenance, progenitor cell-fate specification, migration, differentiation, and survival/death of newly born neurons. The enduring quality and stable cytoarchitecture of adult SVZ-OB neurogenesis may make these complex biological processes experimentally more tractable in comparison to studies of embryonic brain development, which is more ephemeral both temporally and spatially. For these reasons, the adult SVZ-OB system has been intensively studied.

In this chapter, we first review some cellular and molecular aspects of the rodent SVZ NSC niche. We then touch upon recent findings about how different OB interneuron phenotypes may be specified in the adult mouse. The adult human SVZ also harbors NSCs that can be cultured (Roy et al. 1999; Sanai et al. 2004). SVZ NSCs have been propagated in both nonadherent and monolayer cultures, and these NSCs can differentiate into neurons, astrocytes, and oligodendrocytes (Gage et al. 1995; Weiss et al. 1996; McKay 1997; Scheffler et al. 2005). Thus, there is considerable interest in the SVZ as a potential reservoir of neural precursors for therapeutics. Lessons learned from the rodent SVZ-OB neurogenic system may allow us to better manipulate human NSCs in culture for therapeutic transplantation and perhaps even mobilize endogenous precursors to repair diseased or injured brain (Lie et al. 2004; Lim et al. 2007).

THE SVZ MAINTAINS A NEUROGENIC NICHE: EVIDENCE FROM TRANSPLANTATION EXPERIMENTS

For stem cells in general, self-renewal and progenitor cell differentiation are controlled by the specialized microenvironment—or "niche"—in which these cells reside. This concept of a stem cell niche originated from studies of blood cell development where it was found that stem cell fate can be modulated by soluble factors, as well as membrane-bound molecules and extracellular matrix (ECM) (Schofield 1978). These soluble and nonsoluble niche signals may be derived from the stem cells themselves, their progenitors, and the neighboring cells (for review, see Watt and Hogan 2000; Fuchs et al. 2004). SVZ neurogenesis is in part determined by such niche signals, as initially evidenced by transplantation experiments. Mouse SVZ cells grafted homotopically to another SVZ give rise to large numbers of OB interneurons for the recipient animal (Lois and Alvarez-Buylla 1994). In contrast, SVZ cells transplanted to nonneurogenic brain regions produce few, if any, neurons (Herrera et al. 1999). Even purified postnatal SVZ neuroblasts appear to give rise to only glial cells when transplanted to the nonneurogenic striatum, suggesting that late lineage SVZ cells are dependent on local environmental cues to differentiate into or survive as neurons (Seidenfaden et al. 2006). Conversely, the SVZ niche appears to instruct appropriate neurogenesis of cultured progenitors from the hippocampus (Suhonen et al. 1996).

In addition to promoting neuronal differentiation, does the SVZ niche provide signals for NSC self-renewal? Although several molecular pathways have been implicated in having roles in SVZ NSC maintenance (as we discuss later), transplantation experiments demonstrating the existence of an SVZ niche for self-renewal are still lacking. It would be interesting to determine if primary uncultured SVZ cells can be serially grafted from one mouse SVZ to another and retain long-term self-renewal and OB interneuron production. It is, of course, possible that SVZ NSCs are not capable of satisfying this rigorous test of self-renewal; it would nevertheless be instructive to clearly demonstrate that grafting to the SVZ is at least favorable for NSC maintenance/self-renewal than transplantation to nonneurogenic brain regions. Can non-SVZ, nonneurogenic brain regions harbor grafted NSCs in a quiescent or slowly self-renewing state? Or, do nonneurogenic brain regions inexorably commit NSCs to terminally differentiated lineages?

CELLULAR COMPOSITION OF THE ADULT SVZ NICHE

The Primary Cell Types

The adult mouse SVZ is composed of four main cell types (A, B, C, and ependymal cells) as defined by morphological, immunocytochemical, and ultrastructural characteristics (Doetsch et al. 1997). Young neuroblasts (type-A cells) are born throughout the entire SVZ, which extends the entire length of the lateral ventricle (Fig. 1A). Type-A cells migrate in chains (Lois et al. 1996; Wichterle et al. 1997) toward the OB through a network of interconnecting paths widely distributed along the lateral ventricle wall (Fig. 1A,C) (Doetsch and Alvarez-Buylla 1996). These paths converge at the anterior SVZ where the confluence of type-A cell chains form the rostral migratory stream (RMS), a restricted path that enters into the core of the OB. In the OB, type-A cells depart from the chains and migrate radially to more superficial layers where they differentiate into granule and periglomerular interneurons that become incorporated into local circuits (Fig. 1B) (Petreanu and Alvarez-Buylla 2002; Carleton et al. 2003).

The chains of type-A cells in the adult mouse brain are ensheathed by SVZ astrocytes (type-B cells) (Figs. 1C and 2B) (Lois et al. 1996; Doetsch

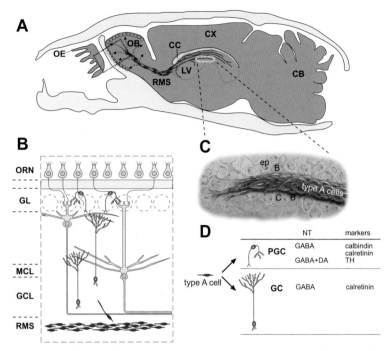

Figure 1. Overview of adult mouse SVZ-OB neurogenesis. (*A*) Sagittal section through mouse head (calvarium is *yellow*). Neuroblasts (type-A cells) born in the subventricular zone (SVZ) of the lateral ventricle (*blue*) migrate through a network of paths (*red*) into the RMS, which enters the olfactory bulb (OB). Cells then leave the RMS (*arrows, dashed lines*) and migrate radially into the OB. Boxed area is shown enlarged in *B*. (*B*) Neuronal layers of OB. Migratory cells depart the RMS and differentiate into granule cells (GCs) or periglomerular cells (PGCs), which reside in the GCL and GL, respectively (type-A cells and differentiated interneurons are *red*). ORNs (*small gray cells*) in the OE project to the GL. (*Gray*) The main projection neurons of the OB (mitral cells and tufted cells). (*C*) Artists' rendition of a chain of migratory type-A cells. These chains are ensheathed by glial cells (type-B cells, *blue*) and are associated with clusters of transit-amplifying cells (type-C cells, *green*). (*D*) Diversity of OB interneurons. Type-A cells differentiate into either PGCs or GCs, which can be distinguished by morphology, NT phenotype, and markers, which are expressed by a subset of the listed cell types. (OE) Olfactory epithelium; (CC) corpus callosum; (RMS) rostral migratory stream; (Cx) cortex; (CE) cerebellum; (ORN) olfactory receptor neuron; (GL) glomerular layer; (MCL) mitral cell layer; (GCL) granule cell layer; (ep) ependymal cell; (NT) neurotransmitter; (TH) tyrosine hydroxylase.

et al. 1997). Type-B cells express the astrocyte marker glial fibrillary acidic protein (GFAP) and have ultrastructural characteristics typical of astrocytes. Although astrocytes have been classically thought of as simply glial "support" elements of the adult brain, some SVZ type-B cells

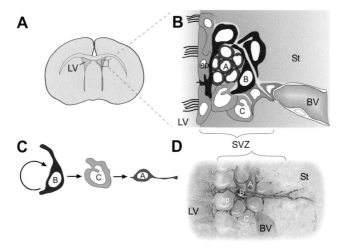

Figure 2. Cellular composition of the SVZ niche. (*A*) Coronal section of adult mouse brain. (*Blue*) Lateral ventricle (LV). Boxed area is shown enlarged in *B*. (*B*) Subventricular zone (SVZ) architecture. Type-B cells (*dark blue*) are the astrocytes that are the SVZ stem cell and also serve as niche cells. Type-C cells (*green*) are rapidly dividing transit-amplifying cells derived from the B cells. Type-C cells give rise to type-A cells (*red*), the migratory neuroblasts. A blood vessel (BV, *pink*) is shown with a perivascular macrophage (*brown*); a basal lamina (*yellow*) extends from the BV and intercalates with SVZ cells. Ciliated ependymal cells (ep, *gray*), microglia (*orange*), and endothelial cells (*white*) may be important niche cells. (*C*) Cell lineage of SVZ neurogenesis; (*D*) artist's conception of the SVZ. The type-B cell shown has a process that makes contact with the ventricle and expresses a single cilium. A similar cell is also shown in *B*.

function as NSCs (Doetsch et al. 1999b; Laywell et al. 2000; Imura et al. 2003; Garcia et al. 2004). These type-B cells are descendant from radial glial cells (Merkle et al. 2004), which in the embryonic brain are multipotent neural precursors (for review, see Merkle and Alvarez-Buylla 2006). In the human brain, a ribbon of GFAP$^+$ astrocytes lines the ventricles (Sanai et al. 2004; Curtis et al. 2007), and some of these human SVZ astrocytes proliferate in vivo and behave as NSCs in vitro (Sanai et al. 2004). Although there is evidence that neurogenesis exists in the adult human OB (Bedard and Parent 2004; Curtis et al. 2007), it does not appear that this neurogenesis is extensive. It has recently been suggested that migration between the SVZ and OB continues in the adult human brain (Curtis et al. 2007); these inferences from autopsy specimens would be extremely interesting if the lineage relationship between human SVZ and OB interneurons could be directly tested.

In addition to serving as the NSC, SVZ astrocytes are also important niche cells. Type-B cells are in direct contact with all other SVZ cell types, including the rapidly dividing transit-amplifying cell as well as the committed migratory neuroblasts (for review, see Alvarez-Buylla et al. 2001). SVZ precursor cells cultured in serum-free medium in direct contact with monolayers of other astrocytes proliferate to form colonies of young neuroblasts (Lim and Alvarez-Buylla 1999). Interestingly, for hippocampal neural precursors, astrocyte-derived soluble and membrane-bound factors also promote neurogenesis (Song et al. 2002), and in the neurogenic region of the hippocampus, astrocyte stem cells form basket-like structures, cradling the newly born neuroblasts (Seri et al. 2004; Seki et al. 2007). Thus, it appears that close proximity and/or contact between astrocytes and NSCs is one important aspect of the neurogenic niche. It is not clear whether SVZ astrocytes are intrinsically different from astrocytes located outside this germinal region. It will be interesting to determine the lineage relationship and molecular characteristics of stem cell astrocytes versus nonstem cell astrocytes.

Distributed along the type-A-cell chains, often interposed between type-B and type-A cells, are clusters of rapidly dividing immature cells (type-C cells) (Doetsch et al. 1997). Type-C cells appear to function as a transit-amplifying intermediate between B and A cells. Thus, the SVZ neurogenic lineage is from type-B cell (SVZ astrocyte) to type-C cell (transit-amplifying intermediate) to type-A cell (migratory neuroblast) (Fig. 2C). Interestingly, similar to type-B cells, type-C cells isolated in vitro can be propagated as NSCs (Doetsch et al. 2002), suggesting that this cell type can be "reprogrammed" by cell culture conditions to acquire the properties of self-renewal and multipotentiality.

The Interdigitation of Type-B Cells within the Ependymal Layer

The lateral ventricle ependyma has often been described as a layer of multiciliated epithelial cells that form a tight barrier between the brain parenchyma and the ventricle lumen. However, electron microscopy (EM) shows that the ependymal layer is not entirely contiguous. In normal mice, a number of type-B cells make direct contact with the ventricle (Doetsch et al. 1999a, 2002; Conover et al. 2000); some type-B cells contact the ventricle by extending a thin cellular process between ependymal cells, and others have a larger luminal surface (illustrated in Fig. 2B,D). Thus, the boundary between the ependymal layer and the

SVZ is blurred by this interdigitation of type-B cells and ependyma; without EM, it can be difficult to distinguish these cell types since they are both very close to the ventricle lumen. The ventricle-contacting type-B cells also possess a single, thin cilium lacking the central pair of microtubules. Similar single cilia with this 9 + 0 microtubule arrangement have been described in embryonic neuroepithelial cells (Sotelo and Trujillo-Cenóz 1958; Stensaas and Stensass 1968) and adult avian brain neuronal precursors (Alvarez-Buylla et al. 1998). It has been suggested that ependymal cells can undergo mitosis and function as stem cells (Johansson et al. 1999), but this has not been supported by other studies (Chiasson et al. 1999; Capela and Temple 2002; Spassky et al. 2005). However, as we describe in the following sections, ependymal cells may be important niche cells.

The SVZ Basal Lamina, Endothelia, Microglia, and Other Potential Niche Components

Mercier et al. (2002) used EM to describe the SVZ vasculature and the associated extravascular basal lamina (BL) which invests itself deeply in the SVZ. Blood vessels that penetrate into the SVZ consist of endothelial cells, pericytes, fibroblasts, and macrophages. The extravascular BL, which is rich in laminin and collagen-1, interdigitates extensively with all SVZ cell types, and there are also many microglial cells in contact with the BL and other SVZ cells. It is possible that the BL concentrates and/or modulates cytokines/growth factors derived from local cells. Perhaps this extensive attachment to the BL is important for type-B-cell maintenance of stem cell properties (Alvarez-Buylla and Lim 2004). Data suggest that endothelial cells and microglial cells are important for the SVZ niche. Endothelial cells cocultured with SVZ explants enhance neuroblast migration and maturation (Leventhal et al. 1999); cultured endothelial cells were also found to secrete soluble factors that stimulate embryonic NSC self-renewal and increase the neuronal production of adult SVZ NSCs (Shen et al. 2004). Recently, Walton et al. (2006) provided in vitro evidence that microglial cells produce soluble factors important for SVZ neurogenesis but not self-renewal. The SVZ niche factors produced by microglial and endothelial cells have not been identified, although there is some evidence that brain-derived neurotrophic factor (BDNF) from endothelial cells (Leventhal et al. 1999) and microglial-derived leukemia inhibitory factor (LIF) (Nakanishi et al. 2007) may be important.

Type-B Cells Are Precursors of Both Oligodendrocytes and OB Interneurons

Most oligodendrocytes develop during embryogenesis and early postnatal life, but some oligodendrocytes are born in the adult brain. In brains with demyelinating lesions, cells from the mouse SVZ migrate to the lesions and generate oligodendrocytes (Nait-Oumesmar et al. 1999; Picard-Riera et al. 2002). Recently, it was demonstrated that SVZ type-B cells also function as precursors of oligodendrocytes in the adult brain (Menn et al. 2006). The fate of GFAP$^+$ type-B cells was followed with injection of RCAS-GFP (replication-competent avian leukosis virus–green fluorescent protein) retrovirus (which can only infect tva receptor-expressing cells) into the SVZ of transgenic mice expressing tva receptor under the control of the GFAP promoter; in these mice, NG2$^+$ oligodendrocyte precursor cells and myelinating oligodendrocytes were GFP$^+$, demonstrating their lineage relationship to type-B cells. Interestingly, after a demyelinating lesion to the corpus callosum, SVZ type-B cells increase oligodendrocyte production fourfold, indicating that the SVZ participates in myelin repair. Whether or not type-B cells in vivo are capable of producing both neurons and oligodendrocytes is not yet known; however, there is at least in vitro evidence for this bipoteniality.

MOLECULAR COMPOSITION OF THE SVZ NICHE

Mitogens, Growth Factors

Fibroblast growth factor-2 (FGF-2) and epidermal growth factor (EGF) are the principal mitogens used to proliferate SVZ NSC in vitro, and they are likely important for proliferation in vivo. It is not clear which SVZ cells express these mitogens in vivo; however, cultured cortical astrocytes express EGF and FGF-2 (Morita et al. 2005), suggesting that SVZ astrocytes may provide these proliferative signals for the SVZ niche. EGF and FGF receptors (EGFR and FGFR) are expressed in the SVZ, and mice null either for FGF-2 (Zheng et al. 2004) or for the EGFR ligand transforming growth factor-α (TGF-α) (Tropepe et al. 1997) have significantly reduced SVZ neurogenesis. Although EGF and FGF are primarily often thought of as mitogens, they may also have roles in cell-fate control of SVZ cells and may even cause NSC dedifferentiation (for review, see Anderson 2001; Raff, 2003).

Other growth factors have been implicated in the regulation of the SVZ (Fig. 3). Intraventricular infusion of BDNF increases SVZ proliferation and OB neurogenesis (Zigova et al. 1998); interestingly, elevated levels

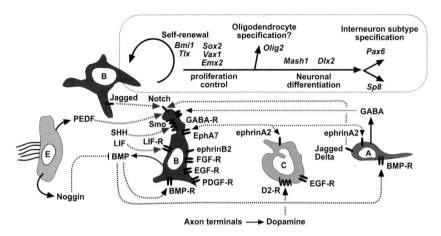

Figure 3. Interactions of selected SVZ niche factors. (*Solid arrows*) Cellular source of secrete factor, when known; (*dotted lines*) molecular interactions, both known and hypothetical; (*blue dotted lines*) interactions that may promote subventricular zone (SVZ) neural stem cell (NSC) self-renewal. Transcription factors that may be involved in self-renewal, proliferation control, and specification of different cell types are in italics in the box above.

of local BDNF expression also appears to promote ectopic neurogenesis (Pencea et al. 2001; Chmielnicki et al. 2004). Intraventricular infusion of ciliary neurotrophic factor (CNTF) (Emsley and Hagg 2003b) increases SVZ proliferation and may increase Notch1 expression (Chojnacki et al. 2003), suggesting a role in NSC self-renewal (see below discussion on Notch). Heparin-binding EGF (HB-EGF) also stimulates SVZ proliferation (Jin et al. 2002b) when administered either into the ventricle or intranasally (Jin et al. 2003). Vascular endothelial growth factor (VEGF), a known angiogenic protein, also stimulates SVZ neurogenesis (Jin et al. 2002a) when infused into the ventricle, suggesting a link between angiogenesis and neurogenesis (Palmer et al. 2000; Louissaint et al. 2002; Greenberg and Jin 2005). Platelet-derived growth factor (PDGF) signaling appears to have a role in balancing neuronal and oligodendrocyte production from the SVZ, and excessive PDGF signaling from ventricular infusion results in hyperplasia with some features of gliomas; a significant subpopulation of cells within the hyperplasias are derived from type-B cells (Jackson et al. 2006).

Shh: SVZ NSC Maintenance Factor?

Sonic hedgehog (Shh) is a classical morphogen in development (Ruiz i Altaba et al. 2003), and although it is a mitogen for a cerebellar granule precursor, it appears to have a distinctive role as an adult NSC mainte-nance factor. To elucidate the role of Shh signaling in NSCs, the Shh core-ceptor Smo (Smoothened) was removed from neural precursors at E12.5 by crossing floxed Smo ($Smo^{n/c}$) with Nestin-Cre (N^{cre}) mice (Machold et al. 2003); postnatally, $Smo^{n/c}$; N^{cre} mice have greatly decreased SVZ cell proliferation and OB neurogenesis. Furthermore, fewer NSCs can be cul-tured from $Smo^{n/c}$;N^{cre} animals. Despite these dramatic effects on adult neurogenesis, the mature brains of $Smo^{n/c}$;N^{cre} mice appear to be of nor-mal size, suggesting that Shh is primarily important for "maintaining" the stem cell population in postnatal and adult brain germinal niches. Although these data are consistent with the role Shh has as an SVZ NSC "maintenance" factor, it has also been shown that Shh acts as a mitogen for SVZ-derived NSC cultures when the EGF concentration is not satu-rating (Palma et al. 2005).

The exact cellular source of Shh for the SVZ has not been deter-mined; however, cells responding to Shh signaling can be inferred by expression of the Gli1 transcription factor (Bai et al. 2002). Gli1$^+$ cells are found in SVZ (Machold et al. 2003) astrocytes and transit-amplifying cells (Palma et al. 2005). Recently, Ahn and Joyner (2005) elegantly showed that Gli1$^+$ SVZ cells behave as NSCs in vivo. These authors engineered mice to express a Cre-estrogen receptor (ER) fusion gene under the con-trol of the Gli1 promoter, restricting Cre-ER expression to cells with Shh signaling. Cre-ER enters the nucleus only in the presence of tamoxifen, thus providing temporal control to Cre-mediated recombination. Gli1–Cre-ER mice crossed to a Cre-reporter line (R26R) thus have LacZ reporter gene expression in a cohort of Shh-signaling cells during the tamoxifen admin-istration. Ahn and Joyner treated Gli1–Cre-ER R26R mice with tamox-ifen, eliminated rapidly dividing cells with administration of an antimitotic, and then followed the fate of LacZ$^+$ cells. Initially, the num-ber of LacZ$^+$ cells in the SVZ increases, possibly representing the expan-sion of type-C cells. During the next year, LacZ$^+$ OB interneurons continue to be generated from the SVZ. Thus, Shh signaling is active in SVZ NSCs, which are relatively quiescent, and these Gli1$^+$ cells can respond to antimitotic insult by increasing proliferation.

BMP Signaling, Modulation by Their Antagonists, Potential Interaction with LIF

Another family of neural morphogens, the bone morphogenetic proteins (BMPs), also regulates adult brain germinal niches. BMP signaling pro-motes astrocyte differentiation of embryonic SVZ-derived precursors at the expense of oligodendrogliogenesis and neurogenesis (Gross et al. 1996). Adult SVZ cells produce BMPs and their receptors (Lim et al. 2000; Peretto et al. 2002). Noggin, a secreted BMP antagonist, is also locally expressed, most strongly in the ependymal cells (Lim et al. 2000; Peretto et al. 2004). This locally derived BMP antagonist may contribute to the neurogenic niche for SVZ stem cells as it promotes neurogenesis both in vitro and in ectopic locations in vivo (Lim et al. 2000). Overexpression of BMP-7 in the SVZ suppresses neurogenesis (Lim et al. 2000), and overexpression of Noggin from the ependyma, suppresses gliogenesis (Chmielnicki et al. 2004). Hence, a "balance" between BMP and their antagonists may control the levels of neurogenesis and gliogenesis from NSCs in adult brain niches.

Are astrocyte-niche cells and astrocyte-stem cells molecularly distinct? Although BMPs and LIF both induce GFAP$^+$ astrocyte differentiation from NSCs, Bonaguidi et al. (2005) demonstrated that BMP- and LIF-induced astrocytes from embryonic SVZ-derived NSCs are distinct populations. BMP-induced GFAP$^+$ cells exit the cell cycle, take on a stellate morphology, and have limited NSC potential. In contrast, LIF treatment generates GFAP$^+$ cells that have a bipolar/tripolar morphology, remain in the cell cycle, express progenitor cell markers, and behave as NSCs in culture. In addition, LIF-treated NSCs have a greater neuronal differentiation potential. In embryonic stem cells, BMPs act in concert with LIF to sustain self-renewal and suppress differentiation (Ying et al. 2003). What is the effect of simultaneous BMP and LIF signaling on adult NSCs? It is possible that the ratio of BMP:LIF signaling on type-B cells determines which is to serve as the niche cell and which is to be the stem cell. Reduction of BMP signaling by Noggin or other BMP antagonists may increase the LIF:BMP signaling ratio, increasing the likelihood that the SVZ astrocyte remains a NSC. Consistent with this hypothesis, overexpression of LIF from an adenoviral vector injected into the brain ventricle promotes SVZ NSC self-renewal, expanding the population of NSCs in vivo while concomitantly decreasing SVZ-OB neurogenesis (Bauer and Patterson 2006).

Roles of PEDF, Notch, and Wnts in NSC Self-renewal

Pigment epithelium-derived factor (PEDF) is another secreted factor of the SVZ niche apparently important for NSC self-renewal. Although PEDF is structurally similar to members of the serpin family of protease inhibitors, PEDF does not inhibit proteases; PEDF has been found to be neurotrophic for a variety of neuronal populations (Becerra et al. 1995). In the SVZ, PEDF is expressed by ependymal cells and endothelial cells. In vitro, PEDF enhances NSC self-renewal without affecting proliferation rate. Intraventricular infusion of PEDF activates slowly dividing type-B cells, whereas infusion of a PEDF blocking reagent reduces type-B-cell proliferation (Ramirez-Castillejo et al. 2006).

In neural development, the Wnt family of secreted signaling molecules has roles in stem cell maintenance, cellular proliferation, differentiation, migration, and axon guidance (Ille and Sommer 2005). Wnt3 has been shown to be a principal regulator of adult hippocampal neurogenesis. Currently, less is known about Wnt signaling in the SVZ; however, Wnt3a and Wnt5a promote proliferation and neuronal differentiation of SVZ-derived NSC cultures (Yu et al. 2006). It is not known which cells in the SVZ express Wnts; however, Wnt5a expression has been described in the postnatal (Shimogori et al. 2004) and adult OB (Lim et al. 2006).

In hematopoiesis, there is evidence that Wnt and Notch signaling are integrated to maintain the stem cell phenotype (Duncan et al. 2005). For NSCs, Notch signaling may suppress neuronal differentiation and maintain precursor cell properties (Gaiano and Fishell 2002). Notch1 and two cognate membrane-bound ligands, Jagged-1 and Delta-1, are expressed in the adult SVZ (Stump et al. 2002; Nyfeler et al. 2005; Givogri et al. 2006). Retroviral induction of activated Notch (ActN) in the embryonic brain promotes radial glial identity and produces dense clusters of SVZ astrocytes postnatally (Gaiano et al. 2000). In postnatal SVZ cells, retrovirally transduced ActN prevents cell migration to the OB, suppresses neuronal differentiation, and decreases proliferation, creating a more "quiescent" cell type (Chambers et al. 2001). More recently, Nyfeler et al. (2005) demonstrated the importance of Jagged1-Notch1 signaling for SVZ prolifertation in vivo; in vitro, soluble Jagged1 promotes SVZ NSC self-renewal and increases their neurogenic potential. In the adult SVZ, Jagged1 is expressed by a subset of GFAP$^+$ astrocytes, and Notch1 is expressed by adjacent clusters of cells; SVZ astrocytes were not found to coexpress Jagged1 and Notch1. One interpretation of these findings is that the Jagged1-expressing astrocytes serve as niche cells for adjacent Notch1-expressing astrocytes. It is intriguing to consider that Jagged1 may be a downstream target of canonical Wnt signaling (Katoh 2006); it is thus possible that Wnt induces Jagged1 expression, which, in turn, signals through Notch1.

In addition to astrocytes, Jagged1 and Delta1 expression is also found in the SVZ neuroblasts, type-A cells (Givogri et al. 2006). This expression pattern suggests a potential feedback regulation of the SVZ niche: Accumulation of newly born type-A cells expressing Jagged1 or Delta1 up-regulate Notch signaling in type-B cells, suppressing differentiation and potentiating self-renewal.

Role of Eph and Ephrins in the SVZ

Holmberg et al. (2005) recently suggested that Ephrin and Eph receptors control SVZ cell numbers through a feedback mechanism. These authors show that ephrin-A2 is expressed by type-A and type-C cells and that the EphA7 receptor is expressed in some type-B cells and ependyma. Ephrin-A2-deficient mice have increased SVZ cell proliferation and OB neurogenesis; disruption of the interaction between ephrin-A2 and EphA7 by infusion of blocking reagents also increases SVZ proliferation and OB neurogenesis. These data indicate that ephrin-A2-EphA7 signaling normally inhibits SVZ proliferation. Holmberg et al. also provide data suggesting that this negative regulation of SVZ proliferation is mediated by reverse signaling through ephrin-A2 in type-A and type-C cells; i.e., type-B cells expressing EphA7 inhibit type-A and type-C cell proliferation which express ephrin-A2. Thus, it is unclear how this functional arrangement of ephrin-A2 and EphA7 in the SVZ can function as a feedback control mechanism for type-A-cell production. Previously, Conover et al. (2000) showed that infusion of EphB2-Fc fusion proteins disrupts SVZ cell migration and increases cell proliferation and type-B-cell number; intriguingly, ventricle-contacting type-B cells increase eightfold. The initial interpretation of this result was that EphB2-Fc activates ephrin-B2 on SVZ type-B cells, increasing their proliferation. However, if as suggested by Holmberg et al. (2005), EphB2-Fc serves primarily as a blocking reagent, thus disinhibiting type-B-cell proliferation, then intercellular signaling between SVZ cells could actually serve as a negative-feedback mechanism controlling type-B-cell numbers, since type-B cells express ephrin-B2 and type-B and/or type-C cells express EphB2.

GABA in a Feedback Control Mechanism of Type-A Cells, and Role of Other Neurotransmitter Molecules

In postnatal SVZ, GABA expression by neuroblasts appears to feedback-inhibit the proliferation of $GFAP^+$ SVZ cells; as more neuroblasts are born, it is possible that extracellular GABA concentrations rise, activating GABA receptors on $GFAP^+$ SVZ cells and decreasing proliferation and neuroblast production (Liu et al. 2005). Other neurotransmitters have been implicated in SVZ neurogenesis control. Serotonin (5-HT) acts through the d-HT2c and 5-HT1a receptors in the SVZ, unregulating cell proliferation and OB neurogenesis (Brezun and Daszuta 1999; Banasr et al. 2004). The role of dopamine in the SVZ niche has been more closely studied and is more controversial. The SVZ receives dopaminergic innervation from the midbrain, with type-C cells being the predominant cell type expressing D2 receptors. Dopaminergic denervation results in decreased SVZ proliferation and OB neurogenesis, and administration of the dopamine precursor levodopa could restore SVZ proliferation to near normal levels. These results suggest that patients with Parkinson's disease may also have decreased neurogenesis (Hoglinger et al. 2004). Contrary to these results, Kippen et al. (2005) found that the D2 receptor blocker haloperidol increases cell SVZ proliferation, expanding the population of NSCs, resulting in increased OB neurogenesis; these authors thus suggest that there is a tonic level of dopamine signaling in the SVZ that normally inhibits, or restrains, neurogenesis. The evidence clearly suggests that dopamine affects cell proliferation, but the controversy is still not resolved.

Nitric oxide (NO) is a free-radical signaling molecule produced by nitrergic neurons that express the neuron-specific form of nitric oxide synthetase (NOS); nitrergic neurons are found in close proximity to the SVZ. Both pharmacological inhibition and genetic disruption of NOS in nitregic neurons increases SVZ-OB neurogenesis, suggesting that NO also serves to restrict adult neurogenesis (Packer et al. 2003; Moreno-Lopez et al. 2004).

The Importance of Cell-intrinsic Factors in the Interpretation of Niche Signals

On the basis of the expression of certain cell-intrinsic factors, the same extracellular, extrinsic signal can be interpreted by a cell in vastly different ways. For instance, although BMPs induce glial differentiation of adult SVZ NSCs (see above) and E17–18 neural precursors (Gross et al. 1996), BMP-signaling promotes neuronal and not glial differentiation of neural precursors from the E13–14 embryo (Li et al. 1998; Mabie et al. 1999). This may be related to the expression of high levels of the Neurogenin1 (Ngn1) transcription factor by E13–14 neural precursors. Ngn1 not only activates genes for neuronal differentiation, but also sequesters the BMP downstream signaling factor SMAD1 from astrocyte differentiation genes. Interestingly, overexpression of Ngn1 can convert BMP into a neuronal differentiation signal from a gliogenic signal (Sun et al. 2001). Along a similar line, in the adult SVZ cellular lineage, BMPs induce astrocyte differentiation in the early precursors (Lim et al. 2000) while inducing cell cycle exit and enhanced survival of late lineage neuroblasts (Coskun and Luskin 2002); this difference of BMP activity may be related to differential expression of BMP receptor subtypes (Lim et al. 2000; Panchision et al. 2001). Clearly, it will be important to study SVZ cell intrinsic factors that modify the effect of niche signals.

TRANSCRIPTION FACTORS FOR SVZ-OB NEUROGENESIS

Genes for Self-renewal and Proliferation Control

What transcription factors are required for adult stem cell self-renewal? Mice null for Bmi1 (Molofsky et al. 2003) or TLX (Shi et al. 2004), both nuclear transcriptional regulators, have impaired self-renewal of adult NSCs. Bmi1 is a member of the polycomb family of chromatin remodeling factors that repress the expression of arrays of genes; Bmi1 appears to promote NSC self-renewal in part by repressing the *Ink4a/arf* locus (Bruggeman et al. 2005; Molofsky et al. 2005); p16Ink4a expression increases with increasing age in the SVZ, leading to senescence of the NSC population (Molofsky et al. 2006). TLX also demonstrates properties of a transcriptional repressor, inhibiting NSC differentiation into astrocytes (Shi et al. 2004). It will be interesting to identify the direct transcriptional targets of Bmi1 and TLX, perhaps on a genome-wide scale by chromatin immunoprecipitation experiments. The homeobox transcription factors Sox2, Vax1, and Emx2 also appear to be involved in SVZ NSC proliferation control. Among many other neurological abnormalities, mice deficient for Sox2 (Ferri et al. 2004) have impaired adult SVZ neurogenesis, and Vax1 null mice develop a very disorganized, abnormal SVZ (Soria et al. 2004). Emx2, which is related to Vax1, has also been implicated in control of SVZ precursor proliferation control (Galli et al. 2002).

Downstream Effectors and Modifiers of Notch Signaling

Numb and Numblike (Numbl) are mammalian homologs of *Drosophila* numb, a protein that antagonizes Notch function in one daughter cell during asymmetric precursor cell division (for review, see Roegiers and Jan 2004). In mice, Numb and Numbl are critical for embryonic neurogenesis (for review, see Johnson 2003). Using a Nestin–Cre-ER transgenic mouse, Kuo et al. (2006) removed Numb/Numbl in postnatal SVZ cells, including ependymal cells, and found that Numb/Numbl has roles in both ependymal wall integrity and SVZ neurogenesis.

The family of Hairy/enhancer of split (Hes) basic helix-loop-helix (bHLH) transcription factors include seven members, and Hes1 and Hes5 are up-regulated by Notch signaling (for review, see Kageyama et al. 2005). Hes5 is expressed in adult SVZ along with the Notch receptor and cognate ligands (Stump et al. 2002). Misexpression of Hes1, Hes3, and Hes5 in embryonic brain inhibits neuronal differentiation and maintains the radial glial phenotype (Ishibashi et al. 1994; Ohtsuka et al. 2001). Conversely, Hes1,Hes5 double-null mice do not propertly maintain radial glial cells and prematurely differentiate into neurons (Ohtsuka et al. 2001); furthermore, Hes1;Hes5 null NSCs cannot be expanded in culture (Ohtsuka et al. 1999). It is possible that Hes5 activation by Notch signaling in the adult SVZ maintains a precursor-like state by repressing prodifferentiation genes. Of note, the premature differentiation of radial glia in Hes-deficient mice correlates with increased expression of proneural bHLH transcription factors including mammalian achaete-scute homolog 1 (Mash1) (Hatakeyama et al. 2004). Furthermore, Hes1 represses Mash1 at the transcriptional level (Sasai et al. 1992) and by protein-protein interaction (Chen et al. 1997).

bHLH Transcription Factors

In the adult SVZ, Mash1 is expressed in a subset of type-B and type-C cells (Kohwi et al. 2005); postnatal mice null for Mash1 have impaired SVZ-OB neurogenesis and oligodendrogliogenesis, suggesting its role in the specification of these cell types (Parras et al. 2004). Another bHLH factor, Olig2, is also expressed in some type-B cells and a subset of type-C cells (Menn et al. 2006); Olig2 is a member of a proneural bHLH subfamily distinct from Mash1. Retroviral vector-driven overexpression of Olig2 in SVZ cells represses the neuronal lineage while promoting oligodendrogliogenesis (Hack et al. 2005). Interestingly, in embryonic development, Olig2 alone also promotes oligodendrocyte development; however, when a proneuronal bHLH factor (Ngn2) is coexpressed, Olig2 promotes motor neuron development (Mizuguchi et al. 2001; Novitch et al. 2001). If a similar mechanism is at play in the adult SVZ, then Olig2—in addition to promoting oligodendrogliogenesis—may also participate in the specification of other cell types when expressed with different combinations of transcription factors.

Generating the Diversity of OB Interneurons

The OB interneuron population is composed of a variety of cellular phenotypes differing in their morphology, OB location, synaptic properties, and neurochemistry (Kosaka et al. 1995, 1998; Kosaka and Kosaka 2005; Parrish-Aungst et al. 2007). OB interneurons can be grossly separated into the granule cells (GCs), located in the granule cell layer (GCL), and the periglomerular cells (PGCs), which are found more peripherally in the glomerular layer (GL) (see Fig. 1B). Mice null for Dlx1/2 homeobox transcription factors have severely impaired genesis of both GCs and PGCs in development (Bulfone et al. 1998). GCs and PGCs are also generated in adulthood. PGCs can be further subclassified based on the expression of calbindin, calretinin, or glutamic acid decarboxylase (GAD) (see Fig. 1C). Most GCs are GABAergic, and, in the rat, about 40% of the PGCs are also GABAergic (Kosaka et al. 1995), and of these, 60–70% are dopaminergic. Recent work has begun to describe the molecular basis of this OB interneuron diversity. Paired box 6 (Pax6), a homeobox transcription factor, both promotes SVZ neurogenesis and directs the generation of dopaminergic PGCs; inhibition of Pax6 by a dominant-negative construct or heterochronic transplantation of embryonic Pax6-null progenitors into the postnatal SVZ results in a severe reduction of dopaminergic PGCs (Hack et al. 2005). The nondopaminergic PGCs, on the other hand, appear to be in part specified by the zinc finger transcription factor Sp8; interestingly, neuroblasts null for Sp8 have increased Pax6 expression (Waclaw et al. 2006). One interpretation of this is that Sp8 normally represses Pax6 expression, shifting the final OB interneuron fate from dopaminergic PGC to nondopaminergic PGC. This notion is supported by the normal expression pattern of Pax6: There are many more Pax6-expressing type-A cells in the RMS than in the OB (Kohwi et al. 2005), suggesting that Pax6 becomes down-regulated in a subpopulation of neuroblasts that reach the OB. Pax6 and Sp8 may thus be the beginnings of a transcriptional combinatorial code for the specification of different PGC interneuron phenotypes. Different populations of GCs can be distinguished by their location in the GCL; some are located in the deep GCL (closer to the center of the OB) and others are more superficial (closer to the mitral cell layer [MCL]). Embryonic Pax6-null progenitors also fail to generate superficial GCs (Kohwi et al. 2005), indicating that Pax6 has a role in determining phenotypes of both GCs and PGCs. Emx1, another homeobox transcription factor, is expressed in cells that become calretinin$^+$, GAD$^+$ superficial GCs and PGCs (Kohwi et al. 2007). Because Emx1 is expressed primarily in the developing pallium (and not the lateral ganglionic eminence, the presumed SVZ primordium), the data suggest that different embryonic anatomical regions contribute to different OB interneuron subtypes. The Er81 transcription factor has an expression pattern similar to that of Sp8, but the function of Er81 in OB interneuron specification is not known (Stenman et al. 2003). It is likely that a complex combinatorial "code" of transcription factors defining OB interneuron diversity will soon emerge.

THE REMARKABLE JOURNEY OF TYPE-A CELLS

After birth in the SVZ, type-A cells migrate a considerable distance (up to several millimeters) through a complex network of paths that converge at the RMS, a restricted path that leads neuroblasts into the OB (see Fig. 1A). This journey is remarkable not only for the great distance, but also for the highly directed nature of the migration. Type-A cells appear to be actively guided rostrally and do not deviate from the restricted path into the OB. By themselves, type-A cells are imbued with an extensive migratory capacity. In vitro, type-A cells form chains of themselves and migrate at relatively high speeds of 120 μm/hr (Wichterle et al. 1997). These neuroblasts move in a stepwise manner, first extending a leading process with a growth cone and then translocating the cell body toward the growth cone tip; this process is repeated, leading to a "saltatory" cellular movement. DCX (doublecortin), a microtubule-associated protein important for neuronal migration in the embryo (Francis et al. 1999), and CRMP-4 (collapsing response mediator protein-4), which is involved in axon guidance (Nacher et al. 2000), are both expressed in type-A cells (Gleeson et al. 1999; Nacher et al. 2000). How DCX and CRMP-4 contribute to the internal molecular machinery of neuroblast migration is not understood. In addition to having a role in SVZ cell proliferation, GABA also reduces the speed of neuroblast migration (Bolteus and Bordey 2004); the intracellular mechanism of how GABA regulates migratory speed is not known, but it appears to involve calcium signaling. More recently, using time-lapse imaging and pharmacological perturbations, Schaar and McConnell (2005) have shown that interactions between microtubules, myosin II, and cell adhesion are required for the saltatory migration of SVZ neuroblasts.

Specific cell surface molecules on type-A cells also confer migratory ability. For instance, type-A cells express PSA-NCAM (polysialic-acid-neural cell adhesion molecule), and there is evidence that PSA and/or PSA-NCAM is critical for SVZ-RMS migration in postnatal mice

(Tomasiewicz et al. 1993; Cremer et al. 1994; Ono et al. 1994; Hu et al. 1996) and is important for efficient chain migration in the adult (Chazal et al. 2000). Certain integrins (Jacques et al. 1998; Murase and Horwitz 2002; Emsley and Hagg 2003a) and ganglioside 9-O-acetyl GD3 (Miyakoshi et al. 2001), a glycolipid on type-A cells, may also be required for chain migration. Tenascin-C is among the several ECM molecules expressed in the RMS, and Tenascin-C interacts with integrins (Yokosaki et al. 1996) as well as 9-O-acetyl GD3 (Probstmeier and Pesheva 1999); however, it is not known whether these particular interactions are essential for migration. Other ECM components of the RMS include chondroitin sulfate proteoglycans (Thomas et al. 1996) and laminin (Murase and Horwitz 2002). Infusion of a laminin peptide or laminin can redirect migration of type-A cells to ectopic locations (Emsley and Hagg 2003a). Thus, the ECM of the migratory paths may create an environment permissive for neuroblast movement.

Perhaps one of the most intriguing processes that occurs in the SVZ-OB is the directional migration of type-A cells. What are the factors that guide the neuroblasts away from the SVZ, rostrally to the OB? It appears that SLIT-ROBO signaling has a crucial role. SLITs are expressed by the septum and choroid plexus (Hu 1999; Li et al. 1999; Nguyen Ba-Charvet et al. 1999), and the Robo-2/Robo-3 receptors are expressed in the SVZ and RMS (Nguyen Ba-Charvet et al. 1999). Secreted SLIT protein is a chemorepulsive factor for type-A cells in vitro (Hu 1999; Wu et al. 1999). A gradient of the SLIT2 chemorepulsive signal—with the highest concentration in the posterior SVZ—can be established by the flow of cerebrospinal fluid (CSF) which parallels the directionality of type-A-cell migration. CSF flow is primarily driven by the ciliary beating of ependymal cells; mice with defective ependymal cilia cannot establish this SLIT gradient and have impaired type-A-cell migration (Sawamoto et al. 2006). Thus, SLIT may be a critical chemorepulsive factor for directional SVZ neuroblast migration, and the gradient of SLIT expression may require proper ependymal polarity, cilia, and CSF flow.

The OB has also been suggested to be a source of chemoattractants for type-A cells. Prokineticin-2 (Ng et al. 2005) and Netrin-1 (Astic et al. 2002; Murase and Horwitz 2002) are both expressed in the OB and can attract type-A cells in vitro; however, other groups have found that Netrin-1 either has no activity or repels type-A cells (Mason et al. 2001; Liu and Rao 2003). In vitro, the OB does not attract SVZ neuroblasts (Hu et al. 1996), and surgical disconnection or removal of the OB does not prevent rostral migration (Jankovski et al. 1998; Kirschenbaum et al.

1999). Thus, although studies have identified molecules that can attract SVZ neuroblalsts, the OB is not essential for most of the directionality of SVZ-OB migration. The polarity of the ependymal layer appears to have an essential role in the orientation of migratory paths within the SVZ.

Upon reaching the OB, type-A cells depart from the tangentially oriented RMS and migrate radially to different layers of the OB. What are the molecular determinates of this change in mode of migration? Reelin, a protein critical for the laminar organization of cortex (Ogawa et al. 1995), is expressed in OB mitral cells and is necessary for the separation of type-A cells from the tangentially oriented chains and initiating radial migration (Hack et al. 2002). Tenascin-R (Saghatelyan et al. 2004) and Prokineticin-2 (Ng et al. 2005) also induce detachment of type-A cells from RMS chains; additionally, these extracellular molecules also appear to attract radially migrating neuroblasts to appropriate OB layers. Interestingly, Tenascin-R expression in the OB is reduced by odor deprivation, suggesting that neuron recruitment may be regulated by patterns of neuronal activity.

CONCLUSION

Significant progress has been made in our understanding of both the SVZ-OB niche and cell-intrinsic molecular program for stem cell maintenance, progenitor cell-fate specification, migration, and terminal differentiation. In addition to harboring proliferative neural lineage cells (type B, C, and A cells), the SVZ is also composed of other cell types (ependymal, microglial, and endothelial) that may serve as niche cells, producing factors important for long-term neurogenesis. Many soluble molecules and signaling pathways have been demonstrated to be critical for the different aspects of SVZ-OB biology, from NSC maintenance to directional migration. Nonsoluble, ECM components have also been demonstrated to have important roles, especially for migration. We are now beginning to appreciate that the adult SVZ not only generates distinct cellular lineages (oligodendroglial and neuronal), but also retains a remarkable capacity to produce a great diversity of OB interneuron phenotypes. The transcriptional factor combinatorial "code" specifying the different OB interneuron subtypes is in its earliest stages. Given that the adult human brain also harbors a population of NSCs in the SVZ, it is likely that certain biological principles discovered in the rodent SVZ-OB system will contribute to our understanding of adult human neural cell genesis, perhaps leading to NSC-based therapies.

In this chapter, we did not touch upon one of the most intriguing questions of SVZ-OB biology: What is the role of adult neurogenesis? Most newly born OB interneurons initially become synaptically integrated into the OB (Belluzzi et al. 2003; Carleton et al. 2003), but only about 50% of adult-born OB interneurons survive for more than 1 month (Petreanu and Alvarez-Buylla 2002). How are these interneurons selected for survival, and what do the surviving neurons do? What are the "reasons" for adult OB neurogenesis? Lledo et al. (2006) review and speculate about some of these questions. Our pursuit of understanding the molecular regulation of SVZ-OB neurogenesis may allow the engineering of precise methods of up-regulating or down-regulating the adult production of specific subpopulations of OB interneurons. Such abilities would be powerful tools for investigations to study the function of adult SVZ-OB neurogenesis. Thus, the SVZ-OB system is an appealing model in which to bridge a rapidly expanding cellular and molecular understanding of neurogenesis with emerging studies on the role of cellular plasticity in neural circuits.

ACKNOWLEDGMENTS

We thank Kenneth Xavier Probst for contributing artwork to Figures 1 and 2. D.A.L. is supported by an award from the AANS research foundation.

REFERENCES

Ahn S. and Joyner A.L. 2005. In vivo analysis of quiescent adult neural stem cells responding to Sonic hedgehog. *Nature* **437:** 894–897.

Alvarez-Buylla A. and Garcia-Verdugo J.M. 2002. Neurogenesis in adult subventricular zone. *J. Neurosci* **22:** 629–634.

Alvarez-Buylla A. and Lim D.A. 2004. For the long run: Maintaining germinal niches in the adult brain. *Neuron* **41:** 683–686.

Alvarez-Buylla A., Garcia-Verdugo J.M., and Tramontin A.D. 2001. A unified hypothesis on the lineage of neural stem cells. *Nat. Rev. Neurosci.* **2:** 287–293.

Alvarez-Buylla A., García-Verdugo J.M., Mateo A., and Merchant-Larios H. 1998. Primary neural precursors and intermitotic nuclear migration in the ventricular zone of adult canaries. *J. Neurosci.* **18:** 1020–1037.

Anderson D.J. 2001. Stem cells and pattern formation in the nervous system: The possible versus the actual. *Neuron* **30:** 19–35.

Astic L., Pellier-Monnin V., Saucier D., Charrier C., and Mehlen P. 2002. Expression of netrin-1 and netrin-1 receptor, DCC, in the rat olfactory nerve pathway during development and axonal regeneration. *Neuroscience* **109:** 643–656.

Bai C.B., Auerbach W., Lee J.S., Stephen D., and Joyner A.L. 2002. Gli2, but not Gli1, is required for initial Shh signaling and ectopic activation of the Shh pathway. *Development* **129:** 4753–4761.

Banasr M., Hery M., Printemps R., and Daszuta A. 2004. Serotonin-induced increases in adult cell proliferation and neurogenesis are mediated through different and common 5-HT receptor subtypes in the dentate gyrus and the subventricular zone. *Neuropsychopharmacology* **29:** 450–460.

Bauer S. and Patterson P.H. 2006. Leukemia inhibitory factor promotes neural stem cell self-renewal in the adult brain. *J. Neurosci* **26:** 12089–12099.

Becerra S.P., Sagasti A., Spinella P., and Notario V. 1995. Pigment epithelium-derived factor behaves like a noninhibitory serpin. Neurotrophic activity does not require the serpin reactive loop. *J. Biol. Chem.* **270:** 25992–25999.

Bedard A. and Parent A. 2004. Evidence of newly generated neurons in the human olfactory bulb. *Brain Res. Dev. Brain Res.* **151:** 159–168.

Belluzzi O., Benedusi M., Ackman J., and LoTurco J.J. 2003. Electrophysiological differentiation of new neurons in the olfactory bulb. *J. Neurosci.* **23:** 10411–10418.

Bolteus A.J. and Bordey A. 2004. GABA release and uptake regulate neuronal precursor migration in the postnatal subventricular zone. *J. Neurosci.* **24:** 7623–7631.

Bonaguidi M.A., McGuire T., Hu M., Kan L., Samanta J., and Kessler J.A. 2005. LIF and BMP signaling generate separate and discrete types of GFAP-expressing cells. *Development* **132:** 5503–5514.

Brezun J.M. and Daszuta A. 1999. Depletion in serotonin decreases neurogenesis in the dentate gyrus and the subventricular zone of adult rats. *Neuroscience* **89:** 999–1002.

Bruggeman S.W., Valk-Lingbeek M.E., van der Stoop P.P., Jacobs J.J., Kieboom K., Tanger E., Hulsman D., Leung C., Arsenijevic Y., Marino S., and van Lohuizen M. 2005. Ink4a and Arf differentially affect cell proliferation and neural stem cell self-renewal in Bmi1-deficient mice. *Genes Dev.* **19:** 1438–1443.

Bulfone A., Wang F., Hevner R., Anderson S., Cutforth T., Chen S., Meneses J., Pedersen R., Axel R., and Rubenstein J.L. 1998. An olfactory sensory map develops in the absence of normal projection neurons or GABAergic interneurons. *Neuron* **21:** 1273–1282.

Capela A. and Temple S. 2002. LeX/ssea-1 is expressed by adult mouse CNS stem cells, identifying them as nonependymal. *Neuron* **35:** 865–875.

Carleton A., Petreanu L.T., Lansford R., Alvarez-Buylla A., and Lledo P.M. 2003. Becoming a new neuron in the adult olfactory bulb. *Nat. Neurosci.* **6:** 507–518.

Chambers C.B., Peng Y., Nguyen H., Gaiano N., Fishell G., and Nye J.S. 2001. Spatiotemporal selectivity of response to Notch1 signals in mammalian forebrain precursors. *Development* **128:** 689–702.

Chazal G., Durbec P., Jankovski A., Rougon G., and Cremer H. 2000. Consequences of neural cell adhesion molecule deficiency on cell migration in the rostral migratory stream of the mouse. *J. Neurosci.* **20:** 1446–1457.

Chen H., Thiagalingam A., Chopra H., Borges M.W., Feder J.N., Nelkin B.D., Baylin S.B., and Ball D.W. 1997. Conservation of the *Drosophila* lateral inhibition pathway in human lung cancer: a hairy-related protein (HES-1) directly represses achaete-scute homolog-1 expression. *Proc. Natl. Acad. Sci.* **94:** 5355–5360.

Chiasson B.J., Tropepe V., Morshead C.M., and van der Kooy D. 1999. Adult mammalian forebrain ependymal and subependymal cells demonstrate proliferative potential, but only subependymal cells have neural stem cell characteristics. *J. Neurosci.* **19:** 4462–4471.

Chmielnicki E., Benraiss A., Economides A.N., and Goldman S.A. 2004. Adenovirally expressed noggin and brain-derived neurotrophic factor cooperate to induce new medium spiny neurons from resident progenitor cells in the adult striatal ventricular zone. *J. Neurosci.* **24:** 2133–2142.

Chojnacki A., Shimazaki T., Gregg C., Weinmaster G., and Weiss S. 2003. Glycoprotein 130 signaling regulates Notch1 expression and activation in the self-renewal of mammalian forebrain neural stem cells. *J. Neurosci.* **23:** 1730–1741.

Conover J.C., Doetsch F., Garcia-Verdugo J.M., Gale N.W., Yancopoulos G.D., and Alvarez-Buylla A. 2000. Disruption of Eph/ephrin signaling affects migration and proliferation in the adult subventricular zone. *Nat. Neurosci.* **3:** 1091–1097.

Coskun V. and Luskin M.B. 2002. Intrinsic and extrinsic regulation of the proliferation and differentiation of cells in the rodent rostral migratory stream. *J. Neurosci. Res.* **69:** 795–802.

Cremer H., Lange R., Christoph A., Plomann M., Vopper G., Roes J., Brown R., Baldwin S., Kraemer P., and Scheff S., et al. 1994. Inactivation of the N-CAM gene in mice results in size reduction of the olfactory bulb and deficits in spatial learning. *Nature* **367:** 455–459.

Curtis M.A., Kam M., Nannmark U., Anderson M.F., Axell M.Z., Wikkelso C., Holtas S., van Roon-Mom W.M., Bjork-Eriksson T., Nordborg C., Frisen J., Dragunow M., Faull R.L., and Eriksson P.S. 2007. Human neuroblasts migrate to the olfactory bulb via a lateral ventricular extension. *Science* **315:** 1243–1249.

Doetsch F. and Alvarez-Buylla A. 1996. Network of tangential pathways for neuronal migration in adult mammalian brain. *Proc. Natl. Acad. Sci.* **93:** 14895–14900.

Doetsch F., Garcia-Verdugo J.M., and Alvarez-Buylla A. 1997. Cellular composition and three-dimensional organization of the subventricular germinal zone in the adult mammalian brain. *J. Neurosci.* **17:** 5046–5061.

———. 1999a. Regeneration of a germinal layer in the adult mammalian brain. *Proc. Natl. Acad. Sci.* **96:** 11619–11624.

Doetsch, F., Caille I., Lim D.A., Garcia-Verdugo J.M., and Alvarez-Buylla A. 1999b. Subventricular zone astrocytes are neural stem cells in the adult mammalian brain. *Cell* **97:** 703–716.

Doetsch F., Petreanu L., Caille I., Garcia-Verdugo J.M., and Alvarez-Buylla A. 2002. EGF converts transit-amplifying neurogenic precursors in the adult brain into multipotent stem cells. *Neuron* **36:** 1021–1034.

Duncan A.W., Rattis F.M., DiMascio L.N., Congdon K.L., Pazianos G., Zhao C., Yoon K., Cook J.M., Willert K., Gaiano N., and Reya T. 2005. Integration of Notch and Wnt signaling in hematopoietic stem cell maintenance. *Nat. Immunol.* **6:** 314–322.

Emsley J.G., and Hagg T. 2003a. alpha6beta1 integrin directs migration of neuronal precursors in adult mouse forebrain. *Exp. Neurol.* **183:** 273–285.

———. 2003b. Endogenous and exogenous ciliary neurotrophic factor enhances forebrain neurogenesis in adult mice. *Exp. Neurol.* **183:** 298–310.

Ferri A.L., Cavallaro M., Braida D., Di Cristofano A., Canta A., Vezzani A., Ottolenghi S., Pandolfi P.P., Sala M., DeBiasi S., and Nicolis S.K. 2004. Sox2 deficiency causes neurodegeneration and impaired neurogenesis in the adult mouse brain. *Development* **131:** 3805–3819.

Francis F., Koulakoff A., Boucher D., Chafey P., Schaar B., Vinet M.C., Friocourt G., McDonnell N., Reiner O., Kahn A., McConnell S.K., Berwald-Netter Y., Denoulet P., and Chelly J. 1999. Doublecortin is a developmentally regulated, microtubule-associated protein expressed in migrating and differentiating neurons. *Neuron* **23:** 247–256.

Fuchs E., Tumbar T., and Guasch G. 2004. Socializing with the neighbors: Stem cells and their niche. *Cell* **116:** 769–778.

Gage F.H., Ray J., and Fisher L.J. 1995. Isolation, characterization, and use of stem cells from the CNS. *Annu. Rev. Neurosci.* **18:** 159–192.

Gaiano N., and Fishell G. 2002. The role of notch in promoting glial and neural stem cell fates. *Annu. Rev. Neurosci.* **25:** 471–490.

Gaiano N. Nye J.S., and Fishell G. 2000. Radial glial identity is promoted by Notch1 signaling in the murine forebrain. *Neuron* **26:** 395–404.

Galli R., Fiocco R., De Filippis L., Muzio L., Gritti A., Mercurio S., Broccoli V., Pellegrini M., Mallamaci A., and Vescovi A.L. 2002. Emx2 regulates the proliferation of stem cells of the adult mammalian central nervous system. *Development* **129:** 1633–1644.

Garcia A.D., Doan N.B., Imura T., Bush T.G., and Sofroniew M.V. 2004. GFAP-expressing progenitors are the principal source of constitutive neurogenesis in adult mouse forebrain. *Nat. Neurosci.* **7:** 1233–1241.

Givogri M.I., de Planell M., Galbiati F., Superchi D., Gritti A., Vescovi A., de Vellis J., and Bongarzone E.R. 2006. Notch signaling in astrocytes and neuroblasts of the adult subventricular zone in health and after cortical injury. *Dev. Neurosci.* **28:** 81–91.

Gleeson J.G., Lin P.T., Flanagan L.A., and Walsh C.A. 1999. Doublecortin is a microtubule-associated protein and is expressed widely by migrating neurons. *Neuron* **23:** 257–271.

Goldman S.A., Kirschenbaum B., Harrison-Restelli C., and Thaler H.T. 1997. Neuronal precursors of the adult rat subependymal zone persist into senescence, with no decline in spatial extent or response to BDNF. *J. Neurobiol.* **32:** 554–566.

Greenberg D.A. and Jin K. 2005. From angiogenesis to neuropathology. *Nature* **438:** 954–959.

Gross R.E., Mehler M.F., Mabie P.C., Zang Z.Y., Santschi L., and Kessler J.A. 1996. Bone morphogenetic proteins promote astroglial lineage commitment by mammalian subventricular zone progenitor cells. *Neuron* **17:** 595–606.

Hack I., Bancila M., Loulier K., Carroll P., and Cremer H. 2002. Reelin is a detachment signal in tangential chain-migration during postnatal neurogenesis. *Nat. Neurosci.* **5:** 939–945.

Hack M.A., Saghatelyan A., de Chevigny A., Pfeifer A., Ashery-Padan R., Lledo P.M., and Gotz M. 2005. Neuronal fate determinants of adult olfactory bulb neurogenesis. *Nat. Neurosci.* **8:** 865–872.

Hatakeyama J., Bessho Y., Katoh K., Ookawara S., Fujioka M., Guillemot F., and Kageyama R. 2004. Hes genes regulate size, shape and histogenesis of the nervous system by control of the timing of neural stem cell differentiation. *Development* **131:** 5539–5550.

Herrera D.G., Garcia-Verdugo J.M., and Alvarez-Buylla A. 1999. Adult-derived neural precursors transplanted into multiple regions in the adult brain. *Ann. Neurol.* **46:** 867–877.

Hoglinger G.U., Rizk P., Muriel M.P., Duyckaerts C., Oertel W.H., Caille I., and Hirsch E.C. 2004. Dopamine depletion impairs precursor cell proliferation in Parkinson disease. *Nat. Neurosci.* **7:** 726–735.

Holmberg J., Armulik A., Senti K.A., Edoff K., Spalding K., Momma S., Cassidy R., Flanagan J.G., and Frisen J. 2005. Ephrin-A2 reverse signaling negatively regulates neural progenitor proliferation and neurogenesis. *Genes Dev.* **19:** 462–471.

Hu H. 1999. Chemorepulsion of neuronal migration by Slit2 in the developing mammalian forebrain. *Neuron* **23:** 703–711.

Hu H., Tomasiewicz H., Magnuson T., and Rutishauser U. 1996. The role of polysialic acid in migration of olfactory bulb interneuron precursors in the subventricular zone. *Neuron* **16:** 735–743.

Ille F. and Sommer L. 2005. Wnt signaling: Multiple functions in neural development. *Cell Mol. Life Sci.* **62:** 1100–1108.

Imura T., Kornblum H.I., and Sofroniew M.V. 2003. The predominant neural stem cell isolated from postnatal and adult forebrain but not early embryonic forebrain expresses GFAP. *J. Neurosci.* **23:** 2824–2832.

Ishibashi M., Moriyoshi K., Sasai Y., Shiota K., Nakanishi S., and Kageyama R. 1994. Persistent expression of helix-loop-helix factor HES-1 prevents mammalian neural differentiation in the central nervous system. *EMBO J.* **13:** 1799–1805.

Jackson E.L., Garcia-Verdugo J.M., Gil-Perotin S., Roy M., Quinones-Hinojosa A., VandenBerg S., and Alvarez-Buylla A. 2006. PDGFR alpha-positive B cells are neural stem cells in the adult SVZ that form glioma-like growths in response to increased PDGF signaling. *Neuron* **51:** 187–199.

Jacques T.S., Relvas J.B., Nishimura S., Pytela R., Edwards G.M., Streuli C.H., and ffrench-Constant C. 1998. Neural precursor cell chain migration and division are regulated through different beta1 integrins. *Development* **125:** 3167–3177.

Jankovski A., Garcia C., Soriano E., and Sotelo C. 1998. Proliferation, migration and differentiation of neuronal progenitor cells in the adult mouse subventricular zone surgically separated from its olfactory bulb. *Eur. J. Neurosci.* **10:** 3853–3868.

Jin K., Zhu Y., Sun Y., Mao X.O., Xie L., and Greenberg D.A. 2002a. Vascular endothelial growth factor (VEGF) stimulates neurogenesis in vitro and in vivo. *Proc. Natl. Acad. Sci.* **99:** 11946–11950.

Jin K., Xie L., Childs J., Sun Y., Mao X.O., Logvinova A., and Greenberg D.A. 2003. Cerebral neurogenesis is induced by intranasal administration of growth factors. *Ann. Neurol.* **53:** 405–409.

Jin K., Mao X.O., Sun Y., Xie L., Jin L., Nishi E., Klagsbrun M., and Greenberg D.A. 2002a,b. Heparin-binding epidermal growth factor-like growth factor: Hypoxia-inducible expression in vitro and stimulation of neurogenesis in vitro and in vivo. *J. Neurosci.* **22:** 5365–5373.

Johansson C.B., Momma S., Clarke D.L., Risling M., Lendahl U., and Frisen J. 1999. Identification of a neural stem cell in the adult mammalian central nervous system. *Cell* **96:** 25–34.

Johnson J.E. 2003. Numb and Numblike control cell number during vertebrate neurogenesis. *Trends Neurosci.* **26:** 395–396.

Kageyama R., Ohtsuka T., Hatakeyama J., and Ohsawa R. 2005. Roles of bHLH genes in neural stem cell differentiation. *Exp. Cell Res.* **306:** 343–348.

Katoh M. 2006. Notch ligand, JAG1, is evolutionarily conserved target of canonical WNT signaling pathway in progenitor cells. *Int. J. Mol. Med.* **17:** 681–685.

Kippin T.E., Kapur S., and van der Kooy D. 2005. Dopamine specifically inhibits forebrain neural stem cell proliferation, suggesting a novel effect of antipsychotic drugs. *J. Neurosci.* **25:** 5815–5823.

Kirschenbaum B., Doetsch F., Lois C., and Alvarez-Buylla A. 1999. Adult subventricular zone neuronal precursors continue to proliferate and migrate in the absence of the olfactory bulb. *J. Neurosci.* **19:** 2171–2180.

Kohwi M., Osumi N., Rubenstein J.L., and Alvarez-Buylla A. 2005. Pax6 is required for making specific subpopulations of granule and periglomerular neurons in the olfactory bulb. *J. Neurosci.* **25:** 6997–7003.

Kohwi M., Petryniak M.A., Long J.E., Ekker M., Obata K., Yanagawa Y., Rubenstein J.L., and Alvarez-Buylla A. 2007. A subpopulation of olfactory bulb GABAergic interneurons is derived from Emx1- and Dlx5/6-expressing progenitors. *J. Neurosci.* **27:** 6878–6891.

Kosaka K., Toida K., Aika Y., and Kosaka T. 1998. How simple is the organization of the olfactory glomerulus?: The heterogeneity of so-called periglomerular cells. *Neurosci. Res.* **30:** 101–110.

Kosaka K., Aika Y., Toida K., Heizmann C.W., Hunziker W., Jacobowitz D.M., Nagatsu I., Streit P., Visser T.J., and Kosaka T. 1995. Chemically defined neuron groups and their subpopulations in the glomerular layer of the rat main olfactory bulb. *Neurosci. Res.* **23:** 73–88.

Kosaka T. and Kosaka K. 2005. Structural organization of the glomerulus in the main olfactory bulb. *Chem. Senses* (suppl. 1) **30:** i107–i108.

Kuhn H.G., Dickinson-Anson H., and Gage F.H. 1996. Neurogenesis in the dentate gyrus of the adult rat: Age-related decrease of neuronal progenitor proliferation. *J. Neurosci.* **16:** 2027–2033.

Kuo C.T., Mirzadeh Z., Soriano-Navarro M., Rasin M., Wang D., Shen J., Sestan N., Garcia-Verdugo J., Alvarez-Buylla A., Jan L.Y., and Jan Y.N. 2006. Postnatal deletion of Numb/Numblike reveals repair and remodeling capacity in the subventricular neurogenic niche. *Cell* **127:** 1253–1264.

Laywell E.D., Rakic P., Kukekov V.G., Holland E.C., and Steindler D.A. 2000. Identification of a multipotent astrocytic stem cell in the immature and adult mouse brain. *Proc. Natl. Acad. Sci.* **97:** 13883–13888.

Leventhal C., Rafii S., Rafii D., Shahar A., and Goldman S.A. 1999. Endothelial trophic support of neuronal production and recruitment from the adult mammalian subependyma. *Mol. Cell. Neurosci.* **13:** 450–464.

Li H.S., Chen J.H., Wu W., Fagaly T., Zhou L., Yuan W., Dupuis S., Jiang Z.H., Nash W., Gick C., Ornitz D.M., Wu J.Y., and Rao Y. 1999. Vertebrate slit, a secreted ligand for the transmembrane protein roundabout, is a repellent for olfactory bulb axons. *Cell* **96:** 807–818.

Li W., Cogswell C.A., and LoTurco J.J. 1998. Neuronal differentiation of precursors in the neocortical ventricular zone is triggered by BMP. *J. Neurosci.* **18:** 8853–8862.

Lie D.C., Song H., Colamarino S.A., Ming G.L., and Gage F.H. 2004. Neurogenesis in the adult brain: New strategies for central nervous system diseases. *Annu. Rev. Pharmacol. Toxicol.* **44:** 399–421.

Lim D.A. and Alvarez-Buylla A. 1999. Interaction between astrocytes and adult subventricular zone precursors stimulates neurogenesis. *Proc. Natl. Acad. Sci.* **96:** 7526–7531.

Lim D.A., Huang Y.C., and Alvarez-Buylla A. 2007. The adult neural stem cell niche: Lessons for future neural cell replacement strategies. *Neurosurg. Clin. N. Am.* **18:** 81–92.

Lim D.A., Tramontin A.D., Trevejo J.M., Herrera D.G., Garcia-Verdugo J.M., and Alvarez-Buylla A. 2000. Noggin antagonizes BMP signaling to create a niche for adult neurogenesis. *Neuron* **28:** 713–726.

Lim D.A., Suarez-Farinas M., Naef F., Hacker C.R., Menn B., Takebayashi H., Magnasco M., Patil N., and Alvarez-Buylla A. 2006. In vivo transcriptional profile analysis reveals RNA splicing and chromatin remodeling as prominent processes for adult neurogenesis. *Mol. Cell. Neurosci.* **31:** 131–148.

Liu G. and Rao Y. 2003. Neuronal migration from the forebrain to the olfactory bulb requires a new attractant persistent in the olfactory bulb. *J. Neurosci.* **23:** 6651–6659.

Liu X., Wang Q., Haydar T.F., and Bordey A. 2005. Nonsynaptic GABA signaling in postnatal subventricular zone controls proliferation of GFAP-expressing progenitors. *Nat. Neurosci.* **8:** 1179–1187.

Lledo P.M., Alonso M., and Grubb M.S. 2006. Adult neurogenesis and functional plasticity in neuronal circuits. *Nat. Rev. Neurosci.* **7:** 179–193.

Lois C. and Alvarez-Buylla A. 1994a. Long-distance neuronal migration in the adult mammalian brain. *Science* **264:** 1145–1148.

Lois C., Garcia-Verdugo J.M., and Alvarez-Buylla A. 1996. Chain migration of neuronal precursors. *Science* **271:** 978–981.

Louissaint A., Jr., Rao S., Leventhal C., and Goldman S.A. 2002. Coordinated interaction of neurogenesis and angiogenesis in the adult songbird brain. *Neuron* **34:** 945–960.

Mabie P.C., Mehler M.F., and Kessler J. A. 1999. Multiple roles of bone morphogenetic protein signaling in the regulation of cortical cell number and phenotype. *J. Neurosci.* **19:** 7077–7088.

Machold R., Hayashi S., Rutlin M., Muzumdar M.D., Nery S., Corbin J.G., Gritli-Linde A., Dellovade T., Porter J.A., Rubin L.L., Dudek H., McMahon A.P., and Fishell G. 2003. Sonic hedgehog is required for progenitor cell maintenance in telencephalic stem cell niches. *Neuron* **39:** 937–950.

Mason H.A., Ito S., and Corfas G. 2001. Extracellular signals that regulate the tangential migration of olfactory bulb neuronal precursors: Inducers, inhibitors, and repellents. *J. Neurosci.* **21:** 7654–7663.

McKay R. 1997. Stem cells in the central nervous system. *Science* **276:** 66–71.

Menn B., Garcia-Verdugo J.M., Yaschine C., Gonzalez-Perez O., Rowitch D., and Alvarez-Buylla A. 2006. Origin of oligodendrocytes in the subventricular zone of the adult brain. *J. Neurosci.* **26:** 7907–7918.

Mercier F., Kitasako J.T., and Hatton G.I. 2002. Anatomy of the brain neurogenic zones revisited: Fractones and the fibroblast/macrophage network. *J. Comp. Neurol.* **451:** 170–188.

Merkle F.T. and Alvarez-Buylla A. 2006. Neural stem cells in mammalian development. *Curr. Opin. Cell. Biol.* **18:** 704–709.

Merkle F.T., Tramontin A.D., Garcia-Verdugo J.M., and Alvarez-Buylla A. 2004. Radial glia give rise to adult neural stem cells in the subventricular zone. *Proc. Natl. Acad. Sci.* **101:** 17528–17532.

Miyakoshi L.M., Mendez-Otero R., and Hedin-Pereira C. 2001. The 9-O-acetyl GD3 gangliosides are expressed by migrating chains of subventricular zone neurons in vitro. *Braz. J. Med. Biol. Res.* **34:** 669–673.

Mizuguchi R., Sugimori M., Takebayashi H., Kosako H., Nagao M., Yoshida S., Nabeshima Y., Shimamura K., and Nakafuku M. 2001. Combinatorial roles of olig2 and neurogenin2 in the coordinated induction of pan-neuronal and subtype-specific properties of motoneurons. *Neuron* **31:** 757–771.

Molofsky A.V., He S., Bydon M., Morrison S.J., and Pardal R. 2005. Bmi-1 promotes neural stem cell self-renewal and neural development but not mouse growth and survival by repressing the p16Ink4a and p19Arf senescence pathways. *Genes Dev.* **19:** 1432–1437.

Molofsky A.V., Pardal R., Iwashita T., Park I.K., Clarke M.F., and Morrison S.J. 2003. Bmi-1 dependence distinguishes neural stem cell self-renewal from progenitor proliferation. *Nature* **425:** 962–967.

Molofsky A.V., Slutsky S.G., Joseph N.M., He S., Pardal R., Krishnamurthy J., Sharpless N.E., and Morrison S.J. 2006. Increasing p16INK4a expression decreases forebrain progenitors and neurogenesis during ageing. *Nature* **443**: 448–452.

Moreno-Lopez B., Romero-Grimaldi C., Noval J.A., Murillo-Carretero M., Matarredona E.R., and Estrada C. 2004. Nitric oxide is a physiological inhibitor of neurogenesis in the adult mouse subventricular zone and olfactory bulb. *J. Neurosci.* **24**: 85–95.

Morita M., Kozuka N., Itofusa R., Yukawa M., and Kudo Y. 2005. Autocrine activation of EGF receptor promotes oscillation of glutamate-induced calcium increase in astrocytes cultured in rat cerebral cortex. *J. Neurochem.* **95**: 871–879.

Murase S. and Horwitz A.F. 2002. Deleted in colorectal carcinoma and differentially expressed integrins mediate the directional migration of neural precursors in the rostral migratory stream. *J. Neurosci.* **22**: 3568–3579.

Nacher J., Rosell D.R., and McEwen B.S. 2000. Widespread expression of rat collapsin response-mediated protein 4 in the telencephalon and other areas of the adult rat central nervous system. *J. Comp. Neurol.* **424**: 628–639.

Nait-Oumesmar B., Decker L., Lachapelle F., Avellana-Adalid V., Bachelin C., and Van Evercooren A.B. 1999. Progenitor cells of the adult mouse subventricular zone proliferate, migrate and differentiate into oligodendrocytes after demyelination. *Eur. J. Neurosci.* **11**: 4357–4366.

Nakanishi M., Niidome T., Matsuda S., Akaike A., Kihara T., and Sugimoto H. 2007. Microglia-derived interleukin-6 and leukaemia inhibitory factor promote astrocytic differentiation of neural stem/progenitor cells. *Eur. J. Neurosci.* **25**: 649–658.

Ng K.L., Li J.D., Cheng M.Y., Leslie F.M., Lee A.G., and Zhou Q.Y. 2005. Dependence of olfactory bulb neurogenesis on prokineticin 2 signaling. *Science* **308**: 1923–1927.

Nguyen Ba-Charvet K.T., Brose K., Marillat V., Kidd T., Goodman C.S., Tessier-Lavigne M., Sotelo C., and Chedotal A. 1999. Slit2-Mediated chemorepulsion and collapse of developing forebrain axons. *Neuron* **22**: 463–473.

Novitch B.G., Chen A.I., and Jessell T.M. 2001. Coordinate regulation of motor neuron subtype identity and pan-neuronal properties by the bHLH repressor Olig2. *Neuron* **31**: 773–789.

Nyfeler Y., Kirch R.D., Mantei N., Leone D.P., Radtke F., Suter U., and Taylor V. 2005. Jagged1 signals in the postnatal subventricular zone are required for neural stem cell self-renewal. *EMBO J.* **24**: 3504–3515.

Ogawa M., Miyata T., Nakajima K., Yagyu K., Seike M., Ikenaka K., Yamamoto H., and Mikoshiba K. 1995. The reeler gene-associated antigen on cajal-retzius neurons is a crucial molecule for laminar organization of cortical neurons. *Neuron* **14**: 899–912.

Ohtsuka T., Sakamoto M., Guillemot F., and Kageyama R. 2001. Roles of the basic helix-loop-helix genes Hes1 and Hes5 in expansion of neural stem cells of the developing brain. *J. Biol. Chem.* **276**: 30467–30474.

Ohtsuka T., Ishibashi M., Gradwohl G., Nakanishi S., Guillemot F., and Kageyama R. 1999. Hes1 and Hes5 as notch effectors in mammalian neuronal differentiation. *EMBO J.* **18**: 2196–2207.

Ono K., Tomasiewicz H., Magnuson T., and Rutishauser U. 1994. N-CAM mutation inhibits tangential neuronal migration and is phenocopied by enzymatic removal of polysialic acid. *Neuron* **13**: 595–609.

Packer M.A., Stasiv Y., Benraiss A., Chmielnicki E., Grinberg A., Westphal H., Goldman S.A., and Enikolopov G. 2003. Nitric oxide negatively regulates mammalian adult neurogenesis. *Proc. Natl. Acad. Sci.* **100**: 9566–9571.

Palma V., Lim D.A., Dahmane N., Sanchez P., Brionne T.C., Herzberg C.D., Gitton Y., Carleton A., Alvarez-Buylla A., and Ruiz i Altaba A. 2005. Sonic hedgehog controls stem cell behavior in the postnatal and adult brain. *Development* **132:** 335–344.

Palmer T.D., Willhoite A.R., and Gage F.H. 2000. Vascular niche for adult hippocampal neurogenesis. *J. Comp. Neurol.* **425:** 479–494.

Panchision D.M., Pickel J.M., Studer L., Lee S.H., Turner P.A., Hazel T.G., and McKay R.D.G. 2001. Sequential actions of BMP receptors control neural precursor cell production and fate. *Genes Dev.* **15:** 2094–2110.

Parent J.M., von dem Bussche N., and Lowenstein D.H. 2006. Prolonged seizures recruit caudal subventricular zone glial progenitors into the injured hippocampus. *Hippocampus* **16:** 321–328.

Parras C.M., Galli R., Britz O., Soares S., Galichet C., Battiste J., Johnson J.E., Nakafuku M., Vescovi A., and Guillemot F. 2004. Mash1 specifies neurons and oligodendrocytes in the postnatal brain. *EMBO J.* **23:** 4495–4505.

Parrish-Aungst S., Shipley M.T., Erdelyi F., Szabo G., and Puche A.C. 2007. Quantitative analysis of neuronal diversity in the mouse olfactory bulb. *J. Comp. Neurol.* **501:** 825–836.

Pencea V., Bingaman K.D., Wiegand S.J., and Luskin M.B. 2001. Infusion of brain-derived neurotrophic factor into the lateral ventricle of the adult rat leads to new neurons in the parenchyma of the striatum, septum, thalamus, and hypothalamus. *J. Neurosci.* **21:** 6706–6717.

Peretto P., Cummings D., Modena C., Behrens M., Venkatraman G., Fasolo A., and Margolis F.L. 2002. BMP mRNA and protein expression in the developing mouse olfactory system. *J. Comp. Neurol.* **451:** 267–278.

Peretto P., Dati C., De Marchis S., Kim H.H., Ukhanova M., Fasolo A., and Margolis F.L. 2004. Expression of the secreted factors noggin and bone morphogenetic proteins in the subependymal layer and olfactory bulb of the adult mouse brain. *Neuroscience* **128:** 685–696.

Petreanu L., and Alvarez-Buylla A. 2002. Maturation and death of adult-born olfactory bulb granule neurons: Role of olfaction. *J. Neurosci.* **22:** 6106–6113.

Picard-Riera N., Decker L., Delarasse C., Goude K., Nait-Oumesmar B., Liblau R., Pham-Dinh D., and Evercooren A.B. 2002. Experimental autoimmune encephalomyelitis mobilizes neural progenitors from the subventricular zone to undergo oligodendrogenesis in adult mice. *Proc. Natl. Acad. Sci.* **99:** 13211–13216.

Probstmeier R. and Pesheva P. 1999. Tenascin-C inhibits beta1 integrin-dependent cell adhesion and neurite outgrowth on fibronectin by a disialoganglioside-mediated signaling mechanism. *Glycobiology* **9:** 101–114.

Raff M. 2003. Adult stem cell plasticity: Fact or artifact? *Annu. Rev. Cell Dev. Biol.* **19:** 1–22.

Ramirez-Castillejo C., Sanchez-Sanchez F., Andreu-Agullo C., Ferron S.R., Aroca-Aguilar J.D., Sanchez P., Mira H., Escribano J., and Farinas I. 2006. Pigment epithelium-derived factor is a niche signal for neural stem cell renewal. *Nat. Neurosci.* **9:** 331–339.

Roegiers F. and Jan Y.N. 2004. Asymmetric cell division. *Curr. Opin. Cell Biol.* **16:** 195–205.

Roy N.S., Wang S., Harrison-Restelli C., Benraiss A., Fraser R.A., Gravel M., Braun P.E., and Goldman S.A. 1999. Identification, isolation, and promoter-defined separation of mitotic oligodendrocyte progenitor cells from the adult human subcortical white matter. *J. Neurosci.* **19:** 9986–9995.

Ruiz i Altaba A., Nguyen V., and Palma V. 2003. The emergent design of the neural tube: Prepattern, SHH morphogen and GLI code. *Curr. Opin. Genet. Dev.* **13:** 513–521.

Saghatelyan A., de Chevigny A., Schachner M., and Lledo P.M. 2004. Tenascin-R mediates activity-dependent recruitment of neuroblasts in the adult mouse forebrain. *Nat. Neurosci.* **7:** 347–356.

Sanai N., Tramontin A.D., Quinones-Hinojosa A., Barbaro N.M., Gupta N., Kunwar S., Lawton M.T., McDermott M.W., Parsa A.T., Manuel-Garcia Verdugo J., Berger M.S., and Alvarez-Buylla A. 2004. Unique astrocyte ribbon in adult human brain contains neural stem cells but lacks chain migration. *Nature* **427:** 740–744.

Sasai Y., Kageyama R., Tagawa Y., Shigemoto R., and Nakanishi S. 1992. Two mammalian helix-loop-helix factors structurally related to *Drosophila* hairy and Enhancer of split. *Genes Dev.* **6:** 2620–2634.

Sawamoto K., Wichterle H., Gonzalez-Perez O., Cholfin J.A., Yamada M., Spassky N., Murcia N.S., Garcia-Verdugo J.M., Marin O., Rubenstein J.L., Tessier-Lavigne M., Okano H., and Alvarez-Buylla A. 2006. New neurons follow the flow of cerebrospinal fluid in the adult brain. *Science* **311:** 629–632.

Schaar B.T. and McConnell S.K. 2005. Cytoskeletal coordination during neuronal migration. *Proc. Natl. Acad. Sci.* **102:** 13652–13657.

Scheffler B., Walton N.M., Lin D.D., Goetz A.K., Enikolopov G., Roper S.N., and Steindler D.A. 2005. Phenotypic and functional characterization of adult brain neuropoiesis. *Proc. Natl. Acad. Sci.* **102:** 9353–9358.

Schofield R. 1978. The relationship between the spleen colony-forming cell and the haemopoietic stem cell. *Blood Cells* **4:** 7–25.

Seidenfaden R., Desoeuvre A., Bosio A., Virard I., and Cremer H. 2006. Glial conversion of SVZ-derived committed neuronal precursors after ectopic grafting into the adult brain. *Mol. Cell. Neurosci.* **32:** 187–198.

Seki T., Namba T., Mochizuki H., and Onodera M. 2007. Clustering, migration, and neurite formation of neural precursor cells in the adult rat hippocampus. *J. Comp. Neurol.* **502:** 275–290.

Seri B., Garcia-Verdugo J.M., Collado-Morente L., McEwen B.S., and Alvarez-Buylla A. 2004. Cell types, lineage, and architecture of the germinal zone in the adult dentate gyrus. *J. Comp. Neurol.* **478:** 359–378.

Shen Q., Goderie S.K., Jin L., Karanth N., Sun Y., Abramova N., Vincent P., Pumiglia K., and Temple S. 2004. Endothelial cells stimulate self-renewal and expand neurogenesis of neural stem cells. *Science* **304:** 1338–1340.

Shi Y., Chichung Lie D., Taupin P., Nakashima K., Ray J., Yu R.T., Gage F.H., and Evans R.M. 2004. Expression and function of orphan nuclear receptor TLX in adult neural stem cells. *Nature* **427:** 78–83.

Shimogori T., VanSant J., Paik E., and Grove E.A. 2004. Members of the Wnt, Fz, and Frp gene families expressed in postnatal mouse cerebral cortex. *J. Comp. Neurol.* **473:** 496–510.

Song H., Stevens C.F., and Gage F.H. 2002. Astroglia induce neurogenesis from adult neural stem cells. *Nature* **417:** 39–44.

Soria J.M., Taglialatela P., Gil-Perotin S., Galli R., Gritti A., Verdugo J.M., and Bertuzzi S. 2004. Defective postnatal neurogenesis and disorganization of the rostral migratory stream in absence of the Vax1 homeobox gene. *J. Neurosci.* **24:** 11171–11181.

Sotelo J.R. and Trujillo-Cenóz O. 1958. Electron microscope study on the development of ciliary components of the neural epithelium of the chick embryo. *Z. Zellforsch.* **49:** 1–12.

Spassky N., Merkle F.T., Flames N., Tramontin A.D., Garcia-Verdugo J.M., and Alvarez-Buylla A. 2005. Adult ependymal cells are postmitotic and are derived from radial glial cells during embryogenesis. *J. Neurosci.* **25:** 10–18.

Stenman J., Toresson H., and Campbell K. 2003. Identification of two distinct progenitor populations in the lateral ganglionic eminence: Implications for striatal and olfactory bulb neurogenesis. *J. Neurosci.* **23:** 167–174.

Stensaas L.J. and Stensass S.S. 1968. Light microscopy of glial cells in turtles and birds. *Z. Zellforsch. Mikrosk. Anat.* **91:** 315–340.

Stump G., Durrer A., Klein A.L., Lutolf S., Suter U., and Taylor V. 2002. Notch1 and its ligands Delta-like and Jagged are expressed and active in distinct cell populations in the postnatal mouse brain. *Mech. Dev.* **114:** 153–159.

Suhonen J.O., Peterson D.A., Ray J., and Gage F.H. 1996. Differentiation of adult hippocampus-derived progenitors into olfactory neurons in vivo. *Nature* **383:** 624–627.

Sun Y., Nadal-Vicens M., Misono S., Lin M.Z., Zubiaga A., Hua X., Fan G., and Greenberg M.E. 2001. Neurogenin promotes neurogenesis and inhibits glial differentiation by independent mechanisms. *Cell* **104:** 365–376.

Thomas L.B., Gates M.A., and Steindler D.A. 1996. Young neurons from the adult subependymal zone proliferate and migrate along an astrocyte, extracellular matrix-rich pathway. *Glia* **17:** 1–14.

Tomasiewicz H., Ono K., Yee D., Thompson C., Goridis C., Rutishauser U., and Magnuson T. 1993. Genetic deletion of a neural cell adhesion molecule variant (N-CAM-180) produces distinct defects in the central nervous system. *Neuron* **11:** 1163–1174.

Tropepe V., Craig C.G., Morshead C.M., and van der Kooy D. 1997. Transforming growth factor-alpha null and senescent mice show decreased neural progenitor cell proliferation in the forebrain subependyma. *J. Neurosci.* **17:** 7850–7859.

Waclaw R.R., Allen Z.J., II, Bell S.M., Erdelyi F., Szabo G., Potter S.S., and Campbell K. 2006. The zinc finger transcription factor Sp8 regulates the generation and diversity of olfactory bulb interneurons. *Neuron* **49:** 503–516.

Walton N.M., Sutter B.M., Laywell E.D., Levkoff L.H., Kearns S.M., Marshall G.P., II, Scheffler B., and Steindler D.A. 2006. Microglia instruct subventricular zone neurogenesis. *Glia* **54:** 815–825.

Watt F.M. and Hogan B.L. 2000. Out of Eden: Stem cells and their niches. *Science* **287:** 1427–1430.

Weiss S., Reynolds B.A., Vescovi A.L., Morshead C., Craig C.G., and Van der Kooy D. 1996. Is there a neural stem cell in the mammalian forebrain? *Trends Neurosci.* **19:** 387–393.

Wichterle H., Garcia-Verdugo J.M., and Alvarez-Buylla A. 1997. Direct evidence for homotypic, glia-independent neuronal migration. *Neuron* **18:** 779–791.

Wu W., Wong K., Chen J., Jiang Z., Dupuis S., Wu J.Y., and Rao Y. 1999. Directional guidance of neuronal migration in the olfactory system by the protein Slit. *Nature* **400:** 331–336.

Ying Q.L., Nichols J., Chambers I., and Smith A. 2003. BMP induction of Id proteins suppresses differentiation and sustains embryonic stem cell self-renewal in collaboration with STAT3. *Cell* **115:** 281–292.

Yokosaki Y., Monis H., Chen J., and Sheppard D. 1996. Differential effects of the integrins alpha9beta1, alphavbeta3, and alphavbeta6 on cell proliferative responses to tenascin.

Roles of the beta subunit extracellular and cytoplasmic domains. *J. Biol. Chem.* **271:** 24144–24150.

Yu J.M., Kim J.H., Song G.S., and Jung J.S. 2006. Increase in proliferation and differentiation of neural progenitor cells isolated from postnatal and adult mice brain by Wnt-3a and Wnt-5a. *Mol. Cell. Biochem.* **288:** 17–28.

Zheng W., Nowakowski R.S., and Vaccarino F.M. 2004. Fibroblast growth factor 2 is required for maintaining the neural stem cell pool in the mouse brain subventricular zone. *Dev. Neurosci.* **26:** 181–196.

Zigova T., Pencea V., Wiegand S.J., and Luskin M.B. 1998. Intraventricular administration of BDNF increases the number of newly generated neurons in the adult olfactory bulb. *Mol. Cell. Neurosci.* **11:** 234–245.

11

Neurogenic Niches in the Adult Mammalian Brain

Dengke K. Ma,[1,2] Guo-li Ming,[1,2,3] Fred H. Gage,[4]
and Hongjun Song[1,2,3]
[1]Institute for Cell Engineering
[2]The Solomon H. Snyder Department of Neuroscience
[3]Department of Neurology
Johns Hopkins University School of Medicine
Baltimore, Maryland 21205
[4]Laboratory of Genetics
Salk Institute for Biological Studies
La Jolla, California 92037

THE MAMMALIAN BRAIN IS A COMPLEX ORGAN composed of trillions of neurons connected with each other in a highly stereotyped yet modifiable manner. Most neurons are born during embryonic development and persist throughout life in the adult brain circuit, in contrast to many other adult tissues, including most from epithelial origins that usually harbor stem cells to maintain homeostatic cellular turnover (Weissman et al. 2001; Li and Xie 2005). The relative stability of neural circuits at the cellular level, especially in higher processing centers of the brain such as the cerebral cortex, was thought to be essential to maintain the ongoing information processing, and any loss or addition to the circuitry component could undermine the cognitive process as a whole (Rakic 1985). Therefore, the discovery of adult neurogenesis—that new neurons are indeed generated in specific regions of adult brains and undergo developmental maturation to become functionally integrated into local neural circuits (Fig. 1a)—came as a surprise (Altman and Das 1965; van Praag et al. 2002). During adult neurogenesis, neural stem cells (NSCs) generate functional neurons through coordinated steps, including cell-fate specification, migration, axonal and dendritic growth, and finally synaptic integration

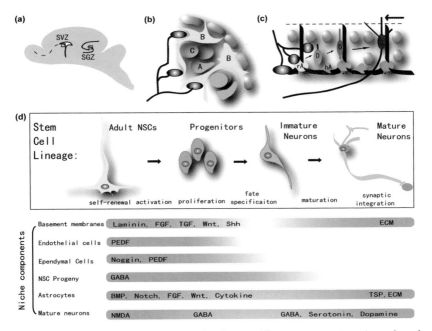

Figure 1. (*a*) A schematic adult mammalian brain with two neurogenic regions: the subventricular zone (SVZ) of the lateral ventricle and the subgranular zone (SGZ) of the
hippocampal dentate gyrus (DG). Newly generated neurons are born locally in the SGZ
to differentiate into dentate granule cells, whereas newborn neurons in the SVZ migrate
along the rostral migratory stream (RMS) (*dashed line*) over long distances to olfactory
bulb (OB). (*b*) Cellular organization of the neurogenic niche in the SVZ. The primary
progenitor type-B cells give rise to transient-amplifying type-C cells and migratory type-
A neuroblasts. Ependymal cells (E cells, *light blue*) line the ventricle and endothelial cells
(ECs, *black white*) locate in proximity to B cells. (*c*) Cellular organization of the neurogenic niche in the SGZ. The primary progenitor radial glia-like astrocytes (rAs) give rise
to transient amplifying type-D cells and new granule neurons. Some horizontal astrocytes line up in the hilus region, and ECs locate in proximity to rAs. (*d*) Distinct roles
of niche cells at different stages of adult neurogenesis from adult neural stem cells
(NSCs). Along the stem cell lineage, adult neurogenesis entails NSC self-renewal, activation, proliferation, fate specification, migration, maturation, and synaptic integration.
Distinct roles of niche components including the extracellular matrix (ECM), ECs, E
cells, astrocytes, and stem cell progeny are shown during different developmental stages
of adult neurogenesis. Known and potential niche molecules are indicated in each bar
with a gradient for varying niche influence.

into the adult brain (Fig. 1d). Since the pioneering studies of Altman in
the early 1960s (Altman 1962), the process of adult neurogenesis has been
unambiguously established in all mammals examined, including humans
(Eriksson et al. 1998; Gage 2000; Lie et al. 2004; Abrous et al. 2005; Ming

and Song 2005; Lledo et al. 2006; Merkle and Alvarez-Buylla 2006). Intriguingly, regions harboring active adult neurogenesis are not ubiquitously present but are instead highly localized to two discrete areas: the subventricular zone (SVZ) of the lateral ventricle and the subgranular zone (SGZ) of the hippocampal dentate gyrus (DG) (Fig. 1a). Although the functional significance of these newborn neurons remains mysterious and is being actively pursued, it has become increasingly clear that neurogenesis in these two neurogenic regions of the adult brain is tightly regulated by their highly specialized microenvironments, termed "neurogenic niches" (Palmer et al. 2000; Alvarez-Buylla and Lim 2004; Ma et al. 2005; Merkle and Alvarez-Buylla 2006).

The concept that somatic stem cells usually reside within specific niches was first suggested from transplantation studies of hematopoietic progenitors in the 1970s (Schofield 1978). Niches are defined by their ability to anatomically house stem cells and functionally control their development in vivo. During the past decade, significant progress has been made in describing the niches at both the cellular and functional levels in several model systems, such as *Drosophila* germ line and mammalian skin, intestine, and bone marrow (Spradling et al. 2001; Li and Xie 2005; Scadden 2006). In the adult brain, much has been learned about the cellular components and molecular elements that characterize the neurogenic niches in the SVZ and SGZ. Astroglia, ependymal cells (E cells), endothelial cells (ECs), the progeny of NSCs, and mature neurons represent five major cellular components of the neurogenic niche. The extracellular matrix (ECM), including the basal membrane, provides a platform for presentation of molecular cues and cellular interaction within the neurogenic niches. We are also beginning to understand the more intricate events and mechanisms by which the neurogenic niches control the full range of NSC development. In addition, recent evidence has implicated critical roles for these niches under several pathological conditions, such as regulation of cancer stem cells (CSC) in tumorigenesis (Vescovi et al. 2006; Calabrese et al. 2007) and repair-like neurogenesis in nonneurogenic regions in response to injuries (Ohab et al. 2006).

In this chapter, we review recent advances in understanding how the neurogenic niche in the adult mammalian brain is organized at the cellular and molecular levels and how the niche components are functionally coordinated to regulate the various steps of neurogenesis, including NSC self-renewal, fate specification, migration, maturation, and synaptic integration into the neural circuit. Emerging roles for the niche and its remodeling under several pathological conditions, such as cancer and stroke, will also be discussed, along with the implications for potential therapeutic interventions.

HISTORICAL ASPECTS AND EMERGENCE
OF THE NEUROGENIC NICHE CONCEPT

It had long been thought that the adult mammalian brain could not generate new neurons, despite pioneering studies by Altman that described the presence of [^{3}H]thymidine-labeled newborn neurons in the postnatal rodent hippocampus (Altman 1962). Due to a lack of functional evidence, these early studies did not stimulate much interest in further investigation. Two lines of evidence that emerged in the 1990s, however, prompted the reexamination of newborn cells in the adult brain. The success of Reynolds and Weiss (1992) in deriving and culturing multipotent neural progenitors from the adult mouse striatum suggested that there might be similar stem cells in vivo that could also give rise to neurons and glia under certain circumstances. This investigation paralleled advances using the newly developed [^{3}H]thymidine replacement, bromodeoxyuridine (BrdU), which can be used in combination with other neuronal cell markers for immunophenotyping of newborn cells (Nowakowski et al. 1989; Rakic 2002). Direct examination of BrdU immunoreactivity and its colocalization with specific neuronal markers provided the first evidence that newly generated cells in the adult rodent brain were indeed neurons (Kuhn et al. 1996). Numerous subsequent studies, along with improved techniques to characterize the newborn cells, unambiguously demonstrated the presence of functional neurogenesis in all mammals examined.

The initial studies by Reynolds and Weiss and subsequent work on progenitors isolated from multiple brain regions also suggested that dormant NSCs are ubiquitously present in many brain regions (Palmer et al. 1997). Once isolated and cultured in vitro, these adult NSCs seemed to be relieved of their in vivo developmental restriction and exhibited the potential to differentiate into neurons (Palmer et al. 1999; Nunes et al. 2003). This notion raised an interesting question: Why do only the neurogenic regions allow functional adult neurogenesis to proceed in vivo? The clues came from studies that transplantation of progenitors isolated from either neurogenic (SVZ of lateral ventricle, SGZ of hippocampus) or nonneurogenic (spinal cord, cerebellum) locations elicited neuronal differentiation only in neurogenic environments in the adult brain (Gage 2000; Shihabuddin et al. 2000; Goh et al. 2003; Emsley et al. 2005). Grafts of hippocampal progenitors to the SVZ even resulted in a novel neuronal subtype (tyrosine-hydroxylase-positive) that was not found in the hippocampus (Suhonen et al. 1996). In addition, injury-induced neurogenesis in nonneurogenic adult cortex was accompanied by the apparent creation of a local neurogenic environment (Magavi et al. 2000). This series of studies provided evidence that local environments, through their unique niche structure and properties, have crucial roles in preserving neurogenesis in specific brain regions to adulthood.

STEM CELL LINEAGE AND FUNCTIONAL ORGANIZATION OF NEUROGENIC NICHES

To describe the neurogenic niche in the SVZ or SGZ, it is necessary to clarify the stem cell lineage within the niches. As with most other somatic stem cells, the primary progenitor in the niche corresponds to the most quiescent NSCs, which divide infrequently and asymmetrically, giving rise to transient amplifying progenitors and subsequent differentiating neurons.

In the SVZ of the lateral ventricle (Fig. 1b), a subset of glial fibrillary acid protein ($GFAP^+$) radial cells that are $S100\beta$-negative, Nestin- and Sox2-positive (collectively called B cells) located underneath the ependymal layer are believed to be the quiescent population and give rise to $Mashl^+$ transient amplifying progenitors (C cells) (Doetsch et al. 1997; Merkle and Alvarez-Buylla 2006). The majority of C cells, in turn, give rise to polysialic-acid-neural cell adhesion molecule ($PSA-NCAM^+$) neuroblast cells (A cells), migrating toward the olfactory bulb (OB) through the rostral migratory stream (RMS) in all mammals, including humans (Merkle and Alvarez-Buylla 2006; Curtis et al. 2007). Within the OB, these new neurons differentiate into two types of interneurons: granular neurons and periglomerular neurons. An oligodendrocyte lineage has also recently been observed differentiating from B and C cells, although it remains unclear whether the neuronal and oligodendroglial lineages share the ancestral origin from the same B cell (Menn et al. 2006). This lineage model from B to C to A cells was initially proposed on the basis of a series of experiments using cell ablation and BrdU/retrovirus-labeling techniques (Doetsch et al. 1999). Recent genetic cell-fate mapping based on the Cre-LoxP system also gave strong support to this model (Garcia et al. 2004). One alternative view is that some of the $CD133^+$ E cells act as the most quiescent stem cell population (Coskun 2006). However, genetic evidence is still needed to unambiguously demonstrate their lineage relationship, if any, to B cells under physiological conditions.

In the SGZ of the DG (Fig. 1c), a similar subset of GFAP$^+$-Sox2$^+$Nestin$^+$ radial glia-like cells are also believed to be the quiescent NSCs (Seri et al. 2004), although self-renewal and multipotentiality of these calls in vivo remain to be demonstrated. These NSCs give rise to transient amplifying progenitors and subsequent doublecortin$^+$ neuroblasts. Unlike the interneuron lineage differentiation in the OB, adult NSCs in the SGZ of the hippocampus give rise to local granule cells, which are presumably glutamatergic excitatory neurons.

The highly hierarchical stem cell lineage in both the SVZ and SGZ requires the precise niche regulation of stem cells from self-renewal, fate specification of NSCs, to maturation and synaptic integration of their neuronal progeny into the existing neural circuit. Thus, neurogenic niches must have evolved to comprise many different components for diverse functionality. The distinct aspects of stem cell lineage in the SVZ and SGZ, on the other hand, also suggest some region-specific roles for neurogenic niches. In the following sections, we describe the major cellular components of neurogenic niches and how these components individually and coordinately contribute to regulation of adult neurogenesis.

Endothelial Component of Neurogenic Niches

ECs form the inner lining of a blood vessel and serve as the interface for metabolite, hormone, and small-molecule exchange between the blood circulation system and the surrounding tissue through the capillary. In the brain, ECs are specialized to form an essential component of the blood brain barrier (BBB), protecting the brain from potentially toxic agents circulating throughout the body. On the basis of these classic roles of ECs, it is not obvious to see how they can directly contribute to de novo neurogenesis in specific regions of the adult brain. However, several key observations made during the initial characterization of adult neurogenesis catalyzed additional investigation of crucial roles of ECs in controlling the first step of adult neurogenesis: stem cell self-renewal.

During early neural development, ECs are among the first type of cells that ventricular zone neural progenitors encounter. In a search for a potential functional interaction between the two, Leventhal et al. (1999) first noticed that coculture of adult rat SVZ explants with ECs generated more neurons than with other cell types. Whether ECs directly interacted with NSCs in the SVZ was not demonstrated. Subsequently, Palmer et al. (2000) observed that dense clusters of dividing cells in the SGZ of DG were anatomically close to vasculature, especially capillaries, suggesting that neurogenesis occurred in an angiogenic niche. In the SVZ, a similar proximity between ECs and LeX⁺ adult NSCs was seen, which led to direct functional characterization of endothelial effects on NSCs from fetal sources (Shen et al. 2004). Remarkably, ECs in coculture stimulate the self-renewal of NSCs, inhibiting their differentiation and enhancing their potential for neuronal differentiation. These effects appeared to be mediated by soluble factors released from ECs and were cell-type-specific, as vascular smooth muscle cells lacked these activities. Although similar effects of ECs on adult NSCs, or the quiescent B cells, remain to be directly demonstrated, this series of studies provided evidence that ECs constitute an angiogenic niche for adult neurogenesis, specifically to promote the self-renewal of adult NSCs. Furthermore, adult NSCs may even "differentiate" into the endothelial lineage when they are cocultured with ECs (Wurmser et al. 2004). This finding highlights the complexity of cellular interactions within the niche and indicates that adult NSCs not only are regulated by their niches, but also, when needed, are able to generate and populate their niche with endothelial niche cells.

What are the molecular identities of these EC-derived factors? How do those factors regulate quiescent NSCs? Initial clues from coculture experiments suggested that Notch signaling became activated by ECs (Shen et al. 2004). Notch signaling is known to be critical in maintaining stem cells in an undifferentiated state (Gaiano and Fishell 2002). Although a recent study suggested that Jagged1, a Notch ligand, is an essential signal for adult SVZ stem cell self-renewal (Nyfeler et al. 2005), it remains to be experimentally verified whether Jagged1 or various other Notch ligands are released from ECs to specifically activate Notch signaling in NSCs. Another study revealed a surprising role of pigment epithelium-derived factor (PEDF) in endothelial control of stem cell self-renewal in the niche of the SVZ (Fig. 1d) (Ramirez-Castillejo et al. 2006). PEDF is highly expressed in the endothelium of the SVZ and promotes the self-renewal of adult NSCs both in vivo and in vitro, whereas a blockade of endogenous PEDF decreases the cycling of adult NSCs. As a secreted glycoprotein, PEDF has well-established antiangiogenic and neurotrophic activities on many different cell types. How does PEDF specifically exert its effect on the self-renewal of adult NSCs? Is the release of PEDF from the endothelium regulated or constitutive? Are effects of PEDF directly on adult NSCs or possibly through other cell types? Answers to these questions, along with the recent identification of the PEDF receptor encoded by Pnpla2 (patatin-like phospholipase domain containing 2) (Notari et al. 2006), should shed light on the molecular mechanisms of how endothelial PEDF controls the self-renewal of adult NSCs.

Astroglial Component of Neurogenic Niches

Astroglia, also called astrocytes with characteristic star shapes, are among the most abundant cell population in the adult brain (Markiewicz and Lukomska 2006). Although they outnumber neurons two to one, the traditional view is that astroglia are mainly supporting cells, contributing to structural and metabolic support, transmitter reuptake, and modulation of synaptic transmission and the BBB. Emerging studies from recent years have revealed that astrocytes actually perform a much wider range of functions than previously appreciated. In a search for mechanisms through which the local environment controls the fate specification of adult NSCs, it was found that astrocytes derived from the hippocampus, but not the spinal cord, promote neurogenesis from adult NSCs under coculture conditions (Song et al. 2002a). These effects by astroglia are mediated by increasing both the proliferation and neuronal fate specification of adult NSCs. Neonatal hippocampal astrocytes appear to be more efficient than their adult counterparts in promoting neurogenesis, consistent with the in vivo observation that neurogenesis decreases with aging. The apparent developmental and regional specificity of astrocytes in their ability to induce neurogenesis from adult NSCs suggests that neuronal production in the adult brain is regulated, at least in part, by distinct properties of local astrocytes, and astrocytes may serve as a component of neurogenic niches to regulate the cell-fate specification of adult NSCs.

It is noteworthy that astrocytes are a highly heterogeneous population. As discussed above, some astrocytes in the SGZ may serve as neurogenic niche cells, whereas a subset of GFAP$^+$ cells in the SVZ and SGZ function as quiescent NSCs. Currently, there are no specific markers to distinguish astrocytic stem cells from a different population of astrocytic niche cells. It also remains an interesting possibility that the NSCs (quiescent B cells) function as niche cells, interacting with each other and leading to asymmetric cell division. In support of this idea, bone morphogenetic proteins (BMPs) and Notch signaling components have been shown to be present in SVZ B cells (Lim et al. 2000; Nyfeler et al. 2005), and activation of BMP or Notch may contribute to the maintenance of astrocytic stem cell state, whereas other neurogenic niche signals (see below) may promote neurogenesis.

The molecular mechanisms underlying astrocyte-induced neurogenesis of adult NSCs have been investigated (Fig. 1d) (Lie et al. 2005; Barkho et al. 2006). The Wnt family genes encode a plethora of secreted glycoproteins that have a fundamental role in neurodevelopment. Wnt-3 is expressed in adult hippocampal astrocytes as well as in the SGZ in vivo.

Overexpression of Wnt-3 is sufficient to increase neurogenesis, and blockade of Wnt signaling dramatically reduces neurogenesis both in vitro and in vivo. Although these studies provide convincing evidence that Wnt signaling is a principal regulator of adult hippocampal neurogenesis, further experiments are required to elucidate the exact in vivo target of Wnt signaling and the underlying mechanisms by which Wnts control the fate specification or neuroblast proliferation in adult neurogenesis. In addition to Wnts, several other secretary factors derived from astrocytes, including specific members of cytokines (Barkho et al. 2006) and noggin-like proteins such as neurogenesin 1 (Ng1) (Ueki et al. 2003), were recently found to promote neuronal differentiation of adult NSCs. Although the mechanisms of cytokines on neurogenesis remain to be investigated, Ng1 acts to prevent the adoption of a glial fate by adult NSCs through its BMP-antagonizing module. The abundant expression of Ng1 by astrocytes in the adult SGZ supports its potential roles in regulating neurogenesis in vivo. Further genetic loss-of-function evidence is needed to substantiate the potential neurogenic function of Ng1 in the niche.

Additional roles for astrocytes in later steps of adult neurogenesis, such as migration, have been demonstrated. In the adult rodent SVZ, a chain of neuroblasts migrates anteriorly along the RMS to the OB through a tunnel formed by astrocytes. Along the RMS, astrocytes first create a physical route for neuronal migration and then communicate with the migrating neurons and regulate their speed of migration (Bolteus and Bordey 2004). Electrophysiological analysis revealed that migrating neuroblasts express functional GABA$_A$ receptors and respond to ambient GABA (γ-amino-n-butyric acid). Surprisingly, the speed of neuroblast migration is regulated by the amount of GABA from the milieu of neuroblasts, which is controlled by astrocytes through their GABA transporters. In the SGZ of adult hippocampus, newly generated neurons migrate for a short distance along the radial processes of astrocytes, reminiscent of the classic mode of radial migration in the developing cortex (Shapiro et al. 2005). These new neurons exhibit extension of dendrites to the upper molecular layer and project axons to the CA3 region of the hippocampus. Immunoelectron microscopic studies reveal that astrocytic radial processes also direct the bidirectional dendritic and axonal outgrowth of newborn neurons. Here, astrocytes function as niche cells by providing a scaffold for the neurite extension of newborn neurons, perhaps supplying nutrients and signaling factors as well.

Adult hippocampal astrocytes also promote synaptic integration and functional maturation of neural progeny of cultured adult hippocampal

NSCs (Song et al. 2002b); whether astrocytes serve similar roles in the SGZ of the hippocampus in vivo requires further studies. The astrocyte-derived synaptogenic factors for newborn neurons also remain to be identified. Recently, secreted molecules, thrombospondin-1 (TSP-1), derived from astrocytes were found to induce synapse assembly by retinal ganglion cells. Given that TSP-4 is intensively expressed in the adult SGZ, it would be interesting to examine whether TSPs also modulate synaptogenesis of newborn neurons from adult NSCs.

Ependymal Component of Neurogenic Niches

E cells line the ventricle wall and function to circulate cerebrospinal fluid (CSF) through the directional cilia beating. Anatomically located between the SVZ and the ventricular cavity of the brain, E cells serve as a natural shield, protecting the SVZ neurogenic niche and simultaneously allowing the niche to sample potential physiological states from CSF. In intimate contact with the quiescent B cells and their progeny, E cells also have instructive roles on the development of adult NSCs within the niche (Fig. 1b).

Noggin belongs to a family of polypeptides that bind to BMP and consequently prevent their activation of BMP receptors (BMPRs). During embryonic development, Noggin functions as a neural inducer to promote neural fate specification from ectoderm by inhibiting BMP activities. In the adult SVZ, Noggin is highly expressed by E cells, whereas BMP and BMPRs are expressed by type-B and -C cells (Lim et al. 2000). This expression pattern immediately suggests its effects on adult NSCs to promote neuronal fate specification. Indeed, purified mouse Noggin protein inhibited BMP signaling and increased neurogenesis from SVZ cells. More interestingly, ectopic Noggin expression promotes neurogenesis from transplanted SVZ cells in the nonneurogenic striatum. These studies strongly implicate the role of Noggin as a potent ependymal-cell-derived neurogenic niche molecule in promoting neurogenesis of adult SVZ NSCs. Whereas loss-of-function evidence of Noggin remains to be shown, it is possible that E cells also provide other soluble or membrane-bound molecules in regulating neurogenesis.

Although the neurogenic niche function of E cells is expected, given their anatomical proximity to the SVZ B cells, a recent study revealed a surprising role for the E cells in directing the orientation of neuroblast migration along the RMS (Sawamoto et al. 2006). To explain the directionality of newborn neurons in the SVZ toward OB, it has been proposed that Slit proteins expressed in septum (Wu et al. 1999) and choroid plexus are the chemorepulsive factors. It was not clear, however, how Slit proteins across the ventricular wall and CSF form a concentration gradient to direct neuroblast migration. Elegant studies using magnetic resonance imaging (MRI) showed that neuroblast migration paralleled the CSF flow, and beating of ependymal cilia is needed for a concentration gradient formation of Slit proteins in directing migration of neuroblasts. These findings highlight another important neurogenic role for E cells in neuronal migration during adult neurogenesis.

In light of their structural and functional roles in the SVZ neurogenic niche, E cells and the ventricular wall are thought to be integral to biogenesis and maintenance of the SVZ stem cell lineage. It therefore came as another surprise that the SVZ and its neurogenic niche could self-repair after extensive damage of the ependymal ventricle by a conditional genetic lesion of Numb/Numblike (Kuo et al. 2006). This finding is reminiscent of endothelial differentiation of adult NSCs as discussed above and suggests the intriguing possibility that SVZ stem cell lineages may have extensive cellular plasticity, producing niche cells to maintain the integrity of the niche under adverse conditions.

Stem Cell Progeny as Components of Neurogenic Niches

Early experiments by Doetsch et al. (1999) observed the proliferation of type-B astrocytes and regeneration of the entire SVZ after elimination of type-C progenitors and type-A neuroblasts. These experiments not only led to the stem cell lineage model described above, but also indicated that local cues or signaling molecules released from stem cell progeny may regulate the self-renewal and fate specification of adult NSCs in a feedback fashion. Liu et al. (2005) found that the spontaneous depolarization of type-A neuroblasts induced GABA release, resulted in tonic $GABA_A$ receptor activation in B cells, and limited progression through the cell cycle. This study suggested a model in which GABA released from neuroblasts provides a feedback mechanism to control the self-renewal of B cells by activating $GABA_A$ receptors. As a result, the reciprocal interaction of NSCs and their progeny may provide a homeostatic mechanism to finely tune adult neurogenesis from the SVZ to OB (Ge et al. 2006a).

Signals from Mature Neurons Contributing to Neurogenic Niches

The hippocampus is important for learning and memory and is endowed with high levels of ongoing neuronal activity. It is also well established that adult hippocampal neurogenesis is regulated by activity, such as exercise, enriched environment, and various behavior-modulating molecules and pathological stimulation (Ming and Song 2005). Thus, adult NSCs in the SGZ may have evolved certain mechanisms to directly sense the neuronal activity elicited from their surrounding mature neurons (Ge et al. 2006a). This theory is supported by evidence that excitatory stimuli were indeed able to act on cultured adult NSCs and profoundly bias the fate toward neuronal specification (Deisseroth et al. 2004). As the chemical messengers of neurons, neurotransmitters may directly mediate the effects of activity from mature neurons on adult neurogenesis. For example, GABA released from interneurons depolarizes neural progenitors and immature neurons. In the SGZ of the adult hippocampus, ambient GABA tonically activates newborn granule cells and facilitates the maturation and synaptic integration of newborn neurons (Tozuka et al. 2005; Ge et al. 2006b). In addition to GABA, another study showed that NMDA (N-methyl-D-aspartate) regulated the competitive survival of newborn neurons in an input- and cell-specific manner, suggesting that new neurons may encode unique information associated with input activity critical for learning and memory (Tashiro et al. 2006). Given the activity-dependent nature of adult neurogenesis, a variety of other neurotransmitters, including serotonin, acetylcholine, and dopamine systems, may also represent information-specific input and exert specific effects on adult neurogenesis (for detail, see Chapter 19).

ECM as a Component of Neurogenic Niches

The ECM is the natural external environment of adult NSCs and thus constitutes an important component of neurogenic niches. Functionally, the ECM can provide support and anchorage for NSCs, sequester and present a variety of growth factors, and regulate intercellular communication. Within the SVZ or SGZ neurogenic niches, the basement membrane formed by ECs or E cells is situated to contact adult NSCs and is particularly enriched in proteoglycans, laminin, and other signaling factors, including epidermal growth factors (EGFs), fibroblast growth factors (FGFs), BMPs, Wnts, and Shh, as important niche molecules (Campos 2005; Labat-Robert and Robert 2005).

Integrins are α-β heterodimeric proteins that are abundant in adult NSCs. As a critical sensor for the ECM, integrin may provide adult NSCs with the capacity to interact with the ECM and regulate the choice of adult NSCs to remain in a quiescent undifferentiated state, to proliferate, or to migrate away from the niche and become differentiated (Campos 2005). The function of integrin as an anchor and sensor for the ECM would be consistent with its important roles in maintenance of a wider variety of somatic stem cells. Conditional genetic ablation of β1-integrin reduces the number of postnatal NSCs and impairs their survival, involving the mitogen-activated protein kinase (MAPK) signaling pathway downstream from β1-integrin (Campos et al. 2004). Further studies have suggested that β1-integrin also affects Notch and EGF signaling pathways. More direct functional evidence will require examination of the role of β1-integrin within NSCs in the SVZ or SGZ by state-of-the-art genetic approaches in animal models.

The adult NSCs and their niches are highly localized, and their developmental regulation is generally independent of adjacent regions, suggesting the presence of mechanisms that restrict or amplify local growth factor signaling. One potential mechanism is the distribution of ECM molecules. For example, tenascin C is highly expressed within the SVZ throughout postnatal and adult life, and genetic ablation of tenascin C alters the number of adult NSCs (Garcion et al. 2004). In addition, tenascin C appears to modulate stem cell sensitivity to FGF-2 and BMP-4 and thus may have an important general role in stem cell niches as a modulator of growth factor signaling.

NICHE REMODELING UNDER PATHOLOGIC CONDITIONS AND ITS IMPLICATIONS FOR THERAPEUTIC INTERVENTIONS

The diverse composition of neurogenic niches and their multiregulatory roles for adult NSC development suggest that niche is not merely a static microenvironment to house NSCs but should also be regarded as a highly dynamic center for complex biochemical signaling and cellular interaction events. This dynamic property further implicates another important feature of neurogenic niches: plasticity for remodeling in response to pathologic conditions.

Stroke causes ischemic neuronal cell death in localized regions of the brain but also, strikingly, induces a regenerative response of the brain in an attempt to repair the damaged area (for detail, see Chapter 26). Early studies suggested that stroke induced an increase in proliferation of endogenous adult NSCs and a number of immature neurons in the SVZ. Surprisingly, many of these newborn neurons were found to migrate toward the damaged area, such as the striatum and cortex, and differentiate into local neurons within 2–4 weeks after stroke (Ohab et al. 2006; Yamashita et al. 2006). Although the number of neurons seems to be quite limited for complete repair of the ischemic region, it is remarkable that the brain appears to have evolved certain mechanisms to direct de novo neurogenesis from endogenous NSCs to repair the damage even distant to the SVZ. It was revealed that neuroblasts migrate in association with reactive astrocytes and blood vessels in response to stroke in the striatum. In the infarcted cortex, immature neurons migrating from the SVZ closely associate with the remodeling vasculature, forming a neurovascular niche for local neurogenesis. Considering the role of neurogenic niches in regulating endogenous NSCs and their neurogenesis in an attempt to self-repair, it is hoped that an understanding of the mechanisms that underlie neurogenic niche regulation may lead to novel therapies that promote neural repair after brain injury.

E cells in the adult brain are viewed as postmitotic and fully differentiated, which is consistent with their roles in maintaining the SVZ neurogenic niche under normal physiologic state. Under certain pathologic conditions, however, E cells can be highly dynamic. Stroke has been shown to acutely stimulate ependymal cell proliferation and, surprisingly, their transformation into radial glia-like cells with morphology, marker expression, and functionality resembling the adult NSCs (Zhang et al. 2007). This transformation correlated with an increased proliferation in the SVZ, and newborn neurons were found to migrate along the fibers of these radial glia-like cells toward the ischemic boundary region. These findings indicate that adult SVZ E cells may harbor a certain cellular plasticity and may dedifferentiate into adult NSCs to direct new neurons toward repair under stroke conditions. The extent of such cellular plasticity in the SVZ, its underlying mechanisms, and how this endogenous regenerative potential can be harnessed for therapeutics remain to be investigated.

Since adult NSCs reside in and under the intimate control of neurogenic niches, dysregulation of neurogenic niches naturally leads to malfunction of adult NSCs. Cancer is now proposed to arise from endogenous stem cells through a series of genetic and epigenetic events; thus, neurogenic niches may be coopted and subverted by cancer stem cells to sustain their uncontrolled growth. Emerging evidence supports such a role of niches in tumorigenesis (Vescovi et al. 2006; Calabrese et al. 2007). Calabrese et al. showed that ECs interact closely with brain cancer cells in vivo and appear to secrete factors that maintain them in a perpetual self-renewal-like state. Increasing the number of ECs or blood vessels in orthotopic brain tumor xenografts accelerated the initiation and growth of tumors. Conversely, depletion of blood vessels from xenografts decreased the self-renewal of tumor stem cells and arrested their growth. This study implicates the endothelial cell niche as an important target in therapeutic interventions for cancer.

SUMMARY

Active adult neurogenesis in the SVZ and SGZ of mammalian brains is now well established. Neurogenic niches in the SVZ and SGZ have emerged as the central regulators of almost every step of neurogenesis from adult NSCs, and they function to preserve adult neurogenesis to adulthood. Tremendous progress has been made during the past few years to understand how the different types of niche cells anatomically compose and functionally organize the neurogenic niches. Less understood are the molecular and cellular mechanisms by which individual niche components control the highly specific developmental decisions made during the distinct stages of adult neurogenesis. Other yet to be identified cell types may also contribute to the niche function, and all the niche components must have evolved to coordinate with each other to orchestrate the complex and precise development of adult NSCs.

Although the neurogenic niche in the SVZ or SGZ serves as a dynamic center for numerous biochemical signaling and intercellular communication events, its remodeling capacity and plasticity under disease conditions are remarkable. The adult brain has utilized the neurogenic niche as a mechanism to initiate endogenous repair for neural injuries. Dysregulation of neurogenic niches may contribute to tumorigenesis from endogenous stem cells and their immature progeny. Understanding how neurogenic niches operate under both normal and pathologic conditions will undoubtedly shed light on the evolutionary implications of adult neurogenesis, as well as guide the principles of therapeutic interventions for neural repair in various neurological disorders.

ACKNOWLEDGMENTS

We thank M.L. Gage for comments. This work was supported by the National Institutes of Health (G.-l.M. and H.S.), Klingenstein fellowship awards in the neurosciences (G.-l.M.), Sloan (G.-l.M.), McKnight (H.S.), and The Robert Packard Center for ALS and MDA (H.S.). D.K.M. is a predoctoral fellow of the American Heart Association.

REFERENCES

Abrous D.N., Koehl M., and Le Moal M. 2005. Adult neurogenesis: From precursors to network and physiology. *Physiol. Rev.* **85:** 523–569.

Altman J. 1962. Are new neurons formed in the brains of adult mammals? *Science* **135:** 1127–1128.

Altman J. and Das G.D. 1965. Post-natal origin of microneurones in the rat brain. *Nature* **207:** 953–956.

Alvarez-Buylla A. and Lim D.A. 2004. For the long run: Maintaining germinal niches in the adult brain. *Neuron* **41:** 683–686.

Barkho B.Z., Song H., Aimone J.B., Smrt R.D., Kuwabara T., Nakashima K., Gage F.H., and Zhao X. 2006. Identification of astrocyte-expressed factors that modulate neural stem/progenitor cell differentiation. *Stem Cells Dev.* **15:** 407–421.

Bolteus A.J. and Bordey A. 2004. GABA release and uptake regulate neuronal precursor migration in the postnatal subventricular zone. *J. Neurosci.* **24:** 7623–7631.

Calabrese C., Poppleton H., Kocak M., Hogg T.L., Fuller C., Hamner B., Oh E.Y., Gaber M.W., Finklestein D., Allen M., Frank A., Bayazitov I.T., Zakharenko S.S., Gajjar A., Davidoff A., and Gilbertson R.J. 2007. A perivascular niche for brain tumor stem cells. *Cancer Cell* **11:** 69–82.

Campos L.S. 2005. Beta1 integrins and neural stem cells: Making sense of the extracellular environment. *BioEssays* **27:** 698–707.

Campos L.S., Leone D.P., Relvas J.B., Brakebusch C., Fassler R., Suter U., and ffrench-Constant C. 2004. Beta1 integrins activate a MAPK signalling pathway in neural stem cells that contributes to their maintenance. *Development* **131:** 3433–3444.

Coskun V., Wu H., Biancotti J.C., Krueger R.C., Kumar S., deVellis J., and Sun Y.E. 2005. CD133 Expressing adult neural stem cells in the postnatal forebrain ependyma. *Abstract Viewer/Itinerary Planner, 2005* (program no. 141.15) Society for Neuroscience, Washington, D.C. (Online.)

Curtis M.A., Kam M., Nannmark U., Anderson M.F., Axell M.Z., Wikkelso C., Holtas S., van Roon-Mom W.M., Bjork-Eriksson T., Nordborg C., Frisen J., Dragunow M., Faull R.L., and Eriksson P.S. 2007. Human neuroblasts migrate to the olfactory bulb via a lateral ventricular extension. *Science* **315:** 1243–1249.

Deisseroth K., Singla S., Toda H., Monje M., Palmer T.D., and Malenka R.C. 2004. Excitation-neurogenesis coupling in adult neural stem/progenitor cells. *Neuron* **42:** 535–552.

Doetsch F., Garcia-Verdugo J.M., and Alvarez-Buylla A. 1997. Cellular composition and three-dimensional organization of the subventricular germinal zone in the adult mammalian brain. *J. Neurosci.* **17:** 5046–5061.

Doetsch F., Caille I., Lim D.A., Garcia-Verdugo J.M., and Alvarez-Buylla A. 1999. Subventricular zone astrocytes are neural stem cells in the adult mammalian brain. *Cell* **97:** 703–716.

Emsley J.G., Mitchell B.D., Kempermann G., and Macklis J.D. 2005. Adult neurogenesis and repair of the adult CNS with neural progenitors, precursors, and stem cells. *Prog. Neurobiol.* **75:** 321–341.

Eriksson P.S., Perfilieva E., Bjork-Eriksson T., Alborn A.M., Nordborg C., Peterson D.A., and Gage F.H. 1998. Neurogenesis in the adult human hippocampus. *Nat. Med.* **4:** 1313–1317.

Gage F.H. 2000. Mammalian neural stem cells. *Science* **287:** 1433–1438.

Gaiano N. and Fishell G. 2002. The role of notch in promoting glial and neural stem cell fates. *Annu. Rev. Neurosci.* **25:** 471–490.

Garcia A.D., Doan N.B., Imura T., Bush T.G., and Sofroniew M.V. 2004. GFAP-expressing progenitors are the principal source of constitutive neurogenesis in adult mouse forebrain. *Nat. Neurosci.* **7:** 1233–1241.

Garcion E., Halilagic A., Faissner A., and ffrench-Constant C. 2004. Generation of an environmental niche for neural stem cell development by the extracellular matrix molecule tenascin C. *Development* **131:** 3423–3432.

Ge S., Pradhan D.A., Ming G.L., and Song H. 2006a. GABA sets the tempo for activity-dependent adult neurogenesis. *Trends Neurosci.* **30:** 1–8.

Ge S., Goh E.L., Sailor K.A., Kitabatake Y., Ming G.L., and Song H. 2006b. GABA regulates synaptic integration of newly generated neurons in the adult brain. *Nature* **439:** 589–593.

Goh E.L., Ma D., Ming G.L., and Song H. 2003. Adult neural stem cells and repair of the adult central nervous system. *J. Hematother. Stem. Cell Res.* **12:** 671–679.

Kuhn H.G., Dickinson-Anson H., and Gage F.H. 1996. Neurogenesis in the dentate gyrus of the adult rat: Age-related decrease of neuronal progenitor proliferation. *J. Neurosci.* **16:** 2027–2033.

Kuo C.T., Mirzadeh Z., Soriano-Navarro M., Rasin M., Wang D., Shen J., Sestan N., Garcia-Verdugo J., Alvarez-Buylla A., Jan L.Y., and Jan Y.N. 2006. Postnatal deletion of Numb/Numblike reveals repair and remodeling capacity in the subventricular neurogenic niche. *Cell* **127:** 1253–1264.

Labat-Robert J. and Robert L. 2005. The extracellular matrix during normal development and neoplastic growth. *Prog. Mol. Subcell. Biol.* **40:** 79–106.

Leventhal C., Rafii S., Rafii D., Shahar A., and Goldman S.A. 1999. Endothelial trophic support of neuronal production and recruitment from the adult mammalian subependyma. *Mol. Cell. Neurosci.* **13:** 450–464.

Li L. and Xie T. 2005. Stem cell niche: structure and function. *Annu. Rev. Cell Dev. Biol.* **21:** 605–631.

Lie D.C., Song H., Colamarino S.A., Ming G.L., and Gage F.H. 2004. Neurogenesis in the adult brain: New strategies for central nervous system diseases. *Annu. Rev. Pharmacol. Toxicol.* **44:** 399–421.

Lie D.C., Colamarino S.A., Song H.J., Desire L., Mira H., Consiglio A., Lein E.S., Jessberger S., Lansford H., Dearie A.R., and Gage F.H. 2005. Wnt signalling regulates adult hippocampal neurogenesis. *Nature* **437:** 1370–1375.

Lim D.A., Tramontin A.D., Trevejo J.M., Herrera D.G., Garcia-Verdugo J.M., and Alvarez-Buylla A. 2000. Noggin antagonizes BMP signaling to create a niche for adult neurogenesis. *Neuron* **28:** 713–726.

Liu X., Wang Q., Haydar T.F., and Bordey A. 2005. Nonsynaptic GABA signaling in postnatal subventricular zone controls proliferation of GFAP-expressing progenitors. *Nat. Neurosci.* **8:** 1179–1187.

Lledo P.M., Alonso M., and Grubb M.S. 2006. Adult neurogenesis and functional plasticity in neuronal circuits. *Nat. Rev. Neurosci.* **7:** 179–193.

Ma D.K., Ming G.L., and Song H. 2005. Glial influences on neural stem cell development: Cellular niches for adult neurogenesis. *Curr. Opin. Neurobiol.* **15:** 514–520.

Magavi S.S., Leavitt B.R., and Macklis J.D. 2000. Induction of neurogenesis in the neocortex of adult mice. *Nature* **405:** 951–955.

Markiewicz I. and Lukomska B. 2006. The role of astrocytes in the physiology and pathology of the central nervous system. *Acta. Neurobiol. Exp.* **66:** 343–358.

Menn B., Garcia-Verdugo J.M., Yaschine C., Gonzalez-Perez O., Rowitch D., and Alvarez-Buylla A. 2006. Origin of oligodendrocytes in the subventricular zone of the adult brain. *J. Neurosci.* **26:** 7907–7918.

Merkle F.T. and Alvarez-Buylla A. 2006. Neural stem cells in mammalian development. *Curr. Opin. Cell Biol.* **18:** 704–709.

Ming G.L. and Song H. 2005. Adult neurogenesis in the mammalian central nervous system. *Annu. Rev. Neurosci.* **28:** 223–250.

Notari L., Baladron V., Aroca-Aguilar J.D., Balko N., Heredia R., Meyer C., Notario P.M., Saravanamuthu S., Nueda M.L., Sanchez-Sanchez F., Escribano J., Laborda J., and Becerra S.P. 2006. Identification of a lipase-linked cell membrane receptor for pigment epithelium-derived factor. *J. Biol. Chem.* **281:** 38022–38037.

Nowakowski R.S., Lewin S.B., and Miller M.W. 1989. Bromodeoxyuridine immunohisto-chemical determination of the lengths of the cell cycle and the DNA-synthetic phase for an anatomically defined population. *J. Neurocytol.* **18:** 311–318.

Nunes M.C., Roy N.S., Keyoung H.M., Goodman R.R., McKhann G., II, Jiang L., Kang J., Nedergaard M., and Goldman S.A. 2003. Identification and isolation of multipotential neural progenitor cells from the subcortical white matter of the adult human brain. *Nat. Med.* **9:** 439–447.

Nyfeler Y., Kirch R.D., Mantei N., Leone D.P., Radtke F., Suter U., and Taylor V. 2005. Jaggedl signals in the postnatal subventricular zone are required for neural stem cell self-renewal. *EMBO J.* **24:** 3504–3515.

Ohab J.J., Fleming S., Blesch A., and Carmichael S.T. 2006. A neurovascular niche for neurogenesis after stroke. *J. Neurosci.* **26:** 13007–13016.

Palmer T. D., Takahashi J., and Gage F.H. 1997. The adult rat hippocampus contains primordial neural stem cells. *Mol. Cell Neurosci.* **8:** 389–404.

Palmer T.D., Markakis E.A., Willhoite A.R., Safar F., and Gage F.H. 1999. Fibroblast growth factor-2 activates a latent neurogenic program in neural stem cells from diverse regions of the adult CNS. *J. Neurosci.* **19:** 8487–8497.

Palmer T.D., Willhoite A.R., and Gage F.H. 2000. Vascular niche for adult hippocampal neurogenesis. *J. Comp. Neurol.* **425:** 479–494.

Rakic P. 1985. Limits of neurogenesis in primates. *Science* **227:** 1054–1056.

———. 2002. Adult neurogenesis in mammals: An identity crisis. *J. Neurosci.* **22:** 614–618.

Ramirez-Castillejo C., Sanchez-Sanchez F., Andreu-Agullo C., Ferron S.R., Aroca-Aguilar J.D., Sanchez P., Mira H., Escribano J., and Farinas I. 2006. Pigment epithelium-derived factor is a niche signal for neural stem cell renewal. *Nat. Neurosci.* **9:** 331–339.

Reynolds B.A. and Weiss S. 1992. Generation of neurons and astrocytes from isolated cells of the adult mammalian central nervous system. *Science* **255:** 1707–1710.

Sawamoto K., Wichterle H., Gonzalez-Perez O., Cholfin J.A., Yamada M., Spassky N., Murcia N.S., Garcia-Verdugo J.M., Marin O., Rubenstein J.L., Tessier-Levigne M., Okano H., and Alvarez-Buylla A. 2006. New neurons follow the flow of cerebrospinal fluid in the adult brain. *Science* **311:** 629–632.

Scadden D.T. 2006. The stem-cell niche as an entity of action. *Nature* **441:** 1075–1079.

Schofield R. 1978. The relationship between the spleen colony-forming cell and the haemopoietic stem cell. *Blood Cells* **4:** 7–25.

Seri B., Garcia-Verdugo J.M., Collado-Morente L., McEwen B.S., and Alvarez-Buylla A. 2004. Cell types, lineage, and architecture of the germinal zone in the adult dentate gyrus. *J. Comp. Neurol.* **478:** 359–378.

Shapiro L.A., Korn M.J., Shan Z., and Ribak C.E. 2005. GFAP-expressing radial glia-like cell bodies are involved in a one-to-one relationship with doublecortin-immunolabeled newborn neurons in the adult dentate gyrus. *Brain Res.* **1040:** 81–91.

Shen Q., Goderie S.K., Jin L., Karanth N., Sun Y., Abramova N., Vincent P., Pumiglia K., and Temple S. 2004. Endothelial cells stimulate self-renewal and expand neurogenesis of neural stem cells. *Science* **304:** 1338–1340.

Shihabuddin L.S., Horner P.J., Ray J., and Gage F.H. 2000. Adult spinal cord stem cells generate neurons after transplantation in the adult dentate gyrus. *J. Neurosci.* **20:** 8727–8735.

Song H., Stevens C.F., and Gage F.H. 2002a. Astroglia induce neurogenesis from adult neural stem cells. *Nature* **417:** 39–44.

————. 2002b. Neural stem cells from adult hippocampus develop essential properties of functional CNS neurons. *Nat. Neurosci.* **5:** 438–445.

Spradling A., Drummond-Barbosa D., and Kai T. 2001. Stem cells find their niche. *Nature* **414:** 98–104.

Suhonen J.O., Peterson D.A., Ray J., and Gage F.H. 1996. Differentiation of adult hippocampus-derived progenitors into olfactory neurons in vivo. *Nature* **383:** 624–627.

Tashiro A., Sandler V.M., Toni N., Zhao C., and Gage F.H. 2006. NMDA-receptor-mediated, cell-specific integration of new neurons in adult dentate gyrus. *Nature* **442:** 929-933.

Tozuka Y., Fukuda S., Namba T., Seki T., and Hisatsune T. 2005. GABAergic excitation promotes neuronal differentiation in adult hippocampal progenitor cells. *Neuron* **47:** 803–815.

Ueki T., Tanaka M., Yamashita K., Mikawa S., Qiu Z., Maragakis N.J., Hevner R.F., Miura N., Sugimura H., and Sato K. 2003. A novel secretory factor, Neurogenesin-1, provides neurogenic environmental cues for neural stem cells in the adult hippocampus. *J. Neurosci.* **23:** 11732–11740.

van Praag H., Schinder A.F., Christie B.R., Toni N., Palmer T.D., and Gage F.H. 2002. Functional neurogenesis in the adult hippocampus. *Nature* **415:** 1030–1034.

Vescovi A.L., Galli R., and Reynolds B.A. 2006. Brain tumour stem cells. *Nat. Rev. Cancer* **6:** 425–436.

Weissman I.L., Anderson D.J., and Gage F. 2001. Stem and progenitor cells: Origins, phenotypes, lineage commitments, and transdifferentiations. *Annu. Rev. Cell Dev. Biol.* **17:** 387–403.

Wu W., Wong K., Chen J., Jiang Z., Dupuis S., Wu J.Y., and Rao Y. 1999. Directional guidance of neuronal migration in the olfactory system by the protein Slit. *Nature* **400:** 331–336.

Wurmser A.E., Nakashima K., Summers R.G., Toni N., D'Amour K.A., Lie D.C., and Gage F.H. 2004. Cell fusion-independent differentiation of neural stem cells to the endothelial lineage. *Nature* **430:** 350–356.

Yamashita T., Ninomiya M., Hernandez Acosta P., Garcia-Verdugo J.M., Sunabori T., Sakaguchi M., Adachi K., Kojima T., Hirota Y., Kawase T., Araki N., Abe K., Okano H., and Sawamoto K. 2006. Subventricular zone-derived neuroblasts migrate and differentiate into mature neurons in the post-stroke adult striatum. *J. Neurosci.* **26:** 6627–6636.

Zhang R.L., Zhang Z.G., Wang Y., Letourneau Y., Liu X.S., Zhang X., Gregg S.R., Wang L., and Chopp M. 2007. Stroke induces ependymal cell transformation into radial glia in the subventricular zone of the adult rodent brain. *J. Cereb. Blood Flow Metab.* **27:** 1201–1212.

12

Adult Neurogenesis: Similarities and Differences in Stem Cell Fate, Proliferation, Migration, and Differentiation in Distinct Forebrain Regions

D. Chichung Lie[1] and Magdalena Götz[2,3]

[1]Institute of Developmental Genetics and
[2]Institute for Stem Cell Research
GSF-National Research Center for Environment and Health
85764 Munich-Neuherberg, Germany
[3]Physiological Genomics
University of Munich
80336 Munich, Germany

SELF-RENEWAL AND PROLIFERATION OF NEURAL STEM CELLS, neuronal fate determination of uncommitted precursors, and migration of neuroblasts are the earliest steps in adult neurogenesis. Self-renewing divisions are required for the maintenance of the stem cell pool, which ensures that neurogenesis continues throughout the lifetime of the organism. Instruction of the stem cell progeny to adopt a neuronal fate is a common feature between the neurogenic niches, yet it is likely that local instructive programs are distinct given that different neuronal phenotypes are generated in neurogenic areas. Finally, immature neurons are born distant from their future location. Thus, migration of the newborn neurons must be tightly regulated to ensure the proper integration of new mature neurons into the neuronal network. In this chapter, we discuss these processes from a functional perspective and summarize current knowledge regarding their cellular and molecular regulation.

Stem cells are defined as cells with the potential to generate differentiated progeny and the potential to undergo unlimited self-renewing divisions (Weissman et al. 2001). In the hematopoietic system, the existence of adult stem cells has been proven through assays, in which a single adult cell and its progeny have been repeatedly challenged to reconstitute the entire hematopoietic system in serial transplantations to lethally irradiated organisms (Weissman et al. 2001). The reconstitution of the entire hematopoietic system demonstrates the multipotentiality of the transplanted cell, whereas their ability to do so in serial transplantations indicates the self-renewal of the initially transplanted cell.

Such stringent stem cell assays are presently not available for the adult CNS. The concept that neural stem cells exist in the adult CNS was initially based on in vitro studies. Cells, which self-renew in vitro in the presence of growth factors and give rise to all neural lineages, have been isolated from multiple regions of the adult brain including the neurogenic subependymal zone (SEZ) and the hippocampus (for review, see Lie et al. 2004). However, the validity of the cell culture assays to reflect in vivo stem cell activity is limited, given that culture conditions can modify the differentiation potential of isolated cells (Kondo and Raff 2000, 2004; Gabay et al. 2003; Hack et al. 2004). In addition, Doetsch et al. (2002a) have provided evidence that transit-amplifying precursor cells give rise to multipotent, self-renewing cells in vitro. More recently, strong evidence has been obtained that a subset of astrocytes are the earliest precursor in the adult neurogenic lineage (for review, see Alvarez-Buylla et al. 2002; Götz 2003). These cells increase their proliferation and reconstitute neurogenesis following pharmacological ablation of highly proliferative precursors in the neurogenic niche and the consequent loss of neurogenesis (Doetsch et al. 1999; Seri et al. 2001). Since there is no apparent change in the number of these specialized astrocytes, it is likely that these cells have undergone self-renewal (Doetsch et al. 1999; Seri et al. 2001). Despite the lack of in vivo proof that a single such astrocyte would be multipotent and could self-renew throughout the lifetime of the organism, we refer to these cells as neural stem cells and not as precursor cells to emphasize that they are at the very beginning of the lineage tree that gives rise to new neurons in the adult brain.

PROLIFERATION IN ADULT NEUROGENESIS

Differences in Proliferative Activity in the Neurogenic Lineage

Neural stem cells are characterized by their ability for extensive proliferation and their undifferentiated state. These properties allow the neural stem cell to generate an immense number of neurons throughout the lifetime of the organism (see, e.g., Ahn and Joyner 2005). In stark contrast to the proliferative activity of neural stem cells during CNS development, proliferation of neural stem cells only accounts for a very low percentage of the total divisions in the neurogenic areas (Morshead et al. 1994, 1998; Doetsch et al. 1999; Filippov et al. 2003; Kronenberg et al. 2003; Kempermann et al. 2004). Indeed, it has been estimated that neural stem cells in the SEZ of the lateral ventricles may divide only once every 2 weeks (Morshead et al. 1994, 1998; Doetsch et al. 1999), a frequency that renders it difficult to distinguish whether the majority of neural stem cells are quiescent or are very slow cycling.

This slow rate of cell division, as well as the potential inheritance of a DNA strand to avoid the accumulation of mutations (Karpowicz et al. 2005), is the basis of the so-called "label-retaining paradigm," a widespread assay for stem cell identification (see, e.g., Potten 2004). Fast-proliferating precursors can be labeled with a single pulse of a DNA-base analogon, such as bromodeoxyuridine (BrdU), but stem cells divide so slowly that BrdU must be continuously applied for 1–2 weeks to be incorporated in a reasonable number of stem cells. Although such a paradigm initially labels all proliferating cells, fast-proliferating cells dilute the label during subsequent occurring divisions if the labeling period is followed by a relatively long label-free period. In contrast, stem cells (and the differentiated postmitotic progeny, which can be readily distinguished from stem cells) maintain the BrdU label because they divide so slowly. Following this label-retaining paradigm, a subset of astrocytes in the adult SEZ and the subgranular zone (SGZ) incorporate BrdU; in contrast, hardly any astrocyte is labeled by a short BrdU pulse. Thus, this paradigm underlines the stem cell nature of astrocytes in these regions (Doetsch et al. 1999; Garcia et al. 2004), but what about the astrocytes not incorporating BrdU? Are they permanently postmitotic or can they be recruited to undergo self-renewal and give rise to differentiated neural progeny as well? If the latter is the case, are these stem cell properties restricted to astrocytes in the two adult neurogenic zones or do most astrocytes scattered throughout the parenchyma have stem cell potential? These questions highlight the problem of distinguishing "normal" astrocytes from "stem cell astrocytes" if there is indeed such a difference (for review, see Mori et al. 2005).

The low proliferative state of stem cells may have a key function, namely, to reduce the risk of accumulation of DNA damage and mutations, which could otherwise lead to stem cell dysfunction and tumorigenesis (Arden 2007; Paik et al. 2007; Tothova et al. 2007). Studies in other adult stem cell systems (Cheng et al. 2000; Lowry et al. 2005) have suggested that increased proliferation or forced cell cycle entry of stem cells can lead to premature depletion of the stem cell pool. Loss of the cell cycle inhibitor p21cip1/waf1 initially leads to increased proliferation in the neurogenic areas during early adulthood (Kippin et al. 2005). At later stages, however, in vivo proliferation and the number of neu-rosphere-forming cells, which can be isolated from these mutants, decrease compared to their normal littermates, suggesting that excessive proliferation of adult neural stem cells leads to exhaustion of the stem cell pool (Kippin et al. 2005). Thus, the low proliferative state may also be essential for long-term maintenance of the adult neural stem cell pool.

These observations raise the question of whether adult neural stem cells can undergo only a limited number of cell divisions, notably, different from single adult hematopoietic stem cells that can generate the hematopoietic system of several animals in serial transplantations (Weissman et al. 2001). It will be interesting to perform similar experiments by serial transplantation of neural stem cells to directly examine whether their proliferation capacity is unlimited and exceeds the life span of a single animal.

Even if neural stem cells can only undergo a limited number of cell divisions, the fact remains that especially in the SEZ/olfactory bulb (OB) system, a significant rate of neurogenesis is sustained throughout the lifetime of the organism. In this regard, it is important to note that stem cells are not the only proliferating cell type in the neurogenic lineage but that several proliferation steps can be distinguished based on the dividing cell type and the speed and number of divisions (Morshead and van der Kooy 1992; Morshead et al. 1994; Doetsch et al. 1999; Kronenberg et al. 2003; Seri et al. 2004). The initial proliferation step is the self-renewing division of the neural stem cell. This may occur in an asymmetric cell division with the stem cell generating two distinct daughter cells or in distinct symmetric cell divisions. In the former case, one of the daughter cells retains stem cell properties and returns to a slowly dividing stem cell state. The other daughter cell differentiates into a transient-amplifying progenitor that divides much faster and has a further limited proliferative capacity. Alternatively, stem cells may divide symmetrically, with some of them generating two fast-dividing transient-amplifying progenitors and others generating two slow-dividing stem cells.

The major burden of precursor expansion is placed on the transient amplifying cells and to a lesser extent on neuroblasts. As mentioned above, these precursors have the ability to rapidly divide but are restricted in the number of divisions that they can undergo in vivo. By shifting the proliferative burden from stem cells to their progeny, the adult brain limits stem cell divisions to infrequent self-renewing divisions and efficiently reduces the risk of stem cell dysfunction and exhaustion while maintaining high levels of neurogenesis.

The proliferation rate of transit-amplifying precursors allows the potent regulation of the number of newly generated neurons. However, transient amplifying cells in different neurogenic areas display different proliferative behavior. Transient-amplifying cells in the SEZ divide several times, with an estimated cell cycle time of 12 hours (Morshead and van der Kooy 1992); in contrast, transient amplifying cells in the SGZ may divide only a few times (Kempermann et al. 2004; Seri et al. 2004) with an approximate cell cycle time of 24 hours (Cameron and McKay 2001) before differentiating into neuroblasts, which after a limited number of divisions, mature into postmitotic neurons. Distinct behavioral modifications increase the proliferation of transient-amplifying progenitors in the SGZ but do not affect the proliferation in the SEZ (Brown et al. 2003b; Kronenberg et al. 2003). Differences in the proliferation rate are also reflected by the number of new neurons that are generated per day in the two adult neurogenic areas. Although it has been estimated that about 80,000 new neurons appear per day in the OB of adult mice, less than 1,000 new neurons are added per day to the dentate gyrus (DG) of young adult mice (Kempermann et al. 1997). Taken together, proliferation and its regulation are likely to be controlled by distinct molecular mechanisms in the two adult neurogenic niches.

Control of Proliferation

Little is known about the mechanisms that control the slow mode of stem cell proliferation in the adult neurogenic zones. Is this due to the low level of proliferation signals or is it controlled by extrinsic factors dampening the rate of proliferation?

In the early postnatal period, release of γ-amino-n-butyric acid (GABA) by stem cell progeny in the SEZ and activation of GABA$_A$ receptors on SEZ stem cells provide a negative feedback signal, which limits stem cell proliferation (Liu et al. 2005). Whether this negative feedback mechanism persists into adulthood to control the proliferative activity of SEZ stem cells is not known. However, ablation of the stem cell progeny in the adult SEZ or SGZ results in the increased proliferation of local neural stem cells, suggesting that quiescence of adult neural stem cells may also be controlled by signals derived from their progeny (Doetsch et al. 1999; Seri et al. 2001, 2004).

What signals control stem cell quiescence and/or their slow mode of proliferation? One can hypothesize that such signals should reduce the rate of proliferation without interfering with the differentiation status of the neural stem cell. It has recently been shown that at least a subpopulation of adult neural stem cells express receptors for transforming growth factor-β (TGF-β) family members (Wachs et al. 2006). TGF-β1 arrests neuronal stem and progenitor cells from the adult SEZ and the hippocampus, which are cultured in the presence of epidermal growth factor (EGF) and fibroblast growth factor-2 (FGF-2), in the G_0/G_1 phase of the cell cycle without affecting their differentiation and self-renewal capacity. This negative effect of TGF-β1 on proliferation is potentially mediated through increased expression of the cyclin-dependent kinase (CDK) inhibitor p21cip1/waf1. In addition, infusion of TGF-β1 into the lateral ventricle strongly reduced cell proliferation in the SEZ and the hippocampus without affecting the differentiation pattern of the remaining newborn cells (Wachs et al. 2006). Thus, TGF-β1-induced signaling may be a physiological signal for the slow rate of neural stem cell proliferation.

Transient-amplifying progenitor cells in the SEZ uniformly express the EGF receptor (EGFR) and respond to infusion of EGF with increased proliferation (Doetsch et al. 2002a). Only a minor proportion of neural stem cells in this area stains positive for the EGFR, and it has been proposed that the expression of the EGFR is an early step in the activation of the quiescent stem cell (Doetsch et al. 2002a). Thus, the suppression of the expression of the EGFR may contribute to allowing the slow mode of stem cell proliferation by quiescence by suppressing the expression of EGFR.

Another class of candidate targets for quiescence signals could be cell cycle regulators, which control the progression of stem cells through the G_1 phase of the cell cycle. Loss of p21cip1/waf1 (Kippin et al. 2005), as well as loss of p53, which in turn leads to decreased p21cip1/waf1 expression (Meletis et al. 2006), results in increased proliferation of neural stem cells in vivo. In contrast, loss of the CDK inhibitor p27Kip1 does not affect the quiescence of stem cells but increases the proliferation of transient-amplifying progenitors (Doetsch et al. 2002b). These distinct roles of CDK inhibitors on specific cell populations in the neurogenic process is analogous to the hematopoietic stem cell system, where p21cip1/waf1 affects stem cell proliferation and p27Kip1 regulates the proliferation of stem cell progeny (Cheng and Scadden 2002). Taken together, characterization of the regulation of stem-cell-specific CDK inhibitors may provide insight into the signaling pathways underlying quiescence of adult neural stem cells.

A number of studies identified stimuli that promote proliferation in the neurogenic zones or the generation of neurospheres in vitro. On the basis of this complexity, these regulators can be divided into regulators on the systems level, on the cellular interaction level, and on the single molecular level (for review, see Ming and Song 2005). It seems likely that stimuli on the systems level such as physical activity alter cellular inter-actions in the neurogenic niche, which then lead to changes in the activ-ity of molecular signaling pathways. Such relationships, however, are far from understood.

Current data indicate that cellular interactions and molecular signals control both stimulation and inhibition of proliferation. The potential existence of a negative feedback signal emanating from neural stem cell progeny to limit proliferation has been described above. Stem cells in the SGZ and the SEZ proliferate in close proximity to endothelial cells (Palmer et al. 2000; Capela and Temple 2002) and astrocytes (Song et al. 2002). Coculture experiments suggest that the interaction of stem cells with astrocytes (Song et al. 2002) and endothelial cells (Shen et al. 2004) stimulates proliferation of stem cells or of their progeny. Further studies have suggested that endothelial cells may increase self-renewing divisions of stem cells potentially through stimulating notch signaling (Shen et al. 2004) and pigment epithelium-derived factor (PEDF)-induced signaling (Ramirez-Castillejo et al. 2006). Indeed, infusion of PEDF into the lateral ventricle increases proliferation of neural stem cells in the SEZ, whereas a blockade of endogenous PEDF decreases the fraction of proliferating neural stem cells (Ramirez-Castillejo et al. 2006). Another factor con-trolling stem cell proliferation may be galectin (Sakaguchi et al. 2006), a lectin binding to a variety of cell surface molecules including the stem cell genes of the integrin family. Similar to PEDF, galectin promotes the number of slowly proliferating cells, suggesting that these factors either recruit additional quiescent stem cells into the slow-proliferation mode or slightly shorten the cell cycle time of slow-proliferating stem cells (Sakaguchi et al. 2006).

Specific populations of the neurogenic lineage show distinct expres-sion patterns of receptors for signaling molecules or distinct patterns of pathway activity in vivo. Platelet-derived growth factor-α receptor (PDGFRα) (Jackson et al. 2006) and Shh signaling (Ahn and Joyner 2005), for example, as well as transient EGFR expression have been local-ized to stem cells (Doetsch et al. 2002a); the bulk of EGFR (Doetsch et al. 2002a), Notch receptor (Nyfeler et al. 2005), dopamine receptor-1-like receptors (Hoglinger et al. 2004), and ephrin A2 (Holmberg et al. 2005) expression appears to be in the transient-amplifying cells, whereas dopamine receptor-2-like receptors (Hoglinger et al. 2004), vascular endothelial growth factor (VEGF) receptor 2 (Jin et al. 2002), and Wnt/ β-catenin-signaling (Lie et al. 2005) have been found in neuroblasts. Although most of these studies cannot fully establish a physiological role of these signals for proliferation of a distinct population, the potentially distinct activity of pathways in subpopulations of the neurogenic lineage is striking in light of the in vivo evidence that the proliferation of each cell type is distinctly regulated (Kronenberg et al. 2003). How these pathways interact and how other prominent candidate signals such as EGF, insulin-like growth factor (IGF), steroids, and neurotransmitters (for an overview, see Ming and Song 2005) contribute to this picture remain to be determined.

Infusion experiments have shown that EGFR signaling appears to be a proliferative signal specific for the transient amplifying cells of the SEZ but not for their counterparts in the SGZ (Kuhn et al. 1997). Interestingly, transit-amplifying precursors in the SEZ express the transcription factor Olig2 (Hack et al. 2004, 2005), which is a downstream target of EGF-mediated signaling and is essential for the high proliferation rate of neurosphere cells in vitro (Gabay et al. 2003; Hack et al. 2004). Blockade of Olig2-mediated transcriptional repression virtually abolishes the transit-amplifying precursor population in the adult SEZ in vivo (Hack et al. 2005), demonstrating that Olig2-mediated repression of target genes is essential for the maintenance of the transit-amplifying precursor state in the adult SEZ and their proliferation. Intriguingly, Olig2 is not expressed in the adult SGZ where transit-amplifying precursors divide much slower than in the SEZ. Similarly, Olig2 is not expressed in a more rostral region close to the OB, where most of the glomerular interneurons of the OB seem to originate (Hack et al. 2005). Indeed, the granule neurons that arise via the hugely expanded population of fast-proliferating Olig2-positive transit-amplifying precursors constitute the majority (80%) of adult generated OB interneurons, whereas the glomerular layer OB interneurons only constitute a small proportion. Taken together, the regulation of the proliferation rate of transit-amplifying precursors through EGF-induced signaling allows the expansion of particular neuronal populations in adult neurogenesis according to functional requirements.

Intracellularly, proliferation signals should influence the cell cycle machinery. Indeed, it has been shown that transgenic mice carrying mutations in genes coding for the positive cell cycle regulator cyclin D2 (Kowalczyk et al. 2004) or for the transcription factor Bmi, which represses the expression of the negative cell cycle regulators p16Ink4a (Molofsky et al. 2003), have decreased proliferation in neurogenic areas.

Interestingly, increasing levels of p16Ink4a in the aging forebrain have been linked to decreased neural stem cell proliferation and senescence (Molofsky et al. 2006), indicating that characterization of the molecular regulation of Bmi and p16Ink4a will provide insight into intracellular pathways regulating stem cell proliferation. Other potential targets may in addition be transcription factors such as Mash1, Dlx (Doetsch et al. 2002a), and Pax6 (Hack et al. 2005), which appear to be expressed within specific proliferating subpopulations of the neurogenic lineage.

Many other interesting questions concerning the regulation of proliferation remain: Do neural stem cells really divide asymmetrically? Can neural stem cells divide symmetrically in vivo? How is asymmetry achieved and regulated? How do neural stem cells escape senescence through telomere shortening? How is the number of cell divisions of stem cell progeny limited? Finally, how is proliferation linked to fate determination?

MAINTENANCE OF THE UNDIFFERENTIATED STATE AND FATE DETERMINATION IN THE NEUROGENIC NICHE

Control of the Undifferentiated State

The undifferentiated state is a prerequisite for the generation of distinct neural lineages. During CNS development, extrinsic signals actively control the undifferentiated state of stem cells. As an example, notch ligands are expressed by neurons and activate Notch receptor signaling in neural stem cells in the developing CNS (Chitnis et al. 1995; de la Pompa et al. 1997; Henrique et al. 1997). Activation of notch signaling promotes the maintenance of a stem-cell-like morphology (Gaiano et al. 2000) and induces the expression of transcriptional effectors of the Hairy enhancer of split family, which repress the expression of differentiation genes (Hatakeyama et al. 2004). These findings illustrate how a feedback loop between stem cells and their progeny maintains the undifferentiated state of the stem cell and controls the number of newly generated differentiated cells. They also highlight the importance of transcriptional regulators, which control the expression of differentiation genes, in the regulation of the undifferentiated state.

The control of the undifferentiated state of adult neural stem cells is less well defined. In vitro, growth factors such as FGF-2 or EGF are necessary to maintain the high proliferation rate of neurosphere cells derived from the adult SEZ or DG, as well as their self-renewal and the capacity to form further neurospheres (for review, see Emsley et al. 2005).

Withdrawal of growth factors in these cultures results in the loss of proliferation and self-renewal and the consequent initiation of differentiation pathways toward neurons or postmitotic astrocytes. Thus, EGF and/or FGF-2 may fulfill several functions in vitro, and the extent to which they perform similar roles in vivo is currently unknown. However, given that only a small percentage of adult neural stem cells in the SEZ express the EGFR (Doetsch et al. 2002a), this strongly indicates that EGF is not necessary for maintenance of the undifferentiated state in vivo, but rather promotes the activation toward a fast-proliferating transit-amplifying state via up-regulation of Olig2 (Doetsch et al. 2002a; Hack et al. 2004, 2005). It is interesting to consider, however, to which extent transit-amplifying precursors are also maintained in an undifferentiated state and to which extent this undifferentiated state is actually different from the undifferentiated state of the stem cell. Notably, stem cells express astroglial identity, whereas transit-amplifying precursors—at least in the SEZ—express proteins and markers such as Olig2, NG2, and PDGFR, which are also expressed by oligodendrocyte precursors (Hack et al. 2005; Menn et al. 2006). In contrast to other stem cells systems, the stem cells in the nervous system are hence not "lineage-marker-negative," even though such a population has recently been identified in the adult brain parenchyma (Buffo et al. 2005). Nevertheless, both stem cells and transit-amplifying precursors are undifferentiated in the sense that they possess the potential to generate neurons, oligodendrocytes, and astrocytes, at least at the population level. It is thus important to identify the molecular factor that allows and regulates such multilineage potential.

Similar to the maintenance of pluripotency in embryonic stem cells (Bernstein et al. 2006; Lee et al. 2006), maintenance of multipotency in adult neural stem cells may require expression of transcriptional regulators and specific chromatin structure (Hsieh and Gage 2005), which suppress the expression of differentiation genes while keeping them accessible for activation. During development, repressor element-1 silencing transcription/neuron-restrictive silencing factor (REST/NRSF) restricts the expression of neuronal genes to neurons through transcriptional repression and epigenetic modifications in nonneuronal cells, including neuronal stem and progenitor cells (Chong et al. 1995; Schoenherr and Anderson 1995; Ballas et al. 2005). REST/NRSF is expressed in the adult DG and in adult hippocampus-derived neural stem cells (Kuwabara et al. 2004). Modification of its transcriptional repressor activity in adult hippocampal neural stem cells is sufficient to induce neuronal gene expression in vitro, arguing that REST/NRSF contributes to maintenance of the undifferentiated state of adult neural stem cells (Kuwabara et al. 2004). Sox2, a transcription factor of the SoxB1 subfamily, and the orphan nuclear receptor TLX are highly expressed in adult neural stem cells in undifferentiated cells of the neurogenic lineage, i.e., adult neural stem cells and transient-amplifying progenitor cells (Zappone et al. 2000; Shi et al. 2004; Steiner et al. 2006). Low expression levels of Sox2 (Ferri et al. 2004) and loss of TLX expression (Shi et al. 2004) lead to the loss of neural stem cells, proliferation, and neurogenesis in the adult mice, indicating that Sox2 and TLX are required for the maintenance of neural stem cells and/or proliferation in the adult neurogenesis zones.

Interestingly, Sox2 expression in neuronal precursor cells of the developing chick spinal cord counteracts the activity of proneuronal transcription factors (Bylund et al. 2003). Moreover, reporter assays in isolated neural stem cells have shown that TLX can repress the expression of genes associated with glial differentiation (Shi et al. 2004). These data suggest that Sox2 and TLX contribute to the maintenance of proliferation in the adult neurogenic zones and promote the undifferentiated state. Interestingly, there is evidence that FGF signaling (Mansukhani et al. 2005; Takemoto et al. 2006) and PEDF signaling (Ramirez-Castillejo et al. 2006) control the expression of Sox2 during development of different organs, raising the possibility that these pathways may regulate an undifferentiated state in vivo. In the future, further characterization of the regulation of Sox2 and TLX expression in adult neural stem cells may yield important clues toward the identity of the extrinsic pathways controlling maintenance of the undifferentiated state.

Fate Determination

Acquisition of a neuronal fate is arguably the most distingushing characteristic of neural stem cell behavior between the neurogenic areas and other regions of the adult CNS and is evidently crucial for the neurogenic process. Recent fate-mapping data from several groups using different experimental paradigms (Ahn and Joyner 2005; Hack et al. 2005; Menn et al. 2006) indicated that stem cells of the adult SEZ in addition to neurons also give rise to new myelinating oligodendrocytes in the corpus callosum, striatum, and fimbria fornix, indicating that mechanisms regulating fate decisions between oligodendrocytes and neurons are operating in the SEZ/OB system. Whether stem cells of the SEZ/OB system generate new astrocytes is less clear as it is difficult or rather impossible to discriminate astrocytes with stem cell properties from "other" astrocytes (see above). It is also a matter of debate whether stem cells in the

SGZ generate all neural lineages. Retrovirus-labeling experiments have shown that cells expressing markers for oligodendrocytes or astrocytes are newly generated in the DG (see, e.g., van Praag et al. 2002). However, there is currently no strong evidence that new glia and neurons are derived from the same or distinct precursors, and it is therefore not clear whether there are competing neural fate programs operating in the hippocampal neurogenic niche.

Intrinsic Control of Neuronal Fate Determination

In contrast, neurogenesis and oligodendrogliogenesis clearly occur in the adult SEZ where manipulation of fate determinants has demonstrated that the neurogenic and oligodendrogliogenic program is competing with each other, thus providing indirect evidence for a common origin of these two lineages. For example, activation of Olig2 in adult SEZ precursors potently interferes with the progression toward the neurogenic lineage and the arrival of new neurons in the OB, whereas expression of Olig2 variants interfering with its normal repressor activity promote neurogenesis and abolish the endogenous low level of oligodendrogliogenesis completely (Hack et al. 2005; Marshall et al. 2005). Similarly, overexpression of the neurogenic fate determinant Pax6 interferes with the endogenous level of oligodendrocyte generation and migration from the SEZ into the corpus callosum (CC) (Hack et al. 2005). Most strikingly, this decision between neuro- and oligodendrogliogenesis is regulated by bone morphogenetic proteins (BMPs) in the adult SEZ (D. Colak et al., unpubl.). Upon inhibition of BMP-mediated signaling monitored by phosphorylation of Smad1,5,8 either by infusion of the extracellular antagonist noggin or by conditional deletion of the central signaling mediator Smad4 in adult neural stem cells, neurogenesis is strongly decreased and oligodendrogliogenesis is potently increased (D. Colak et al., unpubl.). Strikingly, this regulation occurs at the transition between stem cells and the first transit-amplifying precursors, as deletion of Smad4 at later stages of the lineage does not exert any defects in adult OB neurogenesis (D. Colak et al., unpubl.). Taken together, these data suggest a common origin of oligodendrocytes and neurons from precursors of the adult SEZ and demonstrate the necessity to suppress the oligodendroglial lineage program for neurogenesis to occur.

As mentioned above, interneurons destined for the OB may arise not only from the SEZ, but also at further rostral positions in the rostral migratory stream (RMS) (Hack et al. 2005). This conclusion is based on the fact that viral vector injection into the SEZ as a permanent lineage tracer reveals hardly any progeny in the glomerular layer (Hack et al. 2005), although it is also known that interneurons in the glomerular layer are constantly generated in the adult and constitute about 20% of all newborn neurons labeled with BrdU (Winner et al. 2002). Notably, however, injection of the same viral vector into the RMS reveals a prominent contribution to the glomerular neuron lineage, and BrdU birth dating confirmed that new neurons labeled with BrdU arrive in the glomerular layer already a few days after BrdU injection (Hack et al. 2005). Interestingly, a similar local source of progenitors for glomerular neurons also exists during development (Vergano-Vera et al. 2006). Thus, these data imply that three zones of adult neurogenesis persist into adulthood. Notably, two of them, the ones close to the OB and the SGZ, seem not to generate oligodendrocytes, even though they have the potential to do so after culturing in the neurosphere system (Gritti et al. 2002; Emsley et al. 2005).

Most importantly, the neuronal subtypes that are generated in the three neurogenic areas are distinct. Stem cells in the SGZ generate glutamatergic granule neurons of the DG, whereas the generation of GABAergic interneurons is still disputed (Belachew et al. 2003; Liu et al. 2003). Stem cells in the SEZ give rise to the GABAergic granule neurons of the OB that constitute the majority of adult generated neurons (~80%) (Winner et al. 2002). Approximately 20% of the newborn neurons are GABAergic interneurons in the glomerular layer that largely arise close to the OB (Winner et al. 2002). Of these neurons, a small proportion (20%) also contain dopamine as a second neurotransmitter (Winner et al. 2002). Circumstantial evidence obtained from the neuropathological analysis of the OB of Parkinson's patients (Huisman et al. 2004) and the analysis of rodent models for Parkinson's disease (Winner et al. 2004, 2007) suggest that imbalance in the generation of the neuronal subtypes may contribute to olfactory symptoms in Parkinson's patients. These observations emphasize that the differentiation into neuronal subtypes must be tightly controlled in order to support olfactory function.

Indeed, key molecular determinants regulating the dopaminergic neuron phenotype in the OB during development and in adulthood have been identified recently. The transcription factor Pax6 has been demonstrated to be necessary and sufficient for the generation of periglomerular neurons and promotes the acquisition of a dopaminergic fate as monitored by expression of tyrosine hydroxylase, the rate-limiting enzyme for transmitter synthesis (Hack et al. 2005; Kohwi et al. 2005). Overexpression of Pax6 within the RMS results in the complete blockade of granule neuron differentiation and the full conversion to a periglomerular neuron fate

(Hack et al. 2005). Moreover, almost half of these neurons acquire tyrosine hydroxylase immunoreactivity (Hack et al. 2005). Interestingly, the transcription factor Dlx2 exerts the same effect, strongly suggesting that cooperation of these transcription factors specifies dopaminergic neuron fate, similar to their function during development (Andrews et al. 2003).

These data prompt the question of the similarity of fate determination during development and in adult neurogenesis. The above-described similar functions of Pax6 and the basic helix-loop-helix (bHLH) transcription factor Mash1 that is required for neurogenesis and oligodendrogliogenesis both during development and in postnatal stages (Parras et al. 2004) support the concept that similar neural fate determinants act during development and adulthood. Moreover, the apparent maintenance of regional differences in neuronal subtypes generated in the adult brain and the neuronal fate determinants involved in their specification are further reminiscent of similar mechanisms and regionalization during development. Although Gsh2, Olig2, Dlx1,2,5,6, and SP8 are expressed from development into adulthood along the lateral wall of the lateral ventricle (Anderson et al. 1997; Malatesta et al. 2003; Stenman et al. 2003; Waclaw et al. 2004), these transcription factors are absent in the embryonic and adult medial telencephalon, the anlage of the hippocampal formation. These transcription factors are also absent from the adult SGZ, whereas Tbr1, Prox1, and NeuroD are expressed in the developing and adult hippocampus (Pleasure et al. 2000; Hevner et al. 2006). Interestingly, the predominance of Pax6 and Emx1/2 expression in the dorsal region of the telencephalon (Gangemi et al. 2001; Hack et al. 2004, 2005; Hevner et al. 2006; Willaime-Morawek et al. 2006) also corresponds well to the regionalized expression of these transcription factors during development. Thus, the regionalized expression of neurogenic fate determinants persists, with some changes, into adulthood and obviously contributes to the generation of distinct neuronal lineages in distinct adult neurogenic niches.

These data also imply that—in contrast to other stem cell systems—there is no "panneurogenic" fate determination in the CNS either in development or in adulthood. Rather, there may be as many neuronal fate determinants as there are neuronal subtypes in the adult brain. Despite the regional diversty in key neurogenic fate determinants, they exhibit a hierarchy in each region employing a transcripional cascade with earlier and progressively later steps in fate determination. This is best exemplified by the transcriptional cascade of proneural bHLH transcription factors. bHLH transcription factors regulate the commitment to neuronal fate and subsequent terminal neuronal differentiation through the control of cell cycle exit and the expression of neuron-specific proteins. For example, the transcription factor neurogenin2 first acts as an early neuronal fate determinant, but it also regulates later aspects of neuronal differentiation and initiates the expression of bHLH transcription factors later in the cascade, such as Nscl or NeuroD (for review, see Guillemot 2005). Similar cascades can also be found for homeodomain transcription factors, such as Pax6, Tbr2, and Tbr1, that are expressed in this sequence in regions where glutamatergic neurons are generated (Englund et al. 2005; Hevner et al. 2006). Notably, these transcription factors are also expressed in the adult DG, prompting the question of their role in adult neuronal fate specification in this region. Indeed, although much has been learned about the molecular control of intrinsic fate determinants for neurogenesis and neuronal subtype specification in the OB, much less is known in this regard about neurogenesis in the DG. Mice carrying a null mutation for the bHLH transcription factor NeuroD lack the granule cell layer of the DG, demonstrating that NeuroD is essential for the generation of dentate granule neurons in the developing hippocampus (Liu et al. 2000). In the adult DG, immature newborn neurons express NeuroD, suggesting that NeuroD may continue to have an essential role in hippocampal neurogenesis during adulthood (Steiner et al. 2006). Overexpression of NeuroD is sufficient to induce the expression of neuronal markers in adult hippocampal stem cell cultures (Hsieh et al. 2004a). Depolarization of neuronal stem/progenitor cells in culture and GABA-induced depolarization of transient-amplifying cells in hippocampal slice cultures increase the expression of NeuroD in these cell populations (Deisseroth et al. 2004; Tozuka et al. 2005). Moreover, Wnt signaling regulates neuronal fate commitment in the fetal CNS through the control of bHLH transcription factor expression (Hirabayashi et al. 2004), suggesting that GABA and Wnt signaling may converge onto the transcriptional control of proneurogenic bHLH factors to stimulate neuronal fate determination of stem cells in the adult SGZ (Jagasia et al. 2006).

A key issue in fate specification is how different transcriptional regulators may be coordinated for fate changes. The acquisition of a neuronal fate requires the loss of the undifferentiated state at the same time as the initiation of expression of neurogenic fate determinants. Maintenance of the undifferentiated state of neural stem cells relies on transcriptional repression and chromatin modifications at the site of differentiation-associated proteins as well as transcriptional activation of differentiation inhibitors. Consequently, neuronal fate determination is likely to be accompanied by reversal of these mechanisms.

As discussed above, Sox2 and TLX are involved in the maintenance of the undifferentiated state (Ferri et al. 2004; Shi et al. 2004). Both transcription factors are absent in neuronally committed cells, which may be a prerequisite for the expression of neuronal fate determination genes. How the expression and activity of Sox2 and TLX are down-regulated in the process of neuronal fate commitment and whether such regulation is taking place on the transcriptional, translational, or posttranslational level is not known. Similarly, the expression of Olig2 in transit-amplifying precursors needs to be down-regulated for cells to proceed toward neurogenesis (Hack et al. 2004, 2005), and the failure of Olig2 downregulation is at least partly responsible for the failure to elicit neurogenesis in nonneurogenic regions after injury (Buffo et al. 2005).

The control of the transcriptional repressor activity of REST/NRSF (Schoenherr and Anderson 1995) adds an additional layer of complexity to the transcriptional regulation in neuronal fate determination. In cultured adult hippocampal neural stem cells, REST/NRSF represses neuronal gene expression through binding to a defined 21–23-bp sequence named neuron-restrictive silencing element (NRSE) and the recruitment of the mSin3/histone deacetylase (HDAC) 1/2 and the CoREST MeCP2 repressor complexes (Kuwabara et al. 2004). Upon forced differentiation, a noncoding RNA containing the NRSE sequence is expressed and forms double-stranded RNA (dsRNA). NRSE dsRNA binds to the REST/NRSF repressor complex and initiates the switching of cofactors from repressors to activators to drive the expression of neuron-specific genes (Kuwabara et al. 2004). The conversion of the REST/NRSF complex from a transcriptional repressor complex into an transcriptional activator complex may be unique to adult neurogenesis, since neuronal fate determination during development appears to rely on the repression of REST/NRSF expression (Ballas et al. 2005).

The REST/NRSF-dependent mechanism illustrates that cofactors—which regulate the state of the chromatin, for example, through the control of the acetylation and deacetylation state of histone—have a key regulatory role for neuronal fate determination (Hsieh and Gage 2005). During embryonic development, the differentiation of neural stem cells into neurons is paralleled by a transition from "stem cell" chromatin to "neuronal fate" chromatin as evidenced by changes in histone acetylation in genes associated with the neuronal lineage (Ballas et al. 2005). Consistent with the role of histone modifications in fate determination of neural stem cells during development, we have recently unraveled a role for HDAC 2 in adult neurogenesis and neuronal differentiation, in both the SEZ and SGZ (M. Jawerka et al., unpubl.). In addition, treatment of stem cell cultures derived from the adult hippocampus with HDAC inhibitors promotes the expression of neuron-specific proteins including NeuroD in isolated hippocampal neural stem cells and forebrain-derived neural stem cells (Hsieh et al. 2004a; Siebzehnrubl et al. 2007). Curiously, treatment with HDAC inhibitors also potently suppressed the expression of glia-specific proteins (Hsieh et al. 2004a). Thus, the balance of histone deacetylation and acetylation may be especially important to control the progression from an undifferentiated state toward a neuronal fate. Taken together, a number of intrinsic machineries appear to be involved in the transition from an undifferentiated state toward a neuronally committed state. How these machineries are linked to the extrinsic neurogenic signals and how they are coordinated will be an important question to answer in order to understand the process of neuronal fate commitment in adult neurogenesis.

Extrinsic Control of Neuronal Fate Determination and Neuronal Subtype Specification

Cells with the potential to proliferate extensively and to differentiate into neurons and glia have been isolated not only from the adult DG or the SEZ, but also from many regions of the adult CNS (for review, see Lie et al. 2004). Despite their ability to differentiate into neurons and glia in vitro and their proliferative activity in vivo, the endogenous generation of neurons is limited to the DG and the SEZ/OB system. In all other regions, which have been summarized as nonneurogenic or gliogenic regions, neurogenesis fails to occur at significant levels even after brain injury (for review, see Lie et al. 2004). However, these progenitors can react to the same fate determinants that act in adult neurogenesis, such as the transcription factor Pax6. Although no neurogenesis is observed in the adult mouse cortex after stab wound injury upon injection of a control virus expressing green fluorescent protein (GFP) only, about 20% of the infected cells generate young neurons upon injection of a viral vector containing the transcription factor Pax6 (Buffo et al. 2005). Moreover, precursors derived from nonneurogenic regions of the adult CNS give rise to neurons following transplantation into neurogenic areas (Shihabuddin et al. 2000; Lie et al. 2002), demonstrating that at least a subset of precursors in the adult brain still have the potential to react to extrinsic neurogenic signals. Indeed, the adult brain parenchyma contains potent gliogenic signals, as even progenitor cells from the adult neurogenic niches, which already express the early neuron-associated marker polysialic-acid-neural cell adhesion molecule (PSA-NCAM), exclusively give rise to glial cells upon transplantation into normal brain parenchyma (Seidenfaden et al. 2006). These observations demonstrate that environment-derived signals are crucial in controlling the decision between a neuronal fate and a glial fate. Thus, the identification of these regulatory factors is an important aim to allow neuronal stem-cell-mediated repair processes.

One source of local signals in the neurogenic niche are the ependymal cells lining the ventricles that are in intimate contact with neural stem cells in the adult SEZ. It has been proposed that ependymal cells produce the BMP ligands and the antagonist noggin, which acts as a feedback inhibitor to restrict BMP-signaling activity in this region to promote the generation of neurons at the expense of astrocytes (Lim et al. 2000). Our recent data, however, demonstrate that the BMP pathway is active early in the neurogenic lineage of SEZ and that loss of BMP signaling leads to an increased generation of oligodendrocytes at the expense of neurons (D. Colak et al., unpubl.) These data are in line with the more recent model that stem cells in the SEZ face the decision to adopt either a neuronal fate or an oligodendrocytic fate (Menn et al. 2006) and link the BMP pathway to this fate decision. Ependymal cells are not present in the adult DG and activity of the BMP pathway as monitored by phosphorylation of Smad1,5,8 is absent from the SGZ (D. Colak et al., unpubl.), indicating that BMP signaling is specifically regulating neurogenesis in the SEZ.

Another source of crucial neurogenic factors are the astrocytes within the neurogenic niches of both the adult SEZ and the hippocampus. Coculture with astrocytes derived from the neonatal SEZ promotes the generation of colonies of immature neurons from adult SEZ-derived stem/progenitor cells (Lim and Alvarez-Buylla 1999). Similarly, adult hippocampal astrocytes increase neuronal fate commitment of adult-hippocampus-derived stem/progenitor cells in coculture assays (Song et al. 2002). This neurogenic effect is mediated by secreted and membrane-bound factors. Interestingly, astrocytes from gliogenic regions such as the adult spinal cord do not stimulate the neuronal differentiation of neural stem cells, indicating that astrocytes regionally differ in their expression of neurogenic signals (Song et al. 2002). Stem cells in the adult SEZ and SGZ display many characteristics of astrocytes (for review, see Alvarez-Buylla et al. 2002; Götz 2003). Thus, coculture assays may be modeling the interaction of stem cells with their progeny, rather than interaction of stem and progenitor cells with astrocytes outside of the neurogenic lineage. This would indicate that stem cells in the SEZ and SGZ create their own neurogenic microenvironment or that astrocytes come in two variants, as stem cell astrocytes and niche astrocytes.

Several members of the Wnt family of signaling molecules are highly expressed in or close to the adult DG (Shimogori et al. 2004). Wnt3 is expressed by hippocampal astrocytes (Lie et al. 2005). Blocking of Wnt signaling in hippocampal astrocyte/neural stem cell cocultures inhibits the generation of neurons. Conversely, overexpression of Wnt3 in hippocampal neural stem cells strongly increases the generation of immature neurons in vitro through stimulating neuronal fate determination and proliferation of neuroblasts. Most importantly, increased Wnt signaling enhances hippocampal neurogenesis, whereas inhibition of Wnt signaling blocks hippocampal neurogenesis almost completely, demonstrating that Wnt signaling is essential for the maintenance of neurogenesis (Lie et al. 2005).

Neurons have the unique capacity to communicate via electrical activity, and there is an increasing amount of data indicating that increased activity of the hippocampal neural network is associated with enhanced neurogenesis in the DG, whereas inhibition of the neuronal input into the DG reduces neurogenesis (Meltzer et al. 2005; Ge et al. 2007). It has been observed that tetanic stimulation of the DG increases the release of Wnt proteins by neurons, which provides a potential molecular mechanism for the activity dependency of neurogenesis (Chen et al. 2006).

Depolarization of neural stem cells in astrocyte cocultures strongly increases in vitro neurogenesis (Deisseroth et al. 2004). Progenitor cells in the adult SGZ can be depolarized by the neurotransmitter GABA, which is released by inhibitory interneurons of the SGZ (Tozuka et al. 2005; Wang et al. 2005; Ge et al. 2006). $GABA_A$-receptor-mediated depolarization of stem cell progeny enhances expression of neuronal fate determinants (Tozuka et al. 2005) and promotes the development of neuronal morphology (Ge et al. 2006). Pharmacological stimulation of $GABA_A$ receptors in vivo increases the percentage of neurons among newborn cells in the adult DG, further supporting the notion that GABA-induced depolarization is contributing to neuronal fate determination (Tozuka et al. 2005).

Given that GABA release is dependent on hippocampal network activity, it has been hypothesized that GABA acts as an activity sensor in the regulation of hippocampal neurogenesis (Ge et al. 2007).

Control of Glial Cell Fates by Extrinsic Signals

Lineage tracing has shown that stem cells in the adult SEZ generate oligodendrocytes in addition to neurons (Ahn and Joyner 2005; Hack et al. 2005; Menn et al. 2006). Earlier work has shown that PDGF-A-induced signaling through PDGFRα regulates oligodendrocyte production during development (Calver et al. 1998; Fruttiger et al. 1999). PDGF-A is highly expressed in the adult SEZ. SEZ stem cells but not transient-amplifying cells express PDGFRα (Jackson et al. 2006). Importantly, PDGFRα is phosphorylated in a fraction of SEZ stem cells, demonstrating that the PDGF-A/PDGFRα pathway is active under physiological circumstances in vivo. Infusion of PDGF-A into the lateral ventricle stimulates the proliferation of SEZ stem cells and leads to the formation of large nodules containing cells that express Olig2 (Jackson et al. 2006), a bHLH transcription factor crucial for the generation of oligodendrocytes from the adult SEZ as described above (Hack et al. 2005). In contrast, the production of neuroblasts is blocked by stimulation of the PDGF-A/PDGFRα pathway. Importantly, conditional ablation of PDGFRα in adult neural stem cells blocks oligodendrogenesis in the SEZ/OB system (Jackson et al. 2006). Thus, signaling through the PDGF-A/PDGFRα pathway promotes expression of Olig2 and thereby controls the generation of new oligodendrocytes. In addition, these observations strongly suggest that there is no strict distinction between proliferative signals and differentiative signals and that fate determination may already take place during the initial proliferation of the stem cell.

It has also been shown that the insulin-like growth factor (IGF)-signaling pathway can instruct hippocampal neural stem cells to adopt a oligodendrocytic fate in vitro (Hsieh et al. 2004b). Given that there is currently no convincing evidence for the generation of oligodendrocytes in the hippocampal neurogenic niche, it is not clear whether this pathway is of in vivo relevance. However, this observation could suggest that signals are present in the hippocampus that inhibit IGF-induced oligendrogenesis, consistent with the absence of Olig2 in the adult SGZ.

There are two opposing views to the question of whether new astrocytes are generated in the neurogenic areas. Fate-mapping studies have failed to provide evidence that neural stem cells give rise to astrocytes in locations distinct from the stem cell niche (see, e.g., Ahn and Joyner 2005). Therefore, one view is that de novo astrocytogenesis is not taking place. It has, however, been proposed that the astrocytic fate may be a default pathway that is not regulated by determinative bHLH transcription factors but is entered if proneurogenic and prooligenic bHLH transcription factors are not expressed (Kintner 2002). As a consequence, astrocytes could be stimulated to become neurons or oligodendrocytes upon up-regulation of Olig2, which would identify Olig2-positive transit-amplifying precursors as good candidates for precursors of neurons and oligodendrocytes in the adult brain. Indeed, this is supported by the corresponding increase or decrease of neurogenesis and oligodendrogliogenesis as discussed above (Hack et al. 2005).

Adult neural stem cells display many features of astrocytes (Doetsch et al. 1999; Garcia et al. 2004; Raponi et al. 2007). Adult neural stem cells undergo self-renewing divisions in which new stem cells with astrocytic features are generated. Hence, a second view is that during asymmetric divisions of neural stem cells, one daughter cell adopts an astrocytic, i.e., neural stem cell fate.

Such an argument raises the question of whether signals such as notch signaling, BMP signaling, and leukemia inhibitory factor (LIF) signaling, which are potent inducers of astrocytic fate during development and in cultured adult neural stem cells (Nakashima et al. 1999; Gaiano et al. 2000; Tanigaki et al. 2001; Hsieh et al. 2004a), could also be important contributors to adult neural stem cell maintenance. Curiously, BMP and LIF signaling are important in other adult stem cell niches (Fuchs et al. 2004) and embryonic stem cells (Chambers 2004) for maintaining these cells in an undifferentiated state, whereas notch signaling promotes the transition of pluripotent embryonic stem cells into neural stem cells (Lowell et al. 2006). Although there is currently no strong evidence that these signaling pathways participate in the maintenance of adult neural stem cells, it is nevertheless interesting to mention that infusion of LIF into the lateral ventricle appears to promote self-renewing divisions of neural stem cells in vivo (Bauer and Patterson 2006) and that activation of the LIF coreceptor gp130 results in notch activation and increased self-renewal in vitro (Chojnacki et al. 2003).

Adoption of Nonneural Cell Fates

New neurons in the adult CNS are generated in close proximity to blood vessels (Palmer et al. 2000; Capela and Temple 2002). In songbirds, angiogenesis and neurogenesis are closely linked through an intricate cross-regulation involving VEGF and brain-derived neurotrophic factor (BDNF) signaling pathways (Louissaint et al. 2002). In the adult hippocampus, proliferating precursors are located in close proximity to the vasculature and about one third of the newborn cells express endothelial cell markers (Palmer et al. 2000). Moreover, behavioral conditions that promote hippocampal neurogenesis in vivo also appear to promote remodeling of the vasculature in the DG (van Praag et al. 2005; Pereira et al. 2007). In vitro, adult hippocampal neural stem cells give rise to a small percentage of cells expressing endothelial cells markers, which are capable of forming capillary networks (Wurmser et al. 2004). The commitment toward an endothelial fate is rather unexpected because endothelial cells are considered to be mesoderm derivatives as opposed to neuronal cells, which are of ectodermal origin. Although proof is missing that neural stem cells give rise to endothelial cells in vivo, these findings raise the intriguing possibility that neural stem cells can commit to an endothelial fate to modify their vascular niche, which could not only provide signals through cell contact and secretion of signaling molecules, but also allow signals present in the circulation to participate closely in the regulation of neurogenesis. Indeed, complement factors as well as T cells in cooperation with microglia modulate adult neurogenesis (Rahpeymai et al. 2006; Walton et al. 2006; Ziv et al. 2006).

MIGRATION

The correct positioning of neurons is a prerequisite for their functional integration into neuronal networks. During development, two major modes of migration are distinguished: radial migration, which is defined as a migratory mode that depends on the close interaction of migrating neurons and the processes of radial glial cells, and tangential migration, which is defined as a radial-glia-independent migratory mode (for review, see Ayala et al. 2007). The classic radial glia of the developing CNS is no longer existent in the adult brain, even though glial cells with radial morphology have been detected in the adult OB and are a characteristic feature of the adult DG.

In the adult DG, new immature neurons translocate a few cell diameters from the SGZ into the molecular cell layer and come to rest preferentially in the lower third of the granule cell layer (Kempermann et al. 2003; Ahn and Joyner 2005; Esposito et al. 2005; Laplagne et al. 2006). The upper two thirds of the DG appear to be occupied predominantly by granule neurons that were generated during the early postnatal period (Ahn and Joyner 2005; Esposito et al. 2005; Laplagne et al. 2006). Whether this spatial separation is of functional significance remains to be determined. The translocation of the adult-born neurons takes place in close association with the radial-glial-like process of the neural stem cell that spans the entire granule cell layer, suggesting that the radial process may guide the new neurons during their translocation (Seri et al. 2004). One potential mechanism for the translocation is mechanical displacement of immature neurons by cells that have been generated later and are located beneath their "older" cousins. The relative confinement of the adult-born neurons to the lower third of the dentate granule cell layer, the close association with the radial-glia-like process, and the transient expression of doublecortin (Brown et al. 2003a) and the PSA-NCAM (Seki 2002), which are both necessary for proper migration in the SEZ/OB system (Tomasiewicz et al. 1993; Cremer et al. 1994; Ono et al. 1994; Hu et al. 1996; Koizumi et al. 2006), suggest that newborn dentate granule neurons are actively migrating and that the migration process is tightly controlled in a temporal and spatial manner.

A frequently observed phenomenon in human and experimental mesial temporal lobe epilepsy is the appearance of ectopic dentate granule neurons in the hilar region, which is potentially the result of aberrant migration of newborn hippocampal neurons. Curiously, under these pathological circumstances, these migratory newborn neurons use a chain migration-like mode, which under physiological circumstances is only present in the SVZ/OB system (Parent et al. 2006). Reelin, a protein, which is essential for the detachment of immature neurons from the migratory chains in the adult OB (see below), is highly expressed in the adult DG (Heinrich et al. 2006; Gong et al. 2007). Strikingly, prolonged seizures in the rodent hippocampus lead to down-regulation of reelin, causing a dispersion not only of newborn granule neurons, but also virtually of all granule neurons. Thus, the role of reelin in affecting cell migration extends beyond migrating cells even to fully mature neurons in the adult brain that apparently still require specific signals to be kept at their position.

The migration of immature neurons in the SEZ/OB system has been studied more intensively. New neurons are born throughout the entire rostrocaudal extent of the SEZ of the lateral ventricles (Doetsch and Alvarez-Buylla 1996) and migrate through the RMS to the OB (for review, see Ghashghaei et al. 2007). The RMS follows the remnants of the most anterior extension of the lateral ventricle and can be several centimeters in length in humans (Curtis et al. 2007).

In the RMS, immature neurons migrate in chains and slide along each other (Lois et al. 1996). The formation of chains and the sliding of migrating neuroblasts along each other is highly dependent on the presence of the PSA-NCAM on the surface of the migrating neuroblasts. PSA residues allow homophilic interactions between NCAM proteins, which promote the adhesion between neighboring neuroblasts. In vivo, loss of the core protein NCAM or enzymatic removal of the PSA moieties lead to a disruption of migratory chains and reduced migration in the RMS, which ultimately results in a smaller OB (Tomasiewicz et al. 1993; Cremer et al. 1994; Ono et al. 1994; Hu et al. 1996). These findings link PSA-NCAM to the maintenance of chain migration and also demonstrate that chain migration is necessary for proper migration in the adult RMS. Importantly, loss of the PSA residues does not affect the migration of neuroblasts after their detachment from the migratory chains in the OB (Ono et al. 1994), again indicating that PSA-NCAM's major role is during the chain migration in the RMS.

During chain migration, neurons are encapsulated by astroglia (Lois et al. 1996), which appear to have stem cell character and to produce additional immature neurons that join the migrating neurons in the RMS. The glial sheath appears only postnatally and its role for migration is unclear. In some rodents, immature neurons migrate in glia-independent chains through the adult brain parenchyma, indicating that the glial sheath may be dispensable for migration and that immature neurons are guiding each other through the RMS (Luzzati et al. 2003). However, it has been shown that the glial sheath may regulate the migratory speed through regulation of ambient GABA levels (Bolteus and Bordey 2004). Moreover, it has been reported that genetic disruption of the glial sheath leads to an increased number of migrating neuroblasts in ectopic locations, suggesting that the glial sheath may serve as a mechanical boundary that prevents ectopic migration and supports the maintenance of neuroblast chains (Belvindrah et al. 2007). Notably, however, upon brain injury and tumor formation, young neuroblasts can indeed leave the chains and emigrate into the parenchyma (Arvidsson et al. 2002; Glass et al. 2005).

Upon entry into the core of the OB, there is a dramatic change in the mode and directionality of migration. Neurons detach from the chains and switch their direction to migrate radially toward the granule cell layer and the glomerular cell layer and integrate as functional interneurons. Notably, prospective granule or periglomerular neurons have to know their fate by now, as they must recognize whether to stop in the granule layer or proceed to the outer glomerular layer. Consistent with its role in this decision, the transcription factor Pax6 may influence this layer-specific mode of migration (Hack et al. 2005; Kohwi et al. 2005). Interestingly, Tenascin-R and reelin seem to be signals that have a crucial role in the switch from tangential to radial migration (Hack et al. 2002; Saghatelyan et al. 2004). Tenascin-R acts attractive on migrating neuroblasts and seems to attract them normally into the OB. Reelin and Tenascin-R are expressed in the core of the OB, where neuroblasts detach from the chains to subsequently migrate toward the granule cell layer and the periglomerular layer (Hack et al. 2002; Saghatelyan et al. 2004). In mice carrying a mutation for reelin, neuroblasts accumulate at the tip of the RMS and fail to disperse (Hack et al. 2002). Similarly, Tenascin-R mutant mice have reduced density of newborn neurons in the adult OB and show accumulation of neuroblasts in the RMS (Saghatelyan et al. 2004). Interestingly, it was also shown that the expression of Tenascin-R is dependent on the activity of the OB neuronal circuitry, suggesting that Tenascin-R may be a molecular signal that links migration and the activity-dependent integration of new neurons into the olfactory circuitry

(Saghatelyan et al. 2004). Exposure of the RMS to an ectopic source of Tenascin-R leads to premature detachment of neuroblasts from the RMS and directed migration of individualized neuroblasts toward the Tenascin-R source (Saghatelyan et al. 2004). Thus, ectopic expression of Tenascin-R allows the recruitment of adult neuroblasts to ectopic sites (Saghatelyan et al. 2004), thereby attracting them to sites of injury. Indeed, signals from injured brain regions as well as from brain tumors are known to attract adult neuronal progenitors (Magavi et al. 2000; Arvidsson et al. 2002; Glass et al. 2005). Obviously, these signals are crucial to recruit adult progenitors for repair and to restrict tumor growth, a rather exciting novel role of adult neuronal progenitor cells.

Migrating neuroblasts have a dynamic bipolar morphology characterized by a leading process that is extended in the direction of migration and a trailing process (Bellion et al. 2005; Tsai and Gleeson 2005). During the migratory process, neuroblasts undergo repetitive morphological changes: In the first step, the leading process is formed and stabilized. Subsequently, the nucleus is translocated in the direction of the leading process. This step is also termed nucleokinesis. As a final step, the trailing process is retracted and the neuroblast is ready for the next round of leading process extension (Bellion et al. 2005; Tsai and Gleeson 2005).

The leading process and the centrosome are always located at the same area of the migrating neuroblast. The centrosome has a crucial role in the migratory process. It is responsible for the organization of the microtubule network and is linked via this network with the perinuclear microtubule cage (Tanaka et al. 2004; Tsai and Gleeson 2005). Prior to nucleokinesis, the centrosome moves forward into the leading process to determine the subsequent direction of nucleokinesis, which is achieved by the remodeling of the microtubule network. Hence, microtubule network integrity and microtubule dynamics may also be important for proper migration. Indeed, mutation in the microtubule-associated protein doublecortin, which is expressed in migratory neuroblasts of the adult brain, affect the morphology of the leading process, nuclear translocation, and migration speed (Koizumi et al. 2006). Interestingly, doublecortin mutations lead to defects only in long-distance migration (for a recent review, see Ninkovic and Götz 2007).

The migration of neuroblasts in the SEZ/OB system is controlled by cell-cell contact-mediated, substrate-bound, and secreted cues, which together control speed, directionality, and mode of the migratory process.

Neuroblasts express $GABA_A$ receptors ($GABA_AR$) (Bolteus and Bordey 2004) and contain high levels of GABA (Wang et al. 2003).

Application of a $GABA_AR$-specific antagonist significantly reduces the migratory speed of neuroblasts in RMS explants. Notably, the ensheathing astrocytes express the GABA transporter subtype GAT4, and inhibition of astrocytic GABA uptake results in increased ambient GABA concentrations and reduced migratory speed (Bolteus and Bordey 2004). These observations identify GABA as a signal regulating the speed of neuroblast migration in the SEZ/OB system and identify a mechanism of how the glial sheath may participate in the control of migration.

The interplay of chemoattractive and chemorepulsive gradients is thought to have a major role for the directionality of migration. Slit proteins, which act as chemorepulsive signals in axon guidance during development, are highly expressed in the choroid plexus of the lateral ventricles and the septum, which are adjacent to the SEZ, and in migrating neuroblasts themselves (Wu et al. 1999; Marillat et al. 2002; Nguyen-Ba-Charvet et al. 2004). Robo proteins, which are the receptors for Slit proteins, are expressed by neuroblasts in the SEZ and the RMS (Marillat et al. 2002; Nguyen-Ba-Charvet et al. 2004). Studies in explants from the postnatal SEZ and the RMS have indicated that Slit proteins repel newborn neurons in these areas (Wu et al. 1999; Nguyen-Ba-Charvet et al. 2004). Chains of migrating neuroblasts are still present in adult Slit-1 knockout mice. However, these chains are directed toward the corpus callosum, indicating that the Slit/Robo system is necessary to provide orienting cues for migration in the SEZ/OB system (Nguyen-Ba-Charvet et al. 2004). It has been speculated that a caudal-to-rostral Slit gradient may contribute to the directionality of neuroblast migration and that the flow of the cerebrospinal fluid may contribute to the establishment of such a gradient (Sawamoto et al. 2006). This model, however, has been discussed controversially (Götz and Stricker 2006) and awaits further confirmation. The activity of Slit proteins is modulated by interacting pathways and molecules. In vitro studies in the early postnatal SEZ/OB system have shown that the chemorepulsive activity of Slit proteins on neuroblasts may be potentiated or may require a cofactor called migration-inducing activity (MIA). MIA is expressed by glial cells of the early postnatal SEZ and RMS. In vitro, MIA is sufficient to induce nondirectional migration in SEZ explants. In the same study, it was found that Slit proteins by themselves inhibit the migration of neuroblasts and that Slit acts as a chemorepellent only in conjunction with MIA (Mason et al. 2001). Finally, there is strong evidence that Slit proteins bind only weakly to the Robo receptor family and that heparan sulfate is necessary to stabilize their interaction (Hussain et al. 2006). Indeed, removal of heparan sulfate from the surface of SEZ-derived neuroblasts appears to be sufficient to abolish the chemorepulsive activity of Slit (Hu 2001).

Signaling through ErbB4 and Eph receptors, whose ligands, i.e., neuregulins and ephrins, can be present in a membrane-bound form, has also been linked to the control of migration. These receptors are highly expressed on neuroblasts in the RMS, whereas their respective ligands are expressed in close vicinity to the migrating chains (Conover et al. 2000; Anton et al. 2004). Conditional loss of ErbB4 in the SEZ/OB system (Anton et al. 2004; Ghashghaei et al. 2006) and inhibition of the interaction of EphB2 with its ligand (Conover et al. 2000) result in disruption of the migratory chains. The loss of directionality of migration in the ErbB4 mutants indicates that ErbB4 is not only contributing to the organization of the migratory chains, but may also be serving as a receptor for directional cues.

The formation of chains and the sliding of migrating neuroblasts along each other are highly dependent on the presence of the polysialated form of PSA-NCAM on the surface of the migrating neuroblasts. PSA residues allow homophilic interactions between NCAM and proteins, which promote the adhesion between neighboring neuroblasts. In vivo, loss of the core protein NCAM or enzymatic removal of the PSA moieties lead to a disruption of migratory chains and reduced migration in the RMS, which ultimately result in a smaller OB (Tomasiewicz et al. 1993; Cremer et al. 1994; Ono et al. 1994; Hu et al. 1996). These findings link PSA-NCAM to the maintenance of chain migration and also demonstrate that chain migration is necessary for proper migration in the adult RMS. Importantly, loss of the PSA residues does not affect the migration of neuroblasts after their detachment from the migratory chains in the OB (Ono et al. 1994), again indicating that PSA-NCAM has its major role during the chain migration in the RMS.

Finally, it is also important to highlight that extracellular matrix proteins and signaling through their receptors are important for neuroblast migration. Migrating neuroblasts express, for example, $\beta1$-integrins. In vitro, exposure of neuroblasts to the $\beta1$-integrin ligand laminin results in the aggregation of neuroblasts, indicating that laminin-induced $\beta1$-integrin signaling promotes the interaction of neuroblasts. Indeed, loss of $\beta1$-integrin in neuroblasts results in the disruption of neuroblast chains in the RMS, demonstrating that the $\beta1$-integrin signaling system is necessary for the formation and maintenance of chain migration in the RMS (Belvindrah et al. 2007).

How this multitude of extrinsic signals is translated into a coordinated migratory movement remains largely unknown. Given the central role of microtubule rearrangement for nucleokinesis and process exten-sion and the importance of dynamic cell adhesion for the regulation of chain versus individual migration, it is tempting to speculate that the extrinsic signals affect migration predominantly through posttransla-tional mechanisms. However, loss of the transcription factor aristaless-related homeobox gene, which is expressed in migrating neuroblasts, leads to the accumulation of neuroblasts in the RMS (Yoshihara et al. 2005), similar to the phenotype in mice deficient in the serum response factor (Alberti et al. 2005), a transcription factor linking extrinsic signals to the actin cytoskeleton. These observations raise the possibility that dynamic transcriptional regulation through extrinsic signals may also participate in the regulation of the migratory behavior of newborn neu-rons in the adult brain.

Other questions also remain unanswered: In the SEZ/OB neurogenic system, several neuronal subtypes are generated. En route to the OB, neu-roblasts migrate through areas that appear to be potent niches for the specification into specific neuronal subtypes, which raises the question of whether migration may be a necessary step for neuronal subtype speci-fication. It is also not known whether the different neuronal subtypes respond to different migration cues, given that there are subpopulations, which are destined for the granule cell layer and other subpopulations, that will integrate into the glomerular layer. Finally, it will also be inter-esting to determine whether new neurons are migrating toward distinct locations and are integrated into specific local circuits in response to functional demands.

CONCLUSIONS

During the past years, significant progress has been made in our under-standing of how the early steps of neurogenesis are regulated. The emerg-ing picture is that the behavior of stem cells and their progeny is continuously controlled by signals provided by their microenvironment. Some of these signals—predominantly classical developmental mor-phogens and growth factors—have been identified, and we have begun to describe their effects within the neurogenic systems on the cellular level. However, it is becoming increasingly clear that the activity of the CNS is also exerting a major effect on each step of neurogenesis, including the very first events such as stem cell activation, proliferation, and fate determination. To understand how the diversity of signals inter-act to coordinate epigenetic, transcriptional, translational, and posttrans-lational events that are necessary to promote the progression of the stem cell toward a fully mature neuron will be the challenge for the coming years.

REFERENCES

Ahn S. and Joyner A.L. 2005. In vivo analysis of quiescent adult neural stem cells responding to Sonic hedgehog. *Nature* **437:** 894–897.

Alberti S., Krause S.M., Kretz O., Philippar U., Lemberger T., Casanova E., Wiebel F.F., Schwarz H., Frotscher M., Schutz G., and Nordheim A. 2005. Neuronal migration in the murine rostral migratory stream requires serum response factor. *Proc. Natl. Acad. Sci.* **102:** 6148–6153.

Alvarez-Buylla A., Seri B., and Doetsch F. 2002. Identification of neural stem cells in the adult vertebrate brain. *Brain Res. Bull* **57:** 751–758.

Anderson S.A., Qiu M., Bulfone A., Eisenstat D.D., Meneses J., Pedersen R., and Rubenstein J.L. 1997. Mutations of the homeobox genes Dlx-1 and Dlx-2 disrupt the striatal subventricular zone and differentiation of late born striatal neurons. *Neuron* **19:** 27–37.

Andrews G.L., Yun K., Rubenstein J.L., and Mastick G.S. 2003. Dlx transcription factors regulate differentiation of dopaminergic neurons of the ventral thalamus. *Mol. Cell. Neurosci.* **23:** 107–120.

Anton E.S., Ghashghaei H.T., Weber J.L., McCann C., Fischer T.M., Cheung I.D., Gassmann M., Messing A., Klein R., Schwab M.H., Lloyd K.C., and Lai C. 2004. Receptor tyrosine kinase ErbB4 modulates neuroblast migration and placement in the adult forebrain. *Nat. Neurosci.* **7:** 1319–1328.

Arden K.C. 2007. FoxOs in tumor suppression and stem cell maintenance. *Cell* **128:** 235–237.

Arvidsson A., Collin T., Kirik D., Kokaia Z., and Lindvall O. 2002. Neuronal replacement from endogenous precursors in the adult brain after stroke. *Nat. Med.* **8:** 963–970.

Ayala R., Shu T., and Tsai L.H. 2007. Trekking across the brain: The journey of neuronal migration. *Cell* **128:** 29–43.

Ballas N., Grunseich C., Lu D.D., Speh J.C., and Mandel G. 2005. REST and its corepressors mediate plasticity of neuronal gene chromatin throughout neurogenesis. *Cell* **121:** 645–657.

Bauer S. and Patterson P.H. 2006. Leukemia inhibitory factor promotes neural stem cell self-renewal in the adult brain. *J. Neurosci.* **26:** 12089–12099.

Belachew S., Chittajallu R., Aguirre A.A., Yuan X., Kirby M., Anderson S., and Gallo V. 2003. Postnatal NG2 proteoglycan-expressing progenitor cells are intrinsically multipotent and generate functional neurons. *J. Cell Biol.* **161:** 169–186.

Bellion A., Baudoin J.P., Alvarez C., Bornens M., and Metin C. 2005. Nucleokinesis in tangentially migrating neurons comprises two alternating phases: Forward migration of the Golgi/centrosome associated with centrosome splitting and myosin contraction at the rear. *J. Neurosci.* **25:** 5691–5699.

Belvindrah R., Hankel S., Walker J., Patton B.L., and Muller U. 2007. Beta1 integrins control the formation of cell chains in the adult rostral migratory stream. *J. Neurosci.* **27:** 2704–2717.

Bernstein B.E., Mikkelsen T.S., Xie X., Kamal M., Huebert D.J., Cuff J., Fry B., Meissner A., Wernig M., Plath K., Jaenisch R., Wagschal A., Feil R., Schreiber S.L., and Lander

E.S. 2006. A bivalent chromatin structure marks key developmental genes in embryonic stem cells. *Cell* **125**: 315–326.

Bolteus A.J. and Bordey A. 2004. GABA release and uptake regulate neuronal precursor migration in the postnatal subventricular zone. *J. Neurosci.* **24**: 7623–7631.

Brown J.P., Couillard-Despres S., Cooper-Kuhn C.M., Winkler J., Aigner L., and Kuhn H.G. 2003a. Transient expression of doublecortin during adult neurogenesis. *J. Comp. Neurol.* **467**: 1–10.

Brown J., Cooper-Kuhn C.M., Kempermann G., Van Praag H., Winkler J., Gage F.H., and Kuhn H.G. 2003b. Enriched environment and physical activity stimulate hippocampal but not olfactory bulb neurogenesis. *Eur. J. Neurosci.* **17**: 2042–2046.

Buffo A., Vosko M.R., Erturk D., Hamann G.F., Jucker M., Rowitch D., and Götz M. 2005. Expression pattern of the transcription factor Olig2 in response to brain injuries: Implications for neuronal repair. *Proc. Natl. Acad. Sci.* **102**: 18183–18188.

Bylund M., Andersson E., Novitch B.G., and Muhr J. 2003. Vertebrate neurogenesis is counteracted by Sox1-3 activity. *Nat. Neurosci.* **6**: 1162–1168.

Calver A.R., Hall A.C., Yu W.P., Walsh F.S., Heath J.K., Betsholtz C., and Richardson W.D. 1998. Oligodendrocyte population dynamics and the role of PDGF in vivo. *Neuron* **20**: 869–882.

Cameron H.A. and McKay R.D. 2001. Adult neurogenesis produces a large pool of new granule cells in the dentate gyrus. *J. Comp. Neurol.* **435**: 406–417.

Capela A. and Temple S. 2002. LeX/ssea-1 is expressed by adult mouse CNS stem cells, identifying them as nonependymal. *Neuron* **35**: 865–875.

Chambers I. 2004. The molecular basis of pluripotency in mouse embryonic stem cells. *Cloning Stem Cells* **6**: 386–391.

Chen J., Park C.S., and Tang S.J. 2006. Activity-dependent synaptic WNT release regulates hippocampal long-term potentiation. *J. Biol. Chem.* **281**: 11910–11916.

Cheng T. and Scadden D.T. 2002. Cell cycle entry of hematopoietic stem and progenitor cells controlled by distinct cyclin-dependent kinase inhibitors. *Int. J. Hematol.* **75**: 460–465.

Cheng T., Rodrigues N., Shen H., Yang Y., Dombkowski D., Sykes M., and Scadden D.T. 2000. Hematopoietic stem cell quiescence maintained by p21cip1/waf1. *Science* **287**: 1804–1808.

Chitnis A., Henrique D., Lewis J., Ish-Horowicz D., and Kintner C. 1995. Primary neurogenesis in *Xenopus* embryos regulated by a homologue of the *Drosophila* neurogenic gene Delta. *Nature* **375**: 761–766.

Chojnacki A., Shimazaki T., Gregg C., Weinmaster G., and Weiss S. 2003. Glycoprotein 130 signaling regulates Notch1 expression and activation in the self-renewal of mammalian forebrain neural stem cells. *J. Neurosci.* **23**: 1730–1741.

Chong J.A., Tapia-Ramirez J., Kim S., Toledo-Aral J.J., Zheng Y., Boutros M.C., Altshuller Y.M., Frohman M.A., Kraner S.D., and Mandel G. 1995. REST: A mammalian silencer protein that restricts sodium channel gene expression to neurons. *Cell* **80**: 949–957.

Conover J.C., Doetsch F., Garcia-Verdugo J.M., Gale N.W., Yancopoulos G.D., and Alvarez-Buylla A. 2000. Disruption of Eph/ephrin signaling affects migration and proliferation in the adult subventricular zone. *Nat. Neurosci.* **3**: 1091–1097.

Cremer H., Lange R., Christoph A., Plomann M., Vopper G., Roes J., Brown R., Baldwin S., Kraemer P., Scheff S., et al. 1994. Inactivation of the N-CAM gene in mice results in size reduction of the olfactory bulb and deficits in spatial learning. *Nature* **367**: 455–459.

Curtis M.A., Kam M., Nannmark U., Anderson M.F., Axell M.Z., Wikkelso C., Holtas S., van Roon-Mom W.M., Bjork-Eriksson T., Nordborg C., Frisen J., Dragunow M., Faull R.L., and Eriksson P.S. 2007. Human neuroblasts migrate to the olfactory bulb via a lateral ventricular extension. *Science* **315:** 1243–1249.

Deisseroth K., Singla S., Toda H., Monje M., Palmer T.D., and Malenka R.C. 2004. Excitation-neurogenesis coupling in adult neural stem/progenitor cells. *Neuron* **42:** 535–552.

de la Pompa J.L., Wakeham A., Correia K.M., Samper E., Brown S., Aguilera R.J., Nakano T., Honjo T., Mak T.W., Rossant J., and Conlon R.A. 1997. Conservation of the Notch signalling pathway in mammalian neurogenesis. *Development* **124:** 1139–1148.

Doetsch F. and Alvarez-Buylla A. 1996. Network of tangential pathways for neuronal migration in adult mammalian brain. *Proc. Natl. Acad. Sci.* **93:** 14895–14900.

Doetsch F., Caille I., Lim D.A., Garcia-Verdugo J.M., and Alvarez-Buylla A. 1999. Subventricular zone astrocytes are neural stem cells in the adult mammalian brain. *Cell* **97:** 703–716.

Doetsch F., Petreanu L., Caille I., Garcia-Verdugo J.M., and Alvarez-Buylla A. 2002a. EGF converts transit-amplifying neurogenic precursors in the adult brain into multipotent stem cells. *Neuron* **36:** 1021–1034.

Doetsch F., Verdugo J.M., Caille I., Alvarez-Buylla A., Chao M.V., and Casaccia-Bonnefil P. 2002b. Lack of the cell-cycle inhibitor p27Kip1 results in selective increase of transit-amplifying cells for adult neurogenesis. *J. Neurosci.* **22:** 2255–2264.

Emsley J.G., Mitchell B.D., Kempermann G., and Macklis J.D. 2005. Adult neurogenesis and repair of the adult CNS with neural progenitors, precursors, and stem cells. *Prog. Neurobiol.* **75:** 321–341.

Englund C., Fink A., Lau C., Pham D., Daza R.A., Bulfone A., Kowalczyk T., and Hevner R.F. 2005. Pax6, Tbr2, and Tbr1 are expressed sequentially by radial glia, intermediate progenitor cells, and postmitotic neurons in developing neocortex. *J. Neurosci.* **25:** 247–251.

Esposito M.S., Piatti V.C., Laplagne D.A., Morgenstern N.A., Ferrari C.C., Pitossi F.J., and Schinder A.F. 2005. Neuronal differentiation in the adult hippocampus recapitulates embryonic development. *J. Neurosci.* **25:** 10074–10086.

Ferri A.L., Cavallaro M., Braida D., Di Cristofano A., Canta A., Vezzani A., Ottolenghi S., Pandolfi P.P., Sala M., DeBiasi S., and Nicolis S.K. 2004. Sox2 deficiency causes neuro-degeneration and impaired neurogenesis in the adult mouse brain. *Development* **131:** 3805–3819.

Filippov V., Kronenberg G., Pivneva T., Reuter K., Steiner B., Wang L.P., Yamaguchi M., Kettenmann H., and Kempermann G. 2003. Subpopulation of nestin-expressing progenitor cells in the adult murine hippocampus shows electrophysiological and morphological characteristics of astrocytes. *Mol. Cell. Neurosci.* **23:** 373–382.

Fruttiger M., Karlsson L., Hall A.C., Abramsson A., Calver A.R., Bostrom H., Willetts K., Bertold C.H., Heath J.K., Betsholtz C., and Richardson W.D. 1999. Defective oligo-dendrocyte development and severe hypomyelination in PDGF-A knockout mice. *Development* **126:** 457–467.

Fuchs E., Tumbar T., and Guasch G. 2004. Socializing with the neighbors: Stem cells and their niche. *Cell* **116:** 769–778.

Gabay L., Lowell S., Rubin L.L., and Anderson D.J. 2003. Deregulation of dorsoventral patterning by FGF confers trilineage differentiation capacity on CNS stem cells in vitro. *Neuron* **40:** 485–499.

Gaiano N., Nye J.S., and Fishell G. 2000. Radial glial identity is promoted by Notch1 signaling in the murine forebrain. *Neuron* **26:** 395–404.

Gangemi R.M., Daga A., Marubbi D., Rosatto N., Capra M.C., and Corte G. 2001. Emx2 in adult neural precursor cells. *Mech. Dev.* **109:** 323–329.

Garcia A.D., Doan N.B., Imura T., Bush T.G., and Sofroniew M.V. 2004. GFAP-expressing progenitors are the principal source of constitutive neurogenesis in adult mouse forebrain. *Nat. Neurosci.* **7:** 1233–1241.

Ge S., Pradhan D.A., Ming G.L., and Song H. 2007. GABA sets the tempo for activity-dependent adult neurogenesis. *Trends Neurosci.* **30:** 1–8.

Ge S., Goh E.L., Sailor K.A., Kitabatake Y., Ming G.L., and Song H. 2006. GABA regulates synaptic integration of newly generated neurons in the adult brain. *Nature* **439:** 589–593.

Ghashghaei H.T., Lai C., and Anton E.S. 2007. Neuronal migration in the adult brain: Are we there yet? *Nat. Rev. Neurosci.* **8:** 141–151.

Ghashghaei H.T., Weber J., Pevny L., Schmid R., Schwab M.H., Lloyd K.C., Eisenstat D.D., Lai C., and Anton E.S. 2006. The role of neuregulin-ErbB4 interactions on the proliferation and organization of cells in the subventricular zone. *Proc. Natl. Acad. Sci.* **103:** 1930–1935.

Glass R., Synowitz M., Kronenberg G., Walzlein J.H., Markovic D.S., Wang L.P., Gast D., Kiwit J., Kempermann G., and Kettenmann H. 2005. Glioblastoma-induced attraction of endogenous neural precursor cells is associated with improved survival. *J. Neurosci.* **25:** 2637–2646.

Gong C., Wang T.W., Huang H.S., and Parent J.M. 2007. Reelin regulates neuronal progenitor migration in intact and epileptic hippocampus. *J. Neurosci.* **27:** 1803–1811.

Götz M. 2003. Glial cells generate neurons—Master control within CNS regions: Developmental perspectives on neural stem cells. *Neuroscientist* **9:** 379–397.

Götz M. and Stricker S.H. 2006. Go with the flow: Signaling from the ventricle directs neuroblast migration. *Nat. Neurosci.* **9:** 470–472.

Gritti A., Bonfanti L., Doetsch F., Caille I., Alvarez-Buylla A., Lim D.A., Galli R., Verdugo J.M., Herrera D.G., and Vescovi A.L. 2002. Multipotent neural stem cells reside into the rostral extension and olfactory bulb of adult rodents. *J. Neurosci.* **22:** 437–445.

Guillemot F. 2005. Cellular and molecular control of neurogenesis in the mammalian telencephalon. *Curr. Opin. Cell Biol.* **17:** 639–647.

Hack I., Bancila M., Loulier K., Carroll P., and Cremer H. 2002. Reelin is a detachment signal in tangential chain-migration during postnatal neurogenesis. *Nat. Neurosci.* **5:** 939–945.

Hack M.A., Sugimori M., Lundberg C., Nakafuku M., and Götz M. 2004. Regionalization and fate specification in neurospheres: The role of Olig2 and Pax6. *Mol. Cell. Neurosci.* **25:** 664–678.

Hack M.A., Saghatelyan A., de Chevigny A., Pfeifer A., Ashery-Padan R., Lledo P.M., and Götz M. 2005. Neuronal fate determinants of adult olfactory bulb neurogenesis. *Nat. Neurosci.* **8:** 865–872.

Hatakeyama J., Bessho Y., Katoh K., Ookawara S., Fujioka M., Guillemot F., and Kageyama R. 2004. Hes genes regulate size, shape and histogenesis of the nervous system by control of the timing of neural stem cell differentiation. *Development* **131:** 5539–5550.

Heinrich C., Nitta N., Flubacher A., Muller M., Fahrner A., Kirsch M., Freiman T., Suzuki F., Depaulis A., Frotscher M., and Haas C.A. 2006. Reelin deficiency and displacement of mature neurons, but not neurogenesis, underlie the formation of granule cell dispersion in the epileptic hippocampus. *J. Neurosci.* **26:** 4701–4713.

Henrique D., Hirsinger E., Adam J., Le Roux I., Pourquie O., Ish-Horowicz D., and Lewis J. 1997. Maintenance of neuroepithelial progenitor cells by Delta-Notch signalling in the embryonic chick retina. *Curr. Biol.* **7:** 661–670.

Hevner R.F., Hodge R.D., Daza R.A., and Englund C. 2006. Transcription factors in glutamatergic neurogenesis: Conserved programs in neocortex, cerebellum, and adult hippocampus. *Neurosci. Res.* **55:** 223–233.

Hirabayashi Y., Itoh Y., Tabata H., Nakajima K., Akiyama T., Masuyama N., and Gotoh Y. 2004. The Wnt/beta-catenin pathway directs neuronal differentiation of cortical neural precursor cells. *Development* **131:** 2791–2801.

Hoglinger G.U., Rizk P., Muriel M.P., Duyckaerts C., Oertel W.H., Caille I., and Hirsch E.C. 2004. Dopamine depletion impairs precursor cell proliferation in Parkinson disease. *Nat. Neurosci.* **7:** 726–735.

Holmberg J., Armulik A., Senti K.A., Edoff K., Spalding K., Momma S., Cassidy R., Flanagan J.G., and Frisen J. 2005. Ephrin-A2 reverse signaling negatively regulates neural progenitor proliferation and neurogenesis. *Genes Dev.* **19:** 462–471.

Hsieh J., and Gage F.H. 2005. Chromatin remodeling in neural development and plasticity. *Curr. Opin. Cell Biol.* **17:** 664–671.

Hsieh J., Nakashima K., Kuwabara T., Mejia E., and Gage F.H. 2004a. Histone deacetylase inhibition-mediated neuronal differentiation of multipotent adult neural progenitor cells. *Proc. Natl. Acad. Sci.* **101:** 16659–16664.

Hsieh J., Aimone J.B., Kaspar B.K., Kuwabara T., Nakashima K., and Gage F.H. 2004b. IGF-I instructs multipotent adult neural progenitor cells to become oligodendrocytes. *J. Cell. Biol.* **164:** 111–122.

Hu H. 2001. Cell-surface heparan sulfate is involved in the repulsive guidance activities of Slit2 protein. *Nat. Neurosci.* **4:** 695–701.

Hu H., Tomasiewicz H., Magnuson T., and Rutishauser U. 1996. The role of polysialic acid in migration of olfactory bulb interneuron precursors in the subventricular zone. *Neuron* **16:** 735–743.

Huisman E., Uylings H.B., and Hoogland P.V. 2004. A 100% increase of dopaminergic cells in the olfactory bulb may explain hyposmia in Parkinson's disease. *Mov. Disord.* **19:** 687–692.

Hussain S.A., Piper M., Fukuhara N., Strochlic L., Cho G., Howitt J.A., Ahmed Y., Powell A.K., Turnbull J.E., Holt C.E., and Hohenester E. 2006. A molecular mechanism for the heparan sulfate dependence of slit-robo signaling. *J. Biol. Chem.* **281:** 39693–39698.

Jackson E.L., Garcia-Verdugo J.M., Gil-Perotin S., Roy M., Quinones-Hinojosa A., VandenBerg S., and Alvarez-Buylla A. 2006. PDGFR alpha-positive B cells are neural stem cells in the adult SVZ that form glioma-like growths in response to increased PDGF signaling. *Neuron* **51:** 187–199.

Jagasia R., Song H., Gage F.H., and Lie D.C. 2006. New regulators in adult neurogenesis and their potential role for repair. *Trends Mol. Med.* **12:** 400–405.

Jin K., Zhu Y., Sun Y., Mao X.O., Xie L., and Greenberg D.A. 2002. Vascular endothelial growth factor (VEGF) stimulates neurogenesis in vitro and in vivo. *Proc. Natl. Acad. Sci.* **99:** 11946–11950.

Karpowicz P., Morshead C., Kam A., Jervis E., Ramunas J., Cheng V., and van der Kooy D. 2005. Support for the immortal strand hypothesis: Neural stem cells partition DNA asymmetrically in vitro. *J. Cell Biol* **170:** 721–732.

Kempermann G., Kuhn H.G., and Gage F.H. 1997. Genetic influence on neurogenesis in the dentate gyrus of adult mice. *Proc. Natl. Acad. Sci.* **94:** 10409–10414.

Kempermann G., Jessberger S., Steiner B., and Kronenberg G. 2004. Milestones of neuronal development in the adult hippocampus. *Trends Neurosci.* **27:** 447–452.

Kempermann G., Gast D., Kronenberg G., Yamaguchi M., and Gage F.H. 2003. Early determination and long-term persistence of adult-generated new neurons in the hippocampus of mice. *Development* **130:** 391–399.

Kintner C. 2002. Neurogenesis in embryos and in adult neural stem cells. *J Neurosci.* **22:** 639–643.

Kippin T.E., Martens D.J., and van der Kooy D. 2005. p21 loss compromises the relative quiescence of forebrain stem cell proliferation leading to exhaustion of their proliferation capacity. *Genes Dev.* **19:** 756–767.

Kohwi M., Osumi N., Rubenstein J.L., and Alvarez-Buylla A. 2005. Pax6 is required for making specific subpopulations of granule and periglomerular neurons in the olfactory bulb. *J. Neurosci.* **25:** 6997–7003.

Koizumi H., Higginbotham H., Poon T., Tanaka T., Brinkman B.C., and Gleeson J.G. 2006. Doublecortin maintains bipolar shape and nuclear translocation during migration in the adult forebrain. *Nat. Neurosci.* **9:** 779–786.

Kondo T. and Raff M. 2000. Oligodendrocyte precursor cells reprogrammed to become multipotential CNS stem cells. *Science* **289:** 1754–1757.

―――. 2004. Chromatin remodeling and histone modification in the conversion of oligodendrocyte precursors to neural stem cells. *Genes Dev.* **18:** 2963–2972.

Kowalczyk A., Filipkowski R.K., Rylski M., Wilczynski G.M., Konopacki F.A., Jaworski J., Ciemerych M.A., Sicinski P., and Kaczmarek L. 2004. The critical role of cyclin D2 in adult neurogenesis. *J. Cell Biol.* **167:** 209–213.

Kronenberg G., Reuter K., Steiner B., Brandt M.D., Jessberger S., Yamaguchi M., and Kempermann G. 2003. Subpopulations of proliferating cells of the adult hippocampus respond differently to physiologic neurogenic stimuli. *J. Comp. Neurol.* **467:** 455–463.

Kuhn H.G., Winkler J., Kempermann G., Thal L.J., and Gage F.H. 1997. Epidermal growth factor and fibroblast growth factor-2 have different effects on neural progenitors in the adult rat brain. *J. Neurosci.* **17:** 5820–5829.

Kuwabara T., Hsieh J., Nakashima K., Taira K., and Gage F.H. 2004. A small modulatory dsRNA specifies the fate of adult neural stem cells. *Cell* **116:** 779–793.

Laplagne D.A., Esposito M.S., Piatti V.C., Morgenstern N.A., Zhao C., van Praag H., Gage F.H., and Schinder A.F. 2006. Functional convergence of neurons generated in the developing and adult hippocampus. *PLoS Biol.* **4:** e409.

Lee T.I., Jenner R.G., Boyer L.A., Guenther M.G., Levine S.S., Kumar R.M., Chevalier B., Johnstone S.E., Cole M.F., Isono K., Koseki H., Fuchikami T., Abe K., Murray H.L., Zucker J.P., Yuan B., Bell G.W., Herbolsheimer E., Hannett N.M., Sun K., Odom D.T., Otte A.P., Volkert T.L., Bartel D.P., Melton D.A., Gifford D.K., Jaenisch R., and Young R.A. 2006. Control of developmental regulators by Polycomb in human embryonic stem cells. *Cell* **125:** 301–313.

Lie D.C., Song H., Colamarino S.A., Ming G.L., and Gage F.H. 2004. Neurogenesis in the adult brain: New strategies for central nervous system diseases. *Annu. Rev. Pharmacol. Toxicol.* **44:** 399–421.

Lie D.C., Dziewczapolski G., Willhoite A.R., Kaspar B.K., Shults C.W., and Gage F.H. 2002. The adult substantia nigra contains progenitor cells with neurogenic potential. *J. Neurosci.* **22:** 6639–6649.

Lie D.C., Colamarino S.A., Song H.J., Desire L., Mira H., Consiglio A., Lein E.S., Jessberger S., Lansford H., Dearie A.R., and Gage F.H. 2005. Wnt signalling regulates adult hippocampal neurogenesis. *Nature* **437:** 1370–1375.

Lim D.A. and Alvarez-Buylla A. 1999. Interaction between astrocytes and adult subventricular zone precursors stimulates neurogenesis. *Proc. Natl. Acad. Sci.* **96:** 7526–7531.

Lim D.A., Tramontin A.D., Trevejo J.M., Herrera D.G., Garcia-Verdugo J.M., and Alvarez-Buylla A. 2000. Noggin antagonizes BMP signaling to create a niche for adult neurogenesis. *Neuron* **28:** 713–726.

Liu M., Pleasure S.J., Collins A.E., Noebels J.L., Naya F.J., Tsai M.J., and Lowenstein D.H. 2000. Loss of BETA2/NeuroD leads to malformation of the dentate gyrus and epilepsy. *Proc. Natl. Acad. Sci.* **97:** 865–870.

Liu S., Wang J., Zhu D., Fu Y., Lukowiak K., and Lu Y.M. 2003. Generation of functional inhibitory neurons in the adult rat hippocampus. *J. Neurosci.* **23:** 732–736.

Liu X., Wang Q., Haydar T.F., and Bordey A. 2005. Nonsynaptic GABA signaling in postnatal subventricular zone controls proliferation of GFAP-expressing progenitors. *Nat. Neurosci.* **8:** 1179–1187.

Lois C., Garcia-Verdugo J.M., and Alvarez-Buylla A. 1996. Chain migration of neuronal precursors. *Science* **271:** 978–981.

Louissaint A., Jr., Rao S., Leventhal C., and Goldman S.A. 2002. Coordinated interaction of neurogenesis and angiogenesis in the adult songbird brain. *Neuron* **34:** 945–960.

Lowell S., Benchoua A., Heavey B., and Smith A.G. 2006. Notch promotes neural lineage entry by pluripotent embryonic stem cells. *PLoS Biol.* **4:** e121.

Lowry W.E., Blanpain C., Nowak J.A., Guasch G., Lewis L., and Fuchs E. 2005. Defining the impact of beta-catenin/Tcf transactivation on epithelial stem cells. *Genes Dev.* **19:** 1596–1611.

Luzzati F., Peretto P., Aimar P., Ponti G., Fasolo A., and Bonfanti L. 2003. Glia-independent chains of neuroblasts through the subcortical parenchyma of the adult rabbit brain. *Proc. Natl. Acad. Sci.* **100:** 13036–13041.

Magavi S.S., Leavitt B.R., and Macklis J.D. 2000. Induction of neurogenesis in the neocortex of adult mice. *Nature* **405:** 951–955.

Malatesta P., Hack M.A., Hartfuss E., Kettenmann H., Klinkert W., Kirchhoff F., and Götz M. 2003. Neuronal or glial progeny: Regional differences in radial glia fate. *Neuron* **37:** 751–764.

Mansukhani A., Ambrosetti D., Holmes G., Cornivelli L., and Basilico C. 2005. Sox2 induction by FGF and FGFR2 activating mutations inhibits Wnt signaling and osteoblast differentiation. *J. Cell. Biol.* **168:** 1065–1076.

Marillat V., Cases O., Nguyen-Ba-Charvet K.T., Tessier-Lavigne M., Sotelo C., and Chedotal A. 2002. Spatiotemporal expression patterns of slit and robo genes in the rat brain. *J. Comp. Neurol.* **442:** 130–155.

Marshall C.A., Novitch B.G., and Goldman J.E. 2005. Olig2 directs astrocyte and oligodendrocyte formation in postnatal subventricular zone cells. *J. Neurosci.* **25:** 7289–7298.

Mason H.A., Ito S., and Corfas G. 2001. Extracellular signals that regulate the tangential migration of olfactory bulb neuronal precursors: Inducers, inhibitors, and repellents. *J. Neurosci.* **21:** 7654–7663.

Meletis K., Wirta V., Hede S.M., Nister M., Lundeberg J., and Frisen J. 2006. p53 suppresses the self-renewal of adult neural stem cells. *Development* **133:** 363–369.

Meltzer L.A., Yabaluri R., and Deisseroth K. 2005. A role for circuit homeostasis in adult neurogenesis. *Trends Neurosci.* **28:** 653–660.

Menn B., Garcia-Verdugo J.M., Yaschine C., Gonzalez-Perez O., Rowitch D., and Alvarez-Buylla A. 2006. Origin of oligodendrocytes in the subventricular zone of the adult brain. *J. Neurosci.* **26:** 7907–7918.

Ming G.L. and Song H. 2005. Adult neurogenesis in the mammalian central nervous system. *Annu. Rev. Neurosci.* **28:** 223–250.

Molofsky A.V., Pardal R., Iwashita T., Park I.K., Clarke M.F., and Morrison S.J. 2003. Bmi-1 dependence distinguishes neural stem cell self-renewal from progenitor proliferation. *Nature* **425:** 962–967.

Molofsky A.V., Slutsky S.G., Joseph N.M., He S., Pardal R., Krishnamurthy J., Sharpless N.E., and Morrison S.J. 2006. Increasing p16INK4a expression decreases forebrain progenitors and neurogenesis during ageing. *Nature* **443:** 448–452.

Mori T., Buffo A., and Götz M. 2005. The novel roles of glial cells revisited: The contribution of radial glia and astrocytes to neurogenesis. *Curr. Top. Dev. Biol* **69:** 67–99.

Morshead C.M., and van der Kooy D. 1992. Postmitotic death is the fate of constitutively proliferating cells in the subependymal layer of the adult mouse brain. *J. Neurosci.* **12:** 249–256.

Morshead C.M., Craig C.G., and van der Kooy D. 1998. In vivo clonal analyses reveal the properties of endogenous neural stem cell proliferation in the adult mammalian forebrain. *Development* **125:** 2251–2261.

Morshead C.M., Reynolds B.A., Craig C.G., McBurney M.W., Staines W.A., Morassutti D., Weiss S., and van der Kooy D. 1994. Neural stem cells in the adult mammalian forebrain: A relatively quiescent subpopulation of subependymal cells. *Neuron* **13:** 1071–1082.

Nakashima K., Yanagisawa M., Arakawa H., and Taga T. 1999. Astrocyte differentiation mediated by LIF in cooperation with BMP2. *FEBS Lett.* **457:** 43–46.

Nguyen-Ba-Charvet K.T., Picard-Riera N., Tessier-Lavigne M., Baron-Van Evercooren A., Sotelo C., and Chedotal A. 2004. Multiple roles for slits in the control of cell migration in the rostral migratory stream. *J. Neurosci.* **24:** 1497–1506.

Ninkovic J. and Götz M. 2007. Signaling in adult neurogenesis from stem cell niche to neuronal networks. *Curr. Opin. Neurobiol.* **17:** 338–344.

Nyfeler Y., Kirch R.D., Mantei N., Leone D.P., Radtke F., Suter U., and Taylor V. 2005. Jagged1 signals in the postnatal subventricular zone are required for neural stem cell self-renewal. *EMBO J.* **24:** 3504–3515.

Ono K., Tomasiewicz H., Magnuson T., and Rutishauser U. 1994. N-CAM mutation inhibits tangential neuronal migration and is phenocopied by enzymatic removal of polysialic acid. *Neuron* **13:** 595–609.

Paik J.H., Kollipara R., Chu G., Ji H., Xiao Y., Ding Z., Miao L., Tothova Z., Horner J.W., Carrasco D.R., Jiang S., Gilliland D.G., Chin L., Wong W.H., Castrillon D.H., and DePinho R.A. 2007. FoxOs are lineage-restricted redundant tumor suppressors and regulate endothelial cell homeostasis. *Cell* **128:** 309–323.

Palmer T.D., Willhoite A.R., and Gage F.H. 2000. Vascular niche for adult hippocampal neurogenesis. *J. Comp. Neurol.* **425:** 479–494.

Parent J.M., Elliott R.C., Pleasure S.J., Barbaro N.M., and Lowenstein D.H. 2006. Aberrant seizure-induced neurogenesis in experimental temporal lobe epilepsy. *Ann. Neurol.* **59:** 81–91.

Parras C.M., Galli R., Britz O., Soares S., Galichet C., Battiste J., Johnson J.E., Nakafuku M., Vescovi A., and Guillemot F. 2004. Mash1 specifies neurons and oligodendrocytes in the postnatal brain. *EMBO J.* **23:** 4495–4505.

Pereira A.C., Huddleston D.E., Brickman A.M., Sosunov A.A., Hen R., McKhann G.M., Sloan R., Gage F.H., Brown T.R., and Small S.A. 2007. An in vivo correlate of exercise-induced neurogenesis in the adult dentate gyrus. *Proc. Natl. Acad. Sci.* **104**: 5638–5643.

Pleasure S.J., Collins A.E., and Lowenstein D.H. 2000. Unique expression patterns of cell fate molecules delineate sequential stages of dentate gyrus development. *J. Neurosci.* **20**: 6095–6105.

Potten C.S. 2004. Keratinocyte stem cells, label-retaining cells and possible genome protection mechanisms. *J. Investig. Dermatol. Symp. Proc.* **9**: 183–195.

Rahpeymai Y., Hietala M.A., Wilhelmsson U., Fotheringham A., Davies I., Nilsson A.K., Zwirner J., Wetsel R.A., Gerard C., Pekny M., and Pekna M. 2006. Complement: A novel factor in basal and ischemia-induced neurogenesis. *EMBO J.* **25**: 1364–1374.

Ramirez-Castillejo C., Sanchez-Sanchez F., Andreu-Agullo C., Ferron S.R., Aroca-Aguilar J.D., Sanchez P., Mira H., Escribano J., and Farinas I. 2006. Pigment epithelium-derived factor is a niche signal for neural stem cell renewal. *Nat. Neurosci.* **9**: 331–339.

Raponi E., Agenes F., Delphin C., Assard N., Baudier J., Legraverend C., and Deloulme J.C. 2007. S100B expression defines a state in which GFAP-expressing cells lose their neural stem cell potential and acquire a more mature developmental stage. *Glia* **55**: 165–177.

Saghatelyan A., de Chevigny A., Schachner M., and Lledo P.M. 2004. Tenascin-R mediates activity-dependent recruitment of neuroblasts in the adult mouse forebrain. *Nat. Neurosci.* **7**: 347–356.

Sakaguchi M., Shingo T., Shimazaki T., Okano H.J., Shiwa M., Ishibashi S., Oguro H., Ninomiya M., Kadoya T., Horie H., Shibuya A., Mizusawa H., Poirier F., Nakauchi H., Sawamoto K., and Okano H. 2006. A carbohydrate-binding protein, Galectin-1, promotes proliferation of adult neural stem cells. *Proc. Natl. Acad. Sci.* **103**: 7112–7117.

Sawamoto K., Wichterle H., Gonzalez-Perez O., Cholfin J.A., Yamada M., Spassky N., Murcia N.S., Garcia-Verdugo J.M., Marin O., Rubenstein J.L., Tessier-Lavigne M., Okano H., and Alvarez-Buylla A. 2006. New neurons follow the flow of cerebrospinal fluid in the adult brain. *Science* **311**: 629–632.

Schoenherr C.J. and Anderson D.J. 1995. The neuron-restrictive silencer factor (NRSF): A coordinate repressor of multiple neuron-specific genes. *Science* **267**: 1360–1363.

Seidenfaden R., Desoeuvre A., Bosio A., Virard I., and Cremer H. 2006. Glial conversion of SVZ-derived committed neuronal precursors after ectopic grafting into the adult brain. *Mol. Cell. Neurosci.* **32**: 187–198.

Seki T. 2002. Expression patterns of immature neuronal markers PSA-NCAM, CRMP-4 and NeuroD in the hippocampus of young adult and aged rodents. *J. Neurosci. Res.* **70**: 327–334.

Seri B., Garcia-Verdugo J.M., McEwen B.S., and Alvarez-Buylla A. 2001. Astrocytes give rise to new neurons in the adult mammalian hippocampus. *J. Neurosci.* **21**: 7153–7160.

Seri B., Garcia-Verdugo J.M., Collado-Morente L., McEwen B.S., and Alvarez-Buylla A. 2004. Cell types, lineage, and architecture of the germinal zone in the adult dentate gyrus. *J. Comp. Neurol.* **478**: 359–378.

Shen Q., Goderie S.K., Jin L., Karanth N., Sun Y., Abramova N., Vincent P., Pumiglia K., and Temple S. 2004. Endothelial cells stimulate self-renewal and expand neurogenesis of neural stem cells. *Science* **304**: 1338–1340.

Shi Y., Chichung Lie D., Taupin P., Nakashima K., Ray J., Yu R.T., Gage F.H., and Evans R.M. 2004. Expression and function of orphan nuclear receptor TLX in adult neural stem cells. *Nature* **427**: 78–83.

Shihabuddin L.S., Horner P.J., Ray J., and Gage F.H. 2000. Adult spinal cord stem cells generate neurons after transplantation in the adult dentate gyrus. *J. Neurosci.* **20:** 8727–8735.

Shimogori T., VanSant J., Paik E., and Grove E.A. 2004. Members of the Wnt, Fz, and Frp gene families expressed in postnatal mouse cerebral cortex. *J. Comp. Neurol.* **473:** 496–510.

Siebzehnrubl F.A., Buslei R., Eyupoglu I.Y., Seufert S., Hahnen E., and Blumcke I. 2007. Histone deacetylase inhibitors increase neuronal differentiation in adult forebrain precursor cells. *Exp. Brain. Res.* **176:** 672–678.

Song H., Stevens C.F., and Gage F.H. 2002. Astroglia induce neurogenesis from adult neural stem cells. *Nature* **417:** 39–44.

Steiner B., Klempin F., Wang L., Kott M., Kettenmann H., and Kempermann G. 2006. Type-2 cells as link between glial and neuronal lineage in adult hippocampal neurogenesis. *Glia* **54:** 805–814.

Stenman J., Toresson H., and Campbell K. 2003. Identification of two distinct progenitor populations in the lateral ganglionic eminence: Implications for striatal and olfactory bulb neurogenesis. *J. Neurosci.* **23:** 167–174.

Takemoto T., Uchikawa M., Kamachi Y., and Kondoh H. 2006. Convergence of Wnt and FGF signals in the genesis of posterior neural plate through activation of the Sox2 enhancer N-1. *Development* **133:** 297–306.

Tanaka T., Serneo F.F., Higgins C., Gambello M.J., Wynshaw-Boris A., and Gleeson J.G. 2004. Lis1 and doublecortin function with dynein to mediate coupling of the nucleus to the centrosome in neuronal migration. *J. Cell Biol.* **165:** 709–721.

Tanigaki K., Nogaki F., Takahashi J., Tashiro K., Kurooka H., and Honjo T. 2001. Notch1 and Notch3 instructively restrict bFGF-responsive multipotent neural progenitor cells to an astroglial fate. *Neuron* **29:** 45–55.

Tomasiewicz H., Ono K., Yee D., Thompson C., Goridis C., Rutishauser U., and Magnuson T. 1993. Genetic deletion of a neural cell adhesion molecule variant (N-CAM-180) produces distinct defects in the central nervous system. *Neuron* **11:** 1163–1174.

Tothova Z., Kollipara R., Huntly B.J., Lee B.H., Castrillon D.H., Cullen D.E., McDowell E.P., Lazo-Kallanian S., Williams I.R., Sears C., Armstrong S.A., Passegue E., DePinho R.A., and Gilliland D.G. 2007. FoxOs are critical mediators of hematopoietic stem cell resistance to physiologic oxidative stress. *Cell* **128:** 325–339.

Tozuka Y., Fukuda S., Namba T., Seki T., and Hisatsune T. 2005. GABAergic excitation promotes neuronal differentiation in adult hippocampal progenitor cells. *Neuron* **47:** 803–815.

Tsai L.H. and Gleeson J.G. 2005. Nucleokinesis in neuronal migration. *Neuron* **46:** 383–388.

van Praag H., Shubert T., Zhao C., and Gage F.H. 2005. Exercise enhances learning and hippocampal neurogenesis in aged mice. *J. Neurosci.* **25:** 8680–8685.

van Praag H., Schinder A.F., Christie B.R., Toni N., Palmer T.D., and Gage F.H. 2002. Functional neurogenesis in the adult hippocampus. *Nature* **415:** 1030–1034.

Vergano-Vera E., Yusta-Boyo M.J., de Castro F., Bernad A., de Pablo F., and Vicario-Abejon C. 2006. Generation of GABAergic and dopaminergic interneurons from endogenous embryonic olfactory bulb precursor cells. *Development* **133:** 4367–4379.

Wachs F.P., Winner B., Couillard-Despres S., Schiller T., Aigner R., Winkler J., Bogdahn U., and Aigner L. 2006. Transforming growth factor-beta1 is a negative modulator of adult neurogenesis. *J. Neuropathol. Exp. Neurol.* **65:** 358–370.

Waclaw R.R., Wang B., and Campbell K. 2004. The homeobox gene Gsh2 is required for retinoid production in the embryonic mouse telencephalon. *Development* **131:** 4013–4020.

Walton N.M., Sutter B.M., Laywell E.D., Levkoff L.H., Kearns S.M., Marshall G.P., 2nd, Scheffler B., and Steindler D.A. 2006. Microglia instruct subventricular zone neurogenesis. *Glia* **54:** 815–825.

Wang D.D., Krueger D.D., and Bordey A. 2003. GABA depolarizes neuronal progenitors of the postnatal subventricular zone via GABAA receptor activation. *J. Physiol.* **550:** 785–800.

Wang L.P., Kempermann G., and Kettenmann H. 2005. A subpopulation of precursor cells in the mouse dentate gyrus receives synaptic GABAergic input. *Mol. Cell. Neurosci.* **29:** 181–189.

Weissman I.L., Anderson D.J., and Gage F. 2001. Stem and progenitor cells: Origins, phenotypes, lineage commitments, and transdifferentiations. *Annu. Rev. Cell. Dev. Biol.* **17:** 387–403.

Willaime-Morawek S., Seaberg R.M., Batista C., Labbe E., Attisano L., Gorski J.A., Jones K.R., Kam A., Morshead C.M., and van der Kooy D. 2006. Embryonic cortical neural stem cells migrate ventrally and persist as postnatal striatal stem cells. *J. Cell Biol.* **175:** 159–168.

Winner B., Cooper-Kuhn C.M., Aigner R., Winkler J., and Kuhn H.G. 2002. Long-term survival and cell death of newly generated neurons in the adult rat olfactory bulb. *Eur. J. Neurosci.* **16:** 1681–1689.

Winner B., Lie D.C., Rockenstein E., Aigner R., Aigner L., Masliah E., Kuhn H.G., and Winkler J. 2004. Human wild-type alpha-synuclein impairs neurogenesis. *J. Neuropathol. Exp. Neurol.* **63:** 1155–1166.

Winner B., Rockenstein E., Lie D.C., Aigner R., Mante M., Bogdahn U., Couillard-Depres S., Masliah E., and Winkler J. 2007. Mutant alpha-synuclein exacerbates age-related decrease of neurogenesis. *Neurobiol. Aging* (in press)

Wu W., Wong K., Chen J., Jiang Z., Dupuis S., Wu J.Y., and Rao Y. 1999. Directional guidance of neuronal migration in the olfactory system by the protein Slit. *Nature* **400:** 331–336.

Wurmser A.E., Nakashima K., Summers R.G., Toni N., D'Amour K.A., Lie D.C., and Gage F.H. 2004. Cell fusion-independent differentiation of neural stem cells to the endothelial lineage. *Nature* **430:** 350–356.

Yoshihara S., Omichi K., Yanazawa M., Kitamura K., and Yoshihara Y. 2005. Arx homeobox gene is essential for development of mouse olfactory system. *Development* **132:** 751–762.

Zappone M.V., Galli R., Catena R., Meani N., De Biasi S., Mattei E., Tiveron C., Vescovi A.L., Lovell-Badge R., Ottolenghi S., and Nicolis S.K. 2000. Sox2 regulatory sequences direct expression of a (beta)-geo transgene to telencephalic neural stem cells and precursors of the mouse embryo, revealing regionalization of gene expression in CNS stem cells. *Development* **127:** 2367–2382.

Ziv Y., Ron N., Butovsky O., Landa G., Sudai E., Greenberg N., Cohen H., Kipnis J., and Schwartz M. 2006. Immune cells contribute to the maintenance of neurogenesis and spatial learning abilities in adulthood. *Nat. Neurosci.* **9:** 268–275.

13

Proneuronal Genes Drive Neurogenesis on the Road from Development to Adulthood

Elizabeth T. Buchen and Samuel J. Pleasure
Department of Neurology
University of California
San Francisco, California 94143

THE ONGOING PRODUCTION OF NEURONS in selected areas of the adult mammalian brain is tantalizing and has become an active area of research for many investigators. It is exciting to consider the functional importance of adding new neurons to mature circuits, as well as the intricate biological processes regulating their production (Meltzer et al. 2005; Ming and Song 2005; Lledo et al. 2006). Many investigators are also fascinated by the potential of repairing the injured nervous system with adult-generated neurons, those either produced in specialized adult neurogenic niches or induced in regions where little (if any) neurogenesis normally persists, such as the spinal cord or neocortex. Whether the inspiration is systems neuroscience, basic biology, or biomedical applications, the advancement of the field of neurogenesis depends on understanding the underlying molecular mechanisms regulating this process.

When considering this central issue, most investigators have posited that molecular pathways important for development must have similar roles in the adult (Deisseroth et al. 2004; Meltzer et al. 2005; Ming and Song 2005). It is important to realize, however, that the number of studies demonstrating a required and specific role for any developmental regulators in adult neurogenesis is quite small. Adult neurogenesis is contingent on the functioning of the neurogenic niche, which must be produced during development, maintained during postnatal life, and regulated during adulthood. This presents a significant barrier for interpreting most genetic manipulations, as it is virtually impossible to distinguish adult requirements from developmental insults in most studies examining these pathways.

To establish priorities for future studies in this area, it is worth reviewing the literature on the known and suspected roles of some of these developmental factors in adult neurogenesis. This chapter addresses the function of one particular signaling axis known to regulate developmental neurogenesis: the Delta-Notch pathway and the proneuronal genes. We begin by reviewing the role and mechanisms of this pathway during mouse brain development, where it has been extensively studied (for review, see Louvi and Artavanis-Tsakonas 2006). We then explore the possibility that these developmental mechanisms continue to function in adult neurogenesis and discuss recent findings suggesting that this may be the case. We hope to help establish a framework for future studies that aim to definitively determine whether these factors have as central a role in the adult as they do in the embryo.

PRONEURONAL GENES AND CORTICAL DEVELOPMENT

During early embryonic development, the neuroepithelium consists of radially oriented neuroepithelial stem cells, which, by way of symmetric expansile divisions, expand the lateral dimensions of the tube. These divisions produce two daughter cells that are believed to have the same potential as the parent cells (Gotz and Huttner 2005). After about embryonic day 10 (E10), these stem cells begin to produce neurons via asymmetric divisions in the neuroepithelium and begin to express markers of radial glial cells (Gotz and Huttner 2005; Merkle and Alvarez-Buylla 2006). As has been extensively studied, these cells then sequentially produce the various cortical layers as they concomitantly become restricted in their potential; thus, radial glial cells from later gestational ages can no longer produce neurons appropriate for earlier stages (Hevner 2006). At the end of corticogenesis, the radial glial cells produce primarily astrocytes, while some persist as adult progenitor cells in the two neurogenic niches, located in the subventricular zone (SVZ) lining the lateral ventricles and the subgranular zone (SGZ) of the hippocampal dentate gyrus (DG) (Merkle et al. 2004; Merkle and Alvarez-Buylla 2006). Recent studies indicate that oligodendrocytes are produced from the same neuronal stem cells but in a somewhat staged manner: Early-born oligodendrocytes are produced from subcortical sources and migrate tangentially into the cortex, to be replaced at a later time by oligodendrocytes originating

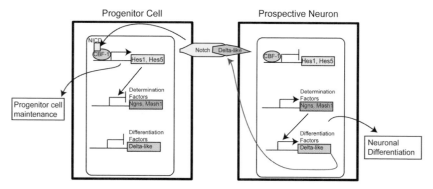

Figure 1. Notch signaling regulates the process of lateral inhibition. A representation of the Notch-mediated interaction between two neuroepithelial cells is shown (data from Kageyama et al. 2005). Activation of the Notch receptor in the "Progenitor Cell" leads to the expression of the Hes genes, which inhibit the neuronal determination genes and thus maintain the cell in an undifferentiated state. Attenuation of Notch signaling in a "Prospective Neuron" permits the expression of the determination factors, which commit the cells to a neuronal fate. The activity of these genes leads to expression of the differentiation factors, which are responsible for neuronal maturation, as well as Notch ligands (e.g., Delta). The expression of Notch ligand in the "Prospective Neuron" further activates Notch signaling in neighboring cells; thus, a cell that commits to a neuronal fate prevents its neighbor from doing the same.

from cortically derived precursors via a group of persisting oligodendrocyte precursor cells (Kessaris et al. 2006; Richardson et al. 2006).

In each of these developmental events, proneuronal genes are central players. These genes code for a set of basic helix-loop-helix (bHLH) transcription factors, originally discovered in *Drosophila* for their role in regulating cell fate and timing during neurogenesis (Chan and Jan 1999). For the purposes of this chapter, we divide the proneuronal genes into two categories: determination factors and differentiation factors. These genes are regulated by Notch signaling and operate in a generally linear cascade (with some exceptions) to control stem cell behavior during neurogenesis (see Figs. 1 and 2).

OVERVIEW OF THE NOTCH-SIGNALING CASCADE

The interaction of Notch with its ligands, Delta or Jagged, leads to intramembrane cleavage of the receptor by γ-secretase, thus releasing the Notch intracellular domain (NICD) (see Fig. 1 for an overview of the Notch-signaling pathway at the cellular level). The NICD then translocates to the nucleus, where it collaborates with a transcriptional complex

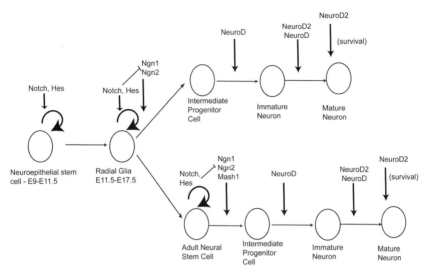

Figure 2. Continuity of the neurogenic lineage and the transcriptional code from embryonic to adult neurogenesis. The proposed lineage of neuronal stem cells from embryonic life into adulthood, indicating known or hypothesized roles of Notch and the proneuronal genes, is shown. Early in cortical development, neuroepithelial stem cells are maintained in an undifferentiated state by Notch signaling and give rise to radial glial cells at the onset of neurogenesis. Radial glial cells undergo differentiative divisions under the control of determination factors, producing neurons, astrocytes, and oligodendrocytes. In both the embryo and the adult, these differentiating cells go through an intermediate expansile phase and are then guided through the maturation process by differentiation factors. Radial glia may also give rise to the astrocyte-like progenitor cells destined to reside in the adult neurogenic niches, likely conserving the neurogenic mechanisms of their predecessors (see text for details).

that includes centromere binding factor 1 (CBF1) and Mastermind (MAML) to drive the expression of a group of inhibitory bHLH factors, the hairy and enhancer of split (Hes) family (Selkoe and Kopan 2003; Hatakeyama et al. 2004; Louvi and Artavanis-Tsakonas 2006). These genes (*Hes1* and *Hes5*) block expression of the determination factors (*Mash1*, *Ngn1*, and *Ngn2*), maintaining progenitors in an undifferentiated, proliferative state. The release of Notch signaling in these cells leads to the derepression of the determination factors, which repress expression of the Hes genes, as well as promote their own expression (Yoon and Gaiano 2005). The determination factors then induce the expression of the differentiation factors (the NeuroD family), which drive the expression of genes necessary for neuronal maturation (Chae et al. 2004).

One of the interesting characteristics of the Notch pathway is that the receptor and known ligands are all membrane-bound, allowing the unique opportunity for directly mediating cell–cell interactions. The Notch pathway capitalizes on these circumstances by incorporating a feedback loop, in which the activation of Notch signaling suppresses the expression of its own ligand; thus, the activation of Notch signaling in one cell decreases Notch activity in its neighbors. This mutually repressive interaction, termed "lateral inhibition," leads to the selection of cells from the epithelium for neurogenesis, inducing them to migrate away and adopt their appropriate nuclear or laminar position (modeled in Fig. 1) (Yoon and Gaiano 2005; Androutsellis-Theotokis et al. 2006; Louvi and Artavanis-Tsakonas 2006). In the embryo, small differences in Notch signaling are created either stochastically or by the asymmetric inheritance of Numb, a cytoplasmic Notch inhibitor, during stem cell divisions, although the mechanism for the latter remains unclear (Zhong et al. 1996; Petersen et al. 2002, 2006; Shen et al. 2002; Zhong 2003; Kuo et al. 2006).

DEVELOPMENTAL PHENOTYPES OF NOTCH-PATHWAY MUTANTS

As discussed above, the Hes genes are the primary effectors of the Notch pathway, blocking the expression of proneuronal genes. Accordingly, loss of function for Notch or Hes genes leads to premature differentiation of neurons in the neural tube and exhaustion of stem cells in the ventricular zone, whereas overexpression of either activated Notch or Hes genes leads to defects in neuronal differentiation (Hatakeyama et al. 2004; Yoon and Gaiano 2005).

When Notch signaling declines, Hes expression follows, thus permitting the expression of *Mash1*, *Ngn1*, and *Ngn2* (Ross et al. 2003; Kageyama et al. 2005). The activity of these determination factors results in the generation of progenitors that are restricted to a neuronal fate and retain a limited mitotic potential: These genes are attributed with coupling cell-fate decisions with proliferation and differentiation. The loss of activity of these genes results in the failure of neuroepithelial cells to generate progenitors committed to a neuronal fate, leading to the precocious generation of glial progenitors, whereas their gain of function drives progenitor cells out of division and promotes neuronal differentiation (Casarosa et al. 1999; Fode et al. 2000; Nieto et al. 2001). Specific determination factors specify individual classes of neurons in distinct anterior–posterior and dorsal–ventral coordinates in the neural tube by collaborating with regionally expressed homeobox genes (Fode et al. 2000; Parras et al. 2002). In recent years, it has become apparent that oligodendrocyte specification is controlled by similar mechanisms, but uses other bHLH family members (Ross et al. 2003).

Determination factors are down-regulated before progenitors exit the cell cycle and, to promote successful neuronal differentiation, must induce the expression of the next class of cells in the proneuronal cascade, the differentiation factors (Fig. 2). This class includes *NeuroD*, *NeuroD2*, *Math2*, and *Math3*, and controls the maturation of immature neurons into specific neuronal subtypes, dependent on the developmental history inherited by the determination factors (Pleasure et al. 2000; Chae et al. 2004). These genes are expressed in late precursors and help drive cell cycle exit and control neuronal subtype characteristics by regulating many aspects of neuronal maturation, including neurite formation, axonal arborization, and survival (Pleasure et al. 2000; Redmond et al. 2000; Schwab et al. 2000; Deisseroth et al. 2004).

Mutations in these genes lead to phenotypes related to the differentiation of particular neuronal subtypes, often with stage specificity. For example, mice lacking functional *NeuroD* fail to produce cerebellar and dentate granule neurons (Liu et al. 2000; Schwab et al. 2000; Chae et al. 2004), whereas mice lacking *NeuroD2* produce dentate granule neurons, but these neurons fail to survive. Accordingly, *NeuroD* is expressed at an earlier stage of granule cell production, but is later down-regulated, leaving *NeuroD2* as the primary differentiation gene expressed in these cells (Pleasure et al. 2000; Olson et al. 2001). Furthermore, *NeuroD2* is highly regulated by calcium-mediated signaling and mediates activity-dependent regulation of differentiation and survival at later stages of brain development (Ince-Dunn et al. 2006).

In summary, the proneuronal genes form the core of a progressive cascade of related regulators of stem cell behavior, cell fate, and differentiation (Fig. 2). During development, at the top of the cascade are the Notch-Delta/Jagged interactions, which are required to maintain stem cell capacity via the action of the antineuronal bHLH genes. The attenuation of Notch signaling activates the determination factors, which drive commitment to a particular neuronal fate and proceed to trigger the differentiation factors. These factors then complete the neurogenic process by driving the expression of a large panoply of cell-type-specific characteristics.

IS THE ROLE OF NOTCH PATHWAY CONSERVED IN ADULT NEUROGENIC ZONES?

Proneuronal genes are key regulators of embryonic neurogenesis, with crucial roles in specifying neuronal fate and controlling maturation. To what extent are these transcriptional cascades conserved in the adult? Our understanding of proneuronal genes in adult neurogenesis has been hampered by a number of obstacles. First, mutations of proneuronal genes (if not lethal) often lead to disruptions in the neurogenic niche, the germinal environment in which adult neurogenesis occurs. It is thus difficult to disentangle aberrant development of the neurogenic niche and its constituents from direct effects on adult neurogenesis. Second, neurogenesis in vitro is problematic, as the culturing conditions hardly recapitulate the in vivo environment of the niche. Third, pharmacological manipulations, where available (e.g., γ-secretase inhibitors that inhibit Notch cleavage and thus inhibit Notch signaling), are likely to be nonspecific and toxic.

In light of these barriers, it is important to recognize that the increasingly detailed findings in developmental neurogenesis form an insightful foundation for investigating whether similar genetic networks operate in the adult. The rationale for this association is that the objective of each process is to produce a functional, integrated neuron from an immature progenitor. Neurons produced in the adult are effectively indistinguishable from those produced in embryos (Laplagne et al. 2006): The equivalence of this final end point, in combination with the parsimony of evolution, indicates that there is a common, underlying molecular mechanism for forming a neuron independent of context (Fig. 2).

Nevertheless, the conservation of molecular mechanisms is not completely incontrovertible: A number of crucial differences between adult and developmental neurogenesis complicate this argument. One consideration is that developmental neurogenesis takes place in a uniform primordium, where lateral inhibition by Notch amplifies stochastic differences to resolve equivalent cells into distinct fates. This mechanism is crucial for regulating the production of appropriately specified precursors, such that sufficient numbers are reserved for later stages, thus permitting sequential neurogenic and gliogenic waves of differentiation.

In the adult, there is no apparent need for coordinating prolific waves of genesis; nevertheless, it is reasonable to apply the same logic of population control offered by Notch-mediated lateral inhibition. In the adult niche, cell-fate coordination among neighbors is necessary for maintaining a sufficient population of undifferentiated progenitors to support neurogenesis for the life of the organism. If this delicate balance is tipped in one direction or the other, the organism suffers from either an exhaustion of the progenitor pool or an insufficient generation of functional neurons and failure to meet the requirement of the organism.

A further difference between developmental and adult neurogenesis that must be appreciated involves the regulatory mechanisms. Developmental

neurogenesis is coordinated by a host of developmentally regulated mor-
phogenic signaling molecules (e.g., Wnts, BMPs, and Shh) produced by
localized signaling centers. Adult neuronal progenitor cells, in contrast,
share an environment with glia, mature neurons, and mature endothelial
cells and are thus exposed to signals from the latter cell types. These cells
have recently come into the spotlight for regulating several aspects of
adult neurogenesis (Palmer et al. 2000; Jin et al. 2002; Song et al. 2002;
Cao et al. 2004; Lie et al. 2005) and may account for the sensitivity of
adult neurogenesis to a number of environmental factors (Ming and Song
2005). It appears, however, that many developmental molecules are rede-
ployed for regulating adult progenitors (Ahn and Joyner 2005; Lie et al.
2005) and may exert their influence by regulating proneuronal genes
(Pozniak and Pleasure 2006). The crucial difference in this system is regulation: Rather than being developmentally regulated, these signals are,
in the adult, under environmental control.

Perhaps the most intriguing difference is that embryonic and adult
progenitor cells have dramatically different capabilities of proliferation
and potency. Embryonic progenitors are capable of self-renewal and produce the entire array of neuronal and glial cell types that constitute the
brain. In the adult, however, only a small subset of neurons is produced,
and progenitor cells appear to be substantially more restricted in their
proliferative potential. These differences can still exist within the context
of a conserved system. It is likely, in fact, that the core genetic program
of proneuronal genes provides a unified mechanism to underlie these differences. Because different proneuronal genes provide different competence and control the acquisition of subtype-specific pathways, the
regulation of these genes by the niche signals may account for the limits
in proliferative and potency potential in the adult.

These aforementioned issues are critical and must be considered
whenever assuming any mechanistic similarities between developmental
and adult neurogenesis. Notwithstanding, as progress in the relatively
nascent field of adult neurogenesis continues to advance, the differences
between adult and developmental neurogenesis are proving to be increasingly superficial. The final end point (Laplagne et al. 2006), as well as the
stages and mechanisms preceding it (Esposito et al. 2005), are effectively
indistinguishable (although small differences may exist [Kee et al. 2007]).
Furthermore, it is now accepted that adult progenitor cells are derived
from the same lineage as those in the embryo (Alvarez-Buylla et al. 2001;
Merkle et al. 2004) and thus that the process leading to the development
of a new neuron in the adult is in a continuum with the related embryonic process (Fig. 2). Given the continuity of this lineage, it remains the

most logical assumption that progenitors use many of the same underlying genetic mechanisms for producing a neuron. As new studies are
published using conditional tools over the next several years, it is likely
that proneuronal genes will prove central to this process.

SPECIFIC IMPLICATIONS FOR ADULT HIPPOCAMPAL NEUROGENESIS

The remainder of this chapter operates under the assumption defended
above: Embryonic neurogenesis is an apt framework against which to
understand the role of proneuronal genes in adult neurogenesis. We
consider the caveats and differences listed above as instructive for establishing the extent to which these genes are conserved in the adult. We
focus on neurogenesis in the SGZ but speculate that similar mechanisms
are involved in the SVZ. However, because different neuronal types are
produced in each location, the specific genes involved are likely to be
different.

Whereas analysis of the Notch pathway and the proneuronal genes in
postnatal DG awaits studies with the appropriate methods, a recent study
did address these questions in the postnatal SVZ (Kuo et al. 2006). In this
study, the investigators used a newly generated Nestin–CreERT line to
manipulate the expression of Notch and the Notch modulators, Numb
and Numblike, in the postnatal SVZ. Interestingly, they found that loss
of Numb modulation of Notch led to injury of the postnatal SVZ and
ependymal cells, and loss of postnatal neurogenesis (Kuo et al. 2006).
This was thought to be caused by deregulated signaling of the Notch
pathway. Supporting this idea, these effects were largely phenocopied
by expression of a dominant-active Notch fragment using the same
conditional approach. The most fascinating aspect of this study was
the regenerative capacity of residual, unrecombined SVZ progenitors
to repopulate the SVZ and largely repair the neurogenic capacity of the
SVZ (Kuo et al. 2006). What is missing from this study, and any others
to date, is the analysis of the actual consequences of manipulation of
the Notch pathway in regulating the production of neurons by progenitors in a cell-autonomous manner; rather, this study was focused on
the role of Notch/Numb in maintaining the appropriate niche for
neurogenesis.

Early Steps and the Likely Role of Notch and Determination Factors

To map specific proneuronal genes to precise stages of the neurogenic
pathway, we must first identify the embryonic correlate of the adult

progenitor, i.e., determine at which point in the sequential phases determination and differentiation begin. We refer to adult hippocampal progenitors as "Type-1 progenitors." These cells are derived from radial astrocytes and are believed to divide slowly or rarely. Adult hippocampal neurogenesis begins when Type-1 progenitors divide to produce intermediate progenitor cells (IPCs), which proliferate to some extent before terminally differentiating and maturing into neurons.

On the basis of the logic of the neurogenic system in embryonic development, the maintenance of the "undifferentiated" state of the Type-1 progenitors is likely to require Notch to repress the expression of certain proneuronal genes and quell differentiation. As in the embryo, there may be stochastic differences in levels of Notch signaling, and thus individual cells may express low levels of determination factors. Positive-feedback loops would then have a major role in amplifying proneuronal gene expression and might select individual progenitors for beginning the neurogenic cascade.

Another intriguing possibility involves the asymmetric inheritance of the Notch receptor or modulators of Notch signaling (e.g., Numb/Numblike), which is a crucial occurrence during developmental neurogenesis. When a Type-1 progenitor divides to produce an IPC, the Notch receptor may remain with the Type-1 progenitor and continue to repress proneuronal genes, while the IPC, newly lacking Notch-mediated repression, begins expressing determination factors. Alternatively, it is also possible that when a Type-1 progenitor divides, one daughter cell is selected to become an IPC based on inherited asymmetry of Notch regulatory molecules (e.g., Numb/Numblike). Testing these propositions in the adult dentate will require the use of conditional Notch alleles (or Numb/Numblike alleles) with temporally and cell-type-restricted Cre lines. Because these reagents have now all been described in the literature (Kuo et al. 2006), these results are likely to be forthcoming.

The determination factors implicated in dentate morphogenesis, and thus most likely to extend their role to the adult dentate, are *Mash1*, *Ngn1*, and *Ngn2* (Pleasure et al. 2000). The expression of *Mash1*, importantly, persists into adulthood, when it becomes restricted to the neurogenic SGZ. In the adult, *Mash1* expression increases under pathological conditions known to increase neurogenesis, such as status epilepticus and transient forebrain ischemia (Elliott et al. 2001; Kawai et al. 2005), suggesting that it contributes to the neurogenesis elicited in these situations. In the near future, studies using conditional alleles for *Mash1*, *Ngn1*, and *Ngn2* will undoubtedly reveal the roles for these factors in driving the differentiation of IPCs toward neuronal fate.

Differentiation Factors

To complete the maturation of a progenitor into a functional neuron, the determination factors activate the expression of the appropriate differentiation factors. These genes control the maturation of the new cells, including both "generic" neuronal characteristics, such as excitability and neurite formation, and subtype-specific attributes, such as the synthesis of appropriate neurotransmitters. In the DG, these genes—*NeuroD* and *NeuroD2*—are likely to be prominent in adult neurogenesis (Pleasure et al. 2000). In future years, it will be quite interesting to determine how the activity of these two transcription factors shapes the specific maturation of granule cells.

NeuroD is required for the proliferation, differentiation, and survival of dentate granule cells in the embryo (Miyata et al. 1999; Liu et al. 2000; Chae et al. 2004). It is expressed at its highest levels during the last stages of IPC proliferation, coincident with the transition into a postmitotic, immature granule neuron (Pleasure et al. 2000). Although the first *NeuroD* mutants died soon after birth, two distinct strategies led to the production of $NeuroD^{-/-}$ mice that survived to adulthood (Miyata et al. 1999; Liu et al. 2000). Concurrent with its role as a differentiation factor, the presumptive granule cells in the *NeuroD*–mutant dentate are generated, indicating unimpaired neuronal determination, but fail to mature, lacking sodium currents and dendritic arborization (Schwab et al. 2000). These studies suggest that adult hippocampal neurogenesis may require *NeuroD* activity (Miyata et al. 1999; Liu et al. 2000), but the massive developmental problems in these mice complicate such interpretations.

In vitro studies with adult progenitors provide further support for *NeuroD* in adult neurogenesis, demonstrating that the increase in neurogenesis following excitatory activity is correlated with an increase in *NeuroD* expression (Deisseroth et al. 2004). Interestingly, excitatory activity did not affect the expression of *Mash1*, indicating that excitation–neurogenesis coupling is more likely to operate at the level of differentiation, not determination.

NeuroD2 is expressed in dentate granule cells after *NeuroD*, both prenatally and in the adult (Pleasure et al. 2000), coinciding with axon elongation and synapse stabilization. *NeuroD2* mutants have a phenotype distinct from that of *NeuroD* mutants (Olson et al. 2001), with significant increases in granule-cell-programmed cell death at the final stages of maturation (Olson et al. 2001). New studies demonstrating that *NeuroD2* is directly regulated by calcium signaling and neuronal activity indicate that *NeuroD2* is a key regulator of the program governing these events in immature neurons (Ince-Dunn et al. 2006). The most compelling model for how these genes are used in the adult incorporates *NeuroD* as regulating events coupled to terminal differentiation, whereas *NeuroD2* is essential for the final maturation. Because of the obvious nature of histologic abnormalities in the hippocampal structure, the dentate is generally the most obvious place where these phenotypes are seen. More careful analysis of the neocortex in *NeuroD2* mutants has revealed defects in activity-dependent developmental events in the early postnatal period (Ince-Dunn et al. 2006).

During the next few years, it will become straightforward to test these propositions and whether they continue to be operative at all ages. In addition to conditional alleles for *NeuroD* and *NeuroD2*, it will be valuable to dissect the distinct transcriptional programs controlled by these two transcription factors. Does *NeuroD* control a set of cues that are specifically involved in terminal differentiation, including cell cycle regulators and modulators of axonal and dendritic morphogenesis? Does *NeuroD2* control a program specific to early stages of synapse formation, maturation, and stabilization? Or, do they control essentially the same program, but differ only in the timing of their expression in the granule cell lineage?

Linking Other Transcription Factors to the Regulation of Postnatal Neurogenesis

As information grows about the functions of other transcription factors, it is likely that these will end up being linked in some ways to the functions of the Notch pathway and the bHLH proteins in regulating postnatal neurogenesis. Recent studies have shown that *Pax6* is likely to be a key regulator of postnatal neurogenesis in the SVZ and DG. In the SVZ, the homeobox gene *Pax6* is an important determinant of cell fate, regulating the production of particular types of olfactory interneurons in a cell-autonomous manner (Hack et al. 2005; Kohwi et al. 2005). Considering also recent studies showing that *Pax6* can cooperate with *Mash1* in the spinal cord in regulating the timing of neuron production (Sugimori et al. 2007), it is quite likely that a similar cascade of events is operating in the SVZ and the DG, where *Pax6* is also expressed in the adult niche (Nacher et al. 2005). Future studies will address the roles of transcription factors such as *Pax6*, and others expressed in dentate and SVZ precursors, for example, Tbr2 (Englund et al. 2005). It is likely that in each case, these factors will work in part by interacting with the underlying neurogenic cascade, controlling the timing of cell fate through the Notch pathway.

FUTURE PROSPECTS

The recent publication of tools allowing temporal and cell-type-restricted manipulation of conditional alleles in the dentate will lead to rapid advances in this area. Both Nestin-CreERT (Kuo et al. 2006) and GFAP-CreERT (Ganat et al. 2006) mice have major promise for dissecting adult neurogenesis when combined with conditional reagents for Notch, *Mash1*, *NeuroD*, and *NeuroD2*. Many of these tools are now available, and these studies are undoubtedly under way. Careful analysis of neurogenesis, with attention paid to the numbers and behavior of progenitor cells, IPCs, and newborn neurons, will reveal whether the numerous developmental roles of proneuronal genes are recapitulated in the adult. In addition, recent studies examining the roles of specific phosphoisoforms of bHLH proteins (Hand et al. 2005; Ge et al. 2006) may yield additional approaches for temporally controlling the function of proneuronal genes.

Investigating the stage-specific roles of proneuronal genes will provide a greater understanding of the granule cell lineage during adulthood. At this point, there is little information about potential heterogeneity, from Type-1 progenitors to postmitotic granule neurons. Studies of the proneuronal genes have already provided valuable stage-specific markers for these cells (Pleasure et al. 2000), but the functions of these genes need to be tested. In addition, elucidating the transcriptional targets of the proneuronal genes will yield major insights into the mechanistic control of neurogenesis and provide novel markers for analyzing the maturation of new neurons.

ACKNOWLEDGMENTS

The authors thank the members of the Pleasure laboratory for helpful discussions; in particular, Grant Li and Christine Pozniak. E.T.B. is supported by a Genentech Predoctoral Fellowship and S.J.P. is supported by the National Institute of Mental Health (NIMH), the National Institute on Drug Abuse (NIDA), and Autism Speaks.

REFERENCES

Ahn S. and Joyner A.L. 2005. In vivo analysis of quiescent adult neural stem cells responding to Sonic hedgehog. *Nature* **437**: 894–897.

Alvarez-Buylla A., Garcia-Verdugo J.M., and Tramontin A.D. 2001. A unified hypothesis on the lineage of neural stem cells. *Nat. Rev. Neurosci.* **2**: 287–293.

Androutsellis-Theotokis A., Leker R.R., Soldner F., Hoeppner D.J., Ravin R., Poser S.W., Rueger M.A., Bae S.K., Kittappa R., and McKay R.D. 2006. Notch signalling regulates stem cell numbers in vitro and in vivo. *Nature* **442**: 823–826.

Cao L., Jiao X., Zuzga D.S., Liu Y., Fong D.M., Young D., and During M.J. 2004. VEGF links hippocampal activity with neurogenesis, learning and memory. *Nat. Genet.* **36:** 827–835.

Casarosa S., Fode C., and Guillemot F. 1999. Mash1 regulates neurogenesis in the ventral telencephalon. *Development* **126:** 525–534.

Chae J.H., Stein G.H., and Lee J.E. 2004. NeuroD: The predicted and the surprising. *Mol. Cell* **18:** 271–288.

Chan Y.M., and Jan Y.N. 1999. Conservation of neurogenic genes and mechanisms. *Curr. Opin. Neurobiol.* **9:** 582–588.

Deisseroth K., Singla S., Toda H., Monje M., Palmer T.D., and Malenka R.C. 2004. Excitation-neurogenesis coupling in adult neural stem/progenitor cells. *Neuron* **42:** 535–552.

Elliott R.C., Khademi S., Pleasure S.J., Parent J.M., and Lowenstein D.H. 2001. Differential regulation of basic helix-loop-helix mRNAs in the dentate gyrus following status epilepticus. *Neuroscience* **106:** 79–88.

Englund C., Fink A., Lau C., Pham D., Daza R.A., Bulfone A., Kowalczyk T., and Hevner R.F. 2005. Pax6, Tbr2, and Tbr1 are expressed sequentially by radial glia, intermediate progenitor cells, and postmitotic neurons in developing neocortex. *J. Neurosci.* **25:** 247–251.

Esposito M.S., Piatti V.C., Laplagne D.A., Morgenstern N.A., Ferrari C.C., Pitossi F.J., and Schinder A.F. 2005. Neuronal differentiation in the adult hippocampus recapitulates embryonic development. *J. Neurosci.* **25:** 10074–10086.

Fode C., Ma Q., Casarosa S., Ang S.L., Anderson D.J., and Guillemot F. 2000. A role for neural determination genes in specifying the dorsoventral identity of telencephalic neurons. *Genes Dev.* **14:** 67–80.

Ganat Y.M., Silbereis J., Cave C., Ngu H., Anderson G.M., Ohkubo Y., Ment L.R., and Vaccarino F.M. 2006. Early postnatal astroglial cells produce multilineage precursors and neural stem cells in vivo. *J. Neurosci.* **26:** 8609–8621.

Ge W., He F., Kim K.J., Blanchi B., Coskun V., Nguyen L., Wu X., Zhao J., Heng J.I., Martinowich K., et al. 2006. Coupling of cell migration with neurogenesis by proneural bHLH factors. *Proc. Natl. Acad. Sci.* **103:** 1319–1324.

Gotz M. and Huttner W.B. 2005. The cell biology of neurogenesis. *Nat. Rev. Mol. Cell Biol.* **6:** 777–788.

Hack M.A., Saghatelyan A., de Chevigny A., Pfeifer A., Ashery-Padan R., Lledo P.M., and Gotz M. 2005. Neuronal fate determinants of adult olfactory bulb neurogenesis. *Nat. Neurosci.* **8:** 865–872.

Hand R., Bortone D., Mattar P., Nguyen L., Heng J.I., Guerrier S., Boutt E., Peters E., Barnes A.P., Parras C., et al. 2005. Phosphorylation of Neurogenin2 specifies the migration properties and the dendritic morphology of pyramidal neurons in the neocortex. *Neuron* **48:** 45–62.

Hatakeyama J., Bessho Y., Katoh K., Ookawara S., Fujioka M., Guillemot F., and Kageyama R. 2004. Hes genes regulate size, shape and histogenesis of the nervous system by control of the timing of neural stem cell differentiation. *Development* **131:** 5539–5550.

Hevner R.F. 2006. From radial glia to pyramidal-projection neuron: Transcription factor cascades in cerebral cortex development. *Mol. Neurobiol.* **33:** 33–50.

Ince-Dunn G., Hall B.J., Hu S.C., Ripley B., Huganir R.L., Olson J.M., Tapscott S.J., and Ghosh A. 2006. Regulation of thalamocortical patterning and synaptic maturation by NeuroD2. *Neuron* **49:** 683–695.

Jin K., Zhu Y., Sun Y., Mao X.O., Xie L., and Greenberg D.A. 2002. Vascular endothelial growth factor (VEGF) stimulates neurogenesis in vitro and in vivo. *Proc. Natl. Acad. Sci.* **99:** 11946–11950.

Kageyama R., Ohtsuka T., Hatakeyama J., and Ohsawa R. 2005. Roles of bHLH genes in neural stem cell differentiation. *Exp. Cell Res.* **306:** 343–348.

Kawai T., Takagi N., Nakahara M., and Takeo S. 2005. Changes in the expression of Hes5 and Mash1 mRNA in the adult rat dentate gyrus after transient forebrain ischemia. *Neurosci. Lett.* **380:** 17–20.

Kee N., Teixeira C.M., Wang A.H., and Frankland P.W. 2007. Preferential incorporation of adult-generated granule cells into spatial memory networks in the dentate gyrus. *Nat. Neurosci.* **10:** 355–362.

Kessaris N., Fogarty M., Iannarelli P., Grist M., Wegner M., and Richardson W.D. 2006. Competing waves of oligodendrocytes in the forebrain and postnatal elimination of an embryonic lineage. *Nat. Neurosci.* **9:** 173–179.

Kohwi M., Osumi N., Rubenstein J.L., and Alvarez-Buylla A. 2005. Pax6 is required for making specific subpopulations of granule and periglomerular neurons in the olfactory bulb. *J. Neurosci.* **25:** 6997–7003.

Kuo C.T., Mirzadeh Z., Soriano-Navarro M., Rasin M., Wang D., Shen J., Sestan N., Garcia-Verdugo J., Alvarez-Buylla A., Jan L.Y., and Jan Y.N. 2006. Postnatal deletion of Numb/Numblike reveals repair and remodeling capacity in the subventricular neurogenic niche. *Cell* **127:** 1253–1264.

Laplagne D.A., Esposito M.S., Piatti V.C., Morgenstern N.A., Zhao C., van Praag H., Gage F.H., and Schinder A.F. 2006. Functional convergence of neurons generated in the developing and adult hippocampus. *PLoS Biol.* **4:** e409.

Lie D.C., Colamarino S.A., Song H.J., Desire L., Mira H., Consiglio A., Lein E.S., Jessberger S., Lansford H., Dearie A.R., and Gage F.H. 2005. Wnt signalling regulates adult hippocampal neurogenesis. *Nature* **437:** 1370–1375.

Liu M. Pleasure S.J., Collins A.E., Noebels J.L., Naya F.J., Tsai M.J., and Lowenstein D.H. 2000. Loss of BETA2/NeuroD leads to malformation of the dentate gyrus and epilepsy. *Proc. Natl. Acad. Sci.* **97:** 865–870.

Lledo P.M., Alonso M., and Grubb M.S. 2006. Adult neurogenesis and functional plasticity in neuronal circuits. *Nat. Rev. Neurosci.* **7:** 179–193.

Louvi A., and Artavanis-Tsakonas S. 2006. Notch signalling in vertebrate neural development. *Nat. Rev. Neurosci.* **7:** 93–102.

Meltzer L.A., Yabaluri R., and Deisseroth K. 2005. A role for circuit homeostasis in adult neurogenesis. *Trends Neurosci.* **28:** 653–660.

Merkle F.T., and Alvarez-Buylla A. 2006. Neural stem cells in mammalian development. *Curr. Opin. Cell Biol.* **18:** 704–709.

Merkle F.T., Tramontin A.D., Garcia-Verdugo J.M., and Alvarez-Buylla A. 2004. Radial glia give rise to adult neural stem cells in the subventricular zone. *Proc. Natl. Acad. Sci.* **101:** 17528–17532.

Ming G.L., and Song H. 2005. Adult neurogenesis in the mammalian central nervous system. *Annu. Rev. Neurosci.* **28:** 223–250.

Miyata T., Maeda T., and Lee J.E. 1999. NeuroD is required for differentiation of the granule cells in the cerebellum and hippocampus. *Genes. Dev.* **13:** 1647–1652.

Nacher J., Varea E., Blasco-Ibanez J.M., Castillo-Gomez E., Crespo C., Martinez-Guijarro F.J., and McEwen B.S. 2005. Expression of the transcription factor Pax 6 in the adult rat dentate gyrus. *J. Neurosci. Res.* **81:** 753–761.

Nieto M., Schuurmans C., Britz O., and Guillemot F. 2001. Neural bHLH genes control the neuronal versus glial fate decision in cortical progenitors. *Neuron* **29:** 401–413.

Olson J.M., Asakura A., Snider L., Hawkes R., Strand A., Stoeck J., Hallahan A., Pritchard J., and Tapscott S.J. 2001. NeuroD2 is necessary for development and survival of central nervous system neurons. *Dev. Biol.* **234:** 174–187.

Palmer T.D., Willhoite A.R., and Gage F.H. 2000. Vascular niche for adult hippocampal neurogenesis. *J. Comp. Neurol.* **425:** 479–494.

Parras C.M., Schuurmans C., Scardigli R., Kim J., Anderson D.J., and Guillemot F. 2002. Divergent functions of the proneural genes Mash1 and Ngn2 in the specification of neuronal subtype identity. *Genes Dev.* **16:** 324–338.

Petersen P.H., Tang H., Zou K., and Zhong W. 2006. The enigma of the numb-Notch relationship during mammalian embryogenesis. *Dev. Neurosci.* **28:** 156–168.

Petersen P.H., Zou K., Hwang J.K., Jan Y.N., and Zhong W. 2002. Progenitor cell maintenance requires numb and numblike during mouse neurogenesis. *Nature* **419:** 929–934.

Pleasure S.J., Collins A.E., and Lowenstein D.H. 2000. Unique expression patterns of cell fate molecules delineate sequential stages of dentate gyrus development. *J. Neurosci.* **20:** 6095–6105.

Pozniak C.D., and Pleasure S.J. 2006. A tale of two signals: Wnt and Hedgehog in dentate neurogenesis. *Sci. STKE:* pe5.

Redmond L., Oh S.R., Hicks C., Weinmaster G., and Ghosh A. 2000. Nuclear Notch1 signaling and the regulation of dendritic development. *Nat. Neurosci.* **3:** 30–40.

Richardson W.D., Kessaris N., and Pringle N. 2006. Oligodendrocyte wars. *Nat. Rev. Neurosci.* **7:** 11–18.

Ross S.E., Greenberg M.E., and Stiles C.D. 2003. Basic helix-loop-helix factors in cortical development. *Neuron* **39:** 13–25.

Schwab M.H., Bartholomae A., Heimrich B., Feldmeyer D., Druffel-Augustin S., Goebbels S., Naya F.J., Zhao S., Frotscher M., Tsai M.J., and Nave K.A. 2000. Neuronal basic helix-loop-helix proteins (NEX and BETA2/Neuro D) regulate terminal granule cell differentiation in the hippocampus. *J. Neurosci.* **20:** 3714–3724.

Selkoe D., and Kopan R. 2003. Notch and Presenilin: Regulated intramembrane proteolysis links development and degeneration. *Annu. Rev. Neurosci.* **26:** 565–597.

Shen Q., Zhong W., Jan Y.N., and Temple S. 2002. Asymmetric Numb distribution is critical for asymmetric cell division of mouse cerebral cortical stem cells and neuroblasts. *Development* **129:** 4843–4853.

Song H., Stevens C.F., and Gage F.H. 2002. Astroglia induce neurogenesis from adult neural stem cells. *Nature* **417:** 39–44.

Sugimori M., Nagao M., Bertrand N., Parras C.M., Guillemot F., and Nakafuku M. 2007. Combinatorial actions of patterning and HLH transcription factors in the spatiotemporal control of neurogenesis and gliogenesis in the developing spinal cord. *Development* **134:** 1617–1629.

Yoon K., and Gaiano N. 2005. Notch signaling in the mammalian central nervous system: Insights from mouse mutants. *Nat. Neurosci.* **8:** 709–715.

Zhong W. 2003. Diversifying neural cells through order of birth and asymmetry of division. *Neuron* **37:** 11–14.

Zhong W., Feder J.N., Jiang M.M., Jan L.Y., and Jan Y.N. 1996. Asymmetric localization of a mammalian numb homolog during mouse cortical neurogenesis. *Neuron* **17:** 43–53.

14

The Balance of Trophic Support and Cell Death in Adult Neurogenesis

H. Georg Kuhn
Center for Brain Repair and Rehabilitation
Department of Neuroscience and Physiology
Göteborg University, Sweden

THE FACT THAT CONTINUOUS PROLIFERATION of stem cells and progenitors, as well as the production of neurons, occurs in the adult CNS raises several basic questions concerning the number of neurons required in a particular system: Can we observe a continued growth of brain regions that sustain neurogenesis? Or does an elimination mechanism exist that keeps the number of cells constant? If so, are the old ones replaced or are the new neurons competing for limited network access? What signals would support their survival and integration and what factors are responsible for their elimination? This chapter addresses these and other questions regarding regulatory mechanisms affecting adult neurogenesis by controlling cell survival.

ARE NEUROGENIC BRAIN REGIONS EXPANDING DESPITE SPACE LIMITATIONS?

This question was initially addressed several decades ago, following the first evidence that adult mammalian neurogenesis exists. Total neuronal cell counts of the olfactory bulb (OB) and dentate gyrus (DG) at different ages revealed that in both regions, a continued growth of the granule cell layer occurs throughout adult life. From 1 month of age, when the developmental production of granule cells can be considered complete, until 1 year of age, the number of DG granule cells doubles in the rat (Bayer 1982; Bayer et al. 1982). A rise in total volume and increased cell density due to reduced cell diameter both contribute to this phenomenon. In the rat OB, a linear growth of the granule cell layer was observed with age (Kaplan et al. 1985), with the number of olfactory granule cells doubling between 3 and 31 months of age.

Considering such substantial growth, we can postulate that new neurons are certainly added to the system. But does this mean that there is no elimination or replacement? Several studies have estimated the amount of new neurons generated in the adult brain. The numbers vary due to differences in thymidine-labeling protocols, species, and strains, but it can be assumed that the young adult rodent DG produces several thousand new granule cells per day. For the OB, the number is even greater and ranges in the tens of thousands per day. From these numbers, it became apparent that the actual growth of the granule cell layers is lower than extrapolated from labeling studies; hence, an elimination mechanism for supernumerous cells was proposed.

APOPTOSIS AS A COMMON ELEMENT OF NEUROGENIC ZONES

Apoptosis, or programmed cell death, has an important role in mammalian brain development, allowing rapid elimination of excess neurons or neurons that have failed to make proper connections (Oppenheim 1991). Cell death can be frequently detected within the neurogenic zones of the adult brain. The cells die through apoptotic mechanisms and can be visualized by terminal deoxynucleotidyl-transferase-mediated dUTP-biotin nick end-labeling (TUNEL) or activated caspase-3 labeling (Biebl et al. 2000; Kuhn et al. 2005). Quantitatively, the highest amount of dying cells is observed in the OB, followed by the rostral migratory stream, subventricular zone (SVZ), and the DG (Biebl et al. 2000). The location of apoptotic profiles indicates that it is intimately connected to the generation of new neurons (see Fig. 1). But are young cells eliminated or are old ones replaced? There are good indicators that a large amount of the dying cells are immature. First, in the DG, for example, cell death is detected largely in the subgranular zone (SGZ), the border between the granule cell layer and hilus, where dividing progenitors reside. Second, apoptotic profiles can be colabeled with immature neuronal markers, such as doublecortin (DCX) (Kuhn et al. 2005). Third, when monitoring cohorts of adult-born bromodeoxyuridine (BrdU)-labeled cells over time, a 50% reduction in the number of progenitors and young neurons is detected over a period of 3 months. These numbers are roughly identical for the rat DG and OB (Winner et al. 2002; Dayer et al. 2003). After this critical 3-month period, adult-born neurons survive largely for the

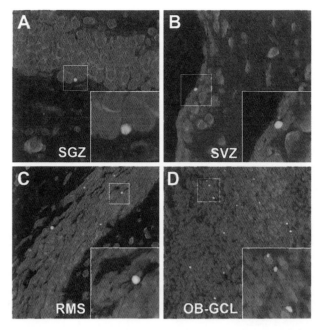

Figure 1. Apoptotic cells can be visualized by TUNEL in sections of the intact adult rodent brain. TUNEL-positive cells (*green*) can be frequently detected in (*A*) the subgranular zone of the dentate gyrus (DG-SGZ), (*B*) the subventricular zone of the lateral ventricle wall (SVZ), (*C*) the rostral migratory stream (RMS), and (*D*) the granule cell layer of the olfactory bulb (OB-GCL). Red counterstain is propidium iodide. (For more details, see Biebl et al. 2000; Cooper-Kuhn and Kuhn 2002; Kuhn et al. 2005.)

rest of the animal's life span. For neurons that are generated during the postnatal period, Cameron and colleagues reported, however, that the number of BrdU-labeled cells continuously declines with age (Dayer et al. 2003). This indicates that mature neurons generated postnatally undergo some degree of cell death. Taken together, we can postulate that a dynamic interplay of cell elimination and addition regulates the neuronal cell numbers in the DG and OB.

NEUROTROPHINS

Neurotrophic substances are able to stimulate the growth and survival of neurons during development and help to maintain adult neurons. They are also capable of inducing increased survival of damaged neurons and regrowth of axons and dendrites. During embryonic and adult neurogenesis, neurotrophic support is an essential component of the signal cascade that controls the generation of neurons from multipotent progenitor cells.

The neurotrophin family—nerve growth factor (NGF), brain-derived neurotrophic factor (BDNF), NT3, NT4/5—is involved in the development and maintenance of a variety of neuronal cell types through multiple cell mechanisms. In the classical view, neurotrophins are released by target structures, acting as neurite-attracting molecules (Lewin and Barde 1996). Upon reaching the target, neurons soon become dependent on retrogradely transported neurotrophin for their survival (Oppenheim 1989). Interestingly, target-independent roles of paracrine- and autocrine-released neurotrophins have also been described previously (Acheson and Lindsay 1996).

For adult neurogenesis, BDNF appears to have a central role in stimulating the differentiation and survival of newly generated neurons in the SVZ, as well as the hippocampus (Zigova et al. 1998; Benraiss et al. 2001; Scharfman et al. 2005; Rossi et al. 2006). Studies on other neurotrophins are less numerous; however, NGF was recently shown to promote neurogenesis following infusion into the lateral ventricle (Frielingsdorf et al. 2006). NT3 also appears to be involved, since adult NT3-deficient mice have impaired granule cell differentiation in the DG, despite normal proliferation (Shimazu et al. 2006).

GROWTH FACTORS

Growth factors bind to specific high-affinity receptors on the surface of responsive cells, inducing signal cascades with the primary effect of activating cellular proliferation and/or differentiation. A prototypic growth factor for neuronal stem cells is fibroblast growth factor-2 (FGF-2). Like many other growth factors, FGF-2 is pleiotrophic, having diverse effects on multiple cell types. Stimulation of neuronal progenitor cells with FGF-2 induces distinct effects at different concentrations. At low concentrations (2 ng/ml), FGF-2 promotes the survival and neuronal differentiation in neuronal progenitor cultures (Walicke et al. 1986), whereas at concentrations above 10 ng/ml, proliferation of progenitor cells can be observed (Ray et al. 1993). The presence of endogenous FGF-2, produced and released by astrocytes (Gomez-Pinilla et al. 1992), makes it an important regulator of adult neurogenesis. Infusion of FGF-2 leads to increased neurogenesis (Kuhn et al. 1997; Wagner et al. 1999), whereas FGF-1-receptor-deficient mice have impaired neurogenesis in the adult hippocampus (Zhao et al. 2007).

Similarly, a variety of growth factors are known to affect adult neurogenic regions. A subset of these peptides, such as epidermal growth factor (EGF), transforming growth factor-β1 (TGF-β1), leukemia-inhibiting factor (LIF), and pigment epithelium-derived factor (PEDF), act only on the proliferation and self-renewal of stem cells (Craig et al. 1996; Kuhn et al. 1997; Jin et al. 2002; Ramirez-Castillejo et al. 2006; Wachs et al. 2006). Other peptide factors, such as FGF-2, vascular endothelial growth factor (VEGF), and pituitary adenylate cyclase-activating polypeptide (PACAP) (Zhu et al. 2003; Schänzer et al. 2004; Mercer et al. 2004; Ohta et al. 2006), have both proliferative and neurotrophic activity. On the other hand, erythropoietin, granulocyte–colony-stimulating factor (G-CSF), and insulin-like growth factor-1 (IGF1) have been identified to increase neurogenesis through their survival-promoting capacity (Aberg et al. 2000; Shingo et al. 2001; Schneider et al. 2005; Lichtenwalner et al. 2006).

STEROID HORMONES

Hormonal signals control many aspects of neuronal development, including cell survival. It is therefore of no surprise that adult neurogenesis is under strong hormonal influence. In summary, stress hormones negatively regulate the amount of new neurons via proliferation, as well as differential survival (for more details, see Chapter 18). Gonadal hormones and neurosteroids, on the other hand, stimulate the generation of new neurons (Karishma and Herbert 2002; Mayo et al. 2005; Galea et al. 2006). Steroid hormones act through nuclear hormone receptors, which bind to promoter regions in key genes for neuronal survival. For example, the neurotrophin BDNF and the Bcl-2 gene family are downstream targets of steroid hormone receptor activation (Almeida et al. 2000; Charalampopoulos et al. 2006; Scharfman and MacLusky 2006; Yao et al. 2007). Moreover, signaling through thyroid hormone and retinoic acid receptors appears to have a strong stimulatory effect on proliferation and survival of neuronal progenitor cells and appears to be important for maintaining postnatal and adult neurogenesis (Ambrogini et al. 2005; Desouza et al. 2005; Lemkine et al. 2005; Wang et al. 2005; Jacobs et al. 2006; Kornyei et al. 2007).

NEUROTRANSMITTERS AND SYNAPTIC ACTIVITY

As immature neurons strive to integrate into functional networks, neuronal communication becomes an important survival-promoting factor. Classical neurotransmitters such as glutamate, acetylcholine, and serotonin, which are released in the vicinity of the neurogenic regions, are involved in regulating adult neurogenesis. This section focuses specifically on the effect of neurotransmitters on the survival of progenitor cells. For a more comprehensive overview, see Chapter 19.

Glutamate

In one of the pioneering studies on the regulation of adult neurogenesis, glutamate was described as suppressing the proliferation and differentiation of granule cells (Cameron et al. 1995). Lesion of the entorhinal cortex, the origin of the perforant path, or treatment with both competitive and noncompetitive N-methyl-D-aspartate (NMDA) receptor antagonists, stimulated the proliferation of granule cell precursors and the production of new granule cells. In contrast, treatment with NMDA receptor agonists inhibited cell proliferation in the DG. Ionotropic and metabotropic glutamate receptors are also capable of modulating adult hippocampal neurogenesis (Bernabeu and Sharp 2000; Yoshimizu and Chaki 2004).

Acetylcholine

Acetylcholine appears to promote the survival of new neurons. Lesions of the basal forebrain cholinergic system lead to significantly lower survival rates in the hippocampus and the OB despite unchanged proliferation of progenitor cells (Cooper-Kuhn et al. 2004; Mohapel et al. 2005). Neuronally committed progenitors express multiple acetylcholine receptor subunits and make contact with cholinergic fibers (Kaneko et al. 2006). Moreover, pharmacological alterations in cholinergic signaling confirmed the positive influence of acetylcholine for hippocampal neurogenesis (Kaneko et al. 2006; Kotani et al. 2006).

Serotonin

Serotonin has a very important role in regulating adult neurogenesis, especially due to the fact that almost all antidepressants, which act through enhancement of serotonergic neurotransmission, stimulate hippocampal granule cell production (for review, see Gould 1999; Jacobs 2002). On the other hand, chronic depletion of serotonin reduces neurogenesis (Brezun and Daszuta 1999). The mechanisms of serotonergic influences are complex. Antidepressants can act via stimulation of progenitor proliferation (Malberg et al. 2000); however, the up-regulation of neurotrophic factors, cAMP response element-binding protein (CREB) and Bcl-2, have also been reported, which points toward a survival- promoting effect (Nibuya et al. 1995, 1996; Chen et al. 2007).

Nonclassical neurotransmitters are also potent regulators of adult neurogenesis. Neuropeptide Y positively regulates hippocampal proliferation (Howell et al. 2005, 2007), whereas substance P and nitric oxide signaling appear to be detrimental to adult neurogenesis by repressing survival signals such as BDNF and pCREB signaling (Morcuende et al. 2003; Packer et al. 2003; Reif et al. 2004; Zhu et al. 2006).

SYNAPTIC ACTIVITY AND SURVIVAL

Synapse formation is viewed as a critical step in survival mechanisms for newly formed neurons, since the new cells can receive target-derived factors that are transported to the soma to influence apoptotic signaling. Surprisingly, neuronal progenitor cells in the hippocampus receive neurotransmitter input at an early stage, when they are still undergoing cell division. GABAergic input to neuronal progenitor cells appears to be the earliest and, at this stage, seems to up-regulate neuronal determination factors, such as NeuroD, which stimulate the cell cycle exit and neuronal differentiation (for more details, see Chapters 15 and 19). Although it has not been directly studied, it is quite likely that mechanisms of synaptic integration induce cell survival in the hippocampus. In the OB neurogenesis system, most dying cells have already matured to the point that they have begun forming dendritic spines and have the potential to receive synaptic input (Petreanu and Alvarez-Buylla 2002). Their survival seems to be very much dependent on sensory input, suggesting a direct role for electrical activity in cell survival (Corotto et al. 1994; Petreanu and Alvarez-Buylla 2002; Rochefort et al. 2002). It appears that electrical activity has no effect on subventricular proliferation, rostral migration, and early differentiation of olfactory neurons; however, once olfactory granule cells reach their target and become synaptically connected, their survival depends on the level of activity they receive. This phase of critical dependence on activity lasts for several months, since a steady decline in the number of new neurons can be observed up to 3 months after the new cells have entered the OB. Thereafter, at least in animals with intact olfactory function, the survival of these cells seems to be secured for the rest of the animal's life span (Winner et al. 2002).

INTRACELLULAR SIGNALING

Trophic support, be it through trophic factors, hormones, neurotransmitters, or other extracellular changes, ultimately needs to influence the activation of intracellular cell death pathways. Apoptotic cell death is characterized by induction of signals that eventually lead to activation of caspases and DNA fragmentation, followed by cell death. Apoptosis can be induced externally through death receptors and intrinsically through control over the integrity of the mitochondrial outer membrane. Cell death receptors, such as Fas, are frequently involved in cell death during inflammatory responses and involve microglia and/or T-cell activation. Although these signals have not been studied in detail with respect to adult neurogenesis, it is unlikely that they have a major role in the intact brain, since inflammatory signals are largely absent (Ekdahl et al. 2003). The intrinsic cell death pathway, which is initiated when cells are subjected to DNA damage or trophic factor deprivation, is regulated by members of the Bcl-2 family. Bcl-2 family proteins are either pro- or antiapoptotic. The major proapoptotic molecules, Bax and Bak, compromise mitochondrial membrane integrity by forming channels that can release apoptogenic signals such as cytochrome c. The antiapoptotic proteins, Bcl-2 and Bcl-X_L, protect the membrane by forming heterodimers with proapoptotic proteins, thus preventing the release of cytochrome c.

In adult neurogenesis, it appears that both pro- and antiapoptotic Bcl-2 family members are involved. Bax-deficient mice have a higher number of neuronal progenitor cells and a higher rate of neurogenesis in the DG, as well as reduced cell death. The lack of apoptosis leads to an increased accumulation of granule cells and a larger DG with age (Sun et al. 2004). Bax-deficient mice, as well as double-knockout mice for Bax and Bak, display a significantly larger pool of neuronal progenitor cells in the SVZ, which can serve as multipotent stem cells in vitro (Lindsten et al. 2003; Shi et al. 2005). The expression of Bcl-2 appears to be intimately linked to neurogenesis in the adult brain with high levels in the SVZ and DG (Bernier and Parent 1998; Bernier et al. 2000). Similar to Bax deficiency, transgenic overexpression of Bcl-2 in neuronal cells results in the increase of hippocampal neurogenesis due to reduced cell death of neuronal progenitor cells (Kuhn et al. 2005).

The survival-promoting effect of neurotrophic factors, hormone, and other extracellular signals is ultimately mediated by interference with apoptosis-inducing signaling pathways. Trophic factors, such as FGF-2,

BDNF, G-CSF, and VEGF, can directly stimulate the expression of antiapoptotic Bcl-2 family proteins (Bryckaert et al. 1999; Desire et al. 2000; Rios-Munoz et al. 2005; Cao et al. 2006; Solaroglu et al. 2006; Milosevic et al. 2007), thereby counteracting Bax and Bak and preventing caspase activation.

CELL DEATH AND PROLIFERATION

An interesting aspect in the regulation of neurogenesis is the question of whether cell death triggers changes in proliferation or vice versa. Manipulations that induce neuronal cell death, such as high levels of cortisol or ischemia, also stimulate cell cycle entry. Moreover, it is known that differentiated neurons, when forced to enter the cell cycle, undergo apoptotic cell death (for review, see Liu and Greene 2001). Nevertheless, is progenitor proliferation a direct result of the "vacant spot"? The most direct evidence contradicting this hypothesis comes from Bax-deficient and Bcl-2-overexpressing mice, where a decrease in cell death did not lead to changes in proliferation (Kuhn et al. 2005; Sun et al. 2004). On the other hand, decreased S-phase entry in mice that are deficient for the cell cycle activator protein E2F1 leads to decreased cell death (Cooper-Kuhn et al. 2002). It is therefore more likely that cell death regulation during adult neurogenesis occurs downstream from proliferative signals.

CONCLUSION

Naturally occurring cell death claims up to 50% of the differentiating neurons in the adult brain. Their survival depends on paracrine or target-derived substances with neurotrophic activity, afferent synaptic activity, cell-cell and cell-matrix interactions, as well as hormonal and other blood-borne signals. It seems that an infinite combination of factors can serve to rescue or influence the survival of immature neurons (Fig. 2). As the cells go through a several-month-long differentiation and integration period, the fine balance of signals can easily be disturbed at many points along the way. The embryonic brain in mammals is largely protected from external stimuli by amniotic fluid and a placental barrier; however, adult neurogenesis takes place in an organism that dynamically interacts with its environment. This might provide a clue to why such an enormous variety of signals and conditions regulate adult neurogenesis.

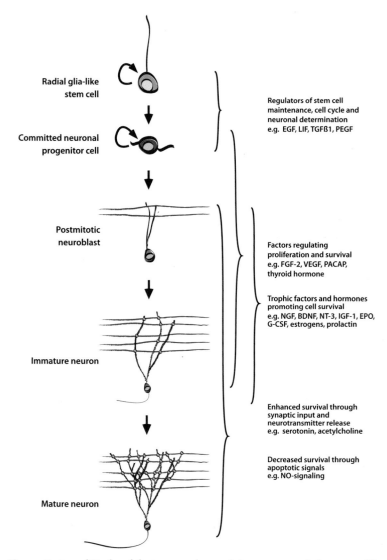

Figure 2. A multitude of factors regulates adult neurogenesis by controlling proliferation, fate determination, and survival of cells. Trophic support is required at all stages during the development from stem cell to neuron; otherwise, depletion of the stem cell pool or apoptotic cell death may significantly reduce the amount of new neurons. Neurotransmitter and synaptic influences are detected at early progenitor stages, indicating a strong influence of the preexisting network on the maturation and survival of neurons in the adult brain.

REFERENCES

Aberg M.A., Aberg N.D., Hedbacker H., Oscarsson J., and Eriksson P.S. 2000. Peripheral infusion of IGF-I selectively induces neurogenesis in the adult rat hippocampus. *J. Neurosci.* **20:** 2896–2903.

Acheson A. and Lindsay R.M. 1996. Non target-derived roles of the neurotrophins. *Philos. Trans. R. Soc. Lond. B. Biol. Sci.* **351:** 417–422.

Almeida O.F., Conde G.L., Crochemore C., Demeneix B.A., Fischer D., Hassan A.H., Meyer M., Holsboer F., and Michaelidis T.M. 2000. Subtle shifts in the ratio between pro- and antiapoptotic molecules after activation of corticosteroid receptors decide neuronal fate. *FASEB J.* **14:** 779–790.

Ambrogini P., Cuppini R., Ferri P., Mancini C., Ciaroni S., Voci A., Gerdoni E., and Gallo G. 2005. Thyroid hormones affect neurogenesis in the dentate gyrus of adult rat. *Neuroendocrinology* **81:** 244–253.

Bayer S.A. 1982. Changes in the total number of dentate granule cells in juvenile and adult rats: A correlated volumetric and 3H-thymidine autoradiographic study. *Exp. Brain Res.* **46:** 315–323.

Bayer S.A., Yackel J.W., and Puri P.S. 1982. Neurons in the rat dentate gyrus granular layer substantially increase during juvenile and adult life. *Science* **216:** 890–892.

Benraiss A., Chmielnicki E., Lerner K., Roh D., and Goldman S.A. 2001. Adenoviral brain-derived neurotrophic factor induces both neostriatal and olfactory neuronal recruitment from endogenous progenitor cells in the adult forebrain. *J. Neurosci.* **21:** 6718–6731.

Bernabeu R. and Sharp F.R. 2000. NMDA and AMPA/kainate glutamate receptors modulate dentate neurogenesis and CA3 synapsin-I in normal and ischemic hippocampus. *J. Cereb. Blood Flow. Metab.* **20:** 1669–1680.

Bernier P.J. and Parent A. 1998. Bcl-2 protein as a marker of neuronal immaturity in postnatal primate brain. *J. Neurosci.* **18:** 2486–2497.

Bernier P.J., Vinet J., Cossette M., and Parent A. 2000. Characterization of the subventricular zone of the adult human brain: Evidence for the involvement of Bcl-2. *Neurosci. Res.* **37:** 67–78.

Biebl M., Cooper C.M., Winkler J., and Kuhn H.G. 2000. Analysis of neurogenesis and programmed cell death reveals a self-renewing capacity in the adult rat brain. *Neurosci. Lett.* **291:** 17–20.

Brezun J.M. and Daszuta A. 1999. Depletion in serotonin decreases neurogenesis in the dentate gyrus and the subventricular zone of adult rats. *Neuroscience* **89:** 999–1002.

Bryckaert M., Guillonneau X., Hecquet C., Courtois Y., and Mascarelli F. 1999. Both FGF1 and bcl-x synthesis are necessary for the reduction of apoptosis in retinal pigmented epithelial cells by FGF2: Role of the extracellular signal-regulated kinase 2. *Oncogene* **18:** 7584–7593.

Cameron H.A., McEwen B.S., and Gould E. 1995. Regulation of adult neurogenesis by excitatory input and NMDA receptor activation in the dentate gyrus. *J. Neurosci.* **15:** 4687–4692.

Cao X.Q., Arai H., Ren Y.R., Oizumi H., Zhang N., Seike S., Furuya T., Yasuda T., Mizuno Y., and Mochizuki H. 2006. Recombinant human granulocyte colony-stimulating factor protects against MPTP-induced dopaminergic cell death in mice by altering Bcl-2/Bax expression levels. *J. Neurochem.* **99:** 861–867.

Charalampopoulos I., Alexaki V.I., Tsatsanis C., Minas V., Dermitzaki E., Lasaridis I., Vardouli L., Stournaras C., Margioris A.N., Castanas E., and Gravanis A. 2006.

Neurosteroids as endogenous inhibitors of neuronal cell apoptosis in aging. *Ann. N.Y. Acad. Sci.* **1088:** 139–152.

Chen S.J., Kao C.L., Chang Y.L., Yen C.J., Shui J.W., Chien C.S., Chen I.L., Tsai T.H., Ku H.H., and Chiou S.H. 2007. Antidepressant administration modulates neural stem cell survival and serotoninergic differentiation through bcl-2. *Curr. Neurovasc. Res.* **4:** 19–29.

Cooper-Kuhn C.M. and Kuhn H.G. 2002. Is it all DNA repair? Methodological considerations for detecting neurogenesis in the adult brain. *Brain Res. Dev. Brain Res.* **134:** 13–21.

Cooper-Kuhn C.M., Winkler J., and Kuhn H.G. 2004. Decreased neurogenesis after cholinergic forebrain lesion in the adult rat. *J. Neurosci. Res.* **77:** 155–165.

Cooper-Kuhn C.M., Vroemen M., Brown J., Ye H., Thompson M.A., Winkler J., and Kuhn H.G. 2002. Impaired adult neurogenesis in mice lacking the transcription factor E2F1. *Mol. Cell. Neurosci.* **21:** 312–323.

Corotto F.S., Henegar J.R., and Maruniak J.A. 1994. Odor deprivation leads to reduced neurogenesis and reduced neuronal survival in the olfactory bulb of the adult mouse. *Neuroscience* **61:** 739–744.

Craig C.G., Tropepe V., Morshead C.M., Reynolds B.A., Weiss S., and van der Kooy D. 1996. In vivo growth factor expansion of endogenous subependymal neural precursor cell populations in the adult mouse brain. *J. Neurosci.* **16:** 2649–2658.

Dayer A.G., Ford A.A., Cleaver K.M., Yassaee M., and Cameron H.A. 2003. Short-term and long-term survival of new neurons in the rat dentate gyrus. *J. Comp. Neurol.* **460:** 563–572.

Desire L., Courtois Y., and Jeanny J.C. 2000. Endogenous and exogenous fibroblast growth factor 2 support survival of chick retinal neurons by control of neuronal neuronal bcl-x(L) and bcl-2 expression through a fibroblast growth factor receptor 1- and ERK-dependent pathway. *J. Neurochem.* **75:** 151–163.

Desouza L.A., Ladiwala U., Daniel S.M., Agashe S., Vaidya R.A., and Vaidya V.A. 2005. Thyroid hormone regulates hippocampal neurogenesis in the adult rat brain. *Mol. Cell. Neurosci.* **29:** 414–426.

Ekdahl C.T., Zhu C., Bonde S., Bahr B.A., Blomgren K., and Lindvall O. 2003. Death mechanisms in status epilepticus-generated neurons and effects of additional seizures on their survival. *Neurobiol. Dis.* **14:** 513–523.

Frielingsdorf H., Simpson D.R., Thal L.J., and Pizzo D.P. 2006. Nerve growth factor promotes survival of new neurons in the adult hippocampus. *Neurobiol. Dis.* **26:** 47–55.

Galea L.A., Spritzer M.D., Barker J.M., and Pawluski J.L. 2006. Gonadal hormone modulation of hippocampal neurogenesis in the adult. *Hippocampus* **16:** 225–232.

Gomez-Pinilla F., Lee J.W., and Cotman C.W. 1992. Basic FGF in adult rat brain: Cellular distribution and response to entorhinal lesion and fimbria-fornix transection. *J. Neurosci.* **12:** 345–355.

Gould E. 1999. Serotonin and hippocampal neurogenesis. *Neuropsychopharmacology* **21:** 46S–51S.

Howell O.W., Doyle K., Goodman J.H., Scharfman H.E., Herzog H., Pringle A., Beck-Sickinger A.G., and Gray W.P. 2005. Neuropeptide Y stimulates neuronal precursor proliferation in the post-natal and adult dentate gyrus. *J. Neurochem.* **93:** 560–570.

Howell O.W., Silva S., Scharfman H.E., Sosunov A.A., Zaben M., Shatya A., McKhann G., 2nd, Herzog H., Laskowski A., and Gray W.P. 2007. Neuropeptide Y is important for basal and seizure-induced precursor cell proliferation in the hippocampus. *Neurobiol. Dis.* **26:** 174–188.

Jacobs B.L. 2002. Adult brain neurogenesis and depression. *Brain Behav. Immun.* **16:** 602–609.

Jacobs S., Lie D.C., DeCicco K.L., Shi Y., DeLuca L.M., Gage F.H., and Evans R.M. 2006. Retinoic acid is required early during adult neurogenesis in the dentate gyrus. *Proc. Natl. Acad. Sci.* **103:** 3902–3907.

Jin K., Mao X.O., Sun Y., Xie L., and Greenberg D.A. 2002. Stem cell factor stimulates neurogenesis in vitro and in vivo. *J. Clin. Invest.* **110:** 311–319.

Kaneko N., Okano H., and Sawamoto K. 2006. Role of the cholinergic system in regulating survival of newborn neurons in the adult mouse dentate gyrus and olfactory bulb. *Genes Cells* **11:** 1145–1159.

Kaplan M.S., McNelly N.A., and Hinds J.W. 1985. Population dynamics of adult-formed granule neurons of the rat olfactory bulb. *J. Comp. Neurol.* **239:** 117–125.

Karishma K.K. and Herbert J. 2002. Dehydroepiandrosterone (DHEA) stimulates neurogenesis in the hippocampus of the rat, promotes survival of newly formed neurons and prevents corticosterone-induced suppression. *Eur. J. Neurosci.* **16:** 445–453.

Kornyei Z., Gocza E., Ruhl R., Orsolits B., Voros E., Szabo B., Vagovits B., and Madarasz E. 2007. Astroglia-derived retinoic acid is a key factor in glia-induced neurogenesis. *FASEB J.* **21:** 2496–2509.

Kotani S., Yamauchi T., Teramoto T., and Ogura H. 2006. Pharmacological evidence of cholinergic involvement in adult hippocampal neurogenesis in rats. *Neuroscience* **142:** 505–514.

Kuhn H.G., Winkler J., Kempermann G., Thal L.J., and Gage F.H. 1997. Epidermal growth factor and fibroblast growth factor-2 have different effects on neural progenitors in the adult rat brain. *J. Neurosci.* **17:** 5820–5829.

Kuhn H.G., Biebl M., Wilhelm D., Li M., Friedlander R.M., and Winkler J. 2005. Increased generation of granule cells in adult Bcl-2-overexpressing mice: A role for cell death during continued hippocampal neurogenesis. *Eur. J. Neurosci.* **22:** 1907–1915.

Lemkine G.F., Raj A., Alfama G., Turque N., Hassani Z., Alegria-Prevot O., Samarut J., Levi G., and Demeneix B.A. 2005. Adult neural stem cell cycling in vivo requires thyroid hormone and its alpha receptor. *FASEB J.* **19:** 863–865.

Lewin G.R. and Barde Y.A. 1996. Physiology of the neurotrophins. *Annu. Rev. Neurosci.* **19:** 289–317.

Lichtenwalner R.J., Forbes M.E., Sonntag W.E., and Riddle D.R. 2006. Adult-onset deficiency in growth hormone and insulin-like growth factor-I decreases survival of dentate granule neurons: Insights into the regulation of adult hippocampal neurogenesis. *J. Neurosci. Res.* **83:** 199–210.

Lindsten T., Golden J.A., Zong W.X., Minarcik J., Harris M.H., and Thompson C.B. 2003. The proapoptotic activities of Bax and Bak limit the size of the neural stem cell pool. *J. Neurosci.* **23:** 11112–11119.

Liu D.X. and Greene L.A. 2001. Neuronal apoptosis at the G1/S cell cycle checkpoint. *Cell Tissue Res.* **305:** 217–228.

Malberg J.E., Eisch A.J., Nestler E.J., and Duman R.S. 2000. Chronic antidepressant treatment increases neurogenesis in adult rat hippocampus. *J. Neurosci.* **20:** 9104–9110.

Mayo W., Lemaire V., Malaterre J., Rodriguez J.J., Cayre M., Stewart M.G., Kharouby M., Rougon G., Le Moal M., Piazza P.V., and Abrous D.N. 2005. Pregnenolone sulfate enhances neurogenesis and PSA-NCAM in young and aged hippocampus. *Neurobiol. Aging* **26:** 103–114.

Mercer A., Ronnholm H., Holmberg J., Lundh H., Heidrich J., Zachrisson O., Ossoinak A., Frisen J., and Patrone C. 2004. PACAP promotes neural stem cell proliferation in adult mouse brain. *J. Neurosci. Res.* **76:** 205–215.

Milosevic J., Maisel M., Wegner F., Leuchtenberger J., Wenger R.H., Gerlach M., Storch A., and Schwarz J. 2007. Lack of hypoxia-inducible factor-1 alpha impairs midbrain neural precursor cells involving vascular endothelial growth factor signaling. *J. Neurosci.* **27:** 412–421.

Mohapel P., Leanza G., Kokaia M., and Lindvall O. 2005. Forebrain acetylcholine regulates adult hippocampal neurogenesis and learning. *Neurobiol. Aging* **26:** 939–946.

Morcuende S., Gadd C.A., Peters M., Moss A., Harris E.A., Sheasby A., Fisher A.S., De Felipe C., Mantyh P.W., Rupniak N.M., Giese K.P., and Hunt S.P. 2003. Increased neurogenesis and brain-derived neurotrophic factor in neurokinin-1 receptor gene knockout mice. *Eur. J. Neurosci.* **18:** 1828–1836.

Nibuya M., Morinobu S., and Duman R.S. 1995. Regulation of BDNF and trkB mRNA in rat brain by chronic electroconvulsive seizure and antidepressant drug treatments. *J. Neurosci.* **15:** 7539–7547.

Nibuya M., Nestler E.J., and Duman R.S. 1996. Chronic antidepressant administration increases the expression of cAMP response element binding protein (CREB) in rat hippocampus. *J. Neurosci.* **16:** 2365–2372.

Ohta S., Gregg C., and Weiss S. 2006. Pituitary adenylate cyclase-activating polypeptide regulates forebrain neural stem cells and neurogenesis in vitro and in vivo. *J. Neurosci. Res.* **84:** 1177–1186.

Oppenheim R.W. 1989. The neurotrophic theory and naturally occurring motoneuron death. *Trends Neurosci.* **12:** 252–255.

———. 1991. Cell death during development of the nervous system. *Annu. Rev. Neurosci.* **14:** 453–501.

Packer M.A., Stasiv Y., Benraiss A., Chmielnicki E., Grinberg A., Westphal H., Goldman S.A., and Enikolopov G. 2003. Nitric oxide negatively regulates mammalian adult neurogenesis. *Proc. Natl. Acad. Sci.* **100:** 9566–9571.

Petreanu L. and Alvarez-Buylla A. 2002. Maturation and death of adult-born olfactory bulb granule neurons: Role of olfaction. *J. Neurosci.* **22:** 6106–6113.

Ramirez-Castillejo C., Sanchez-Sanchez F., Andreu-Agullo C., Ferron S.R., Aroca-Aguilar J.D., Sanchez P., Mira H., Escribano J., and Farinas I. 2006. Pigment epithelium-derived factor is a niche signal for neural stem cell renewal. *Nat. Neurosci.* **9:** 331–339.

Ray J., Peterson D.A., Schinstine M., and Gage F.H. 1993. Proliferation, differentiation, and long-term culture of primary hippocampal neurons. *Proc. Natl. Acad. Sci.* **90:** 3602–3606.

Reif A., Schmitt A., Fritzen S., Chourbaji S., Bartsch C., Urani A., Wycislo M., Mossner R., Sommer C., Gass P., and Lesch K.P. 2004. Differential effect of endothelial nitric oxide synthase (NOS-III) on the regulation of adult neurogenesis and behaviour. *Eur. J. Neurosci.* **20:** 885–895.

Rios-Munoz W., Soto I., Duprey-Diaz M.V., Blagburn J., and Blanco R.E. 2005. Fibroblast growth factor 2 applied to the optic nerve after axotomy increases Bcl-2 and decreases Bax in ganglion cells by activating the extracellular signal-regulated kinase signaling pathway. *J. Neurochem.* **93:** 1422–1433.

Rochefort C., Gheusi G., Vincent J.D., and Lledo P.M. 2002. Enriched odor exposure increases the number of newborn neurons in the adult olfactory bulb and improves odor memory. *J. Neurosci.* **22:** 2679–2689.

Rossi C., Angelucci A., Costantin L., Braschi C., Mazzantini M., Babbini F., Fabbri M.E., Tessarollo L., Maffei L., Berardi N., and Caleo M. 2006. Brain-derived neurotrophic factor (BDNF) is required for the enhancement of hippocampal neurogenesis following environmental enrichment. *Eur. J. Neurosci.* **24:** 1850–1856.

Schänzer A., Wachs F.P., Wilhelm D., Acker T., Cooper-Kuhn C.M., Beck H., Winkler J., Aigner L., Plate K.H., and Kuhn H.G. 2004. Direct stimulation of adult neural stem cells in vitro and neurogenesis in vivo by vascular endothelial growth factor. *Brain Pathol.* **14:** 237–248.

Scharfman H., Goodman J., Macleod A., Phani S., Antonelli C., and Croll S. 2005. Increased neurogenesis and the ectopic granule cells after intrahippocampal BDNF infusion in adult rats. *Exp. Neurol.* **192:** 348–356.

Scharfman H.E. and MacLusky N.J. 2006. Estrogen and brain-derived neurotrophic factor (BDNF) in hippocampus: Complexity of steroid hormone-growth factor interactions in the adult CNS. *Front. Neuroendocrinol.* **27:** 415–435.

Schneider A., Kruger C., Steigleder T., Weber D., Pitzer C., Laage R., Aronowski J., Maurer M.H., Gassler N., Mier W., Hasselblatt M., Kollmar R., Schwab S., Sommer C., Bach A., Kuhn H.G., and Schabitz W.R. 2005. The hematopoietic factor G-CSF is a neuronal ligand that counteracts programmed cell death and drives neurogenesis. *J. Clin. Invest.* **115:** 2083–2098.

Shi J., Parada L.F., and Kernie S.G. 2005. Bax limits adult neural stem cell persistence through caspase and IP3 receptor activation. *Cell Death Differ.* **12:** 1601–1612.

Shimazu K., Zhao M., Sakata K., Akbarian S., Bates B., Jaenisch R., and Lu B. 2006. NT-3 facilitates hippocampal plasticity and learning and memory by regulating neurogenesis. *Learn. Mem.* **13:** 307–315.

Shingo T., Sorokan S.T., Shimazaki T., and Weiss S. 2001. Erythropoietin regulates the in vitro and in vivo production of neuronal progenitors by mammalian forebrain neural stem cells. *J. Neurosci.* **21:** 9733–9743.

Solaroglu I., Tsubokawa T., Cahill J., and Zhang J.H. 2006. Anti-apoptotic effect of granulocyte-colony stimulating factor after focal cerebral ischemia in the rat. *Neuroscience* **143:** 965–974.

Sun W., Winseck A., Vinsant S., Park O.H., Kim H., and Oppenheim R.W. 2004. Programmed cell death of adult-generated hippocampal neurons is mediated by the proapoptotic gene Bax. *J. Neurosci.* **24:** 11205–11213.

Wachs F.P., Winner B., Couillard-Despres S., Schiller T., Aigner R., Winkler J., Bogdahn U., and Aigner L. 2006. Transforming growth factor-beta1 is a negative modulator of adult neurogenesis. *J. Neuropathol. Exp. Neurol.* **65:** 358–370.

Wagner J.P., Black I.B., and DiCicco-Bloom E. 1999. Stimulation of neonatal and adult brain neurogenesis by subcutaneous injection of basic fibroblast growth factor. *J. Neurosci.* **19:** 6006–6016.

Walicke P., Cowan W.M., Ueno N., Baird A., and Guillemin R. 1986. Fibroblast growth factor promotes survival of dissociated hippocampal neurons and enhances neurite extension. *Proc. Natl. Acad. Sci.* **83:** 3012–3016.

Wang T.W., Zhang H., and Parent J.M. 2005. Retinoic acid regulates postnatal neurogenesis in the murine subventricular zone-olfactory bulb pathway. *Development* **132:** 2721–2732.

Winner B., Cooper-Kuhn C.M., Aigner R., Winkler J., and Kuhn H.G. 2002. Long-term survival and cell death of newly generated neurons in the adult rat olfactory bulb. *Eur. J. Neurosci.* **16:** 1681–1689.

Yao M., Nguyen T.V., and Pike C.J. 2007. Estrogen regulates Bcl-w and Bim expression: Role in protection against beta-amyloid peptide-induced neuronal death. *J. Neurosci.* **27:** 1422–1433.

Yoshimizu T. and Chaki S. 2004. Increased cell proliferation in the adult mouse hippocampus following chronic administration of group II metabotropic glutamate receptor antagonist, MGS0039. *Biochem. Biophys. Res. Commun.* **315:** 493–496.

Zhao M., Li D., Shimazu K., Zhou Y.X., Lu B. and Deng C.X. 2007. Fibroblast growth factor receptor-1 is required for long-term potentiation, memory consolidation, and neurogenesis. *Biol. Psychiatry* (in press).

Zhu X.J., Hua Y., Jiang J., Zhou Q.G., Luo C.X., Han X., Lu Y.M., and Zhu D.Y. 2006. Neuronal nitric oxide synthase-derived nitric oxide inhibits neurogenesis in the adult dentate gyrus by down-regulating cyclic AMP response element binding protein phosphorylation. *Neuroscience* **141:** 827–836.

Zhu Y., Jin K., Mao X.O., and Greenberg D.A. 2003. Vascular endothelial growth factor promotes proliferation of cortical neuron precursors by regulating E2F expression. *FASEB J.* **17:** 186–193.

Zigova T., Pencea V., Wiegand S.J., and Luskin M.B. 1998. Intraventricular administration of BDNF increases the number of newly generated neurons in the adult olfactory bulb. *Mol. Cell. Neurosci.* **11:** 234–245.

15

Maturation and Functional Integration of New Granule Cells into the Adult Hippocampus

Josef Bischofberger
Institute of Physiology
University of Freiburg
Germany

Alejandro F. Schinder
Leloir Institute
Buenos Aires, Argentina

THE HIPPOCAMPUS, LOCATED WITHIN THE MEDIAL TEMPORAL LOBE of the cerebral cortex, is critically important for the formation of semantic and episodic memory (Squire et al. 2004). As with other cortical circuits, the hippocampal network (Fig. 1) is highly dynamic and has the capacity to modify its connectivity by changing the number and strength of synaptic contacts in an activity-dependent manner. Synaptic connections can be added, strengthened, weakened, or eliminated in response to neuronal activity, a phenomenon called synaptic plasticity. The plasticity of specific hippocampal synapses has a significant role in memory formation and learning of hippocampus-dependent tasks (Nakazawa et al. 2004; Whitlock et al. 2006). The dentate gyrus (DG) of the adult hippocampus has the additional capacity of modifying the circuitry by the addition of new neurons. Thus, network remodeling is not limited to synapses, but also includes the incorporation of new functional units (neurons) that provide an additional dimension of plasticity to the existing hippocampal circuitry (Schinder and Gage 2004; Song et al. 2005; Lledo et al. 2006; Piatti et al. 2006).

Adult Neurogenesis ©2008 Cold Spring Harbor Laboratory Press 978-087969-784-6

299

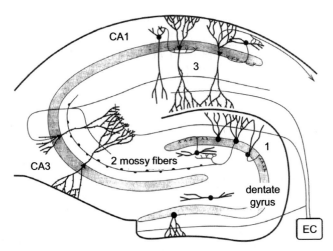

Figure 1. Synaptic circuits of the hippocampus. A transversal slice through the hippocampus shows the different subregions of the dentate gyrus, CA3, and CA1. The principal cells located in densely packed cell layers are synaptically connected via the so-called trisynaptic circuit (synapses 1–3). Inhibitory synapses are formed by interneurons that are mainly located outside the prinicipal cell layers and widely distributed in a mosaic-like fashion. (Modified, with permission from Springen Science and Business Media, from Bischofberger et al. 2006.)

The biological significance of adult hippocampal neurogenesis depends on the extent to which adult-born neurons can participate in signal processing in the hippocampal network. The impact of new neurons on the adult neuronal circuitry will be highly determined by how they become engaged in network activity and how their intrinsic properties and connectivities compare to those of existing dentate granule cells (GCs) that were generated during development. To list some possibilities, adult-born neurons could be continuously added to replace dying cells or to increase their number to adjust the size of the neural network in an activity-dependent manner. Alternatively, adult-born neurons might have different intrinsic functional properties as compared to the older preexisting cells or process incoming information in a unique way due to a specific network connectivity. As described in depth in this chapter, a combination of these possibilities seems to apply to adult-born neurons during specific stages of their development and neuronal maturation.

SYNAPTIC CIRCUITS OF THE HIPPOCAMPUS

Similarly to what occurs in the neocortex, GABAergic interneurons and glutamatergic principal cells are the primary components of the hippocampal network. GABAergic interneurons are distributed across the entire hippocampus in a mosaic-like pattern. In contrast, cell bodies of glutamatergic neurons are arranged in well-defined layers that can even be distinguished in transversal sections with low magnification. There are three types of glutamatergic principal neurons whose distinctive anatomical properties define the subregions of the hippocampus called DG, CA3, and CA1 (Fig. 1). The GCs of the DG receive excitatory glutamatergic synaptic inputs (synapse 1) from the superficial layers of the entorhinal cortex. GCs project onto the proximal dendrites of CA3 pyramidal cells via the so-called mossy fiber tract (synapse 2). They send axons to the CA1 pyramidal cells (synapse 3), which, in turn, project to the subiculum and back to the deep layers of the entorhinal cortex. In addition to this trisynaptic circuit, there is also a direct projection from the entorhinal cortex to the CA3 and CA1 pyramidal cells, providing parallel synaptic pathways for hippocampal information processing (Johnston and Amaral 1998). Remarkably, only DG GCs can be generated in the adult hippocampus.

The core processing unit of the hippocampal circuitry is thought to be the autoassociative network of the CA3 region (Fig. 1). Memory items might be represented by cell assemblies generated by interconnected CA3 pyramidal cells, whose synaptic contacts are formed during learning and reactivated during retrieval of episodic memory (Nakazawa et al. 2004). Due to the associative collaterals, the learned entorhinal input patterns will activate the corresponding cell assembly even when the input is partially disrupted, a process called pattern completion (Nakazawa et al. 2004). This is why we recognize our bedroom even when the bed is missing.

In addition to the direct entorhinal afferents, CA3 pyramidal cells also receive synaptic inputs from GCs of the DG via mossy fibers (Fig. 1), which appear to have some remarkable functional properties (Bischofberger et al. 2006). Most importantly, mossy fiber synapses can operate within a large dynamic range and with high efficacy, such that short spike trains in a single presynaptic GC can effectively discharge CA3 pyramidal cells (Henze et al. 2002). The average firing rate of GCs in vivo is usually relatively low, and only a small fraction of GCs is active during exploration of an environment (~2% of GCs as compared to 40% of pyramidal cells) (Chawla et al. 2005), consistent with a sparse coding of information. Nevertheless, the GCs might fire short high-frequency bursts during distinct behavioral tasks (Jung and McNaugton 1993). The functional properties and the specific synaptic connectivity of the GCs might help to generate unique activity patterns within the CA3 network, even for similar but nevertheless distinct entorhinal inputs, a process called pattern separation (Leutgeb et al. 2007).

Consistent with the sparse coding concept, GCs outnumber by far the number of neurons in both the input and the output regions. In young adult rats, the numbers of GCs were estimated to be about 1 million, whereas entorhinal layer II and CA3 pyramidal cells can reach 225,000 and 330,000 per hippocampus, respectively (Amaral et al. 1990). Most GCs are added after birth, which is due to the fact that new GCs are continuously generated, even in the adult nervous system.

Because newly generated neurons must be incorporated into the existing network in a meaningful way without disrupting hippocampal function, their development and functional integration must be tightly controlled. As described below, these problems in the adult hippocampus are partly solved by mechanisms that are similar to those operating during the critical periods described for early postnatal development.

STEM-CELL PROLIFERATION AND NEURONAL DIFFERENTIATION

Adult neurogenesis can be divided into three major steps that are termed proliferation, neuronal determination, and maturation (Fig. 2). These different developmental stages are regulated by distinct physiological processes. The hippocampal neural stem cells (NSCs) are located in the subgranular zone, adjacent to the granule cell layer (GCL) of the DG, and share several features of glial cells (Filippov et al. 2003). Similarly to the radial glia that gives rise to glutamatergic neurons during cortical embryonic development, they express the intermediate filament proteins nestin and GFAP (glial fibrillary acidic protein) (Filippov et al. 2003; Rakic 2003). Furthermore, they possess a relatively thick apical process, projecting radially through the GCL into the molecular layer where they ramify into many small branches (Encinas et al. 2006).

The electrophysiological properties of NSCs were studied using transgenic animals expressing the enhanced green fluorescent protein (EGFP) under the control of the nestin promoter (Filippov et al. 2003). They express a high density of voltage-independent potassium channels typical of glial cells. As a consequence, they have a low electrical input resistance of about 70 MΩ (Fukuda et al. 2003). In addition, NSCs lack voltage-gated sodium channels and are therefore not electrically excitable, similar to normal glial cells.

NSC proliferation was investigated by injection of 5-bromo-2'-deoxyuridine (BrdU), which is inserted into the DNA of dividing cells during the S phase of the cell cycle. Afterward, it can be detected by

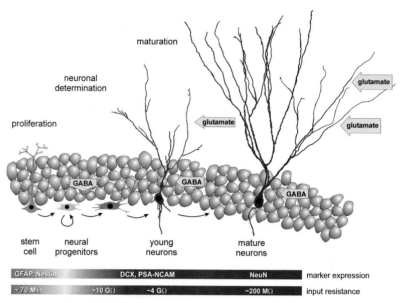

Figure 2. Development of newly generated granule cells (GCs). Neural stem cells (*green*) give rise to neural progenitor cells (*yellow*) that continue to divide for a few more days. After the last cell cycle, the young neurons (*orange, red*) rapidly grow and differentiate during a 6-week maturation period until the new neurons resemble fully mature hippocampal GCs (*blue*).

immunohistochemistry, resulting in the labeling of all cells that were dividing at the time of BrdU injection. After a single BrdU pulse, approximately 2–10% of NSCs are labeled, indicating that their mitotic activity is usually relatively low (Filippov et al. 2003; Encinas et al. 2006). The NSCs give rise to neural progenitors that lack the apical process and rapidly down-regulate GFAP while maintaining nestin expression for some time. Because the progenitors continue to divide for a short period of approximately 3–5 days with high mitotic activity, this population is also called "transiently amplifying cells." Accordingly, immunostaining against Ki-67, which is expressed during most of the cell cycle (G_1, S, and G_2), was shown to label about 80% of neural progenitor cells but only about 25% of NSCs (Tozuka et al. 2005). Overall, a remarkable number of new cells can be generated, which was estimated to be about 9000 per day in the hippocampus of young adult rats (Cameron and McKay 2001).

Because neural progenitor cells start losing the glial potassium channels, they show a significantly higher input resistance (>500 MΩ) as compared to the NSCs (~70 MΩ) (Fukuda et al. 2003). Furthermore, they already express neurotransmitter receptors such as γ-amino-n-butyric acid A (GABA$_A$) receptors (Wang et al. 2005). These GABA receptors can apparently be activated by extrasynaptic GABA due to spillover or even by GABAergic synaptic contacts (Wang et al. 2005; Ge et al. 2006). GABA$_A$ receptors appear to be very important for neuronal differentiation of daughter cells. Typically, a large fraction of about 70–90% of the cells will differentiate into neurons. The young cells have a relatively depolarized chloride reversal potential (E$_{Cl}$ = −35 mV), due to the expression of the Na$^+$-K$^+$-Cl$^-$ cotransporter NKCC1 (Tozuka et al. 2005). The GABA release from GABAergic interneurons of the DG therefore leads to depolarization and to Ca^{2+} influx via voltage-gated Ca^{2+} channels. Most likely, candidate interneurons are DG basket cells, which show dense axonal ramification within the GCL and fire high-frequency action potentials during hippocampal θ-γ oscillations (Jonas et al. 2004). The Ca^{2+} signals, in turn, are important for the expression of the early neuronal transcription factor NeuroD that determines a neuronal fate (Tozuka et al. 2005). As a consequence, progenitor cells express neuronal voltage-gated sodium channels as well as other early neuronal proteins as, for example, the microtubule-associated protein doublecortin (DCX; Fig. 2) (Brown et al. 2003) or the polysialic-acid–neural cell adhesion molecule (PSA-NCAM) (Seki 2002; Fukuda et al. 2003). Therefore, neuronal activity in the DG will determine the proportion of newly generated cells differentiating into neurons.

In addition to GABA receptors, the progenitors were also reported to express functional AMPA (α-amino-3-hydroxy-5-methyl-4-isoazole) and NMDA (N-methyl-D-aspartate) receptors (Deisseroth et al. 2004; Wang et al. 2005). Although it is unclear if and how these receptors are activated in vivo, there is also evidence for NMDA-receptor-mediated induction of NeuroD expression (Deisseroth et al. 2004). These data show that the neuronal progenitor cells are electrically excitable, in strong contrast to the stem cells. Furthermore, they already express neurotransmitter receptors, supporting the activity-dependent regulation of adult neurogenesis.

MATURATION OF YOUNG NEURONS

After birth, young neurons go through a relatively long period where intrinsic membrane properties and synaptic connectivity develop. Although synapse formation and functional maturation is largely finished by the sixth week, the size of synaptic spines can still increase up to about 2–4 months (Espósito et al 2005; Zhao et al. 2006).

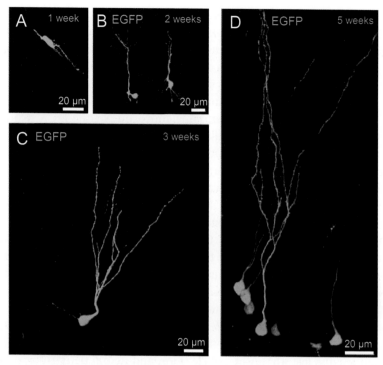

Figure 3. Morphological development of enhanced green fluorescent protein (EGFP)-labeled newly generated granule cells (GCs). Confocal images show GCs at 1 week (*A*), 2 weeks (*B*), 3 weeks (*C*), and 5 weeks (*D*) after cell division. The EGFP-tagged retrovirus was delivered to the dentate gyrus of adult (7 weeks old) mice, and the animals were sacrificed at the indicated times after surgery. (Images by V.C. Piatti and N.A. Morgenstern.)

During the early stages (≤3 weeks), they continue to express DCX and PSA-NCAM as well as other typical embryonic proteins, which might be important for migration and morphological differentiation (Seki 2002; Brown et al. 2003). Figure 3 shows young neurons in the DG of 2-month-old adult mice, which were labeled with EGFP using a replication-deficient retrovirus. Because oncoretroviral DNA cannot pass the nuclear membrane, such retroviruses can only infect dividing cells during the M phase of the cell cycle when nuclear membranes become disassembled. Therefore, a modified mouse leukemia virus has been used to selectively transduce fluorescent proteins into dividing cells of the subgranular zone by stereotactic injection (van Praag et al. 2002). Because EGFP is inserted in the DNA of dividing progenitor cells, their progeny will be fluorescently labeled in a specific fashion. Fluorescent markers are expressed throughout maturation, and the entire morphological and functional development of new neurons can therefore be analyzed in living brain tissue with a reasonable temporal precision of a few days (Fig. 3) (Espósito et al. 2005; Zhao et al. 2006). Alternatively, the total population of young neurons can be labeled in transgenic animals that express EGFP or DsRed under the control of the POMC (pro-opiomelanocortin) or the DCX promoters (Overstreet et al. 2004; Couillard-Despres et al. 2006).

During the first few days after mitosis, the dendrites start growing and the axon rapidly extends toward the CA3 region (Hasting and Gould 1999; Zhao et al. 2006). The apical dendrites of young neurons reach the molecular layer already after a few days. After 2 weeks, the axon has grown far into the CA3 region and the dendrites reach approximately the middle of the molecular layer (Zhao et al. 2006). Initially, they also generate basal dendrites similar to that of pyramidal cells, which might represent a default growth pattern of glutamatergic neurons (Fig. 3B). However, during the next 2 weeks, basal dendrites are retracted and apical dendrites further extend toward the hippocampal fissure, until the new neurons show the typical morphology of a fully differentiated mature GC (Fig. 3D) (van Praag et al. 2002; Espósito et al 2005).

During the last phases of proliferation and neural determination, all of the glial potassium channels are lost. Therefore, the first DCX- and PSA-NCAM-expressing postmitotic neurons show a remarkably high electrical input resistance of about 10 GΩ (Fig. 4) (Schmidt-Hieber et al. 2004; Couillard-Despres et al. 2006). With progressing maturation, the input resistance decreases again, which might be due to cell growth as well as to an increase in the specific membrane conductance mediated by neuronal leakage channels (Ambrogini et al. 2004). The density of voltage-gated channels rapidly increases, so that the neurons can fire large-amplitude-overshooting action potentials (APs) already within 2–3 weeks of birth when they are still positive for DCX (Fig. 4D–F) (Couillard-Despres et al. 2006). Furthermore, young neurons express voltage-gated Ca^{2+} channels, similarly to neuronal progenitor cells. They express a relatively high density of T-type channels, with a particular low activation threshold (Schmidt-Hieber et al. 2004). These channels can generate low-threshold Ca^{2+} spikes or boost excitatory potentials to elicit full-amplitude sodium action potentials. As a result, APs can be generated with very small excitatory currents in the range of a few picoamperes (Fig. 4F) (Couillard-Despres et al. 2006). In contrast, in mature GCs, much larger currents are necessary to elicit an AP (Fig. 4A–C) (Schmidt-Hieber et al. 2004; van Praag et al. 2002).

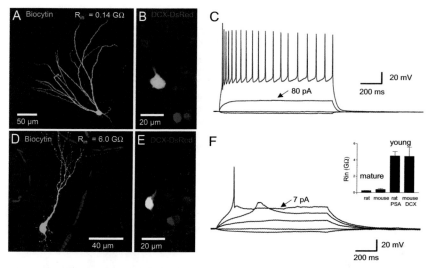

Figure 4. Enhanced excitability of young granule cells (GCs) (*A*) Mature GC that was filled with biocytin during whole-cell patch-clamp recording. Cells were recorded in the dentate gyrus of a 2-month-old transgenic mouse expressing the red fluorescent protein DsRed under the control of the doublecortin (DCX) promoter (*B, red*). (*C*) Firing pattern of the cell in *A*. (*D,E,F*) DCX-DsRed expressing young GC with a high input resistance (6 GΩ) generates isolated low-threshold spikes and overshooting action potentials (APs) with small excitatory current injections (*arrow*). (*Inset*) Input resistances of mature and young GCs identified by expression of polysialic-acid-neural cell adhesion molecule, (PSA-NCAM) and DCX-DsRed. (Reprinted, with permission, from Couillard-Despres et al. 2006.)

Similar to its role for progenitor cells, GABA appears to be also important for the development of the young neurons. After about 1 week, the young cells already receive GABAergic synaptic inputs (Fig. 5A) (Espósito et al. 2005; Ge et al. 2006). These early GABAergic synaptic events show a relatively slow time course which might be due to the expression of specific high-affinity GABA$_A$ receptor subunits (Overstreet-Wadiche et al. 2005). In addition, the early synaptic contacts might target preferentially the small processes that ramify within the GCL and the inner molecular layer, leading to significant dendritic filtering (Espósito et al. 2005).

Developing neurons express high levels of the Na^+-K^+-Cl^- cotransporter NKCC1, which maintains a high concentration of intracellular Cl^- and renders a Cl^- reversal potential of about −30 to −50 mV within the first week after mitosis (Ge et al. 2006; Karten et al 2006). As a result, GABAergic synaptic inputs onto young neurons are depolarizing, similar

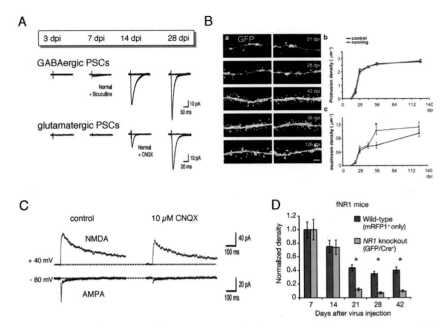

Figure 5. Development of GABAergic and glutamatergic synaptic inputs. (*A*) GABAergic and glutamatergic synaptic currents in enhanced green fluorescent protein (EGFP) labeled newly generated granule cells (GCs). (Reprinted, with permission, from Macmillan Publishers Ltd., from Ge et al. 2006.) (*B*) Spine density at different time points after GC birth in control cages (*blue*) and cages including running wheels (*red*). (Reprinted, with permission, from Zhao et al. 2006.) (*C*) Excitatory α-amino-3-hydroxy-5-methyl-4-isoazole (AMPA) and N-methyl-D-aspartate (NMDA)-receptor-mediated synaptic currents in a young GC, recorded at membrane potentials indicated at the left. AMPA receptors were blocked by the competitive antagonist CNQX (C. Schmidt-Hieber and J. Bischofberger, unpubl.). (*D*) Density of newly generated GCs at several time points after cell birth with NMDA-receptor-deficient (*green*) or wild-type postsynaptic receptors (*red*). (Reprinted, with permission, from Macmillan Publishers Ltd., from Tashiro et al. 2006.)

to those described during early postnatal development (Ben-Ari 2002). After about 2 weeks, the expression of NKCC1 decreases and the K^+-Cl^- cotransporter KCC2 is up-regulated, leading to a decreased level of intracellular Cl^- concentration. As a consequence, GABA becomes inhibitory.

GABAergic excitation appears to be very important for neuronal maturation. By using retroviral overexpression of short-hairpin RNA against NKCC1, the Cl^- reversal potential of young neurons was shifted toward negative values, abrogating GABA-mediated excitation and leading to a dramatic slowdown of dendritic growth and synapse formation (Ge et al. 2006). $GABA_A$ receptors might therefore act as sensors for the surrounding network activity to support activity-dependent growth during this early phase of neuronal maturation (Ge et al. 2007). Consistent with this view, the time course of the development of newly generated GCs in the adult hippocampus can vary considerably depending on environmental conditions or pathophysiological situations (Overstreet-Wadiche et al. 2006a,b).

The first glutamatergic synapses are formed about 2 weeks after birth (Fig. 5A) (Ge et al. 2006). The first synaptic spines appear shortly after and rapidly grow during the second half of the first month to reach a density of about 2 spines/μm at the age of 6 weeks (Fig. 5B) (Zhao et al. 2006). Assuming a total dendritic length of about 3000 μm (Schmidt-Hieber et al. 2004), this would correspond to a total number of about 6000 newly formed synapses. Figure 5C shows excitatory synaptic currents (ESCs) in young GCs after stimulation of entorhinal fibres. The young cells show a relatively large NMDA:AMPA receptor ratio, which might contribute to the generation and stabilization of the new excitatory synaptic contacts. It was shown that NMDA receptor deletion in floxed NR1 mice by retroviral expression of the Cre recombinase dramatically reduces survival rates of the young neurons during the first 2–3 weeks after birth (Fig. 5D) (Tashiro et al. 2006). Synaptic activation of NMDA receptors might also contribute to the enhanced survival of newly generated GCs after exposure to an enriched environment and hippocampus-dependent learning (Gould et al. 1999; Leuner et al. 2004; Tashiro et al. 2007). Taken together, these data indicate that young neurons form glutamatergic synaptic contacts at about 2 weeks after birth, which seems to be important for survival. In contrast, cells without excitatory synaptic connections are eliminated (Fig. 5D). After young neurons have survived through a critical period of about 4 weeks, they can persist for many months and years (Dayer et al. 2003; Kempermann et al. 2003; Zhao et al. 2006).

Maturation of the young neurons within the adult hippocampus follows a sequence that appears to be remarkably similar to that of neuronal development in postnatal animals. During the first 2 weeks, GABAergic excitation supports dendritic growth that appears to be dependent on general hippocampal activity. This process is followed by another 4-week period with GABAergic inhibition together with the formation of several thousands of new excitatory synaptic contacts.

FUNCTIONAL PROPERTIES OF MATURE ADULT-BORN NEURONS

At the age of about 6–8 weeks, new neurons display overall morphological and functional characteristics of fully mature GCs (van Praag et al. 2002; Espósito et al. 2005; Zhao et al. 2006). They display a round cell body localized inside the GCL with complex spiny dendrites reaching the hippocampal fissure and an axon that projects through the hilus toward CA3. They exhibit an input resistance of about 250 MΩ and fire repetitive action potentials with after-hyperpolarization and frequency-dependent adaptation in response to depolarizing current steps (Laplagne et al. 2006). Adult-born GCs are completely integrated into the hippocampal circuitry, as they receive functional inhibitory (GABAergic) and excitatory (glutamatergic) inputs, with the latter formed onto dendritic spines.

To understand whether all mature GCs ultimately belong to the same functional population or whether adult-born neurons maintain a distinctive functional phenotype, double retroviral labeling was used to distinguish neurons generated in the developing DG versus adult DG in acute hippocampal slices (Fig. 6) (Laplagne et al. 2006). A retroviral construct expressing EGFP was delivered to the DG to label progenitor cells that were dividing during early postnatal development (Fig. 6A, green). A second retroviral injection was then performed during adulthood to express a red fluorescent protein (RFP) in adult-born neurons of the same hippocampus (Fig. 6A, red). Adult-born mature GCs receive excitatory glutamatergic afferents from the entorhinal cortex through the medial and lateral perforant paths (Fig. 6C). They also receive GABAergic synaptic inputs of persiomatic origin (arising most likely from basket cells) as well as of dendritic origin. Despite the broad range of functional parameters that were quantitatively assessed, GCs generated in the developing and adult DG displayed a striking similarity. The convergence of connectivity and function was reflected by the strength, kinetics, and short-term plasticity of evoked glutamatergic responses (Fig. 6C,D) and also in the conductance, reversal potential, and kinetics of both evoked and spontaneous GABAergic responses originating at the soma and dendrites.

To investigate whether adult-born neurons can integrate excitatory synaptic inputs to generate a spike, action currents were recorded in the cell-attached configuration. MPP (medial perforant path) stimulation elicited spikes whose probability of occurrence increased as a function of the stimulus strength. In addition, simultaneous dual recordings from neurons born in the developing and adult hippocampus rendered a similar spiking probability for a given stimulus strength, demonstrating that they can integrate excitatory inputs in an equivalent fashion (Fig. 6E). Therefore, developmental and adult neurogenesis seem to produce neurons that will ultimately belong to the same functional population.

Although afferent connectivity of adult-born GCs has been thoroughly characterized in both young and mature neurons, no functional

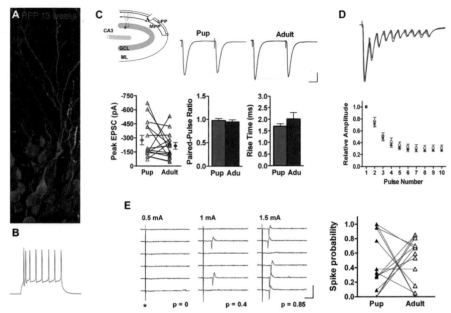

Figure 6. Functional properties of newly generated mature neurons. (*A*) Simultaneous retroviral labeling of neurons born in the dentate gyrus of pups (enhanced green fluorescent protein [EGFP], *green*) and adult mice (mRFP1, *red*). (*B*) Spiking of an adult-born neuron elicited by a depolarizing current step. (*C*) Excitatory postsynaptic currents (EPSCs) in neurons labeled in pups (*green traces and symbols*) and adult mice (*red traces and symbols*) recorded in response to paired-pulse stimulation of the medial perforant path. Bars, 25 msec, 50 pA. (*D*) EPSCs in response to high-frequency stimulation (10 pulses, 50 Hz). Traces are normalized to the first EPSC amplitude. (*Bottom*) EPSC amplitudes normalized to the first event. No difference was found between pup and adult responses. (*E*) Action currents in cell-attached configuration recorded from an adult-born granule cell (GC) in response to increasing stimulus strengths. Spiking probability (*p*) is shown below the traces. Bars, 10 msec, 50 pA. (*Right*) Firing behavior of neurons born in pup and adult brain during simultaneous paired experiments showed no significant differences. Bicuculline was present in all recordings. Neurons were about 18 weeks old (pup) and approximately 13 weeks old (adult). (Modified from Laplagne et al. 2006.)

studies have so far addressed their postsynaptic connectivity. This is due to the experimental challenge of identifying and recording presynaptic adult-born neurons connected to postsynaptic target cells in hippocampal slices. Whether adult-born GCs release glutamate or other neurotransmitters and whether presynaptic function is similar to that of preexisting neurons are all highly relevant problems that await further investigation.

SYNAPTIC PLASTICITY IN YOUNG AND MATURE GRANULE CELLS

The strength of the hippocampal synaptic connections can be adjusted in an activity-dependent manner by long-term potentiation (LTP) and long-term depression (LTD). As in other brain regions, the induction of LTP in DG GCs appears to be dependent on the activation of postsynaptic NMDA receptors, which are only effectively opened after both binding of glutamate and sufficient postsynaptic depolarization to relieve voltage-dependent Mg^{2+} block (Dan and Poo 2006; Lin et al. 2006).

Although LTP in rats and mice can be induced throughout the whole life span of the animals, the contribution of NMDA receptors to synaptic transmission onto GCs and the capacity for bidirectional plasticity decrease during postnatal development (Trommer et al. 1995). Furthermore, the threshold for LTP induction was shown to increase in hippocampal pyramidal cells and GCs during maturation and aging (Paulsen and Sejnowski 2000; Burke and Barnes 2006). These data indicate that the properties of synaptic plasticity in young and old hippocampal neurons differ markedly. Therefore, adult neurogenesis not only might increase the number of neurons, but also might generate a population of young GCs with juvenile properties such as, for example, a higher degree of synaptic plasticity.

As mentioned earlier, the firing of GCs is tightly controlled by inhibitory interneurons. On the other hand, the Cl^- concentration in the young neurons, and thus the reversal potential for $GABA_A$ receptors, slowly decreases during the initial 6 weeks of maturation. Therefore, the action of GABA is initially excitatory and becomes inhibitory only at mature developmental stages. As a consequence, the induction of LTP in synapses formed onto young GCs is only slightly affected by the application of the $GABA_A$ receptor blocker picrotoxin (Wang et al. 2000). Furthermore, there is also a strong difference between young and old neurons in the threshold for LTP induction, even in the absence of GABAergic transmission (Schmidt-Hieber et al. 2004). Figure 7 shows that the EPSP amplitude in PSA-NCAM-positive young GCs (Fig. 7A,B, right) is potentiated by pairing of a relatively weak presynaptic stimulation with single postsynaptic APs (TBS1), showing that the young neurons have a low threshold for the induction of LTP. In contrast, old GCs in the adult hippocampus only express LTP after very strong stimulations (TBS2), indicating that synapses formed onto old neurons are much more stable.

To assess the contribution of young cells to overall synaptic plasticity in the dentate, the dividing progenitor cells were killed by low-dose

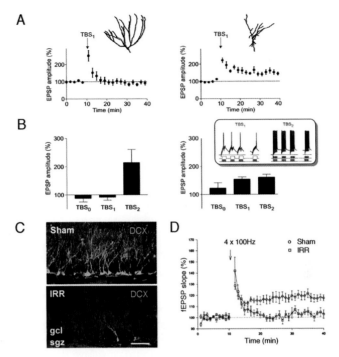

Figure 7. Enhanced synaptic plasticity in young hippocampal granule cells (GCs). (*A,B*) The EPSP amplitude in polysialic-acid-neural cell adhesion molecule (PSA-NCAM)-positive young (*right*) but not in old GCs (*left*) is increased after an induction protocol with single postsynaptic action potentials (APs) (TBS1). Strong stimulation paradigms (TBS2) also induce long-term potentiation (LTP) in old GCs. (Reprinted, with permission from Macmillan Publishers Ltd., from Schmidt-Hieber et al. 2004.) (*C*) Brief γ-irradiation destroys dividing stem cells leading to nearly complete absence of young GCs 4 weeks later. (Reprinted, with permission, from Snyder et al. 2005.) (*D*) This treatment strongly reduced LTP of synaptic field potentials. (Reprinted, with permission, from Snyder et al. 2001.)

γ-irradiation (Snyder et al. 2001). Using this treatment, it is possible to effectively eliminate all of the young DCX-positive neurons with minimal side effects (Fig. 7C) (Wojtowicz 2006, see also Chapter 21). LTP induction in the presence of normal artificial cerebrospinal fluid (ACSF) revealed a remarkable contribution of the 1–4-week-old cells to the overall potentiation of the synaptic field potentials (Fig. 7D).

The late phase of LTP and the stable memory formation depend on gene transcription and expression of new proteins. The first to come are the so-called immediate-early genes (IEGs), including Zif268 and Arc/Arg3.1 which are expressed within several minutes after LTP

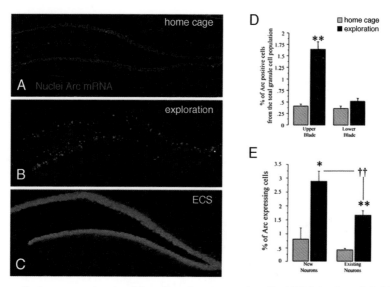

Figure 8. Arc expression in old and newborn granule cells (GCs) in the adult hippocampus. (*A,B,C*) Arc mRNA in the hippocampus was detected by fluorescence in situ hybridization (FISH, *red*), and nuclei were counterstained with propidium iodide (*blue*). Rats were either taken from the home cages (*A*), put into a new environment for spatial exploration (*B*), or treated with electroconvulsive shocks (ECS, *C*) and sacrificed immediately afterward. (Reprinted, with permission, from Chawla et al. 2005.) (*D,E*) Newly generated GCs, identified by 5-bromo-2′-deoxyuridine (BrdU) insertion, are significantly more likely to be activated during spatial exploration as compared to preexisting old GCs. (Reprinted, with permission, from Ramirez-Amaya et al. 2006.)

induction or spatial exploration (Fig. 8) (Guzowski et al. 2000; Bozon et al. 2002; Chawla et al. 2005). Consistent with the slice recordings described above, a significant fraction of the 2-week-old cells express Zif268 after LTP induction in vivo (Bruel-Jungerman et al. 2006). A more physiological stimulus such as spatial exploration in a novel environment also induces the expression of Zif268 and other IEGs in adult-born neurons at 2–5 weeks after birth (Jessberger and Kempermann 2003). Actually, it was shown that Arc expression during spatial exploration and learning occurs even more likely in newly generated GCs as compared to neighboring old cells (Fig. 8E) (Ramirez-Amaya et al. 2006; Kee et al. 2007). Similarly, newly generated 3-week-old GCs in the olfactory bulb preferentially increase c-*fos* and Zif268 expression during a novel odor-learning task (Magavi et al. 2005). Taken together, these data indicate that newly generated GCs in the adult hippocampus not only show a low threshold for synaptic plasticity, but might also contribute significantly to hippocampal information processing.

CONCLUDING REMARKS

Neurogenesis in the adult hippocampus is precisely controlled by hippocampal network activity. Neural progenitor cells follow a specific sequence of fate determination and neuronal maturation lasting for about 1–2 months. During this period, several thousands of new excitatory and inhibitory synapses are formed. Similar to postnatal development, GABA facilitates functional maturation, and a substantial amount of newly generated neurons die most likely due to the lack of proper synaptic inputs. Newborn neurons that survived the first month will persist for many months and years and can be activated during hippocampus-dependent behavioral tasks, supporting the notion that they contribute to hippocampal information processing.

During the maturation period, the newborn young GCs show distinct physiological properties, including enhanced excitability and a low threshold for synaptic plasticity. This might indicate a specific role for the young neurons within the hippocampal network. Alternatively, the increased synaptic plasticity could act as a mechanism to facilitate functional integration of new neurons that are already partially connected but not yet participating in information processing. In contrast, the newborn mature neurons are remarkably similar to the preexisting GCs, indicating that they would be suited to replace dying GCs and to adjust the total number of neurons within the DG. This might be important to continuously support sparse coding of information even with an increasing memory load.

A recent provocative hypothesis proposes a critical role for young GCs with high excitability in the association of new memories that occurred within a restricted temporal window (Aimone et al. 2006) (see Chapter 22). In this case, the high excitability of young neurons might provide a mechanism of low-threshold activation. On the other hand, the DG might contribute to the separation of similar hippocampal input patterns to facilitate the unique representation of new memories (Leutgeb et al. 2007). Therefore, the enhanced associative synaptic plasticity in newly generated young cells might be helpful for categorization of new hippocampal input patterns to facilitate pattern separation within the hippocampal network (Bischofberger 2007).

Despite the large progress achieved during the past years, many questions still remain open. What are the molecular mechanisms of plasticity and stability in young and old GCs? When do they establish functional outputs onto CA3 pyramidal cells? What are the functional properties of the presynaptic terminals? Do they release glutamate or other neurotransmitters? Do they show presynaptic LTP as typically described for hippocampal mossy fiber boutons? Addressing these questions will be necessary to understand the mechanisms and significance of functional integration of newly generated GCs in the adult hippocampus.

REFERENCES

Aimone J.B., Wiles J., and Gage F.H. 2006. Potential role for adult neurogenesis in the encoding of time in new memories. *Nat. Neurosci.* **9:** 723–727.

Amaral D.G., Ishizuka N., and Claiborne B. 1990. Neurons, numbers and the hippocampal network. *Prog. Brain Res.* **83:** 1–11.

Ambrogini P., Lattanzi D., Ciuffoli S., Agostini D., Bertini L., Stocchi V., Santi S., and Cuppini R. 2004. Morpho-functional characterization of neuronal cells at different stages of maturation in granule cell layer of adult rat dentate gyrus. *Brain Res.* **1017:** 21–31.

Ben-Ari Y. 2002. Excitatory actions of GABA during development: The nature of the nurture. *Nat. Rev. Neurosci.* **3:** 728–739.

Bischofbeger J. 2007. Young and excitable: New neurons in memory networks. *Nat. Neurosci.* **10:** 273–275.

Bischofberger J., Engel D., Frotscher M., and Jonas P. 2006. Timing and efficacy of transmitter release at mossy fiber synapses in the hippocampal network. *Eur. J. Physiol.* **453:** 361–372.

Bozon B., Davis S., and Laroche S. 2002. Regulated transcription of the immediate-early gene Zif268: Mechanisms and gene dosage-dependent function in synaptic plasticity and memory formation. *Hippocampus* **12:** 570–577.

Brown J.P., Couillard-Despres S., Cooper-Kuhn C.M., Winkler J., Aigner L., and Kuhn H.G. 2003. Transient expression of doublecortin during adult neurogenesis. *J. Comp. Neurol.* **467:** 1–10.

Bruel-Jungerman E., Davis S., Rampon C., and Laroche S. 2006. Long-term potentiation enhances neurogenesis in the adult dentate gyrus. *J. Neurosci.* **26:** 5888–5893.

Burke S.N. and Barnes C.A. 2006. Neural plasticity in the ageing brain. *Nat. Rev. Neurosci.* **7:** 30–40.

Cameron H.A. and McKay R.D. 2001. Adult neurogenesis produces a large pool of new granule cells in the dentate gyrus. *J. Comp. Neurol.* **435:** 406–417.

Chawla M.K., Guzowski J.F., Ramirez-Amaya V., Lipa P., Hoffman K.L., Marriott L.K., Worley P.F., McNaughton B.L., and Barnes C.A. 2005. Sparse, environmentally selective expression of Arc RNA in the upper blade of the rodent fascia dentata by brief spatial experience. *Hippocampus* **15:** 579–586.

Couillard-Despres S., Winner B., Karl C., Lindemann G., Schmid P., Aigner R., Laemke J., Bogdahn U., Winkler J., Bischofberger J., and Aigner L. 2006. Transgene expression in neuronal precursors: Watching young neurons in the old brain. *Eur. J. Neurosci.* **24:** 1535–1545.

Dan Y. and Poo M.M. 2006. Spike timing-dependent plasticity: From synapse to perception. *Physiol. Rev.* **86:** 1033–1048.

Dayer A.G., Ford A.A., Cleaver K.M., Yassaee M., and Cameron H.A. 2003. Short-term and long-term survival of new neurons in the rat dentate gyrus. *J. Comp. Neurol.* **460**: 563–572.

Deisseroth K., Singla S., Toda H., Monje M., Palmer T.D., and Malenka R.C. 2004. Excitation-neurogenesis coupling in adult neural stem/progenitor cells. *Neuron* **42**: 535–552.

Encinas J.M., Vaahtokari A., and Enikolopov G. 2006. Fluoxetine targets early progenitor cells in the adult brain. *Proc. Natl. Acad. Sci.* **103**: 8233–8238.

Espósito M.S., Piatti V.C., Laplagne D.A., Morgenstern N.A., Ferrari C.C., Pitossi F.J., and Schinder A.F. 2005. Neuronal differentiation in the adult hippocampus recapitulates embryonic development. *J. Neurosci.* **25**: 10074–10086.

Filippov V., Kronenberg G., Pivneva T., Reuter K., Steiner B., Wang L.P., Yamaguchi M., Kettenmann H., and Kempermann G. 2003. Subpopulation of nestin-expressing progenitor cells in the adult murine hippocampus shows electrophysiological and morphological characteristics of astrocytes. *Mol. Cell. Neurosci.* **23**: 373–382.

Fukuda S., Kato F., Tozuka Y., Yamaguchi M., Miyamoto Y., and Hisatsune T. 2003. Two distinct subpopulations of nestin-positive cells in adult mouse dentate gyrus. *J. Neurosci.* **23**: 9357–9366.

Ge S., Pradhan D.A., Ming G.L., and Song H. 2007. GABA sets the tempo for activity-dependent adult neurogenesis. *Trends Neurosci.* **30**: 1–8.

Ge S., Goh E.L., Sailor K.A., Kitabatake Y., Ming G., and Song H. 2006. GABA regulates synaptic integration of newly generated neurons in the adult brain. *Nature* **439**: 589–593.

Gould E., Beylin A., Tanapat P., Reeves A., and Shors T.J. 1999. Learning enhances adult neurogenesis in the hippocampal formation. *Nat. Neurosci.* **2**: 260–265.

Guzowski J.F., Lyford G.L., Stevenson G.D., Houston F.P., McGaugh J.L., Worley P.F., and Barnes C.A. 2000. Inhibition of activity-dependent arc protein expression in the rat hippocampus impairs the maintenance of long-term potentiation and the consolidation of long-term memory. *J. Neurosci.* **20**: 3993–4001.

Hastings N.B. and Gould E. 1999. Rapid extension of axons into the CA3 region by adult-generated granule cells. *J. Comp. Neurol.* **413**: 146–154.

Henze D.A., Wittner L., and Buzsaki G. 2002. Single granule cells reliably discharge targets in the hippocampal CA3 network in vivo. *Nat. Neurosci.* **5**: 790–795.

Jessberger S. and Kempermann G. 2003. Adult-born hippocampal neurons mature into activity-dependent responsiveness. *Eur. J. Neurosci.* **18**: 2707–2712.

Johnston D. and Amaral D.G. 1998. Hippocampus. In *The synaptic organization of the brain* (ed. G. M. Shepherd), pp. 417–458. Oxford University Press, New York.

Jonas P., Bischofberger J., Fricker D., and Miles R. 2004. Interneuron diversity series: Fast in, fast out—Temporal and spatial signal processing in hippocampal interneurons. *Trends Neurosci.* **27**: 30–40.

Jung M.W. and McNaughton B.L. 1993. Spatial selectivity of unit activity in the hippocampal granular layer. *Hippocampus* **3**: 165–182.

Karten Y.J., Jones M.A., Jeurling S.I., and Cameron H.A. 2006. GABAergic signaling in young granule cells in the adult rat and mouse dentate gyrus. *Hippocampus* **16**: 312–320.

Kee N., Teixeira C.M., Wang A.H., and Frankland P.W. 2007. Preferential recruitment of adult generated granule cells into spatial memory networks in the dentate gyrus. *Nat. Neurosci.* **10**: 355–362.

Kempermann G., Gast D., Kronenberg G., Yamaguchi M., and Gage F.H. 2003. Early determination and long-term persistence of adult-generated new neurons in the hippocampus of mice. *Development* **130:** 391–399.

Laplagne D.A., Esposito M.S., Piatti V.C., Morgenstern N.A., Zhao C., van Praag H., Gage F.H., and Schinder A.F. 2006. Functional convergence of neurons generated in the developing and adult hippocampus. *PloS. Biol.* **4:**e409.

Leuner B., Mendolia-Loffredo S., Kozorovitskiy Y., Samburg D., Gould E., and Shors T.J. 2004. Learning enhances the survival of new neurons beyond the time when the hippocampus is required for memory. *J. Neurosci.* **24:** 7477–7481.

Leutgeb J.K., Leutgeb S., Moser M.B., and Moser E.I. 2007. Pattern separation in the dentate gyrus and CA3 of the hippocampus. *Science* **315:** 961–966.

Lin Y.W., Yang H.W., Wang H.J., Gong C.L., Chiu T.H., and Min M.Y. 2006. Spike-timing-dependent plasticity at resting and conditioned lateral perforant path synapses on granule cells in the dentate gyrus: Different roles of *N*-methyl-D-aspartate and group I metabotropic glutamate receptors. *Eur. J. Neurosci.* **23:** 2362–2374.

Lledo P.M., Alonso M., and Grubb M.S. 2006. Adult neurogenesis and functional plasticity in neuronal circuits. *Nat. Rev. Neurosci.* **7:** 179–193.

Magavi S.S.P., Mitchell B.D., Szentirmai O., Carter B.S., and Macklis J.D. 2005. Adult-born and preexisting olfactory granule neurons undergo distinct experience-dependent modifications of their olfactory responses in vivo. *J. Neurosci.* **25:** 10729–10739.

Nakazawa K., McHugh T.J., Wilson M.A., and Tonegawa S. 2004. NMDA receptors, place cells and hippocampal spatial memory. *Nat. Rev. Neurosci.* **5:** 361–372.

Overstreet-Wadiche L.S., Hentges S.T., Bumaschny V.F., de Souza P.S., Smart J.L., Santangelo A.M., Low M.J., Westbrook G.L., and Rubinstein M. 2004. A transgenic marker for newly born granule cells in dentate gyrus. *J. Neurosci.* **24:** 3251–3259.

Overstreet-Wadiche L.S., Bensen A.L., and Westbrook G.L. 2006a. Delayed development of adult-generated granule cells in dentate gyrus. *J. Neurosci.* **26:** 2326–2334.

Overstreet-Wadiche L.S., Bromberg D.A., Bensen A.L., and Westbrook G.L. 2005. GABAergic signaling to newborn neurons in dentate gyrus. *J. Neurophysiol.* **94:** 4528–4532.

———. 2006b. Seizures accelerate functional integration of adult-generated granule cells. *J. Neurosci.* **26:** 4095–4103.

Paulsen O. and Sejnowsky T.J. 2000. Natural patterns of activity and long-term synaptic plasticity. *Curr. Opin. Neurobiol.* **10:** 172–179.

Piatti V.C., Esposito M.S., and Schinder A.F. 2006. The timing of neuronal development in adult hippocampal neurogenesis. *Neuroscientist* **12:** 463–468.

Ramirez-Amaya V., Marrone D.F., Gage F.H., Worley P.F., and Barnes C.A. 2006. Integration of new neurons into functional neural networks. *J. Neurosci.* **26:** 12237–12241.

Rakic P. 2003. Developmental and evolutionary adaptations of cortical radial glia. *Cereb. Cortex* **13:** 541–549.

Rolls E. and Treves A. 1998. *Neural networks and brain function.* Oxford University Press, United Kingdom.

Schinder A.F. and Gage F.H. 2004. A hypothesis about the role of adult neurogenesis in hippocampal function. *Physiology* **19:** 253–261.

Schmidt-Hieber C., Jonas P., and Bischofberger J. 2004. Enhanced synaptic plasticity in newly generated granule cells of the adult hippocampus. *Nature* **429:** 184–187.

Seki T. 2002. Expression patterns of immature neuronal markers PSA-NCAM, CRMP-4 and NeuroD in the hippocampus of young adult and aged rodents. *J. Neurosci. Res.* **70:** 327–334.

Snyder J.S., Kee N., and Wojtowicz J.M. 2001. Effects of adult neurogenesis on synaptic plasticity in the rat dentate gyrus. *J. Neurophysiol.* **85:** 2423–2431.

Snyder J.S., Hong N.S., McDonald R.J., and Wojtowicz J.M. 2005. A role for adult neurogenesis in spatial long-term memory. *Neuroscience* **130:** 843–852.

Song H., Kempermann G., Overstreet-Wadiche L., Zhao C., Schinder A.F., and Bischofberger J. 2005. New neurons in the adult mammalian brain: Synaptogenesis and functional integration. *J. Neurosci.* **25:** 10366–10368.

Squire L.R., Stark C.E., and Clark R.E. 2004. The medial temporal lobe. *Ann. Rev. Neurosci.* **27:** 279–306.

Tashiro A., Makino H., and Gage F.H. 2007. Experience-specific functional modification of the dentate gyrus through adult neurogenesis: A critical period during an immature stage. *J. Neurosci.* **27:** 3252–3259.

Tashiro A., Sandler V.M., Toni N., Zhao C., and Gage F.H. 2006. NMDA-receptor-mediated, cell-specific integration of new neurons in adult dentate gyrus. *Nature* **442:** 929–933.

Tozuka Y., Fukuda S., Namba T., Seki T., and Hisatsune T. 2005. GABAergic excitation promotes neuronal differentiation in adult hippocampal progenitor cells. *Neuron* **47:** 803–815.

Trommer B.L., Kennelly J.J., Colley P.A., Overstreet L.S., Slater N.T., and Pasternak J.F. 1995. AP5 blocks LTP in developing rat dentate gyrus and unmasks LTD. *Exp. Neurol.* **131:** 83–92.

van Praag H., Schinder A.F., Christie B.R., Toni N., Palmer T.D., and Gage F.H. 2002. Functional neurogenesis in the adult hippocampus. *Nature* **415:** 1030–1034.

Wang L.P., Kempermann G., and Kettenmann H. 2005. A subpopulation of precursor cells in the mouse dentate gyrus receives synaptic GABAergic input. *Mol. Cell. Neurosci.* **29:** 181–189.

Wang S., Scott B.W., and Wojtowicz J.M. 2000. Heterogenous properties of dentate granule neurons in the adult rat. *J. Neurobiol.* **42:** 248–257.

Whitlock J.R., Heynen A.J., Shuler M.G., and Bear M.F. 2006. Learning induces long-term potentiation in the hippocampus. *Science* **313:** 1093–1097.

Winocur G., Wojtowicz J.M., Sekeres M., Snyder J.S., and Wang S. 2006. Inhibition of neurogenesis interferes with hippocampus dependent memory function. *Hippocampus* **16:** 296–304.

Wojtowicz J.M. 2006. Irradiation as an experimental tool in studies of adult neurogenesis. *Hippocampus* **16:** 261–266.

Zhao C., Teng E.M., Summers R.G., Ming G., and Gage F.H. 2006. Distinct morphological stages of dentate granule neuron maturation in the adult mouse hippocampus. *J. Neurosci.* **26:** 3–11.

16

Genetics and Epigenetics
in Adult Neurogenesis

Jenny Hsieh[1] and Jay W. Schneider[2]
[1]Department of Molecular Biology and Cecil H. and Ida Green
Center for Reproductive Biology Sciences
[2]Department of Internal Medicine, Division of Cardiology
University of Texas
Southwestern Medical Center
Dallas, Texas 75390

CHROMATIN STRUCTURE AND FUNCTION ARE DYNAMICALLY regulated in stem cells of the brain, which serve as an important paradigm for understanding the regulatory mechanisms that transduce physiological and pathophysiological signals to the stem cell genome. In the adult vertebrate brain, the production of newborn neurons from stem cells (neurogenesis) takes place in discrete proliferation zones (niches), such as the subventricular zone (SVZ) of the lateral ventricle and the subgranular zone (SGZ) of the dentate gyrus of the hippocampus (Gage 2000). A variety of signals, ranging from excitation due to locally released neurotransmitters to systemic factors or drugs that cross the blood-brain barrier, converge upon clusters of neuronal stem/progenitor cells (NSCs) residing in these niches, which are intimately associated with the cerebral microvasculature. Balanced control of self-renewal, differentiation, and survival of NSCs produces new neurons and glial cells necessary for functional homeostasis of the brain and also has an important role in brain function such as memory and learning. Moreover, as potential cancer stem cells, NSCs are suspected to be the root of brain malignancies such as glioblastoma multiforme. To become neurons, NSCs require coordinated changes in the pattern of gene expression, primarily regulated at the level of gene transcription. Epigenetic chromatin remodeling has emerged as a fundamental higher-order mechanism for fine-tuning and coordinating gene expression during neurogenesis. Important aspects of brain function such as synaptic plasticity are also governed by chromatin-remodeling enzymes, cell-type-specific transcriptional regulators, and small regulatory noncoding RNAs. Thus, signaling to the genome through diverse epigenetic regulatory mechanisms is critical to neurogenesis and brain function during development and throughout life.

Although it is a relatively new concept in neuroscience, the importance of epigenetic mechanisms has been appreciated for many years in the developmental and cancer biology fields. Indeed, drugs that act on epigenetic mechanisms are already in cancer clinical trials. Classical epigenetic processes, such as X-chromosome inactivation, imprinting, and gene silencing, have been studied in traditional model systems such as plants, invertebrates, and mice. At this time, epigenetic mechanisms are considered equally important to sequence-specific DNA-binding transcription factors as regulators of gene expression. Among the most commonly employed definitions of epigenetics are meiotically and mitotically heritable changes in patterns of gene expression that are not encoded in the primary DNA sequence itself, essentially a term to describe alterations that lead to new cellular phenotypes without a change in genotype. In this sense, epigenetic changes in chromatin organization and biochemical modifications allow genotypically identical cells to behave phenotypically different. Much of the basic cellular chromatin and transcriptional machinery has been elucidated from biochemical approaches and the use of simple genetic model organisms. However, a full understanding of the signaling circuitry and transcriptional regulatory mechanisms in complex organs such as the mammalian brain remains sketchy. This chapter presents key aspects of the epigenetic regulation of adult neurogenesis and the convergence of epigenetic mechanisms that govern the neuronal genome in both normal and disease states.

CHROMATIN STRUCTURE AND HISTONE MODIFICATIONS

Chromatin is a nucleoprotein complex composed of nucleosome repeats of 147 bp of DNA sequence physically wrapped around two copies each of the histonal proteins, H2A, H2B, H3, and H4 (Luger and Richmond 1998). One of the most exciting breakthroughs in chromatin biology in this last decade is the discovery that the amino-terminal tails of core histones are subject to a variety of covalent modifications such as acetylation and methylation and that DNA site- or domain-specific histone modifications ("the histone code") control the activation or repression of the associated genes (Jenuwein and Allis 2001). Some histone marks such as acetylation of Lys-9 and Lys-14, di- or trimethylation of Lys-4, and phosphorylation of Ser-10 on histone H3 are signatures of actively expressed chromatin. Other marks, such as di- or trimethylation of Lys-9 on histone H3 are associated with silent chromatin domains. The histone code is specified by chromatin-modifying enzymes such as histone acetyltransferases (HATs), histone deacetylases (HDACs), and histone methyltransferases (HMTs), which are targeted to specific chromatin loci through direct association with sequence-specific DNA-binding proteins in large, multicomponent complexes. Many of the components of these complexes, even the chromatin-modifying enzymes themselves, are signal-responsive, resulting in a complex regulatory hierarchy for control of the genome.

The HATs and HDACs are the most extensively studied chromatin-modifying enzymes. HATs induce histone acetylation that usually results in relaxation of the nucleosomes and increased transcription, whereas HDACs catalyze the reverse reaction; in the deacetylated state, nucleosomes are more condensed, preventing access of transcriptional activators to their target sites which results in transcriptional repression. In terms of the brain, global changes in histone modifications have been observed in different regions, particularly after traumatic brain injury, electroconvulsive seizures, or drugs of abuse such as cocaine (Crosio et al. 2003; Tsankova et al. 2004; Kumar et al. 2005; Gao et al. 2006; Sng et al. 2006). Although these findings are intriguing, the role of cell-type-specific histone-modifying enzymes in neural development and function remains to be determined. One class of tissue-specific HDACs (class II) has a well-documented signal-responsive role in controlling cardiac muscle cell hypertrophy (Nakagawa et al. 2006); this class of enzyme is also highly enriched in the developing and adult brain (Thiagalingam et al. 2003), yet little is known regarding the specific function of class II HDACs in neurobiology. One important clue regarding the role of class I and class II HDACs comes from small-molecule HDAC inhibitors (HDACi) such as trichostatin A or valproic acid (VPA) (Grozinger and Schreiber 2002). HDACi can induce neuronal differentiation in embryonic (Hao et al. 2004) and adult neuronal progenitor cells (Hsieh et al. 2004). In our own work, we found that VPA treatment in rats decreased NSC proliferation in the SGZ (Hsieh et al. 2004) and, in seizure animals, attenuated the aberrant neurogenesis associated with seizures and, most importantly, improved cognitive function (Jessberger et al. 2007). These results suggest that pharmacological inhibition of HDAC activity might become an effective clinical strategy for treating seizure disorders and epilepsy. Moreover, due to their neuroprotective effects, HDACi epigenetic drugs have shown therapeutic promise in humans or animal models of Huntington's disease (Butler and Bates 2006), amyotrophic lateral sclerosis (Petri et al. 2006), Friedreich's ataxia (Herman et al. 2006), spinal muscular atrophy (Hahnen et al. 2006), ischemic brain injury (Faraco et al. 2006), seizure disorders (Eyal et al. 2004), and even brain cancer cells (Komata et al. 2005; van den Boom et al. 2006). The future application of HDACi for treating CNS disorders remains an exciting possibility and, combined with our increasing understanding of NSCs in the adult brain, can be a powerful approach to combating neurological disease. These global studies of histone modifications and HDACi in neurological disorders and brain function are intriguing; however, detailed studies are needed to understand how specific genes are regulated to control neuronal stem-cell-fate decisions.

CHROMATIN COMPLEXES CONTROLLING NEURONAL CELL FATE

One of the best-characterized mechanisms for regulating HDAC activity in the cell involves signal-responsive export of the active class II enzyme from the nucleus and derepression of gene expression (Grozinger and Schreiber 2000; McKinsey et al. 2000). A second, perhaps even more common, mechanism involves the recruitment of HDACs and multiprotein chromatin-remodeling complexes to gene-specific promoters by association with DNA-binding proteins (Grozinger and Schreiber 2002). During cortical development, NSCs first undergo limited expansion through rounds of symmetric divisions and then undergo neurogenesis, mainly through asymmetric divisions (Temple 2001). Toward the end of neurogenesis, cortical progenitors switch back to symmetric divisions and give rise to astrocytes and oligodendrocytes. Cell-fate specification toward neuronal or glial lineages involves the reciprocal regulation of several gene batteries. For example, the genes responsible for the specification of neuronal cells must be repressed or silenced when the developing NSC diverges toward astrocytic or oligodendrocytic cell fate, and, vice versa, gliogenic genes must be inhibited during neurogenesis. This is coordinated through both transcriptional and epigenetic regulatory mechanisms that determine the potential of the cell to respond to extrinsic signals (Fig. 1). Many elegant studies collectively describe the existence of transcriptional codes involving the spatial and temporal expression of cell-type-specific transcriptional activators and repressors that control neuronal subtype specification and patterning in the developing CNS (Bertrand et al. 2002; Ross et al. 2003). Many of these factors have been studied in terms of their regulation of individual target genes during neural specification; however, a unifying mechanism that controls the

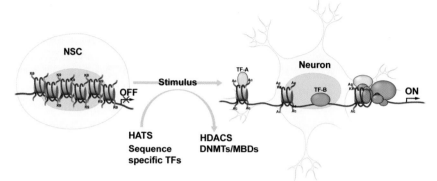

Figure 1. Sequence-specific transcription factors work in concert with the chromatin machinery to direct neuronal stem cells (NSCs) toward neuronal lineage differentiation. In the stem cell state, repressive machinery such as HDACs/DNMTs/MBDs maintain neuronal gene repression (OFF) through one set of epigenetic marks, such as histone H3 Lys-K9 methylation and DNA cytosine methylation. Stimulation by environmental factors and/or stress signals can induce adult neurogenesis by de-repressing or activating neuronal gene expression (ON) through the epigenetic machinery transducing these signals to the genome. (NSC) Neuronal stem cell; (HATs) histone acetyltransferases; (TFs) transcription factors; (HDACs) histone deacetylases; (DNMTs) DNA methyltransferases; (MBDs) methyl-DNA binding proteins.

process of neuronal cell fate specification through maturation has not been found.

A candidate master regulator for neurogenesis and gliogenesis is the transcriptional regulator NRSF (neuron-restrictive silencing factor, also called REST) (Chong et al. 1995; Schoenherr and Anderson 1995). Initially described as a repressor of neuronal genes in nonneuronal cells, NRSF is a zinc finger protein that binds to a conserved 21–23-bp motif known as NRSE (neuron-restrictive silencing element, also called RE1). The NRSE sequence is found scattered throughout the genome, many in the regulatory regions of neuron-specific genes, including neuronal growth factors, ion channels, neurotransmitter receptors, and guidance/migration molecules, among other important neurogenic genes. The precise mechanism by which NRSF mediates transcriptional regulation of its target genes remains a mystery; in certain gene contexts, NRSF acts as a repressor, whereas in other contexts, it acts as an activator of gene transcription. Emerging evidence suggests that NRSF functions by recruiting other corepressor complexes, such as mSin3A/B (Naruse et al. 1999), N-CoR (Jepsen et al. 2000), CtBP (Garriga-Canut et al. 2006), or CoREST (Ballas et al. 2001), to specific RE1 sites in the genome. Importantly, these regulatory complexes also recruit chromatin-modifying class I HDAC enzymes to the genome. Recently, Nakagawa et al. (2006) demonstrated direct interaction between NRSF and class II HDACs in ventricular cardiac muscle cells in vivo and in vitro (Nakagawa et al. 2006). Bioinformatics analyses have identified hundreds to thousands of NRSE sites throughout the mammalian genome (Bruce et al. 2004), supporting the idea that NRSF functions as a master regulator of neuronal gene expression in NSCs as well as in mature neurons (Sun et al. 2005; Greenway et al. 2006). Recently, NRSF was found to be important for triggering adult hippocampal neurogenesis from NSCs by acting as an activator of neuronal gene expression (Kuwabara et al. 2004). The mechanism of NRSF action in adult NSCs appeared to involve a novel non-coding double-stranded RNA (discussed in another section). To add to the intrigue, NRSF function has also been implicated in various pathological states in the brain as well as outside the brain, such as in different kinds of neuronal and neuroendocrine cancer cells (Coulson 2005), Huntington's disease (Zuccato et al. 2003), ischemic insults (Calderone et al. 2003), epilepsy (Garriga-Canut et al. 2006), and cardiovascular disease (Cheong et al. 2005).

In addition to neuronal gene regulation by NRSF and recruitment of histone-modifying enzymes, recent evidence suggests that members of the SWI/SNF and Polycomb family of chromatin-remodeling complexes have important roles in neural development and cancer and stem cell biology (Cui et al. 2006; Lee et al. 2006). Chromatin-remodeling complexes such as SWI/SNF use ATP hydrolysis to disrupt histone-DNA associations and contain different combinations of proteins that include both conserved and nonconserved components (Becker and Horz 2002). They also interact with HATs or HDACs and/or sequence-specific transcription factors to activate or repress target genes. Typically, SWI/SNF complexes contain either the ATPase Brahma (Brm) or Brahma-related protein 1 (Brg-1). Chromodomain proteins, which possess the ATPase domain found in SWI/SNF proteins, are also recruited by Polycomb repressors and are involved in homeotic gene silencing to control tissue patterning. Recently, Brm, Brg-1, and the Polycomb repressor protein Bmi-1 have all been implicated in some aspect of neural development and function. The SRY-related HMG-box-containing protein (Sox-2) is a key transcription factor in NSCs, and the recruitment of Brm for its own transcriptional regulation appears to be linked to its role in stem cell maintenance (Kondo and Raff 2004). Brg-1 is also required for *Xenopus* neurogenesis; loss of Brg-1 correlated with increased neuronal progenitor proliferation and an expansion of Sox2-positive cells in embryos at later stages of neurogenesis. Brg-1 function appears to be linked to the regulation of two proneural basic helix-loop-helix (bHLH) proteins: Neurogenin (Ngn)-related 1 and NeuroD (Seo et al. 2005). Moreover, the Polycomb group proteins are emerging to have fundamental roles in stem cell maintenance and cancer. Bmi-1 is required for the self-renewal of stem cells in the CNS and PNS; mice lacking Bmi-1 experience neurological defects, presumably due to a postnatal depletion of their NSCs (van der Lugt et al. 1994; Molofsky et al. 2003, 2005). Together, these data indicate the important role of chromatin-remodeling factors in brain development and function; the exact role of ATP-dependent chromatin-remodeling complexes remains to be determined.

CHROMATIN REMODELING AND ADULT NEUROGENESIS

In addition to development, transcriptional coactivators and chromatin-remodeling factors are also emerging to be important for adult neurogenesis. Querkopf (Qkf)-deficient mice, a MYST family histone acetyltransferase, have fewer NSCs and fewer migrating neuroblasts in the rostral migratory stream of the adult SVZ (Merson et al. 2006). Furthermore, neurospheres derived from Qkf mutant mice contain fewer cells with the potential to generate secondary neurospheres and form fewer neurons, suggesting that Qkf mutant stem cells have a defect in self-renewal and lineage specification in vitro. Under similar conditions, the nuclear corepressor (N-CoR) is also required to suppress astrocyte differentiation and promote neuronal differentiation (Hermanson et al. 2002), suggesting commonalities in the transcriptional machinery regulating neuronal lineage differentiation. Many of the HDAC proteins themselves have been characterized exclusively with respect to heart function; virtually nothing is known regarding their roles in the brain, particularly during adult neurogenesis. Interestingly, a study profiling the gene expression patterns associated with SVZ neurogenesis revealed an up-regulation of chromatin modifiers belonging to the SWI/SNF and Polycomb/Trithorax family of proteins (Lim et al. 2006), raising the possibility that chromatin remodeling may be important for the maintenance of the stem cell state by keeping lineage specification genes repressed, but primed to receive a specific extrinsic signal. Another important epigenetic modification implicated in adult neurogenesis is methylation of CpG dinucleotides (discussed in the next section).

THE ROLE OF DNA METHYLATION IN THE BRAIN

A classically studied epigenetic modification is DNA methylation where mammalian DNA can be covalently modified through methylation of the carbon at the fifth position on the pyrimidine ring of the cytosine residue, which is usually found at symmetrical CpG dinucleotides. DNA methylation is a major epigenetic mechanism for the establishment of parental-specific imprints during gametogenesis and gene silencing of the inactivated X chromosome and retrotransposons (Jaenisch and Bird 2003). DNA methylation is mediated by cellular methyltransferases, Dnmt3a and Dnmt3b, which add methyl groups de novo onto unmethylated DNA. When cells divide, the methyltransferase Dnmt1 preferentially recognizes hemimethylated DNA and maintains DNA methylation. Interestingly, both classes of methyltransferases have been shown to participate in various stages of neural fate and neurogenesis. During the initial specification of neurons and glia (Feng et al. 2005), as well as during later stages of neuronal maturation and function (Levenson et al. 2006), Dnmt3a and Dnmt3b are important. Dnmt1 is also very important in the brain and involved in JAK-STAT signaling to control the timing of when precursor cells switch from neurogenesis to gliogenesis during development (Fan et al. 2001, 2005). Site-specific methylation of DNA has been shown to be important for astrocyte differentiation and activation of the glial fibrillary acidic protein (GFAP) promoter. Activation of the GFAP promoter requires binding of the signal transducer and activator of transcription 3 (STAT3) in response to leukemia inhibitory factor (LIF) signaling (Ross et al. 2003). Early cortical progenitors are refractory to astrocyte fates, even in the presence of LIF and STAT3 activation, likely due to methylation of the STAT3-binding site, which prevents STAT3 binding to the GFAP promoter (Takizawa et al. 2001). At later stages, this site becomes demethylated, and gliogenesis can proceed. A similar case was observed in the STAT3-binding site of the S100β promoter, a calcium-binding protein expressed in astrocytes (Namihira et al. 2004). This directly demonstrates the role of a specific epigenetic mark on neuronal cell-fate specification.

In addition to DNA methylation, histone methylation has also been found to be important for the regulation of the GFAP gene in cortical progenitors. Using chromatin immunoprecipitation (ChIP) assays, Song and Ghosh (2004) showed that there was increased histone H3 Lys-9 methylation, a mark associated with gene silencing, at the STAT3-binding site within GFAP at progenitor stages, when GFAP was repressed. Stimulation with basic fibroblast growth factor-2 (FGF-2) potentiated the ability of CNTF (ciliary neurotrophic factor) to induce astrocyte differentiation. Strikingly, FGF-2 induction caused Lys-9 to become demethylated, but Lys-4, the active mark, to become hypermethylated, resulting in STAT3 binding and activation of GFAP. In contrast, mature neurons had low levels of both Lys-9 and Lys-4 methylation. Still remaining is the question of how GFAP or astrocytic genes are repressed in neurons and whether there are factors that actively promote gliogenesis, without necessarily blocking neurogenesis. In fact, a recent study identified two nuclear factor proteins (NF1A and NF1B) for the initial specification of gliogenesis (Deneen et al. 2006). These studies highlight the diverse transcriptional and epigenetic mechanisms that control lineage-specific gene expression for the generation of different neuronal cell types during development and throughout adulthood.

The world of epigenetic and histone modifications is becoming increasingly complex and fascinating. Recently, histone arginine methylation has been shown to regulate the pluripotent state of cells in the inner cell mass and, intriguingly, may be among the first set of "marks" that contribute to the pluripotent state (Torres-Padilla et al. 2007). These types of studies will likely extend to adult neurogenesis, where stem cells still retain multipotency but are no longer pluripotent. Understanding both the epigenetic and genetic determinants of plasticity is important for determining how to utilize and manipulate stem cells for regenerative medicine.

EPIGENETICS AND NEURODEVELOPMENTAL DISORDERS

There are many examples of how epigenetic regulation controls lineage-specific gene expression during development and in adult neurogenesis; however, epigenetic control mechanisms also have an important functional role in mature postmitotic neurons in the brain. As described in the previous section, DNA methylation can mediate silencing by interfering with transcription factor binding to target genes by methylation at CpG sites. Another mechanism by which DNA methylation leads to transcriptional gene silencing is the binding of methyl-CpGs by methyl-CpG-binding proteins (MBDs), which further recruit HDAC repressor complexes, leading to the formation of repressive chromatin. MBDs comprise a family of proteins, two of which (MeCP2 and MBD1), and possibly more, are turning out to have important roles in the nervous system. MeCP2 is found highly enriched in postmitotic neurons (Zoghbi 2003), and mutations in MeCP2 have been linked to the neurological disorder Rett syndrome (Guy et al. 2001; Jung et al. 2003). Rett syndrome is characterized by normal development until 1 year of age, is most commonly seen in females, and is followed by a rapid deterioration, involving loss of speech and motor skills, microcephaly, seizures, autism, ataxia, and stereotypical hand wringing (Rett 1966). Although the exact role of MeCp2 in neurons is unclear, mounting evidence suggests that MeCP2 is a key regulator of the essential brain protein, brain-derived neurotrophic factor (BDNF) (West et al. 2002), but may regulate many more target genes yet to be uncovered. Interestingly, exposure of neurons to elevated levels of extracellular potassium resulted in the phosphorylation of MeCP2 and transcriptional derepression of the BDNF promoter (Chen et al. 2003; Martinowich et al. 2003). Phosphorylation of MeCP2 at a specific amino acid residue, S421, has recently been shown to occur selectively in the nervous system and is important for regulating dendritic growth and spine maturation and to explain, at least in part, the neuron-specific pathology of Rett syndrome (Zhou et al. 2006). Other MBD proteins may also have roles in the brain, particularly during adult neurogenesis. MBD1 knockout mice have diminished hippocampal neurogenesis, defects in spatial learning, and a reduction in long-term potentiation in the dentate gyrus of the hippocampus (Zhao et al. 2003).

Another commonly inherited mental retardation syndrome is Fragile X. Patients have excessive expansion of the CGG sequence from the 5'UTR (untranslated region) of the *Fmr1* gene, resulting in transcriptional silencing of the gene. When the CGG repeats exceed 200, the region is hypermethylated, preventing the binding of transcriptional machinery. Treatments with DNA demethylating drugs result in decreased histone H3 methylation at the Lys-9 residue, increased H3 and H4 acetylation, and decreased methylation at histone H3 Lys-4, resulting in the activation of *Fmr1* (Tabolacci et al. 2005). This finding strongly suggests a role for chromatin remodeling in the silencing of *Fmr1*. It appears that MeCP2 and Brm can associate and jointly silence the *Fmr1* gene. RNA interference (RNAi) knockdown of MeCP2 and Brm resulted in increased mRNA transcription of *Fmr1*. These emerging studies suggest that epigenetic factors contribute to neurological disorders such as Rett syndrome and Fragile X mental retardation and represent a new avenue of therapeutic targets.

EPIGENETIC CONTROL OF NEURAL PLASTICITY, LEARNING, AND MEMORY

Epigenetic mechanisms have also been suggested to underlie certain forms of memory and neural plasticity, especially after stress signals, some of which are physiological and others of which are pathological (Fig. 2). Experimental evidence has implicated histone modifications and chromatin-remodeling factors in animal models of drug addiction, as well as certain types of learning (Colvis et al. 2005). Since drug addiction has

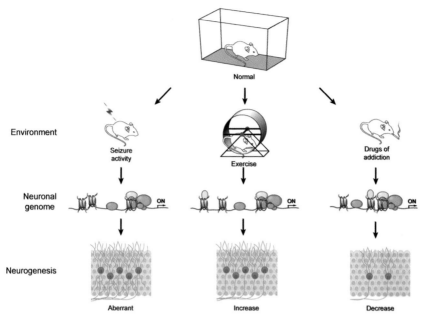

Figure 2. Signaling to the genome through diverse epigenetic regulatory mechanisms is critical to neurogenesis and brain function. Environmental or behavioral stimulation, both physiological (such as voluntary exercise) and pathological (such as seizure induction or drugs of addiction), can lead to global changes in histone modifications and chromatin remodeling of target genes controlling the neuronal genome. These context-dependent alterations in chromatin structure and transcriptional activity could produce acute changes in neuronal stem cell proliferation, differentiation, and survival of cells (neurogenesis), as well as cause a long-term sustained effect. The epigenetic machinery could then serve as a key mediator acting at the cellular and molecular levels, sensing perturbations in the environment and triggering long-lasting changes in gene expression associated with various disease or pathological states.

a long-term sustained effect, evidence is leading to alterations in brain cell chromatin structure as the possible underlying mechanism to mediate long-lasting changes in gene expression associated with the drug-addicted state. c-*fos* and ΔFosB, members of the Fos family of transcription factors, are implicated in behavioral responses and synaptic plasticity induced by abused drugs. Interestingly, cocaine can induce H4 acetylation and H3 phosphoacetylation at the c-*fos* gene promoter in the striatum after acute administration but is desensitized after repeated exposure to cocaine (Kumar et al. 2005). As for the FosB promoter, H3 acetylation is still observed under chronic cocaine treatment. ΔFosB, a product of the *fosB* gene, also known to accumulate under chronic cocaine exposure, increases the binding of Brg-1 (a component of the SWI-SNF complex mentioned earlier) with cyclin-dependent kinase 5, a gene implicated in neurogenesis, synaptic plasticity, learning, and memory (Nikolic et al. 1996; Angelo et al. 2006). This evidence suggests that ΔFosB can recruit chromatin-remodeling factors such as HATs and Brg-1-containing complexes to its downstream targets, therefore regulating the expression of key genes involved in sustaining the long-term effects of drugs of abuse.

Epigenetic modifications have also been linked to changes in neural plasticity and long-term memory (Kandel 2001). Recently, two different mouse models for Rubinstein-Taybi syndrome (RTS), a well-defined congenital syndrome composed of mental retardation, postnatal growth deficiency, microcephaly, specific facial characteristics, broad thumbs, and big toes (Rubinstein and Taybi 1963), which is genetically linked to the CREB-binding protein (CBP) HAT (Petrij et al. 1995), have been used to demonstrate that histone acetylation is required for long-term potentiation, learning, and memory (Alarcon et al. 2004; Korzus et al. 2004). CBP heterozygotes exhibited normal levels of short-term memory; however, their long-term memory, determined as contextual and cued fear conditioning and novel object recognition tasks, was significantly abnormal. Interestingly, when the HDAC inhibitor suberoylanilide hydroxamic acid was delivered intraventricularly into CBP heterozygotes, the mice showed improvement in contextual fear conditioning. In a second mouse model bearing a hippocampal CA1- and dentate-gyrus-specific inducible form of mutant CBP lacking HAT activity, there were long-term memory deficits, although short-term memory appeared to be intact. Similarly, when an HDAC inhibitor trichostatin A was used, or when the transgene was shut off, the memory deficits could be rescued, suggesting that pharmacological manipulation of HAT activity might be a potential therapeutic approach to treat RTS. Perhaps more importantly, the findings that genes such as CBP and MeCP2, which have such global effects on gene expression, have such specific phenotypic consequences in the brain, suggest that epigenetic mechanisms may represent an untapped area for understanding and treating cognitive disorders. A more comprehensive discussion and review of neurodevelopmental and cognitive disorders linked to transcriptional regulators have been described and discussed elsewhere (Hong et al. 2005).

REGULATORY SMALL RNAs IN THE BRAIN

The discovery of small noncoding RNAs has revolutionized our understanding of the mechanisms that regulate gene expression in all cells

(Novina and Sharp 2004), including those of the mammalian brain (Cao et al. 2006). Their small size and sequence complementarity allow extreme versatility to target mRNAs for repression or activation of gene expression or modify chromatin structures of targeted genes. Double-stranded RNA was originally found to silence genes by targeting their mRNA intermediates for degradation, but during the past several years, further mechanisms of action of small RNAs have been discovered and potential new classes of such RNAs have been revealed. Small RNAs fall into at least four major classes. Small interfering RNAs (siRNAs) and microRNAs (miRNAs) both silence gene expression; siRNAs usually target mRNAs for degradation by sequence complementarity to a single site, whereas miRNAs recognize numerous partially complementary binding sites that act synergistically within target mRNAs found to be abundantly distributed in the postnatal brain, suggesting that they have important roles in neuronal function. Of the approximately 250 miRNAs in mammals, it would appear that at least 125 miRNAs are expressed in the brain at various levels (Krichevsky et al. 2003; Sempere et al. 2004). Some of these miRNAs found in brain are miR9, miR29, miR124, miR125a, miR125b, miR127, miR128, miR132, miR137, miR138, and miR139, among the ever-growing list. Recently, a CREB-induced miRNA (miR132) was found to regulate neuronal morphogenesis in cultured cortical neurons (Vo et al. 2005). Another miRNA implicated in neuronal function is miR134, which is dendritically localized and involved in the control of spine size (Schratt et al. 2006).

We recently demonstrated that a class of small RNAs could participate with chromatin-remodeling factors to regulate adult NSCs from the rodent hippocampus. In this study, a small noncoding RNA matching the binding sequence of NRSF, mentioned above, called NRSE small modulatory RNA (or just smRNA), could effectively trigger neurogenesis (Kuwabara et al. 2004). This smRNA appeared to mediate its function in an entirely novel way, unlike siRNAs or miRNAs: It interacted with the NRSF protein and induced neuron-specific gene expression. Clearly, this is just the beginning as we will see many more studies regarding brain-enriched small RNAs both in normal physiology and in disease.

CONCLUSIONS

The genome is no longer considered a static and privileged storage depot of genetic information; rather, it is now recognized to be a highly dynamic entity that undergoes profound structural and functional changes in response to extrinsic signals. Epigenetic and transcriptional regulation of the neuronal genome during adult neurogenesis has emerged as one of the best and most useful paradigms to study the role of chromatin modification in cell-fate specification. After cell-fate decisions are triggered, the next step is to determine how master regulators of the neuronal genome interact with their cognate binding partners at sequence-specific binding sites in a context-dependent manner to regulate downstream events controlling neuronal differentiation, maturation, and synaptic plasticity in an ever-changing environment. The fact that many epigenetic processes in the cell can act in a reversible and gene-specific manner allows the precise control of gene expression that is critical for brain development and function. Moreover, epigenetic processes in developing and mature neuronal cells appear to be targeted by a number of pathophysiological stimuli, including seizure activity and drugs such as cocaine. Although these studies are good starting points for studying epigenetic and transcriptional regulation of neuronal target genes, much work is still needed to fully understand the role of master regulators of the neuronal genome such as NRSF. Indeed, small chemical compounds and/or ribonucleic acids would provide excellent tools to begin gaining insight into these regulatory mechanisms and could be the starting points for novel neurological drugs.

ACKNOWLEDGMENTS

We thank Sean Goetsch for graphics. J.H. received funding from the Esther A. & Joseph Klingenstein Fund, the Ellison Medical Foundation, the Advanced Research Program, Texas Higher Education Coordinating Board, the Welch Foundation, and the University of Texas Southwestern (UTSW) President's Research Council. J.W.S. received funding from the Haberecht Wildhare-Idea Program and the Reynolds Foundation at UTSW.

REFERENCES

Alarcon J.M., Malleret G., Touzani K., Vronskaya S., Ishii S., Kandel E.R., and Barco A. 2004. Chromatin acetylation, memory, and LTP are impaired in CBP+/− mice: A model for the cognitive deficit in Rubinstein-Taybi syndrome and its amelioration. *Neuron* **42:** 947–959.

Angelo M., Plattner F., and Giese K.P. 2006. Cyclin-dependent kinase 5 in synaptic plasticity, learning and memory. *J. Neurochem.* **99:** 353–370.

Ballas N., Battaglioli E., Atouf F., Andres M.E., Chenoweth J., Anderson M.E., Burger C., Moniwa M., Davie J.R., Bowers W.J., et al. 2001. Regulation of neuronal traits by a novel transcriptional complex. *Neuron* **31:** 353–365.

Becker P.B. and Horz W. 2002. ATP-dependent nucleosome remodeling. *Annu. Rev. Biochem.* **71:** 247–273.

Bertrand N., Castro D.S., and Guillemot F. 2002. Proneural genes and the specification of neural cell types. *Nat. Rev. Neurosci.* **3:** 517–530.

Bruce A.W., Donaldson I.J., Wood I.C., Yerbury S.A., Sadowski M.I., Chapman M., Gottgens B., and Buckley N.J. 2004. Genome-wide analysis of repressor element 1 silencing transcription factor/neuron-restrictive silencing factor (REST/NRSF) target genes. *Proc. Natl. Acad. Sci.* **101:** 10458–10463.

Butler R. and Bates G.P. 2006. Histone deacetylase inhibitors as therapeutics for polyglutamine disorders. *Nat. Rev. Neurosci.* **7:** 784–796.

Calderone A., Jover T., Noh K.M., Tanaka H., Yokota H., Lin Y., Grooms S.Y., Regis R., Bennett M.V., and Zukin R.S. 2003. Ischemic insults derepress the gene silencer REST in neurons destined to die. *J. Neurosci.* **23:** 2112–2121.

Cao X., Yeo G., Muotri A.R., Kuwabara T., and Gage F.H. 2006. Noncoding RNAs in the mammalian central nervous system. *Annu. Rev. Neurosci.* **29:** 77–103.

Chen W.G., Chang Q., Lin Y., Meissner A., West A.E., Griffith E.C., Jaenisch R., and Greenberg M.E. 2003. Derepression of BDNF transcription involves calcium-dependent phosphorylation of MeCP2. *Science* **302:** 885–889.

Cheong A., Bingham A.J., Li J., Kumar B., Sukumar P., Munsch C., Buckley N.J., Neylon C.B., Porter K.E., Beech D.J., and Wood I.C. 2005. Downregulated REST transcription factor is a switch enabling critical potassium channel expression and cell proliferation. *Mol. Cell* **20:** 45–52.

Chong J.A., Tapia-Ramirez J., Kim S., Toledo-Aral J.J., Zheng Y., Boutros M.C., Altshuller Y.M., Frohman M.A., Kraner S.D., and Mandel G. 1995. REST: A mammalian silencer protein that restricts sodium channel gene expression to neurons. *Cell* **80:** 949–957.

Colvis C.M., Pollock J.D., Goodman R.H., Impey S., Dunn J., Mandel G., Champagne F.A., Mayford M., Korzus E., Kumar A., et al. 2005. Epigenetic mechanisms and gene networks in the nervous system. *J. Neurosci.* **25:** 10379–10389.

Coulson J.M. 2005. Transcriptional regulation: Cancer, neurons and the REST. *Curr. Biol.* **15:** R665–668.

Crosio C., Heitz E., Allis C.D., Borrelli E., and Sassone-Corsi P. 2003. Chromatin remodeling and neuronal response: Multiple signaling pathways induce specific histone H3 modifications and early gene expression in hippocampal neurons. *J. Cell. Sci.* **116:** 4905–4914.

Cui H., Ma J., Ding J., Li T., Alam G., and Ding H.F. 2006. Bmi-1 regulates the differentiation and clonogenic self-renewal of I-type neuroblastoma cells in a concentration-dependent manner. *J. Biol. Chem.* **281:** 34696–34704.

Deneen B., Ho R., Lukaszewicz A., Hochstim C.J., Gronostajski R.M., and Anderson D.J. 2006. The transcription factor NFIA controls the onset of gliogenesis in the developing spinal cord. *Neuron* **52:** 953–968.

Eyal S., Yagen B., Sobol E., Altschuler Y., Shmuel M., and Bialer M. 2004. The activity of antiepileptic drugs as histone deacetylase inhibitors. *Epilepsia* **45:** 737–744.

Fan G., Martinowich K., Chin M.H., He F., Fouse S.D., Hutnick L., Hattori D., Ge W., Shen Y., Wu H., et al. 2005. DNA methylation controls the timing of astrogliogenesis through regulation of JAK-STAT signaling. *Development* **132:** 3345–3356.

Fan G., Beard C., Chen R.Z., Csankovszki G., Sun Y., Siniaia M., Biniszkiewicz D., Bates B., Lee P.P., Kuhn R., et al. 2001. DNA hypomethylation perturbs the function and survival of CNS neurons in postnatal animals. *J. Neurosci.* **21:** 788–797.

Faraco G., Pancani T., Formentini L., Mascagni P., Fossati G., Leoni F., Moroni F., and Chiarugi A. 2006. Pharmacological inhibition of histone deacetylases by suberoylanilide hydroxamic acid specifically alters gene expression and reduces ischemic injury in the mouse brain. *Mol. Pharmacol.* **70:** 1876–1884.

Feng J., Chang H., Li E., and Fan G. 2005. Dynamic expression of de novo DNA methyltransferases Dnmt3a and Dnmt3b in the central nervous system. *J. Neurosci. Res.* **79:** 734–746.

Gage F.H. 2000. Mammalian neural stem cells. *Science* **287:** 1433–1438.

Gao W.M., Chadha M.S., Kline A.E., Clark R.S., Kochanek P.M., Dixon C.E., and Jenkins L.W. 2006. Immunohistochemical analysis of histone H3 acetylation and methylation—Evidence for altered epigenetic signaling following traumatic brain injury in immature rats. *Brain Res.* **1070:** 31–34.

Garriga-Canut M., Schoenike B., Qazi R., Bergendahl K., Daley T.J., Pfender R.M., Morrison J.F., Ockuly J., Stafstrom C., Sutula T., and Roopra A. 2006. 2-Deoxy-D-glucose reduces epilepsy progression by NRSF-CtBP-dependent metabolic regulation of chromatin structure. *Nat. Neurosci.* **9:** 1382–1387.

Greenway D.J., Street M., Jeffries A., and Buckley N.J. 2006. REST maintains a repressive chromatin environment in embryonic hippocampal neural stem cells. *Stem Cells* **25:** 354–363.

Grozinger C.M. and Schreiber S.L. 2000. Regulation of histone deacetylase 4 and 5 and transcriptional activity by 14-3-3-dependent cellular localization. *Proc. Natl. Acad. Sci.* **97:** 7835–7840.

———. 2002. Deacetylase enzymes: Biological functions and the use of small-molecule inhibitors. *Chem. Biol.* **9:** 3–16.

Guy J., Hendrich B., Holmes M., Martin J.E., and Bird A. 2001. A mouse Mecp2-null mutation causes neurological symptoms that mimic Rett syndrome. *Nat. Genet.* **27:** 322–326.

Hahnen E., Eyupoglu I.Y., Brichta L., Haastert K., Trankle C., Siebzehnrubl F.A., Riessland M., Holker I., Claus P., Romstock J., et al. 2006. In vitro and ex vivo evaluation of second-generation histone deacetylase inhibitors for the treatment of spinal muscular atrophy. *J. Neurochem.* **98:** 193–202.

Hao Y., Creson T., Zhang L., Li P., Du F., Yuan P., Gould T., Manji H., and Chen G. 2004. Mood stabilizer valproate promotes ERK pathway-dependent cortical neuronal growth and neurogenesis. *J. Neurosci.* **24:** 6590–6599.

Herman D., Jenssen K., Burnett R., Soragni E., Perlman S.L., and Gottesfeld J.M. 2006. Histone deacetylase inhibitors reverse gene silencing in Friedreich's ataxia. *Nat. Chem. Biol.* **2:** 551–558.

Hermanson O., Jepsen K., and Rosenfeld M.G. 2002. N-CoR controls differentiation of neural stem cells into astrocytes. *Nature* **419:** 934–939.

Hong E.J., West A.E., and Greenberg M.E. 2005. Transcriptional control of cognitive development. *Curr. Opin. Neurobiol.* **15:** 21–28.

Hsieh J., Nakashima K., Kuwabara T., Mejia E., and Gage F.H. 2004. Histone deacetylase inhibition-mediated neuronal differentiation of multipotent adult neural progenitor cells. *Proc. Natl. Acad. Sci.* **101:** 16659–16664.

Jaenisch R. and Bird A. 2003. Epigenetic regulation of gene expression: How the genome integrates intrinsic and environmental signals. *Nat. Genet.* (suppl.) **33:** 245–254.

Jenuwein T. and Allis C.D. 2001. Translating the histone code. *Science* **293:** 1074–1080.

Jepsen K., Hermanson O., Onami T.M., Gleiberman A.S., Lunyak V., McEvilly R.J., Kurokawa R., Kumar V., Liu F., Seto E., et al. 2000. Combinatorial roles of the nuclear receptor corepressor in transcription and development. *Cell* **102:** 753–763.

Jessberger S., Nakashima K., Clemenson G.D., Jr., Mejia E., Mathews E., Ure K., Ogawa S., Sinton C.M., Gage F.H., and Hsieh J. 2007. Epigenetic modulation of seizure-induced neurogenesis and cognitive decline. *J. Neurosci.* **27**: 5967–5975.

Jung B.P., Jugloff D.G., Zhang G., Logan R., Brown S., and Eubanks J.H. 2003. The expression of methyl CpG binding factor MeCP2 correlates with cellular differentiation in the developing rat brain and in cultured cells. *J. Neurobiol.* **55**: 86–96.

Kandel E.R. 2001. The molecular biology of memory storage: A dialog between genes and synapses. *Biosci. Rep.* **21**: 565–611.

Komata T., Kanzawa T., Nashimoto T., Aoki H., Endo S., Kon T., Takahashi H., Kondo S., and Tanaka R. 2005. Histone deacetylase inhibitors, *N*-butyric acid and trichostatin A, induce caspase-8- but not caspase-9-dependent apoptosis in human malignant glioma cells. *Int. J. Oncol.* **26**: 1345–1352.

Kondo T. and Raff M. 2004. Chromatin remodeling and histone modification in the conversion of oligodendrocyte precursors to neural stem cells. *Genes Dev.* **18**: 2963–2972.

Korzus E., Rosenfeld M.G., and Mayford M. 2004. CBP histone acetyltransferase activity is a critical component of memory consolidation. *Neuron* **42**: 961–972.

Krichevsky A.M., King K.S., Donahue C.P., Khrapko K., and Kosik K.S. 2003. A microRNA array reveals extensive regulation of microRNAs during brain development. *RNA* **9**: 1274–1281.

Kumar A., Choi K.H., Renthal W., Tsankova N.M., Theobald D.E., Truong H.T., Russo S.J., Laplant Q., Sasaki T.S., Whistler K.N., et al. 2005. Chromatin remodeling is a key mechanism underlying cocaine-induced plasticity in striatum. *Neuron* **48**: 303–314.

Kuwabara T., Hsieh J., Nakashima K., Taira K., and Gage F.H. 2004. A small modulatory dsRNA specifies the fate of adult neural stem cells. *Cell* **116**: 779–793.

Lee T.I., Jenner R.G., Boyer L.A., Guenther M.G., Levine S.S., Kumar R.M., Chevalier B., Johnstone S.E., Cole M.F., Isono K., et al. 2006. Control of developmental regulators by Polycomb in human embryonic stem cells. *Cell* **125**: 301–313.

Levenson J.M., Roth T.L., Lubin F.D., Miller C.A., Huang I.C., Desai P., Malone L.M., and Sweatt J.D. 2006. Evidence that DNA (cytosine-5) methyltransferase regulates synaptic plasticity in the hippocampus. *J. Biol. Chem.* **281**: 15763–15773.

Lim D.A., Suarez-Farinas M., Naef F., Hacker C.R., Menn B., Takebayashi H., Magnasco M., Patil N., and Alvarez-Buylla A. 2006. In vivo transcriptional profile analysis reveals RNA splicing and chromatin remodeling as prominent processes for adult neurogenesis. *Mol. Cell. Neurosci.* **31**: 131–148.

Luger K. and Richmond T.J. 1998. The histone tails of the nucleosome. *Curr. Opin. Genet. Dev.* **8**: 140–146.

Martinowich K., Hattori D., Wu H., Fouse S., He F., Hu Y., Fan G., and Sun Y.E. 2003. DNA methylation-related chromatin remodeling in activity-dependent BDNF gene regulation. *Science* **302**: 890–893.

McKinsey T.A., Zhang C.L., Lu J., and Olson E.N. 2000. Signal-dependent nuclear export of a histone deacetylase regulates muscle differentiation. *Nature* **408**: 106–111.

Merson T.D., Dixon M.P., Collin C., Rietze R.L., Bartlett P.F., Thomas T., and Voss A.K. 2006. The transcriptional coactivator Querkopf controls adult neurogenesis. *J. Neurosci.* **26**: 11359–11370.

Molofsky A.V., He S., Bydon M., Morrison S.J., and Pardal R. 2005. Bmi-1 promotes neural stem cell self-renewal and neural development but not mouse growth and survival by repressing the p16Ink4a and p19Arf senescence pathways. *Genes Dev.* **19**: 1432–1437.

Molofsky A.V., Pardal R., Iwashita T., Park I.K., Clarke M.F., and Morrison S.J. 2003. Bmi-1 dependence distinguishes neural stem cell self-renewal from progenitor proliferation. *Nature* **425:** 962–967.

Nakagawa Y., Kuwahara K., Harada M., Takahashi N., Yasuno S., Adachi Y., Kawakami R., Nakanishi M., Tanimoto K., Usami S., et al. 2006. Class II HDACs mediate CaMK-dependent signaling to NRSF in ventricular myocytes. *J. Mol. Cell. Cardiol.* **41:** 1010–1022.

Namihira M., Nakashima K., and Taga T. 2004. Developmental stage dependent regulation of DNA methylation and chromatin modification in an immature astrocyte specific gene promoter. *FEBS Lett.* **572:** 184–188.

Naruse Y., Aoki T., Kojima T., and Mori N. 1999. Neural restrictive silencer factor recruits mSin3 and histone deacetylase complex to repress neuron-specific target genes. *Proc. Natl. Acad. Sci.* **96:** 13691–13696.

Nikolic M., Dudek H., Kwon Y.T., Ramos Y.F., and Tsai L.H. 1996. The cdk5/p35 kinase is essential for neurite outgrowth during neuronal differentiation. *Genes Dev.* **10:** 816–825.

Novina C.D. and Sharp P.A. 2004. The RNAi revolution. *Nature* **430:** 161–164.

Petri S., Kiaei M., Kipiani K., Chen J., Calingasan N.Y., Crow J.P., and Beal M.F. 2006. Additive neuroprotective effects of a histone deacetylase inhibitor and a catalytic antioxidant in a transgenic mouse model of amyotrophic lateral sclerosis. *Neurobiol. Dis.* **22:** 40–49.

Petrij F., Giles R.H., Dauwerse H.G., Saris J.J., Hennekam R.C., Masuno M., Tommerup N., van Ommen G.J., Goodman R.H., Peters D.J., et al. 1995. Rubinstein-Taybi syndrome caused by mutations in the transcriptional co-activator CBP. *Nature* **376:** 348–351.

Rett A. 1966. On an unusual brain atrophy syndrome in hyperammonemia in childhood. *Wiener medizinische Wochenschrift (1946)* **116:** 723–726.

Ross S.E., Greenberg M.E., and Stiles C.D. 2003. Basic helix-loop-helix factors in cortical development. *Neuron* **39:** 13–25.

Rubinstein J.H. and Taybi H. 1963. Broad thumbs and toes and facial abnormalities. A possible mental retardation syndrome. *Am. J. Dis. Children (1960)* **105:** 588–608.

Schoenherr C.J. and Anderson D.J. 1995. The neuron-restrictive silencer factor (NRSF): A coordinate repressor of multiple neuron-specific genes. *Science* **267:** 1360–1363.

Schratt G.M., Tuebing F., Nigh E.A., Kane C.G., Sabatini M.E., Kiebler M., and Greenberg M.E. 2006. A brain-specific microRNA regulates dendritic spine development. *Nature* **439:** 283–289.

Sempere L.F., Freemantle S., Pitha-Rowe I., Moss E., Dmitrovsky E., and Ambros V. 2004. Expression profiling of mammalian microRNAs uncovers a subset of brain-expressed microRNAs with possible roles in murine and human neuronal differentiation. *Genome Biol.* **5:** R13.

Seo S., Richardson G.A., and Kroll K.L. 2005. The SWI/SNF chromatin remodeling protein Brg1 is required for vertebrate neurogenesis and mediates transactivation of Ngn and NeuroD. *Development* **132:** 105–115.

Sng J.C., Taniura H., and Yoneda Y. 2006. Histone modifications in kainate-induced status epilepticus. *Eur. J. Neurosci.* **23:** 1269–1282.

Song M.R. and Ghosh A. 2004. FGF2-induced chromatin remodeling regulates CNTF-mediated gene expression and astrocyte differentiation. *Nat. Neurosci.* **7:** 229–235.

Sun Y.M., Greenway D.J., Johnson R., Street M., Belyaev N.D., Deuchars J., Bee T., Wilde S., and Buckley N.J. 2005. Distinct profiles of REST interactions with its target genes at different stages of neuronal development. *Mol. Biol. Cell* **16:** 5630–5638.

Tabolacci E., Pietrobono R., Moscato U., Oostra B.A., Chiurazzi P., and Neri G. 2005. Differential epigenetic modifications in the FMR1 gene of the fragile X syndrome after reactivating pharmacological treatments. *Eur. J. Hum. Genet.* **13:** 641–648.

Takizawa T., Nakashima K., Namihira M., Ochiai W., Uemura A., Yanagisawa M., Fujita N., Nakao M., and Taga T. 2001. DNA methylation is a critical cell-intrinsic determinant of astrocyte differentiation in the fetal brain. *Dev. Cell* **1:** 749–758.

Temple S. 2001. The development of neural stem cells. *Nature* **414:** 112–117.

Thiagalingam S., Cheng K.H., Lee H.J., Mineva N., Thiagalingam A., and Ponte J.F. 2003. Histone deacetylases: Unique players in shaping the epigenetic histone code. *Ann. N.Y. Acad. Sci.* **983:** 84–100.

Torres-Padilla M.E., Parfitt D.E., Kouzarides T., and Zernicka-Goetz M. 2007. Histone arginine methylation regulates pluripotency in the early mouse embryo. *Nature* **445:** 214–218.

Tsankova N.M., Kumar A., and Nestler E.J. 2004. Histone modifications at gene promoter regions in rat hippocampus after acute and chronic electroconvulsive seizures. *J. Neurosci.* **24:** 5603–5610.

van den Boom J., Wolter M., Blaschke B., Knobbe C.B., and Reifenberger G. 2006. Identification of novel genes associated with astrocytoma progression using suppression subtractive hybridization and real-time reverse transcription-polymerase chain reaction. *Int. J. Cancer* **119:** 2330–2338.

van der Lugt N.M., Domen J., Linders K., van Roon M., Robanus-Maandag E., te Riele H., van der Valk M., Deschamps J., Sofroniew M., van Lohuizen M., et al. 1994. Posterior transformation, neurological abnormalities, and severe hematopoietic defects in mice with a targeted deletion of the bmi-1 proto-oncogene. *Genes Dev.* **8:** 757–769.

Vo N., Klein M.E., Varlamova O., Keller D.M., Yamamoto T., Goodman R.H., and Impey S. 2005. A cAMP-response element binding protein-induced microRNA regulates neuronal morphogenesis. *Proc. Natl. Acad. Sci.* **102:** 16426–16431.

West A.E., Griffith E.C., and Greenberg M.E. 2002. Regulation of transcription factors by neuronal activity. *Nat. Rev. Neurosci.* **3:** 921–931.

Zhao X., Ueba T., Christie B.R., Barkho B., McConnell M.J., Nakashima K., Lein E.S., Eadie B.D., Willhoite A.R., Muotri A.R., et al. 2003. Mice lacking methyl-CpG binding protein 1 have deficits in adult neurogenesis and hippocampal function. *Proc. Natl. Acad. Sci.* **100:** 6777–6782.

Zhou Z., Hong E.J., Cohen S., Zhao W.N., Ho H.Y., Schmidt L., Chen W.G., Lin Y., Savner E., Griffith E.C., et al. 2006. Brain-specific phosphorylation of MeCP2 regulates activity-dependent Bdnf transcription, dendritic growth, and spine maturation. *Neuron* **52:** 255–269.

Zoghbi H.Y. 2003. Postnatal neurodevelopmental disorders: Meeting at the synapse? *Science* **302:** 826–830.

Zuccato C., Tartari M., Crotti A., Goffredo D., Valenza M., Conti L., Cataudella T., Leavitt B.R., Hayden M.R., Timmusk T., et al. 2003. Huntingtin interacts with REST/NRSF to modulate the transcription of NRSE-controlled neuronal genes. *Nat. Genet.* **35:** 76–83.

17

Activity Dependency and Aging in the Regulation of Adult Neurogenesis

Gerd Kempermann
Genomics of Regeneration
Center for Regenerative Therapies Dresden
01307 Dresden, Germany

AGE AND ACTIVITY MIGHT BE CONSIDERED THE TWO antagonistic key regulators of adult neurogenesis. Whereas adult neurogenesis declines with age, different kinds of "activities" positively regulate adult neurogenesis. An interaction between these two mechanisms exists. Aging influences aging, and activity affects aging processes.

Aging is a principal determinant of life and as such cuts across all biological, psychological, and sociological research. The very essence of aging is difficult to conceptualize, because it is a uniquely omnipresent variable. Aging can thus only be addressed in an transdisciplinary approach, especially if the consequences of aging on complex brain functions are to be studied. In the context of neurogenesis research, "aging" has so far been largely equaled with the biology of long timescales. Implicit in this understanding is that age-dependent changes essentially reflect a unidirectional development in that everything builds on what has occurred before. In this sense, aging can also be seen as continued or lifelong development. This idea has limitations but is instructive with regard to adult neurogenesis because adult neurogenesis is neuronal development under the conditions of the adult brain. The age-related alterations of adult neurogenesis themselves have quantitative and qualitative components. So far, most research has focused on the quantitative aspects. But there can be little doubt that qualitative changes do not simply follow quantitative changes, for example, in cell or synapse numbers but emerge on a systems level and above, when an organism ages.

The observation that adult neurogenesis is regulated by activity relates to this idea. From a behavioral to a synaptic level, activity increases adult neurogenesis. This regulation does not seem to occur in an all-or-nothing fashion but rather influences different stages of neuronal development differently.

To date, the biology of aging has largely focused on aspects of molecular and cellular biology and has looked at entire organisms only if, for example, longevity was being studied. With respect to adult neurogenesis, only one multilevel experiment has been conducted, and even in that study, only three timepoints were investigated (Driscoll et al. 2006). A more complete theory of adult neurogenesis across the life span will require close reference to concepts on many levels of research, including systems biology, behavioral and cognitive neuroscience, and developmental and cognitive psychology.

Recent years have brought fundamentally new insight in how brain function and structure are linked. This mutual interaction is one key meaning of "plasticity." The term "plasticity," despite all its importance for modern neurobiology and psychology, remains elusive. But the concept is important because it states that one cannot see cognitive function independent of the underlying structure (and also vice versa). In this context, adult neurogenesis has become a topic of particular attraction (Lledo et al. 2006).

The observation that adult neurogenesis is regulated by activity cannot be considered independent of a functional theory (Cecchi et al. 2001; Kempermann 2002; Chambers et al. 2004; Deisseroth et al. 2004; Schinder and Gage 2004; Becker 2005; Lehmann et al. 2005; Aimone et al. 2006; Wiskott et al. 2006). If "activity" means "functionally relevant activity" (which remains to be proven), the regulation becomes part of the function. In other words, the functional theory should incorporate the idea of the reciprocal effect of function back on neurogenesis as much as a concept of "regulation" cannot do without an idea of what the entire process is leading to functionally.

AGE EFFECTS ON ADULT NEUROGENESIS

Even Altman's first description of adult hippocampal neurogenesis in 1965 contained a paragraph on the noticeable decrease in cell genesis within months after birth (Fig. 1) (Altman and Das 1965a). Kaplan later reported that adult neurogenesis would occur in old rodents but did not include a detailed quantitative analysis (Kaplan and Bell 1984; Kaplan 1985). The first studies to specifically look at age effects on adult neurogenesis came

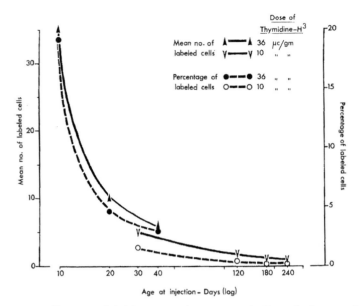

Figure 1. Age effects on adult hippocampal neurogenesis. Even the inaugural paper of the field, Altman and Das's original description of adult hippocampal neurogenesis contained data revealing how strongly postnatal and adult neurogenesis decrease with age. Today, absolute numbers can be assigned to the different time-points, revealing that throughout most adult life, hippocampal neurogenesis ranges on a minute scale. (Reprinted, with permission, from Altman and Das 1965 © John Wiley & Sons.)

from Seki and Arai (1995) and Kuhn et al. (1996), the analogous report for mice came in 1998 (Kempermann et al. 1998b). The same year saw the groundbreaking study of Eriksson (1998) on neurogenesis in aging humans; he presented evidence for adult hippocampal neurogenesis in individuals as old as in their early 70s. Since then, adult neurogenesis is accepted as a lifelong process, although the production of new neurons in the senescent brain ranges on a minute scale. The big question is: What can the functional contribution of so few new neurons be?

Information on age effects on neurogenesis in the olfactory system has only more recently received considerable attention. Although Tropepe et al. (1997) reported that the aging subventricular zone (SVZ) contained fewer precursor cells than in younger rats, only in 2004 was a direct in vivo study published (Enwere et al. 2004). Luo et al. (2006) finally gave a complete account of the structural and neurogenic changes in the aging SVZ in 2006. Ex vivo, precursor cells from the adult SVZ showed longer cell cycle times with increasing age (Tropepe et al. 1997), but this was questioned by other investigators (Luo et al. 2006).

In both neurogenic systems, aging seems to result in a reduced number of precursor cells, most notably a decline in the number of transiently amplifying cells, and a decrease in cell proliferation (Seki 2002; Garcia et al. 2004; Luo et al. 2006). If this is also true for the more quiescent stem cells, from which the progenitor cells might reappear if adequately stimulated, is not quite clear. Maslov et al. (2004) reported a twofold decrease in the number of putative stem cells in the SVZ (based on the expression of G_1-phase-specific proliferation marker Mcm2 that presumably allows a distinction between the relative quiescent and the more proliferative transient amplifying cells) and a corresponding reduction in the yield of neurospheres that could be obtained from the aging SVZ. Luo et al. (2006) did not find a decrease in astrocyte-like cells in the SVZ or even an increased number of astrocyte touching the ventricular lumen. It is exactly this contact that has previously been identified as a hallmark of the putative stem cells in the SVZ (Doetsch et al. 1997).

For rodents, the decrease in adult hippocampal neurogenesis appears to be almost hyperbolical, with a very strong decline early in life and a rather low but persistent level for the remainder of the life span. Barker et al. (2005) suggested that in other rodent species, the decrease might be more linear. This remains to be confirmed in larger comparative experiments but indicates that genetic factors determine the impact of aging on adult neurogenesis.

Although aging is often called a strong (or even the strongest) negative regulator of adult neurogenesis, the term "regulator" is actually problematic in the context of aging. It is undisputed that neurogenesis decreases with age; in old age, adult neurogenesis occurs at only a small fraction of the level in early adulthood, but the decline does not seem to be "regulated" but rather to be the by-product of many age-related changes of other kinds. This range of impact makes aging a particularly complex variable.

MECHANISMS UNDERLYING THE AGE-RELATED DECLINE IN ADULT NEUROGENESIS

Besides the presence or absence of the precursor cells, the decline of neurogenesis in the senescent brain might be due to a loss of extrinsic signals, a reduced responsiveness of the aging precursor cells to normal signaling, or both. Senescence of the precursor cells in the neurogenic zones has to date only been investigated in one study: An increase of p16INK4a, a cyclin-dependent kinase inhibitor linked to cellular senescence, was associated with the age-related loss of precursor cells in the SVZ but not in the hippocampus (Molofsky et al. 2006). This regional discrepancy with respect to cellular senescence might be due to the different cellular compositions and stem cell features in the two neurogenic regions (Seaberg and van der Kooy 2002) but requires further work.

One characteristic of stem cells is the expression of telomerase that prevents telomere shortening under sustained proliferation (Lee et al. 1998). Telomerase activity has been linked to the regulation of adult neurogenesis, but no study focusing specifically on age effects has been published (Caporaso et al. 2003).

Adult neurogenesis requires a specific microenvironment, the so-called niche, that provides the signals needed to maintain and control the proliferation and differentiation of the precursor cells (Palmer et al. 2000; Mercier et al. 2002; Seki 2003). Precursor cells and their niche form a functional unit. The neurogenic niche consists of the precursor cells and their progeny, glia and endothelial cells, possibly immune cells and macrophages, and an extracellular matrix, surrounded by a shared basal membrane. Within the neurogenic niche, cell-cell contacts, paracrine signaling of neurotransmitters, neurotrophic factors, and growth factors as well as synaptic contacts control neurogenesis. It has been proposed that the decline in precursor cell activity in the aging hippocampus is a result of increased precursor cell quiescence due to changing niche properties (Hattiangady and Shetty 2006).

Changes in growth factor levels might underlie changes in neurogenesis as a consequence of disease or aging. Neurotrophic factors, growth factors, and their receptors are abundant during development and decline with age (Wise 2003; Shetty et al. 2005; Chadashvili and Peterson 2006). Growth factors have a strong influence on precursor cell proliferation and neuronal differentiation and thus might be important mediators of cellular plasticity in aging (Cameron et al. 1998; Kwon 2002; Hattiangady et al. 2005; Shetty et al. 2005). The aged mouse brain retains the capacity to respond to the neurogenesis-stimulating effects of growth factors (Lichtenwalner et al. 2001; Jin et al. 2003; Sun et al. 2005).

Although some factors decrease with age, others such as the cell cycle regulator transforming growth factor-β1 (TGF-β1) might increase. Overexpression of TGF-β1 in old mice inhibited the proliferation of early precursor cells (Buckwalter et al. 2006).

The age-related increase in serum glucocorticoids has been proposed as the main culprit for the age-dependent loss of new neurons (Cameron and McKay 1999). We discuss them below, when a synthesis is attempted between the findings of age dependency and activity effects on adult neurogenesis.

PRINCIPLES OF ACTIVITY-DEPENDENT REGULATION OF NEUROGENESIS

Brain development originates from a genetic program that unfolds during the fetal period and early postnatally and becomes increasingly influenced by regulatory influences from the outside world. Whereas the general pattern of brain anatomy is genetically controlled and not affected by activity, the details of network formation, especially on the synaptic level, are activity-dependent.

Similarly, the general flow of adult neurogenesis is determined genetically. The general pattern of neurogenic regulation brings about the idea that regulation of adult neurogenesis might be divided into a less specific phase of precursor cell proliferation and a specific phase of selective survival (Kempermann et al. 1997a, 2004b, 2006; Winner et al. 2002; Kronenberg et al. 2003). In a genetic study, we have found that most of the net neurogenic regulation is determined by the control of survival and not of the expansion phase (Kempermann et al. 2006). The fine regulation on the quantitative and presumably the qualitative level, however, occurs in coordination with parameters of activity. Both cell proliferation and survival are influenced by or even depend on activity.

With regard to activity on a behavioral level, one can grossly distinguish between physical exercise and cognitive stimulation (Fig. 2). Physical exercise robustly increases the proliferation of precursor cells in the hippocampus (but not in the olfactory bulb [Brown et al. 2003]), thereby expanding the population of progenitor cells that is available for further neuronal maturation and functional integration (van Praag et al. 1999; Kronenberg et al. 2003). It has sometimes been postulated that the particular type of physical activity might influence the results, because in studies on the effects of water maze training on neurogenesis, a yoked control that swam for the same time as the groups in the learning test had no signs of increased neurogenesis (van Praag et al. 1999; Ehninger and Kempermann 2006). However, in those studies, swimming lasted only a few minutes per day, so that from these data, the conclusion that swimming would be less efficient than running is not justified (Ra et al. 2002). Voluntary wheel running and forced exercise in treadmill paradigms thus remain the best-studied paradigms of physical exercise in rodents with respect to their effects on adult neurogenesis (van Praag et al. 1999; Trejo et al. 2001; Ra et al. 2002; Kronenberg et al. 2003, 2005;

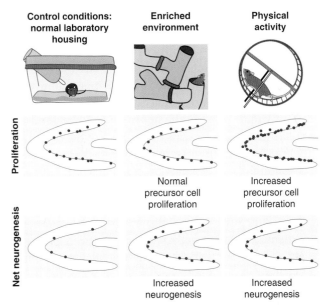

Figure 2. Different types of activity differently affect adult hippocampal neurogenesis. Whereas, presumably, nonspecific stimuli such as physical activity increase the proliferation of precursor cells in the dentate gyrus (DG), more specific stimuli such as exposure to a complex environment (environmental enrichment) preferentially increases the survival of newborn cells. The figure is a schematic rendering of data from van Praag et al. (1999).

Kim et al. 2004; Naylor et al. 2005). The effects of voluntary wheel running on hippocampal neurogenesis were even transmissible from the exercising pregnant mouse to her offspring (Bick-Sander et al. 2006).

In contrast, stimuli that supposedly are more specific to hippocampal function have no or limited effect on cell proliferation but recruit new neurons for long-term survival (Fig. 2). This distinction, however, is strain-dependent and thus presumably again subject to genetic determination (Kempermann et al. 1998a). We have used environmental enrichment to demonstrate this survival-promoting effect (Kempermann et al. 1997b, 2002; Kempermann and Gage 1999; Kronenberg et al. 2003); other investigators have shown such results with specific stimuli such as learning tasks (Gould et al. 1999; Dobrossy et al. 2003; Leuner et al. 2004; Hairston et al. 2005). In addition, the induction of long-term potentiation, the putative electrophysiological correlate of learning, increased neurogenesis in the adult hippocampus (Bruel-Jungerman et al. 2006). Olfactory enrichment had analogous effects on precursor cells migrating to the olfactory bulb from the SVZ but no parallel effect in the hippocampus (Rochefort et al. 2002).

The activity-dependent recruitment for long-term survival mostly takes place once the cells have exited from the cell cycle and have made first synaptic contacts (Brandt et al. 2003; Kempermann et al. 2004b). Environmental enrichment during this critical period in the development of a new neuron would provide the appropriate behavioral stimulus for the selective recruitment. In rodents, this time window appears to be relatively narrow and might amount to not more than 2–3 weeks (Gould et al. 1999; Greenough et al. 1999; Dobrossy et al. 2003; Leuner et al. 2004; Hairston et al. 2005; Ehninger and Kempermann 2006). The steep decline in survival after exit from the cell cycle suggests that most of the recruitment actually occurs very early. This implies that, quantitatively, the activity-dependent selection occurs well before the cells are finally integrated into the existing network. Although dendrites are extended early after exit from the cell cycle (Plumpe et al. 2006), dendritic spines first appear only 2–3 weeks later (Zhao et al. 2006). It seems that the regulatory synaptic drive that reaches the cells even on a progenitor level is GABAergic and responsible for the subsequent neurodevelopmental events (Tozuka et al. 2005; Wang et al. 2005; Ge et al. 2006).

From the fact that in the course of their individual development, new neurons in the adult hippocampus go through a phase of increased synaptic sensitivity (and a reduced threshold for the induction of long-term potentiation) (Wang et al. 2000; Schmidt-Hieber et al. 2004), it has been inferred that during this particular phase, the new cells might also serve a particular function, presumably distinct from their lasting function once they are fully integrated into the network. On the other hand, one might argue that the increased synaptic sensitivity serves the preferential recruitment of the new cells into the existing network, rather than a function of its own. The hypothesis is that transient and persistent functions are linked and that the former serves the latter (Kempermann et al. 2004a).

In long-term experiments, both environmental enrichment and voluntary wheel running had additional effects that had not been apparent in more acute settings. In both paradigms, survival-promoting effects acted upon the precursor cells themselves and not only on their progeny (Kempermann and Gage 1999; Kempermann et al. 2002; Kronenberg et al. 2006). Long-term wheel running had both proproliferative and survival-promoting effects, the latter resulting in an increased potential for neurogenesis that went unused in the absence of additional specific (cognitive) stimuli (Kronenberg et al. 2006).

Similar to voluntary physical exercise, chemically induced seizures as a model of pansynaptic activation led to an increase in cell proliferation in the adult hippocampus (Bengzon et al. 1997; Parent et al. 1997; Parent and Lowenstein 2002). Similarly, there is a nonspecific response in cell proliferation in cases of various types of pathology throughout the brain (for review, see Dietrich and Kempermann 2006). All of these might lead to an initial increase in net neurogenesis, but arguably, this increase will not last. Acute effects of nonspecific stimuli "wear off" (Kralic et al. 2005; Kronenberg et al. 2006). From a functional perspective, this seems to be plausible: Nonspecific, proproliferative stimuli would lead to a transient increase in a potential for new neurons, but additional survival-promoting stimuli, presumably more specific to hippocampal function, would be necessary. Net regulation of adult neurogenesis is thus dependent on the activity-dependent control of two regulatory stages: expansion and selective survival. Transient expansion of the precursor pool apparently occurs under many circumstances of increased "activity" in a very general sense. That nonspecific activity has any lasting effect at all seems surprising. Why should physical activity have any functionally relevant consequences on the hippocampal network? For a rodent, "cognition" is inseparable from locomotion. An animal discovers the world by moving in it. Learning is to a large degree spatial learning and navigation. Only with the development of language did cognition independent of locomotion become possible. If adult neurogenesis thus allows an activity-dependent adaptation of the neuronal network in the hippocampus, locomotor activity would reflect movement in the environment. Increased locomotion might (not alone, but prominently) be a sign of increased exploration. Increased activity in the sense of locomotion might thus generally be indicative of situations of increased likelihood of novel and complex cognitive stimuli. Physical activity would thus not by itself provide specific hippocampus-relevant stimuli that induce net neurogenesis but be associated with a greater chance to encounter specific relevant stimuli. Only in the lab situation can we separate these two conditions.

MECHANISMS OF ACTIVITY-DEPENDENT REGULATION

Precursor cells in the neurogenic niches of the adult brain might be able to directly sense activity and translate this stimulus into the initiation of a program of neuronal development. Deisseroth et al. (2004) reported that even short-lasting and mild depolarization of hippocampal precursor cells in vitro caused long-lasting increases in the number of newly generated neurons. In proliferating cells, a fast up-regulation of NeuroD was found, accompanied by a down-regulation of transcription factors favoring glial development (Deisseroth et al. 2004). The effect was dependent on NMDA (*N*-methyl-D-aspartate) receptor signaling and mediated by calcium signaling. The potential relevance of NMDA receptors for the control of adult neurogenesis had been noted early (Cameron et al. 1995; Nacher and McEwen 2006) and might provide an interesting conceptual link between synaptic plasticity and adult neurogenesis as an example of cellular plasticity.

Given the possibility that systemic factors might be involved in the activity-dependent control of adult neurogenesis, one has tried early to identify circulating factors that would mediate the up-regulating effects of exercise on neurogenesis. For both insulin-like growth factor-1 (IGF-1) and vascular endothelial growth factor (VEGF), the argument has been made that either factor could serve as sole mediator between exercise and neurogenesis (Trejo et al. 2001; Fabel et al. 2003). Blocking IGF-1 or VEGF prevented the exercise-induced increase in adult hippocampal neurogenesis. Because both factors have been implied as important for neurogenesis in general, the interpretation of these results is ambiguous; rather, they show that both factors are necessary for the regulation of neurogenesis but do not prove that they are sufficient. In fact, the two claims are mutually exclusive.

Fibroblast growth factor-2 (FGF-2), which is necessary to grow hippocampal precursor cells in vitro (Palmer et al. 1995), is another key candidate for mediating activity-dependent effects across the life span (Jin et al. 2003), because FGF-2 can act systemically (Wagner et al. 1999), is proneurogenic in the adult brain (Kuhn et al. 1997), declines with age (Shetty et al. 2005), and is up-regulated in activity paradigms (Bick-Sander et al. 2006), and FGF-2 receptor expression is prominent in the neurogenic zones of the adult and aged rodent brain (Chadashvili and Peterson 2006).

Hormones constitute the key regulatory system that has been investigated as a potential link between age effects on neurogenesis and its activity-dependent regulation. To date, most information is available on the effects of glucocorticoids. Acute stress decreases the proliferation of precursor cells and seems to do so through the action of corticosterone (Mirescu and Gould 2006). Stress elevates glucocorticoid levels and stimulates glutamate release in the hippocampus. This leads to a down-regulation of precursor cell proliferation in the adult dentate gyrus (DG) (Gould et al. 1997; Gould and Tanapat 1999). The stress-induced effects of glucocorticoids have been linked to age-related cognitive problems in humans (Heffelfinger and Newcomer 2001; Magri et al. 2006). In animal models, the effects of chronic stress and chronically high levels of corticosterone on adult neurogenesis have been rather difficult to assess, and the data do not yet provide a clear picture. A number of chronic stress models did not show changes in adult neurogenesis (Heine et al. 2004b, 2005). In addition, the activity-based models of increased levels of adult neurogenesis, voluntary wheel running, and environmental enrichment are associated not with lower serum levels of corticosterone but with higher levels (van Praag et al. 1999). Nevertheless, adult neurogenesis in tree shrews was more vulnerable to chronic psychosocial stress in old animals than in the young ones (Simon et al. 2005).

Age-related cognitive decline as well as the age-dependent decrease in adult neurogenesis might be due to chronically increased glucocorticoid levels even in the absence of stressful experience, because a dysregulation of the HPA (hypothalamic-pituitary-adrenal) system might lead to increased serum corticosterone levels (Koehl et al. 1999; Sapolsky 1999). The age-related reduction in adult neurogenesis might be a result of the increased corticosteroid levels in older age (Cameron and McKay 1999), although some conflicting data exist (Heine et al. 2004a). Corticosteroid receptor expression increases on precursor cells in older age, possibly making them more sensitive to corticosterone action (Garcia et al. 2004). Interestingly, the expression pattern did not change under voluntary wheel running or environmental enrichment. Even a brief treatment with the glucocorticoid receptor antagonist mifepristone increased hippocampal neurogenesis and improved cognition in old rats (Mayer et al. 2006). The prevention of glucocorticoid-mediated effects in aging rats maintained higher neurogenesis levels (Cameron and McKay 1999; Montaron et al. 2006). Because both environmental enrichment and physical activity are associated with increased rather than decreased levels of corticosteroids, however, the interaction might be more complex than these data alone suggest. Not unexpectedly, one study found that acute stress, for example, interfered with the up-regulating effect of wheel running on adult neurogenesis (Stranahan et al. 2006). Activity-dependent control thus stands in regulatory competition with other factors not only on the behavioral level, but also on the mechanistic level.

Sex hormones, both female and male, can influence adult neurogenesis (Galea et al. 2006). In the hippocampus of female rodents, cell proliferation peaks during the estrogen-high proestrous phase (Tanapat et al. 1999). Exogenous application of estradiol led only to a transient increase in cell proliferation, presumably because a compensating feedback mechanisms might exist (Ormerod et al. 2003). Estrogen has neuroprotective properties and estrogen deficiency during menopause can be associated with cognitive dysfunction. Estrogen substitution therapy in postmenopausal women might have a positive effect on these problems, but no consistent link between hormone replacement therapy and cognitive performance has been found (Hogervorst et al. 2002; Low and Anstey 2006). In songbirds, a fundamental link between testosterone signaling and the control of adult neurogenesis has been described previously (Goldman and Nottebohm 1983; Louissaint et al. 2002).

In mammals, sex hormones presumably have a largely modulatory role on neurogenesis. Estrogens and IGF-1 interact in the promotion of neuronal survival (Garcia-Segura et al. 2000). Furthermore, estrogen might act through a serotonin-mediated pathway to stimulate precursor cell proliferation in the adult DG (Banasr et al. 2001).

PATHOLOGICAL AGING, NEURODEGENERATION, AND ADULT NEUROGENESIS

In pathological aging, changes to cognitive functions exceed the level that is considered to be normal and specific pathological events can be detected. Neurodegenerative disease is defined by an age-dependent progressive loss of neurons due to the accumulation of misfolded proteins, e.g., α-synuclein in Parkinson's disease and amyloid-β in Alzheimer's disease. In a wider sense, neurodegenerative disease includes secondary degeneration such as in inflammatory disorders or ischemia.

Enthusiasm about adult neurogenesis has led to the speculative idea that neurodegeneration might only represent the flip side of neuronal development and regeneration. Neurodegenerative disorders might rather be a consequence of a lack of neuroregeneration than an increase in degeneration per se. The underlying assumption is that under normal conditions, degeneration and regeneration, cell death, and birth were in an equilibrium. Under pathological conditions, this equilibrium might become disturbed and lead to the manifestations of the disease. There is very little evidence that this concept applies in a fundamental sense, although adult neurogenesis responds to a wide range of pathologies, and stem cell biology has raised new hopes for regenerative medicine, especially with respect to aging-related disorders, because stem cells embody the potential for regeneration (Lie et al. 2004; Dietrich and Kempermann 2006; Kempermann 2006; Lindvall and Kokaia 2006; Rando 2006). In contrast, it might be the physiological function of the new neurons that is influenced or impaired by aging and that, if adequately preserved, adult neurogenesis might contribute to the compensation of age-dependent functional losses. In the course of neurodegenerative disorders, there is an increase in a specific degenerative pathology, not just a reduction in regeneration.

Despite being exceptional and rare in old age, adult neurogenesis might be of profound functional significance because it occurs at a strategic bottleneck location in the hippocampus (Kempermann 2002). A failure of adult hippocampal neurogenesis might contribute to different types of age-related neurological and psychiatric disorders (Steiner et al. 2006), most notably major depression (Jacobs et al. 2000), neurodegenerative disorders such as Alzheimer's disease (AD) (Jin et al. 2004), and the cognitive decline associated with Parkinson's disease (PD) (Hoglinger et al. 2004).

In addition, adult neurogenesis might represent the tip of the iceberg of a much larger spectrum of cellular plasticity. The recent increase in knowledge about NG2 cells that might turn out to constitute a previously unrecognized entity of cells with precursor cell properties indicates that in terms of cellular plasticity, the adult brain contains a still largely uncharted terrain (Mallon et al. 2002; Nishiyama et al. 2002). The fact that in the adult brain, cellular regeneration after damage is scarce might have misled us to believe that it was absent in general. There are thus aspects of activity-dependent plasticity related to neuronal precursor cell biology that are independent of neurogenesis in the canonical neurogenic regions of the adult brain.

THE INTERACTION BETWEEN EFFECTS OF AGING AND ACTIVITY ON ADULT NEUROGENESIS: THE NEURAL RESERVE HYPOTHESIS

The cognitive or neural reserve theory proposes that neuronal networks that are strengthened by constant cognitive challenges can withstand degeneration and functional loss to a greater extent than less trained networks (Satz 1993; Stern 2002). Cognitive reserve thus describes the ability of the brain to recruit compensatory neuronal networks in the case of degeneration or age-related loss. The cognitive reserve hypothesis might explain the large interindividual differences in coping with the symptoms of dementia in neurodegenerative diseases and the variable relationship between neuropathological findings and clinical manifestation of the disease. Cellular plasticity and adult neurogenesis might contribute to these effects. In fact, the neural reserve might (among other aspects) represent the interaction between age and activity effects in the control of neurogenesis.

The neural reserve hypothesis can be tested in the animal model of environmental enrichment. Environmental enrichment leads to an increase in adult hippocampal neurogenesis, and this effect is seen even in old animals, where the baseline level of adult neurogenesis is very low (Kempermann et al. 1998b). Long-term environmental enrichment for the second half of the life of a mouse led to a sustained induction of adult neurogenesis (in contrast to effects that might be seen only after acute exposure), suggesting that living in a challenging environment has lasting consequences on the potential for cellular hippocampal plasticity (Kempermann et al. 2002).

In line with the results discussed in the previous paragraphs, adult hippocampal neurogenesis might thus contribute to a structural or neural reserve that if appropriately trained early in life might provide a compensatory buffer of brain plasticity in the face of increasing neurodegeneration or nonpathological age-related functional losses.

Environmental enrichment, for example, had positive effects on symptoms and the cause of disease in murine models of AD (Arendash et al. 2004; Wolf et al. 2006). Despite stable Aβ plaque, load enrichment improved learning and memory performance in aged APP (amyloid precursor protein) transgenic mice. The effect was not apparent in APP transgenic mice with access to a running wheel. The enhancement of cognitive function was paralleled by a hippocampus-specific increase of neurotrophic factors and increased neurogenesis (Wolf et al. 2006).

Despite such suggestive findings, there is still only limited information on the activity-dependent parameters that help to prevent the age-dependent decrease in adult neurogenesis and maintain cellular plasticity. A number of "usual suspects" such as corticosterone (Cameron and McKay 1999; Montaron et al. 2006), IGF-1 (Lichtenwalner et al. 2001), or exercise (Kronenberg et al. 2006) dominate the discussion, yet very little is known about the intracellular signaling pathways that mediate and modulate those effects.

The cAMP response element-binding protein (CREB) is a key factor controlling neuronal survival, learning, and memory and of plasticity in the nervous system. Fittingly, the hippocampal expression of CREB decreases with age (Kudo et al. 2005). CREB is expressed by polysialilated neural cell adhesion molecule (PSA-NCAM)-positive progenitor cells and their postmitotic progeny in the adult hippocampus (Nakagawa et al. 2002a), and the activation of CREB increases many aspects of adult neurogenesis, including proliferation, survival, and dendritic maturation (Nakagawa et al. 2002b; Fujioka et al. 2004). However, intracellular signaling from numerous pathways converges on CREB, so that CREB action can only be understood in the context of larger networks of regulatory mechanisms. In a related context, a comprehensive model of transcriptional mechanisms that distinguish successful from unsuccessful aging in the hippocampus have been proposed (Lund et al. 2004) but remains to be applied to adult neurogenesis.

SUMMARY

Age and activity might be considered the two antagonistic key regulators of adult neurogenesis. Adult neurogenesis decreases with age but remains present, albeit at a very low level, even in oldest individuals. Activity, be it physical or cognitive, increases adult neurogenesis and thereby seems to counteract age effects. It is thus proposed that activity-dependent regulation of adult neurogenesis might contribute to some sort of "neural reserve," the brain's ability to compensate functional loss associated with aging or neurodegeneration. Activity can have nonspecific and specific effects on adult neurogenesis. Mechanistically, nonspecific stimuli that largely affect precursor cell stages might be related by the local microenvironment, whereas more specific survival-promoting effects take place on later stages of neuronal development and require the synaptic integration of the new cell and its particular synaptic plasticity.

REFERENCES

Aimone J.B., Wiles J., and Gage F.H. 2006. Potential role for adult neurogenesis in the encoding of time in new memories. *Nat. Neurosci.* **9:** 723–727.

Altman J. and Das G.D. 1965a. Autoradiographic and histologic evidence of postnatal neurogenesis in rats. *J. Comp. Neurol.* **124:** 319–335.

Arendash G.W., Garcia M.F., Costa D.A., Cracchiolo J.R., Wefes I.M., and Potter H. 2004. Environmental enrichment improves cognition in aged Alzheimer's transgenic mice despite stable beta-amyloid deposition. *Neuroreport* **15:** 1751–1754.

Banasr M., Hery M., Brezun J.M., and Daszuta A. 2001. Serotonin mediates oestrogen stimulation of cell proliferation in the adult dentate gyrus. *Eur. J. Neurosci.* **14:** 1417–1424.

Barker J.M., Wojtowicz J.M., and Boonstra R. 2005. Where's my dinner? Adult neurogenesis in free-living food-storing rodents. *Genes Brain Behav.* **4:** 89–98.

Becker S. 2005. A computational principle for hippocampal learning and neurogenesis. *Hippocampus* **15:** 722–738.

Bengzon J., Kokaia Z., Elmér E., Nanobashvili A., Kokaia M., and Lindvall O. 1997. Apoptosis and proliferation of dentate gyrus neurons after single and intermittent limbic seizures. *Proc. Natl. Acad. Sci.* **94:** 10432–10437.

Bick-Sander A., Steiner B., Wolf S.A., Babu H., and Kempermann G. 2006. Running in pregnancy transiently increases postnatal hippocampal neurogenesis in the offspring. *Proc. Natl. Acad. Sci.* **103:** 3852–3857.

Brandt M.D., Jessberger S., Steiner B., Kronenberg G., Reuter K., Bick-Sander A., Von der Behrens W., and Kempermann G. 2003. Transient calretinin-expression defines early postmitotic step of neuronal differentiation in adult hippocampal neurogenesis of mice. *Mol. Cell. Neurosci.* **24:** 603–613.

Brown J., Cooper-Kuhn C.M., Kempermann G., Van Praag H., Winkler J., Gage F.H., and Kuhn H.G. 2003. Enriched environment and physical activity stimulate hippocampal but not olfactory bulb neurogenesis. *Eur. J. Neurosci.* **17:** 2042–2046.

Bruel-Jungerman E., Davis S., Rampon C., and Laroche S. 2006. Long-term potentiation enhances neurogenesis in the adult dentate gyrus. *J. Neurosci.* **26:** 5888–5893.

Buckwalter M.S., Yamane M., Coleman B.S., Ormerod B.K., Chin J.T., Palmer T., and Wyss-Coray T. 2006. Chronically increased transforming growth factor-beta1 strongly inhibits hippocampal neurogenesis in aged mice. *Am. J. Pathol.* **169:** 154–164.

Cameron H.A. and McKay R.D. 1999. Restoring production of hippocampal neurons in old age. *Nat. Neurosci.* **2:** 894–897.

Cameron H.A., Hazel T.G., and McKay R.D. 1998. Regulation of neurogenesis by growth factors and neurotransmitters. *J. Neurobiol.* **36:** 287–306.

Cameron H.A., McEwen B.S., and Gould E. 1995. Regulation of adult neurogenesis by excitatory input and NMDA receptor activation in the dentate gyrus. *J. Neurosci.* **15:** 4687–4692.

Caporaso G.L., Lim D.A., Alvarez-Buylla A., and Chao M.V. 2003. Telomerase activity in the subventricular zone of adult mice. *Mol. Cell. Neurosci.* **23:** 693–702.

Cecchi G.A., Petreanu L.T., Alvarez-Buylla A., and Magnasco M.O. 2001. Unsupervised learning and adaptation in a model of adult neurogenesis. *J. Comput. Neurosci.* **11:** 175–182.

Chadashvili T. and Peterson D.A. 2006. Cytoarchitecture of fibroblast growth factor receptor 2 (FGFR-2) immunoreactivity in astrocytes of neurogenic and non-neurogenic regions of the young adult and aged rat brain. *J. Comp. Neurol.* **498:** 1–15.

Chambers R.A., Potenza M.N., Hoffman R.E., and Miranker W. 2004. Simulated apoptosis/neurogenesis regulates learning and memory capabilities of adaptive neural networks. *Neuropsychopharmacology* **29:** 747–758.

Deisseroth K., Singla S., Toda H., Monje M., Palmer T.D., and Malenka R.C. 2004. Excitation-neurogenesis coupling in adult neural stem/progenitor cells. *Neuron* **42:** 535–552.

Dietrich J. and Kempermann G. 2006. Role of endogenous neural stem cells in neurological disease and brain repair. *Adv. Exp. Med. Biol.* **557:** 191–220.

Dobrossy M.D., Drapeau E., Aurousseau C., Le Moal M., Piazza P.V., and Abrous D.N. 2003. Differential effects of learning on neurogenesis: Learning increases or decreases the number of newly born cells depending on their birth date. *Mol. Psychiatry* **8:** 974–982.

Doetsch F., Garcia-Verdugo J.M., and Alvarez-Buylla A. 1997. Cellular composition and three-dimensional organization of the subventricular germinal zone in the adult mammalian brain. *J. Neurosci.* **17:** 5046–5061.

Driscoll I., Howard S.R., Stone J.C., Monfils M.H., Tomanek B., Brooks W.M., and Sutherland R.J. 2006. The aging hippocampus: A multi-level analysis in the rat. *Neuroscience* **139:** 1173–1185.

Ehninger D. and Kempermann G. 2006. Paradoxical effects of learning the Morris water maze on adult hippocampal neurogenesis in mice may be explained by a combination of stress and physical activity. *Genes Brain Behav.* **5:** 29–39.

Enwere E., Shingo T., Gregg C., Fujikawa H., Ohta S., and Weiss S. 2004. Aging results in reduced epidermal growth factor receptor signaling, diminished olfactory neurogenesis, and deficits in fine olfactory discrimination. *J. Neurosci.* **24:** 8354–8365.

Eriksson P.S., Perfilieva E., Bjork-Eriksson T., Alborn A.M., Nordborg C., Peterson D.A., and Gage F.H. 1998. Neurogenesis in the adult human hippocampus. *Nat. Med.* **4:** 1313–1317.

Fabel K., Tam B., Kaufer D., Baiker A., Simmons N., Kuo C.J., and Palmer T.D. 2003. VEGF is necessary for exercise-induced adult hippocampal neurogenesis. *Eur. J. Neurosci.* **18:** 2803–2812.

Fujioka T., Fujioka A., and Duman R.S. 2004. Activation of cAMP signaling facilitates the morphological maturation of newborn neurons in adult hippocampus. *J. Neurosci.* **24:** 319–328.

Galea L.A., Spritzer M.D., Barker J.M., and Pawluski J.L. 2006. Gonadal hormone modulation of hippocampal neurogenesis in the adult. *Hippocampus* **16:** 225–232.

Garcia A., Steiner B., Kronenberg G., Bick-Sander A., and Kempermann G. 2004. Age-dependent expression of glucocorticoid- and mineralocorticoid receptors on neural precursor cell populations in the adult murine hippocampus. *Aging Cell* **3:** 363–371.

Garcia-Segura L.M., Cardona-Gomez G.P., Chowen J.A., and Azcoitia I. 2000. Insulin-like growth factor-I receptors and estrogen receptors interact in the promotion of neuronal survival and neuroprotection. *J. Neurocytol.* **29:** 425–437.

Ge S., Goh E.L., Sailor K.A., Kitabatake Y., Ming G.L., and Song H. 2006. GABA regulates synaptic integration of newly generated neurons in the adult brain. *Nature* **439:** 589–593.

Goldman S.A. and Nottebohm F. 1983. Neuronal production, migration and differentiation in a vocal control nucleus of the adult female canary brain. *Proc. Natl. Acad. Sci.* **80:** 2390–2394.

Gould E. and Tanapat P. 1999. Stress and hippocampal neurogenesis. *Biol. Psychiatry* **46:** 1472–1479.

Gould E., Beylin A., Tanapat P., Reeves A., and Shors T.J. 1999. Learning enhances adult neurogenesis in the hippoampal formation. *Nat. Neurosci.* **2:** 260–265.

Gould E., McEwen B.S., Tanapat P., Galea L.A., and Fuchs E. 1997. Neurogenesis in the dentate gyrus of the adult tree shrew is regulated by psychosocial stress and NMDA receptor activation. *J. Neurosci.* **17:** 2492–2498.

Greenough W.T., Cohen N.J., and Juraska J.M. 1999. New neurons in old brains: Learning to survive? *Nat. Neurosci.* **2:** 203–205.

Hairston I.S., Little M.T., Scanlon M.D., Barakat M.T., Palmer T.D., Sapolsky R.M., and Heller H.C. 2005. Sleep restriction suppresses neurogenesis induced by hippocampus-dependent learning. *J. Neurophysiol.* **94:** 4224–4233.

Hattiangady B. and Shetty A.K. 2006. Aging does not alter the number or phenotype of putative stem/progenitor cells in the neurogenic region of the hippocampus. *Neurobiol. Aging* (in press).

Hattiangady B., Rao M.S., Shetty G.A., and Shetty A.K. 2005. Brain-derived neurotrophic factor, phosphorylated cyclic AMP response element binding protein and neuropeptide Y decline as early as middle age in the dentate gyrus and CA1 and CA3 subfields of the hippocampus. *Exp. Neurol.* **195:** 353–371.

Heffelfinger A.K. and Newcomer J.W. 2001. Glucocorticoid effects on memory function over the human life span. *Dev. Psychopathol.* **13:** 491–513.

Heine V.M., Maslam S., Joels M., and Lucassen P.J. 2004a. Prominent decline of newborn cell proliferation, differentiation, and apoptosis in the aging dentate gyrus, in absence of an age-related hypothalamus-pituitary-adrenal axis activation. *Neurobiol. Aging* **25:** 361–375.

Heine V.M., Maslam S., Zareno J., Joels M., and Lucassen P.J. 2004b. Suppressed proliferation and apoptotic changes in the rat dentate gyrus after acute and chronic stress are reversible. *Eur. J. Neurosci.* **19:** 131–144.

Heine V.M., Zareno J., Maslam S., Joels M., and Lucassen P.J. 2005. Chronic stress in the adult dentate gyrus reduces cell proliferation near the vasculature and VEGF and Flk-1 protein expression. *Eur. J. Neurosci.* **21:** 1304–1314.

Hogervorst E., Yaffe K., Richards M., and Huppert F. 2002. Hormone replacement therapy for cognitive function in postmenopausal women. *Cochrane Database Syst. Rev.* **3:** CD003122.

Hoglinger G.U., Rizk P., Muriel M.P., Duyckaerts C., Oertel W.H., Caille I., and Hirsch E.C. 2004. Dopamine depletion impairs precursor cell proliferation in Parkinson disease. *Nat. Neurosci.* **7:** 726–735.

Jacobs B.L., Praag H., and Gage F.H. 2000. Adult brain neurogenesis and psychiatry: A novel theory of depression. *Mol. Psychiatry* **5:** 262–269.

Jin K., Galvan V., Xie L., Mao X.O., Gorostiza O.F., Bredesen D.E., and Greenberg D.A. 2004. Enhanced neurogenesis in Alzheimer's disease transgenic (PDGF-APPSw,Ind) mice. *Proc. Natl. Acad. Sci.* **101:** 13363–13367.

Jin K., Sun Y., Xie L., Batteur S., Mao X.O., Smelick C., Logvinova A., and Greenberg D.A. 2003. Neurogenesis and aging: FGF-2 and HB-EGF restore neurogenesis in hippocampus and subventricular zone of aged mice. *Aging Cell* **2:** 175–183.

Kaplan M.S. 1985. Formation and turnover of neurons in young and senescent animals: An electronmicroscopic and morphometric analysis. *Ann. N.Y. Acad. Sci.* **457:** 173–192.

Kaplan M.S. and Bell D.H. 1984. Mitotic neuroblasts in the 9-day-old and 11-month-old rodent hippocampus. *J. Neurosci.* **4:** 1429–1441.

Kempermann G. 2002. Why new neurons? Possible functions for adult hippocampal neurogenesis. *J. Neurosci.* **22:** 635–638.

———. 2006. *Adult neurogenesis—Stem cells and neuronal development in the adult brain.* Oxford University Press, New York.

Kempermann G. and Gage F.H. 1999. Experience-dependent regulation of adult hippocampal neurogenesis: Effects of long-term stimulation and stimulus withdrawal. *Hippocampus* **9:** 321–332.

Kempermann G., Brandon E.P., and Gage F.H. 1998a. Environmental stimulation of 129/SvJ mice causes increased cell proliferation and neurogenesis in the adult dentate gyrus. *Curr. Biol.* **8:** 939–942.

Kempermann G., Gast D., and Gage F.H. 2002. Neuroplasticity in old age: Sustained fivefold induction of hippocampal neurogenesis by long-term environmental enrichment. *Ann. Neurol.* **52:** 135–143.

Kempermann G., Kuhn H.G., and Gage F.H. 1997a. Genetic influence on neurogenesis in the dentate gyrus of adult mice. *Proc. Natl. Acad. Sci.* **94:** 10409–10414.

———. 1997b. More hippocampal neurons in adult mice living in an enriched environment. *Nature* **386:** 493–495.

———. 1998b. Experience-induced neurogenesis in the senescent dentate gyrus. *J. Neurosci.* **18:** 3206–3212.

Kempermann G., Wiskott L., and Gage F.H. 2004a. Functional significance of adult neurogenesis. *Curr. Opin. Neurobiol.* **14:** 186–191.

Kempermann G., Jessberger S., Steiner B., and Kronenberg G. 2004b. Milestones of neuronal development in the adult hippocampus. *Trends Neurosci.* **27:** 447–452.

Kempermann G., Chesler E.J., Lu L., Williams R.W., and Gage F.H. 2006. Natural variation and genetic covariance in adult hippocampal neurogenesis. *Proc. Natl. Acad. Sci.* **103:** 780–785.

Kim Y.P., Kim H., Shin M.S., Chang H.K., Jang M.H., Shin M.C., Lee S.J., Lee H.H., Yoon J.H., Jeong I.G., and Kim C.J. 2004. Age-dependence of the effect of treadmill exercise on cell proliferation in the dentate gyrus of rats. *Neurosci. Lett.* **355:** 152–154.

Koehl M., Darnaudery M., Dulluc J., Van Reeth O., Le Moal M., and Maccari S. 1999. Prenatal stress alters circadian activity of hypothalamo-pituitary-adrenal axis and hippocampal corticosteroid receptors in adult rats of both gender. *J. Neurobiol.* **40:** 302–315.

Kralic J.E., Ledergerber D.A., and Fritschy J.M. 2005. Disruption of the neurogenic potential of the dentate gyrus in a mouse model of temporal lobe epilepsy with focal seizures. *Eur. J. Neurosci.* **22:** 1916–1927.

Kronenberg G., Bick-Sander A., Bunk E., Wolf C., Ehninger D., and Kempermann G. 2005. Physical exercise prevents age-related decline in precursor cell activity in the mouse dentate gyrus. *Neurobiol. Aging* **27:** 1505–1513.

———. 2006. Physical exercise prevents age-related decline in precursor cell activity in the mouse dentate gyrus. *Neurobiol. Aging* **27:** 1505–1513.

Kronenberg G., Reuter K., Steiner B., Brandt M.D., Jessberger S., Yamaguchi M., and Kempermann G. 2003. Subpopulations of proliferating cells of the adult hippocampus respond differently to physiologic neurogenic stimuli. *J. Comp. Neurol.* **467:** 455–463.

Kudo K., Wati H., Qiao C., Arita J., and Kanba S. 2005. Age-related disturbance of memory and CREB phosphorylation in CA1 area of hippocampus of rats. *Brain Res.* **1054:** 30–37.

Kuhn H.G., Dickinson-Anson H., and Gage F.H. 1996. Neurogenesis in the dentate gyrus of the adult rat: Age-related decrease of neuronal progenitor proliferation. *J. Neurosci.* **16:** 2027–2033.

Kuhn H.G., Winkler J., Kempermann G., Thal L.J., and Gage F.H. 1997. Epidermal growth factor and fibroblast growth factor-2 have different effects on neural progenitors in the adult rat brain. *J. Neurosci.* **17:** 5820–5829.

Kwon Y.K. 2002. Effect of neurotrophic factors on neuronal stem cell death. *J. Biochem. Mol. Biol.* **35:** 87–93.

Lee H.W., Blasco M.A., Gottlieb G.J., Horner J.W., 2nd, Greider C.W., and DePinho R.A. 1998. Essential role of mouse telomerase in highly proliferative organs. *Nature* **392:** 569–574.

Lehmann K., Butz M., and Teuchert-Noodt G. 2005. Offer and demand: Proliferation and survival of neurons in the dentate gyrus. *Eur. J. Neurosci.* **21:** 3205–3216.

Leuner B., Mendolia-Loffredo S., Kozorovitskiy Y., Samburg D., Gould E., and Shors T.J. 2004. Learning enhances the survival of new neurons beyond the time when the hippocampus is required for memory. *J. Neurosci.* **24:** 7477–7481.

Lichtenwalner R.J., Forbes M.E., Bennett S.A., Lynch C.D., Sonntag W.E., and Riddle D.R. 2001. Intracerebroventricular infusion of insulin-like growth factor-I ameliorates the age-related decline in hippocampal neurogenesis. *Neuroscience* **107:** 603–613.

Lie D.C., Song H., Colamarino S.A., Ming G.L., and Gage F.H. 2004. Neurogenesis in the adult brain: New strategies for central nervous system diseases. *Annu. Rev. Pharmacol. Toxicol.* **44:** 399–421.

Lindvall O. and Kokaia Z. 2006. Stem cells for the treatment of neurological disorders. *Nature* **441:** 1094–1096.

Lledo P.M., Alonso M., and Grubb M.S. 2006. Adult neurogenesis and functional plasticity in neuronal circuits. *Nat. Rev. Neurosci.* **7:** 179–193.

Louissaint A., Jr., Rao S., Leventhal C., and Goldman S.A. 2002. Coordinated interaction of neurogenesis and angiogenesis in the adult songbird brain. *Neuron* **34:** 945–960.

Low L.F. and Anstey K.J. 2006. Hormone replacement therapy and cognitive performance in postmenopausal women—A review by cognitive domain. *Neurosci. Biobehav. Rev.* **30:** 66–84.

Lund P.K., Hoyt E.C., Bizon J., Smith D.R., Haberman R., Helm K., and Gallagher M. 2004. Transcriptional mechanisms of hippocampal aging. *Exp. Gerontol.* **39:** 1613–1622.

Luo J., Daniels S.B., Lennington J.B., Notti R.Q., and Conover J.C. 2006. The aging neurogenic subventricular zone. *Aging Cell* **5:** 139–152.

Magri F., Cravello L., Barili L., Sarra S., Cinchetti W., Salmoiraghi F., Micale G., and Ferrari E. 2006. Stress and dementia: The role of the hypothalamicpituitary-adrenal axis. *Aging Clin. Exp. Res.* **18:** 167–170.

Mallon B.S., Shick H.E., Kidd G.J., and Macklin W.B. 2002. Proteolipid promoter activity distinguishes two populations of NG2-positive cells throughout neonatal cortical development. *J. Neurosci.* **22:** 876–885.

Maslov A.Y., Barone T.A., Plunkett R.J., and Pruitt S.C. 2004. Neural stem cell detection, characterization, and age-related changes in the subventricular zone of mice. *J. Neurosci.* **24:** 1726–1733.

Mayer J.L., Klumpers L., Maslam S., de Kloet E.R., Joels M., and Lucassen P.J. 2006. Brief treatment with the glucocorticoid receptor antagonist mifepristone normalises the corticosterone-induced reduction of adult hippocampal neurogenesis. *J. Neuroendocrinol.* **18:** 629–631.

Mercier F., Kitasako J.T., and Hatton G.I. 2002. Anatomy of the brain neurogenic zones revisited: Fractones and the fibroblast/macrophage network. *J. Comp. Neurol.* **451:** 170–188.

Mirescu C. and Gould E. 2006. Stress and adult neurogenesis. *Hippocampus* **16:** 233–238.

Molofsky A.V., Slutsky S.G., Joseph N.M., He S., Pardal R., Krishnamurthy J., Sharpless N.E., and Morrison S.J. 2006. Increasing p16INK4a expression decreases forebrain progenitors and neurogenesis during ageing. *Nature* **443:** 448–452.

Montaron M.F., Drapeau E., Dupret D., Kitchener P., Aurousseau C., Le Moal M., Piazza P.V., and Abrous D.N. 2006. Lifelong corticosterone level determines age-related decline in neurogenesis and memory. *Neurobiol. Aging* **27:** 645–654.

Nacher J. and McEwen B.S. 2006. The role of N-methyl-D-asparate receptors in neurogenesis. *Hippocampus* **16:** 267–270.

Nakagawa S., Kim J.E., Lee R., Chen J., Fujioka T., Malberg J., Tsuji S., and Duman R.S. 2002a. Localization of phosphorylated cAMP response element-binding protein in immature neurons of adult hippocampus. *J. Neurosci.* **22:** 9868–9876.

Nakagawa S., Kim J.E., Lee R., Malberg J.E., Chen J., Steffen C., Zhang Y.J., Nestler E.J., and Duman R.S. 2002b. Regulation of neurogenesis in adult mouse hippocampus by cAMP and the cAMP response element-binding protein. *J. Neurosci.* **22:** 3673–3682.

Naylor A.S., Persson A.I., Eriksson P.S., Jonsdottir I.H., and Thorlin T. 2005. Extended voluntary running inhibits exercise-induced adult hippocampal progenitor proliferation in the spontaneously hypertensive rat. *J. Neurophysiol.* **93:** 2406–2414.

Nishiyama A., Watanabe M., Yang Z., and Bu J. 2002. Identity, distribution, and development of polydendrocytes: NG2-expressing glial cells. *J. Neurocytol.* **31:** 437–455.

Ormerod B.K., Lee T.T., and Galea L.A. 2003. Estradiol initially enhances but subsequently suppresses (via adrenal steroids) granule cell proliferation in the dentate gyrus of adult female rats. *J. Neurobiol.* **55:** 247–260.

Palmer T.D., Ray J., and Gage F.H. 1995. FGF-2-responsive neuronal progenitors reside in proliferative and quiescent regions of the adult rodent brain. *Mol. Cell. Neurosci.* **6:** 474–486.

Palmer T.D., Willhoite A.R., and Gage F.H. 2000. Vascular niche for adult hippocampal neurogenesis. *J. Comp. Neurol.* **425:** 479–494.

Parent J.M. and Lowenstein D.H. 2002. Seizure-induced neurogenesis: Are more new neurons good for an adult brain? *Prog. Brain Res.* **135:** 121–131.

Parent J.M., Yu T.W., Leibowitz R.T., Geschwind D.H., Sloviter R.S., and Lowenstein D.H. 1997. Dentate granule cell neurogenesis is increased by seizures and contributes to aberrant network reorganization in the adult rat hippocampus. *J. Neurosci.* **17:** 3727–3738.

Plumpe T., Ehninger D., Steiner B., Klempin F., Jessberger S., Brandt M., Romer B., Rodriguez G.R., Kronenberg G., and Kempermann G. 2006. Variability of doublecortin-associated dendrite maturation in adult hippocampal neurogenesis is independent of the regulation of precursor cell proliferation. *BMC Neurosci.* **7:** 77.

Ra S.M., Kim H., Jang M.H., Shin M.C., Lee T.H., Lim B.V., Kim C.J., Kim E.H., Kim K.M., and Kim S.S. 2002. Treadmill running and swimming increase cell proliferation in the hippocampal dentate gyrus of rats. *Neurosci. Lett.* **333:** 123–126.

Rando T.A. 2006. Stem cells, ageing and the quest for immortality. *Nature* **441:** 1080–1086.

Rochefort C., Gheusi G., Vincent J.D., and Lledo P.M. 2002. Enriched odor exposure increases the number of newborn neurons in the adult olfactory bulb and improves odor memory. *J. Neurosci.* **22:** 2679–2689.

Sapolsky R.M. 1999. Glucocorticoids, stress, and their adverse neurological effects: Relevance to aging. *Exp. Gerontol.* **34:** 721–732.

Satz P. 1993. Brain reserve capacity on symptom onset after brain injury: A formulation and review of evidence for threshold theory. *Neuropsychology* **7:** 273–295.

Schinder A.F. and Gage F.H. 2004. A hypothesis about the role of adult neurogenesis in hippocampal function. *Physiology* **19:** 253–261.

Schmidt-Hieber C., Jonas P., and Bischofberger J. 2004. Enhanced synaptic plasticity in newly generated granule cells of the adult hippocampus. *Nature* **429:** 184–187.

Seaberg R.M. and van der Kooy D. 2002. Adult rodent neurogenic regions: The ventricular subependyma contains neural stem cells, but the dentate gyrus contains restricted progenitors. *J. Neurosci.* **22:** 1784–1793.

Seki T. 2002. Expression patterns of immature neuronal markers PSA-NCAM, CRMP-4 and NeuroD in the hippocampus of young adult and aged rodents. *J. Neurosci. Res.* **70:** 327–334.

———. 2003. Microenvironmental elements supporting adult hippocampal neurogenesis. *Anat. Sci. Int.* **78:** 69–78.

Seki T. and Arai Y. 1995. Age-related production of new granule cells in the adult dentate gyrus. *Neuroreport* **6:** 2479–2482.

Shetty A.K., Hattiangady B., and Shetty G.A. 2005. Stem/progenitor cell proliferation factors FGF-2, IGF-1, and VEGF exhibit early decline during the course of aging in the hippocampus: role of astrocytes. *Glia* **51:** 173–186.

Simon M., Czeh B., and Fuchs E. 2005. Age-dependent susceptibility of adult hippocampal cell proliferation to chronic psychosocial stress. *Brain Res.* **1049:** 244–248.

Steiner B., Wolf S.A., and Kempermann G. 2006. Adult neurogenesis and neurodegenerative disorders. *Regen. Medicine* **1:** 15–28.

Stern Y. 2002. What is cognitive reserve? Theory and research application of the reserve concept. *J. Int. Neuropsychol. Soc.* **8:** 448–460.

Stranahan A.M., Khalil D., and Gould E. 2006. Social isolation delays the positive effects of running on adult neurogenesis. *Nat. Neurosci.* **9:** 526–533.

Sun L.Y., Evans M.S., Hsieh J., Panici J., and Bartke A. 2005. Increased neurogenesis in dentate gyrus of long-lived Ames dwarf mice. *Endocrinology* **146:** 1138–1144.

Tanapat P., Hastings N.B., Reeves A.J., and Gould E. 1999. Estrogen stimulates a transient increase in the number of new neurons in the dentate gyrus of the adult female rat. *J. Neurosci.* **19:** 5792–5801.

Tozuka Y., Fukuda S., Namba T., Seki T., and Hisatsune T. 2005. GABAergic excitation promotes neuronal differentiation in adult hippocampal progenitor cells. *Neuron* **47:** 803–815.

Trejo J.L., Carro E., and Torres-Aleman I. 2001. Circulating insulin-like growth factor I mediates exercise-induced increases in the number of new neurons in the adult hippocampus. *J. Neurosci.* **21:** 1628–1634.

Tropepe V., Craig C.G., Morshead C.M., and van der Kooy D. 1997. Transforming growth factor-alpha null and senescent mice show decreased neural progenitor cell proliferation in the forebrain subependyma. *J. Neurosci.* **17:** 7850–7859.

van Praag H., Kempermann G., and Gage F.H. 1999. Running increases cell proliferation and neurogenesis in the adult mouse dentate gyrus. *Nat. Neurosci.* **2:** 266–270.

Wagner J.P., Black I.B., and DiCicco-Bloom E. 1999. Stimulation of neonatal and adult brain neurogenesis by subcutaneous injection of basic fibroblast growth factor. *J. Neurosci.* **19:** 6006–6016.

Wang L.P., Kempermann G., and Kettenmann H. 2005. A subpopulation of precursor cells in the mouse dentate gyrus receives synaptic GABAergic input. *Mol. Cell. Neurosci.* **29:** 181–189.

Wang S., Scott B.W., and Wojtowicz J.M. 2000. Heterogeneous properties of dentate granule neurons in the adult rat. *J. Neurobiol.* **42:** 248–257.

Winner B., Cooper-Kuhn C.M., Aigner R., Winkler J., and Kuhn H.G. 2002. Long-term survival and cell death of newly generated neurons in the adult rat olfactory bulb. *Eur. J. Neurosci.* **16:** 1681–1689.

Wise P.M. 2003. Creating new neurons in old brains. *Sci. Aging Knowledge Environ.* **2003:** PE13.

Wiskott L., Rasch M.J., and Kempermann G. 2006. A functional hypothesis for adult hippocampal neurogenesis: Avoidance of catastrophic interference in the dentate gyrus. *Hippocampus* **16:** 329–343.

Wolf S.A., Kronenberg G., Lehmann K., Blankenship A., Overall R., Staufenbiel M., and Kempermann G. 2006. Cognitive and physical activity differently modulate disease progression in the amyloid precursor protein (APP)-23 model of Alzheimer's disease. *Biol. Psychiatry* **60:** 1314–1323.

Zhao C., Teng E.M., Summers R.G., Jr., Ming G.L., and Gage F.H. 2006. Distinct morphological stages of dentate granule neuron maturation in the adult mouse hippocampus. *J. Neurosci.* **26:** 3–11.

18

Regulation of Hippocampal Neurogenesis by Systemic Factors Including Stress, Glucocorticoids, Sleep, and Inflammation

Paul J. Lucassen and Charlotte A. Oomen
Centre for Neuroscience
Swammerdam Institute of Life Sciences
University of Amsterdam
Amsterdam, The Netherlands

Anne-Marie van Dam
Free University Medical Center
Department of Anatomy and Neurosciences
Amsterdam, The Netherlands

Boldizsár Czéh
Clinical Neurobiology Laboratory
German Primate Center
Göttingen, Germany

THIS CHAPTER SUMMARIZES AND DISCUSSES the regulation of adult neurogenesis and hippocampal cellular plasticity by systemic factors. We focus on the role of stress, glucocorticoids, and related factors such as sleep deprivation and inflammation.

THE CONCEPT OF STRESS

Ever present as stress may be in the modern Western society, it represents an old, yet essential, alarm system for an organism. By definition, stress systems are activated whenever a discrepancy occurs between an

Adult Neurogenesis ©2008 Cold Spring Harbor Laboratory Press 978-087969-784-6

organism's expectations and the reality it encounters, particularly when it involves a threat to the organism's homeostasis, well-being, or health.

Lack of information, loss of control, unpredictability, and uncertainty when faced with predator threat in animals or psychosocial demands in humans can all produce stress signals. The same holds for perturbations of a physical or biological nature, such as food shortage, injury, or inflammation. Various sensory and cognitive signals converge to activate a stress response that triggers several adaptive processes in the body and brain aimed to restore homeostasis.

THE STRESS RESPONSE

In mammals, the stress response develops in a stereotypic manner through three phases: (1) an initial alarm reaction, (2) resistance, and, only after prolonged exposure, (3) exhaustion. The first phase largely involves activation of the sympathoadrenal system through the rapid release of epinephrine and norepinephrine from the adrenal medulla; these hormones elevate basal metabolic rate and increase blood flow to vital organs such as the heart and muscles.

At a later stage, the limbic hypothalamus-pituitary-adrenal (HPA) system is activated, i.e., a classic neuroendocrine circuit in which limbic and hypothalamic brain structures integrate emotional, cognitive, neuroendocrine, and autonomic inputs that together determine the magnitude and duration of the organism's behavioral, neural, and hormonal responses to stress.

HPA Axis and Glucocorticoids

Stress-induced activation of the HPA axis involves the production of corticotropin-releasing hormone (CRH) in parvocellular neurons of the hypothalamic paraventricular nucleus (PVN) (Fig. 1). This induces adrenocorticotropic hormone (ACTH) release from the anterior pituitary gland, which in turn causes the release of glucocorticoids (GCs) (cortisol in primates, corticosterone in rodents) from the adrenal cortex into the general circulation. When stress is prolonged, the stimulatory effect on ACTH release is potentiated through coexpression of vasopressin (AVP) by the parvocellular CRH neurons.

Glucocorticoid Actions

Upon their release, GCs exert a wide range of effects. In the periphery, GCs mobilize energy by raising glucose levels. They further affect carbohydrate

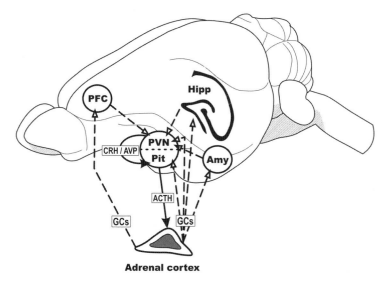

Adrenal cortex

Figure 1. Schematic drawing of the hypothalamus-pituitary-adrenal (HPA) axis and its inputs. The paraventricular nucleus (PVN) of the hypothalamus contains neuroendocrine neurons that secrete vasopressin (AVP) and corticotropin-releasing hormone (CRH). These peptides stimulate the pituitary gland to secrete adrenocorticotropic hormone (ACTH) which acts on the adrenal cortices that produce glucocorticoid (GC) hormones. These steroids exert negative feedback by binding to glucocorticoid receptors in the hypothalamus and pituitary to suppress CRH and ACTH production. Anatomical connections between limbic areas such as the prefrontal cortex, amygdala, hippocampus, and hypothalamus regulate HPA axis activation and are sensitive to GCs as well. (*Solid line arrows*) Stimulatory effects; (*dashed arrows*) mainly inhibitory effects. (Amy) Amygdala; (GCs) glucocorticoids; (Hipp) hippocampus; (PFC) prefrontal cortex; (Pit) pituitary gland.

and lipid metabolism and can have catabolic actions on muscle and bone tissues. They also exert "permissive" actions that enhance sympathoadrenal activity, among others. GCs respond more slowly than the fast-acting catecholamines and enable a continued sympathoadrenal activity (resistance). If stress becomes chronic, an imbalance may occur and GCs can exert deleterious effects by inducing muscle wasting, hyperglycemia (diabetes mellitus), gastrointestinal ulceration, and atrophy of the immune system (exhaustion phase).

Feedback Regulation

Although fast GC effects have also been reported (Karst et al. 2005), most GC effects involve slow genomic actions following activation of the mineralocorticoid receptor (MR) and glucocorticoid receptor (GR). The MR has a high affinity for aldosterone and corticosterone. It is highly expressed in the hippocampus, lateral septum, and amygdala. In contrast, the GR has a tenfold lower affinity for corticosterone and is ubiquitously distributed, with enrichments in the hippocampus, PVN, and pituitary, i.e., the main feedback sites through which GCs regulate their own release (de Kloet et al. 2005; Han et al. 2005). In addition, other brain regions such as the amygdala and prefrontal cortex modulate HPA feedback and (re)activity (Fig. 1).

Due to differences in affinity, the degree of MR and GR occupation depends on circulating GC levels, which fluctuate over the course of the day. Under rest, circulating GC levels are low and activate mainly the MR while occupying only a small fraction of the GRs. Only after stress, or at the circadian peak prior to the onset of the activity period, do GRs become activated. Changes in the degree of MR and GR activation influence gene expression that may result in persistent changes in electrophysiological properties of the hippocampal network. The variable MR/GR ratio is particularly relevant for neurons that express both receptor types, i.e., CA1 pyramidal neurons and dentate gyrus (DG) granular neurons. In the CA3 region, very few GRs are present. Thus, not only is the hippocampus very sensitive to circulating GC levels, it is also important in emotional processing and key aspects of learning and memory where adult neurogenesis has also been implicated (Kempermann and Gage 2002; de Kloet et al. 2005; Aimone et al. 2006; Leuner et al. 2006; Saxe et al. 2006).

Short-term exposure to stress induces behavioral adaptation and is considered harmless. Prolonged exposure to stress, however, may induce alterations in HPA feedback that can lead to overexposure of the brain and body to aberrant GC levels that form a risk factor for disease (de Kloet et al. 2005). Even though feedback is largely mediated through the GR, chronic stress may also alter function of the MR that is implicated in tonic inhibitory control of the HPA axis and modulates adult neurogenesis (Gass et al. 2000; Gesing et al. 2001; Fischer et al. 2002).

STRESS-INDUCED STRUCTURAL PLASTICITY IN THE HIPPOCAMPUS

Stress Effects on Hippocampal Function, Volume, Cell Number, and Apoptosis

Although in earlier studies, chronic stress was reported to induce hippocampal neuronal loss particularly in the CA3 subregion (Sapolsky et al. 1990), more recent studies failed to find evidence that stress induces massive loss of the principal hippocampal cells (Sousa et al. 1998; Lucassen and de Kloet 2001; Lucassen et al. 2001; Muller et al. 2001; O'Brien et al. 2001; Heine et al. 2004c) as reviewed recently (Lucassen et al. 2006). A wide range of studies have used state-of-the-art methodologies to address the consequences of chronic stress and hypercortisolism for hippocampal structure and function. GC excess is generally associated with deleterious alterations in hippocampal excitability, long-term potentiation, and hippocampus-related memory performance. However, positive effects of stress have also been described (Bartolomucci et al. 2002; Kim and Diamond 2002; Joels et al. 2006, 2007; Roozendaal et al. 2006).

When stress is prolonged, reductions in neuropil and hippocampal volume have been reported (Czéh et al. 2001; Sheline 2003; Fuchs et al. 2004). Magnetic resonance imaging (MRI) and morphometrical studies in tree shrews, for example, show that chronic psychosocial stress results in a mild reduction in hippocampal volume, i.e., about 10% (Czéh et al. 2001). Effects of stress on hippocampal volume and hippocampal cell number are relatively mild and subregion-specific and appear to occur already shortly after stressor onset, but prior to cognitive disturbances (Fuchs et al. 2006). As major loss in the CA and DG neuronal layers is not responsible, hippocampal volume changes after stress must be derived from other factors as discussed elsewhere (Czéh and Lucassen 2007).

Stress Effects on Dendritic Atrophy, Spine, and Synaptic Changes

Structural substrates for the stress-induced functional alterations are generally thought to involve axonal changes, synapse loss, alterations in postsynaptic densities, and dendritic reorganization. Chronic stress or corticosterone application initially induces atrophy of the apical dendrites of CA3c and to a lesser extent of CA1 cells and DG granule cells (Watanabe et al. 1992; Stewart et al. 2005; Donohue et al. 2006; Fuchs et al. 2006). Furthermore, alterations in CA3 synapses and in the morphology of their mossy fiber terminals have been described (Magarinos et al. 1997; Sousa et al. 2000; Sandi et al. 2003; Stewart et al. 2005; Tata et al. 2006). This synaptic remodeling may also involve cortical areas, where changes in cell adhesion molecule expression, attention, spatial memory, and fear conditioning often occur in parallel. The latter dendritic changes appear to occur early and last for long periods of time (Sandi 2004; Brown et al. 2005; Izquierdo et al. 2006; Radley et al. 2006).

STRESS AND ADULT NEUROGENESIS

Stress and GCs are among the most potent inhibitors of adult neurogenesis in the DG (Mirescu and Gould 2006) but not the subventricular zone (SVZ) or olfactory bulb (Brown et al. 2003). Adult neurogenesis has been observed in various species (Amrein et al. 2004), including primates (Gould et al. 2001) and humans (Eriksson et al. 1998), and effects of stress have so far been described in at least four different animal species—mice, rats, tree shrews, and marmoset monkeys—and appear to generalize across different paradigms regarding the direction of the effect. Psychosocial (Gould et al. 1997; Tanapat et al. 2001; Falconer and Galea 2003; Dong et al. 2004; Heine et al. 2004c; Oomen et al. 2007) or physical stressors (Malberg and Duman 2003; Pham et al. 2003; Vollmayr et al. 2003; Bain et al. 2004; Heine et al. 2004c; Nacher et al. 2004; Mirescu and Gould 2006) all seem to inhibit one or more phases of the adult neurogenesis process.

The phase that is affected appears to depend on the experimental design and the nature and timing of the stressor (see below). Of note, a considerable proportion of the adult-generated hippocampal cells die within a few days by apoptosis (Dayer et al. 2003), a process also regulated by stress and antidepressants (Lucassen et al. 2004, 2006). This close correlation between cell birth and death implies a continuous turnover among the newborn cells in the DG. Under stressful conditions, whether or not cell death among the older cells also increases and thereby stimulates a compensatory neurogenic effect awaits further research.

Importantly, although a temporary change in DG turnover rate following stress exposure may not directly influence total DG cell number, stress will thus change the overall composition, average age, identity, and projections of the DG cells, and may thereby affect the properties of the hippocampal memory circuit.

Glucocorticoids as Mediators

Effects of stress on cell proliferation and adult neurogenesis are largely mediated by GCs (Gould et al. 1991; Cameron and Gould 1994; Ambrogini et al. 2002; Montaron et al. 2003; Mayer et al. 2006), although other steroids have also been implicated (Mayo et al. 2001, 2005; Karishma and Herbert 2002; Suzuki et al. 2004; Ambrogini et al. 2005; Galea et al. 2006; Montero-Pedrazuela et al. 2006). Removal of the circulating adrenal steroids by adrenalectomy increases cell proliferation and adult neurogenesis in young and aged rodents (Cameron and Gould 1994; Cameron and

McKay 1999; Krugers et al. 2007). Increased GC levels inhibit not only proliferation, but also survival and differentiation of the newly formed cells (see Fig. 2) (Ambrogini et al. 2002; Montaron et al. 2003; Wong and Herbert 2004, 2006; Mayer et al. 2006). In support of this, manipulations that interfere with HPA axis activity, such as pharmacological blockade of corticotropin-releasing factor 1 (CRF1) and vasopressin (V1b) receptors (Alonso et al. 2004) or the GR, normalize stress or corticosterone-induced reductions in adult neurogenesis (see Fig. 2) (Wong and Herbert 2005; Mayer et al. 2006; Oomen et al. 2007).

Methodological Considerations

As outlined elsewhere, the study of adult neurogenesis requires a careful experimental design. Small differences in the concentration, timing, and survival after BrdU (bromodeoxyuridine) injection, for example, or the use of different immunocytochemical markers to identify adult neurogenesis can explain considerable discrepancies between otherwise seemingly similar paradigms (Gould et al. 1999; Greenough et al. 1999; van Praag et al. 1999a).

When studying effects of stress on adult neurogenesis, additional psychological variables, interindividual and gender differences in the behavioral reactivity to stress, and the coping ability of the animal will affect the impact of a stressor and hence the effect on adult neurogenesis. As handling itself already changes neuroendocrine characteristics of the animal, the inclusion of a handled control group not exposed to stress as well as naive controls is critical. In addition, given the circadian rhythm of endogenous corticosterone, the time of day and the state of the stressed animal at sacrifice are important (Goergen et al. 2002; Holmes et al. 2004; Kochman et al. 2006).

Seemingly irrelevant conditions such as the recent cleaning of its cage, individual housing, pain, or transport to the laboratory may represent a heterotypic stressor for an animal and exacerbate the negative consequences of stress (Lu et al. 2003; Duric and McCarson 2006). Altered GC levels may subsequently affect behavior or the response of running on adult neurogenesis (Stranahan et al. 2006). Conversely, the hippocampal activation required for water maze or fear learning, the associated rise in stress hormone levels, or the amount of physical exercise during such tasks can also influence adult neurogenesis (Dobrossy et al. 2003; Namestkova et al. 2005; Olariu et al. 2005; Pham et al. 2005; Bruel-Jungerman et al. 2006; Ehninger and Kempermann 2006).

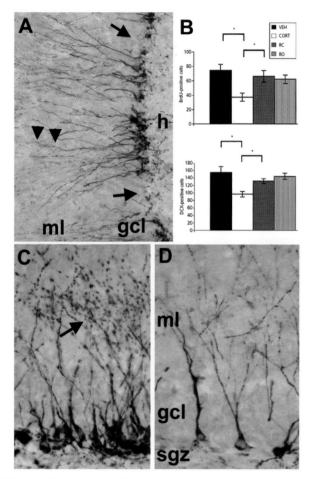

Figure 2. (*A*) Doublecortin (DCX) immunostaining of immature neurons in the adult hippocampal dentate gyrus (DG). DCX-positive somata are located in the subgranular zone at the border between the granule cell layer (GCL) and the hilus (h) with extensions (*arrowheads*) passing through the GCL and running into the molecular layer (ml). The presence of empty regions or "gaps" (*arrows*), where no DCX-positive cells are found, is more frequent in stressed animals. (*B*) Graph depicting quantification of BrdU- and DCX-positive cell numbers in the hippocampus of rats treated with vehicle (veh), corticosterone (cort), or corticosterone plus the GR antagonist mifepristone (RC), and with mifepristone alone (RO). The significant reduction in both BrdU- (21-day-old cells) and DCX-positive cell numbers after 21 days of corticosterone treatment is normalized by mifepristone treatment for the last 4 days, whereas the drug alone has no effect (Mayer et al. 2006). Similar results are obtained after chronic stress (C.A. Oomen et al. 2007). (*C,D*) Details of the different individual morphological patterns of DCX-positive cells showing strong immunostaining of long and complex processes extending into the ml (*arrow*) (*C*) versus regions with fewer cells and often shorter and less complex dendritic extensions (*D*). In chronically stressed animals, the latter type is more prominent.

Considerable interindividual differences in hippocampal cell proliferation can occur in resident-intruder paradigms that relate to inborn differences in behavioral traits such as defensive behavior and reaction to novelty or in biological traits such as HPA reactivity. Traits in reactivity may correlate with differences in corticosterone secretion and with basal adult neurogenesis rates (Lemaire et al. 1999; Kozorovitskiy and Gould 2004; Abrous et al. 2005; Mitra et al. 2006) that can be influenced by early life stressors (Caldji et al. 2000; Seckl 2001; Neumann et al. 2005).

Many studies on adult neurogenesis have employed stressors that are both uncontrollable and unpredictable. Hierarchy-based paradigms are clear examples of uncontrollable stress (Kozorovitskiy and Gould 2004; Fuchs et al. 2006) as is the "learned helplessness" model (Malberg and Duman 2003). Predictability and controllability, however, strongly influence the negative consequences of stress (Shors et al. 1989, 2007). An uncontrollable stressor generally induces stronger HPA axis activation than controllable stressors. Chronic restraint is, for example, an uncontrollable and strong, but very predictable, stressor, and when applied for prolonged periods, chronic restraint causes adaptation of the stress response (Pham et al. 2003). Active avoidance learning, in contrast, is stressful but controllable and does not affect DG proliferation (Van der Borght et al. 2005).

Social isolation stress and restraint stress further have a different impact on male versus female animals (Shors et al. 2001; Westenbroek et al. 2004). In male, but not female, rats, exposure to predator odor reduces newborn cell number (Tanapat et al. 2001; Falconer and Galea 2003; Mirescu et al. 2004). These sex differences, and related ones in other hippocampal parameters, may be attributed to the effects of estradiol (Tanapat et al. 1999; Ormerod and Galea 2001; Ormerod et al. 2003, 2004; Lagace et al. 2007; Shors et al. 2007). In contrast, androgens do not seem to affect proliferation in stressed male rats (Kambo and Galea 2006). In addition, Galea and McEwen (1999) have found seasonal effects on adult neurogenesis associated with fluctuations in steroid levels, whereas Lavenex et al. (2000) have not. In conclusion, the study of stress effects on adult neurogenesis requires properly controlled and timed experimental designs. Differences therein may explain some of the present discrepancies in the literature.

Acute Stress and Adult Neurogenesis

With few exceptions (Pham et al. 2003; Nacher et al. 2004; Thomas et al. 2006), most studies in rats report reductions in the proliferation of adult-generated cells after a short exposure to a diversity of stressors, e.g., predator odor, forced swim, or immobilization (Falconer and Galea

2003; Malberg and Duman 2003; Heine et al. 2004c; Hill et al. 2006b). In other species, a single exposure to a dominant male inhibits the number of dividing cells in the DG of subordinate tree shrews (Gould et al. 1997) and marmoset monkeys (Gould et al. 1998). Of interest, the effect of such stressors appear to be transient, at least in rats, as they are normalized again after 24 hours of recovery (Heine et al. 2004c), thus confirming the idea that the influence of acute stress is short-lived. Alternatively, diminished DG proliferation resulting in decreased immature neuron production may be followed by a period of enhanced cell survival, leaving the ultimate number of new neurons unchanged (Gould and Tanapat 1999; Fuchs et al. 2006).

The differential effects of acute stress on proliferation (Heine et al. 2004c; Nacher et al. 2004; Thomas et al. 2006) may relate to the experimental design. When BrdU injections are applied concurrently with exposure to the predator odor TMT (trimethylthiazoline), no effect is seen on proliferation studied 2 hours later, despite elevated corticosterone plasma levels and behavioral evidence of stress (Thomas et al. 2006). Although similar discrepancies between stress and DG proliferation have been observed (Nacher et al. 2004; Krugers et al. 2007), these differ from studies in which animals were exposed to the same stressor but sacrificed 24 hours later. In this case, TMT was found to significantly suppress DG cell proliferation, albeit with a clear difference between the sexes (Falconer and Galea 2003; Hill et al. 2006a).

Chronic and Repeated Stress

Repeated and prolonged (3 weeks) but not acute (2 or 6 hours) restraint stress decreased cell proliferation in rats (Pham et al. 2003; Xu et al. 2006) but up-regulated PSA-NCAM (polysialic-acid-linked neural cell adhesion molecule) expression (Nacher et al. 2004). When using a 3–4-week chronic unpredictable stress paradigm, structural parameters such as cell number and volume of the granule cell layer (GCL) were not affected, but cell proliferation (Heine et al. 2004c) and adult neurogenesis were reduced (Oomen et al. 2007). In chronically stressed tree shrews, cytogenesis in the DG was also reduced (Czéh et al. 2001). In a different paradigm of chronic mild stress in rodents, cell proliferation was reduced (Alonso et al. 2004; Jayatissa et al. 2006), whereas others only found the survival (and not proliferation) phase to be reduced (Lee et al. 2006).

In the "learned helplessness" model of chronic uncontrollable stress (Malberg and Duman 2003; Henn and Vollmayr 2005), mice or rats are exposed to a series of inescapable stressors that induce behavioral despair and disable the animal's ability to escape other stressors days later (Malberg and Duman 2003). Considerable interindividual variation exists in the behavioral responses to "learned helplessness," but the effect on adult neurogenesis was the same in rats that developed the syndrome and those that did not (Vollmayr et al. 2003). Prolonged stress or hypercortisolism can further reduce survival and neuronal differentiation of newborn cells (Wong and Herbert 2004, 2006; Mayer et al. 2006; Oomen et al. 2007). These effects are mediated through the serotonin 1A receptor as well as through GRs (Abrous et al. 2005; Huang and Herbert 2005; Wong and Herbert 2005).

In contrast to the abundant data available on rats, neurogenesis studies on mice have only recently become available. Although one study reports a stimulatory effect of acute restraint stress on proliferation in mice (Bain et al. 2004), most studies report reductions in cell proliferation and/or adult neurogenesis following acute or prolonged stressors (Alonso et al. 2004; Ehninger and Kempermann 2006; Mitra et al. 2006; Yap et al. 2006; Veenema et al. 2007). Social stress in a resident-intruder mouse model also reduced hippocampal cell proliferation (Mitra et al. 2006) but did not affect proliferation in the amygdala, thus suggesting that stress effects on neurogenesis are comparable in rats and mice.

Stress and Aging

Aging is accompanied by a strong reduction in the rate of adult neurogenesis, and in at least some, but not all, rat strains, this is paralleled by an age-related dysregulation of the HPA axis and elevated corticosterone levels (Kuhn et al. 1996; Heine et al. 2004a; Rao et al. 2005, 2006). Although some studies suggest that the extent of lifetime exposure to GCs determines the extent of age-related decline in hippocampal neurogenesis, others do not support a simple inverse relationship between chronic plasma GC levels and age-dependent granule cell neurogenesis (Heine et al. 2004a; Brunson et al. 2005; Montaron et al. 2006). Chronic stress does appear to have an age-dependent effect on the suppression of DG proliferation, in such a way that middle-aged and old animals are generally more vulnerable to the inhibitory effects of stress (Simon et al. 2005).

Perinatal Stress

In humans, early life stressors are among the strongest predisposing factors for disorders such as major depression. In rodents, many of the effects of stress are much more pronounced when studied in juvenile or perinatal animals. The juvenile and early perinatal phase, in particular, appears to offer a time window in which maternal factors and corticosterone can induce "organizing" effects on the rate of adult neurogenesis, stress reactivity, and behavior in the offspring that can last throughout life and correlate with hippocampal function (Caldji et al. 2000; Seckl 2001; Fenoglio et al. 2006). Restraint stress of pregnant rats or macaques, for example, affects HPA axis activity and induces lifelong reductions in DG proliferation in the offspring for life (Lemaire et al. 2000; Coe et al. 2003; Bosch et al. 2006). In rats, this suppression in adult neurogenesis was paralleled by hippocampal learning impairments and could be normalized by postnatal handling and stimulation of the pups (Lemaire et al. 2006). A chronic stress regime in pregnant mice induces increased anxiety-related behavior, prolonged HPA axis dysfunctioning, and sleep disturbances in the adult offspring. Likewise, repeated maternal separation during the early postnatal period inhibits cell proliferation and the production of immature neurons in the DG of adult offspring (Mirescu et al. 2004); other investigators could not find such differences (Greisen et al. 2005). Changes in cell death after maternal separation have also been observed (Zhang et al. 2002). Isolation stress induced by individually housing mice from weaning onward was further shown to decrease hippocampal proliferation at 6 months of age (Dong et al. 2004).

Sleep Deprivation as a Stressor

In addition to exercise and stress, sleep can potently modulate adult neurogenesis. Of interest, one night of sleep deprivation (SD) can improve mood in depressed patients for at least 1 day. In animals, acute SD upregulates adult hippocampal neurogenesis. Together with the stimulatory effects of many antidepressive drugs, it has been proposed that a neurogenic response to sleep could be implicated in the beneficial effects elicited in depressed patients (Grassi Zucconi et al. 2006).

Other investigators, however, have found reductions in proliferation after 24 hours of SD (Roman et al. 2005) and also in the number of new neurons in the DG of adult rats with longer periods of SD (Guzman-Marin et al. 2005; Roman et al. 2005; Tung et al. 2005; Mirescu et al. 2006). Similarly, sleep restriction is known to selectively impair spatial learning, whereas performance in a nonspatial task was improved. SD also impaired hippocampus-dependent learning in object-recognition tasks (Palchykova et al. 2006) and abolished learning-induced adult neurogenesis (Hairston et al. 2005).

Given the altered stress responses and GC levels associated with SD (Meerlo et al. 2002; Mirescu et al. 2006; Sgoifo et al. 2006) that may become worse when SD is prolonged, a lack of sleep is considered stressful. SD inhibits adult neurogenesis at a time when circulating levels of corticosterone are elevated. Reducing corticosterone prevents SD-induced reductions in proliferation, thus suggesting that SD inhibits adult neurogenesis by acting as a stressor. Indeed, SD has been associated with adverse consequences for cognition and susceptability to depressive symptoms. In contrast, other effects of SD occurring later in time do not seem to be GC-dependent (Mirescu et al. 2006).

Inflammation as a Stressor: Microglia

The HPA axis is activated not only after stress, but also during disease processes. Particularly when fever and inflammatory processes are involved, cells of the immune system produce proinflammatory cytokines such as interleukins. Interleukin-1β (IL-1β) and interleukin-6 (IL-6) elicit a variety of (patho)physiological reactions during inflammation and injury that together coordinate the "nonspecific symptoms of sickness." These symptoms include the activation of the HPA axis through enhanced ACTH release, which results in a rise in GC levels. Elevated GC levels have a strong immunosuppressive effect that prevents the immune system from overshooting. Thus, a clear bidirectional communication exists between the immune and neuroendocrine system.

In addition, interleukins are produced within the brain under conditions such as ischemia, Alzheimer's disease, multiple sclerosis, and epilepsy and in disease models such as adjuvant arthritis, which is associated with GC overexposure. In most situations, microglial cells are the main source of these interleukins. Activated microglial cells and the subsequent production of inflammatory mediators are generally considered detrimental for neuronal viability under pathological conditions. A role for microglia and interleukins has also been implicated in processes such as brain plasticity following stress exposure (Bartolomucci et al. 2003). Hence, neuroinflammation, defined by an activation of microglia and the presence of proinflammatory mediators, forms a stressor for the brain that may affect adult neurogenesis. Moreover, synthetic GCs such as prednisone and dexamethasone are often prescribed for the treatment of inflammatory disorders. They not only block prostaglandin and cytokine synthesis and hamper leukocyte infiltration in affected tissues, but also inhibit adult neurogenesis, thus suggesting an indirect, GC-mediated, inhibitory effect of neuroinflammation on neurogenesis.

Inflammation and cytokine expression can affect adult neurogenesis directly (Vallieres et al. 2002; Monje et al. 2003), whereas immune modulators such as transforming growth factor-β (TGF-β) (Wachs et al. 2006) have a (concentration-dependent) neurogenic potential in the adult brain (Battista et al. 2006). Other proinflammatory cytokines such as tumor necrosis factor-α (TNF-α) (Iosif et al. 2006) or interferon-α decrease adult neurogenesis through modulation of IL-1 (Kaneko et al. 2006). In addition, impairment of IL-1β action prevents the attenuated rate of adult neurogenesis in response to stress. This supports the idea that proinflammatory mediators in the brain can have a role in restricting adult neurogenesis. On the other hand, increased neurogenesis has been shown to occur inside inflammatory lesions in the spinal cord of an experimental multiple sclerosis model (Danilov et al. 2006), and complement activation products can promote basal levels of adult neurogenesis (Rahpeymai et al. 2006). Local cues in the brain may thus determine whether adult neurogenesis is impaired or stimulated.

Conversely, selective factors capable of affecting cell genesis can, in parallel, also influence microglial activation. As part of the neuroinflammatory response, activated microglia modulates the neurogenic niche and can be detrimental to the survival of new hippocampal neurons. Dependent on whether they are activated by IL-4 or by IFN-γ, microglial cells can differentially induce oligodendrogenesis and neurogenesis, respectively, from adult stem/progenitor cells (Butovsky et al. 2006).

Reducing neuroinflammation, i.e., by proinflammatory mediators and/or microglia activation, by different classes of drugs including COX-2 (cyclooxygenase) inhibitors, indomethacin, or the antibiotic minocycline, can restore or increase adult neurogenesis in multiple sclerosis models or after different types of hippocampal insults, cerebral ischemia, or radiation (Ekdahl et al. 2003; Monje et al. 2003; Aharoni et al. 2005; Hoehn et al. 2005). Of interest, recently identified central nervous system (CNS)-specific autoimmune T cells seem to act in concert with resident microglia and underlie, in part, hippocampal plasticity and learning through an effect on progenitor cell proliferation (Ziv et al. 2006).

The general picture emerging from these studies is that inflammatory responses in the brain influence adult neurogenesis, which may be mediated through elevated GC levels. In addition to HPA changes and alterations in neurogenesis, inflammatory mediators have also been implicated in the pathophysiology of depression (Hayley et al. 2005; Pace et al. 2006).

Putative Regulatory Mechanisms

Direct Mechanisms

On the basis of the different properties of MR and GR (de Kloet et al. 2005), the suppressing effects of GCs on adult neurogenesis are most likely mediated through GRs. Stress-induced reductions in the number of proliferating cells could result from a GR-mediated loss of progenitor cells or from a cell cycle arrest. Indeed, reduced proliferation after acute stress in the subgranular zone (SGZ) was paralleled by increased numbers of apoptotic cells, but as the phenotype of these apoptotic cells could not be established, it remains unclear here whether new cells or mature cells are preferentially affected by stress. After 3 weeks of chronic stress, proliferation and apoptosis were both reduced, indicating that the turnover of the DG had slowed down (Heine et al. 2004c). Of interest, chronic stress not only affected DG turnover, but concomitantly elevated levels of p27Kip1, an endogenous cell cycle inhibitor, in the SGZ, thus suggesting that stress may prevent progenitors from reentering the cell cycle (Heine et al. 2004b).

The regulation of individual precursor cells by stress may occur by different means: directly through specific receptors presumably located on the precursor cell themselves, indirectly through changes in neighboring cells, or through changes in the local environment that involve alterations in growth factor levels (Kuhn et al. 1997; Heine et al. 2005). Given the variations in sensitivity of progenitor cells versus mature neurons, changes in the local environment may differentially influence GR/MR ratio and thereby affect different stages of the adult neurogenesis process. In young mice, only a quarter of the dividing cells express GR and little or no MR. After 4 weeks, the majority of the newborn cells express both GR and MR (Cameron et al. 1993; Garcia et al. 2004). Newborn cells express low and variable levels of GR; therefore, the effect of stress is dependent on the age of the individual newborn cell.

Indirect Mechanisms

Glutamate/NMDA receptors. Indirect mechanisms, in contrast, are more likely to impact the entire population of progenitor cells in the DG. Stress increases glutamate release throughout the hippocampus (Abraham et al. 1998), and several lines of evidence suggest that enhanced excitatory neurotransmission affects dendritic remodeling and reduces adult neurogenesis. In both developing and adult animals, NMDA (*N*-methyl-D-aspartate) receptor activation reduces the number of new cells in the DG, whereas a blockade of NMDA receptors or a lesion of the entorhinal cortex, which provides the major excitatory input to the DG, has the opposite effect on cell proliferation (Cameron et al. 1998; Nacher et al. 2003). Recent studies indicate that NR1 and NR2B are expressed in some proliferating cells and in glial fibrillary acidic protein (GFAP)-positive cells but are rarely found in granule cells younger than 60 hours and are not expressed in transiently amplifying progenitors (type 2–3 cells). These results demonstrate that adult hippocampal neurogenesis may be regulated by NMDA receptors present in precursor cells and in differentiating granule neurons (Nacher et al. 2007).

Other glutamate receptors have also been studied with respect to stress including the AMPA (α-amino-3-hydroxy-5-methyl-4-isoazole) receptor. Differential changes occur in the assembly of AMPA receptor subunits and isoforms, as well as in the glial glutamate transporters, that are stressor- and hippocampal-subregion-dependent and may underlie neuroplastic changes induced by stress (Fuchs et al. 2004; Reagan et al. 2004). Of interest, selective AMPA receptor potentiators were found to increase adult hippocampal proliferation (Bai et al. 2003).

Growth factors. Additional factors mediating the effects of corticosterone include insulin-like growth factor-1 (IGF-1), brain-derived neurotrophic factor (BDNF), and epidermal growth factor (EGF) (Kuhn et al. 1997; Aberg et al. 2000). Stress and GCs also influence vascular dynamics and may, at least in part, exert their effects on neurogenesis by altering endothelial or smooth-muscle status. For example, GCs can down-regulate vascular endothelial growth factor (VEGF) receptors as well as decrease the expression of VEGF in models of CNS tumor angiogenesis and vascular cell interaction during angiogenesis. Pericytes are particularly interesting in this respect as they have a key role in endothelial activation during angiogenesis. Interestingly, pericytes respond to and produce many of the factors that have a role in both angiogenesis and adult neurogenesis such as IGF-1, VEGF, platelet-derived growth factor (PDGF), and fibroblast growth factor-2 (FGF-2) (Jin et al. 2002; Heine et al. 2005).

Vasculature/glia. The proximity of the precursors to blood vessels and the response of mammalian neural precursors to endothelial BDNF (Leventhal et al. 1999; Palmer et al. 2000) suggest that precursors communicate intimately with cells of the vasculature. However, there is no evidence that the small capillaries lack a blood-brain barrier formed by pericytes and astroglial end-feet. This suggests that vascular effects may be "translated" by other cell types within the neurogenic niche (see Chapter 11). Not withstanding their proposed role as stem cells, astrocytes are commonly known for their role in the formation of the blood-brain barrier. Astrocytic end-feet envelop the capillary bed and selectively transport proteins and metabolites to and from the brain parenchyma. Recent work demonstrates that cultured astrocytes support the proliferation and survival of developing neurons and can stimulate the production of neurons from SVZ precursors. Taken together, this suggests that astrocytes are a prime candidate for mediating vascular cues in the context of stress. Moreover, astrocytes themselves possess GRs and are significantly affected by stress (Czéh et al. 2006).

PARADOXICAL FINDINGS AND REMAINING ISSUES

Correlations between Glucocorticoid Levels and Neurogenesis

In contrast to the general picture emerging from the literature on the general inhibitory effects of stress on adult neurogenesis, there are also exceptions. Certainly not all physiological processes that elevate GCs result in a decrease in adult neurogenesis, and an overlapping time window is not always present. Despite the fact that rises in GC levels are often paralleled by an inhibition of adult neurogenesis and by decreases in memory function, increases in learning and memory parameters can also occur after stress. Issues of timing and context specificity appear to be critical here regarding specific phases of memory, as discussed in detail elsewhere (Joels et al. 2006, 2007).

In addition, several examples of a persistent and long-lasting inhibition of adult neurogenesis exist, despite normalized GC levels. Early life stressors decrease neurogenesis or proliferation not only at the time of stress (Tanapat et al. 1998; Zhang et al. 2002), but also into adulthood when basal levels of adrenal steroids are normal again (Lemaire et al. 2000; Mirescu et al. 2004). Similar findings were shown in the "learned helplessness" paradigm for adult animals. Inescapable shock (IS) exposure decreased DG proliferation. When the analysis was conducted 9 days after IS, a significant decrease in proliferation persisted, despite the fact that no difference was present in corticosterone levels. This indicates that reductions in proliferation are not directly linked to increased levels of adrenal steroids (Malberg and Duman 2003).

These and other findings (Nacher et al. 2004; Van der Borght et al. 2005; Thomas et al. 2006) suggest that although GCs may be involved in the initial suppression of cell proliferation, and particularly in early life, they are not always necessary for the maintenance of this effect. Alternatively, it is possible that GCs continue to be involved in reducing cell proliferation but that this influence is not detectable by merely examining total peripheral levels of GCs; changes in mediators of steroid action such as corticosteroid-binding globulin (CBG) or 11-Beta-hydroxysteroid dehydrogenase (11BOHSD) levels may have been induced.

Exercise and Learning

Paradoxically, living in an enriched environment (Moncek et al. 2004) and training on specific learning paradigms (Leuner et al. 2004) increase circulating GC levels. Yet, such conditions are often but not always (Boekhoorn et al. 2006b) associated with an enhancement of adult hippocampal and SVZ neurogenesis (Brown et al. 2003) by enhancing survival of new cells (Kempermann et al. 1997, 2002; Nilsson et al. 1999). Apparently, the positive influences of these experiences on adult neurogenesis are so strong that either they override the negative effects of elevated glucocorticoids or they take place in an entirely different context.

Another example of an unexpected relationship between circulating GC levels and cell proliferation occurs with physical activity. Running is a well-known stimulus for adult neurogenesis (van Praag et al. 1999b) that is accompanied by pronounced increases in circulating GCs. Previously sedentary animals elevate their GC levels shortly after running, a response that does not adapt as the animals become more physically fit. This suggests that the positive effects on adult neurogenesis may not be as profound as those observed after negative stressors.

Another interesting contradiction exists in the direction of the effect of running on adult neurogenesis when different durations of voluntary running are taken into account. In rats, short-term voluntary running for 9 days potently stimulated dentate neurogenesis. In contrast, long-term running for 24 days induced a strong down-regulation of progenitor proliferation rate to 50% of nonrunning controls. This finding was paralleled by the development of a strong activation of the HPA axis while the vasculature and the opioid system are also involved (Hagg et al. 2004; Persson et al. 2004; Naylor et al. 2005). Athough exercise is generally associated with beneficial changes, also in depression (Ernst et al. 2006), prolonged running can develop into a stressor overruling its positive effects on adult neurogenesis. Together with other data showing that the effect of exercise (Brown et al. 2003; Olson et al. 2006) is modulated by both circadian phase and the amount of daily exercise (Holmes et al. 2004), these data suggest that running-induced factors that stimulate neurogenesis can only be effective when HPA axis activation is minimal.

Stress-induced Changes in Adult Neurogenesis: Relevant for Depression?

Stress and elevated GC levels have important roles in various clinical disorders. Alterations of HPA axis (re)activity are common in major depression and posttraumatic stress disorder (PTSD), among others (de Kloet et al. 2005; Swaab et al. 2005). The set point of HPA axis activity appears to be programmed by genotype, but it can be altered by developmental influences or chronic stress early or later in life (Kendler et al. 1999; Seckl 2001). Many studies have implicated adult neurogenesis in depression and antidepressant drug action, although exceptions also exist (Henn and Vollmayr 2004; Feldmann et al. 2006; Czéh and Lucassen 2007). Moreover, apoptosis, glia, dendritic complexity, spine density, various plasticity-related proteins, and hippocampal volume are also influenced by these drugs (Norrholm and Ouimet 2001; Lucassen et al. 2004; Czéh et al. 2006; Sairanen et al. 2007). For further details, see other chapters in this volume.

Restoration of Stress-induced Reductions in Neurogenesis

The demonstration that chronic, but not acute, treatment with antidepressant drugs or electroconvulsive stimulation can stimulate neurogenesis in the adult DG (Malberg et al. 2000) has been followed quickly by compelling experimental evidence (generated exclusively from preclinical studies) that stress-induced reductions of adult DG neurogenesis can be reversed by antidepressant treatment (Dranovsky and Hen 2006; Warner-Schmidt and Duman 2006). Examples include the selective serotonin reuptake inhibitors (SSRIs) fluoxetine (Malberg and Duman 2003) and escitalopram or the tricyclic antidepressants (TCAs) such as clomipramine and imipramine, and tianeptine (Czéh et al. 2001), a unique compound described as a selective serotonin reuptake enhancer. It should be noted that beside these classic antidepressant drugs, atypical antipsychotics such as quetiapine and venlafaxine or GR antagonists such as mifepristone are also effective in preventing chronic stress-induced suppression in adult cytogenesis and neurogenesis (Mayer et al. 2006; Oomen et al. 2007). In addition to the chronic stress paradigms, olfactory bulbectomy, another model of depression, results in a significant reduction of dentate cell proliferation that can be normalized by imipramine treatment (Jaako-Movits et al. 2006).

These findings have strengthened the concept that alterations in adult cytogenesis and neurogenesis after stress may contribute to the pathogenesis of depression (Jacobs et al. 2000; Duman 2004). Together with the demonstration that the behavioral effects of antidepressants require adult dentate neurogenesis (Santarelli et al. 2003), the promotion of hippocampal neurogenesis has been regarded as a promising strategy for identifying new antidepressant drug targets. Accordingly, several potential antidepressant compounds, such as selective neurokinin-1 (NK1) receptor antagonists (Czéh et al. 2005), corticotropin-releasing factor (CRF1), and vasopressin (V1b) receptor antagonist (Alonso et al. 2004), have been tested in chronic stress paradigms. All of these novel compounds could prevent or normalize the inhibitory effects of stress on adult cell proliferation. So far, only repetitive transcranial magnetic stimulation (rTMS), a possible therapeutic intervention for major depression, appeared to be completely ineffective in reversing the stress-induced reduction of neurogenesis (Czéh et al. 2002). However, rTMS has a limited brain penetrance and may not target the hippocampus.

It should be emphasized that the functional significance of the newly generated neurons in the etiology of depression, or in the hippocampal volume changes, is still subject to debate (Henn and Vollmayr 2004; Czéh and Lucassen 2007). Convincing evidence for a central role of adult neurogenesis in depression would require its direct examination in the brains of depressed patients. Only one study (Reif et al. 2006) has so far examined cell proliferation in the postmortem hippocampus of depressed subjects, and although they confirm its very rare occurrence in human brain (Eriksson et al. 1998; Boekhoorn et al. 2006a), no changes were found in depression. Future studies in which adult neurogenesis can be visualized in live subjects using, for example, positron emission tomography imaging might address these issues (Shapiro et al. 2006).

We conclude that altered adult neurogenesis may contribute to the etiology of stress-related disorders such as depression, to antidepressant drug action, and to the cognitive symptoms occurring in depression.

SUMMARY/CONCLUDING REMARKS

In summary, there is now compelling experimental evidence that stress, through both hormonal and environmental components, can potently inhibit different phases of neurogenesis in the adult DG. Yet, the functional impact of this process on cognition or behavior remains unclear. A reduced rate of adult neurogenesis is most commonly interpreted as indicative of impaired hippocampal plasticity. This reflects an adaptive response as stress-induced reductions in neurogenesis can normalize again, for example, after appropiate recovery periods, exercise, or antidepressant drug treatment. On the basis of the assumption that adult neurogenesis is implicated in various learning processes, reductions in neurogenesis and in DG turnover will change the overall composition, average age, and identity of the DG cells, and thereby affect the properties of the hippocampal memory circuit, which may, over time, contribute to the cognitive impairment in stress-related human disorders. Another possibility is that the slower neuronal renewal reflects an adaptive response by which the system tries to "protect" itself from unwanted memories during stressful life events. Whether stress-induced reductions in adult neurogenesis occur in humans awaits further investigation.

ACKNOWLEDGMENTS

We thank Ms. A. Thompson (SILS-CNS, UvA) for her corrections to the English language. We are grateful to J. Müller-Keuker (DPZ Gottingen) for her help in the preparation of the HPA-axis figure. P.J.L. is supported by the VolkswagenStiftung Germany, the EU, and the Nederlandse HersenStichting.

REFERENCES

Aberg M.A., Aberg N.D., Hedbacker H., Oscarsson J., and Eriksson P.S. 2000. Peripheral infusion of IGF-I selectively induces neurogenesis in the adult rat hippocampus. *J. Neurosci.* **20:** 2896–2903.

Abraham I., Juhasz G., Kekesi K.A., and Kovacs K.J. 1998. Corticosterone peak is responsible for stress-induced elevation of glutamate in the hippocampus. *Stress* **2:** 171–181.

Abrous D.N., Koehl M., and Le Moal M. 2005. Adult neurogenesis: From precursors to network and physiology. *Physiol. Rev.* **85:** 523–569.

Aharoni R., Arnon R., and Eilam R. 2005. Neurogenesis and neuroprotection induced by peripheral immunomodulatory treatment of experimental autoimmune encephalomyelitis. *J. Neurosci.* **25:** 8217–8228.

Aimone J.B., Wiles J., and Gage F.H. 2006. Potential role for adult neurogenesis in the encoding of time in new memories. *Nat. Neurosci.* **9:** 723–727.

Alonso R., Griebel G., Pavone G., Stemmelin J., Le Fur G., and Soubrie P. 2004. Blockade of CRF(1) or V(1b) receptors reverses stress-induced suppression of neurogenesis in a mouse model of depression. *Mol. Psychiatry* **9:** 278–286.

Ambrogini P., Orsini L., Mancini C., Ferri P., Barbanti I., and Cuppini R. 2002. Persistently high corticosterone levels but not normal circadian fluctuations of the hormone affect cell proliferation in the adult rat dentate gyrus. *Neuroendocrinology* **76:** 366–372.

Ambrogini P., Cuppini R., Ferri P., Mancini C., Ciaroni S., Voci A., Gerdoni E., and Gallo G. 2005. Thyroid hormones affect neurogenesis in the dentate gyrus of adult rat. *Neuroendocrinology* **81:** 244–253.

Amrein I., Slomianka L., Poletaeva I.I., Bologova N.V., and Lipp H.P. 2004. Marked species and age-dependent differences in cell proliferation and neurogenesis in the hippocampus of wild-living rodents. *Hippocampus* **14:** 1000–1010.

Bai F., Bergeron M., and Nelson D.L. 2003. Chronic AMPA receptor potentiator (LY451646) treatment increases cell proliferation in adult rat hippocampus. *Neuropharmacology* **44:** 1013–1021.

Bain M.J., Dwyer S.M., and Rusak B. 2004. Restraint stress affects hippocampal cell proliferation differently in rats and mice. *Neurosci. Lett.* **368:** 7–10.

Bartolomucci A., de Biurrun G., Czeh B., van Kampen M., and Fuchs E. 2002. Selective enhancement of spatial learning under chronic psychosocial stress. *Eur. J. Neurosci.* **15:** 1863–1866.

Bartolomucci A., Palanza P., Parmigiani S., Pederzani T., Merlot E., Neveu P.J., and Dantzer R. 2003. Chronic psychosocial stress down-regulates central cytokines mRNA. *Brain Res. Bull.* **62:** 173–178.

Battista D., Ferrari C.C., Gage F.H., and Pitossi F.J. 2006. Neurogenic niche modulation by activated microglia: Transforming growth factor beta increases neurogenesis in the adult dentate gyrus. *Eur. J. Neurosci.* **23:** 83–93.

Boekhoorn K., Joels M., and Lucassen P.J. 2006a. Increased proliferation reflects glial and vascular-associated changes, but not neurogenesis in the presenile Alzheimer hippocampus. *Neurobiol. Dis.* **24:** 1–14.

Boekhoorn K., Terwel D., Biemans B., Borghgraef P., Wiegert O., Ramakers G.J., de Vos K., Krugers H., Tomiyama T., Mori H., Joels M., van Leuven F., and Lucassen P.J. 2006b. Improved long-term potentiation and memory in young tau-P301L transgenic mice before onset of hyperphosphorylation and tauopathy. *J. Neurosci.* **26:** 3514–3523.

Bosch O.J., Kromer S.A., and Neumann I.D. 2006. Prenatal stress: Opposite effects on anxiety and hypothalamic expression of vasopressin and corticotropin-releasing hormone in rats selectively bred for high and low anxiety. *Eur. J. Neurosci.* **23:** 541–551.

Brown J., Cooper-Kuhn C.M., Kempermann G., van Praag H., Winkler J., Gage F.H., and Kuhn H.G. 2003. Enriched environment and physical activity stimulate hippocampal but not olfactory bulb neurogenesis. *Eur. J. Neurosci.* **17:** 2042–2046.

Brown S.M., Henning S., and Wellman C.L. 2005. Mild, short-term stress alters dendritic morphology in rat medial prefrontal cortex. *Cereb. Cortex* **15:** 1714–1722.

Bruel-Jungerman E., Davis S., Rampon C., and Laroche S. 2006. Long-term potentiation enhances neurogenesis in the adult dentate gyrus. *J. Neurosci.* **26:** 5888–5893.

Brunson K.L., Baram T.Z., and Bender R.A. 2005. Hippocampal neurogenesis is not enhanced by lifelong reduction of glucocorticoid levels. *Hippocampus* **15:** 491–501.

Butovsky O., Ziv Y., Schwartz A., Landa G., Talpalar A.E., Pluchino S., Martino G., and Schwartz M. 2006. Microglia activated by IL-4 or IFN-gamma differentially induce neurogenesis and oligodendrogenesis from adult stem/progenitor cells. *Mol. Cell. Neurosci.* **31:** 149–160.

Caldji C., Diorio J., and Meaney M.J. 2000. Variations in maternal care in infancy regulate the development of stress reactivity. *Biol. Psychiatry* **48:** 1164–1174.

Cameron H.A. and Gould E. 1994. Adult neurogenesis is regulated by adrenal steroids in the dentate gyrus. *Neuroscience* **61:** 203–209.

Cameron H.A. and McKay R.D. 1999. Restoring production of hippocampal neurons in old age. *Nat. Neurosci.* **2:** 894–897.

Cameron H.A., Tanapat P., and Gould E. 1998. Adrenal steroids and N-methyl-D-aspartate receptor activation regulate neurogenesis in the dentate gyrus of adult rats through a common pathway. *Neuroscience* **82:** 349–354.

Cameron H.A., Woolley C.S., McEwen B.S., and Gould E. 1993. Differentiation of newly born neurons and glia in the dentate gyrus of the adult rat. *Neuroscience* **56:** 337–344.

Coe C.L., Kramer M., Czeh B., Gould E., Reeves A.J., Kirschbaum C., and Fuchs E. 2003. Prenatal stress diminishes neurogenesis in the dentate gyrus of juvenile rhesus monkeys. *Biol. Psychiatry* **54:** 1025–1034.

Czéh B. and Lucassen P.J. 2007. Hippocampal volume changes in depression: Are neurogenesis, gliogenesis and apoptosis causally involved? *Eur. Arch. Gen. Psychiatry Clin. Med.* (in press).

Czéh B., Simon M., Schmelting B., Hiemke C., and Fuchs E. 2006. Astroglial plasticity in the hippocampus is affected by chronic psychosocial stress and concomitant fluoxetine treatment. *Neuropsychopharmacology* **31:** 1616–1626.

Czéh B., Michaelis T., Watanabe T., Frahm J., de Biurrun G., van Kampen M., Bartolomucci A., and Fuchs E. 2001. Stress-induced changes in cerebral metabolites, hippocampal volume, and cell proliferation are prevented by antidepressant treatment with tianeptine. *Proc. Natl. Acad. Sci.* **98:** 12796–12801.

Czéh B., Pudovkina O., van der Hart M.G., Simon M., Heilbronner U., Michaelis T., Watanabe T., Frahm J., and Fuchs E. 2005. Examining SLV-323, a novel NK1 receptor antagonist, in a chronic psychosocial stress model for depression. *Psychopharmacology* **180:** 548–557.

Czéh B., Welt T., Fischer A.K., Erhardt A., Schmitt W., Muller M.B., Toschi N., Fuchs E., and Keck M.E. 2002. Chronic psychosocial stress and concomitant repetitive transcranial magnetic stimulation: Effects on stress hormone levels and adult hippocampal neurogenesis. *Biol. Psychiatry* **52:** 1057–1065.

Danilov A.I., Covacu R., Moe M.C., Langmoen I.A., Johansson C.B., Olsson T., and Brundin L. 2006. Neurogenesis in the adult spinal cord in an experimental model of multiple sclerosis. *Eur. J. Neurosci.* **23:** 394–400.

Dayer A.G., Ford A.A., Cleaver K.M., Yassaee M., and Cameron H.A. 2003. Short-term and long-term survival of new neurons in the rat dentate gyrus. *J. Comp. Neurol.* **460:** 563–572.

de Kloet E.R., Joels M., and Holsboer F. 2005. Stress and the brain: From adaptation to disease. *Nat. Rev. Neurosci.* **6:** 463–475.

Dobrossy M.D., Drapeau E., Aurousseau C., Le Moal M., Piazza P.V., and Abrous D.N. 2003. Differential effects of learning on neurogenesis: Learning increases or decreases the number of newly born cells depending on their birth date. *Mol. Psychiatry* **8:** 974–982.

Dong H., Goico B., Martin M., Csernansky C.A., Bertchume A., and Csernansky J.G. 2004. Modulation of hippocampal cell proliferation, memory, and amyloid plaque deposition in APPsw (Tg2576) mutant mice by isolation stress. *Neuroscience* **127:** 601–609.

Donohue H.S., Gabbott P.L., Davies H.A., Rodriguez J.J., Cordero M.I., Sandi C., Medvedev N.I., Popov V.I., Colyer F.M., Peddie C.J., and Stewart M.G. 2006. Chronic restraint stress induces changes in synapse morphology in stratum lacunosum-moleculare CA1 rat hippocampus: A stereological and three-dimensional ultrastructural study. *Neuroscience* **140:** 597–606.

Dranovsky A. and Hen R. 2006. Hippocampal neurogenesis: Regulation by stress and antidepressants. *Biol. Psychiatry* **59:** 1136–1143.

Duman R.S. 2004. Depression: A case of neuronal life and death? *Biol. Psychiatry* **56:** 140–145.

Duric V. and McCarson K.E. 2006. Persistent pain produces stress-like alterations in hippocampal neurogenesis and gene expression. *J. Pain* **7:** 544–555.

Ehninger D. and Kempermann G. 2006. Paradoxical effects of learning the Morris water maze on adult hippocampal neurogenesis in mice may be explained by a combination of stress and physical activity. *Genes Brain Behav.* **5:** 29–39.

Ekdahl C.T., Claasen J.H., Bonde S., Kokaia Z., and Lindvall O. 2003. Inflammation is detrimental for neurogenesis in adult brain. *Proc. Natl. Acad. Sci.* **100:** 13632–13637.

Eriksson P.S., Perfilieva E., Bjork-Eriksson T., Alborn A.M., Nordborg C., Peterson D.A., and Gage F.H. 1998. Neurogenesis in the adult human hippocampus. *Nat. Med.* **4:** 1313–1317.

Ernst C., Olson A.K., Pinel J.P., Lam R.W., and Christie B.R. 2006. Antidepressant effects of exercise: Evidence for an adult-neurogenesis hypothesis? *J. Psychiatry Neurosci.* **31:** 84–92.

Falconer E.M. and Galea L.A. 2003. Sex differences in cell proliferation, cell death and defensive behavior following acute predator odor stress in adult rats. *Brain Res.* **975:** 22–36.

Feldmann R.E., Jr., Sawa A., and Seilder G.H. 2007. Causality of stem cell based neurogenesis and depression—To be or not to be, is that the question? *J. Psychiatr. Res.* **41:** 713–723.

Fenoglio K.A., Brunson K.L., and Baram T.Z. 2006. Hippocampal neuroplasticity induced by early-life stress: Functional and molecular aspects. *Front. Neuroendocrinol.* **27:** 180–192.

Fischer A.K., von Rosenstiel P., Fuchs E., Goula D., Almeida O.F., and Czéh B. 2002. The prototypic mineralocorticoid receptor agonist aldosterone influences neurogenesis in the dentate gyrus of the adrenalectomized rat. *Brain Res.* **947:** 290–293.

Fuchs E., Flugge G., and Czéh B. 2006. Remodeling of neuronal networks by stress. *Front Biosci.* **11:** 2746–2758.

Fuchs E., Czéh B., Kole M.H., Michaelis T., and Lucassen P.J. 2004. Alterations of neuroplasticity in depression: The hippocampus and beyond. *Eur. Neuropsychopharmacol.* (suppl. 5) **14:** S481–S490.

Galea L.A. and McEwen B.S. 1999. Sex and seasonal differences in the rate of cell proliferation in the dentate gyrus of adult wild meadow voles. *Neuroscience* **89:**955–964.

Galea L.A., Spritzer M.D., Barker J.M., and Pawluski J.L. 2006. Gonadal hormone modulation of hippocampal neurogenesis in the adult. *Hippocampus* **16:** 225–232.

Garcia A., Steiner B., Kronenberg G., Bick-Sander A., and Kempermann G. 2004. Age-dependent expression of glucocorticoid- and mineralocorticoid receptors on neural precursor cell populations in the adult murine hippocampus. *Aging Cell* **3:** 363–371.

Gass P., Kretz O., Wolfer D.P., Berger S., Tronche F., Reichardt H.M., Kellendonk C., Lipp H.P., Schmid W., and Schutz G. 2000. Genetic disruption of mineralocorticoid receptor leads to impaired neurogenesis and granule cell degeneration in the hippocampus of adult mice. *EMBO Rep.* **1:** 447–451.

Gesing A., Bilang-Bleuel A., Droste S.K., Linthorst A.C., Holsboer F., and Reul J.M. 2001. Psychological stress increases hippocampal mineralocorticoid receptor levels: Involvement of corticotropin-releasing hormone. *J. Neurosci.* **21:** 4822–4829.

Goergen E.M., Bagay L.A., Rehm K., Benton J.L., and Beltz B.S. 2002. Circadian control of neurogenesis. *J. Neurobiol.* **53:** 90–95.

Gould E. and Tanapat P. 1999. Stress and hippocampal neurogenesis. *Biol. Psychiatry* **46:** 1472–1479.

Gould E., Woolley C.S., and McEwen B.S. 1991. Adrenal steroids regulate postnatal development of the rat dentate gyrus. I. Effects of glucocorticoids on cell death. *J. Comp. Neurol.* **313:** 479–485.

Gould E., Vail N., Wagers M., and Gross C.G. 2001. Adult-generated hippocampal and neocortical neurons in macaques have a transient existence. *Proc. Natl. Acad. Sci.* **98:** 10910–10917.

Gould E., Beylin A., Tanapat P., Reeves A., and Shors T.J. 1999. Learning enhances adult neurogenesis in the hippocampal formation. *Nat. Neurosci.* **2:** 260–265.

Gould E., McEwen B.S., Tanapat P., Galea L.A., and Fuchs E. 1997. Neurogenesis in the dentate gyrus of the adult tree shrew is regulated by psychosocial stress and NMDA receptor activation. *J. Neurosci.* **17:** 2492–2498.

Gould E., Tanapat P., McEwen B.S., Flugge G., and Fuchs E. 1998. Proliferation of granule cell precursors in the dentate gyrus of adult monkeys is diminished by stress. *Proc. Natl. Acad. Sci.* **95:** 3168–3171.

Grassi Zucconi G., Cipriani S., Balgkouranidou I., and Scattoni R. 2006. "One night" sleep deprivation stimulates hippocampal neurogenesis. *Brain Res. Bull.* **69:** 375–381.

Greenough W.T., Cohen N.J., and Juraska J.M. 1999. New neurons in old brains: Learning to survive? *Nat. Neurosci.* **2:** 203–205.

Greisen M.H., Altar C.A., Bolwig T.G., Whitehead R., and Wortwein G. 2005. Increased adult hippocampal brain-derived neurotrophic factor and normal levels of neurogenesis in maternal separation rats. *J. Neurosci. Res.* **79:** 772–778.

Guzman-Marin R., Suntsova N., Methippara M., Greiffenstein R., Szymusiak R., and McGinty D. 2005. Sleep deprivation suppresses neurogenesis in the adult hippocampus of rats. *Eur. J. Neurosci.* **22:** 2111–2116.

Hagg U., Andersson I., Naylor A.S., Gronros J., Jonsdottir I.H., Bergstrom G., and Gan L.M. 2004. Voluntary physical exercise-induced vascular effects in spontaneously hypertensive rats. *Clin. Sci.* **107:** 571–581.

Hairston I.S., Little M.T., Scanlon M.D., Barakat M.T., Palmer T.D., Sapolsky R.M., and Heller H.C. 2005. Sleep restriction suppresses neurogenesis induced by hippocampus-dependent learning. *J. Neurophysiol.* **94:** 4224–4233.

Han F., Ozawa H., Matsuda K., Nishi M., and Kawata M. 2005. Colocalization of mineralocorticoid receptor and glucocorticoid receptor in the hippocampus and hypothalamus. *Neurosci. Res.* **51:** 371–381.

Hayley S., Poulter M.O., Merali Z., and Anisman H. 2005. The pathogenesis of clinical depression: Stressor- and cytokine-induced alterations of neuroplasticity. *Neuroscience* **135:** 659–678.

Heine V.M., Maslam S., Joels M., and Lucassen P.J. 2004a. Prominent decline of newborn cell proliferation, differentiation, and apoptosis in the aging dentate gyrus, in absence of an age-related hypothalamus-pituitary-adrenal axis activation. *Neurobiol. Aging* **25:** 361–375.

———. 2004b. Increased P27KIP1 protein expression in the dentate gyrus of chronically stressed rats indicates G1 arrest involvement. *Neuroscience* **129:** 593–601.

Heine V.M., Maslam S., Zareno J., Joels M., and Lucassen P.J. 2004c. Suppressed proliferation and apoptotic changes in the rat dentate gyrus after acute and chronic stress are reversible. *Eur. J. Neurosci.* **19:** 131–144.

Heine V.M., Zareno J., Maslam S., Joels M., and Lucassen P.J. 2005. Chronic stress in the adult dentate gyrus reduces cell proliferation near the vasculature and VEGF and Flk-1 protein expression. *Eur. J. Neurosci.* **21:** 1304–1314.

Henn F.A. and Vollmayr B. 2004. Neurogenesis and depression: Etiology or epiphenomenon? *Biol. Psychiatry* **56:** 146–150.

———. 2005. Stress models of depression: Forming genetically vulnerable strains. *Neurosci. Biobehav. Rev.* **29:** 799–804.

Hill M.N., Kambo J.S., Sun J.C., Gorzalka B.B., and Galea L.A. 2006a. Endocannabinoids modulate stress-induced suppression of hippocampal cell proliferation and activation of defensive behaviours. *Eur. J. Neurosci.* **24:** 1845–1849.

Hill M.N., Ho W.S., Sinopoli K.J., Viau V., Hillard C.J., and Gorzalka B.B. 2006b. Involvement of the endocannabinoid system in the ability of long-term tricyclic antidepressant treatment to suppress stress-induced activation of the hypothalamic-pituitary-adrenal axis. *Neuropsychopharmacology* **31:** 2591–2599.

Hoehn B.D., Palmer T.D., and Steinberg G.K. 2005. Neurogenesis in rats after focal cerebral ischemia is enhanced by indomethacin. *Stroke* **36:** 2718–2724.

Holmes M.M., Galea L.A., Mistlberger R.E., and Kempermann G. 2004. Adult hippocampal neurogenesis and voluntary running activity: Circadian and dose-dependent effects. *J. Neurosci. Res.* **76:** 216–222.

Huang G.J. and Herbert J. 2005. Serotonin modulates the suppressive effects of corticosterone on proliferating progenitor cells in the dentate gyrus of the hippocampus in the adult rat. *Neuropsychopharmacology* **30:** 231–241.

Iosif R.E., Ekdahl C.T., Ahlenius H., Pronk C.J., Bonde S., Kokaia Z., Jacobsen S.E., and Lindvall O. 2006. Tumor necrosis factor receptor 1 is a negative regulator of progenitor proliferation in adult hippocampal neurogenesis. *J. Neurosci.* **26:** 9703–9712.

Izquierdo A., Wellman C.L., and Holmes A. 2006. Brief uncontrollable stress causes dendritic retraction in infralimbic cortex and resistance to fear extinction in mice. *J. Neurosci.* **26:** 5733–5738.

Jaako-Movits K., Zharkovsky T., Pedersen M., and Zharkovsky A. 2006. Decreased hippocampal neurogenesis following olfactory bulbectomy is reversed by repeated citalopram administration. *Cell. Mol. Neurobiol.* **26:** 1557–1568.

Jacobs B.L., Praag H., and Gage F.H. 2000. Adult brain neurogenesis and psychiatry: A novel theory of depression. *Mol. Psychiatry* **5:** 262–269.

Jayatissa M.N., Bisgaard C., Tingstrom A., Papp M., and Wiborg O. 2006. Hippocampal cytogenesis correlates to escitalopram-mediated recovery in a chronic mild stress rat model of depression. *Neuropsychopharmacology* **31:** 2395–2404.

Jin K., Zhu Y., Sun Y., Mao X.O., Xie L., and Greenberg D.A. 2002. Vascular endothelial growth factor (VEGF) stimulates neurogenesis in vitro and in vivo. *Proc. Natl. Acad. Sci.* **99:** 11946–11950.

Joels M., Karst H., Krugers H.J., and Lucassen P.J. 2007. Chronic stress; implications for neuron morphology, function and neurogenesis. *Front. Neuroendocrinol.* (in press).

Joels M., Pu Z., Wiegert O., Oitzl M.S., and Krugers H.J. 2006. Learning under stress: How does it work? *Trends Cogn. Sci.* **10:** 152–158.

Kambo J.S. and Galea L.A. 2006. Activational levels of androgens influence risk assessment behaviour but do not influence stress-induced suppression in hippocampal cell proliferation in adult male rats. *Behav. Brain Res.* **175:** 263–270.

Kaneko N., Kudo K., Mabuchi T., Takemoto K., Fujimaki K., Wati H., Iguchi H., Tezuka H., and Kanba S. 2006. Suppression of cell proliferation by interferon-alpha through interleukin-1 production in adult rat dentate gyrus. *Neuropsychopharmacology* **31:** 2619–2626.

Karishma K.K. and Herbert J. 2002. Dehydroepiandrosterone (DHEA) stimulates neurogenesis in the hippocampus of the rat, promotes survival of newly formed neurons and prevents corticosterone-induced suppression. *Eur. J. Neurosci.* **16:** 445–453.

Karst H., Berger S., Turiault M., Tronche F., Schutz G., and Joels M. 2005. Mineralocorticoid receptors are indispensable for nongenomic modulation of hippocampal glutamate transmission by corticosterone. *Proc. Natl. Acad. Sci.* **102:** 19204–19207.

Kempermann G. and Gage F.H. 2002. Genetic determinants of adult hippocampal neurogenesis correlate with acquisition, but not probe trial performance, in the water maze task. *Eur. J. Neurosci.* **16:** 129–136.

Kempermann G., Gast D., and Gage F.H. 2002. Neuroplasticity in old age: Sustained fivefold induction of hippocampal neurogenesis by long-term environmental enrichment. *Ann. Neurol.* **52:** 135–143.

Kempermann G., Kuhn H.G., and Gage F.H. 1997. More hippocampal neurons in adult mice living in an enriched environment. *Nature* **386:** 493–495.

Kendler K.S., Karkowski L.M., and Prescott C.A. 1999. Causal relationship between stressful life events and the onset of major depression. *Am. J. Psychiatry* **156:** 837–841.

Kim J.J. and Diamond D.M. 2002. The stressed hippocampus, synaptic plasticity and lost memories. *Nat. Rev. Neurosci.* **3:** 453–462.

Kochman L.J., Weber E.T., Fornal C.A., and Jacobs B.L. 2006. Circadian variation in mouse hippocampal cell proliferation. *Neurosci. Lett.* **406:** 256–259.

Kozorovitskiy Y. and Gould E. 2004. Dominance hierarchy influences adult neurogenesis in the dentate gyrus. *J. Neurosci.* **24:** 6755–6759.

Krugers H.J., van der Linden S., van Olst E., Alfarez D.N., Maslam S., Lucassen P.J., and Joels M. 2007. Dissociation between apoptosis, neurogenesis, and synaptic potentiation in the dentate gyrus of adrenalectomized rats. *Synapse* **61:** 221–230.

Kuhn H.G., Dickinson-Anson H., and Gage F.H. 1996. Neurogenesis in the dentate gyrus of the adult rat: Age-related decrease of neuronal progenitor proliferation. *J. Neurosci.* **16:** 2027–2033.

Kuhn H.G., Winkler J., Kempermann G., Thal L.J., and Gage F.H. 1997. Epidermal growth factor and fibroblast growth factor-2 have different effects on neural progenitors in the adult rat brain. *J. Neurosci.* **17:** 5820–5829.

Lagace D.C., Fischer S.J., and Eisch A.J. 2007. Gender and endogenous levels of estradiol do not influence adult hippocampal neurogenesis in mice. *Hippocampus* **17:** 175–180.

Lavenex P., Steele M.A., and Jacobs L.F. 2000. The seasonal pattern of cell proliferation and neuron number in the dentate gyrus of wild adult eastern grey squirrels. *Eur. J. Neurosci.* **12:** 643–648.

Lee K.J., Kim S.J., Kim S.W., Choi S.H., Shin Y.C., Park S.H., Moon B.H., Cho E., Lee M.S., Choi S.H., Chun B.G., and Shin K.H. 2006. Chronic mild stress decreases survival, but not proliferation, of new-born cells in adult rat hippocampus. *Exp. Mol. Med.* **38:** 44–54.

Lemaire V., Aurousseau C., Le Moal M., and Abrous D.N. 1999. Behavioural trait of reactivity to novelty is related to hippocampal neurogenesis. *Eur. J. Neurosci.* **11:** 4006–4014.

Lemaire V., Koehl M., Le Moal M., and Abrous D.N. 2000. Prenatal stress produces learning deficits associated with an inhibition of neurogenesis in the hippocampus. *Proc. Natl. Acad. Sci.* **97:** 11032–11037.

Lemaire V., Lamarque S., Le Moal M., Piazza P.V., and Abrous D.N. 2006. Postnatal stimulation of the pups counteracts prenatal stress-induced deficits in hippocampal neurogenesis. *Biol. Psychiatry* **59:** 786–792.

Leuner B., Gould E., and Shors T.J. 2006. Is there a link between adult neurogenesis and learning? *Hippocampus* **16:** 216–224.

Leuner B., Mendolia-Loffredo S., Kozorovitskiy Y., Samburg D., Gould E., and Shors T.J. 2004. Learning enhances the survival of new neurons beyond the time when the hippocampus is required for memory. *J. Neurosci.* **24:** 7477–7481.

Leventhal C., Rafii S., Rafii D., Shahar A., and Goldman S.A. 1999. Endothelial trophic support of neuronal production and recruitment from the adult mammalian subependyma. *Mol. Cell. Neurosci.* **13:** 450–464.

Lu L., Bao G., Chen H., Xia P., Fan X., Zhang J., Pei G., and Ma L. 2003. Modification of hippocampal neurogenesis and neuroplasticity by social environments. *Exp. Neurol.* **183:** 600–609.

Lucassen P.J. and de Kloet E.R. 2001. Glucocorticoids and the aging brain; cause or consequence? In *Functional neurobiology of aging* (ed. P. Hof and C. Mobbs), pp. 883–905. Academic Press, New York.

Lucassen P.J., Fuchs E., and Czéh B. 2004. Antidepressant treatment with tianeptine reduces apoptosis in the hippocampal dentate gyrus and temporal cortex. *Biol. Psychiatry* **55:** 789–796.

Lucassen P.J., Muller M.B., Holsboer F., Bauer J., Holtrop A., Wouda J., Hoogendijk W.J., De Kloet E.R., and Swaab D.F. 2001. Hippocampal apoptosis in major depression is a minor event and absent from subareas at risk for glucocorticoid overexposure. *Am. J. Pathol.* **158:** 453–468.

Lucassen P.J., Heine V.M., Muller M.B., van der Beek E.M., Wiegant V.M., De Kloet E.R., Joels M., Fuchs E., Swaab D.F., and Czéh B. 2006. Stress, depression and hippocampal apoptosis. *CNS Neurol. Disord. Drug Targets* **5:** 531–546.

Magarinos A.M., Verdugo J.M., and McEwen B.S. 1997. Chronic stress alters synaptic terminal structure in hippocampus. *Proc. Natl. Acad. Sci.* **94:** 14002–14008.

Malberg J.E. and Duman R.S. 2003. Cell proliferation in adult hippocampus is decreased by inescapable stress: Reversal by fluoxetine treatment. *Neuropsychopharmacology* **28:** 1562–1571.

Malberg J.E., Eisch A.J., Nestler E.J., and Duman R.S. 2000. Chronic antidepressant treatment increases neurogenesis in adult rat hippocampus. *J. Neurosci.* **20:** 9104–9110.

Mayer J.L., Klumpers L., Maslam S., de Kloet E.R., Joels M., and Lucassen P.J. 2006. Brief treatment with the glucocorticoid receptor antagonist mifepristone normalises the corticosterone-induced reduction of adult hippocampal neurogenesis. *J. Neuroendocrinol.* **18:** 629–631.

Mayo W., Le Moal M., and Abrous D.N. 2001. Pregnenolone sulfate and aging of cognitive functions: Behavioral, neurochemical, and morphological investigations. *Horm. Behav.* **40:** 215–217.

Mayo W., Lemaire V., Malaterre J., Rodriguez J.J., Cayre M., Stewart M.G., Kharouby M., Rougon G., Le Moal M., Piazza P.V., and Abrous D.N. 2005. Pregnenolone sulfate enhances neurogenesis and PSA-NCAM in young and aged hippocampus. *Neurobiol. Aging* **26:** 103–114.

Meerlo P., Koehl M., van der Borght K., and Turek F.W. 2002. Sleep restriction alters the hypothalamic-pituitary-adrenal response to stress. *J. Neuroendocrinol.* **14:** 397–402.

Mirescu C. and Gould E. 2006. Stress and adult neurogenesis. *Hippocampus* **16:** 233–238.

Mirescu C., Peters J.D., and Gould E. 2004. Early life experience alters response of adult neurogenesis to stress. *Nat. Neurosci.* **7:** 841–846.

Mirescu C., Peters J.D., Noiman L., and Gould E. 2006. Sleep deprivation inhibits adult neurogenesis in the hippocampus by elevating glucocorticoids. *Proc. Natl. Acad. Sci.* **103:** 19170–19175.

Mitra R., Sundlass K., Parker K.J., Schatzberg A.F., and Lyons D.M. 2006. Social stress-related behavior affects hippocampal cell proliferation in mice. *Physiol. Behav.* **89:** 123–127.

Moncek F., Duncko R., Johansson B.B., and Jezova D. 2004. Effect of environmental enrichment on stress related systems in rats. *J. Neuroendocrinol.* **16:** 423–431.

Monje M.L., Toda H., and Palmer T.D. 2003. Inflammatory blockade restores adult hippocampal neurogenesis. *Science* **302:** 1760–1765.

Montaron M.F., Piazza P.V., Aurousseau C., Urani A., Le Moal M., and Abrous D.N. 2003. Implication of corticosteroid receptors in the regulation of hippocampal structural plasticity. *Eur. J. Neurosci.* **18:** 3105–3111.

Montaron M.F., Drapeau E., Dupret D., Kitchener P., Aurousseau C., Le Moal M., Piazza P.V., and Abrous D.N. 2006. Lifelong corticosterone level determines age-related decline in neurogenesis and memory. *Neurobiol. Aging* **27:** 645–654.

Montero-Pedrazuela A., Venero C., Lavado-Autric R., Fernandez-Lamo I., Garcia-Verdugo J.M., Bernal J., and Guadano-Ferraz A. 2006. Modulation of adult hippocampal neurogenesis by thyroid hormones: Implications in depressive-like behavior. *Mol. Psychiatry* **11:** 361–371.

Muller M.B., Lucassen P.J., Yassouridis A., Hoogendijk W.J., Holsboer F., and Swaab D.F. 2001. Neither major depression nor glucocorticoid treatment affects the cellular integrity of the human hippocampus. *Eur. J. Neurosci.* **14:** 1603–1612.

Nacher J., Gomez-Climent M.A., and McEwen B. 2004. Chronic non-invasive glucocorticoid administration decreases polysialylated neural cell adhesion molecule expression in the adult rat dentate gyrus. *Neurosci. Lett.* **370:** 40–44.

Nacher J., Alonso-Llosa G., Rosell D.R., and McEwen B.S. 2003. NMDA receptor antagonist treatment increases the production of new neurons in the aged rat hippocampus. *Neurobiol. Aging* **24:** 273–284.

Nacher J., Varea E., Miguel Blasco-Ibanez J., Gomez-Climent M.A., Castillo-Gomez E., Crespo C., Martinez-Guijarro F.J., and McEwen B.S. 2007. N-methyl-D-aspartate receptor expression during adult neurogenesis in the rat dentate gyrus. *Neuroscience* **144:** 855–864.

Namestkova K., Simonova Z., and Sykova E. 2005. Decreased proliferation in the adult rat hippocampus after exposure to the Morris water maze and its reversal by fluoxetine. *Behav. Brain Res.* **163:** 26–32.

Naylor A.S., Persson A.I., Eriksson P.S., Jonsdottir I.H., and Thorlin T. 2005. Extended voluntary running inhibits exercise-induced adult hippocampal progenitor proliferation in the spontaneously hypertensive rat. *J. Neurophysiol.* **93:** 2406–2414.

Neumann I.D., Wigger A., Kromer S., Frank E., Landgraf R., and Bosch O.J. 2005. Differential effects of periodic maternal separation on adult stress coping in a rat model of extremes in trait anxiety. *Neuroscience* **132:** 867–877.

Nilsson M., Perfilieva E., Johansson U., Orwar O., and Eriksson P.S. 1999. Enriched environment increases neurogenesis in the adult rat dentate gyrus and improves spatial memory. *J. Neurobiol.* **39:** 569–578.

Norrholm S.D. and Ouimet C.C. 2001. Altered dendritic spine density in animal models of depression and in response to antidepressant treatment. *Synapse* **42:** 151–163.

O'Brien J., Thomas A., Ballard C., Brown A., Ferrier N., Jaros E., and Perry R. 2001. Cognitive impairment in depression is not associated with neuropathologic evidence of increased vascular or Alzheimer-type pathology. *Biol. Psychiatry* **49:** 130–136.

Olariu A., Cleaver K.M., Shore L.E., Brewer M.D., and Cameron H.A. 2005. A natural form of learning can increase and decrease the survival of new neurons in the dentate gyrus. *Hippocampus* **15:** 750–762.

Olson A.K., Eadie B.D., Ernst C., and Christie B.R. 2006. Environmental enrichment and voluntary exercise massively increase neurogenesis in the adult hippocampus via dissociable pathways. *Hippocampus* **16**: 250–260.

Oomen C.A., Mayer J.L., Joels M., de Kloet E.R., and Lucassen P.J. 2007. Brief treatment with the GR antagonist mifepristone normalizes the reduction in neurogenesis after chronic stress. *J. Neurosci.* (in press).

Ormerod B.K. and Galea L.A. 2001. Reproductive status influences cell proliferation and cell survival in the dentate gyrus of adult female meadow voles: A possible regulatory role for estradiol. *Neuroscience* **102**: 369–379.

Ormerod B.K., Lee T.T., and Galea L.A. 2003. Estradiol initially enhances but subsequently suppresses (via adrenal steroids) granule cell proliferation in the dentate gyrus of adult female rats. *J. Neurobiol.* **55**: 247–260.

———. 2004. Estradiol enhances neurogenesis in the dentate gyri of adult male meadow voles by increasing the survival of young granule neurons. *Neuroscience* **128**: 645–654.

Pace T.W., Mletzko T.C., Alagbe O., Musselman D.L., Nemeroff C.B., Miller A.H., and Heim C.M. 2006. Increased stress-induced inflammatory responses in male patients with major depression and increased early life stress. *Am. J. Psychiatry* **163**: 1630–1633.

Palchykova S., Winsky-Sommerer R., Meerlo P., Durr R., and Tobler I. 2006. Sleep deprivation impairs object recognition in mice. *Neurobiol. Learn. Mem.* **85**: 263–271.

Palmer T.D., Willhoite A.R., and Gage F.H. 2000. Vascular niche for adult hippocampal neurogenesis. *J. Comp. Neurol.* **425**: 479–494.

Persson A.I., Naylor A.S., Jonsdottir I.H., Nyberg F., Eriksson P.S., and Thorlin T. 2004. Differential regulation of hippocampal progenitor proliferation by opioid receptor antagonists in running and non-running spontaneously hypertensive rats. *Eur. J. Neurosci.* **19**: 1847–1855.

Pham K., McEwen B.S., Ledoux J.E., and Nader K. 2005. Fear learning transiently impairs hippocampal cell proliferation. *Neuroscience* **130**: 17–24.

Pham K., Nacher J., Hof P.R., and McEwen B.S. 2003. Repeated restraint stress suppresses neurogenesis and induces biphasic PSA-NCAM expression in the adult rat dentate gyrus. *Eur. J. Neurosci.* **17**: 879–886.

Radley J.J., Rocher A.B., Miller M., Janssen W.G., Liston C., Hof P.R., McEwen B.S., and Morrison J.H. 2006. Repeated stress induces dendritic spine loss in the rat medial prefrontal cortex. *Cereb. Cortex* **16**: 313–320.

Rahpeymai Y., Hietala M.A., Wilhelmsson U., Fotheringham A., Davies I., Nilsson A.K., Zwirner J., Wetsel R.A., Gerard C., Pekny M., and Pekna M. 2006. Complement: A novel factor in basal and ischemia-induced neurogenesis. *EMBO J.* **25**: 1364–1374.

Rao M.S., Hattiangady B., and Shetty A.K. 2006. The window and mechanisms of major age-related decline in the production of new neurons within the dentate gyrus of the hippocampus. *Aging Cell* **5**: 545–558.

Rao M.S., Hattiangady B., Abdel-Rahman A., Stanley D.P., and Shetty A.K. 2005. Newly born cells in the ageing dentate gyrus display normal migration, survival and neuronal fate choice but endure retarded early maturation. *Eur. J. Neurosci.* **21**: 464–476.

Reagan L.P., Rosell D.R., Wood G.E., Spedding M., Munoz C., Rothstein J., and McEwen B.S. 2004. Chronic restraint stress up-regulates GLT-1 mRNA and protein expression in the rat hippocampus: Reversal by tianeptine. *Proc. Natl. Acad. Sci.* **101**: 2179–2184.

Reif A., Fritzen S., Finger M., Strobel A., Lauer M., Schmitt A., and Lesch K.P. 2006. Neural stem cell proliferation is decreased in schizophrenia, but not in depression. *Mol. Psychiatry* **11**: 514–522.

Roman V., Van der Borght K., Leemburg S.A., Van der Zee E.A., and Meerlo P. 2005. Sleep restriction by forced activity reduces hippocampal cell proliferation. *Brain Res.* **1065:** 53–59.

Roozendaal B., Okuda S., Van der Zee E.A., and McGaugh J.L. 2006. Glucocorticoid enhancement of memory requires arousal-induced noradrenergic activation in the basolateral amygdala. *Proc. Natl. Acad. Sci.* **103:** 6741–6746.

Sairanen M., O'Leary O.F., Knuuttila J.E., and Castren E. 2007. Chronic antidepressant treatment selectively increases expression of plasticity-related proteins in the hippocampus and medial prefrontal cortex of the rat. *Neuroscience* **144:** 368–374.

Sandi C. 2004. Stress, cognitive impairment and cell adhesion molecules. *Nat. Rev. Neurosci.* **5:** 917–930.

Sandi C., Davies H.A., Cordero M.I., Rodriguez J.J., Popov V.I., and Stewart M.G. 2003. Rapid reversal of stress induced loss of synapses in CA3 of rat hippocampus following water maze training. *Eur. J. Neurosci.* **17:** 2447–2456.

Santarelli L., Saxe M., Gross C., Surget A., Battaglia F., Dulawa S., Weisstaub N., Lee J., Duman R., Arancio O., Belzung C., and Hen R. 2003. Requirement of hippocampal neurogenesis for the behavioral effects of antidepressants. *Science* **301:** 805–809.

Sapolsky R.M., Uno H., Rebert C.S., and Finch C.E. 1990. Hippocampal damage associated with prolonged glucocorticoid exposure in primates. *J. Neurosci.* **10:** 2897–2902.

Saxe M.D., Battaglia F., Wang J.W., Malleret G., David D.J., Monckton J.E., Garcia A.D., Sofroniew M.V., Kandel E.R., Santarelli L., Hen R., and Drew M.R. 2006. Ablation of hippocampal neurogenesis impairs contextual fear conditioning and synaptic plasticity in the dentate gyrus. *Proc. Natl. Acad. Sci.* **103:** 17501–17506.

Seckl J.R. 2001. Glucocorticoid programming of the fetus; adult phenotypes and molecular mechanisms. *Mol. Cell. Endocrinol.* **185:** 61–71.

Sgoifo A., Buwalda B., Roos M., Costoli T., Merati G., and Meerlo P. 2006. Effects of sleep deprivation on cardiac autonomic and pituitary-adrenocortical stress reactivity in rats. *Psychoneuroendocrinology* **31:** 197–208.

Shapiro E.M., Gonzalez-Perez O., Manuel Garcia-Verdugo J., Alvarez-Buylla A., and Koretsky A.P. 2006. Magnetic resonance imaging of the migration of neuronal precursors generated in the adult rodent brain. *Neuroimage* **32:** 1150–1157.

Sheline Y.I. 2003. Neuroimaging studies of mood disorder effects on the brain. *Biol. Psychiatry* **54:** 338–352.

Shors T.J., Chua C., and Falduto J. 2001. Sex differences and opposite effects of stress on dendritic spine density in the male versus female hippocampus. *J. Neurosci.* **21:** 6292–6297.

Shors T.J., Seib T.B., Levine S., and Thompson R.F. 1989. Inescapable versus escapable shock modulates long-term potentiation in the rat hippocampus. *Science* **244:** 224–226.

Shors T.J., Mathew J., Sisti H.M., Edgecomb C., Beckoff S., and Dalla C. 2007. Neurogenesis and helplessness are mediated by controllability in males but not in females. *Biol. Psychiatry* (in press).

Simon M., Czeh B., and Fuchs E. 2005. Age-dependent susceptibility of adult hippocampal cell proliferation to chronic psychosocial stress. *Brain Res.* **1049:** 244–248.

Sousa N., Almeida O.F., Holsboer F., Paula-Barbosa M.M., and Madeira M.D. 1998. Maintenance of hippocampal cell numbers in young and aged rats submitted to chronic unpredictable stress. Comparison with the effects of corticosterone treatment. *Stress* **2:** 237–249.

Sousa N., Lukoyanov N.V., Madeira M.D., Almeida O.F. and Paula-Barbosa M.M. 2000. Reorganization of the morphology of hippocampal neurites and synapses after stress-induced damage correlates with behavioral improvement. *Neuroscience* **97:** 253–266.

Stewart M.G., Davies H.A., Sandi C., Kraev I.V., Rogachevsky V.V., Peddie C.J., Rodriguez J.J., Cordero M.I., Donohue H.S., Gabbott P.L., and Popov V.I. 2005. Stress suppresses and learning induces plasticity in CA3 of rat hippocampus: A three-dimensional ultrastructural study of thorny excrescences and their postsynaptic densities. *Neuroscience* **131:** 43–54.

Stranahan A.M., Khalil D., and Gould E. 2006. Social isolation delays the positive effects of running on adult neurogenesis. *Nat. Neurosci.* **9:** 526–533.

Suzuki M., Wright L.S., Marwah P., Lardy H.A., and Svendsen C.N. 2004. Mitotic and neurogenic effects of dehydroepiandrosterone (DHEA) on human neural stem cell cultures derived from the fetal cortex. *Proc. Natl. Acad. Sci.* **101:** 3202–3207.

Swaab D.F., Bao A.M., and Lucassen P.J. 2005. The stress system in the human brain in depression and neurodegeneration. *Ageing Res. Rev.* **4:** 141–194.

Tanapat P., Galea L.A., and Gould E. 1998. Stress inhibits the proliferation of granule cell precursors in the developing dentate gyrus. *Int. J. Dev. Neurosci.* **16:** 235–239.

Tanapat P., Hastings N.B., Reeves A.J., and Gould E. 1999. Estrogen stimulates a transient increase in the number of new neurons in the dentate gyrus of the adult female rat. *J. Neurosci.* **19:** 5792–5801.

Tanapat P., Hastings N.B., Rydel T.A., Galea L.A., and Gould E. 2001. Exposure to fox odor inhibits cell proliferation in the hippocampus of adult rats via an adrenal hormone-dependent mechanism. *J. Comp. Neurol.* **437:** 496–504.

Tata D.A., Marciano V.A., and Anderson B.J. 2006. Synapse loss from chronically elevated glucocorticoids: Relationship to neuropil volume and cell number in hippocampal area CA3. *J. Comp. Neurol.* **498:** 363–374.

Thomas R.M., Urban J.H., and Peterson D.A. 2006. Acute exposure to predator odor elicits a robust increase in corticosterone and a decrease in activity without altering proliferation in the adult rat hippocampus. *Exp. Neurol.* **201:** 308–315.

Tung A., Takase L., Fornal C., and Jacobs B. 2005. Effects of sleep deprivation and recovery sleep upon cell proliferation in adult rat dentate gyrus. *Neuroscience* **134:** 721–723.

Vallieres L., Campbell I.L., Gage F.H., and Sawchenko P.E. 2002. Reduced hippocampal neurogenesis in adult transgenic mice with chronic astrocytic production of interleukin-6. *J. Neurosci.* **22:** 486–492.

Van der Borght K., Meerlo P., Luiten P.G., Eggen B.J., and Van der Zee E.A. 2005. Effects of active shock avoidance learning on hippocampal neurogenesis and plasma levels of corticosterone. *Behav. Brain Res.* **157:** 23–30.

van Praag H., Kempermann G., and Gage F.H. 1999a. Running increases cell proliferation and neurogenesis in the adult mouse dentate gyrus. *Nat. Neurosci.* **2:** 266–270.

van Praag H., Christie B.R., Sejnowski T.J., and Gage F.H. 1999b. Running enhances neurogenesis, learning, and long-term potentiation in mice. *Proc. Natl. Acad. Sci.* **96:** 13427–13431.

Veenema A.H., de Kloet E.R., de Wilde M.C., Roelofs A.J., Kawata M., Buwalda B., Neumann I.D., Koolhass J.M., and Lucassen P.J. 2007. Differential effects of stress on adult hippocampal cell proliferation in low and high aggressive mice. *J. Neuroendocrinol.* **19:** 489–498.

Vollmayr B., Simonis C., Weber S., Gass P., and Henn F. 2003. Reduced cell proliferation in the dentate gyrus is not correlated with the development of learned helplessness. *Biol. Psychiatry* **54:** 1035–1040.

Wachs F.P., Winner B., Couillard-Despres S., Schiller T., Aigner R., Winkler J., Bogdahn U., and Aigner L. 2006. Transforming growth factor-beta1 is a negative modulator of adult neurogenesis. *J. Neuropathol. Exp. Neurol.* **65:** 358–370.

Warner-Schmidt J.L. and Duman R.S. 2006. Hippocampal neurogenesis: Opposing effects of stress and antidepressant treatment. *Hippocampus* **16:** 239–249.

Watanabe Y., Gould E., and McEwen B.S. 1992. Stress induces atrophy of apical dendrites of hippocampal CA3 pyramidal neurons. *Brain Res.* **588:** 341–345.

Westenbroek C., Den Boer J.A., Veenhuis M., and Ter Horst G.J. 2004. Chronic stress and social housing differentially affect neurogenesis in male and female rats. *Brain Res. Bull.* **64:** 303–308.

Wong E.Y. and Herbert J. 2004. The corticoid environment: A determining factor for neural progenitors' survival in the adult hippocampus. *Eur. J. Neurosci.* **20:** 2491–2498.

———. 2005. Roles of mineralocorticoid and glucocorticoid receptors in the regulation of progenitor proliferation in the adult hippocampus. *Eur. J. Neurosci.* **22:** 785–792.

———. 2006. Raised circulating corticosterone inhibits neuronal differentiation of progenitor cells in the adult hippocampus. *Neuroscience* **137:** 83–92.

Xu H., Chen Z., He J., Haimanot S., Li X., Dyck L., and Li X.M. 2006. Synergetic effects of quetiapine and venlafaxine in preventing the chronic restraint stress-induced decrease in cell proliferation and BDNF expression in rat hippocampus. *Hippocampus* **16:** 551–559.

Yap J.J., Takase L.F., Kochman L.J., Fornal C.A., Miczek K.A., and Jacobs B.L. 2006. Repeated brief social defeat episodes in mice: Effects on cell proliferation in the dentate gyrus. *Behav. Brain Res.* **172:** 344–350.

Zhang L.X., Levine S., Dent G., Zhan Y., Xing G., Okimoto D., Kathleen Gordon M., Post R.M., and Smith M.A. 2002. Maternal deprivation increases cell death in the infant rat brain. *Brain Res. Dev. Brain Res.* **133:** 1–11.

Ziv Y., Ron N., Butovsky O., Landa G., Sudai E., Greenberg N., Cohen H., Kipnis J., and Schwartz M. 2006. Immune cells contribute to the maintenance of neurogenesis and spatial learning abilities in adulthood. *Nat. Neurosci.* **9:** 268–275.

19

Regulation of Adult Neurogenesis by Neurotransmitters

Mi-Hyeon Jang, Hongjun Song, and Guo-li Ming
Institute for Cell Engineering, Department of Neurology
The Solomon H. Snyder Department of Neuroscience
Johns Hopkins University School of Medicine
Baltimore, Maryland 21205

ACTIVE ADULT NEUROGENESIS OCCURS FROM NEURONAL progenitor cells (Npcs) in discrete regions of the adult mammalian central nervous system (CNS) (Abrous et al. 2005; Ming and Song 2005; Lledo et al. 2006). The generation of nascent neurons from NPcs in the intact adult CNS is restricted to the subventricular zone (SVZ) of the lateral ventricle and the subgranular zone (SGZ) of the hippocampal dentate gyrus (DG) (Alvarez-Buylla and Lim 2004). Outside of these two regions, proliferating NPcs normally generate only glia cells, but they appear to be able to give rise to neurons after insults (Emsley et al. 2005). Accumulative evidence suggests that continuous neuronal production in the adult brain under physiological conditions is involved in specific brain functions, such as olfaction, learning, and memory (Kempermann et al. 2004a). On the other hand, neural production of NPcs under pathological conditions may contribute to brain repair (Emsley et al. 2005). Functional integration of nascent neurons is achieved by progression through sequential developmental steps that resemble embryonic and fetal neurogenesis, from proliferation and fate specification of NPcs, to differentiation, migration, axonal/dendritic development, and synaptic integration of newborn neurons (Ming and Song 2005). In contrast to developing neurogenesis, adult neurogenesis arises from a significantly different environment and proceeds with concurrent activities of mature neurons within the existing circuit.

Adult neurogenesis, a striking form of structural plasticity in the intact adult CNS, is dynamically regulated by many physiological and pathological stimuli (Abrous et al. 2005; Ming and Song 2005). For example, environmental enrichment (Kempermann et al. 1997), electroconvulsive shock stimulation (Scott et al. 2000), and seizures (Parent et al. 1997; Overstreet-Wadiche et al. 2006b) dramatically influence adult neurogenesis. Neuronal activities, and neurotransmitters in particular, are major mediators for such modulation. Emerging evidence suggests that neuronal activities regulate multiple steps of adult neurogenesis through both synaptic and nonsynaptic mechanisms (Ge et al. 2007). Classic neurotransmitters (glutamate, γ-amino butyric acid [GABA], acetylcholine [ACh], serotonin, dopamine, glycine, and norepinephrine) and nonclassic neurotransmitters (nitric oxide [NO], carbon monoxide [CO], neuropeptides) released from excitatory and inhibitory neurons in response to neuronal activation can diffuse from the synaptic cleft into the extrasynaptic space and are thus present as a tonic source to mediate and/or modulate the process of adult neurogenesis (Farrant and Nusser 2005). Besides these neuronal sources, some neurotransmitters are re-uptaken by surrounding glia cells and then released from these nonneuronal cells to contribute to the ambient level of neurotransmitters (Halassa et al. 2007). In addition to these nonsynaptic mechanisms, neurotransmitters released from mature neurons also directly activate progeny of NPCs once they receive functional synaptic inputs (Ge et al. 2007). Many of these transmitters appear to regulate the early phases of adult neurogenesis, including the proliferation and differentiation of NPCs, whereas others can affect almost the whole spectrum of developmental steps of adult neurogenesis (Figs. 1 and 2). In this chapter, we review current findings on how classic and nonclassic neurotransmitters regulate various steps of adult neurogenesis mainly in neurogenic regions. For a comparison of the roles of neurotransmitters in regulating neurogenesis during embryonic stages, see comprehensive reviews by Barker et al. (1998), Ben-Ari et al. (1997), and Owens and Kriegstein (2002).

γ-AMINO BUTYRIC ACID

GABA was first identified as a major inhibitory neurotransmitter in the mammalian brain in the 1950s (Roberts and Frankel 1950; Owens and Kriegstein 2002). As a classic neurotransmitter, GABA exerts its physiological effect by altering the chloride ion (Cl^-) permeability of GABA receptor channels. Because of the differences in cytoplasmic Cl^- concentrations, GABA causes depolarization or hyperpolarization of immature and

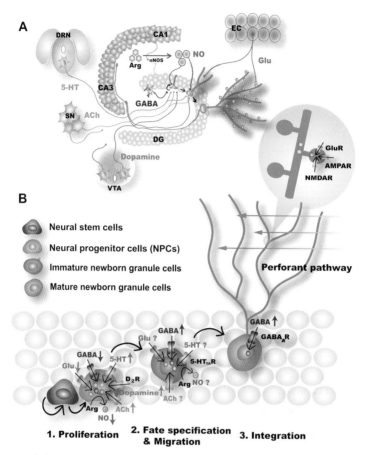

Figure 1. Adult neurogenesis in the dentate gyrus of the hippocampus and its regulation by various neurotransmitters. (*A*) Generation of newborn neurons from NPCs located in the SGZ can be regulated by various neurotransmitters, such as glutamate (Glu) from the entorhinal cortical inputs (EC), GABA from the local interneurons, acetylcholine (ACh) from axons originating from the septal nucleus (SN), dopamine from axons originating from the ventral tegmental area (VTA), serotonin (5-HT) axons originating from the dorsal raphe nucleus (DRN), and nitric oxide (NO) synthesized by the neuronal nitric oxide synthase (nNOS). (*B*) Sequential steps involved in generating functional and integrated newborn granule cells in the dentate gyrus (DG). (*1*) Proliferation of NPCs; (*2*) neuronal fate specification and migration; (*3*) synaptic integration of newborn granule cells into the existing hippocampal circuitry. Various neurotransmitters can potentially regulate all of these steps.

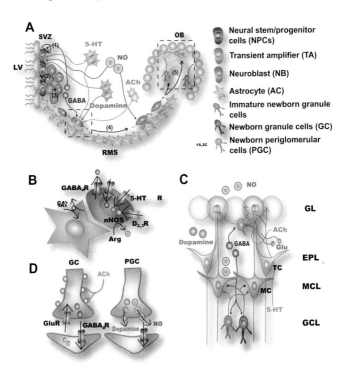

Figure 2. Adult neurogenesis in the SVZ/OB system and its regulation by various neurotransmitters. (*A*) Sequential steps involved in generating functional and integrated new olfactory interneurons from the NPCs. (*1*) Proliferation of NPCs in the SVZ; (*2*) NPCs giving rise to transient amplifying cells; (*3*) differentiation into neuroblasts; (*4*) tangential migration of neuroblasts in chains through the rostralmigratory stream (RMS); (*5*) synaptic integration of new interneurons in the olfactory bulb (OB). The boxed regions outlined are shown in more detail in *B* and *C*. NPCs and their neuronal progeny are regulated by various neurotransmitters, including glutamate and GABA from the mitral/tufted cells, nitric oxide (NO), and dopamine from the periglomerular cells, 5-HT from axons originating from the dorsal raphe, ACh from axons originating from the septal nucleus of basal forebrain. (LV) Lateral ventricle. (*B*) Cell-cell interactions through various receptors of neurotransmitters in the adult RMS. (*C*) Integration of differentiated interneurons into the OB. Immature granule cells (GC) in the granule cell layer (GCL) and periglomerular cells (PGC) in the glomerular layer (GL) are regulated by neurotransmitters. (TC) Tufted cells; (MC) mitral cells; (EPL) external plexiform layer; (MCL) mitral cell layer. (*D*) Cellular distribution of neurotransmitters and their receptors in the GC and PGC.

mature neurons, respectively (Bordey 2006; Ge et al. 2007). Three major types of GABA receptors (GABARs) have been identified: the ionotropic $GABA_A$ and $GABA_C$ receptors and the metabotropic $GABA_B$ receptor (Chebib and Johnston 1999).

Electrophysiological analysis has shown the existence of functional $GABA_A Rs$ in NPCs and their neuronal progeny in the adult brain (Wang et al. 2003a; Wang et al. 2005). GABA, released by local interneurons, has emerged as a crucial player in regulating multiple developmental steps of adult neurogenesis (Ge et al. 2007).

Role of GABA in Hippocampal Neurogenesis

In the SGZ of the adult DG, neurogenesis involves a series of differentiation steps from radial glia-like stem/progenitor (type-1) cells, to transient amplifying neuronal progenitor (type-2) cells, to postmitotic dentate granule cells (DGCs; Fig. 1A) (Kempermann et al. 2004b). In this review, both type-1 and type-2 cells will be collectively referred to as NPCs. Electrophysiological recording with retrovirus-mediated birth dating showed that newborn cells are activated by ambient GABA before receiving any functional synaptic inputs in the adult hippocampus (Ge et al. 2006). Recordings in adult transgenic mice that express green fluorescent protein (GFP) under the control of nestin promoter have further shown that type-2 cells in the SGZ already exhibit both spontaneous and evoked GABAergic synaptic transmission (Tozuka et al. 2005; Wang et al. 2005). Interestingly, application of a $GABA_A R$ antagonist in vivo substantially increases the proliferation of type-2 NPCs, whereas $GABA_A R$ agonist treatment exhibits the opposite effect (Table 1) (Tozuka et al. 2005).

GABA also regulates NPC differentiation during adult hippocampal neurogenesis. In a slice culture preparation, activation of $GABA_A Rs$ in new neurons induces the expression of transcription factor NeuroD, a positive regulator of neuronal differentiation (Tozuka et al. 2005). This effect of GABA is likely due to the depolarizing action of GABA on these immature neurons from high expression of Na^+-K^+-$2Cl^-$ cotransporter NKCC1 that accumulates cytoplasmic Cl^- (Tozuka et al. 2005; Ge et al. 2006). The role of GABA-induced depolarization in hippocampal neurogenesis in vivo was examined using retrovirus-mediated expression of short-hairpin RNAs to knock down the expression of endogenous NKCC1 specifically in NPCs and their progeny (Ge et al. 2006). Interestingly, a lack of GABA-induced depolarization in newborn neurons leads to significant defects in their dendritic development and synapse formation in the adult brain. In another study, using pro-opiomelanocortin (POMC)-GFP transgenic mice to identify newborn DGCs, it was suggested that the much delayed DGC maturation in adult compared to neonates is partially due to a reduced depolarizing GABA-mediated network activity in adults (Overstreet-Wadiche et al. 2006a). Taken together, these studies

Table 1. Effects of manipulations of neurotransmitter systems on adult neurogenesis

Neurotransmitters	Dentate gyrus of the hippocampus		
	proliferation	migration/differentiation	synaptic integration
Glutamate	↓ NMDAR activation (Cameron et al. 1995; Gould et al. 1997; Kitayama et al. 2003; Nacher et al. 2001, 2003; Poulsen et al. 2005) ↓ mGluRs activation (Yoshimizu and Chaki 2004) ↓ AMPA/kainate receptors activation (Bernabeu and Sharp 2000; Poulsen et al. 2004) ↓ mGluR5 KO/mGluR2/3 inhibition (Di Giorgi-Gerevini et al. 2005) ↑ AMPAR activation (Bai et al. 2003) ↑ kainate receptors activation (Gray and Sundstrom 1998)	↑ NMDAR inhibition (Cameron et al. 1995; Nacher et al. 2001, 2003) ↑ NMDARs activation (Kitayama et al. 2003; Tashiro et al. 2006)	
GABA	↓ $GABA_AR$ (Tozuka et al. 2005)	↑ $GABA_AR$ (Tozuka et al. 2005)	↑GABA via NKCC1 (Ge et al. 2006)
Acetylcholine	↓ ACh inhibition by forebrain injury (Mohapel et al. 2005) ↓ nAChR β2 subunit KO (Harrist et al. 2004) → pharmacological ACh enhancement (Kaneko et al. 2006)	↑ pharmacological ACh enhancement (Kaneko et al. 2006; Kotani et al. 2006) ↓ ACh inhibition by forebrain injury (Cooper-Kuhn et al. 2004; Mohapel et al. 2005)	
Serotonin	↑ 5-HT synthesis inhibition (Jha et al. 2006) ↑ $5\text{-}HT_{1A}$ receptor (Radley and Jacobs 2002; Santarelli et al. 2003, Banasr et al. 2004; Encinas et al. 2006)	↓ 5-HT deletion (Brezun and Daszuta 1999) ↑ $5\text{-}HT_{1A}$ receptor (Banasr et al. 2004)	

Neurotransmitters	proliferation	SVZ/OB	
		migration/differentiation	synaptic integration
	↑ SSRIs (Malberg et al. 2000; Lee et al. 2001; Malberg and Duman 2003; Kodama et al. 2004; Encinas et al. 2006) → 5-HT depletion (Jha et al. 2006)	→ SSRIs (Malberg et al. 2000; Santarelli et al. 2003; Encinas et al. 2006) → 5-HT level depletion (Jha et al. 2006)	
Dopamine	↓ dopaminergic denervation (Hoglinger et al. 2004)		
Nitric oxide	↑ nNOS inhibition with L-NAME (Packer et al. 2003; Park et al. 2003) ↑ nNOS inhibition with 7-NI (Park et al. 2003; Zhu et al. 2006) ↑ nNOS KO (Packer et al. 2003; Zhu et al. 2006) ↑ NO donor (Zhang et al. 2001; Lu et al. 2003) → nNOS inhibition with L-NAME/7-NI (Moreno-Lopez et al. 2004)		
Glutamate	↓ mGluR2, 3,5 activation (Di Giorgi-Gerevini et al. 2005) ↑ AMPA (Xu et al. 2005) → NMDA (Kitayama et al. 2003)		
GABA	↓ GABA$_A$R (Liu et al. 2005)	↓ GABA$_A$R (Bolteus and Bordey 2004)	↑ GABA$_A$R (Gascon et al. 2006)
Acetylcholine	→ nAChR β2 subunit KO (Mechawar et al. 2004) → ACh inhibition by forebrain injury (Kaneko et al. 2006)	↑ pharmacological ACh enhancement (Kaneko et al. 2006) ↓ ACh inhibition by forebrain injury (Cooper-Kuhn et al. 2004)	

(continued)

Table 1. (*continued*)

Neurotransmitters	SVZ/OB		
	proliferation	migration/differentiation	synaptic integration
Serotonin	↑5-HT deletion (Brezun and Daszuta 1999) ↑5-HT$_{1A}$ receptor (Banasr et al. 2004) ↓5-HT$_{2A/2C}$ receptor (Banasr et al. 2004; Wang et al. 2004; Green et al. 2006) ↑5-HT$_{2C}$ receptor (Banasr et al. 2004) →SSRI (Malberg et al. 2000; Santarelli et al. 2003; Kodama et al. 2004; Encinas et al. 2006)	↑5-HT$_{1A}$ agonists (Banasr et al. 2004) ↑5-HT$_{2C}$ agonists (Banasr et al. 2004)	
Dopamine	↑D$_2$ receptor (Hoglinger et al. 2004) ↑D$_3$ receptor (Baker et al. 2004; Coronas et al. 2004; Hoglinger et al. 2004; Van Kampen et al. 2004) ↓Dopaminergic denervation (Baker et al. 2004; Hoglinger et al. 2004; Freundlieb et al. 2006) →D$_1$ receptor (Hoglinger et al. 2004)	↑D$_2$ receptor (Feron et al. 1999) ↑D$_3$ receptor (Van Kampen et al. 2004) ↓dopaminergic denervation (Hoglinger et al. 2004)	
Nitric oxide	↑nNOS inhibition with L-NAME (Cheng et al. 2003; Packer et al. 2003; Moreno-Lopez et al. 2004; Torroglosa et al. 2007) ↑nNOS inhibition with 7-NI (Moreno-Lopez et al. 2004) ↑nNOS KO (Packer et al. 2003) ↓NO donor (Matarredona et al. 2004)	↓nNOS inhibition with L-NAME (Cheng et al. 2003; Moreno-Lopez et al. 2004) →nNOS inhibition with L-NAME (Moreno-Lopez et al. 2004) →nNOS inhibition with L-NAME on differentiation (Packer et al. 2003)	↓nNOS KO (Chen et al. 2004)

established that GABA-induced depolarization promotes the differentiation, maturation, and functional integration of newly generated DGCs in the adult brain (Fig. 1B).

Role of GABA in SVZ Neurogenesis

In the adult SVZ, a subset of glial fibrillary acidic protein (GFAP)-expressing cells exhibit some NPC properties and give rise to neuroblasts, which then migrate through the rostral migratory stream (RMS) to the olfactory bulb (OB) and differentiate into either granule cells or periglomerular cells, two types of interneurons (Fig. 2A) (Alvarez-Buylla and Lim 2004). Interestingly, GABA is released by migrating neuroblasts while NPCs of the postnatal SVZ express functional GABA$_A$Rs (Liu et al. 2005). Similar to the situation in the embryonic SVZ (LoTurco et al. 1995; Haydar et al. 2000), activation of GABA$_A$Rs inhibits proliferation of postnatal GFAP-expressing NPCs in a slice culture preparation (Liu et al. 2005). Thus, GABA released from neuroblasts may provide a negative-feedback mechanism to control the proliferation of NPCs through tonic activation of GABA$_A$Rs. Whether GABA also regulates the differentiation of NPCs into neurons in the adult SVZ has yet to be determined.

Tangentially migrating neuroblasts from the SVZ maintain expression of functional GABA$_A$Rs while they are still lacking synaptic inputs (Fig. 2A,B) (Carleton et al. 2003). Electrophysiological recordings have shown that GABA depolarizes newborn neurons in the SVZ and RMS (Wang et al. 2003b). Interestingly, inhibition of GABA$_A$Rs by an antagonist bicuculline speeds up the migration (Bolteus and Bordey 2004), suggesting that endogenous GABA reduces the velocity of neuronal migration during adult SVZ neurogenesis. Once these migrating cells arrive at the OB, growth of the dendritic structure is an essential step that leads to the synaptic integration of these newborn neurons into the existing circuitry. In contrast to the negative role of GABA on early neurogenic steps (Gascon et al. 2006), activation of GABA$_A$Rs promotes growth cone motility and stimulates the initiation of dendritic growth of SVZ-derived newborn neurons in vitro (Fig. 2C) (Vutskits et al. 2006). Thus, similarly as during adult hippocampal neurogenesis, GABA may promote dendritic development of newly generated olfactory interneurons.

In the SVZ and SGZ, immature cells are tonically activated by ambient GABA before receiving any functional synaptic inputs (Liu et al. 2005; Ge et al. 2006). Within the DG, principal DGCs and interneurons form extensive recurrent connections (Freund and Buzsaki 1996). The levels of ambient GABA, regulated by interneuron activities, may serve as a general indicator of the dynamic neuronal network activity and regulate behaviors of NPCs and their progeny at early stages of development. GABAergic synaptic inputs, formed at the progenitor stage well before the formation of functional glutamatergic inputs (Overstreet Wadiche et al. 2005; Wang et al. 2005; Ge et al. 2006; Karten et al. 2006), can further regulate newborn neurons in an input-specific fashion. Together, GABA signaling serves as the major activity-dependent regulator for early phases of adult neurogenesis.

GLUTAMATE

Glutamate was discovered as the first major excitatory neurotransmitter in the CNS in the 1960s (Curtis and Watkins 1961). Disregulation of glutamate has been associated with many neurological disorders, such as amyotrophic lateral sclerosis, Alzheimer's disease, Huntington's disease, Parkinson's disease, ischemia, and epilepsy (Lipton and Rosenberg 1994; Maragakis and Rothstein 2004). Glutamate receptors are classified into ionotropic glutamate receptors (iGluRs) and metabotropic glutamate receptors (mGluRs) (Kew and Kemp 2005). According to their preferred agonists, the iGluRs include N-methyl-D-aspartic acid receptors (NMDARs), α-amino-3-hydroxy-5-methyl-4-isoxazolepropionic acid receptors (AMPARs), and kainate receptors. Similar to GABA, glutamate signaling also regulates adult neurogenesis via different mechanisms. First, glutamate escaped from the synaptic cleft and released from astrocytes may tonically activate the receptors on these immature cells and regulate their development (Montana et al. 2004). Second, recent electrophysiological studies have shown that newborn neurons from the SGZ and SVZ begin to receive functional glutamatergic synaptic inputs 2 weeks after they are born in the adult brain (Belluzzi et al. 2003; Carleton et al. 2003; Esposito et al. 2005; Ge et al. 2006). Therefore, glutamate appears to exhibit a direct regulatory role on later phases of adult neurogenesis. Third, as the major excitatory neurotransmitter in the adult brain, glutamate activates a variety of mature neurons, which in turn may release factors to regulate adult neurogenesis.

Role of Glutamate in Hippocampal Neurogenesis

Mature DGCs receive dense glutamatergic inputs from the entorhinal cortex via perforant path fibers (Fig. 1A) (Collingridge and Lester 1989).

Glutamate has been shown to influence the proliferation and differentiation of NPCs as well as the migration and survival of their neuronal progeny (Table 1) (Cameron et al. 1995; Lujan et al. 2005; Suzuki et al. 2006). The role of glutamate in regulating NPC proliferation in the adult brain is controversial. Inhibition of NMDARs with various antagonists induces NPC proliferation in the DG in vivo (Gould et al. 1994; Cameron et al. 1995; Nacher et al. 2001, 2003; Kitayama et al. 2003) and in vitro (Poulsen et al. 2005). Similar results have been obtained by direct application of mGluR antagonists in adult mouse hippocampus (Yoshimizu and Chaki 2004). However, knock out of the mGluR5 subunit or systemic injection of mGluR2/3 antagonists suppresses NPC proliferation in the adult DG (Di Giorgi-Gerevini et al. 2005). Opposite effects of glutamate via AMPARs on NPC proliferation have also been reported (Table 1). For example, blockade of AMPARs (Poulsen et al. 2005) or AMPA/kainate receptors (Bernabeu and Sharp 2000) increases NPC proliferation in some experimental paradigms, whereas application of the AMPAR potentiator (Bai et al. 2003) or kainic acid, an agonist of kainate receptors (Gray and Sundstrom 1998), also increases NPC proliferation in the adult hippocampus in other studies. The discrepancy between these results may be due to differences in antagonists applied, different GluR subtypes involved, and different experimental approaches used in each study. More importantly, globally manipulating the signaling of the main excitatory neurotransmitter in the brain will certainly affect neuronal activity of the existing circuitry; thus, effects of glutamate on NPCs are likely to be complex and may be indirect.

Glutamate also regulates the differentiation of NPCs and survival of newborn DGCs (Fig. 1B). Antagonists of NMDARs have been shown to increase the number of new immature DGCs expressing polysialic-acid-neural cell adhesion molecule (PSA-NCAM) or doublecortin (DCX), two immature neuronal markers, in the adult hippocampus (Nacher et al. 2001, 2003). Furthermore, blockade of NMDARs triggers neuronal differentiation as measured by colabeling of bromodeoxyuridine (BrdU) and neuron-specific enolase (NSE), a neuronal marker (Cameron et al. 1995). In contrast, activation of NMDARs of cultured NPCs derived from the adult hippocampus increases the expression of NeuroD while decreasing the expression of glial fate genes such as *Hes1* and *Id2* (Deisseroth et al. 2004). Moreover, direct application of NMDA promotes neuronal differentiation of hippocampal NPCs in vitro (Kitayama et al. 2004). The differences in these in vivo and in vitro findings may be explained by cell autonomous and noncell autonomous effects of complex glutamate signaling. Therefore, manipulating glutamate signaling specifically in NPCs may provide insight into the regulatory role of glutamate signaling in adult neurogenesis. Indeed, a recent study with "single-cell knockout" of the NMDAR subunit NR1 specifically in proliferating NPCs demonstrated that survival of newborn DGCs is regulated by NMDAR signaling in the adult hippocampus (Tashiro et al. 2006).

Role of Glutamate in SVZ Neurogenesis

Much less is known about the role of glutamate in regulating adult SVZ neurogenesis. Administration of mGluR2/3 or 5 subunit antagonists as well as deletion of the mGluR5 subunit in mice result in an enhanced proliferation of NPCs in the adult SVZ (Di Giorgi-Gerevini et al. 2005). On the other hand, NMDA fails to affect either NPC proliferation or the expression of nestin and proliferating cell nuclear antigen (PCNA) in the adult mouse SVZ and OB in another study (Kitayama et al. 2003). It is currently unknown whether glutamate also regulates neuronal differentiation and migration in the adult SVZ. During embryonic development, elevated glutamate, via NMDAR activation, stimulates radial migration of neurons from SVZ/VZ to the developing cortex (Behar et al. 1999); thus, glutamate may also regulate the motility of migratory neurons in the adult brain.

Taken together, these studies have revealed the importance and complexity of glutamate signaling in regulating different steps of adult hippocampal and SVZ neurogenesis. Future studies with more sophisticated genetic approaches are needed to sort out the cell autonomous and non-cell autonomous roles of glutamate signaling in regulating adult neurogenesis and may provide insight into the puzzling opposite effects of glutamate under various conditions.

ACETYLCHOLINE

ACh was first identified as a neurotransmitter in 1921 (Eiden 1998). ACh-mediated innervation regulates many neuronal processes, such as transmitter release, cell excitability, and neuronal integration (Changeux et al. 1998). There are two main classes of acetylcholine receptors (AChRs): nicotinic AChRs (nAChRs), a family of ligand-gated ion channels, and metabotropic muscarinic AChRs (mAChRs), a family of G-protein-coupled receptors (Caulfield and Birdsall 1998; Changeux et al. 1998). Cholinergic fibers innervate newborn neurons in both adult hippocampus and olfactory system.

Role of ACh in Hippocampal Neurogenesis

Cholinergic innervation from the medial septal nucleus and the vertical limb of the diagonal band projects to the hippocampus (Fig. 1) (Mesulam et al. 1983). Multiple types of AChR subunits are expressed in NPCs (Mohapel et al. 2005) and their neuronal progeny in the adult DG (Kaneko et al. 2006; Kotani et al. 2006). Accumulating evidence suggests that ACh signaling regulates NPC proliferation in the adult hippocampus (Table 1). For example, depletion of ACh by forebrain injury reduces NPC proliferation, whereas systemic administration of cholinergic agonists has opposite effects (Mohapel et al. 2005). Similarly, knockout of the β2 subunit of the nAChR in mice results in decreased proliferation of NPCs in the adult SGZ (Harrist et al. 2004). Application of donepezil, a long-lasting inhibitor of acetylcholinesterase, however, does not lead to significant changes in the NPC proliferation (Kaneko et al. 2006). Experiments with pharmacological application (Kaneko et al. 2006) and disturbance of cholinergic inputs by forebrain injury (Cooper-Kuhn et al. 2004) also suggest a role of ACh in promoting the survival and neuronal differentiation of NPCs. Whether ACh also regulates the synaptic integration of newborn DGCs has yet to be determined.

Role of ACh in SVZ Neurogenesis

In the olfactory system, ACh appears to regulate late developmental steps of newborn granule neurons derived from NPCs (Fig. 2A,C). Knock out of β2-nAChRs in mice or depletion of ACh by forebrain injury does not effect NPC proliferation in the SVZ and RMS (Mechawar et al. 2004; Kaneko et al. 2006). Various types of AChR subunits are expressed in newborn immature neurons when they migrate out of the SVZ (Kaneko et al. 2006). Enhanced ACh levels promote neuronal differentiation and survival of newborn neurons in OB without affecting migration and neuronal fate specification (Cooper-Kuhn et al. 2004; Mechawar et al. 2004; Kaneko et al. 2006).

In summary, ACh serves as a positive regulator of adult neurogenesis in both the SVZ/OB and SGZ/hippocampus (Cooper-Kuhn et al. 2004; Mohapel et al. 2005; Kaneko et al. 2006; Kotani et al. 2006). It remains largely unclear whether observed effects of ACh is through direct/indirect and synaptic/nonsynaptic mechanisms.

DOPAMINE

Dopamine, one of the major neurotransmitters involved in Parkinson's disease, influences neurogenesis in the adult CNS via regulation of NPC proliferation (Table 1). Dopamine receptors are classified into two groups according to the structural homology and shared second-messenger cascades: the D_1-like subclass including the D_1 and D_5 receptors and the D_2-like subclass including the D_2, D_3, and D_4 receptors (Missale et al. 1998; Borta and Hoglinger 2007). Dopamine synthesis involves tyrosine hydroxylase (TH) as the rate-limiting enzyme, and expression of TH is commonly used as a marker of dopaminergic neurons. Dopaminergic neurons are localized mainly in the pars compacta of the substantia nigra and the ventral tegmental area, whereas subtypes of the dopamine receptor are widely expressed in the hippocampus (Amenta et al. 2001) and SVZ (Diaz et al. 1997). Activation of dopamine receptors has been shown to modulate neurogenesis in both the developing (Ohtani et al. 2003) and adult brain (Table 1) (Van Kampen et al. 2004).

Role of Dopamine in Hippocampal Neurogenesis

Dopaminergic fibers from the ventral tegmental area are found in the vicinity of NPCs in the adult SGZ (Fig. 1). It has been shown that the proliferation of NPCs is decreased by dopaminergic denervation induced by 1-methyl 4-phenyl 1,2,3,6-tetrahydropyridine (MPTP) or 6-hydroxy-dopamine (6-OHDA), two neurotoxins commonly used to induce dopaminergic neuron degeneration in animal models (Hoglinger et al. 2004). Enhanced dopaminergic neurotransmission due to cocaine administration, however, decreases NPC proliferation without significantly affecting either their maturation or their differentiation (Yamaguchi et al. 2004; Dominguez-Escriba et al. 2006). These seemingly contradictory results on cell proliferation in the SGZ might be caused by differences in the activation of receptor subtypes and/or experimental approaches. For example, the effect of dopaminergic denervation by MPTP or 6-OHDA is mainly associated with the D_2 receptor (Hoglinger et al. 2004), whereas that of cocaine involves the D_1 receptor (Harvey 2004). Furthermore, cocaine induces complex changes in the CNS neurochemistry and thereby also influences neurogenesis via mechanisms independent of the dopaminergic system.

Role of Dopamine in SVZ Neurogenesis

The SVZ receives significant axonal innervations from the substantia nigra, the main source of dopamine in the brain (Fig. 2A) (Prakash and Wurst 2006; Riquelme et al. 2007). Accumulating evidence indicates that dopamine stimulates precursor cell proliferation mainly through the D_2-like receptor class (Baker et al. 2004; Freundlieb et al. 2006). Activation of D_2 or D_3 receptors by pharmacological manipulations promotes the proliferation of adult SVZ cells (Coronas et al. 2004; Van Kampen et al. 2004), whereas activation of other receptors, such as D_1 receptors, does not (Fig. 2C) (Hoglinger et al. 2004). In addition, ablation of dopaminergic innervation using neurotoxins MPTP and 6-OHDA reduces SVZ cell proliferation, an effect alleviated by administration of a D_3 receptor agonist (Baker et al. 2004; Hoglinger et al. 2004). Dopamine also appears to influence neuronal migration and differentiation in the SVZ. For example, dopaminergic denervation decreases neuroblast migration and the number of differentiated new neurons in the OB (Hoglinger et al. 2004). Furthermore, activation of D_2 or D_3 receptors has been shown to enhance the maturation and differentiation of NPCs into neurons in the SVZ-RMS and OB (Feron et al. 1999; Van Kampen et al. 2004).

Role of Dopamine in Nonneurogenic Regions

A combination of in vitro and in vivo experiments suggests that the substantia nigra of adult mammals contains multipotent NPCs with the capacity to differentiate into new neurons (Lie et al. 2002; Zhao et al. 2003). Dopamine, through the activation of D_2 or D_3 receptors, appears to regulate this population of NPCs in the substantia nigra (Zhao et al. 2003) and in the striatum (Van Kampen et al. 2004). Chronic D_3 receptor stimulation induces cell proliferation and appears to promote the expression of dopaminergic neuronal subtype markers in some of these nascent cells (Van Kampen and Robertson 2005). In a more recent study using a 6-OHDA animal model of Parkinson's disease, the number of TH, BrdU, and NeuN colabeled cells appeared to be significantly increased following administration of a D_3 receptor agonist in the lesioned substantia nigra (Van Kampen and Eckman 2006). Several other groups, however, failed to observe any significant number of new TH-expressing neurons in the substantia nigra under normal or similar pathological conditions (Lie et al. 2003; Cooper and Isacson 2004). Residential NPCs have been shown to exist in various brain regions and to be reactivated by specific injuries (Emsley et al. 2005). Clearly, future studies are needed to further clarify the existence of endogenous and injury-induced adult neurogenesis in the substantia nigra and its regulation by dopamine.

SEROTONIN

Serotonin (5-hydroxytryptamine, 5-HT) is an important neurotransmitter and neuromodulator in the mammalian CNS. Impaired serotonin transmission has been implicated in cognitive and various psychiatric disorders, such as generalized anxiety disorders, depression, and schizophrenia (Kempermann 2002). Serotoninergic neurons are located in the dorsal raphe nuclei (DRN) and medial raphe nuclei (MRN) and contain tryptophan hydroxylase (TPH), the rate-limiting enzyme for serotonin synthesis. Several subtypes of 5-HT receptors are known, including $5\text{-}HT_{1A,1B}$, $5\text{-}HT_{2A,2B,2C}$, $5\text{-}HT_{3A}$, $5\text{-}HT_4$, $5\text{-}HT_{5A}$, and $5\text{-}HT_7$. Of these, the involvement of $5\text{-}HT_{1A,1B}$ and $5\text{-}HT_{2A,2C}$ in adult neurogenesis has been examined most extensively to date, mainly through systemic administration of various agonists and antagonists (Table 1). Whether $5\text{-}HT_{3A}$, $5\text{-}HT_4$, $5\text{-}HT_{5A}$, and $5\text{-}HT_7$ are also involved in regulating adult neurogenesis remains largely unknown (Radley and Jacobs, 2002; Banasr et al. 2004).

Role of Serotonin in Hippocampal Neurogenesis

In the DG, serotoninergic axons from the DRN terminate in the molecular layer, whereas those from the MRN project mainly to the hilus (Fig. 1) (Stanfield and Trice 1988). It has been shown that an increase in the level of 5-HT enhances the proliferation of NPCs through activation of the $5\text{-}HT_{1A}$ receptor (Santarelli et al. 2003), whereas inhibition of $5\text{-}HT_{1A}$ receptor results in reduced NPC proliferation (Radley and Jacobs 2002). Selective serotonin reuptake inhibitors (SSRIs), such as fluoxetine (Prozac), are the most commonly used antidepressants and function to increase the level of 5-HT by selectively blocking the presynaptic reuptake of serotonin. Administration of SSRIs has been shown to regulate the proliferation of NPCs in the adult hippocampus under both physiological (Malberg et al. 2000) and pathological conditions (Lee et al. 2001; Malberg and Duman 2003). Serotonin affects neurogenesis from the early progenitor stage (transient amplifying cells) to the differentiation phase in the adult DG (Encinas et al. 2006). Depletion of 5-HT by blocking 5-hydroxytryptamine synthesis or selective lesions of serotoninergic neurons suppresses the differentiation of NPCs (Brezun and Daszuta 1999), whereas elevation of 5-HT enhances neuronal differentiation of NPCs (Banasr et al. 2004). In some other studies, however, an increase in 5-HT levels induced proliferation but not differentiation of NPCs (Malberg et al. 2000; Santarelli et al. 2003). Likewise, selective 5-HT depletion was reported to have no effects on proliferation, survival, and differentiation of SGZ NPCs in the adult hippocampus (Jha et al. 2006). There are several possible explanations for the discrepancy, such as differences in animal strains, gender, methods, and pathological conditions. In addition to a direct regulatory role of serotonin in the proliferation and differentiation of NPCs, activation of the 5-HT$_{1A}$ receptor also initiates neurotrophic events in modulating neurite outgrowth and neuronal survival by activating various signal transduction cascades (Fricker et al. 2005).

Role of Serotonin in SVZ Neurogenesis

5-HT, released by serotoninergic neurons projecting into the ventricles via the supraependymal nerve plexus, also appears to regulate the proliferation of NPCs in SVZ (Fig. 2). Depletion of 5-HT by TPH inhibitors or selective lesion of 5-HT neurons inhibits NPC proliferation (Brezun and Daszuta 1999). Interestingly, activation of different 5-HT receptor subtypes exhibits differential effects on the proliferation of NPCs. Administration of agonists for 5-HT$_{1A}$ and 5-HT$_{2C}$ and antagonists for 5-HT$_{2A/2C}$ significantly enhances the proliferation of NPCs in the SVZ, whereas 5-HT$_{1B}$ agonists decrease NPC proliferation without affecting their differentiation into neuronal and glial phenotypes (Banasr et al. 2004). Consistent with these results, olanzapine, a potent 5-HT$_{2A/2C}$ antagonist that has been used for treating patients with schizophrenia and manic episodes associated with bipolar I disorder, was shown to increase NPC proliferation in the SVZ (Wang et al. 2004; Green et al. 2006). Whether serotonin regulates integration of new neurons awaits further investigation.

NITRIC OXIDE

NO, a gaseous radical synthesized from L-arginine by NO synthase (NOS), was first identified as a neurotransmitter and biological messenger molecule in the CNS in 1988 and has since been implicated in many physiological and pathological processes in the brain (Garthwaite et al. 1988). Three different isoforms of NOS have been identified in mammals: neuronal NOS (nNOS), inducible NOS (iNOS), and endothelial NOS (eNOS). Of these, nNOS is the principal isotype expressed in the brain, including regions where adult neurogenesis occurs, such as hippocampal DG (Islam et al. 2003), forebrain SVZ, RMS (Moreno-Lopez et al. 2000), and OB (Chen et al. 2004). nNOS acts as a modulator of neuronal differentiation, survival, and synaptic plasticity (Gibbs 2003; Hoglinger et al. 2004). The other isoforms, eNOS and iNOS, are also found in other cell types in the CNS (Heneka and Feinstein, 2001; Dinerman et al. 1994), and eNOS has been suggested to regulate adult hippocampal neurogenesis (Reif et al. 2004).

Role of NO in Hippocampal Neurogenesis

Although still controversial, most studies suggest that NO negatively regulates hippocampal NPC proliferation (Fig. 1; Table 1). For example, pharmacological inhibition of nNOS by N^{ω}-nitro-L-arginine methyl ester (L-NAME) or 7-nitroindazole (7-NI), as well as deletion of the nNOS gene, enhances NPC proliferation in the SGZ of the adult rodent hippocampus (Packer et al. 2003; Park et al. 2003; Zhu et al. 2006). eNOS-deficient mice, however, exhibit a significant reduction in SGZ NPC proliferation (Reif et al. 2004). In one study, inhibition of nNOS exhibited no effects on NPC proliferation in the hippocampus, but produced a dose- and time-dependent increase in NPC proliferation in the olfactory system (Moreno-Lopez et al. 2004). In contrast to the loss-of-function studies, gain-of-function analysis with NO donor administration appears to promote NPC proliferation (Zhang et al. 2001; Lu et al. 2003).

nNOS knockout and pharmacological studies suggest that NO might also regulate the differentiation of NPCs and negatively influence the migration and survival of newborn neurons in the DG (Zhu et al. 2006). On the other hand, administration of DETA/NONOate, an NO donor, significantly enhanced the proliferation, survival, migration, and differentiation of NPCs in the hippocampus after brain injury (Lu et al. 2003). Since DETA/NONOate also stimulates the secretion of vascular endothelial growth factor (VEGF) (Zhang et al. 2003), which is known to promote neurogenesis (Sun and Guo 2005), the enhanced neurogenesis observed in the DG following systemic administration of NO donors may involve indirect mechanisms.

Role of NO in SVZ Neurogenesis

NO also appears to negatively regulate SVZ NPCs under physiological conditions (Fig. 2) (Cheng et al. 2003; Matarredona et al. 2004; Moreno-Lopez et al. 2004; Torroglosa et al. 2007). Inhibition of nNOS through administration of L-NAME and 7-NI enhances the proliferation of NPCs (Moreno-Lopez et al. 2004; Packer et al. 2003), whereas administration of an NO donor leads to a decrease in proliferation without increasing apoptosis (Matarredona et al. 2004). Similar results were obtained when these agents were administrated directly into the cerebral ventricle (Cheng et al. 2003) or using continuous-release osmotic minipumps over a period of 7 days (Packer et al. 2003). Consistent with results from these pharmacological experiments, knockout of the nNOS gene in mice also resulted in increased NPC proliferation in the SVZ, including quiescent cells with astrocyte-like properties (Packer et al. 2003).

No effect of NO on cell survival and migration was observed following systemic (Moreno-Lopez et al. 2004) or local administration (Packer et al. 2003) of NOS inhibitors. Yet, nNOS inhibition has been shown to cause a delay in neuronal differentiation of newly arriving neuroblasts at the OB (Moreno-Lopez et al. 2004). It has subsequently been reported that NO appears to be necessary for glomerular organization and contributes to the preservation of synaptic integrity in the adult OB (Chen et al. 2004).

SUMMARY

Significant progress has been made in recent years to document differential regulatory roles of various neurotransmitters in regulating distinct steps of adult neurogenesis in the SGZ/hippocampus and SVZ/OB. These studies also demonstrate that experimental manipulations of neurotransmitter signaling could lead to cell autonomous/direct effects on NPCs and their progeny, as well as noncell autonomous/indirect effects through modulation of mature neurons within the existing neuronal networks. Many of the controversies arisen from early studies can be addressed in the future with specific genetic manipulation of NPCs and their progeny in the adult brain, as in the case of examining GABA (Ge et al. 2006) and NMDAR (Tashiro et al. 2006) signaling. Adult neurogenesis is well known to be regulated by different physiological and pathological stimuli that change the neural network activity and therefore the level of neurotransmitter release (Abrous et al. 2005; Ming and Song 2005). There are many questions that remain to be answered. First, we do not yet have a complete picture of how different neurotransmitter systems mediate the effects of these external stimuli on different steps of adult neurogenesis. Second, we know very little about the downstream signaling mechanisms underlying differential effects of neurotransmitters on NPC proliferation and development. Third, we know even less about the roles of different neurotransmitter systems in regulating neurogenesis in nonneurogenic regions after insults or degenerative neurological diseases.

Adult neurogenesis contributes to both plasticity and regenerative capacity of the adult brain, and potential future therapeutic applications may very well be based on the manipulation of such regenerative capacity of the mature CNS (Lie et al. 2004; Ming and Song 2005; Sailor et al. 2006). During the past decade, significant progress has been made to decipher the basic mechanisms regulating the processes involved in adult neurogenesis (Abrous et al. 2005; Ming and Song 2005; Lledo et al. 2006). A better understanding of such mechanisms, including neurotransmitter signaling, may provide further insights into functions of persistent neurogenesis in healthy brains. Ultimately, this knowledge may provide the foundation for therapeutic interventions for neurodegenerative diseases.

ACKNOWLEDGMENTS

We thank Rusty Gage, Gerd Kempermann, and Dedrick Jordan for comments. Research in our laboratories is supported by the National Institutes of Health (NIH), March of Dimes, Klingenstein fellowship Award in the Neuroscience, Whitehall Foundation, and Sloan Scholar Award (to G.-l.M.) and by the NIH, McKnight Scholar Award, the Rett Syndrome Research Foundation, and The Packard Center for ALS at Johns Hopkins and MDA's Wings Over Wall Street (to H.S.).

REFERENCES

Abrous D.N., Koehl M., and Le Moal M. 2005. Adult neurogenesis: From precursors to network and physiology. *Physiol. Rev.* **85:** 523–569.

Alvarez-Buylla A. and Lim D.A. 2004. For the long run: Maintaining germinal niches in the adult brain. *Neuron* **41:** 683–686.

Amenta F., Mignini F., Ricci A., Sabbatini M., Tomassoni D., and Tayebati S.K. 2001. Age-related changes of dopamine receptors in the rat hippocampus: A light microscope autoradiography study. *Mech. Ageing Dev.* **122:** 2071–2083.

Bai F., Bergeron M., and Nelson D.L. 2003. Chronic AMPA receptor potentiator (LY451646) treatment increases cell proliferation in adult rat hippocampus. *Neuropharmacology* **44:** 1013–1021.

Baker S.A., Baker K.A., and Hagg T. 2004. Dopaminergic nigrostriatal projections regulate neural precursor proliferation in the adult mouse subventricular zone. *Eur. J. Neurosci.* **20:** 575–579.

Banasr M., Hery M., Printemps R., and Daszuta A. 2004. Serotonin-induced increases in adult cell proliferation and neurogenesis are mediated through different and common 5-HT receptor subtypes in the dentate gyrus and the subventricular zone. *Neuropsychopharmacology* **29:** 450–460.

Barker J.L., Behar T., Li Y.X., Liu Q.Y., Ma W., Maric D., Maric I., Schaffner A.E., Serafini R., Smith S.V., et al. 1998. GABAergic cells and signals in CNS development. *Persp. Dev. Neurobiol.* **5:** 305–322.

Behar T.N., Scott C.A., Greene C.L., Wen X., Smith S.V., Maric D., Liu Q.Y., Colton C.A., and Barker J.L. 1999. Glutamate acting at NMDA receptors stimulates embryonic cortical neuronal migration. *J. Neurosci.* **19:** 4449–4461.

Belluzzi O., Benedusi M., Ackman J., and LoTurco J.J. 2003. Electrophysiological differentiation of new neurons in the olfactory bulb. *J. Neurosci.* **23:** 10411–10418.

Ben-Ari Y., Khazipov R., Leinekugel X., Caillard O., and Gaiarsa J.L. 1997. GABAA, NMDA and AMPA receptors: A developmentally regulated "menage a trois." *Trends Neurosci.* **20:** 523–529.

Bernabeu R. and Sharp F.R. 2000. NMDA and AMPA/kainate glutamate receptors modulate dentate neurogenesis and CA3 synapsin-I in normal and ischemic hippocampus. *J. Cereb. Blood Flow Metab.* **20:** 1669–1680.

Bolteus A.J. and Bordey A. 2004. GABA release and uptake regulate neuronal precursor migration in the postnatal subventricular zone. *J. Neurosci.* **24:** 7623–7631.

Bordey A. 2006. Adult neurogenesis: Basic concepts of signaling. *Cell Cycle* **5:** 722–728.

Borta A. and Hoglinger G.U. 2007. Dopamine and adult neurogenesis. *J. Neurochem.* **100:** 587–595.

Brezun J.M. and Daszuta A. 1999. Depletion in serotonin decreases neurogenesis in the dentate gyrus and the subventricular zone of adult rats. *Neuroscience* **89:** 999–1002.

Cameron H.A., McEwen B.S., and Gould E. 1995. Regulation of adult neurogenesis by excitatory input and NMDA receptor activation in the dentate gyrus. *J. Neurosci.* **15:** 4687–4692.

Carleton A., Petreanu L.T., Lansford R., Alvarez-Buylla A., and Lledo P.M. 2003. Becoming a new neuron in the adult olfactory bulb. *Nat. Neurosci.* **6:** 507–518.

Caulfield M.P. and Birdsall N.J. 1998. International Union of Pharmacology. XVII. Classification of muscarinic acetylcholine receptors. *Pharmacol. Rev.* **50:** 279–290.

Changeux J.P., Bertrand D., Corringer P.J., Dehaene S., Edelstein S., Lena C., Le Novere N., Marubio L., Picciotto M., and Zoli M. 1998. Brain nicotinic receptors: Structure and regulation, role in learning and reinforcement. *Brain Res. Brain Res. Rev.* **26:** 198–216.

Chebib M. and Johnston G.A. 1999. The "ABC" of GABA receptors: A brief review. *Clin. Exp. Pharmacol. Physiol.* **26:** 937–940.

Chen J., Tu Y., Moon C., Matarazzo V., Palmer A.M., and Ronnett G.V. 2004. The localization of neuronal nitric oxide synthase may influence its role in neuronal precursor proliferation and synaptic maintenance. *Dev. Biol.* **269:** 165–182.

Cheng A., Wang S., Cai J., Rao M.S., and Mattson M.P. 2003. Nitric oxide acts in a positive feedback loop with BDNF to regulate neural progenitor cell proliferation and differentiation in the mammalian brain. *Dev. Biol.* **258:** 319–333.

Collingridge G.L. and Lester R.A. 1989. Excitatory amino acid receptors in the vertebrate central nervous system. *Pharmacol. Rev.* **41:** 143–210.

Cooper O. and Isacson O. 2004. Intrastriatal transforming growth factor alpha delivery to a model of Parkinson's disease induces proliferation and migration of endogenous adult neural progenitor cells without differentiation into dopaminergic neurons. *J. Neurosci.* **24:** 8924–8931.

Cooper-Kuhn C.M., Winkler J., and Kuhn H.G. 2004. Decreased neurogenesis after cholinergic forebrain lesion in the adult rat. *J. Neurosci. Res.* **77:** 155–165.

Coronas V., Bantubungi K., Fombonne J., Krantic S., Schiffmann S.N., and Roger M. 2004. Dopamine D3 receptor stimulation promotes the proliferation of cells derived from the post-natal subventricular zone. *J. Neurochem.* **91:** 1292–1301.

Curtis D.R. and Watkins J.C. 1961. Analogues of glutamic and gamma-amino-n-butyric acids having potent actions on mammalian neurones. *Nature* 191: 1010–1011.

Deisseroth K., Singla S., Toda H., Monje M., Palmer T.D., and Malenka R.C. 2004. Excitation-neurogenesis coupling in adult neural stem/progenitor cells. *Neuron* 42: 535–552.

Di Giorgi-Gerevini V., Melchiorri D., Battaglia G., Ricci-Vitiani L., Ciceroni C., Busceti C.L., Biagioni F., Iacovelli L., Canudas A.M., Parati E., De Maria R., and Nicoletti F. 2005. Endogenous activation of metabotropic glutamate receptors supports the proliferation and survival of neural progenitor cells. *Cell Death Differ.* 12: 1124–1133.

Diaz J., Ridray S., Mignon V., Griffon N., Schwartz J.C., and Sokoloff P. 1997. Selective expression of dopamine D3 receptor mRNA in proliferative zones during embryonic development of the rat brain. *J. Neurosci.* 17: 4282–4292.

Dinerman J.L., Dawson T.M., Schell M.J., Snowman A., and Snyder S.H. 1994. Endothelial nitric oxide synthase localized to hippocampal pyramidal cells: Implications for synaptic plasticity. *Proc. Natl. Acad. Sci.* 91: 4214–4218.

Dominguez-Escriba L., Hernandez-Rabaza V., Soriano-Navarro M., Barcia J.A., Romero F.J., Garcia-Verdugo J.M., and Canales J.J. 2006. Chronic cocaine exposure impairs progenitor proliferation but spares survival and maturation of neural precursors in adult rat dentate gyrus. *Eur. J. Neurosci.* 24: 586–594.

Eiden L.E. 1998. The cholinergic gene locus. *J. Neurochem.* 70: 2227–2240.

Emsley J.G., Mitchell B.D., Kempermann G., and Macklis J.D. 2005. Adult neurogenesis and repair of the adult CNS with neural progenitors, precursors, and stem cells. *Prog. Neurobiol.* 75: 321–341.

Encinas J.M., Vaahtokari A., and Enikolopov G. 2006. Fluoxetine targets early progenitor cells in the adult brain. *Proc. Natl. Acad. Sci.* 103: 8233–8238.

Esposito M.S., Piatti V.C., Laplagne D.A., Morgenstern N.A., Ferrari C.C., Pitossi F.J., and Schinder A.F. 2005. Neuronal differentiation in the adult hippocampus recapitulates embryonic development. *J. Neurosci.* 25: 10074–10086.

Farrant M. and Nusser Z. 2005. Variations on an inhibitory theme: Phasic and tonic activation of GABA(A) receptors. *Nat. Rev. Neurosci.* 6: 215–229.

Feron F., Vincent A., and Mackay-Sim A. 1999. Dopamine promotes differentiation of olfactory neuron in vitro. *Brain Res.* 845: 252–259.

Freund T.F. and Buzsaki G. 1996. Interneurons of the hippocampus. *Hippocampus* 6: 347–470.

Freundlieb N., Francois C., Tande D., Oertel W.H., Hirsch E.C., and Hoglinger G.U. 2006. Dopaminergic substantia nigra neurons project topographically organized to the subventricular zone and stimulate precursor cell proliferation in aged primates. *J. Neurosci.* 26: 2321–2325.

Fricker A.D., Rios C., Devi L.A., and Gomes I. 2005. Serotonin receptor activation leads to neurite outgrowth and neuronal survival. *Brain Res. Mol. Brain Res.* 138: 228–235.

Garthwaite J., Charles S.L., and Chess-Williams R. 1988. Endothelium-derived relaxing factor release on activation of NMDA receptors suggests role as intercellular messenger in the brain. *Nature* 336: 385–388.

Gascon E., Dayer A.G., Sauvain M.O., Potter G., Jenny B., De Roo M., Zgraggen E., Demaurex N., Muller D., and Kiss J.Z. 2006. GABA regulates dendritic growth by stabilizing lamellipodia in newly generated interneurons of the olfactory bulb. *J. Neurosci.* 26: 12956–12966.

Ge S., Pradhan D.A., Ming G.L., and Song H. 2007. GABA sets the tempo for activity-dependent adult neurogenesis. *Trends Neurosci.* 30: 1–8.

Ge S., Goh E.L., Sailor K.A., Kitabatake Y., Ming G.L., and Song H. 2006. GABA regulates synaptic integration of newly generated neurons in the adult brain. *Nature* **439:** 589–593.

Gibbs S.M. 2003. Regulation of neuronal proliferation and differentiation by nitric oxide. *Mol. Neurobiol.* **27:** 107–120.

Gould E., Cameron H.A., and McEwen B.S. 1994. Blockade of NMDA receptors increases cell death and birth in the developing rat dentate gyrus. *J. Comp. Neurol.* **340:** 551–565.

Gould E., McEwen B.S., Tanapat P., Galea L.A., and Fuchs E. 1997. Neurogenesis in the dentate gyrus of the adult tree shrew is regulated by psychosocial stress and NMDA receptor activation. *J. Neurosci.* **17:** 2492–2498.

Gray W.P. and Sundstrom L.E. 1998. Kainic acid increases the proliferation of granule cell progenitors in the dentate gyrus of the adult rat. *Brain Res.* **790:** 52–59.

Green W., Patil P., Marsden C.A., Bennett G.W., and Wigmore P.M. 2006. Treatment with olanzapine increases cell proliferation in the subventricular zone and prefrontal cortex. *Brain Res.* **1070:** 242–245.

Halassa M.M., Fellin T., and Haydon P.G. 2007. The tripartite synapse: Roles for gliotransmission in health and disease. *Trends Mol. Med.* **13:** 54–63.

Harrist A., Beech R.D., King S.L., Zanardi A., Cleary M.A., Caldarone B.J., Eisch A., Zoli M., and Picciotto M.R. 2004. Alteration of hippocampal cell proliferation in mice lacking the beta 2 subunit of the neuronal nicotinic acetylcholine receptor. *Synapse* **54:** 200–206.

Harvey J.A. 2004. Cocaine effects on the developing brain: Current status. *Neurosci. Biobehav. Rev.* **27:** 751–764.

Haydar T.F., Wang F., Schwartz M.L., and Rakic P. 2000. Differential modulation of proliferation in the neocortical ventricular and subventricular zones. *J. Neurosci.* **20:** 5764–5774.

Heneka M.T. and Feinstein D.L. 2001. Expression and function of inducible nitric oxide synthase in neurons. *J. Neuroimmunol.* **114:** 8–18.

Hoglinger G.U., Rizk P., Muriel M.P., Duyckaerts C., Oertel W.H., Caille I., and Hirsch E.C. 2004. Dopamine depletion impairs precursor cell proliferation in Parkinson disease. *Nat. Neurosci.* **7:** 726–735.

Islam A.T., Kuraoka A., and Kawabuchi M. 2003. Morphological basis of nitric oxide production and its correlation with the polysialylated precursor cells in the dentate gyrus of the adult guinea pig hippocampus. *Anat. Sci. Int.* **78:** 98–103.

Jha S., Rajendran R., Davda J., and Vaidya V.A. 2006. Selective serotonin depletion does not regulate hippocampal neurogenesis in the adult rat brain: Differential effects of p-chlorophenylalanine and 5,7-dihydroxytryptamine. *Brain Res.* **1075:** 48–59.

Kaneko N., Okano H., and Sawamoto K. 2006. Role of the cholinergic system in regulating survival of newborn neurons in the adult mouse dentate gyrus and olfactory bulb. *Genes Cells* **11:** 1145–1159.

Karten Y.J., Jones M.A., Jeurling S.I., and Cameron H.A. 2006. GABAergic signaling in young granule cells in the adult rat and mouse dentate gyrus. *Hippocampus* **16:** 312–320.

Kempermann G. 2002. Regulation of adult hippocampal neurogenesis—Implications for novel theories of major depression. *Bipolar Disord.* **4:** 17–33.

Kempermann G., Kuhn H.G., and Gage F.H. 1997. More hippocampal neurons in adult mice living in an enriched environment. *Nature* **386:** 493–495.

Kempermann G., Wiskott L., and Gage F.H. 2004a. Functional significance of adult neurogenesis. *Curr. Opin. Neurobiol.* **14:** 186–191.

Kempermann G., Jessberger S., Steiner B., and Kronenberg G. 2004b. Milestones of neuronal development in the adult hippocampus. *Trends Neurosci.* **27:** 447–452.

Kew J.N. and Kemp J.A. 2005. Ionotropic and metabotropic glutamate receptor structure and pharmacology. *Psychopharmacology* **179:** 4–29.

Kitayama T., Yoneyama M., and Yoneda Y. 2003. Possible regulation by N-methyl-D-aspartate receptors of proliferative progenitor cells expressed in adult mouse hippocampal dentate gyrus. *J. Neurochem.* **84:** 767–780.

Kitayama T., Yoneyama M., Tamaki K., and Yoneda Y. 2004. Regulation of neuronal differentiation by N-methyl-D-aspartate receptors expressed in neural progenitor cells isolated from adult mouse hippocampus. *J. Neurosci. Res.* **76:** 599–612.

Kodama M., Fujioka T., and Duman R.S. 2004. Chronic olanzapine or fluoxetine administration increases cell proliferation in hippocampus and prefrontal cortex of adult rat. *Biol. Psychiatry* **56:** 570–580.

Kotani S., Yamauchi T., Teramoto T., and Ogura H. 2006. Pharmacological evidence of cholinergic involvement in adult hippocampal neurogenesis in rats. *Neuroscience* **142:** 505–514.

Lee H.J., Kim J.W., Yim S.V., Kim M.J., Kim S.A., Kim Y.J., Kim C.J., and Chung J.H. 2001. Fluoxetine enhances cell proliferation and prevents apoptosis in dentate gyrus of maternally separated rats. *Mol. Psychiatry* **6:** 618–752.

Lie D.C., Song H., Colamarino S.A., Ming G.L., and Gage F.H. 2004. Neurogenesis in the adult brain: New strategies for central nervous system diseases. *Annu. Rev. Pharmacol. Toxicol.* **44:** 399–421.

Lie D.C., Dziewczapolski G., Willhoite A.R., Kaspar B.K., Shults C.W., and Gage F.H. 2002. The adult substantia nigra contains progenitor cells with neurogenic potential. *J. Neurosci.* **22:** 6639–6649.

Lie D.C., Lansford H., Dearie A., Song H., Colamarino S.A., Kaspar B., and Gage F.H. 2003. Dopaminergic neurons derived from adult neural stem cells. *Soc. Neurosci. Abs.* **670.8.**

Lipton S.A. and Rosenberg P.A. 1994. Excitatory amino acids as a final common pathway for neurologic disorders. *N. Engl. J. Med.* **330:** 613–622.

Liu X., Wang Q., Haydar T.F., and Bordey A. 2005. Nonsynaptic GABA signaling in postnatal subventricular zone controls proliferation of GFAP-expressing progenitors. *Nat. Neurosci.* **8:** 1179–1187.

Lledo P.M., Alonso M., and Grubb M.S. 2006. Adult neurogenesis and functional plasticity in neuronal circuits. *Nat. Rev. Neurosci.* **7:** 179–193.

LoTurco J.J., Owens D.F., Heath M.J., Davis M.B., and Kriegstein A.R. 1995. GABA and glutamate depolarize cortical progenitor cells and inhibit DNA synthesis. *Neuron* **15:** 1287–1298.

Lu D., Mahmood A., Zhang R., and Copp M. 2003. Upregulation of neurogenesis and reduction in functional deficits following administration of DEtA/NONOate, a nitric oxide donor, after traumatic brain injury in rats. *J. Neurosurg.* **99:** 351–361.

Lujan R., Shigemoto R., and Lopez-Bendito G. 2005. Glutamate and GABA receptor signalling in the developing brain. *Neuroscience* **130:** 567–580.

Malberg J.E. and Duman R.S. 2003. Cell proliferation in adult hippocampus is decreased by inescapable stress: Reversal by fluoxetine treatment. *Neuropsychopharmacology* **28:** 1562–1571.

Malberg J.E., Eisch A.J., Nestler E.J., and Duman R.S. 2000. Chronic antidepressant treatment increases neurogenesis in adult rat hippocampus. *J. Neurosci.* **20:** 9104–9110.

Maragakis N.J., and Rothstein J.D. 2004. Glutamate transporters: Animal models to neurologic disease. *Neurobiol. Dis.* **15:** 461–473.

Matarredona E.R., Murillo-Carretero M., Moreno-Lopez B., and Estrada C. 2004. Nitric oxide synthesis inhibition increases proliferation of neural precursors isolated from the postnatal mouse subventricular zone. *Brain Res.* **995:** 274–284.

Mechawar N., Saghatelyan A., Grailhe R., Scoriels L., Gheusi G., Gabellec M.M., Lledo P.M., and Changeux J.P. 2004. Nicotinic receptors regulate the survival of newborn neurons in the adult olfactory bulb. *Proc. Natl. Acad. Sci.* **101:** 9822–9826.

Mesulam M.M., Mufson E.J., Wainer B.H., and Levey A.I. 1983. Central cholinergic pathways in the rat: An overview based on an alternative nomenclature (Ch1-Ch6). *Neuroscience* **10:** 1185–1201.

Ming G.L. and Song H. 2005. Adult neurogenesis in the mammalian central nervous system. *Annu. Rev. Neurosci.* **28:** 223–250.

Missale C., Nash S.R., Robinson S.W., Jaber M., and Caron M.G. 1998. Dopamine receptors: From structure to function. *Physiol. Rev.* **78:** 189–225.

Mohapel P., Leanza G., Kokaia M., and Lindvall O. 2005. Forebrain acetylcholine regulates adult hippocampal neurogenesis and learning. *Neurobiol. Aging* **26:** 939–946.

Montana V., Ni Y., Sunjara V., Hua X., and Parpura V. 2004. Vesicular glutamate transporter-dependent glutamate release from astrocytes. *J. Neurosci.* **24:** 2633–2642.

Moreno-Lopez B., Noval J.A., Gonzalez-Bonet L.G., and Estrada C. 2000. Morphological bases for a role of nitric oxide in adult neurogenesis. *Brain Res.* **869:** 244–250.

Moreno-Lopez B., Romero-Grimaldi C., Noval J.A., Murillo-Carretero M., Matarredona E.R., and Estrada C. 2004. Nitric oxide is a physiological inhibitor of neurogenesis in the adult mouse subventricular zone and olfactory bulb. *J. Neurosci.* **24:** 85–95.

Nacher J., Alonso-Llosa G., Rosell D.R., and McEwen B.S. 2003. NMDA receptor antagonist treatment increases the production of new neurons in the aged rat hippocampus. *Neurobiol. Aging* **24:** 273–284.

Nacher J., Rosell D.R., Alonso-Llosa G., and McEwen B.S. 2001. NMDA receptor antagonist treatment induces a long-lasting increase in the number of proliferating cells, PSA-NCAM-immunoreactive granule neurons and radial glia in the adult rat dentate gyrus. *Eur. J. Neurosci.* **13:** 512–520.

Ohtani N., Goto T., Waeber C., and Bhide P.G. 2003. Dopamine modulates cell cycle in the lateral ganglionic eminence. *J. Neurosci.* **23:** 2840–2850.

Overstreet-Wadiche L.S., Bensen A.L., and Westbrook G.L. 2006a. Delayed development of adult-generated granule cells in dentate gyrus. *J. Neurosci.* **26:** 2326–2334.

Overstreet-Wadiche L., Bromberg D.A., Bensen A.L., and Westbrook G.L. 2005. GABAergic signaling to newborn neurons in dentate gyrus. *J. Neurophysiol.* **94:** 4528–4532.

———. 2006b. Seizures accelerate functional integration of adult-generated granule cells. *J. Neurosci.* **26:** 4095–4103.

Owens D.F. and Kriegstein A.R. 2002. Is there more to GABA than synaptic inhibition? *Nat. Rev. Neurosci.* **3:** 715–727.

Packer M.A., Stasiv Y., Benraiss A., Chmielnicki E., Grinberg A., Westphal H., Goldman S.A., and Enikolopov G. 2003. Nitric oxide negatively regulates mammalian adult neurogenesis. *Proc. Natl. Acad. Sci.* **100:** 9566–9571.

Parent J.M., Yu T.W., Leibowitz R.T., Geschwind D.H., Sloviter R.S., and Lowenstein D.H. 1997. Dentate granule cell neurogenesis is increased by seizures and contributes to aberrant network reorganization in the adult rat hippocampus. *J. Neurosci.* **17:** 3727–3738.

Park C., Sohn Y., Shin K.S., Kim J., Ahn H., and Huh Y. 2003. The chronic inhibition of nitric oxide synthase enhances cell proliferation in the adult rat hippocampus. *Neurosci. Lett.* **339:** 9–12.

Poulsen C.F., Simeone T.A., Maar T.E., Smith-Swintosky V., White H.S., and Schousboe A. 2004. Modulation by topiramate of AMPA and kainate mediated calcium influx in cultured cerebral cortical, hippocampal and cerebellar neurons. *Neurochem. Res.* **29:** 275–282.

Poulsen F.R., Blaabjerg M., Montero M., and Zimmer J. 2005. Glutamate receptor antagonists and growth factors modulate dentate granule cell neurogenesis in organotypic, rat hippocampal slice cultures. *Brain Res.* **1051:** 35–49.

Prakash N. and Wurst W. 2006. Development of dopaminergic neurons in the mammalian brain. *Cell. Mol. Life. Sci.* **63:** 187–206.

Radley J.J. and Jacobs B.L. 2002. 5-HT1A receptor antagonist administration decreases cell proliferation in the dentate gyrus. *Brain Res.* **955:** 264–267.

Reif A., Schmitt A., Fritzen S., Chourbaji S., Bartsch C., Urani A., Wycislo M., Mossner R., Sommer C., Gass P., and Lesch K.P. 2004. Differential effect of endothelial nitric oxide synthase (NOS-III) on the regulation of adult neurogenesis and behaviour. *Eur. J. Neurosci.* **20:** 885–895.

Riquelme P.A., Drapeau E., and Doetsch F. 2007. Brain micro-ecologies: Neural stem cell niches in the adult mammalian brain. *Philos. Trans. R. Soc. Lond. B. Biol. Sci.* (in press).

Roberts E. and Frankel S. 1950. gamma-Aminobutyric acid in brain: Its formation from glutamic acid. *J. Biol. Chem.* **187:** 55–63.

Sailor K.A., Ming G.L., and Song H. 2006. Neurogenesis as a potential therapeutic strategy for neurodegenerative diseases. *Expert Opin. Biol. Ther.* **6:** 879–890.

Santarelli L., Saxe M., Gross C., Surget A., Battaglia F., Dulawa S., Weisstaub N., Lee J., Duman R., Arancio O., Belzang C., and Hen R. 2003. Requirement of hippocampal neurogenesis for the behavioral effects of antidepressants. *Science* **301:** 805–809.

Scott B.W., Wojtowicz J.M., and Burnham W.M. 2000. Neurogenesis in the dentate gyrus of the rat following electroconvulsive shock seizures. *Exp. Neurol.* **165:** 231–236.

Stanfield B.B. and Trice J.E. 1988. Evidence that granule cells generated in the dentate gyrus of adult rats extend axonal projections. *Exp. Brain Res.* **72:** 399–406.

Sun F.Y. and Guo X. 2005. Molecular and cellular mechanisms of neuroprotection by vascular endothelial growth factor. *J. Neurosci. Res.* **79:** 180–184.

Suzuki M., Nelson A.D., Eickstaedt J.B., Wallace K., Wright L.S., and Svendsen C.N. 2006. Glutamate enhances proliferation and neurogenesis in human neural progenitor cell cultures derived from the fetal cortex. *Eur. J. Neurosci.* **24:** 645–653.

Tashiro A., Sandler V.M., Toni N., Zhao C., and Gage F.H. 2006. NMDA-receptor-mediated, cell-specific integration of new neurons in adult dentate gyrus. *Nature* **442:** 929–933.

Torroglosa A., Murillo-Carretero M., Romero-Grimaldi C., Matarredona E.R., Campos-Caro A., and Estrada C. 2007. Nitric oxide decreases subventricular zone stem cell proliferation by inhibition of epidermal growth factor receptor and phosphoinositide-3-kinase/ Akt pathway. *Stem Cells* **25:** 88–97.

Tozuka Y., Fukuda S., Namba T., Seki T., and Hisatsune T. 2005. GABAergic excitation promotes neuronal differentiation in adult hippocampal progenitor cells. *Neuron* **47:** 803–815.

Van Kampen J.M. and Eckman C.B. 2006. Dopamine D3 receptor agonist delivery to a model of Parkinson's disease restores the nigrostriatal pathway and improves locomotor behavior. *J. Neurosci.* **26:** 7272–7280.

Van Kampen J.M. and Robertson H.A. 2005. A possible role for dopamine D3 receptor stimulation in the induction of neurogenesis in the adult rat substantia nigra. *Neuroscience* **136:** 381–386.

Van Kampen J.M., Hagg T., and Robertson H.A. 2004. Induction of neurogenesis in the adult rat subventricular zone and neostriatum following dopamine D receptor stimulation. *Eur. J. Neurosci.* **19:** 2377–2387.

Vutskits L., Gascon E., Tassonyi E., and Kiss J.Z. 2006. Effect of ketamine on dendritic arbor development and survival of immature GABAergic neurons in vitro. *Toxicol. Sci.* **91:** 540–549.

Wang D.D., Krueger D.D., and Bordey A. 2003a. Biophysical properties and ionic signature of neuronal progenitors of the postnatal subventricular zone in situ. *J. Neurophysiol.* **90:** 2291–2302.

———. 2003b. GABA depolarizes neuronal progenitors of the postnatal subventricular zone via GABAA receptor activation. *J. Physiol.* **550:** 785–800.

Wang H.D., Dunnavant F.D., Jarman T., and Deutch A.Y. 2004. Effects of antipsychotic drugs on neurogenesis in the forebrain of the adult rat. *Neuropsychopharmacology* **29:** 1230–1238.

Wang L.P., Kempermann G., and Kettenmann H. 2005. A subpopulation of precursor cells in the mouse dentate gyrus receives synaptic GABAergic input. *Mol. Cell. Neurosci.* **29:** 181–189.

Xu G., Ong J., Liu Y.Q., Silverstein F.S., and Barks J.D. 2005. Subventricular zone proliferation after alpha-amino-3-hydroxy-5-methyl-4-isoxazolepropionic acid receptor-mediated neonatal brain injury. *Dev. Neurosci.* **27:** 228–234.

Yamaguchi M., Suzuki T., Seki T., Namba T., Juan R., Arai H., Hori T., and Asada T. 2004. Repetitive cocaine administration decreases neurogenesis in adult rat hippocampus. *Ann. N.Y. Acad. Sci.* **1025:** 351–362.

Yoshimizu T. and Chaki S. 2004. Increased cell proliferation in the adult mouse hippocampus following chronic administration of group II metabotropic glutamate receptor antagonist, MGS0039. *Biochem. Biophys. Res. Commun.* **315:** 493–496.

Zhang R., Wang L., Zhang L., Chen J., Zhu Z., Zhang Z., and Chopp M. 2003. Nitric oxide enhances angiogenesis via the synthesis of vascular endothelial growth factor and cGMP after stroke in the rat. *Circ. Res.* **92:** 308–313.

Zhang R., Zhang L., Zhang Z., Wang Y., Lu M., Lapointe M., and Chopp M. 2001. A nitric oxide donor induces neurogenesis and reduces functional deficits after stroke in rats. *Ann. Neurol.* **50:** 602–611.

Zhao M., Momma S., Delfani K., Carlen M., Cassidy R.M., Johansson C.B., Brismar H., Shupliakov O., Frisen J., and Janson A.M. 2003. Evidence for neurogenesis in the adult mammalian substantia nigra. *Proc. Natl. Acad. Sci.* **100:** 7925–7930.

Zhu X.J., Hua Y., Jiang J., Zhou Q.G., Luo C.X., Han X., Lu Y.M., and Zhu D.Y. 2006. Neuronal nitric oxide synthase-derived nitric oxide inhibits neurogenesis in the adult dentate gyrus by down-regulating cyclic AMP response element binding protein phosphorylation. *Neuroscience* **141:** 827–836.

20

Adult Neurogenesis in the Olfactory Bulb

Pierre-Marie Lledo
Pasteur Institute, Head of the Laboratory of Perception and Memory
Director of the CNRS Unit Genes, Synapses & Cognition
Paris Cedex 15, France

Most organisms rely on an olfactory system to detect and analyze chemical cues from the external world in the context of essential behavior. From worms to vertebrates, chemicals are detected by odorant receptors expressed by olfactory sensory neurons, which send an axon to the primary processing center—the olfactory bulb, in vertebrates. Within this relay, sensory neurons form excitatory synapses with projection neurons and with inhibitory interneurons. Thus, due to complex synaptic interactions in the olfactory bulb circuit, the output of a given projection neuron is determined not only by the sensory input, but also by the activity of local inhibitory interneurons that are concerned by adult neurogenesis throughout life. Recent studies have provided clues about how these new neurons incorporate into preexisting networks, how they survive or die once integrated into proper microcircuits, and how basic network functions are maintained despite the continual renewal of a large percentage of neurons. We know that external influences modulate the process of late neurogenesis at various stages. Thus, this process is probably flexible, allowing brain performance to be optimized for its environment. But optimized how? And why?

This chapter describes the adaptation of new interneuron production to experience-induced plasticity. In particular, how the survival of newly generated neurons is highly sensitive not only to the level of sensory inputs, but also to the behavioral context is discussed. Also discussed is how neurogenesis may finely tune the functioning of the neural network, optimizing the processing of sensory information. Adult neurogenesis maintains continual turnover of bulbar interneurons that can be shaped by environmental cues and/or internal states. In this way, adult neurogenesis is associated with improvement of various cognitive performances. This chapter brings together recent descriptions of the properties of newly generated neurons in the olfactory bulb and emerging principles concerning their functions. It is now being realized that new cells have a role much more complex than that of simple gatekeepers inhibiting the olfactory bulb network. This chapter, concentrating exclusively on the mammalian olfactory system, highlights adult neurogenesis as part of the brain's arsenal for dealing with an environment and with internal states that are constantly changing.

STABILITY VERSUS FLEXIBILITY: A QUANDARY FOR ADULT NEURONAL CIRCUITS

It has become clear during the past decades that new neurons are continually generated in the adult brain. Two regions—the olfactory bulb and the dentate gyrus (DG) of the hippocampus—receive and continually integrate newly generated neurons throughout adult life (Ming and Song 2005). The molecular and cellular mechanisms by which these cells are generated, and by which they migrate to their future targets, are areas of intense research that will doubtless provide critical information for clinical approaches attempting to regenerate damaged or degraded portions of neural tissue. Old neuronal circuits have a continually renewing pool of neurons, raising a series of fundamental questions concerning the role that neurogenesis has in the normal functioning of neuronal circuits (Kempermann et al. 2004). It is not understood why neurogenesis persists in various parts of the adult brain, but not in other parts, and whether this process is a recapitulation of embryogenesis or a unique feature of the adult forebrain. Neurogenesis is restricted, apparently, to two specific regions under normal conditions, but we do not know why or how these regions balance the need for plasticity with that to maintain information processing networks that are already functional. Is neurogenesis in the adult brain a constant restorative process or is it flexible, producing different numbers of neurons in different regions in response to environmental stimuli or internal needs? We also ask whether new neurons are generated in the adult brain to perform particular tasks not possible for mature neurons or whether they are generated as flexible units that can undertake whatever role their target structure most requires.

A key conclusion drawn from the discovery of neuronal production in the mature brain is that adult neurogenesis is another means by which the brain alters its structural organization and therefore functional circuitry. Flexibility in morphology and function exists in the nervous system on various levels, from genes to neuronal compartments, including the addition and removal of synaptic contacts. What more does neuronal turnover bring to the brain's ability to adapt than synaptic plasticity alone? A host of recent evidence shows that this renovation at the cell level is not merely restorative; instead, adult neurogenesis is a dynamic form of neuronal plasticity that enables the brain to adapt itself to an environment and to internal states that are constantly evolving. In contrast, the addition of newly generated neurons to adult circuitry raises questions about stability. In particular, how do new cells integrate into an already functioning system without disturbing its existing properties? The brain must preserve the circuitry and synaptic organization required to constantly process information so that behavior is maintained. At the same time, it must adapt its circuitry in response to learning, injuries, and environmental changes. This dilemma between stability and plasticity (Abrous et al. 2005) is a matter of passionate debate that is far from being resolved. The exact functional implications of adult neurogenesis are not known. Studies suggest that newly generated neurons in the adult olfactory bulb and hippocampus are critical for various aspects of learning, memory, and sensory performance. This chapter presents more general issues concerning the plasticity conferred to adult circuits by newly generated cells and those that are replaceable.

One might wonder whether neurogenesis acts directly, in a specific manner, on neuronal information processing or whether it acts indirectly by preparing host circuits for general experiences and increased general challenges. It is noteworthy that common concepts about the functional consequences of neurogenesis may emerge from exploration of functional differences between the olfactory bulb and the hippocampus, as well as between new cell generation in developing and mature brain. Possible reasons for the marked spatial restriction of neurogenesis in the adult brain should be provided by comparing early and late neurogenesis and by determining the differences of these processes from those in other parts of the central nervous system.

PLASTICITY OF OLFACTORY BULB MICROCIRCUITS: WHAT IS IT FOR?

The interneuron populations of the olfactory bulb have a key role in inducing and maintaining synchronization of projection neurons (Lledo and Lagier 2006). Thus, it is not surprising that their functional regulation is complex and occurs by two major ways. First, extensive systems of centrifugal fibers innervating the olfactory bulb primarily contact local interneurons (Luskin and Price 1983). These centrifugal systems (Halasz and Shepherd 1983) comprise excitatory inputs from the piriform cortex, the telencephalic basal ganglion, the magnocellular basal forebrain nuclei, and the brain stem locus cœruleus (Shipley and Ennis 1996), as well as inhibitory inputs from the nucleus of the horizontal limb of the diagonal band (Kunze et al. 1992). These fibers provide the contextual information required for sensory formatting of odor representations by early olfactory circuits. Therefore, they are important in early stages of olfactory information processing. Second, in addition to extensive regulation of interneuron activity by neuromodulators, neuronal networks of the olfactory bulb are influenced by adult neurogenesis. Periglomerular cells (GABAergic and/or dopaminergic interneurons) and granule cells (GABAergic interneurons only) are continually produced throughout the life span of mammals (Altman 1969). Despite the elimination of some, more than 30,000 newly generated interneurons reach the olfactory bulb circuit every day (Petreanu and Alvarez-Buylla 2002). These two distinct levels of regulation (i.e., neuromodulation and neurogenesis), although sometimes intermingled because neuromodulators also regulate survival of newly generated cells (see, e.g., Bauer et al. 2003; Mechawar et al. 2004), result in a high degree of plasticity in olfactory bulb microcircuits. Thus, they facilitate the modulation of all steps involved in odor sensing, ranging from detection and discrimination to learning and memory.

It has become clear that newly generated interneurons are integrated into the local inhibitory network of the olfactory bulb throughout life. However, the effect of this integration on the function of the network is still unclear (Lledo et al. 2006). Most newly generated cells that reach the olfactory bulb are granule cells (~94%). The remaining cell population is made up of periglomerular cells (~4%) and astrocytes (<2%). Granule cells make up the largest percentage of neurons in the olfactory bulb (Shepherd et al. 2004), and each granule cell contacts several hundred relay neurons which, in turn, contact many pyramidal cells of the piriform cortex. Thus, occurring in the early part of the olfactory pathway, bulbar neurogenesis is ideal for amplifying the neurogenic effect. Perhaps one of the major functions of adult neurogenesis, acting in concert with centrifugal fibers, is to improve plasticity of neuronal networks. This hypothesis is supported by a recent finding that sensory deprivation substantially decreases the number, the dendritic length, and the spine density of newly generated interneurons, leaving all preexisting interneurons intact (Saghatelyan et al. 2005). Thus, newly generated interneurons enable the olfactory bulb circuit to adapt to novel sensory challenges (Lledo and Saghatelyan 2005).

It is noteworthy that bulbar neurogenesis occurs in a neuronal network in which sensory afferent inputs are subjected to continual replacement as well. Mature olfactory sensory neurons have a limited life span—about 60 days in rodents—tightly regulated by environmental factors. Bulbar neurogenesis might therefore be a mechanism for modulation of sensory information processing in the brain, in response to changing sensory afferent neurons, which are highly sensitive to the complexity of the odorant space. More experiments are needed to elucidate whether sensory neurogenesis and bulbar neurogenesis are tightly concerted and whether sensory neuron generation influences changes in bulbar microcircuits.

NEWBORN NEURONS IN THE ADULT OLFACTORY BULB: USE THEM OR LOSE THEM!

Understanding the functional effect of adult neurogenesis is currently one of the major challenges in *Neurosciences*. As discussed above, the recruitment of new neurons into adult circuits is one process in a large repertoire of neuroadaptive responses (Abrous et al. 2005) that strongly depends on the overall activity of the adult brain. In sharp contrast to the situation of the developing brain, in adult neurogenesis, newly generated neurons integrate into already functioning circuits. In the olfactory bulb, the newcomers originate from the periventricular region, called the subventricular zone (SVZ), migrate along a rostral migratory stream (RMS), and differentiate into local interneurons before integrating into functional circuitry (Fig. 1) (Belluzzi et al. 2003; Carleton et al. 2003). We now know that recruitment of newly generated neurons into bulbar circuits correlates strongly with the complexity of the olfactory environment (Petreanu and Alvarez-Buylla 2002). Yet, we do not know how new neurons are selected locally to integrate and survive in the appropriate microcircuit. Understanding how this selective process affects microcircuit function is a major challenge.

At first glance, adult neurogenesis seems to have no purpose as only a subset of newly generated interneurons is selected to survive, whereas the majority is eliminated. Bromodeoxyuridine (BrdU) and [^{3}H]thymidine-labeling experiments in rodent olfactory bulbs showed that nearly half of new granule cells are eliminated within a few weeks (Petreanu and Alvarez-Buylla 2002; Winner et al. 2002). The survival rate of newly generated neurons changes substantially with age. Whereas the entire population of newly generated granule cells survives until adulthood in early postnatal neurogenesis, newly generated neurons have a shorter life span

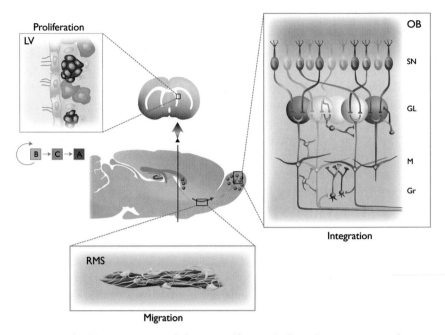

Figure 1. The hippocampus and the SVZ-olfactory bulb pathway represent the two constitutive neurogenic zones in the adult central nervous system (CNS). (*Center panel*) Sagittal section of the forebrain; (*arrow*) tangential migration of neuroblasts (*purple dots*) toward the olfactory bulb (OB). Neurogenesis refers to the replacement of glutamatergic neurons in the hippocampus and of bulbar local interneurons in the OB. (*Right panel*) Wiring of the OB. Each sensory neuron (SN) expresses only one of the 1000 odorant receptor genes, and the axons from all cells expressing that particular receptor converge onto one or a few glomeruli (GL) in the olfactory bulb. The nearly 2000 glomeruli in the rat OB are spherical knots of neuropil, ~50–100 μm in diameter, which contain the incoming axons of sensory neurons and the apical dendrites of the main input-output neuron of the OB, the mitral cell (M). Mitral axons leaving the OB project widely to higher brain structures including the piriform cortex, hippocampus, and amygdala. Processing of the olfactory message into the bulb occurs through two populations of interneurons: periglomerular cells and granule cells (Gr). (*Left panel*) Neurogenic niche. Here, proliferation in the subventricular zone (SVZ) takes place in the medial wall of the lateral ventricle (LV), where stem cells (in *green*, type-B cells) divide to generate transit amplifying cells (in *brown*, type-C cells), which in turn give rise to neuroblasts (in *purple*, type-A cells) that migrate in the rostral migratory stream (RMS) (*bottom panel*) to their final destination in the OB, where they differentiate into bulbar interneurons.

in older animals (Lemasson et al. 2005). This difference in the survival rates suggests that the functional implications of neurogenesis change with age. During early postnatal periods, new neurons are continually added to the developing olfactory bulb and therefore must be considered to be long-lasting building blocks. During adulthood, new neurons appear to be labile, replaceable functional units required for coping with an ever-changing external world and/or internal states.

Sensory activity is one of the major factors regulating the life span of adult-generated neurons. Sensory deprivation by naris occlusion reduces the number of granule cells through increased apoptosis (see, e.g., Corotto et al. 1994). Conversely, reopening the naris following early occlusion (Cummings and Brunjes 1997) and olfactory enrichment in adults (Rochefort et al. 2002) promote the survival of newly formed bulbar interneurons. Thus, the death of newly generated neurons appears to be an important effector of input-dependent structural plasticity in the olfactory bulb. Consequently, the rate of bulbar neurogenesis is determined not only by the constitutive rate of proliferation in the neurogenic niche, but also by the cell death that occurs in the targeted tissue. It is possible that primary olfactory inputs indirectly regulate the early steps of neurogenesis, as apoptosis also occurs in the SVZ (Morshead and van der Kooy 1992). It has been reported that proliferation rates in the SVZ changed during olfactory deprivation in adults (Corotto et al. 1994), by surgical disruption of the RMS (Jankovski et al. 1998), and ablation of the olfactory bulb (Kirschenbaum et al. 1999). Mandairon et al. (2003) have shown that cell death and proliferation increased in the SVZ, RMS, and olfactory bulb during the first 2 weeks following olfactory deafferentation by axotomy and returned to control values 1 month after deafferentation. These changes closely followed the temporal pattern of olfactory sensory neuron death and regeneration. These data strongly suggest that cell apoptosis occurring throughout the SVZ-olfactory bulb system influences all stages of adult neurogenesis, from cell proliferation to integration of neurons into networks, and is tightly controlled by the degree of sensory inputs (i.e., animal's experiences). Thus, apoptosis of newly generated neurons is a key mechanism for experience-induced changes in adult neurogenesis.

Survival of newly generated neurons is critically regulated by the degree of sensory inputs, only when occurring during a precise window of time (Yamaguchi and Mori 2005). Sensory-experience-dependent neuronal plasticity, occurring during a specific time window only, is a crucial mechanism for establishing a finely tuned neuronal circuit in the developing brain. For example, deprivation of visual input from one eye shifts the response property of binocular zone neurons in the visual cortex preferentially toward the undeprived eye input. This ocular dominance plasticity occurs during a sensitive period after the birth of animals, called the *critical period*, during which monocular deprivation shifts ocular dominance (Hensch 2005). Yamaguchi and Mori (2005) examined whether there is a critical period during which olfactory sensory experience more efficiently influences the survival of newly generated neurons in the adult, by applying a naris cauterization method that has been used to reveal sensory-experience-dependent changes in the olfactory bulb circuit. These authors identified a critical time window during which sensory experience strongly influenced new granule cell survival in the adult mouse olfactory bulb. This period corresponded to days 14–28 after the generation of cells when the newcomers start receiving synaptic contacts, thus suggesting that new granule cells are susceptible to cell death in a sensory-experience-dependent manner during the period of synapse formation.

ADULT NEUROGENESIS: A NOVEL NEURONAL BASIS FOR EXPERIENCE-INDUCED REWIRING

Adult neurogenesis is a biological process that is conserved throughout evolution; thus, its adaptive functions should be explored. What are the advantages of bulbar neurogenesis in terms of adaptivity for an organism? Is it involved in survival, reproduction, or fitness? The physiological processes underlying adult neurogenesis have been well-studied. Nevertheless, other than investigations reported in seminal papers in birds and chickadees, very few experiments have addressed its possible role in adaptation. Due to seasonal variation in the availability of food, some bird species store it in various locations. Winter survival of these food-storing birds critically depends on their cache-recovery ability, but lesions of the hippocampus impair their spatial memory (Sherry and Vaccarino 1989). These species have larger hippocampus volumes than nonstoring birds (Healy and Krebs 1993). Large numbers of newly generated neurons are recruited to the hippocampal complex of the free-ranging adult black-capped chickadees during the fall, when the birds are storing food. This behavioral context strongly suggests that the recruitment of newly generated neurons is important for survival.

Recent studies have begun to explore adult neurogenesis in more relevant ethological contexts. For instance, Shingo et al. (2003) investigated bulbar neurogenesis during pregnancy and lactation in the mouse. They reported that the rate of neurogenesis increased in the first trimester of gestation, during the first week of lactation, and following mating. Of the various stimulating maternal hormones, prolactin (PRL) appeared to be a key factor triggering bulbar neurogenesis in the dams. There was a larger number of BrdU-positive cells in the SVZ and in the olfactory bulb in mice chronically administered PRL than in controls. There were significantly fewer neuroblasts in the SVZ of mated females heterozygous for the PRL receptor than in controls. This up-regulated neurogenesis might be involved in maternal care at parturition and during lactation. Further behavioral investigations in lactating females are needed to assess whether olfactory offspring discrimination is also enhanced.

Two other speculations about adaptive functions arise from the observation of olfactory neurogenesis during early pregnancy and lactation. The first is that an increase in the density of newly generated neurons in the accessory olfactory bulb is required for the female's olfactory memory of a stud male's pheromones after mating. The initial week of pregnancy is similar to the time period during which female mice form an olfactory memory to pheromones of a stud male after mating. Such memory depends on the accessory olfactory bulb. Adult neurogenesis has also been shown to occur in the accessory olfactory bulb, although changes in this region during early pregnancy have not been examined. The second speculation is that the increase in olfactory neurogenesis during early gestation and lactation facilitates the ability of females to assess the relatedness of potential nesting partners. House mice nest communally in the wild and in the laboratory under seminatural conditions (Manning et al. 1992). In communal nests, several females give birth and nurse offspring and nonoffspring indiscriminately. Female mice nest communally with other females possessing a similar major histocompatibility complex; more generally, communal nesting partners tend to be kin (Manning et al. 1992). Even if mothers recognized their own pups, it is unlikely that in mixed-litter nests, lactating females would nurse their pups and block access to nonoffspring. In terms of adaptive fitness, it is possible that mother mice reduce the cost of sharing milk with nonoffspring by nesting with closely related conspecifics (Hayes 2000).

It is possible that the functional consequences of neurogenesis can be observed even in nonpregnant mice. This assumption has been tested by studies in transgenic mice exhibiting low levels of adult neurogenesis. For example, neural cell adhesion molecule (NCAM) mutant mice, shown by quantification of bulbar neurogenesis to have 40% fewer newly formed interneurons than wild-type animals, have been used for behavioral analyses. This low level of neurogenesis was accompanied by impaired odor discrimination (Gheusi et al. 2000), revealing a specific role for newly formed interneurons in the downstream coding of olfactory information. This finding is supported by observations in another transgenic line with impaired γ-amino-n-butyric acid (GABA$_A$) receptor-mediated synaptic inhibition (Nusser et al. 2001; Lagier et al. 2007) and by theoretical analyses (Cecchi et al. 2001; Bathelier et al. 2006). Together, these data indicate that a critical level of inhibition, mediated by the activation of $GABA_A$ receptors located on the dendrites of principal neurons of the olfactory bulb, is necessary for normal olfactory processing.

IS ADULT NEUROGENESIS A REAL BENEFIT FOR LEARNING?

The olfactory bulb is involved in consolidating processes associated with long-term odor memory. Thus, it is possible that changes in the number of GABAergic interneurons may regulate both olfactory perception and memory. This hypothesis has been investigated by studies aiming to determine whether a change in the number of newly formed bulbar interneurons alters a specific olfactory memory. In these experiments, animals were reared in a complex olfactory environment as both physical and intellectual activities positively influence brain structure and function (Rosenzweig and Bennett 1996). Following a 40-day period in enriched olfactory environments, mice produced twice as many new interneurons as control mice. This difference was specific to the olfactory bulb, with no changes observed in the DG. Animals with higher levels of neurogenesis retain learned olfactory information for longer periods of time than controls (Rochefort et al. 2002). In particular, animals reared in enriched olfactory environments recognize familiar odors more readily and in a more sustained manner than animals reared under standard conditions. Although the consequences of an enriched environment on synaptic efficacy and/or projection neuron activity remain to be explored, these results are consistent with a correlation between size of the newly formed interneuron population and memory capacity. Nevertheless, most of the findings about adult neurogenesis have been observed in laboratory animals kept under artificial conditions that are very different from their natural habitat. Even the odor-enriched environments used in experiments are arguably much poorer in terms of sensory stimulation than conditions in the wild.

Various studies have demonstrated that changes in the level of sensory activity regulate the number of newly generated neurons by adjusting the level of cell death. Therefore, one may ask whether all newly generated neurons are beneficial for memory or whether cell elimination is also required? So far, the hippocampus represents the model that has provided the most compelling information to this quest. Despite the growing number of studies aimed at characterizing the functional implications of adult neurogenesis, conflicting results, rather than consensual opinions, have emerged on this issue. For instance, Gould et al. (1999) demonstrated increased survival of newly generated hippocampal neurons in rats subjected to spatial learning tasks. In contrast, van Praag et al. (1999) did not observe any changes in neurogenesis following similar spatial training in mice.

The complex nature of adult neurogenesis has been confirmed by more recent studies simultaneously reporting lower and higher levels of neurogenesis in response to water maze tasks (see, e.g., Ambrogini et al. 2004; Ehninger and Kempermann 2006). As already pointed out by Döbrössy et al. (2003), these discrepancies may be caused by the interval between the generation of new neurons and their exposure to memory tasks. It is noteworthy that the functions of newly generated neurons may evolve during their sequential maturation.

In the olfactory bulb, this hypothesis has been recently addressed (Alonso et al. 2006). In the context of odorant discrimination learning, it has been shown that the degree of sensory input not only controls the number of newly generated neurons by preventing their death, but also regulates their spatial distribution. Learning between two odorants has been shown to reduce the elimination of newly generated neurons by apoptosis that usually occurs between 15 and 30 days after birth (Petreanu and Alvarez-Buylla 2002; Winner et al. 2002; Giachino et al. 2005; Yamaguchi and Mori 2005). Surprisingly, this effect was not linked to the activated region of the olfactory bulb associated with the reward. In contrast, survival rates were significantly higher in loci where the activation driven by the nonreinforced cue was highest. It is thus possible that odor information without reward (i.e., negative information) has a more dominant role in survival of newly generated neurons than does positive information in associative conditioning. Whatever the mechanism, a learning period of only 6 days is clearly sufficient for substantially better survival rates of newly generated neurons in some regions and not in others.

Why does the newly generated neuron population grow in some restricted areas, but not in others? It is possible that adjusting the distribution of newly generated neurons in a learning context enhances the contrast between spatial activation patterns produced by two odorants. This supports the hypothesis that small differences in regional pattern activity can be used for odorant discrimination as long as animals are subjected to the motivational and experimental consequences of differential reinforcement. The most recent results indicate that sensory inputs in a learning context are important for the local control of the survival of new neurons in behaviorally relevant networks in the adult olfactory bulb. They also indicate that sensory experience shapes olfactory bulb microcircuits beyond the simple modulation of existing synaptic connections.

THE HUMAN CASE

One of the oldest beliefs about human perception is that we have a poor sense of smell. At first glance, this general belief seems to be grounded in scientific evidence, as recent human genetic studies showed a decline in the number of functional olfactory receptor genes during primate evolution to humans (Gilad et al. 2004). According to this view, the use of an arboreal habitat and the adoption of an erect posture during human evolution have led to the gradual increase in the importance of vision and decrease in the importance of smell. However, several overlooked human features including the structure of the nasal cavity, retronasal smell, olfactory brain areas, and language call for reassessing the status of olfaction in humans (Shepherd 2004). After all, the mismatching of genetic prediction and actual functions should not surprise us. As systems biology has taught us, there is not a one-to-one relationship between the number of genes and complex behavior. Thus, a decreased number of genes coding for olfactory receptors does not necessarily correlate with a loss of olfactory capacity. Behavior is influenced by multiple factors. This is true for the olfactory system in which olfactory receptor genes do not necessarily reflect olfactory capacities.

If human smell perception is therefore better than we thought, it may have conserved some of its original traits. Previous findings that human olfactory bulbs incorporate newly generated neurons (Liu and Martin 2003; Bedard and Parent 2004) support this assumption. Supporting these results is the very recent observation of neuroblasts migrating to the olfactory bulb via a lateral ventricular extension in humans (Curtis et al. 2007). Newly generated neurons in humans may modulate olfactory discrimination in a manner similar to that observed in rodents. This may come as a surprise to many people, although not to those who make their living by their noses, including enologists, perfumers, and food scientists. Anyone who has taken part in a wine tasting, or observed professional testing of food flavors or perfumes, knows that the human sense of smell has extraordinary capacities for discrimination.

Time has seen the fall of a century-old dogma and the introduction of the new field of adult human neurogenesis. The priority for investigators in this field has shifted from documentation of the existence of newly generated neurons to understanding their functions (Rakic 2004). Many basic mechanisms regulating adult neurogenesis are still unknown. New neurons are generated continually in many regions of the adult central nervous system in nonmammalian vertebrates (Zupanc 2001); thus, comparisons of adult neurogenesis in a broad range of species should yield new insights into the evolution and functions of this process. A central question in the field of neurogenesis in the human olfactory system has yet to be answered: What are all the functions of this well-conserved biological process in humans?

Active neurogenesis in adult humans shows that the mature central nervous system has the capacity to regenerate, raising hopes that the central nervous system, damaged by injury or degenerative neuronal diseases, can be repaired. Understanding the basic mechanisms regulating adult neurogenesis under normal and pathological conditions will provide the foundation for cell replacement therapy, with either endogenous adult neural stem cells or transplanted cells from various sources. Technical advances have propelled progress in the field of adult neurogenesis. Aided by novel technologies, including live imaging of migrating newly generated neurons in living subjects, the best is yet to come.

CONCLUDING REMARKS

The sense of smell has had an essential role during mammalian evolution. Odor representation is dynamic and highly complex and may require unique plasticity mechanisms. Neuronal production, migration, and the replacement of mature interneurons in the adult olfactory bulb represent some of these adaptive mechanisms.

We have seen that relay neurons of the mammalian olfactory bulb circuit readily synchronize their firing and generate fast oscillations in response to olfactory nerve inputs. Due to its basic architecture and its synaptic organization, the local inhibitory network provides the olfactory bulb with a unique form of inhibition that is crucial to induce and distribute rhythms to downstream olfactory structures. Inhibitory interneurons are involved in both the temporal and spatial organization of olfactory bulb outputs: As transducers, they transform the spatial dimension of the sensory information reaching the olfactory bulb into spatiotemporal patterns. Continuous neurogenesis is therefore a mechanism by which the spatiotemporal coding is finely tuned. Such a mechanism may act together with experience-induced changes in synaptic weights, adding to the known range of adaptive properties of the adult olfactory bulb.

This chapter has tackled two of the main problems related to the functional significance of adult neurogenesis. Both are fundamental issues. First, we have seen the role of adult neurogenesis in the context of olfactory bulb functions. Assuming that the bulb processes odor information before relaying it to the olfactory cortex, it can be hypothesized that adult neurogenesis enables the olfactory bulb to adjust the degree of sensory information processing appropriately. Second, the existence of a pool of juvenile neurons, enabling the system to adapt to similar future situations, raises the possibility that adult neurogenesis acts post hoc to provide a structural basis for neuronal plasticity and learning. Thus, research on adult neurogenesis in the olfactory system not only is interesting in itself, but also provides new avenues of exploration to increase our understanding of adult circuits adaptation.

As a result of unprecedented developments in methods for examining the structure, development, function, and neurochemistry of olfactory system circuits, research in olfactory plasticity has progressed substantially in recent years. Nevertheless, applying new technologies, including those of molecular biology, neurophysiology, and functional imaging should help us to unravel the mysteries of how neurogenesis shapes olfactory coding. After digging deeply into the fundamental basis on how the olfactory system processes sensory information, we are now starting to release exquisite fragrances of awareness.

SUMMARY

With a growing number of studies confirming that neural stem cells reside in the adult central nervous system, the function of newly generated neuronal cells in the adult brain is the source of intense research and debate. Adult neurogenesis is modulated by a wide variety of physiopathological conditions and environmental stimuli, offering the possibility that newly generated neuronal cells might be functionally associated with the response to these processes. Pioneering studies have proposed that newly generated neuronal cells support various brain functions, including learning, memory, and mood. However, more recent studies have begun to challenge these ideas. It is possible that adult neurogenesis alters the olfactory bulb and hippocampus at the cellular, network, and system levels. Computational models suggest that cell turnover might be especially beneficial for the learning of new information. In the olfactory bulb, it is clear that the number of new neurons increases after a learning task because newly generated neurons have longer life spans. Quite surprisingly, simple exposure to olfactory cues does not alter neurogenesis. Only in a learning context does sensory activation control the total number of newly generated neurons and refine their precise locations. These findings suggest that associative learning based on sensory stimuli can modulate adult neurogenesis.

Definitive experiments to demonstrate the function(s) of adult neurogenesis await the development of techniques that can specifically eliminate this process in very limited areas. In particular, precisely localized "loss-of-function" experiments could help to determine the role of adult neurogenesis. For this, a focal and nongenetic approach represented by irradiation to selectively eliminate dividing cells might be helpful (Fig. 2). γ-rays and X-rays damage DNA so extensively during mitosis that endogenous repair mechanisms are overwhelmed and apoptotic elimination is induced (Tada et al. 2000). Studies in animal models have shown that exposure to therapeutic doses of radiation leads to ablation of adult neurogenesis but not gliogenesis (Tada et al. 2000; Santarelli et al. 2003). This dramatic reduction of adult neurogenesis results from the combined effects of acute cell death, decreased proliferation, and neuronal differentiation of the neural progenitors. γ-ray or X-ray irradiation induces apoptosis in dividing cells, sparing postmitotic cells such as mature neurons. Therefore, irradiation is a very flexible technique that is not invasive and requires no surgery. It is an ideal tool to target adult neurogenic zones without damaging neighboring tissue. Long-term reductions in neurogenesis within the subgranular zone (SGZ) of the DG have previously been reported after low-dose irradiation of the heads of adult rodents (Tada et al. 2000; Santarelli et al. 2003). Experiments showing that neuronal precursors are more sensitive to radiation than glial precursors (Mizumatsu et al. 2003; Snyder et al. 2005) indicate that the effects of irradiation are very selective for depletion of new neurons in the brain.

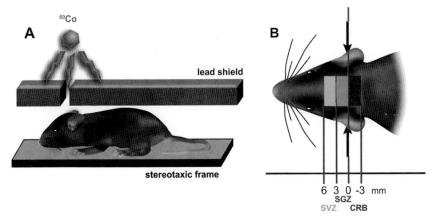

Figure 2. γ-ray treatment ablation of cell proliferation. (A) Mice are anesthetized, placed in a stereotaxic frame, and exposed to cranial irradiation. (B) A lead shield protects the mouse body while exposing the SVZ, SGZ, or cerebellum (CRB) to γ-rays; 5 Grays are delivered on days 1, 3, and 5, before behavioral tests beginning on day 28.

For instance, two to three fractionated, low doses of γ-rays or X-rays, repeated at 24–48 hours, are less damaging to the brain tissue than a single large dose (for review, see Wojtowicz 2006). A local irradiation of the SVZ by shielding the body and most of the head should not directly affect the olfactory system, the hippocampus, or other tissues. Experiments using irradiation are now in progress to study the functional consequences of specific neurogenesis ablation in the SVZ for the sense of smell (Fig. 2). Whatever the results gained, we already have numerous lines of correlative and experimental evidence that olfactory bulb neurogenesis is important for controlling sensory discrimination and learning.

ACKNOWLEDGMENTS

This work was supported by the *Fédération pour la Recherche sur le Cerveau,* the *Fondation pour la Recherche Médicale,* and the *Agence Nationale de la Recherche* (ANR-05-Neur-028-01). I thank Cecile Moreau for help with illustrations and I apologize to those authors whose references, although relevant to this subject, have not been included in this review due to space constraints.

REFERENCES

Abrous D.N., Koehl M., and Le Moal M. 2005. Adult neurogenesis: From precursors to network and physiology. *Physiol. Rev.* **85:** 523–569.

Alonso M., Viollet C., Gabellec M.-M., Meas-Yadid V., Olivo-Marin J.-C., and Lledo P.-M. 2006. Olfactory discrimination learning increases the survival of adult-born neurons in the olfactory bulb. *J. Neurosci.* **26:** 10508–10513.

Altman J. 1969. Autoradiographic and histological studies of postnatal neurogenesis. IV. Cell proliferation and migration in the anterior forebrain, with special reference to persisting neurogenesis in the olfactory bulb. *J. Comp. Neurol.* **137:** 433–457.

Ambrogini P., Orsini L., Mancini C., Ferri P., Ciaroni S., and Cuppini R. 2004. Learning may reduce neurogenesis in adult rat dentate gyrus. *Neurosci. Lett.* **359:** 13–16.

Bathelier B., Lagier S., Faure P., and Lledo P.-M. 2006. Circuit properties generating gamma oscillations in the mammalian olfactory bulb. *J. Neurophysiol.* **95:** 2678–2691.

Bauer S., Moyse E., Jourdan F., Colpaert F., Martel J.C., and Marien M. 2003. Effects of the alpha 2-adrenoreceptor antagonist dexefaroxan on neurogenesis in the olfactory bulb of the adult rat *in vivo:* Selective protection against neuronal death. *Neuroscience* **117:** 281–291.

Bedard A. and Parent A. 2004. Evidence of newly generated neurons in the human olfactory bulb. *Brain Res. Dev. Brain Res.* **151:** 159–168.

Belluzzi O., Benedusi M., Ackman J., and LoTurco J.J. 2003. Electrophysiological differentiation of new neurons in the olfactory bulb. *J. Neurosci.* **23:** 10411–10418.

Carleton A., Petreanu L.T., Lansford R., Alvarez-Buylla A., and Lledo P.-M. 2003. Becoming a new neuron in the adult olfactory bulb. *Nat. Neurosci.* **6:** 507–518.

Cecchi G.A., Petreanu L.T., Alvarez-Buylla A., and Magnasco M.O. 2001. Unsupervised learning and adaptation in a model of adult neurogenesis. *J. Comput. Neurosci.* **11:** 175–182.

Corotto F.S., Henegar J.R., and Maruniak J.A. 1994. Odor deprivation leads to reduced neurogenesis and reduced neuronal survival in the olfactory bulb of the adult mouse. *Neuroscience* **61:** 739–744.

Cummings D.M. and Brunjes P.C. 1997. The effects of variable periods of functional deprivation on olfactory bulb development in rats. *Exp. Neurol.* **148:** 360–366.

Curtis M.A., Kam M., Nannmark U., Anderson M.F., Axell M.Z., Wikkelso C., Holtas S., van Roon-Mom W.M., Bjork-Eriksson T., Nordborg C., Frisen J., Dragunow M., Faull R.L., and Eriksson P.S. 2007. Human neuroblasts migrate to the olfactory bulb via a lateral ventricular extension. *Science* **315:** 1243–1249.

Döbrössy M.D., Drapeau E., Aurousseau C., Le Moal M., Piazza P.V., and Abrous D.N. 2003. Differential effects of learning on neurogenesis: Learning increases or decreases the number of newly born cells depending on their birth date. *Mol. Psychiatry* **8:** 974–982.

Ehninger D. and Kempermann G. 2006. Paradoxical effects of learning the Morris water maze on adult hippocampal neurogenesis in mice may be explained by a combination of stress and physical activity. *Genes Brain Behav.* **5:** 29–39.

Gheusi G., Cremer H., McLean H., Chazal G., Vincent J.-D., and Lledo P.-M. 2000. Importance of newly generated neurons in the adult olfactory bulb for odor discrimination. *Proc. Natl. Acad. Sci.* **97:** 1823–1828.

Giachino C., De Marchis S., Giampietro C., Parlato R., Perroteau I., Schutz G., Fasolo A., and Peretto P. 2005. cAMP response element-binding protein regulates differentiation and survival of newborn neurons in the olfactory bulb. *J. Neurosci.* **25:** 10105–10108.

Gilad Y., Wiebe V., Przeworski M., Lancet D., and Pääbo S. 2004. Loss of olfactory receptor genes coincides with the acquisition of full trichromatic vision in primates. *PLoS Biol.* **2:** e5.

Gould E., Beylin A., Tanapat P., Reeves A., and Shors T.J. 1999. Learning enhances adult neurogenesis in the hippocampal formation. *Nat. Neurosci.* **2:** 260–265.

Halasz N. and Shepherd G.M. 1983. Neurochemistry of the vertebrate olfactory bulb. *Neuroscience* **10:** 579–619.

Hayes L.D. 2000. To nest communally or not to nest communally: A review of rodent communal nesting and nursing. *Anim. Behav.* **59:** 677–688.

Healy S.D. and Krebs J.R. 1993. Development of hippocampal specialisation in a food-storing bird. *Behav. Brain Res.* **53:** 127–131.

Hensch T.K. 2005. Critical period plasticity in local cortical circuits. *Nat. Rev. Neurosci.* **6:** 877–888.

Jankovski A., Garcia C., Soriano E., and Sotelo C. 1998. Proliferation, migration and differentiation of neuronal progenitor cells in the adult mouse subventricular zone surgically separated from its olfactory bulb. *Eur. J. Neurosci.* **10:** 3853–3868.

Kempermann G., Wiskott L., and Gage F.H. 2004. Functional significance of adult neurogenesis. *Curr. Opin. Neurobiol.* **14:** 186–191.

Kirschenbaum B., Doetsch F., Lois C., and Alvarez-Buylla A. 1999. Adult subventricular zone neuronal precursors continue to proliferate and migrate in the absence of the olfactory bulb. *J. Neurosci.* **19:** 2171–2180.

Kunze W.A., Shafton A.D., Kem R.E., and McKenzie J.S. 1992. Intracellular responses of olfactory bulb granule cells to stimulating the horizontal diagonal band nucleus. *Neuroscience* **48:** 363–369.

Lagier S., Panzanelli P., Russo R., Sassoè-Pognetto M., Fritschy J.M., and Lledo P.-M. 2007. Tuning gamma oscillation in the olfactory bulb: insights from alpha1 knock-out mice. *Proc. Natl. Acad. Sci.* **104:** 7259–7264.

Lemasson M., Saghatelyan A., Olivo-Marin J.C., and Lledo P.-M. 2005. Neonatal and adult neurogenesis provide two distinct populations of granule cells in the mouse olfactory bulb. *J. Neurosci.* **25:** 6816–6825.

Liu Z. and Martin L.J. 2003. Olfactory bulb core is a rich source of neural progenitor and stem cells in adult rodent and human. *J. Comp. Neurol.* **459:** 368–391.

Lledo P.-M. and Lagier S. 2006. Local interneurons transduce spatial coding into temporal patterning in the mammalian olfactory bulb. *Semin. Cell Dev. Biol.* **17:** 443–453.

Lledo P.-M. and Saghatelyan A. 2005. Integrating new neurons into the adult olfactory bulb: Joining the network, life/death decisions, and the effects of sensory experience. *Trends Neurosci.* **28:** 248–254.

Lledo P.-M., Grubb M., and Alonso M. 2006. Adult neurogenesis and functional plasticity in neuronal circuits. *Nat. Rev. Neurosci.* **7:** 179–193.

Luskin M.B. and Price J.L. 1983. The topographic organization of associational fibers of the olfactory system in the rat, including centrifugal fibers to the olfactory bulb. *J. Comp. Neurol.* **216:** 264–291.

Mandairon N., Jourdan F., and Didier A. 2003. Deprivation of sensory inputs to the olfactory bulb up-regulates cell death and proliferation in the subventricular zone of adult mice. *Neuroscience* **119:** 507–516.

Manning C.J., Wakeland E.K., and Potts W.K. 1992. Communal nesting patterns in mice implicate MHC genes in kin recognition. *Nature* **360:** 581–583.

Mechawar N., Saghatelyan A., Grailhe R., Scoriels L., Gheusi G., Gabellec M.-M., Lledo P.-M., and Changeux J.-P. 2004. Nicotinic receptors regulate the survival of newborn neurons in the adult olfactory bulb. *Proc. Natl. Acad. Sci.* **101:** 9822–9826.

Ming G.L. and Song H. 2005. Adult neurogenesis in the mammalian central nervous system. *Annu. Rev. Neurosci.* **28:** 223–250.

Mizumatsu S., Monje M.L., Morhardt D.R., Rola R., Palmer T.D., and Fike J.R. 2003. Extreme sensitivity of adult neurogenesis to low doses of X-irradiation. *Cancer Res.* **63:** 4021–4027.

Morshead C.M. and van der Kooy D. 1992. Postmitotic death is the fate of constitutively proliferating cells in the subependymal layer of the adult mouse brain. *J. Neurosci.* **12:** 249–256.

Nusser Z., Kay L.M., Laurent G., Homanics G.E., and Mody I. 2001. Disruption of GABA$_A$ receptors on GABAergic interneurons leads to increased oscillatory power in the olfactory bulb network. *J. Neurophysiol.* **86:** 2823–2833.

Petreanu L. and Alvarez–Buylla A. 2002. Maturation and death of adult-born olfactory bulb granule neurons: Role of olfaction. *J. Neurosci.* **22:** 6106–6113.

Rakic P. 2004. Neuroscience: Immigration denied. *Nature* **427:** 685–686.

Rochefort R., Gheusi G., Vincent J.-D., and Lledo P.-M. 2002. Enriched odor exposure increases the number of newborn neurons in the adult olfactory bulb and improves odor memory. *J. Neurosci.* **22:** 2679–2689.

Rosenzweig M.R. and Bennett E.L. 1996. Psychobiology of plasticity: Effects of training and experience on brain and behavior. *Behav. Brain Res.* **78:** 57–65.

Saghatelyan A., Roux P., Migliore M., Rochefort C., Charneau P., Shepherd G., and Lledo P.-M. 2005. Activity-dependent adjustments of the inhibitory network in the adult olfactory bulb following early postnatal deprivation. *Neuron* **46:** 103–116.

Santarelli L., Saxe M., Gross C., Surget A., Battaglia F., Dulawa S., Weisstaub N., Lee J., Duman R., Arancio O., Belzung C., and Hen R. 2003. Requirement of hippocampal neurogenesis for the behavioral effects of antidepressants. *Science* **301:** 805–809.

Sherry D.F. and Vaccarino A.L. 1989. Hippocampus and memory for food caches in black-capped chickadees. *Behav. Neurosci.* **103:** 308–318.

Shepherd G.M. 2004. The human sense of smell: Are we better than we think? *PLoS Biol.* **2:** E146.

Shepherd G.M., Wei R.C., and Greer C.A. 2004. Olfactory bulb. In *The synaptic organization of the brain*, 5th edition (ed. G.M. Shepherd), pp. 165–216. Oxford University Press, Oxford, United Kingdom.

Shingo T., Gregg C., Enwere E., Fujikawa H., Hassam R., Geary C., Cross J., and Weiss S. 2003. Pregnancy-stimulated neurogenesis in the adult female forebrain mediated by prolactin. *Science* **299:** 117–120.

Shipley M.T. and Ennis M. 1996. Functional organization of olfactory system. *J. Neurobiol.* **30:** 123–176.

Snyder J.S., Hong N.S., McDonald R.J., and Wojtowicz J.M. 2005. A role for adult neurogenesis in spatial long-term memory. *Neuroscience* **130:** 843–852.

Tada E., Parent J.M., Lowenstein D.H., and Fike J.R. 2000. X-irradiation causes a prolonged reduction in cell proliferation in the dentate gyrus of adult rats. *Neuroscience* **99:** 33–41.

van Praag H., Christie B.R., Sejnowski T.J., and Gage F.H. 1999. Running enhances neurogenesis, learning, and long-term potentiation in mice. *Proc. Natl. Acad. Sci.* **96:** 13427–13431.

Winner B., Cooper-Kuhn C.M., Aigner R., Winkler J., and Kuhn H.G. 2002. Long-term survival and cell death of newly generated neurons in the adult rat olfactory bulb. *Eur. J. Neurosci.* **16:** 1681–1689.

Wojtowicz J.M. 2006. Irradiation as an experimental tool in studies of adult neurogenesis. *Hippocampus* **16:** 261–266.

Yamaguchi M. and Mori K. 2005. Critical period for sensory experience-dependent survival of newly generated granule cells in the adult mouse olfactory bulb. *Proc. Natl. Acad. Sci.* **102:** 9697–9702.

Zupanc G.K. 2001. A comparative approach towards the understanding of adult neurogenesis. *Brain Behav. Evol.* **58:** 246–249.

21

Neurogenesis and Hippocampal Memory System

D. Nora Abrous
Neurogenesis and Pathophysiology Laboratory
Bordeaux Neuroscience Research Center
Inserm 862, Bordeaux, France
University of Bordeaux, Bordeaux, France

J. Martin Wojtowicz
Department of Physiology
University of Toronto, Medical Sciences Building
Toronto, Ontario, M5S 1A8, Canada

WHEN DISCUSSING A BRAIN FUNCTION SUCH AS MEMORY, one should relate it to brain plasticity. One definition of plasticity is an alternative way of performing the same function. Anecdotal evidence suggests that the human brain can perform amazing memory feats in unexpected, alternative ways. For example, the established ability of savants (individuals with partial brain damage) to memorize events, sequences of numbers, letters, or musical notes, and to perform arithmetical calculations, suggests that compensatory rewiring of brain circuits after injury can affect learning. Which particular form of brain plasticity could be responsible for such astounding learning abilities as those seen in Kim Peek ("Rain Man") and Daniel Tammet ("Brainman"), two individuals diagnosed as autistic savants (www.savantsyndrome.com)? In this chapter, we describe a radical form of plasticity, adult neurogenesis, in hippocampal formation (HF). The discovery of adult neurogenesis (production of new neurons in adult brain) has radically changed our ideas of how the brain can adapt to physiological and environmental challenges. The process of neuronal production is highly regulated and is involved in hippocampal functions under physiological conditions. In some cases, neurogenesis can

Adult Neurogenesis ©2008 Cold Spring Harbor Laboratory Press 978-087969-784-6

respond to hippocampus-related pathologies such as epilepsy, ischemia, mood disorders, and addiction. Understanding neurogenesis, along with other forms of brain plasticity, may help us to understand normal memory and perhaps the enhanced memory such as that seen in individuals with the Savant Syndrome (Treffert and Christensen 2005).

LESIONS OF THE NEUROGENIC REGION

The HF is part of an integrated network involved in learning and memory (Eichenbaum 2000, 2001; Morris 2006). However, the precise functional role of the HF is still a matter of debate, this entity being involved in cognitive mapping, relational memory, configural associations, episodic-like memories, etc. (for review, see Morris 2006). Given this complexity, the contribution of the different hippocampal subregions also remains a matter of dispute. However, to understand the role of adult neurogenesis, one may ask what can be expected from lesions of the whole dentate gyrus (DG). Borrowing from a recent article outlining a neurobiological theory of hippocampal function, we suggest that the DG is not equally involved in all types of hippocampus-dependent memories (Morris 2006). This follows from a parallel arrangement of the two main inputs into the hippocampus, the medial and lateral entorhinal cortices projecting to CA1 and CA3/DG (Fig. 1) (Witter 1993). A lesion in the DG would be expected to disrupt only one of these inputs originating from layer II in the entorhinal cortices. At the functional level, various forms of memory can be disrupted by colchicine lesions of the DG, i.e., encoding reference spatial memory (Nanry et al. 1989; Xavier et al. 1999; Jeltsch et al. 2001; Nakayama and Sawada 2002), retrieval of reference spatial memory (Nanry et al. 1989), and working memory (Xavier et al. 1999). However, the role of DG is not absolute, and some spatial-related tasks, such as odor-place or object-place paired associations, are spared (Gilbert and Kesner 2003). Furthermore, the distinction between the roles of DG and CA1 becomes subtle when behavioral tasks involve temporal or spatial pattern separation (Gilbert et al. 2001). All such experiments involving lesions must take into consideration that in cases when one region is damaged, the other pathway may compensate for the loss. In support of this explanation, the CA1 place cells, which are corollary of spatial learning, can be formed despite selective lesions of the DG (McNaughton et al. 1989; Morris 2006). However, this evidence does not disprove an argument for the involvement of the DG-CA3 pathway, together with the CA3 collateral system, in optimal, normal memory processes (Nakazawa et al. 2002). In summary, taking into account the complexity of hippocampal

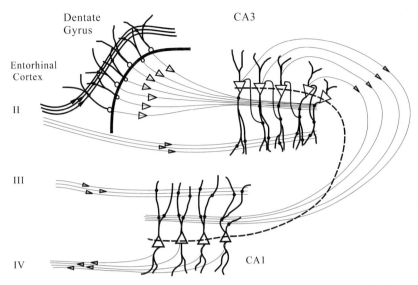

Figure 1. A diagram of cortico-hippocampal connections emphasizing parallel inputs into the hippocampus. The classical trisynaptic pathway from entorhinal cortex via dentate gyrus (DG) is accompanied by the direct pathways to CA3 and CA1.

functioning and considering that newly born granule neurons within DG are a small minority (perhaps 10%) of the total population, the behavioral consequences of their depletion cannot be easily predicted (Snyder et al. 2001). However, given their strategic location in the gateway of the HF, they may have a pivotal role.

A TWO-WAY INTERRELATIONSHIP BETWEEN NEUROGENESIS AND LEARNING

There are at least two ways in which neurogenesis and learning have been linked: (1) The rate of neurogenesis determines learning performance and memory and (2) learning tasks enhance neurogenesis by enhancing survival of adult-born neurons at a particular stage of their development.

The Rate of Neurogenesis Determines Learning and Memory Abilities

Three categories of correlative evidence suggest involvement of adult-born neurons in memory: (1) conditions that enhance neurogenesis also improve learning, (2) conditions that decrease neurogenesis also impair learning, and (3) quantitative correlation between neurogenesis and learning performances. We only focus on those studies in which memory abilities and neurogenesis have been conducted in the same animals (within-subjects designs).

Conditions Increasing Neurogenesis Enhance Learning Abilities

Most studies evaluating effects of an enriched environment and/or physical activity support the contention that within a physiological adaptive range, the addition of adult-born neurons may be beneficial for adult brain functioning, particularly for spatial memory. Exposure of mice to an enriched environment or to a running wheel increases both the spatial memory ability in the water maze and the number of adult-born neurons in DG (Kempermann et al. 1997, 1998a; van Praag et al. 1999b). This beneficial effect of environmental experience also ameliorates reduced brain functioning during aging. Indeed, both neurogenesis and spatial memory performances are improved in senescent or middle-aged mice raised in an enriched environment (Kempermann et al. 1998b, 2002) or in senescent mice housed with access to a running wheel (van Praag et al. 2005). Furthermore, exposure of rats to an enriched environment during the prenatal period, the early postnatal period (Koo et al. 2003), and adulthood (Nilsson et al. 1999) also increased both neurogenesis and spatial learning abilities in the water maze.

Another set of experiments supports the hypothesis that improvement in memory abilities depends on increased production of new neurons. First, lowering corticosterone secretion from midlife for the rest of the animal's life increases behavioral performance in the water maze and also hippocampal neurogenesis at the point of senescence (Montaron et al. 2006). Second, treatment of adult rats with the cognitive enhancer ginseng enhances memory performances in the contextual fear-conditioning paradigm and increases the number of bromodeoxyuridine (BrdU)-labeled cells (Qiao et al. 2005).

Conditions Decreasing Neurogenesis Impair Learning

On the basis of environmental, lesioning, pharmacological, and genetic approaches, it has been suggested that a reduction of hippocampal neurogenesis leads to reduced memory abilities. First, stressful events during the prenatal period caused a lifelong reduction of neurogenesis and disrupted spatial working memory in the Y maze and/or spatial reference memory in the water maze (Lemaire et al. 2000; Koo et al. 2003). Second, olfactory bulbectomy and lead exposure during the first three postnatal weeks induced contextual memory deficits that are associated with a decreased neurogenesis (Jaako-Movits and Zharkovsky 2005; Jaako-Movits et al. 2005). Third, FoxG1 haploinsufficiency induced impairments in contextual fear conditioning and decreased hippocampal neurogenesis (Shen et al. 2006). Fourth, mice with mutation of a member of the methylated DNA-binding protein, methyl-CpG-binding protein 1, exhibited increased genomic instability, spatial memory deficits, and reduced neurogenesis (Zhao et al. 2003). Fifth, genetically based variations of neurogenesis in mice have been related to differences in spatial memory, i.e., poor learning is associated with low levels of neurogenesis (Kempermann and Gage 2002). Sixth, neurotrophin 3 (NT3) conditional KO mice, in which the NT3 gene is selectively deleted in the brain throughout development, exhibit deficits in spatial memory and impairs neurogenesis (Shimazu et al. 2006). Finally, lesion of the cholinergic septohippocampal pathway impairs spatial learning and reduces neurogenesis (Mohapel et al. 2005).

Quantitative Correlation between Learning Abilities and Neurogenesis

A quantitative relationship between learning and the number of newly generated neurons has been shown in senescent rats by characterizing spatial memory abilities in the water maze (Drapeau et al. 2003). Animals with preserved spatial memory (i.e., aged-unimpaired rats) exhibited 1 month after training a higher level of cell proliferation and a higher number of new neurons in comparison to rats with spatial memory impairments (i.e., aged-impaired rats). Two studies, however, failed to demonstrate such correlation (Bizon and Gallagher 2003; Merrill et al. 2003). Various experimental differences in the BrdU-labeling protocol, the number of subjects, the rat strain, and the gender of the animals could explain this apparent controversy.

Summary

There is a general consensus that neurogenesis is related to learning and memory abilities. However, it is not yet clear how this is accomplished. In young animals that are capable of high rates of neurogenesis and hence high rates of neuronal turnover, new neurons could participate in day-to-day memory acquisitions as implied by the experiments described below. However, in older animals, the rates may be too low for such function, and gradual structural changes in network connectivity are more likely. Such adaptive restructuring of the hippocampus may be species-dependent and probably occurs on a much larger scale in animals living in a natural environment instead of contrived laboratory conditions. Furthermore, hippocampal function (improved or decreased memory) is not solely due to neurogenesis (high or low rate of production). Modifications in synaptic and structural plasticity at the level of the dendrites and spines, for example, in vascularization and in metabolic adaptation, occurring within and outside the HF could change memory independently of neurogenesis. Experimental dissection of these factors will be required to establish a causal role of new neurons in memory.

Influence of Learning on Neurogenesis

Training on learning tasks requiring the HF has been shown to exert a complex influence on distinct steps of neurogenesis: cell survival, cell proliferation, and cell death. We review the influence of learning on each of these steps separately, keeping in mind that these might be intermingled processes.

Effect on Cell Survival

The effect of associative (trace eye-blink conditioning) or spatial learning has been examined on *survival* of newly born cells. To this end, animals received a single BrdU injection and were submitted to the tasks 1 week later. The 1-week delay allows for cell differentiation into neurons but not to their full maturation since at this time, the axons and dendritic trees are still under development (Piatti et al. 2006). Animals sacrificed immediately following trace eye-blink conditioning showed an increase in the number of BrdU-labeled cells (Gould et al. 1999). One week after completion of the task, most newly born cells colabeled with neuronal markers (Gould et al. 1999). The learning-induced increase in neurogenesis was maintained for 2 months after completion of the task (Leuner et al. 2004). The behavioral performances positively correlated with the number of surviving cells, indicating that learning, and not training, rescued the adult-born cells (Leuner et al. 2004). The observed enhancement of cell survival was specific to hippocampus-dependent associative learning, as neurogenesis remained unchanged in rats following delay-eye-blink conditioning (classically attributed to the cerebellum; Gould et al. 1999) or to active shock avoidance (another classical Pavlovian conditioning task that does not strictly depend on the HF; Van der Borght et al. 2005a). The degree of task difficulty also seems to be an important factor for obtaining a learning-promoting effect on cell survival. Indeed, establishing a contextual conditioned stimulus representation acquired in a single training trial is not sufficient to change the survival of cells born 10 days before exposure to the task (Pham et al. 2005).

A positive effect of learning on *cell survival* has also been described in the water maze task. The number of newly born cells, labeled with BrdU 1 week before exposure to the task, was increased following 4 days of training (Gould et al. 1999; Hairston et al. 2005). The observed enhancement of cell survival was specific to spatial learning, as neurogenesis remained unchanged in rats exposed to the task without a platform, but producing the same amount of motor responses (stress group). Furthermore, learning-induced up-regulation of neurogenesis has been specifically attributed to hippocampal functioning, as training on a cued test in the water maze, a hippocampus-independent type of learning, does not modify neurogenesis (Gould et al. 1999). Most recent evidence shows that the survival-promoting effects of training are strongest in the animals that learn the task well (Sisti and Shors 2006). Several studies did not observe a survival-promoting effect in the water maze (Ambrogini et al. 2000; Van der Borght et al. 2005b). These discrepancies may be due to restricting the counting of BrdU cell numbers to the dorsal DG (Ambrogini et al. 2000), and/or the platform location being changed during the course of training (Van der Borght et al. 2005b), and/or differences in the number of daily trials (Snyder et al. 2005).

Effect on Cell Proliferation

Recent evidence indicates that spatial learning also induces the *proliferation* of neural precursors. During learning of the water-maze task, two phases can be distinguished: an early phase, during which performance improves rapidly, and a late phase, during which performance stabilizes and reaches an asymptotic level. Döbrössy et al. (2003) have shown that the early phase of learning does not modify proliferation, whereas the late phase does. Indeed, when animals are injected with BrdU during the early phase and sacrificed at the end of this phase, learning does not modify BrdU-immunoreactive cell numbers. In contrast, when animals are injected with BrdU during the late phase of learning and sacrificed 1 day later, the number of BrdU-labeled cells is increased (Lemaire et al. 2000; Döbrössy et al. 2003). Learning-induced cell proliferation was not correlated to learning performances (Döbrössy et al. 2003), suggesting that these newly born cells do not directly sustain ongoing learning. Their function remains to be determined. The learning-induced increase in the genesis of cells born contingently with the late phase of learning is long-lasting, persisting for at least 5 weeks after the animals had acquired the task (Döbrössy et al. 2003). However, some controversy exists, since cell proliferation, as measured with Ki-67, was not influenced following 4 days of training and decreased following 14 days in the water maze (Mohapel et al. 2006).

Effect on Cell Death

More surprisingly, spatial learning also decreased the number of newborn cells (Döbrössy et al. 2003; Ambrogini et al. 2004). This decline in BrdU cell numbers was not related to stress/and or physical activity since (1) animals were habituated to the pool before training in order to diminish its stressful component (but see Ehninger and Kempermann 2006) and (2) no change in BrdU cell numbers was observed in yoked animals exposed to the pool but without the platform. Indeed, in animals injected with BrdU during the early phase and sacrificed at the end of the late phase, the number of BrdU-labeled cells decreased. This decrease was specifically induced by the late phase of learning and not by the passage of time (Döbrössy et al. 2003). Even more surprisingly, the decline in newly born cells was correlated with spatial abilities, as rats with the lowest number of BrdU-labeled cells (and most likely the highest rates of cell death) had the best memory performances (Döbrössy et al. 2003). This observation indicates that the decline in newly born cells is "involved" in memory and that learning, not training, decreased the number of adult-born cells. The decline in BrdU cell number most likely resulted from a cell death process, since spatial learning, but not cue training, increased cell death as evaluated with the TUNEL (terminal deoxynucleotidyl-transferease-mediated dUTP-biotin nick end-labeling) technique (Ambrogini et al. 2004). These intriguing results, together with the fact that learning also increased the number of newly born cells, may explain why no changes in BrdU cell numbers were observed in animals injected with BrdU during the entire period of training (van Praag et al. 1999a; Döbrössy et al. 2003).

Summary

Altogether, the reported results suggest a complex chain of changes in neurogenesis that accompanies spatial learning. The first step, occurring during learning of the task, is characterized by an increase in the survival of cells that have been produced before the learning experience. Remarkably, the survival-promoting effect of hippocampus-dependent learning seems to occur during the second week of cellular development. During this "critical period," cells are still immature but already differentiated as neurons and express immature neuronal markers and physiological neuronal characteristics (Kempermann et al. 2004; Piatti et al. 2006). One hypothesis is that these cells have reached an adequate developmental period to be "stabilized" by activity-dependent stimuli generated in the course of learning. This is supported by electrophysiological evidence showing that enhanced synaptic activity (obtained by the stimulation of perforant path) enhances cell survival (Bruel-Jungerman et al. 2006). An activity-dependent process might promote neurogenesis either by GABAergic depolarization (Esposito et al. 2005; Overstreet et al. 2005; Tozuka et al. 2005; Wang et al. 2005; Ge et al. 2006; Overstreet-Wadiche et al. 2006) or by glutamate activation of the L-type (Deisseroth et al. 2004) or the T-type Ca^{2+} channels (Schmidt-Hieber et al. 2004). The observation that reduction of learning-induced increase in cell survival by sleep restriction selectively impaired spatial learning (Hairston et al. 2005) suggests that the neurons rescued by learning participated in memory processes. However, at least one fundamental question remains: What is the function for these surviving neurons? One possibility is outlined by Becker and Wojtowicz (2007), who proposed that clusters of newly born neurons induce formation of functional neuronal assemblies in CA3. These assemblies could represent the memory traces available for further consolidation and ultimately retrieval.

The second, even less understood, step occurring during the late phase of learning is characterized by a decrease in the number of newly produced cells, likely due to the elimination of immature newly generated cells that have not been selected (stabilized) by learning. The elimination of more immature cells may be necessary for the survival of older newly generated cells "by making room." Indeed, newborn neurons are certainly competing for available resources to survive, and the death of one population could facilitate the survival and the integration of the remaining older ones. According to the selective stabilization theory (Changeux et al. 1973), regressive events, including cell death, will stabilize particular networks among others, thereby sculpting the circuits that are crucial for a given function. Yet another scenario has been proposed, where the selection of immature neurons is balanced by cell death of more mature newborn neurons (Ambrogini et al. 2000, 2004). Thus, although both scenarios rely on an active selection process, they differ in the age of the newly born cells that are selected or killed by learning.

A third step also takes place during the late phase of learning and is characterized by an increase in cell proliferation. It may constitute a compensatory mechanism to the selection process and consequently replace the pool of neurons that encounter death.

IS NEUROGENESIS CAUSALLY RELATED TO MEMORY ABILITIES?

One of the most important challenges in the field of hippocampal neurogenesis is to demonstrate a causal relationship between memory and neurogenesis. So far, only three methods have been used to target the neural progenitor cells in the DG. Two of the methods have been borrowed from cancer research, in particular from cancer treatment using high-energy irradiation and chemotherapy. These methods take advantage of the known sensitivity of rapidly dividing cells to γ-rays and to various chemical agents that disrupt the mitotic cell cycle. The most recent approach consists of development of inducible glial fibrillary acidic protein (GFAP)-thymidine kinase (TK) transgenic (TG) mice (Garcia et al. 2004). Selectivity and possible side effects of these methods have been discussed elsewhere; here, we only deal with the interpretation of the results and not with various possible shortcomings of the methods (Monje and Palmer 2003; Wojtowicz 2006).

Analysis of the literature indicates that the road to a conclusion on the existence of a causal relationship between memory and neurogenesis is long. Indeed, treatment with the antimitotic agent MAM (methylazoxymethanol) has been shown to disrupt trace eye-blink conditioning and trace fear conditioning (Shors et al. 2001, 2002). In contrast, no deficits were observed in two other hippocampus-dependent memory tasks, i.e., contextual fear conditioning and spatial memory in the water maze (Shors et al. 2002). In the case of irradiation treatment, contextual fear conditioning (Winocur et al. 2006), place recognition in a T maze (Madsen et al. 2003), spatial learning in the Barnes maze test (Raber et al. 2004), delayed nonmatching to sample in the water maze (especially when the delay between sample and test trials is long; Winocur et al. 2006), and long-term retention in the water maze (Snyder et al. 2005) have all been demonstrated to be altered by irradiation. In contrast, spatial learning in the water maze has been reported to remain unchanged following irradiation treatment (Madsen et al. 2003; Raber et al. 2004; Snyder et al. 2005; Meshi et al. 2006; Saxe et al. 2006), with the exception of one study reporting altered spatial learning following cranial irradiation (Fan et al. 2007). Ablating neurogenesis in GFAP-TK TG mice using a subcutaneous infusion of ganciclovir led to deficits in contextual fear conditioning

(Saxe et al. 2006). The same studies revealed that nonhippocampal tasks were not affected.

To summarize, the impact of ablating neurogenesis has been more extensively examined using contextual fear conditioning and the Morris water maze test. It appears that contextual fear memories are altered in some cases (Saxe et al. 2006; Winocur et al. 2006) but not in others (Shors et al. 2002). The data concerning the water maze are more conflicting given that spatial memory deficits were described in two studies (Snyder et al. 2005; Fan et al. 2007), but not in others (Shors et al. 2002; Madsen et al. 2003; Raber et al. 2004; Meshi et al. 2006; Saxe et al. 2006; Winocur et al. 2006). The controversy might be linked to the method used to ablate neurogenesis, the duration of the treatments before the testing procedure (and thus the age of the adult-born neurons at the time of behavioral testing), the species used (mice, rats, gerbils), the testing procedure used (configuration of the tests, training schedule, etc.), the type of memory examined (configural, relational, working), "task complexity," and/or the memory phase examined (encoding, consolidation, retrieval).

An important aspect of this work is its guidance by theoretical and computational models that can incorporate data into conceptual frameworks and formulate predictions for future experiments. A number of such theoretical models have been put forward (Becker 2005; Meltzer et al. 2005; Wiskott et al. 2006). Such interactions between experimental and theoretical approaches should prove fruitful in advancing neurogenesis research and also in its integration within the mainstream neuroscience.

CONCLUSION AND PERSPECTIVES

On the basis of evidence from several correlative studies, it has been proposed that neurogenesis is involved in hippocampus-dependent memory and in particular in spatial memory. This proposal is reinforced by the observation that, reciprocally, learning influences the rate of production and the number of surviving neurons. On the other hand, the experiments involving lesions of the neurogenic region suggest no significant role for new neurons in spatial learning. This is in contrast to the results obtained with complete DG lesions that can impair acquisition of spatial reference memory. These data may be explained by the existence of a compensatory mechanism that can sustain the apparently normal function with either the mature granule cells or other neurons of the Ammon's Horn via the direct projections from the entorhinal cortex (see

Fig. 1). However, the door is still open to new experiments. For example, theoretical considerations suggest that the dependence of spatial learning on new neurons may be particularly strong when several similar learning tasks are presented in sequence (Becker 2005). This and other computational predictions can be addressed with existing experimental methods. Although the present state of knowledge in this area is confusing, it provides us with certain directions for future progress.

1. Experiments should continue taking the correlative and causative approaches into account. These approaches are complementary and should ultimately lead to a unified picture of how neurogenesis participates in learning.
2. A variety of behavioral tasks should be used since neurogenesis is likely to have different roles in different behavioral circumstances.
3. Relating cells of certain ages (e.g., 1–2 weeks and 2–3 weeks old) to specific learning functions should be a priority.
4. Determining which plastic properties of new neurons are relevant to learning. Do the changes occur at the perforant path or mossy fiber synapses?
5. Development of new transgenic models to ablate adult-born neurons with emphasis on targeting different development stages of neurons (i.e., eliminating neurons at stages of proliferation, differentiation, and maturation).
6. Development of new drugs to kill neurons or to stimulate their production.

SUMMARY

The discovery of a continuous renewal of neurons in the adult mammalian brain has been a long process, one of the most controversial of modern neuroscience. The existence of de novo production of neurons in the adult hippocampus, a structure involved in memory, has stimulated research on their potential role in the physiology and pathophysiology of the hippocampus. Here, we have reviewed the current knowledge on the putative role of adult hippocampal neurogenesis in memory. In particular, we have illustrated that there is a two-way interrelationship between neurogenesis and learning: The rate of neurogenesis determines learning and memory and, reciprocally, learning influences the rate of neurogenesis. However, we have also highlighted that the state of our knowledge on the causal role of newborn neurons in memory is still controversial.

ACKNOWLEDGMENTS

D.N.A. is supported by INSERM, France. J.M.W. is supported by NSERC and CIHR, Canada.

REFERENCES

Ambrogini P., Orsini L., Mancini C., Ferri P., Ciaroni S., and Cuppini R. 2004. Learning may reduce neurogenesis in adult rat dentate gyrus. *Neurosci. Lett.* **359:** 13–16.

Ambrogini P., Cuppini R., Cuppini C., Ciaroni S., Cecchini T., Ferri P., Sartini S., and Del Grande P. 2000. Spatial learning affects immature granule cell survival in adult rat dentate gyrus. *Neurosci. Lett.* **286:** 21–24.

Becker S. 2005. A computational principle for hippocampal learning and neurogenesis. *Hippocampus* **15:** 722–738.

Becker S. and Wojtowicz J.M. 2007. A model of hippocampal neurogenesis in memory and mood disorders. *Trends Cogn. Sci.* **11:** 70–76.

Bizon J.L. and Gallagher M. 2003. Production of new cells in the rat dentate gyrus over the lifespan: Relation to cognitive decline. *Eur. J. Neurosci.* **18:** 215–219.

Bruel-Jungerman E., Davis S., Rampon C., and Laroche S. 2006. Long-term potentiation enhances neurogenesis in the adult dentate gyrus. *J. Neurosci.* **26:** 5888–5893.

Changeux J.P., Courrege P., and Danchin A. 1973. A theory of the epigenesis of neuronal networks by selective stabilization of synapses. *Proc. Natl. Acad. Sci.* **70:** 2974–2978.

Deisseroth K., Singla S., Toda H., Monje M., Palmer T.D., and Malenka R.C. 2004. Excitation-neurogenesis coupling in adult neural stem/progenitor cells. *Neuron* **42:** 535–552.

Döbrössy M.D.E., Aurousseau C., Le Moal M., Piazza P.V., and Abrous D.N. 2003. Differential effects of learning on neurogenesis: Learning increases or decreases the number of newly born cells depending on their birth date. *Mol. Psychiatry* **8:** 974–982.

Drapeau E., Mayo W., Aurousseau C., Le Moal M., Piazza P.V., and Abrous D.N. 2003. Spatial memory performances of aged rats in the water maze predict levels of hippocampal neurogenesis. *Proc. Natl. Acad. Sci.* **100:** 14385–14390.

Ehninger D. and Kempermann G. 2006. Paradoxical effects of learning the Morris water maze on adult hippocampal neurogenesis in mice may be explained by a combination of stress and physical activity. *Genes Brain Behav.* **5:** 29–39.

Eichenbaum H. 2000. A cortical-hippocampal system for declarative memory. *Nat. Rev. Neurosci.* **1:** 41–50.

———. 2001. The hippocampus and declarative memory: Cognitive mechanisms and neural codes. *Behav. Brain Res.* **127:** 199–207.

Esposito M.S., Piatti V.C., Laplagne D.A., Morgenstern N.A., Ferrari C.C., Pitossi F.J., and Schinder A.F. 2005. Neuronal differentiation in the adult hippocampus recapitulates embryonic development. *J. Neurosci.* **25:** 10074–10086.

Fan Y., Liu E., Weinstein P.R., Fike J.R., and Liu J. 2007. Environmental enrichment enhances neurogenesis and improve functional outcome after cranial irradiation. *Eur. J. Neurosci.* **25:** 38–46.

Garcia A.D., Doan N.B., Imura T., Bush T.G., and Sofroniew M.V. 2004. GFAP-expressing progenitors are the principal source of constitutive neurogenesis in adult mouse forebrain. *Nat. Neurosci.* **7:** 1233–1241.

Ge S., Goh E.L., Sailor K.A., Kitabatake Y., Ming G.L., and Song H. 2006. GABA regulates synaptic integration of newly generated neurons in the adult brain. *Nature* **439**: 589–593.

Gilbert P.E. and Kesner R.P. 2003. Localization of function within the dorsal hippocampus: The role of the CA3 subregion in paired-associate learning. *Behav. Neurosci.* **117**: 1385–1394.

Gilbert P.E., Kesner R.P., and Lee I. 2001. Dissociating hippocampal subregions: Double dissociation between dentate gyrus and CA1. *Hippocampus* **11**: 626–636.

Gould E., Beylin A., Tanapat P., Reeves A., and Shors T.J. 1999. Learning enhances adult neurogenesis in the hippocampal formation. *Nat. Neurosci.* **2**: 260–265.

Hairston I.S., Little M.T., Scanlon M.D., Barakat M.T., Palmer T.D., Sapolsky R.M., and Heller H.C. 2005. Sleep restriction suppresses neurogenesis induced by hippocampus-dependent learning. *J. Neurophysiol.* **94**: 4224–4233.

Jaako-Movits K. and Zharkovsky A. 2005. Impaired fear memory and decreased hippocampal neurogenesis following olfactory bulbectomy in rats. *Eur. J. Neurosci.* **22**: 2871–2878.

Jaako-Movits K., Zharkovsky T., Romantchik O., Jurgenson M., Merisalu E., Heidmets L.T., and Zharkovsky A. 2005. Developmental lead exposure impairs contextual fear conditioning and reduces adult hippocampal neurogenesis in the rat brain. *Int. J. Dev. Neurosci.* **23**: 627–635.

Jeltsch H., Bertrand F., Lazarus C., and Cassel J.C. 2001. Cognitive performances and locomotor activity following dentate granule cell damage in rats: Role of lesion extent and type of memory tested. *Neurobiol. Learn. Mem.* **76**: 81–105.

Kempermann G. and Gage F.H. 2002. Genetic influence on phenotypic differentiation in adult hippocampal neurogenesis. *Dev. Brain Res.* **134**: 1–12.

Kempermann G., Brandon E.P., and Gage F.H. 1998a. Environmental stimulation of 129/SvJ mice causes increased cell proliferation and neurogenesis in the adult dentate gyrus. *Curr. Biol.* **8**: 939–942.

Kempermann G., Gast D., and Gage F.H. 2002. Neuroplasticity in old age: Sustained five-fold induction of hippocampal neurogenesis by long-term environmental enrichment. *Ann. Neurol.* **52**: 135–143.

Kempermann G., Kuhn H.G., and Gage F.H. 1997. More hippocampal neurons in adult mice living in an enriched environment. *Nature* **386**: 493–495.

———. 1998b. Experience-induced neurogenesis in the senescent dentate gyrus. *J. Neurosci.* **18**: 3206–3212.

Kempermann G., Jessberger S., Steiner B., and Kronenberg G. 2004. Milestones of neuronal development in the adult hippocampus. *Trends Neurosci.* **27**: 447–452.

Koo J.W., Park C.H., Choi S.H., Kim N.J., Kim H.S., Choe J.C., and Suh Y.H. 2003. The postnatal environment can counteract prenatal effects on cognitive ability, cell proliferation, and synaptic protein expression. *FASEB J.* **17**: 1556–1558.

Lemaire V., Koehl M., Le Moal M., and Abrous D.N. 2000. Prenatal stress produces learning deficits associated with an inhibition of neurogenesis in the hippocampus. *Proc. Natl. Acad. Sci.* **97**: 11032–11037.

Leuner B., Mendolia-Loffredo S., Kozorovitskiy Y., Samburg D., Gould E., and Shors T.J. 2004. Learning enhances the survival of new neurons beyond the time when the hippocampus is required for memory. *J. Neurosci.* **24**: 7477–7481.

Madsen T.M., Kristjansen P.E., Bolwig T.G., and Wortwein G. 2003. Arrested neuronal proliferation and impaired hippocampal function following fractionated brain irradiation in the adult rat. *Neuroscience* **119**: 635–642.

McNaughton B.L., Barnes C.A., Meltzer J., and Sutherland R.J. 1989. Hippocampal granule cells are necessary for normal spatial learning but not for spatially-selective pyramidal cell discharge. *Exp. Brain Res.* **76:** 485–496.

Meltzer L.A., Yabaluri R., and Deisseroth K. 2005. A role for circuit homeostasis in adult neurogenesis. *Trends Neurosci.* **28:** 653–660.

Merrill D.A., Karim R., Darraq M., Chiba A., and Tuszynski M.H. 2003. Hippocampal cell genesis does not correlate with spatial learning ability in aged rats. *J. Comp. Neurol.* **459:** 201–207.

Meshi D., Drew M.R., Saxe M., Ansorge M.S., David D., Santarelli L., Malapani C., Moore H., and Hen R. 2006. Hippocampal neurogenesis is not required for behavioral effects of environmental enrichment. *Nat. Neurosci.* **9:** 729–731.

Mohapel P., Leanza G., Kokaia M., and Lindvall O. 2005. Forebrain acetylcholine regulates adult hippocampal neurogenesis and learning. *Neurobiol. Aging* **26:** 939–946.

Mohapel P., Mundt-Petersen K., Brundin P., and Frielingsdorf H. 2006. Working memory training decreases hippocampal neurogenesis. *Neuroscience* **142:** 609–613.

Monje M.L. and Palmer T. 2003. Radiation injury and neurogenesis. *Curr. Opin. Neurol.* **16:** 129–134.

Montaron M.F., Drapeau E., Dupret D., Kitchener P., Aurousseau C., Le Moal M., Piazza P.V., and Abrous D.N. 2006. Lifelong corticosterone level determines age-related decline in neurogenesis and memory. *Neurobiol. Aging* **27:** 654.

Morris R.G. 2006. Elements of a neurobiological theory of hippocampal function: The role of synaptic plasticity, synaptic tagging and schemas. *Eur. J. Neurosci.* **23:** 2829–2846.

Nakayama T. and Sawada T. 2002. Involvement of microtubule integrity in memory impairment caused by colchicine. *Pharmacol. Biochem. Behav.* **71:** 119–138.

Nakazawa K., Quirk M.C., Chitwood R.A., Watanabe M., Yeckel M.F., Sun L.D., Kato A., Carr C.A., Johnston D., Wilson M.A., and Tonegawa S. 2002. Requirement for hippocampal CA3 NMDA receptors in associative memory recall. *Science* **297:** 211–218.

Nanry K.P., Mundy W.R., and Tilson H.A. 1989. Colchicine-induced alterations of reference memory in rats: Role of spatial versus non-spatial task components. *Behav. Brain Res.* **35:** 45–53.

Nilsson M., Perfilieva E., Johansson U., Orwar O., and Eriksson P.S. 1999. Enriched environment increases neurogenesis in the adult rat dentate gyrus and improves spatial memory. *J. Neurobiol.* **39:** 569–578.

Overstreet W.L., Bromberg D.A., Bensen A.L., and Westbrook G.L. 2005. GABAergic signaling to newborn neurons in dentate gyrus. *J. Neurophysiol.* **94:** 4528–4532.

Overstreet-Wadiche L.S., Bensen A.L., and Westbrook G.L. 2006. Delayed development of adult-generated granule cells in dentate gyrus. *J. Neurosci.* **26:** 2326–2334.

Pham K., McEwen B.S., LeDoux J.E., and Nader K. 2005. Fear learning transiently impairs hippocampal cell proliferation. *Neuroscience* **130:** 17–24.

Piatti V.C., Esposito M.S., and Schinder A.F. 2006. The timing of neuronal development in adult hippocampal neurogenesis. *Neuroscientist* **12:** 463–468.

Qiao C., Den R., Kudo K., Yamada K., Takemoto K., Wati H., and Kanba S. 2005. Ginseng enhances contextual fear conditioning and neurogenesis in rats. *Neurosci. Res.* **51:** 31–38.

Raber J., Rola R., LeFevour A., Morhardt D., Curley J., Mizumatsu S., VandenBerg S.R., and Fike J.R. 2004. Radiation-induced cognitive impairments are associated with changes in indicators of hippocampal neurogenesis. *Radiat. Res.* **162:** 39–47.

Saxe M.D., Battaglia F., Wang J.W., Malleret G., David D.J., Monckton J.E., Garcia A.D., Sofroniew M.V., Kandel E.R., Santarelli L., Hen R., and Drew M.R. 2006. Ablation of hippocampal neurogenesis impairs contextual fear conditioning and synaptic plasticity in the dentate gyrus. *Proc. Natl. Acad. Sci.* **103:** 17501–17506.

Schmidt-Hieber C., Jonas P., and Bischofberger J. 2004. Enhanced synaptic plasticity in newly generated granule cells of the adult hippocampus. *Nature* **429:** 184–187.

Shen L., Nam H.S., Song P., Moore H., and Anderson S.A. 2006. FoxG1 haploinsufficiency results in impaired neurogenesis in the postnatal hippocampus and contextual memory deficits. *Hippocampus* **16:** 875–890.

Shimazu K., Zhao M., Sakata K., Akbarian S., Bates B., Jaenisch R., and Lu B. 2006. NT-3 facilitates hippocampal plasticity and learning and memory by regulating neurogenesis. *Learn. Mem.* **13:** 307–315.

Shors T.J., Townsend D.A., Zhao M., Kozorovitskiy Y., and Gould E. 2002. Neurogenesis may relate to some but not all types of hippocampal-dependent learning. *Hippocampus* **12:** 578–584.

Shors T.J., Miesegaes G., Beylin A., Zhao M., Rydel T., and Gould E. 2001. Neurogenesis in the adult is involved in the formation of trace memories. *Nature* **410:** 372–376.

Sisti H.M., Glass A.L., and Shors T.J. 2007. Neurogenesis and the spacing effect: Learning over time enhances memory and the survival of new neurons. *Learn. Mem.* **14:** 368–375.

Snyder J.S., Kee N., and Wojtowicz J.M. 2001. Effects of adult neurogenesis on synaptic plasticity in the rat dentate gyrus. *J. Neurophysiol.* **85:** 2423–2431.

Snyder J.S., Hong N.S., McDonald R.J., and Wojtowicz J.M. 2005. A role for adult neurogenesis in spatial long-term memory. *Neuroscience* **130:** 843–852.

Tozuka Y., Fukuda S., Namba T., Seki T., and Hisatsune T. 2005. GABAergic excitation promotes neuronal differentiation in adult hippocampal progenitor cells. *Neuron* **47:** 803–815.

Treffert D.A. and Christensen D.D. 2005. Inside the mind of a savant. *Sci. Am.* **293:** 108–113.

Van der Borght K., Meerlo P., Luiten P.G., Eggen B.J., and Van der Zee E.A. 2005a. Effects of active shock avoidance learning on hippocampal neurogenesis and plasma levels of corticosterone. *Behav. Brain Res.* **157:** 23–30.

Van der Borght K., Wallinga A.E., Luiten P.G., Eggen B.J., and Van der Zee E.A. 2005b. Morris water maze learning in two rat strains increases the expression of the polysialylated form of the neural cell adhesion molecule in the dentate gyrus but has no effect on hippocampal neurogenesis. *Behav. Neurosci.* **119:** 926–932.

van Praag H., Kempermann G., and Gage F.H. 1999a. Running increases cell proliferation and neurogenesis in the adult mouse dentate gyrus. *Nat. Neurosci.* **2:** 266–270.

van Praag H., Christie B.R., Sejnowski T.J., and Gage F.H. 1999b. Running enhances neurogenesis, learning, and long-term potentiation in mice. *Proc. Natl. Acad. Sci.* **96:** 13427–13431.

van Praag H., Shubert T., Zhao C., and Gage F.H. 2005. Exercise enhances learning and hippocampal neurogenesis in aged mice. *J. Neurosci.* **25:** 8680–8685.

Wang L.P., Kempermann G., and Kettenmann H. 2005. A subpopulation of precursor cells in the mouse dentate gyrus receives synaptic GABAergic input. *Mol. Cell. Neurosci.* **29:** 181–189.

Winocur G., Wojtowicz J.M., Sekeres M., Snyder J.S., and Wang S. 2006. Inhibition of neurogenesis interferes with hippocampus-dependent memory function. *Hippocampus* **16:** 296–304.

Wiskott L., Rasch M.J., and Kempermann G. 2006. A functional hypothesis for adult hippocampal neurogenesis: Avoidance of catastrophic interference in the dentate gyrus. *Hippocampus* **16:** 329–343.

Witter M.P. 1993. Organization of the entorhinal-hippocampal system: A review of current anatomical data. *Hippocampus* (spec. no.) **3:** 33–44.

Wojtowicz J.M. 2006. Irradiation as an experimental tool in studies of adult neurogenesis. *Hippocampus* **16:** 261–266.

Xavier G.F., Oliveira-Filho F.J.B., and Santos A.M.G. 1999. Dentate gyrus-selective colchicine lesion and disruption of performance in spatial tasks: Difficulties in place "strategy" because of a lack of flexibility in the use of environmental cues? *Hippocampus* **9:** 668–681.

Zhao X., Ueba T., Christie B.R., Barkho B., McConnell M.J., Nakashima K., Lein E.S., Eadie B.D., Willhoite A.R., Muotri A.R., Summers R.G., Chun J., Lee K.F., and Gage F.H. 2003. Mice lacking methyl-CpG binding protein 1 have deficits in adult neurogenesis and hippocampal function. *Proc. Natl. Acad. Sci.* **100:** 6777–6782.

22

Computational Modeling of Adult Neurogenesis

James B. Aimone
Laboratory of Genetics
The Salk Institute for Biological Studies
La Jolla, California 92037

Laurenz Wiskott
Institute for Theoretical Biology
Humboldt-Universität zu Berlin
D-10115 Berlin, Germany

ONE OF THE MOST INTRIGUING DIFFERENCES between adult and developmental neurogenesis is that in the adult brain, new neurons are integrating into already-developed, functioning circuits. Newborn neurons develop highly complex neuronal morphology—an impressive feat, considering that the extracellular signaling environment (thought to be important during development) is considerably different in the adult. Adult neurogenesis has been observed in most animal species, both in the normal course of life and in response to injury in many nonmammals. The fact that adult neurogenesis is essentially limited to two regions in mammalian brains suggests that the addition of new neurons to these regions (the olfactory bulb [OB] and dentate gyrus [DG]) is of particular importance.

Although the function of regenerative neurogenesis is self-evident, the purpose for lifelong neurogenesis remains unclear. There are several reasons why taking a computational modeling approach has potential. One is that any effect of adding new neurons will first be manifested computationally in the network and will only then be observed behaviorally. Modeling can permit the observation of an effect that otherwise would go unseen in standard behavioral assays. This provides a framework by which new predictions can be made that can be specifically tested experimentally. Furthermore, a well-developed computational model or theory can be altered in a manner that is impractical or impossible in animal models, such as increasing the rate of neurogenesis by tenfold or studying the effects of neurogenesis in nonneurogenic areas. Finally, modeling the computational aspects of a system often helps focus future experiments, bringing unknown parameters that need further biological study to the forefront. This chapter discusses research into the computational implications of adding new neurons to existing biological networks. Computational models have been most prominent in mammalian adult neurogenesis, but theoretical circuit functions have been suggested in other classes as well, including birds and fish, and there is a history of "neurogenesis" in nonbiological, artificial neural-network theory.

Although the details of adult neurogenesis vary from one system to another (which is discussed at length elsewhere in this book), several aspects of neurogenesis are consistent across species and important in understanding the computational implication of new cells. The first is that they are functional neurons. In the mammalian hippocampus, OB, and the bird song system these neurons have convincingly been shown to be functionally integrated into their circuit (Paton and Nottebohm 1984, van Praag et al. 2002; Carleton et al. 2003). Furthermore, most evidence suggests that they integrate into the network in a manner anatomically and physiologically indistinguishable from more mature neurons (Esposito et al. 2005).

A second observation critical to understanding function is that whereas neuronal addition is coupled with cell death in most cases, the relationship is more complex than a simple one-for-one replacement. At least in some systems, cell birth and death are likely sufficiently decoupled to be considered separate processes, providing the neurogenic regions with fluctuating numbers of neurons over time, whereas in other systems, periods of neuron birth and death are phasic. Furthermore, in most instances, both proliferation rates and newborn neuron survival are regulated by a wide range of external and behavioral factors (for review, see Ming and Song 2005).

"NEUROGENESIS" IN ARTIFICIAL NEURAL NETWORKS

Although adult neurogenesis is a relatively novel biological concept, it has existed in some form in the artificial neural-network literature for several decades. Neurogenesis has been described in both supervised and unsupervised networks. In the former case, the network is told exactly how to respond to a particular input. In the latter, it does not receive detailed guidance and has to learn based on a rather general objective. A particularly interesting example for supervised algorithms that successively add new units to improve the performance (for review, see Bishop 1995) is the cascade-correlation learning architecture (Fahlman and Lebiere 1990). It is a feedforward network with input units, output units, and hidden units, which carry out most of the computation from input to output. There are no lateral or feedback connections. The network starts with no hidden unit, and all connections go directly from input to output. After having trained this reduced network on a particular learning task, a single hidden unit is added and the whole network is trained again with only new connections of this hidden unit modified and all old units and connections kept fixed. Then the first hidden unit is kept fixed while a second hidden unit is added and trained. This way, the network is extended by single hidden units and retrained until network performance is satisfactory. Fahlman and Lebiere (1990) have shown much faster learning compared to standard feedforward networks trained with the popular backpropagation algorithm.

Clustering (finding groups of similar patterns in a data set) and vector quantization (tiling up the data space in a reasonable way that reflects the distribution of the data) are two related examples of unsupervised learning. The adaptive resonance theory (ART) network (Carpenter and Grossberg 1987; Carpenter et al. 1991) finds clusters of similar patterns and represents each one with a single unit. If patterns have to be represented that are distinct from all patterns seen before, the ART network assumes there is a new cluster and adds a unit to represent it. Growing neural networks performing vector quantization have been extensively studied by Fritzke (1994). For a more detailed discussion of neurogenesis in artificial neural networks, see Wiskott et al. (2006).

MAMMALIAN NEUROGENESIS: THE DENTATE GYRUS

As has been observed in a wide range of mammals, including humans, the DG structure in the hippocampus experiences a substantial level of adult neurogenesis throughout life (Eriksson et al. 1998). DG neurogenesis is of substantial computational interest not only because of its potential role in human episodic memory formation, but also because the hippocampus has historically been a focus of both neural-network modeling and biological studies. The hippocampus is a region of the brain considered to be important for short-term memory formation and spatial learning. Behavioral and computational work has dissociated the functions of its different subregions (for review, see Rolls and Kesner 2006). A simple view of the hippocampus shows an ideal circuit for computational modeling: the DG as the input layer, which is thought to have a preprocessing role, such as sparsification and pattern separation; the CA3 as an associative network, which is considered ideal for memory formation because of its many feedback (recurrent) connections; and the CA1 as the output layer, possibly inverting the preprocessing by the DG and/or performing pattern-completion, integrating over its CA3 inputs. Along with numerous subcortical structures, such as the septum, the entorhinal cortex (EC) is the primary cortical source and target of hippocampal projections. In reality, the hippocampus is considerably more complex: Feedback circuitry is found in the DG as well, a wide range of interneurons are present in each layer, and the EC projects to the CA3 and CA1 directly, as well as through the DG (Fig. 1).

A wide range of computational models have been generated for the hippocampus, and, more recently, a number of models appeared addressing neurogenesis in the DG. They range from conceptual to anatomically precise, with each providing unique insights into the potential role of new neurons in cognition.

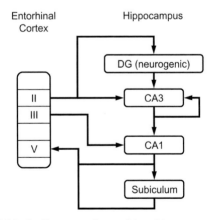

Figure 1. Simplified block diagram of entorhino–hippocampal circuitry. Principal excitatory projections are shown with each region also having significant internal inhibitory (and in some cases, excitatory) components as well. The granule cell projection neurons of the dentate gyrus (DG) are the only population of hippocampal neurons that are born throughout life.

Small-world Network Elaboration

Small-world networks have been proposed as an intermediate between fully connected networks, where each unit is linked to every other unit but the number of connections is physically prohibitive, and locally connected networks, where physical connectivity is minimized but at the expense of time delays and information loss between distant parts of the network. In a small-world network, a small percentage of neurons project to distant neurons, dramatically reducing distance between neurons while only slightly increasing connection density.

Manev and Manev (2005) take the idea that the cortex functions as a small-world network and suggest that adding new neurons throughout life optimizes this connectivity by placing new units between regions that were previously distant, forming a new, shorter path. This model is entirely conceptual and does not specifically develop the use of small-world networks in the hippocampal circuit. Although specific relevance to hippocampal neurogenesis is unclear, these investigators point to interesting ramifications of the concept within the context of therapeutic neurogenesis in typically nonneurogenic regions, where the random addition of new neurons would serve to increase small-world connectivity.

Proliferation and Homeostasis of Dentate Activity

Unlike other neurogenesis models that focus on the information processing of a network with newborn neurons, Lehmann et al. (2005) and Butz et al. (2006) discuss the homeostatic implications of adding new neurons to a network. Their approach builds on the concept that a network uses the gain and loss of synapses to respond to different strengths of inputs. According to their hypothesis, increased levels of EC inputs would normally result in a decrease of excitatory synapses. The addition of new neurons in the DG would provide another source of this plasticity, helping to balance the network. The drawback to this process is that adding too many new neurons results in a destabilization of the network.

This idea was modeled in Butz et al. (2006) by using a recurrent network that stabilizes itself by gaining or losing synapses in response to network perturbations. The effect of new neurons is most evident under conditions in which the external input is altered signficantly. In systems without neurogenesis, a burst of high-excitatory input results in a compensatory decay of synapses from most units. However, by using neurogenesis and apoptosis, the network becomes significantly more robust to such perturbations. In their model, the combination of eliminating the neurons most affected by the shift of excitation and adding new neurons better suited for the current network conditions provides a strong stabilizing force. Importantly, this buffering effect of new neurons is rate-dependent: At high levels of neurogenesis, the network's ability to settle into a homeostatic state was greatly perturbed.

Hidden-layer Turnover

Several models simulate the effects of neurogenesis by replacing neurons in the hidden layer of a three-layer neural network. The first such model, by Chambers et al. (2004), has been elaborated in two subsequent studies (Crick and Miranker 2006; Chambers and Conroy 2007) and investigates the ability of the network to learn different alphabets. Around the same time, a second model by Deisseroth et al. (2004) was proposed that is architecturally similar but with several significant differences, which are outlined below.

The Chambers model uses a backpropagation algorithm to train a three-layer feedforward network to learn first the Roman and then the Greek alphabet. The network structure uses a hidden layer (considered the DG) in the middle to transform the images of letters, represented by a 5×7 grid of units (EC), into a 1-out-of-N code of their identity in an output layer (CA3) with 26 units. Their most significant finding is that replacing units in the modeled DG layer facilitates the learning of new letters while accelerating clearance of old memories. Various permutations of turnover were tested, with gradual replacement of neurons during training leading to the best acquisition of the second (Greek) alphabet.

The Deisseroth model (part of a larger paper investigating the role of N-methyl-D-aspartate [NMDA] in proliferation) is similar to the original Chambers model, although there are several differences. Most notable are that it uses Hebbian learning rather than backpropagation, the units in this model are binary threshold units, and the network is considerably larger (500 neurons per layer) than that used in the Chambers and Crick studies (10–32 neurons per DG layer). Nevertheless, the results are similar to the results of the Chambers study—replacement of neurons in the middle layer cleared older memories and increased learning rates for new memories, confirming that the neuronal replacement effects seen are robust to network size, input structure, and learning rules.

Two subsequent studies have built on the approach of the Chambers study. Crick and Miranker developed the model further by adding lateral-inhibitory connections, using Hebbian learning (as in the Deisseroth model), and adding selective apoptosis. Their work included a thorough exploration of the parameter space, altering layer size, connectivity, and learning extent, ultimately showing that the neurogenesis effects seen in the original study are robust to network architecture. In addition, they discuss the rationale for having directed rather than random apoptosis. Recently, Chambers and Conroy (2007) used the Chambers model to test how dynamic regulation of neurogenesis levels can help learn new alphabets of different levels of novelty. After initial training on the Roman alphabet, the network was trained on one of three alphabets (Roman, Cyrillic, or Hebrew). They showed that learning a very different alphabet, such as Hebrew, can only be accomplished effectively with high levels of neurogenesis, whereas learning less novel information necessitates only low levels of replacement.

The results from the Chambers and Deiserroth studies consistently show the benefit of neuronal replacement for learning in a three-layer pattern-recognition network. After a series of patterns have been learned, the units in the hidden layer become specialized to certain features of the input patterns, with many of their weights assuming large values. Replacing them with new units resets their connection strengths to smaller random levels, which do not represent stored information. Training these naïve units is substantially easier than training an already-trained unit in the hidden layer. Thus, neuronal replacement increases learning of new patterns at the expense of old patterns.

Unique Coding in the Hippocampus

The Becker model is a full hippocampus model (EC→DG→CA3→ CA1→EC) that investigates the role of both neurogenesis and a particular learning rule in hippocampal memory formation (Becker 2005). Although the circuit modeled is considerably more complex, the memory capacity results of her model are consistent with the earlier studies of neurogenesis, with the critical exception that increasing neurogenesis does not cause loss of memory. This is because unlike previous models, which involved the DG in both storage and retrieval, the Becker model only involves the DG in the memory storage phase. Retrieval is initiated by the direct EC to CA3 projection.

The Becker model investigates two conditions—the effect of DG size on hippocampal memory and the effect of neuronal turnover. Models of hippocampus with sufficiently large DGs were able to show near-perfect recall, but had difficulty dissociating similar patterns. In the neurogenesis case, neurons were replaced in the DG layer between the presentation of input patterns to be learned. With neurogenesis, the network performed much better in dissociating similar patterns, although the optimal performance in this model was at 100% replacement between learning epochs. Neurogenesis occurs at much lower rates than was tested in this model, so it remains to be seen how these effects are manifested at more biologically relevant levels of neurogenesis.

In a recent follow-up paper, Becker and Wojtowicz (2007) extended this model conceptually by proposing that neurons born at the same time have similar but not identical inputs, allowing associations of different aspects of an environment to be encoded simultaneously. This function is then associated with affective disorders such as depression, in which neurogenesis levels are thought to be possibly misregulated. According to their prediction and model, the memory deficits seen during depression are a result of decreased neurogenesis, which impairs the association of positive context to new memories. In addition, they suggest that new neurons may be functional in clusters, with neurons born at the same time encoding varied aspects of new information.

Reduction of Catastrophic Interference

As in previous models, Wiskott et al. (2006) assume that DG is performing a preprocessing of input patterns for storage in CA3 and that DG is modeled as a middle layer in a three-layer network. In contrast to previous models, however, neurogenesis is not assumed to be a turnover process but, rather, accumulative. The general idea is that to optimize the encoding process, the DG adapts to the statistics of the input patterns. If the input statistics change for any reason, such as when an animal moves into a new environment, the DG's representation ceases to be optimal and should readapt. In general, this would change the previous encoding and disrupt cued recall of previously stored patterns, an effect referred to as catastrophic interference. The hypothesis is that this catastrophic interference is avoided by keeping old units in the DG, keeping their weights fixed, and adding new plastic units to encode the different features of the new input patterns. Thus, if the input statistics change, the encoding is extended rather than changed. In a simple model simulation, Wiskott et al. (2006) showed that adding new units can indeed help to adapt to new input statistics without catastrophic interference.

Time in Memories

The model postulated by Aimone et al. (2006) is similar to the Wiskott and Becker models in that it focuses on the DG's role in memory encoding.

However, instead of concentrating on the role of new neurons in improving adaptation to new input patterns, this hypothesis focuses exclusively on the involvement of immature neurons in the encoding of temporal relationships between events. Immature neurons have unique physiological properties, possibly making them relatively more active in the DG circuitry. They propose that during memory formation, new neurons will preferentially be involved in the encoding of events. Because of the sparse activity of the DG, separate memories will tend to activate a distinct set of mature neurons but will involve many of the same, less selective immature neurons. As a result, events close in time use overlapping sets of new neurons, thus linking the memories, whereas events far in time will be encoded with separate sets of neurons.

A computational model has been developed for this hypothesis, with results consistent with those predicted by the hypothesis (presented by Aimone et al., *Soc. Neurosci. Abstr.* 419.10, 2006). In this model, the full DG circuit was simulated (including the excitatory and inhibitory feedback neurons of the hilus), with the projection to CA3 treated as the output. In their simulations, the presence of new neurons increases the overlap between mossy fiber inputs to the CA3 for two events close in time, although events distinct in time (~2 weeks apart) are encoded separately.

Other Models and Theories of Hippocampal Neurogenesis

We have summarized above the principal computational studies regarding hippocampal neurogenesis, although there are several theories and unpublished studies which deserve to be mentioned. Early work by Gould et al. (1999) suggested a possible role in the storage of memories that will soon be forgotten. According to this hypothesis, the fact that the vast majority of new neurons die within several weeks is indicative of their function being complete—the memory is no longer needed, so they disappear. A theoretical review by Schinder and Gage (2004) describes how functional neurogenesis would complement the synaptic plasticity and memory function of the older neurons. In this context, new neurons would provide both more synapses for learning and added processing abilities. Finally, a model by V.I. Weisz and P.F. Argibay was recently presented (*Soc. Neurosci. Abstr.* 101.30, 2006) that simulates neurogenesis in the full hippocampus circuit, as with the Becker model, but uses neuronal addition rather than replacement. Their conclusions suggest that neurogenesis increases hippocampal capacity and facilitates retrieval.

In summary, these models are based on different assumptions and with different levels of biological detail. A key difference between the models by Chambers et al. (2004), Deisseroth et al. (2004), and Becker (2005), on the one hand, and Manev and Manev (2005), Wiskott et al. (2006), and Aimone et al. (2006), on the other hand, is that the former assume neuronal turnover, whereas the latter assume addition of new neurons. We believe that there is more evidence for neuronal addition (Bayer et al. 1982; Boss et al. 1985), but this issue does not seem to be completely resolved. It is clear, however, that adult neurogenesis decreases with age, a fact only addressed by the model of Wiskott et al. (2006). In addition, many of these models are simple and are conceptually driven, rather than explicitly based on hippocampal circuitry. Two exceptions are those by Aimone et al. (2006) and Becker (2005). The latter is the only model for neurogenesis that takes direct perforant path connections to CA3 and CA1 into account. Finally, the Lehmann et al. (2005) and Butz et al. (2006) studies are unique in that they address dynamic stability issues, rather than functional aspects.

MAMMALIAN NEUROGENESIS: THE OLFACTORY BULB

The second mammalian region that has lifelong neurogenesis is the OB and olfactory epithelium (for details, see Chapter 20). Unlike the DG, where only one class of cells is continuously generated, several different types of neurons in the olfactory region are renewed throughout life (Fig. 2). New OB neurons are born in the subventricular zone (SVZ) and migrate to their ultimate integration site. Olfactory receptor neurons, which reside outside the cribriform plate in the olfactory epithelium inside the nasal cavity, are generated from local progenitor populations and appear to undergo a complete turnover process.

The OB has a highly ordered structure, which has made it amenable to modeling (Lledo et al. 2005). The primary sensory neurons are the olfactory receptor neurons (ORNs). Each ORN expresses a single olfactory receptor out of the hundreds of possibilities encoded by its DNA. The ORNs project their axons through the cribriform plate into the OB and converge with other similar ORNs in the appropriate glomerulus. The glomeruli are clusters of synapses between the ORN axons and OB neuron dendrites, and each is specific for a single olfactory receptor. Although the glomeruli are specific to a particular receptor, it must be noted that the receptors themselves are not very specific.

The primary neurons that receive input within the glomeruli are the mitral and tufted cells. Generally, mitral cells receive input from a single glomerulus, whereas tufted cells may have dendrites in several glomeruli. This apparently simple convergence from ORNs to mitral and tufted cells

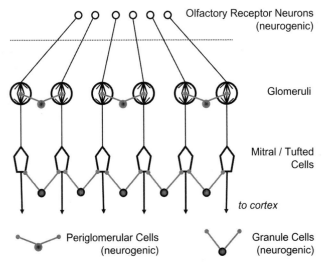

Figure 2. Basic circuit architecture of the olfactory bulb (OB). Olfactory receptor neurons (ORNs), the primary sensory neurons for smell, project to bundles of neuropil, known as glomeruli, where they synapse onto mitral and tufted cell neurons, which relay information to the olfactory cortex. Inhibitory neurons provide lateral inhibition between individual glomeruli (periglomerular cells) and between mitral/tufted cells (granule cells). ORNs, periglomerular cells, and granule cells are all neurogenic groups and are continuously replaced throughout adulthood.

is confounded by the presence of large amounts of lateral communication between glomeruli. There are several populations of interneurons that provide lateral inhibition between principal neurons, with one class, the periglomerular cells, experiencing adult neurogenesis. Periglomerular neurons typically provide inhibition to one or several neighboring glomeruli. The second level of lateral inhibition in the circuit resides outside the glomeruli, between the output tufted and mitral cells. These cells, known as granule cells, are also neurogenic and provide both lateral and feedback inhibition onto the principal neurons. Granule cells are believed to link neurons from many different glomeruli.

As in most sensory systems, the morphology and physiology of olfactory neurons are unique. For instance, dendro–dendritic synapses are prevalent in the bulbar system, particularly in the glomeruli where mitral dendrites release glutamate onto interneuron dendritic spines, which in turn release GABA back onto the mitral cell. Furthermore, the dendritic architecture of mitral and tufted cells is such that lateral inhibition by the granule cells likely affects local spike propagation, rather than spike generation. The result of this microcircuitry is a complex interaction between excitation and inhibition regulating the dynamics and is not well understood.

The models of OB function are diverse, and several general computational theories have been proposed for what the OB is doing. Many are related to the "center-surround" lateral-inhibition model often suggested for visual processing. Multiple glomeruli can be activated in response to a single odorant, because the one chemical will often activate many different types of receptor neurons. According to this lateral-inhibition idea, the activation of one glomerulus will lead to the inhibition of the others, thus increasing the contrast in the output of the OB. Other computational approaches center on the dynamical properties of the network, which has been explored in detail in insect olfactory systems but has also been considered for vertebrate olfaction (Laurent 2002). Interneurons, like periglomerular cells and granule cells, have a large role in these models because of their widespread effects on the network. Although numerous computational models of the OB exist, few models have explored the role of neurogenesis in this circuit. One model that explicitly models OB neurogenesis is that of Cecchi et al. (2001). The Cecchi model is a two-layer network using mitral and tufted cells as the output units and the granule cells providing lateral inhibition. Although anatomically based, it is limited in scope, looking exclusively at neurogenesis and survival, and does not account for the full complexity of the olfactory system. Nevertheless, the Cecchi model is interesting for several reasons. First, unlike most DG models, the Cecchi model uses the feature that newborn granule cell survival is experience-dependent. Second, it shows that the integration of these new cells is effective at increasing mitral cell discrimination of inputs.

The Cecchi model shows that continuously replacing interneurons and only keeping those that are useful is sufficient for increasing discrimination. However, it remains unclear how necessary neurogenesis is for these results because the olfactory system has learning and more complex interneuron connectivity. In a later study, Cecchi and Magnasco (2005) briefly address this issue of necessity, suggesting that neurogenesis, unlike learning, will permit the network to adjust its size to optimally process the inputs.

AVIAN NEUROGENESIS: THE SONG SYSTEM

Adult neurogenesis in songbirds occurs in many areas of the forebrain, including the hippocampus (for review, see Chapter 28; Doetsch and Scharff 2001; Brainard and Doupe 2002; Nottebohm 2002). The region most thoroughly studied in birds has been the song system. Songbirds usually learn their songs from their conspecifics, called tutors, in two (in some species overlapping) phases (for review, see Prather and Mooney 2004; Brenowitz and Beecher 2005; and the introductions of the theoretical papers cited below). First, they memorize the tutor song, and then they learn to reproduce it from memory. The learned song is initially variable and plastic and later crystallizes or becomes stereotyped.

The song system (Fig. 3) comprises a considerable number of nuclei, but only a few are considered in computational studies—most importantly, the HVC region (formerly known as "higher vocal center"). It receives highly processed auditory input and projects to the robust nucleus of the archipallium (RA) and the anterior forebrain pathway (AFP). The RA and subsequent nuclei form the motor-output pathway that controls the syrinx, where the actual sound is produced. Neurons in the HVC fire sparse bursts that seem to represent whole syllables, whereas the activity in the RA is much more distributed and seems to represent more detailed motor controls. The AFP is apparently involved in sensory-motor integration, matching the produced song with the stored template song. Lesions within the AFP primarily affect song learning and maintenance and do not greatly affect song production in adult birds.

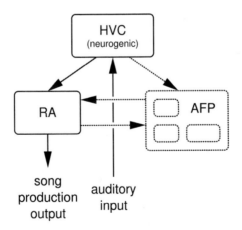

Figure 3. Simplified block diagram of the bird-song system. The higher vocal center (HVC) region of the bird brain receives input from auditory brain regions and projects to both the anterior forebrain pathway (AFP) (containing several nuclei), which is involved in song learning (indicated by dashed lines), and the robust nucleus of the archipallium (RA), which is part of the song vocalization pathway. The group of neurons in the HVC that project to the RA experiences neurogenesis.

In the HVC, neurogenesis appears to be a turnover process that only involves the neurons projecting to the RA. In adult canaries, neuronal turnover shows a seasonal pattern correlated with phases of song plasticity and recrystallization. However, seasonal neurogenesis also occurs in birds that do not change their song. Survival of new cells depends on the bird singing and auditory input.

There are several computational models of the bird song system. Margoliash and Bankes (1993) have trained a time-delayed neural network with backpropagation to reproduce neural activities in the auditory pathway. Doya and Sejnowski (1995; 1998) have shown that a simple model of the song production system (HVC→RA→ . . . →syrinx) can be trained with reinforcement learning to produce songs that are similar to stored template songs. Fry (1996) has extended this model to also include perceptual learning and classification of song syllables with a Kohonen network, which is assumed to be required for tutor-song memorization. In a similar framework, Troyer and Doupe (2000a) have investigated how learning of a memorized song can take place despite the large time delay of 100 msec between the initiation of song production and auditory feedback in the RA (Troyer and Doupe 2000a), and how sequences of syllables can be learned (Troyer and Doupe 2000b). Abarbanel and colleagues have presented much more detailed models comprising conductance-based Hodgkin–Huxley neurons. Their models focus on dynamical aspects of how HVC activity translates into auditory features in the bird's song (Abarbanel et al. 2004a) and how the HVC–RA synapses can be stabilized within the HVC–RA–AFP circuitry (Abarbanel et al. 2004b,c).

We are not aware of any computational model for adult neurogenesis in the bird-song system and there seems to be little speculation about the functional role of neurogenesis in this system in general. Nottebohm (2002) has put forward the idea that "*neural replacement provides, in the absence of net brain growth, a mechanism for adding new, unlearned neurons to existing circuits and with it, perhaps, a chance to rejuvenate the brain's learning potential.*" This is similar to ideas also developed for the hippocampus (Chambers et al. 2004; Deisseroth et al. 2004; Crick and Miranker 2006) and applies to the seasonal changes in canaries. However, as in the hippocampus case, the question arises of whether plasticity could not be regained more easily by other means and whether there may be a reason for neurogenesis that is more computational. For instance, as in the cascade-correlation learning architecture (Fahlman and Lebiere 1990), successively adding new neurons during the learning of a new song might be an efficient strategy to guide learning by slowly introducing more and more degrees of freedom to the system. This might result in faster learning than if all neurons, i.e. all degrees of freedom, were in place right from the beginning. Nevertheless, both hypotheses face the question of why neurogenesis occurs in songbirds even when they do not change their song, for example, zebra finches, which maintain the same song throughout life.

FISH AND REPTILIAN NEUROGENESIS

Adult neurogenesis has been observed in other vertebrates as well, including reptiles and fish (for review, see Garcia-Verdugo et al. 2002; Zupanc 2006). Neurogenesis in animals other than birds and mammals is discussed in greater detail in Chapter 27. Proliferation and neuronal differentiation in these animals are more pronounced than in mammals or birds, and neurogenesis has been shown to occur in many more regions than in the mammalian brain. Interestingly, however, the OB and areas thought to be equivalent to the mammalian DG are still among the most neurogenic regions.

Despite its ubiquity, the neurogenesis process in these species is not well understood, and its functions remain even less clear than in birds and mammals. Neurogenesis research in lower vertebrates is mainly focused on neuronal regeneration, which is significantly greater in the brains of these animals (Lopez-Garcia et al. 2002). Perhaps the case of nonregenerative neurogenesis best studied with regard to function is the fish cerebellum. The vast majority of new neurons in fish brains are born in the molecular layer of the cerebellum and integrate into the local circuits, with new neurons believed to often survive for the life of the fish.

According to Zupanc (2006), the large capacity for neurogenesis seen in the cerebellum is attributable to the fact that fish can continue to grow throughout life. The cerebellum in fish, just as in other vertebrates, is thought to be responsible for proprioception and mapping the sensory and motor space of the animal. Unlike mammals, the postnatal growth of fish is coupled with developing new muscle fibers and sensory neurons. This increase in motor and sensory information requires a concurrent increase in cerebellar processing capability. To our knowledge, the Zupanc hypothesis has not been investigated computationally, although it is likely that the cerebellum would greatly benefit from neurogenesis in the event of such an increase in input dimensionality.

CONCLUSION

In conclusion, there are several important things to note. For one, despite the successes of the above models in noting how neuronal addition or replacement can alter neural-network function, the precise role of new neurons in any of the brain circuits described above remains elusive. Nevertheless, general ideas about how these neurons are involved in cognition are beginning to emerge. First, the computational studies above help to dispel the notion that neurogenesis is an evolutionary artifact without any real effect on the processing of information. This is particularly true in the "uninteresting" case that neurogenesis is simply a replacement of older cells, as this would lead to substantial disruption of old memories in favor of newer ones. Second, models of both the DG and OB show that the inclusion of new neurons can greatly facilitate the encoding of novel information. Finally, it remains possible that the new neurons are providing additional information to the system, by virtue of their maturation, survival, or some other factor.

The current breakthrough in our realization of the neurogenesis process is an exciting development in our understanding of the computational potential of the brain. Although neurogenesis is not as ubiquitous as synaptic plasticity, the capability to change the neuronal population, in conjunction with the ability to alter the connections between neurons, provides the brain with the potential to modify itself—to learn—at timescales ranging from milliseconds to minutes to weeks. Undoubtedly, our understanding of the added possibilities is just beginning, and although the models above describe a wide range of functions for neurogenesis, the full story behind this process is not yet realized.

ACKNOWLEDGMENTS

We thank Constance Scharff for helpful comments on the section about avian neurogenesis, and Lindsey Aimone for editorial comments. J.B.A. is supported by the Neuroplasticity of Aging Training Grant (National Institutes of Health) and the Temporal Dynamics of Learning Center (National Science Foundation). L.W. has been supported by the Volkswagen Foundation.

REFERENCES

Abarbanel H.D.I., Gibb L., Mindlin G.B., and Talathi S. 2004a. Mapping neural architectures onto acoustic features of birdsong. *J. Neurophysiol.* **92:** 96–110.
Abarbanel H.D.I., Gibb L., Mindlin G.B., Rabinovich M.I., and Talathi S. 2004b. Spike

timing and synaptic plasticity in the premotor pathway of birdsong. *Biol. Cybern.* **91:** 159–167.

Abarbanel H.D.I., Talathi S.S., Mindlin G., Rabinovich M., and Gibb L. 2004c. Dynamical model of birdsong maintenance and control. *Phys. Rev. E* **70:** 1–16.

Aimone J.B., Wiles J., and Gage F.H. 2006. Potential role for adult neurogenesis in the encoding of time in new memories. *Nat. Neurosci.* **9:** 723–727.

Bayer S.A., Yackel J.W., and Puri P.S. 1982. Neurons in the rat dentate gyrus granular layer substantially increase during juvenile and adult life. *Science* **216:** 890–892.

Becker S. 2005. A computational principle for hippocampal learning and neurogenesis. *Hippocampus* **15:** 722–738.

Becker S. and Wojtowicz J.M. 2007. A model of hippocampal neurogenesis in memory and mood disorders. *Trends Cognit. Sci.* **11:** 70–76.

Bishop C.M. 1995. *Neural networks for pattern recognition.* Oxford University Press, Oxford, United Kingdom.

Boss B.D., Peterson G.M., and Cowan W.M. 1985. On the number of neurons in the dentate gyrus of the rat. *Brain Res.* **338:** 144–150.

Brainard M.S. and Doupe A.J. 2002. What songbirds teach us about learning. *Nature* **417:** 351–358.

Brenowitz E.A. and Beecher M.D. 2005. Song learning in birds: Diversity and plasticity, opportunities and challenges. *Trends Neurosci.* **28:** 127–132.

Butz M., Lehmann K., Dammasch I.E., and Teuchert-Noodt G. 2006. A theoretical network model to analyse neurogenesis and synaptogenesis in the dentate gyrus. *Neural Net.* **19:** 1490–1505.

Carpenter G.A. and Grossberg S. 1987. ART 2: Self-organization of stable category recognition codes for analog input patterns. *Appl. Optics* **26:** 4919–4930.

Carpenter G.A., Grossberg S., and Rosen D.B. 1991. ART 2-A: An adaptive resonance algorithm for rapid category learning and recognition. *Neural Netw.* **4:** 493–504.

Cecchi G.A. and Magnasco M.O. 2005. Computational Models of Neurogenesis. *Physica A* **356:** 43–47.

Cecchi G.A., Petreanu L.T., Alvarez-Buylla A., and Magnasco M.O. 2001. Unsupervised learning and adaptation in a model of adult neurogenesis. *J. Comput. Neurosci.* **11:** 175–182.

Chambers R.A. and Conroy S.K. 2007. Network modeling of adult neurogenesis: Shifting rates of neuronal turnover optimally gears network learning according to novelty gradient. *J. Cogn. Neurosci.* **19:** 1–12.

Chambers R.A., Potenza M.N., Hoffman R.E., and Miranker W. 2004. Simulated apoptosis/neurogenesis regulates learning and memory capabilities of adaptive neural networks. *Neuropsychopharm.* **29:** 747–758.

Crick C. and Miranker W. 2006. Apoptosis, neurogenesis, and information content in Hebbian networks. *Biol. Cybern.* **94:** 9–19.

Deisseroth K., Singla S., Toda H., Monje M., Palmer T.D., and Malenka R.C. 2004. Excitation-neurogenesis coupling in adult neural stem/progenitor cells. *Neuron* **42:** 535–552.

Doetsch F. and Scharff C. 2001. Challenges for brain repair: Insights from adult neurogenesis in birds and mammals. *Brain Behav. Evol.* **58:** 306–322.

Doya K. and Sejnowski T. 1995. A novel reinforcement model of birdsong vocalization learning. In *Advances in neural information processing systems 7* (ed. G. Tesauro et al.), pp. 101–108. MIT Press, Cambridge, Massachusetts.

———. 1998. A computational model of birdsong learning by auditory experience and auditory feedback. In *Central auditory processing and neural modeling* (ed. Poon and Brugge), Plenum Press, New York.

Fahlman S.E. and Lebiere C. 1990. The cascade-correlation learning architecture. In *Advances in neural information processing systems 2* (ed. D. Touretzky), pp. 524–532. Morgan-Kaufmann, San Francisco, California.

Fritzke B. 1994. Growing cell structures—A self-organizing network for unsupervised and supervised learning. *Neural Netw.* **7:** 1441–1460.

Fry C.L. 1996. How perception guides production in birdsong learning. In *Advances in neural information processing systems 8* (ed. D.S. Touretzky, et al.), pp. 110–116. MIT Press, Cambridge, Massachusetts.

Garcia-Verdugo J.M., Ferron S., Flames N., Collado L., Desfilis E., and Font E. 2002. The proliferative ventricular zone in adult vertebrates: A comparative study using reptiles, birds, and mammals. *Brain Res. Bull.* **57:** 765–775.

Gould E., Tanapaat P., Hastings N.B., and Shors T.J. 1999. Neurogenesis in adulthood: A possible role in learning. *Trends Cogn. Sci.* **3:** 186–192.

Kirn J., O'Loughlin B., Kasparian S., and Nottebohm F. 1994. Cell death and neuronal recruitment in the high vocal center of adult male canaries are temporally related to changes in song. *Proc. Natl. Acad. Sci.* **91:** 7836–7858.

Laurent G. 2002. Olfactory network dynamics and the coding of multidimensional signals. *Nat. Rev. Neurosci.* **3:** 884–895.

Lehmann K., Butz M., and Teuchert-Noodt G. 2005. Offer and demand: Proliferation and survival of neurons in the dentate gyrus. *Eur. J. Neurosci.* **21:** 3205–3216.

Lledo P.M, Gheusi G., and Vincent J.D. 2005. Information processing in the mammalian olfactory system. *Physiol. Rev.* **85:** 281–317.

Lopez-Garcia C., Molowny A., Nacher J., Ponsada X., Sancho-Bielsa F., and Alonso-Llosa G. 2002. The lizard cerebral cortex as a model to study neuronal regeneration. *An. Acad. Bras. Cienc.* **74:** 85–104.

Manev R. and Manev H. 2005. The meaning of mammalian adult neurogenesis and the function of newly added neurons: the "small-world" network. *Med. Hypotheses* **64:** 114–117.

Margoliash D. and Bankes S.C. 1993. Computations in the ascending auditory pathway in songbirds related to song learning. *Am. Zool.* **33:** 94–103.

Ming G.L. and Song H. 2005. Adult neurogenesis in the mammalian central nervous system. *Annu. Rev. Neurosci.* **28:** 223–250.

Nottebohm F. 2002. Neural replacement in adult brain. *Brain Res. Bull.* **57:** 737–749.

Paton J.A. and Nottebohm F.N. 1984. Neurons generated in the adult brain are recruited into functional circuits. *Science* **225:** 1046–1048.

Prather J.F. and Mooney R. 2004. Neural correlates of learned song in the avian forebrain: Simultaneous representation of self and others. *Curr. Opin. Neurobiol.* **14:** 496–502.

Rolls E.T. and Kesner R.P. 2006. A computational theory of hippocampal function, and empirical tests of the theory. *Prog. Neurobiol.* **79:** 1–48.

Schinder A.F. and Gage F.H. 2004. A hypothesis about the role of adult neurogenesis in hippocampal function. *Physiology* **19:** 253–261.

Troyer T.W. and Doupe A.J. 2000a. An associational model of birdsong sensorimotor learning I. Efference copy and the learning of song syllables *J. Neurophysiol.* **84:** 1204–1223.

———. 2000b. An associational model of birdsong sensorimotor learning II. Temporal hierarchies and the learning of song sequence. *J. Neurophysiol.* **84:** 1224–1239.

van Praag H., Schinder A.F., Christie B.R., Toni N., Palmer T.D., and Gage F.H. 2002. Functional neurogenesis in the adult hippocampus. *Nature* **415:** 1030–1034.

Wiskott L., Rasch M.J., and Kempermann G. 2006. A functional hypothesis for adult hippocampal neurogenesis: Avoidance of catastrophic interference in the dentate gyrus. *Hippocampus* **16:** 329–343.

Zupanc G.K.H. 2006. Neurogenesis and neuronal regeneration in the adult fish brain. *J. Comp. Physiol. A* **192:** 649–670.

23

Hippocampal Neurogenesis: Depression and Antidepressant Responses

Amar Sahay and Rene Hen
Department of Psychiatry
Center for Neurobiology and Behavior
Division of Integrative Neuroscience
Columbia University, New York, New York 10032

Ronald S. Duman
Division of Molecular Psychiatry
Departments of Psychiatry and Pharmacology
Yale University School of Medicine
New Haven, Connecticut 06510

BASIC RESEARCH AND CLINICAL STUDIES HAVE PROVIDED evidence that stress and depression can result in structural alterations in limbic brain regions implicated in mood disorders, including atrophy and loss of neurons and glia. These studies also demonstrate that antidepressant (AD) treatments block or reverse these effects. Several mechanisms contribute to the structural alterations and loss of cells in response to stress and depression, but one of intense interest is the involvement of neurogenesis in the adult hippocampal formation. Basic research studies consistently demonstrate that stress and AD treatment exert opposing actions on neurogenesis in the hippocampal dentate gyrus (DG).

The study of adult hippocampal neurogenesis has revealed it to be a robust phenomenon that is capable of conferring previously unrecognized forms of plasticity to the DG. The progression from neuronal stem cell to mature dentate granule neuron can be divided into discrete stages, each of which is defined by distinct physiological and morphological

Adult Neurogenesis ©2008 Cold Spring Harbor Laboratory Press 978-087969-784-6

properties (Esposito et al. 2005; Song et al. 2005) and is influenced by a plethora of factors comprising growth factors, neurotrophins, and chemokines (Lledo et al. 2006). These factors act in concert with network activity to regulate the balance between proliferation, differentiation, and survival of neuronal stem cells in vivo. It is through this general mechanism that levels of adult hippocampal neurogenesis change in response to aversive and enriching experiences, such as stress and learning, respectively, and the physiological state of the organism. Recent studies relying on experimental approaches that ablate adult hippocampal neurogenesis in rodents have shed some light on the contribution of baseline and increased neurogenesis to emotional reactivity and learning.

EVIDENCE OF NEURONAL ATROPHY IN DEPRESSION

Brain-imaging studies have provided evidence that the volume of certain limbic brain structures is reduced in depressed subjects. One of the most highly studied regions is the hippocampus. Although the hippocampal formation is largely recognized for its role in learning and memory, it has also been implicated in the actions of stress, which can precipitate or exacerbate depression. The hippocampus expresses high levels of adrenal-glucocorticoid receptors and contributes to negative feedback of the hypothalamic-pituitary-adrenal (HPA) axis, and exposure to stress causes atrophy of CA3 pyramidal cells and decreased proliferation of DG granule cells (McEwen 2001; Sapolsky 2001, 2003; Gold and Chrousos 2002; McEwen 2005). In addition to disruption of the HPA axis, learning, and memory, dysfunction of the hippocampus could underlie altered anxiety and cognition via connections with the amygdala and prefrontal cortex (Warner-Schmidt and Duman 2006). In addition, the hippocampus influences the activity of mesolimbic dopamine neurons and could thereby alter motivation and reward in a manner that leads to anhedonia, another hallmark symptom of depression (Lisman and Grace 2005).

On the basis of this large body of work, magnetic resonance imaging studies of the hippocampus of depressed subjects have been conducted, and there are more than a dozen papers demonstrating that the volume of the hippocampus is decreased in depressed subjects (Sheline et al. 1996, 1999; Bremner et al. 2000; MacQueen et al. 2003; for review, see Warner-Schmidt and Duman 2006). There are also reports that AD treatment can reverse or block hippocampal atrophy (Sheline et al. 2003; Vermetten et al. 2003). Postmortem studies of the hippocampus report a decrease in neuronal cell body size (Stockmeier et al. 2004) or no change (Muller et al. 2001), although the latter was based on a qualitative assessment. In addition to the hippocampus, reduced volume or atrophy of the prefrontal cortex has also been reported (Drevets et al. 1997). Moreover, postmortem studies of the prefrontal cortex consistently report reductions in the number of glia and size of neurons (Ongur et al. 1998; Rajkowska et al. 1999). Together, these studies provide convincing evidence for atrophy and loss of cells in two regions of the brain implicated in depression.

OPPOSING ACTIONS OF STRESS AND ANTIDEPRESSANTS ON ADULT NEUROGENESIS

There are a number of cellular mechanisms that could underlie the atrophy and cell loss caused by stress and depression, including atrophy of neuronal processes and neuropil, degeneration or loss of neurons and glia, and decreased proliferation of neurons and glia. It is possible that all of these processes contribute to decreased volume of the hippocampus as well as the prefrontal cortex in depression and that ADs could act by reversing or blocking these effects. We focus here on studies demonstrating that stress and ADs exert opposing actions on neurogenesis in the adult hippocampus.

Stress exerts many effects on neuronal function and behavior, but one of the most consistent cellular effects that has been reported is a decrease in the proliferation of newborn neurons in the DG of the hippocampal formation. These studies demonstrate that hippocampal neurogenesis is decreased by exposure to stress and that the type of stress can be emotional or physical, including social/intruder stress (Gould et al. 1997; Czeh et al. 2001), predatory odor (Galea et al. 2001), acute or chronic restraint (Pham et al. 2003; Vollmayr et al. 2003; Rosenbrock et al. 2005), foot shock (Malberg and Duman 2003; Vollmayr et al. 2003), and chronic mild stress (Alonso et al. 2004; Banasr et al. 2007). Administration of corticosterone also decreases neurogenesis in the adult hippocampal DG, indicating that this adrenal steroid could mediate the effects of stress (Cameron et al. 1998).

In contrast, one of the most consistent actions of ADs is to increase neurogenesis in the DG. Increased hippocampal neurogenesis is reported with different classes of ADs, including 5-HT (serotonin) and NE (norepinephrine) reuptake inhibitors, monoamine oxidase inhibitors, electroconvulsive seizures, and atypical ADs (Madsen et al. 2000; Malberg et al. 2000; Manev et al. 2001; Nakagawa et al. 2002; Santarelli et al. 2003; J. Warner-Schmidt and R.S. Duman, in prep; for review, see Duman 2004). The actions of chemical ADs are dependent on chronic administration, consistent with the time course for the therapeutic actions of these agents. The more robust and rapid induction of neurogenesis by ECS (electroconvulsive shock) is notable given the superior therapeutic efficacy of this treatment modality. ADs increase the proliferation as well as survival of newborn neurons, but they do not alter the differentiation of newborn cells into neurons and glia (Malberg et al. 2000; Nakagawa et al. 2002). Atypical antipsychotic agents that are used for add-on therapy for treating depressed subjects also increase neurogenesis (Kodama et al. 2004). Typical antipsychotic drugs do not influence neurogenesis, and drugs of abuse decrease neurogenesis, demonstrating the pharmacological specificity of AD induction of newborn neurons.

These findings provide strong evidence for a role of adult hippocampal neurogenesis in the actions of stress and ADs. However, studies to determine the functional consequences of altered neurogenesis are required to directly test this hypothesis. Sophisticated approaches to address this question are discussed in the following sections.

EXPERIMENTAL APPROACHES TO PROBE THE ROLE OF ADULT HIPPOCAMPAL NEUROGENESIS IN BEHAVIOR

Preclinical studies have proven to be tremendously informative in bridging the "causality gap" between adult hippocampal neurogenesis and behavior. To date, three different loss-of-function approaches have been used to ablate adult hippocampal neurogenesis. These approaches rely on either a pharmacological intervention (MAM), X-ray and γ-ray irradiation, or a pharmacogenetic approach—glial fibrillary acidic protein–thymidine kinase (GFAP-TK) and Nestin-TK—to target the dividing stem cells in the subgranular zone (SGZ). Briefly, MAM or methylazoxymethanol acetate is a methylating agent that blocks neurogenesis by interfering with cell division (Shors et al. 2001). Although some studies have succeeded in using MAM effectively, the interpretation of the findings is encumbered by the deleterious side effects associated with MAM.

A second method used to abolish adult hippocampal neurogenesis is by focal application of ionizing radiation to the animal's head or only at a narrow region that includes the hippocampus, as performed in our laboratory (Santarelli et al. 2003; Wojtowicz 2006). Exposure to low-dose irradiation is sufficient to arrest neurogenesis completely and irreversibly. Electrophysiological recordings reveal that mature neurons within the hippocampal formation (Snyder et al. 2001; Santarelli et al. 2003; Saxe et al. 2006) are unaffected by the procedure. Furthermore, many hippocampus-dependent behaviors are unchanged by irradiation, indicating a relatively high degree of specificity of the focal hippocampus-directed X-ray irradiation protocol (Saxe et al. 2006). One limitation inherent to this method is the increased inflammatory response elicited by irradiation, which, although transient, may have long-lasting effects on the stem cell niche (Monje et al. 2003).

Finally, a third and alternative method for arresting neurogenesis is a transgenic mouse model in which herpes simplex virus–thymidine kinase (HSV-TK) is expressed under the control of the GFAP promoter (Garcia et al. 2004). GFAP is expressed in neuronal stem cells located in the SGZ and subventricular zone (SVZ) (Doetsch et al. 1999; Seri et al. 2001; Filippov et al. 2003). The presence of HSV-TK in dividing cells confers sensitivity to the cytotoxic effects of the antiviral drug gancyclovir. Recently, a Nestin-TK mouse line has also been developed that allows for ablation of Nestin-positive nonquiescent stem cells using gancyclovir (Parent et al. 2006). The TK system does not completely abolish neurogenesis, as quiescent stem cells are unaffected. None of the above-described approaches spare gliogenesis in the adult SGZ.

Despite limitations inherent to the irradiation and TK-dependent systems, considerable advances have been made using these strategies to understand the role of baseline and increased neurogenesis in behavior. It remains essential, however, to replicate findings using more specific approaches. To this end, experiments are under way in our laboratory using inducible genetic systems in combination with the doublecortin promoter to selectively manipulate adult neurogenesis without affecting glial cell production (A. Sahay and R. Hen, unpubl.).

ASSESSING THE ROLE OF BASELINE NEUROGENESIS IN EMOTIONAL REACTIVITY AND LEARNING

Using X-ray irradiation and GFAP-TK, our laboratory has investigated the contribution of baseline-adult hippocampal neurogenesis to a number of different behavioral paradigms that measure emotional reactivity and hippocampus-dependent learning. To date, there is no evidence implicating adult hippocampal neurogenesis in the pathogenesis of anxiety or depression-like behavior. Ablation of hippocampal neurogenesis does not affect anxiety-related behavior as assessed in conflict-based tests such as the open field, light-dark choice, and elevated plus-maze (Saxe et al. 2006). Hippocampal irradiation also fails to affect behavior in two other anxiety tests that are also used to screen for ADs, novelty-suppressed feeding, and novelty-induced hypophagia (Santarelli et al. 2003; Meshi et al. 2006). Finally, blocking neurogenesis does not affect behavior in the forced-swim test and the chronic unpredictable stress paradigm, two tests used to assess AD-like behavior (Santarelli et al. 2003; Holick et al. 2007) Taken together, these studies indicate that abolishing hippocampal neurogenesis *in adult wild-type* animals does not result in altered anxiety or depression-like behavior.

Studies on hippocampus-dependent learning, in contrast to emotional reactivity, have revealed specific requirements for adult-generated dentate granule cells. Reducing or blocking SGZ neurogenesis in rats or mice has been reported to cause impairments in a hippocampus-dependent form of classical conditioning (trace conditioning) (Shors et al. 2001), long-term spatial memory (Snyder et al. 2005; Saxe et al. 2006), working memory (Winocur et al. 2006; Saxe et al. 2007), and contextual fear conditioning (Saxe et al. 2006; Winocur et al. 2006). Recently, there has also been one report of improved performance in a specific working memory paradigm after neurogenesis ablation (Saxe et al. 2007). Although there are some inconsistencies in these findings, which may reflect differential behavioral requirements for adult hippocampal neurogenesis in mice and rats or variations germane to methodology, it is clear that neurogenesis contributes to adult DG function in a behaviorally relevant way. Perhaps the most significant of these observations is the requirement of neurogenesis for contextual fear conditioning, which has been demonstrated in both rats and mice, using three different methods for arresting neurogenesis (Saxe et al. 2006; Winocur et al. 2006). Studies from our laboratory using both focally X-ray-irradiated mice and the GFAP-TK system have shown that blocking neurogenesis prior to training reduces the amount of contextual fear expressed when rodents are reexposed to the training context (Saxe et al. 2006). Importantly, blocking neurogenesis does not impair fear conditioning to a discrete tone stimulus, indicating that shock sensitivity and motor control of the fear response are not impaired. The impairment of contextual fear conditioning may thus reflect a requirement of neurogenesis in encoding contextual information, consistent with the well-recognized role of the DG in pattern separation.

Our studies in mice using focal X-ray irradiation indicate that hippocampal neurogenesis is not required for other hippocampus-dependent behaviors such as spatial learning, assessed in both the Morris water maze and the Y-maze (Saxe et al. 2006). Notably, blockade of neurogenesis does not result in spatial long-term memory deficits, as previously reported for the rat using whole-head irradiation (Snyder et al. 2005).

Investigations in our laboratory into a role for adult-generated dentate granule cells in hippocampus-dependent working memory tasks have revealed, unexpectedly, an enhancement in working memory in both

X-ray-irradiated and GFAP-TK mice (Saxe et al. 2007). Mice lacking hippocampal neurogenesis performed better than controls in a delayed nonmatching-to-place task on a radial eight-arm maze, but only when the intertrial interval exceeded 15 seconds (consistent with when the hippocampus becomes engaged in working memory tests) and when there was interference from earlier trials. Thus, it appears that with sufficient memory loading and presentation of repetitive information, neurogenesis may handicap the ability of the DG to ignore conflicting information from previous trials. Whether this is true for all working memory paradigms is not clear. For example, a recent study using a non-spatial working memory task showed that whole-brain-irradiated rats perform worse as a function of delay (Winocur et al. 2006).

Much work is still required to understand how and under what conditions neurogenesis is required for specific kinds of learning. Studies are needed to assess the differential contribution of adult-generated neurons in different stages of learning such as acquisition, consolidation, and retrieval. Finally, it is conceivable that adult-generated neurons have different functions depending on their state of maturation. Intriguingly, one report using rats has even suggested a functional role for 10-day-old neurons in learning (Shors et al. 2001). We predict that the novel genetic approaches mentioned earlier will confer the flexibility and specificity needed to test these different hypotheses.

BEHAVIORAL EFFECTS ASSOCIATED WITH INCREASED NEUROGENESIS

It is widely recognized that all of the major classes of ADs are associated with a several-week delay in the onset of therapeutic efficacy. One form of structural plasticity within the hippocampus that is consistent with the delayed onset of ADs is the birth and subsequent integration of newly generated neurons in the adult DG (Esposito et al. 2005; Overstreet-Wadiche and Westbrook 2006). Moreover, almost all ADs known to date including serotonin and norepinephrine selective reuptake inhibitors (SSRIs and NSRIs), tricyclics, monoamine oxidase inhibitors, phosphodiesterase inhibitors, and ECS therapy increase neurogenesis (Madsen et al. 2000; Malberg et al. 2000; Manev et al. 2001; Nakagawa et al. 2002). Therefore, the idea that ADs may work through enhancing neurogenesis has received abundant attention, and recent preclinical studies have tested this hypothesis (Santarelli et al. 2003; Jiang et al. 2005; Meltzer et al. 2006).

Our laboratory first demonstrated that neurogenesis is required for the effects of both imipramine, a classic tricyclic AD, and fluoxetine (SSRI) in two mouse behavioral screens for AD activity: the novelty-suppressed feeding (NSF) test and a chronic stress procedure (Santarelli et al. 2003). Importantly, this study has since been replicated in our laboratory using the GFAP-TK mice (unpublished results). More recently, and in a series of experiments conducted by another laboratory, the synthetic cannabinoid HU210 was shown to have AD-like behavioral effects in the NSF paradigm, which, interestingly, was lost following X-ray irradiation (Jiang et al. 2005). In addition, a recent preliminary report (Meltzer et al. 2006) showed that irradiation blocks the behavioral effects of fluoxetine in the forced-swim test. Thus, a neurogenic dependence for the behavioral effects of ADs has been revealed for three different drugs in three different AD screens and using two different ways to ablate neurogenesis. Together, these studies identify a previously unrecognized role for neurogenesis in the adult hippocampus in mediating the behavioral effects of SSRIs.

The above work suggests that enhancing neurogenesis confers AD-like behavioral responses in several tests of chronic AD action. Given the pleiotropic effects of ADs and the limitations inherent to the AD screens (Nestler et al. 2002; Duman and Monteggia 2006; Warner-Schmidt and Duman 2006), it is likely that some clinically important features of these treatments are neurogenesis-independent. Three studies from our laboratory confirm this notion and suggest that the net effects of increasing neurogenesis on behavior are likely to be influenced by factors such as levels of network activity and the genetic makeup of the organism (Scharfman and Hen 2007). One study examined the effects of AD drugs on behavior and DG neurogenesis in the BALB/c mouse strain, a strain that exhibits high anxiety in behavioral tests (Holick et al. 2007). In this strain, chronic fluoxetine treatment reduced anxiety-like and depressive behavior in the novelty-induced hypophagia and forced-swim paradigms but failed to increase neuronal proliferation. Not surprisingly, these behavioral effects were unaffected by hippocampal irradiation. Thus, in BALB/c mice, the behavioral effects of ADs are mediated through a neurogenesis-independent mechanism. A second study has found that environmental enrichment and exercise, which are known to increase neurogenesis, improved learning and had anxiolytic effects independently of neurogenesis (Meshi et al. 2006). Finally, a third study showed that the anxiolytic and AD-like effects of a novel melanin-concentrating hormone receptor antagonist were also independent of its neurogenic effects (David et al. 2007).

These studies indicate that AD-like effects can be achieved through at least two different pathways, one neurogenesis-dependent and one neurogenesis-independent (Fig. 1). Moreover, they illustrate that the require-

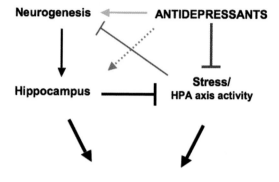

Behavioral Effects

Figure 1. Schematic showing two potential neural substrates of the antidepressant (AD) response. Current monoaminergic AD treatments may work by increasing hippocampal neurogenesis and/or dampening hypothalamic-pituitary-adrenal (HPA) reactivity. The neurogenic effects of ADs are likely to be mediated through neurotrophin-dependent mechanisms. AD drugs can also modulate hippocampal functions independent of neurogenesis.

ment for neurogenesis to mediate the behavioral effects of ADs is revealed only in certain biological contexts and not others. An outstanding and unanswered question in this regard is whether increasing neurogenesis is *sufficient* to confer the behavioral effects of ADs. Addressing the sufficiency hypothesis is central to our endeavor to understand the contribution of increasing neurogenesis to different behaviors and assessment of its potential as a target for novel nonmonoamine-based therapeutics.

NEUROGENESIS-DEPENDENT AND -INDEPENDENT MECHANISMS OF AD ACTION

Considerable efforts have been directed at dissecting the molecular mediators of the neurogenic effects of ADs with an emphasis on the serotonin and neurotrophin systems. The serotonin system is composed of a plethora of different 5-HT receptors that are expressed in a variety of cell types, making it likely that SSRIs act directly and indirectly, through regulation of network properties, on neurogenesis. A recent study has shed some light on this subject, showing that only a specific proliferative cell type, the transiently amplifying neuronal progenitor (ANP) (Filippov et al. 2003; Tozuka et al. 2005; Encinas et al. 2006), directly responds to fluoxetine. Moreover, the study confirmed that once exposure to fluoxetine ends, the rate of progenitor cell division is restored to baseline and that the increase in ANPs translates into a net increase in the number of new neurons. Thus, it appears that expansion of the neuronal progenitor pool may catalyze the AD-induced neurogenic response. The identity of 5-HT receptors required for the expansion of ANPs is unknown. Signaling through multiple postsynaptic 5-HT receptors can affect adult hippocampal neurogenesis (Radley and Jacobs 2002; Santarelli et al. 2003; Banasr et al. 2004). Indirect evidence for the 5-HT1A receptors as mediators of the neurogenic effects of ADs comes from a study done in our laboratory that showed a failure of 5-HT1A KO mice to respond to the neurogenic and behavioral effects of ADs (Santarelli et al. 2003). Whether this lack of response to ADs reflects an acute requirement for 5-HT1A is not known. Characterization of 5-HT receptor expression in the SGZ and within mature cell types of the hippocampal formation is under way in our laboratory. These studies combined with cell-type-restricted manipulation of specific 5-HT receptors will help identify mechanisms recruited by SSRIs for their neurogenic and behavioral effects.

The complex pathophysiology of MDD (major depressive disorder) underscores the possibility that new AD treatments may rely on modulation of neural circuits within different brain regions such as the HPA axis, prefrontal cortex, nucleus accumbens, and the hippocampus (Fig. 1). The prominent role of the HPA axis in MDD has fueled considerable investigation into strategies that dampen increased HPA reactivity. These efforts have resulted in identification of CRF (corticotropin-releasing factor) and vasopressin system antagonists with AD potential (Berton and Nestler 2006). Other examples of novel candidate AD drugs include melatonin receptor agonists (agomelatine), CB1 receptor agonists and antagonists, and galanin receptor-ligands (Berton and Nestler 2006). Future studies will reveal how disparate brain regions affected by these different signaling systems relate to the behavioral responses associated with successful AD treatments.

AD INDUCTION OF ADULT NEUROGENESIS: ROLE OF NEUROTROPHIC FACTORS

Previous studies have demonstrated that AD treatment increases the expression of a number of neurotrophic/growth factors in the hippocampus, including brain-derived neurotrophic factor (BDNF) (Nibuya et al. 1995, 1996), vascular endothelial growth factor (VEGF) (Newton et al. 2003; Warner-Schmidt and Duman 2007), and insulin-like growth factor-1 (IGF-1) (Khawaja et al. 2004). These studies raise the possibility that induction of trophic factor expression could contribute to up-regulation of neurogenesis (Fig. 2). Infusions of BDNF are reported to increase neurogenesis in nonhippocampal regions (Pencea et al. 2001), and a role for this neurotrophic factor in the actions of AD has been demonstrated recently (Sairanen et al. 2005). However, this study finds a specific role for BDNF in the survival of newborn neurons but not AD induction of cell proliferation. These findings are supported by studies demonstrating that heterozygous deletion of BDNF or overexpression of a dominant-negative mutant of the BDNF receptor, TrkB, decreases the survival of newborn neurons and blocks the effects of AD treatment on survival but does not influence AD-induced proliferation.

More recent studies have demonstrated an essential role for VEGF in the proliferative actions of AD treatment (Warner-Schmidt and Duman 2007). These studies demonstrate that blockade of VEGF-Flk-1 receptor signaling blocks the induction of neurogenesis by chemical ADs, including 5-HT and NSRIs, and by ECS. In contrast, infusion of VEGF (intracerebral ventricular) increases hippocampal neurogenesis, consistent with a previous qualitative report (Jin et al. 2002). The functional conse-

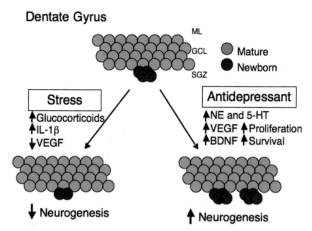

Figure 2. Schematic showing mechanisms underlying the actions of stress and antidepressant (AD) treatment on adult hippocampal neurogenesis. Stress exposure activates the hypothalamic-pituitary-adrenal (HPA) axis, increases levels of adrenal glucocorticoids and interleukin-1b (IL-1b), and decreases the expression of vascular endothelial growth factor (VEGF), effects that contribute to decreased neurogenesis. Chronic AD treatment increases levels of monoamines (5-HT and norepinephrine [NE]), leading to increased levels of VEGF and brain-derived neurotrophic factor (BDNF), which underlie AD induction of neuronal proliferation and survival, respectively. (ML) Molecular layer; (GCL) granule cell layer; (SVZ) subventricular zone.

quences of VEGF regulation of neurogenesis were also examined in behavioral models that link AD treatment and neurogenesis, novelty-suppressed feeding and chronic unpredictable stress. These studies demonstrate that blockade of VEGF-Flk-1 signaling also blocks the effects of chronic AD treatment in both of these models (Warner-Schmidt and Duman 2007). We also found that inhibition of VEGF signaling blocked the effects of ADs in two other behavioral models, the forced-swim test and learned helplessness paradigm that only require short-term AD treatment. The actions of ADs in these models are neurogenesis-independent and indicate that VEGF acts through other mechanisms, possibly at the level of synaptic function or activity to produce an AD response.

As with the 5-HT system, much work is still needed to define precisely how these two neurotrophic factors are recruited in vivo by ADs. Given the large repertoire of biological functions subserved by neurotrophins and growth factors, it is very likely that the enhancement in growth factor signaling following AD treatment results in changes in levels of neurogenesis, structural and electrophysiological properties of synapses, and even alterations in the vasculature. The use of cell-type-specific conditional knockouts will help define the contribution of the neurogenic and nonneurogenic effects of VEGF and BDNF to the AD-mediated behavioral response.

FACTORS UNDERLYING STRESS INHIBITION OF ADULT NEUROGENESIS

Although activation of the HPA axis and elevated adrenal glucocorticoid levels have been implicated in the down-regulation of neurogenesis resulting from exposure to stress, the specific mechanisms have not been identified (Fig. 2). There is a report that decreased neurogenesis resulting from stress or corticosterone can be blocked by administration of an NMDA (N-methyl-D-aspartate) receptor antagonist, indicating that glutamate transmission or overflow is increased by stress (Cameron and Gould 1996). It is also possible that decreased expression of neurotrophic/ growth factors, such as VEGF, could contribute to the actions of stress (Heine et al. 2005). Although this could explain, in part, a reduction in neurogenesis resulting from long-term stress exposure, it is not likely to account for the rapid effects of acute stress.

Recent studies have demonstrated a role for the cytokine interleukin-1b (IL-1b) in the actions of stress (Fig. 2). Infusions of IL-1b (intracerebral ventricular) decrease neurogenesis in the DG, but not SVZ, and infusions of an IL-1b antagonist (IL-1Ra) block the effects of either acute restraint or foot-shock stress (Koo and Duman 2006). Preliminary studies demonstrate that continued IL-1Ra infusions also block the effects of long-term exposure to unpredictable stress (J. Koo and R.S. Duman, unpubl.).

Immunohistochemical studies demonstrate that neuronal progenitors in vivo and in vitro express the primary IL-1b receptor, IL-R1, and incubation of progenitor cells in vitro with IL-1b decreases cell proliferation (Koo and Duman 2006). Studies are currently under way to determine whether blockade of IL-1b blocks the effects of stress and whether IL-1Ra infusions produce AD effects in behavioral models of depression.

CORTICAL CELL PROLIFERATION: INFLUENCE OF STRESS AND AD TREATMENT

Under physiological conditions, neurogenesis is not typically observed in cortical brain regions, although very low levels of GABAergic cell proliferation (<0.1% of proliferating cells) have been reported (Dayer et al. 2005). However, proliferation of nonneuronal cells, including oligodendrocytes, astrocytes, microglia, and endothelial cells, occurs throughout the cerebral cortex as well as other structures of the adult brain. On the basis of the consistent reports of decreased glia in depressed subjects (Ongur et al. 1998; Rajkowska et al. 1999; Cotter et al. 2001; Uranova et al. 2004), we hypothesized that decreased cell proliferation could account, in part, for the reduction in glia observed in depression. This hypothesis is confirmed by studies demonstrating that chronic unpredictable stress (15 days) decreases the proliferation of oligodendrocytes and endothelial cells, as well as a population of unidentified cells in the infralimbic cortex of rats (Banasr et al. 2007). The reduction in levels of oligodendrocytes is consistent with a postmortem study demonstrating a loss of this cell type in depressed subjects (Uranova et al. 2004). Decreased cell proliferation is also observed in cingulate and motor cortices, but not in striatum. In contrast to the effects of stress, chronic AD treatment increases cell proliferation in the infralimbic cortex and reverses the effects of stress (Kodama et al. 2004; Madsen et al. 2005; Banasr et al. 2007). This includes a chemical AD (fluoxetine) and ECS, as well as an atypical antipsychotic agent (olanzapine) shown to have AD efficacy.

Taken together, this work extends studies of neurogenesis in the DG to cell proliferation in cortical brain regions implicated in depression and AD action. The functional consequences of glial reductions in response to stress have not been determined, but they could contribute to the atrophy of neurons observed in the prefrontal cortex of depressed subjects (Rajkowska et al. 1999) and the reduction in apical dendrites observed in rodent stress models (Radley et al. 2004, 2006). However, it is also possible that loss of glia occurs in conjunction with or results from stress-induced neuronal atrophy. Conversely, increased cell proliferation in response to chronic AD could contribute to a reversal of neuronal atrophy and to the behavioral actions of AD treatments, particularly those that are dependent on the prefrontal cortex. Studies are currently under way to develop selective cell ablation approaches similar to those described for hippocampal neurogenesis to test the function of cortical cell proliferation.

CONCLUDING REMARKS

Preclinical studies have revealed candidate molecular pathways and neural substrates to target for new AD treatments. Adult hippocampal neurogenesis (and the DG) has gained considerable traction as a target for AD action, although considerable work is still needed to understand the biological contexts in which adult hippocampal neurogenesis is found necessary for the actions of ADs. As stated earlier, it remains to be determined whether increasing adult hippocampal neurogenesis is sufficient to confer AD-like behavioral responses. At a systems level, we must ask (and show) how introducing new functional units in the adult DG affects hippocampal function as well as downstream structures such as the amygdala and the prefrontal cortex. From an etiological standpoint, we know much less about pathogenic mechanisms of depression. Studies to date have not identified a link between reduced neurogenesis and depression, suggesting that different neural substrates may be involved in the pathogenesis of this illness. However, these studies have been conducted in adult wild-type animals, and it is likely that developmental and genetic interactions are critical determinants of depression. If neurogenesis is to be evaluated as a causal factor in depression, then we need to ask whether altering neurogenesis and DG function during early development, and when combined with a genetic predisposition or environmental insult, results in pathophysiology in adulthood.

ACKNOWLEDGMENTS

This work was supported by NARSAD (A.S, R.H., and R.S.D.) and NIMH (R.H. and R.S.D.).

REFERENCES

Alonso R., Griebel G., Pavone G., Stemmelin J., Le Fur G., and Soubrie P. 2004. Blockade of CRF1 or V1B receptors reverses stress-induced suppression of neurogenesis in a mouse model of depression. *Mol. Psychiatry* **9:** 278–286.

Banasr M., Hery M., Printemps R., and Daszuta A. 2004. Serotonin-induced increases in adult cell proliferation and neurogenesis are mediated through different and common 5-HT receptor subtypes in the dentate gyrus and the subventricular zone. *Neuropsychopharmacology* **29:** 450–460.

Banasr M., Valentine G.W., Li X.Y., Gourley S., Taylor J., and Duman R.S. 2007. Chronic stress decreases cell proliferation in adult cerebral cortex of rat: Reversal by antidepressant treatment. *Biol. Psychiatry* (in press).

Berton O. and Nestler E.J. 2006. New approaches to antidepressant drug discovery: Beyond monoamines. *Nat. Rev. Neurosci.* **7:** 137–151.

Bremner J.D., Vermetten E., and Mazure C.M. 2000. Development and preliminary psychometric properties of an instrument for the measurement of childhood trauma: The early trauma inventory. *Depression Anxiety* **12:** 1–12.

Cameron H. and Gould E. 1996. Distinct populations of cells in the adult dentate gyrus undergo mitosis or apoptosis in response to adrenalectomy. *J. Comp. Neurol.* **369:** 56–63.

Cameron H., Tanapat P., and Gould E. 1998. Adrenal steroids and N-Methyl-D-Aspartate receptor activation regulate neurogenesis in the dentate gyrus of adult rats through a common pathway. *Neuroscience* **82:** 349–354.

Cotter D., Mackay D., Landau S., Kerwin R., and Everall I. 2001. Reduced glial cell density and neuronal size in the anterior cingulate cortex in major depressive disorder. *Arch. Gen. Psychiat.* **58:** 545–553.

Czeh B., Michaelis T., Watanabe T., Frahm J., de Biurrun G., van Kampen M., Bartololomucci A., and Fuchs E. 2001. Stress-induced changes in cerebral metabolites, hippocampal volume, and cell proliferation are prevented by antidepressant treatment with tianeptine. *Proc. Natl. Acad. Sci.* **98:** 12796–12801.

David D.J., Klemenhagen K.C., Holick K.A., Saxe M.D., Mendez I., Santarelli L., Craig D., Zong H., Swanson C., Hegde L.G., et al. 2007. Efficacy of the MCHR1 antagonist N-[3-(1-{[4-(3,4-difluorophenoxy)phenyl]methyl}(4-piperidyl))-4-methylphen yl]-2-methylpropanamide (SNAP 94847) in mouse models of anxiety and depression following acute and chronic administration is independent of hippocampal neurogenesis. *J. Pharmacol. Exp. Ther.* **321:** 237–248.

Dayer A., Cleaver K.M., Abouantoun T., and Cameron H.A. 2005. New GABAergic interneurons in the adult neocortex and striatum are generated from different precursors. *J. Cell Biol.* **168:** 415–427.

Doetsch F., Caille I., Lim D.A., Garcia-Verdugo J.M., and Alvarez-Buylla A. 1999. Subventricular zone astrocytes are neural stem cells in the adult mammalian brain. *Cell* **97:** 703–716.

Drevets W.C., Price J.L., Simpson J.R., Todd R.D., Reich T., Vannier M., and Raichle M.E. 1997. Subgenual prefrontal cortex abnormalities in mood disorders. *Nature* **386:** 824–827.

Duman R. 2004. Depression: A case of neuronal life and death? *Biol. Psychiatry* **56:** 140–145.

Duman R.S. and Monteggia L.M. 2006. A neurotrophic model for stress-related mood disorders. *Biol. Psychiatry* **59:** 1116–1127.

Encinas J.M., Vaahtokari A., and Enikolopov G. 2006. Fluoxetine targets early progenitor cells in the adult brain. *Proc. Natl. Acad. Sci.* **103:** 8233–8238.

Esposito M.S., Piatti V.C., Laplagne D.A., Morgenstern N.A., Ferrari C.C., Pitossi F.J., and Schinder A.F. 2005. Neuronal differentiation in the adult hippocampus recapitulates embryonic development. *J. Neurosci.* **25:** 10074–10086.

Filippov V., Kronenberg G., Pivneva T., Reuter K., Steiner B., Wang L.P., Yamaguchi M., Kettenmann H., and Kempermann G. 2003. Subpopulation of nestin-expressing progenitor cells in the adult murine hippocampus shows electrophysiological and morphological characteristics of astrocytes. *Mol. Cell. Neurosci.* **23:** 373–382.

Galea L., Wide J.K., and Barr A.M. 2001. Estradiol alleviates depressive-like symptoms in a novel animal model of post-partum depression. *Behav. Brain Res.* **122:** 1–9.

Garcia A.D., Doan N.B., Imura T., Bush T.G., and Sofroniew M.V. 2004. GFAP-expressing progenitors are the principal source of constitutive neurogenesis in adult mouse forebrain. *Nat. Neurosci.* **7:** 1233–1241.

Gold P. and Chrousos G.P. 2002. Organization of the stress system and its dysregulation in melancholic and atypical depression: High vs low CRH/NE states. *Mol. Psychiat.* **7:** 254–275.

Gould E., McEwen B.S., Tanapat P., Galea L.A.M., and Fuchs E. 1997. Neurogenesis in the dentate gyrus of the adult tree shrew is regulated by psychosocial stress and NMDA receptor activation. *J. Neurosci.* **17:** 2492–2498.

Heine V., Zareno J., Maslam S., Joels M., and Lucassen P.J. 2005. Chronic stress in the adult dentate gyrus reduces cell proliferation near the vasculature and VEGF and Flk-1 protein expression. *Eur. J. Neurosci.* **21:** 1304–1314.

Holick K.A., Lee D.C., Hen R., and Dulawa S.C. 2007. Behavioral effects of chronic fluoxetine in BALB/cJ mice do not require adult hippocampal neurogenesis or the serotonin 1A receptor. *Neuropsychopharmacology* (in press).

Jiang W., Zhang Y., Xiao L., Van Cleemput J., Ji S.P., Bai G., and Zhang X. 2005. Cannabinoids promote embryonic and adult hippocampus neurogenesis and produce anxiolytic- and antidepressant-like effects. *J. Clin. Invest.* **115:** 3104–3116.

Jin K., Zhu Y., Yunjuan S., Mao X.O., Xie L., and Greenberg D.A. 2002. Vascular endothelial growth factor (VEGF) stimulates neurogenisis in vitro and in vivo. *Proc. Natl. Acad. Sci.* **99:** 11946–11950.

Khawaja X., Xu J., Liang J-J., and Barrett J.E. 2004. Proteomic analysis of protein changes developing in rat hippocampus after chronic antidepressant treatment: Implications for depressive disorders and future therapies. *J. Neurosci.* **75:** 451–460.

Kodama M., Fujioka T., and Duman R.S. 2004. Chronic olanzapine or fluoxetine treatment increases cell proliferation in rat hippocampus and frontal cortex. *J. Biol. Psychol.* **56:** 570–580.

Koo J. and Duman R.S. 2006. Interleukin-1b mediates the inhibitory effects of acute stress on hippocampal cell proliferation. *Soc. Neurosci. Abstr.* 289.9/0038.

Lisman J. and Grace A.A. 2005. The hippocampal-VTA loop: Controlling the entry of information into long-term memory. *Neuron* **46:** 703–713.

Lledo P.M., Alonso M., and Grubb M.S. 2006. Adult neurogenesis and functional plasticity in neuronal circuits. *Nat. Rev. Neurosci.* **7:** 179–193.

MacQueen G., Campbell S., McEwen B.S., Macdonald K., Amano S., Joffe R.T., Nahmias C., and Young L.T. 2003. Course of illness, hippocampal function, and hippocampal volume in major depression. *Proc. Natl. Acad. Sci.* **100:** 1387–1392.

Madsen T., Yeh D.D., and Duman R.S. 2005. Electroconvulsive seizure treatment increases cell proliferation in rat frontal cortex. *Neuropsychopharmacology* **30:** 27–34.

Madsen T., Treschow A., Bengzon J., Bolwig T.G., Lindvall O., and Tingström A. 2000. Increased neurogenesis in a model of electroconvulsive therapy. *Biol. Psychiatry* **47:** 1043–1049.

Malberg J. and Duman R.S. 2003. Cell proliferation in adult hippocampus is decreased by inescapable stress: Reversal by fluoxetine treatment. *Neuropsychopharmacology* **28:** 1562–1571.

Malberg J., Eisch A.J., Nestler E.J., and Duman R.S. 2000. Chronic antidepressant treatment increases neurogenesis in adult hippocampus. *J. Neurosci.* **20:** 9104–9110.

Manev H., Uz T., Smalheiser N.R., and Manev R. 2001. Antidepressants alter cell proliferation in the adult brain in vivo and in neural cultures in vitro. *Eur. J. Pharmacol.* **411:** 67–70.

McEwen B. 2001. Estrogens effects on the brain: Multiple sites and molecular mechanisms. *J. Appl. Physiol.* **91:** 2785–2801.

———. 2005. Glucocorticoids, depression, and mood disorders: Structural remodeling in the brain. *Metabolism* **54:** 20–23.

Meltzer L.A., Roy M., Parente V., and Deisseroth K. 2006. Hippocampal neurogenesis is required for lasting antidepressant effects of fluoxetine. *Soc. Neurosci. Abstr.* 418.2/A27.

Meshi D., Drew M.R., Saxe M., Ansorge M.S., David D., Santarelli L., Malapani C., Moore H., and Hen R. 2006. Hippocampal neurogenesis is not required for behavioral effects of environmental enrichment. *Nat. Neurosci.* **9:** 729–731.

Monje M.L., Toda H., and Palmer T.D. 2003. Inflammatory blockade restores adult hippocampal neurogenesis. *Science* **302:** 1760–1765.

Muller M., Lucassen P.J., Yassouridis A., Hoogendijk J.G., Holsboer F., and Swabb D.F. 2001. Neither major depression nor glucocorticoid treatment affects the cellular integrity of the human hippocampus. *Eur. J. Neurosci.* **14:** 1603–1612.

Nakagawa S., Kim J.E., Lee R., Malberg J.E., Chen J., Steffen C., Zhang Y.J., Nestler E.J., and Duman R.S. 2002. Regulation of neurogenesis in adult mouse hippocampus by cAMP and the cAMP response element-binding protein. *J. Neurosci.* **22:** 3673–3682.

Nestler E.J., Barrot M., DiLeone R.J., Eisch A.J., Gold S.J., and Monteggia L.M. 2002. Neurobiology of depression. *Neuron* **34:** 13–25.

Newton S., Collier E., Hunsberger J., Adams D., Salvanayagam E., and Duman R.S. 2003. Gene profile of electroconvulsive seizures: Induction of neurogenic and angiogenic factors. *J. Neurosci.* **23:** 10841–10851.

Nibuya M., Morinobu, S., and Duman, R.S. 1995. Regulation of BDNF and trkB mRNA in rat brain by chronic electroconvulsive seizure and antidepressant drug treatments. *J. Neurosci.* **15:** 7539–7547.

Nibuya M., Nestler E.J., and Duman R.S. 1996. Chronic antidepressant administration increases the expression of cAMP response element binding protein (CREB) in rat hippocampus. *J. Neurosci.* **16:** 2365–2372.

Ongur D., Drevets W.C., and Price J.L. 1998. Glial reduction in the subgenual prefrontal cortex in mood disorders. *Proc. Natl. Acad. Sci.* **95:** 13290–13295.

Overstreet-Wadiche L.S. and Westbrook G.L. 2006. Functional maturation of adult-generated granule cells. *Hippocampus* **16:** 208–215.

Parent J., Lichtenwalner R.J., Velander A.J., Fuller C.L., and Burant C.F. 2006. Conditional suicide gene ablation of neural progenitors in the adult mouse forebrain subventricular zone. *Soc. Neurosci. Abstr.* 419.2/B16.

Pencea V., Bingaman K.D., Wiegand S.J., and Luskin M.B. 2001. Infusion of brain-derived neurotrophic factor into the lateral ventricle of the adult rat leads to new neurons in the parenchyma of the striatum, septum, thalamus, and hypothalamus. *J. Neurosci.* **21:** 6706–6717.

Pham K., Nacher J., Hof P.R., and McEwen B.S. 2003. Repeated restraint stress suppresses neurogenesis and reduces biphasic PSA-NCAM expression in the adult rat dentate gyrus. *Eur. J. Neurosci.* **17:** 879–886.

Radley J.J. and Jacobs B.L. 2002. 5-HT1A receptor antagonist administration decreases cell proliferation in the dentate gyrus. *Brain Res.* **955:** 264–267.

Radley J.J., Rocher A.B., Miller M., Janssen W.G., Liton C., Hof P.R., McEwen B.S., and Morrison J.H. 2006. Repeated stress induces dendrtic spine loss in the rat medial prefrontal cortex. *Cereb. Cortex* **16:** 313–320.

Radley J.J., Sisti H.M., Hao J., Rocher A.B., McCall T., Hof P.R., McEwen B.S., and Morrison J.H. 2004. Chronic behavioral stress induces apical dendritic reorganization in pyramidal neurons of the medial prefrontal cortex. *Neuroscience* **125:** 1–6.

Rajkowska G., Miguel-Hidalgo J.J., Wei J., Dilley G., Pittman S.D., Meltzer H.Y., Overholser J.C., Roth B.L., and Stockmeier C.A. 1999. Morphometric evidence for neuronal and glial prefrontal cell pathology in major depression. *Biol. Psychiatry* **45:** 1085–1098.

Rosenbrock H., Koros E., Bloching A., Podhorna J., and Borsini F. 2005. Effect of chronic intermittent restraint stress on hippocampal expression of marker proteins for synaptic plasticity and progenitor cell proliferation in rats. *Brain Res.* **1040:** 55–63.

Sairanen M., Lucas G., Ernfors P., Castren M., and Casren E. 2005. Brain-derived neurotrophic factor and antidepressant drugs have different but coordinated effects on neuronal turnover, proliferation, and survival in the adult dentate gyrus. *J. Neurosci.* **25:** 1089–1094.

Santarelli L., Saxe M., Gross C., Surget A., Battaglia F., Dulawa S., Weisstaub N., Lee J., Duman R., Arancio O., et al. 2003. Requirement of hippocampal neurogenesis for the behavioral effects of antidepressants. *Science* **301:** 805–809.

Sapolsky R. 2001. Depression, antidepressants, and the shrinking hippocampus. *Proc. Natl. Acad. Sci.* **98:** 12320–12322.

———. 2003. Neuroprotective gene therapy against acute neurological insults. *Nat. Rev. Neurosci.* **4:** 61–69.

Saxe M., Malleret G., Vronskaya S., Drew M., Santarelli L., Kandel E.R., and Hen R. 2007. Paradoxical consequences of ablating hippocampal neurogenesis. *Proc. Natl. Acad. Sci.* **104:** 4642–4646.

Saxe M.D., Battaglia F., Wang J.W., Malleret G., David D.J., Monckton J.E., Garcia A.D., Sofroniew M.V., Kandel E.R., Santarelli L., Hen R., and Drew M.R. 2006. Ablation of hippocampal neurogenesis impairs contextual fear conditioning and synaptic plasticity in the dentate gyrus. *Proc. Natl. Acad. Sci.* **103:** 17501–17506.

Scharfman H.E. and Hen R. 2007. Neuroscience. Is more neurogenesis always better? *Science* **315:** 336–338.

Seri B., Garcia-Verdugo J.M., McEwen B.S., and Alvarez-Buylla A. 2001. Astrocytes give rise to new neurons in the adult mammalian hippocampus. *J. Neurosci.* **21:** 7153–7160.

Sheline Y., Gado M.H., and Kraemer H.C. 2003. Untreated depression and hippocampal volume loss. *Am. J. Psychol.* **160:** 1516–1518.

Sheline Y., Sanghavi M., Mintun M.A., and Gado M.H. 1999. Depression duration but not age predicts hippocampal volume loss in medically healthy women with recurrent major depression. *J. Neurosci.* **19:** 5034–5043.

Sheline Y., Wany P., Gado M.H., Csernansky J.G., and Vannier M.W. 1996. Hippocampal atrophy in recurrent major depression. *Proc. Natl. Acad. Sci.* **93:** 3908–3913.

Shors T.J., Miesegaes G., Beylin A., Zhao M., Rydel T., and Gould E. 2001. Neurogenesis in the adult is involved in the formation of trace memories. *Nature* **410:** 372–376.

Snyder J.S., Kee N., and Wojtowicz J.M. 2001. Effects of adult neurogenesis on synaptic plasticity in the rat dentate gyrus. *J. Neurophysiol.* **85:** 2423–2431.

Snyder J.S., Hong N.S., McDonald R.J., and Wojtowicz J.M. 2005. A role for adult neurogenesis in spatial long-term memory. *Neuroscience* **130:** 843–852.

Song H., Kempermann G., Overstreet Wadiche L., Zhao C., Schinder A.F., and Bischofberger J. 2005. New neurons in the adult mammalian brain: Synaptogenesis and functional integration. *J. Neurosci.* **25:** 10366–10368.

Stockmeier C., Mahajan G.J., Konick L.C., Overholser J.C., Jurjus G.J., Meltzer H.Y., Uylings H.B.M., Friedman L., and Rajkowska G. 2004. Cellular changes in the postmortem hippocampus in major depression. *Biol. Psychiatry* **56:** 640–650.

Tozuka Y., Fukuda S., Namba T., Seki T., and Hisatsune T. 2005. GABAergic excitation promotes neuronal differentiation in adult hippocampal progenitor cells. *Neuron* **47:** 803–815.

Uranova N., Vostrikov V.M., Orlovskaya D.D., and Rachmanova V.I. 2004. Oligodendroglial density in the prefrontal cortex in schizophrenia and mood disorders: A study for the Stanley Neuropathology Consortium. *Schizophr. Res.* **67:** 269–275.

Vermetten E., Vythilingam M., Southwick S.M., Charney D.S., and Bremner J.D. 2003. Long-term treatment with paroxetine increases verbal declarative memory and hippocampal volume in posttraumatic stress disorder. *Biol. Psychiatry* **54:** 693–702.

Vollmayr B., Simonis C., Weber S., Gass P., and Henn F. 2003. Reduced cell proliferation in the dentate gyrus is not correlated with the development of learned helplessness. *Biol. Psychiatry* **54:** 1035–1040.

Warner-Schmidt J. and Duman R.S. 2006. Hippocampal neurogenesis: Opposing effects of stress and antidepressant treatment. *Hippocampus* **16:** 239–249.

Winocur G., Wojtowicz J.M., Sekeres M., Snyder J.S., and Wang S. 2006. Inhibition of neurogenesis interferes with hippocampus-dependent memory function. *Hippocampus* **16:** 296–304.

Wojtowicz J.M. 2006. Irradiation as an experimental tool in studies of adult neurogenesis. *Hippocampus* **16:** 261–266.

24

Adult Neurogenesis in Neurodegenerative Diseases

Patrik Brundin,[1] Jürgen Winkler,[2,3] and Eliezer Masliah[3,4]
[1]Neuronal Survival Unit, Department of Experimental Medical Science
Wallenberg Neuroscience Center
Lund University, 221 84 Lund, Sweden
[2]Department of Neurology, University of Regensburg
D-93053 Regensburg, Germany
[3]Department of Neurosciences and [4]Pathology
University of California
San Diego, La Jolla, California 92093

THE NEURODEGENERATIVE DISORDERS PARKINSON'S DISEASE (PD), Huntington's disease (HD), Alzheimer's disease (AD), and human immunodeficiency virus (HIV)-associated cognitive impairment (HACI) all present with a gradual loss of relatively well-defined neuronal populations. Under all of these conditions, progression is slow. In some cases, the neuropathology is relatively restricted, leaving significant parts of the nervous system unaffected. They have therefore become interesting targets for restorative therapies. One of the most exciting ideas for repair is the concept that one might be able to harness the adult brain's endogenous capacity for cell renewal. Thus, it might be possible to direct newborn cells in the adult brain to migrate to the regions affected by the disease and there differentiate into the specific types of neurons that succumb due to the disease (Jordan et al. 2006). This concept is based on the realization that the adult mammalian brain also has the capacity to generate new neurons.

The interaction between neurogenesis in the adult brain and neurodegenerative disease can also be viewed from another angle. It is conceivable that failure of a normal reparative process, i.e., adult neurogenesis, contributes to the development of the disease. Taken to its extreme, this idea has even led to the hypothesis that the symptoms in some neu-rodegenerative diseases may partly be the consequence of reduced adult neurogenesis, resulting in a failed replacement of dying neurons (Armstrong and Barker 2001). The objectives of this chapter are to describe neurogenesis in the adult brain and to determine to what extent it is affected in animal models of PD, HD, AD, and HACI. In addition, we also describe a small number of cases where neurogenesis in human brains affected by these diseases has been studied. We also briefly discuss to what extent it seems possible that enhancing neurogenesis could be used as a therapeutic strategy in these disorders.

NEUROGENESIS IN THE ADULT BRAIN

In the normal adult mammalian brain, neurogenesis is known to occur in two locations: in the subgranular zone (SGZ) of the hippocampal dentate gyrus (DG) and the subventricular zone (SVZ) adjacent to the lateral ven-tricles (Ming and Song 2005; Kempermann 2006). Newborn cells in the hippocampus can differentiate into granule neurons and incorporate into the adult hippocampal circuitry (Laplagne et al. 2006). Importantly, there is evidence that neurogenesis also takes place in the hippocampus of humans (Eriksson et al. 1998). Hippocampal neurogenesis has been asso-ciated with learning and memory functions (Aimone et al. 2006). There is, however, no conclusive evidence regarding its impact on these processes, and the precise functional role of hippocampal neurogenesis in the adult brain remains enigmatic (Scharfman and Hen 2007).

In rodents, it is clear that newborn cells arising in the SVZ normally migrate along the rostral migratory stream to the olfactory bulb (OB) where they differentiate into GABAergic and dopaminergic neurons (Ming and Song 2005; Kempermann 2006). In the human brain, the exis-tence of an equivalent to the rostral migratory stream has been debated (Sanai et al. 2004). A recent study provided strong evidence that also in humans, the cells born in the SVZ migrate along a specific pathway that terminates in the OB (Curtis et al. 2007).

Factors Regulating Neurogenesis

Neurogenesis involves three crucial steps (Ming and Song 2005; Kempermann 2006). Asymmetric cell division of a stem cell results in one daughter stem cell and one cell that can develop into a neuron. The sec-ond step entails the migration of the newborn cell to its final and appro-priate destination in the brain. The third and final step involves maturation of the cells into a neuron that forms both efferent and afferent connections with the brain. Importantly, each of the three steps may be significantly altered by disease in the adult brain. Factors that normally can regulate neurogenesis are particularly interesting when discussing how disease states may influence neurogenesis. These factors are discussed in great detail elsewhere in this volume (see chapters 12 and 19). Briefly, studies in experimental animals have defined several environmental conditions (e.g., environmental enrichment, exercise, learning tasks, and stress) (Kempermann et al. 1998; Gould et al. 1999; Gould and Tanapat 1999; van Praag et al. 1999; Mohapel et al. 2006) and pharmacological treatments (e.g., drugs affecting the serotonergic system, steroids, and "cognitive enhancers") (Santarelli et al. 2003; Longo et al. 2006; Scharfman and Hen 2007) that either enhance or inhibit the rate of hippocampal neurogenesis. Certain acute experimental brain lesions and pathological conditions, such as stroke, brain trauma, and seizures, are also known to increase cell proliferation in the brain (Parent 2003; Kokaia et al. 2006; Lichtenwalner and Parent 2006; Miles and Kernie 2006). Some of these newborn cells differentiate into neurons, and in some, but not all, cases, there is evidence that they integrate into the circuitry and contribute to brain function. For example, hippocampal granule cells that are generated in response to seizures integrate into the circuitry and appear to contribute to information processing in the hippocampus (Jakubs et al. 2006). In such situations, the newborn granule neurons are similar to those cells that are normally generated in the adult hippocampus.

On the other hand, under certain circumstances, brain lesions may promote migration of neuronal precursors from the SVZ to regions where they are not normally found and therefore might be less suited to integrate within the existing circuitry (Romanko et al. 2004). Thus, it is still not clear whether the increased numbers of neurons that are generated from cells originating in the SVZ following stroke functionally integrate into the damaged brain and promote recovery of symptoms (Kokaia et al. 2006; Lichtenwalner and Parent 2006; Miles and Kernie 2006). Several studies have reported that neurogenesis can be boosted following brain lesions by administration of different growth factors (Fallon et al. 2000; Nakatomi et al. 2002; Teramoto et al. 2003; Mohapel et al. 2005b; Kobayashi et al. 2006). With this in mind, it is conceivable that it will be possible to develop treatments in the future that improve functional integration of newborn cells in brains damaged due to disease or injury. Therefore, it is very important to understand to what extent neurogenesis is affected in neurodegenerative diseases, to try to understand the molecular mechanisms that underlie the changes, and to evaluate whether neurogenesis in the adult brain may represent a target for future therapies.

ALTERATIONS IN NEUROGENESIS IN PARKINSON'S DISEASE

Parkinson's Disease and Related Disorders

The most common movement disorder in the world is PD. It belongs to an heterogeneous group of disorders with parkinsonism, dysautonomia, and dementia; denominated Lewy body disease (LBD), which includes dementia with Lewy bodies; PD dementia; and idiopathic PD (McKeith et al. 1996, 2004; Lippa et al. 2007). Both sporadic and familial forms of LBD are characterized by progressive degeneration of neurons in the nigrostriatal system, although earlier in the degenerative process, there is synapse loss and axonal injury (Galvin et al. 1999; Hashimoto et al. 2003; Maguire-Zeiss and Federoff 2003). In addition to the dopaminergic cells, other neuronal groups degenerate, including the cholinergic cells in the nucleus basalis, adrenergic neurons in the locus ceruleus, autonomic ganglia, amygdala, hippocampus, OB (Jellinger 1991; Braak and Braak 2000; Braak et al. 2003, 2004; Jellinger and Attems 2006), and temporal and cingulate cortex (Harding and Halliday 2001). The central neuropathological feature in LBD is the formation of Lewy bodies and neurites (Spillantini et al. 1997; Wakabayashi et al. 1997; Takeda et al. 1998; Trojanowski and Lee 1998; Shults 2006). These inclusions contain α-synuclein, ubiquitin, and synaptic and cytoskeletal components. Mutations and multiplications (Polymeropoulos et al. 1997; Kruger et al. 1998; Singleton et al. 2003; Johnson et al. 2004) in α-synuclein have been linked to rare familial forms of parkinsonism, and α-synuclein overexpression in flies and rodents results in neuropathological alterations reminiscent of LBD (Hashimoto et al. 2003).

Disease Mechanisms and Treatment Strategies

Similar to the amyloid β (Aβ) and huntingtin proteins, abnormal oligomerization and accumulation of α-synuclein in synapses and neurites are proposed as central mechanisms leading to neurodegeneration of dopaminergic and nondopaminergic neurons in PD and related disorders (Hashimoto et al. 2004). Therefore, strategies aimed at reducing α-synuclein oligomerization and aggregation are suggested as potential therapies. Current treatment strategies are mainly limited to symptomatic approaches aimed at increasing dopamine levels in the degenerating nigrostriatal system. Considerable attention has been placed on developing therapies based on cell replacement. Cell therapies such as transplantation of dopamine-producing fetal cells provided the proof of principle that transplanted neurons can survive, innervate the patient's brain, and elicit beneficial effects (Hagell and Brundin 2001). However, limited access to suitable donor tissue, variability in the outcome, and adverse side effects (graft-related dyskinesias) in some patients discourage this therapeutic option (Bjorklund et al. 2003). The pool of endogenous neuronal stem cells of the adult brain provides an alternative and attractive cell source for cell-based therapies. New developments encourage research to determine the possible impact of changes in neuronal stem and progenitor cells on the symptoms of PD. Therefore, several animal models of PD have been examined in detail concerning the potential and modulation of endogenous neurogenesis.

Changes in Adult Neurogenesis in Genetic Models of Parkinson's Disease

As mentioned above, new neurons are normally added to the OB and the hippocampus throughout a lifetime: Proliferating neuronal precursor cells (NPCs) residing in the basal forebrain SVZ constantly give rise to neuronal precursors; these cells migrate along the rostral migratory stream toward the OB where they are functionally integrated into olfactory circuitries (Carlen et al. 2002; Carleton et al. 2003). Accumulation of α-synuclein in the OB is of particular clinical relevance because olfactory dysfunction is one early symptom in PD (Berendse et al. 2001; Sobel et al. 2001) and is correlated with pathological changes in this region.

Recently, using a transgenic mouse line that expresses high levels of wild-type human α-synuclein (WTS), we reported that increased levels of WTS affected adult neurogenesis. Both the OB and the SGZ are areas where α-synuclein accumulates in WTS mice (Fig. 1). In WTS mice, expression of human α-synuclein led to the death of immature neurons in the brain of adult animals, indicating that overexpression of human α-synuclein has an impact on adult neurogenesis (Winner et al. 2004). A second study examined SVZ/OB neurogenesis in aging mice expressing WTS. Aging WTS animals also generated significantly fewer new neurons than their nontransgenic littermates. Moreover, increased numbers of terminal deoxynucleotidyl-transferase-mediated dUTP-biotin nick end-labeling (TUNEL)-positive profiles were detected in the OB in aging WTS mice (Winner et al. 2007).

α-synuclein mutations lead to an early onset and accelerated course of the disease in mice. Reduction of neurogenesis was more pronounced

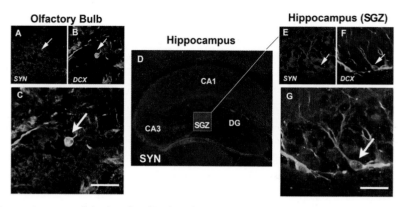

Figure 1. α-synuclein is colocalized with neuronal precursor markers in the brains of transgenic mice. The laser-scanning confocal image in the FITC channel is double-cortin (DCX) and that in the red channel is synuclein (SYN) from 6-month-old transgenic mice. (*A–C*) Colocalization of SYN and DCX in NPCs in the glomerular layer of the olfactory bulb (OB). (*D*) Low-magnification view of the hippocampus from a SYN transgenic mouse. Note the extensive accumulation of SYN in the neuropil and neuronal cell bodies in the CA3-CA1 regions of the hippocampus. (*E–G*) Colocalization of SYN and DCX in NPCs in the dentate gyrus (DG) subgranular zone (SGZ). (Modified, with permission from the Journal of Neuropathology and Experimental Neurology, from Winner et al. 2004.)

in A53T α-synuclein transgenic mice as compared to WTS mice. Here, additional mechanisms contributed to the decreased neurogenesis measured in the A53T mice: Both a decrease in cell proliferation and an increase in apoptotic profiles within the SVZ were observed (Winner et al. 2007). Taken together, these data indicate that the accumulation of toxic species of α-synuclein has a profound influence on the generation and survival of neuronal stem and progenitor cells. Moreover, α-synuclein aggregates contribute to neurodegeneration by interfering with neurogenesis in the hippocampus and OB; however, the mechanisms are not yet clear. A better understanding of the molecular pathways involved could lead to therapies that might block synuclein toxicity and stimulate endogenous neurogenesis.

Changes in Neurogenesis in Neurotoxin-induced Models of Parkinson's Disease

Several neurotoxin-induced animal models of PD have revealed decreased proliferation and/or survival of newly generated neurons in regions of neurogenesis. In this context, in 6-hydroxydopamine (6-OHDA) lesion models, which exhibit cell death in the dopaminergic system, a decrease in SVZ proliferation was noted by several groups (Baker et al. 2004; Hoglinger et al. 2004; Winner et al. 2006). Similar findings were also reported for a mouse treated with MPTP (1-methyl-4-phenyl-1,2,3,6-tetrahydropyridine) (Hoglinger et al. 2004). A recent study also supports the hypothesis that the proliferation rate is also affected in patients with PD, as is suggested by the decrease in the number of mitotic cells in the SVZ as well as in the hippocampus (Hoglinger et al. 2004). Interestingly, toxin-induced models, besides demonstrating a reduction in proliferation, show an increase in dopaminergic neurogenesis in the OB glomerular layer, which has been described for the MPTP (Yamada et al. 2004) as well as for the 6-OHDA model (Winner et al. 2006). This finding is of particular interest, as an increase in dopaminergic olfactory neurons has been described in the OB of PD patients (Huisman et al. 2004) and parallels these experimental findings.

Can Neurogenesis in the Substantia Nigra Provide a Therapy for Parkinson's Disease?

The idea that neurogenesis might take place in the adult substantia nigra has stimulated several research groups. If newborn cells can migrate into the adult substantia nigra and differentiate into neurons, the possibility that they may adopt a dopaminergic phenotype and extend axons to the striatum is very exciting. Promoting such a process may provide the basis for a novel therapy for PD. Since the SVZ and and SGZ are classically considered the neurogenic regions in the adult mammalian brain, then the fundamental question is: Can neurogenesis normally take place in the adult substantia nigra or is such a process stimulated by damage or disease? Yoshimi et al. (2005) reported that there were polysialic-acid-neural cell adhesion molecule (PSA-NCAM)-positive cells in the substantia nigra of PD patients and suggested that they may be newly generated cells undergoing maturation into neurons. There are two problems with this interpretation. First, PSA-NCAM can also be expressed both by neurons undergoing plastic changes and damage and by glia (Oumesmar et al. 1995; Emery et al. 2003; Bonfanti 2006). Therefore, expression of PSA-NCAM is not proof of newborn neurons. Second, the PSA-NCAM-positive cells were located in pars reticulata, not pars compacta (where dopaminergic neurons reside), in the patients' brains. In conclusion, there is still no unequivocal evidence that neurogenesis takes place in the substantia nigra pars compacta of humans. If it does at all, the basal rate is clearly not sufficient to compensate for the loss of dopaminergic neurons that occurs in PD.

The technical problems associated with studying adult neurogenesis in the human brain mean that it is more practical to examine whether neurogenesis can take place in the substantia nigra of experimental animals. Several studies have addressed specifically the question of whether new dopaminergic neurons are generated in rodents when the brain is intact, subjected to a lesion, or stimulated by drug or growth factor treatment. Borta and Hoglinger (2007) recently reviewed these studies in detail and with great care; therefore, we will only deal with them briefly here. Thus, one highly publicized study claimed that a significant number of new dopaminergic neurons are generated in the substantia nigra of intact mice and that the rate is increased in response to a toxin-induced lesion of the nigra (Zhao et al. 2003). We and other investigators have not been able to replicate this finding (Kay and Blum 2000; Mao et al. 2001; Lie et al. 2002; Frielingsdorf et al. 2004; Mohapel et al. 2005b; Reimers et al. 2006; Steiner et al. 2006). This may in part be due to differences in the criteria used to define cells as double-labeled by bromodeoxyuridine (BrdU) and tyrosine hydroxylase (in this context, a marker for dopaminergic neurons). Undoubtedly, newborn cells can populate the adult substantia nigra, but the majority of them appear to differentiate into different types of glial cells (Mao et al. 2001; Lie et al. 2002; Chen et al. 2005; Steiner et al. 2006). Similarly, groups studying the effects of different growth factors, e.g., glial-cell-derived neurotrophic factor (GDNF), transforming growth factor-β (TGF-β), platelet-derived growth factor (PDGF), liver growth factor, and brain-derived neurotrophic factor (BDNF), on neurogenesis in rats with nigral lesions have also not been able to detect newly generated dopaminergic neurons (Cooper and Isacson 2004; Chen et al. 2005; Mohapel et al. 2005b; Reimers et al. 2006). Three recent studies have once again suggested that neurogenesis may occur in the adult substantia nigra and, in one study, even restore lesion-induced behavioral deficits (Van Kampen and Robertson 2005; Shan et al. 2006; Van Kampen and Eckman 2006). In their comprehensive review, Borta and Hoglinger (2007) present compelling evidence that there are technical shortcomings (e.g., related to features of the BrdU staining) in each of these three studies, the details of which extend beyond the scope of this chapter. In conclusion, the current evidence that neurogenesis takes place in the substantia nigra is weak. Instead, studies that suggest the contrary are increasing in number.

NEUROGENESIS IN HUNTINGTON'S DISEASE

Recent research has shown that changes in adult neurogenesis occur both in animal models of HD and in brains obtained postmortem from HD patients. It is technically difficult to precisely determine changes in cell genesis in patients' brains, as it is not possible to label the newborn cells with mitotic markers. One is limited to the use of cell cycle markers, such as proliferating cell nuclear antigen (PCNA). In the SVZ of HD patient brains, the numbers of PCNA-immunopositive neurons are increased with increasing disease stage (Curtis et al. 2003, 2005b), and this is matched by similar findings using western blotting to assess the levels of PCNA (Curtis et al. 2003). In double-labeling immunofluorescence experiments, approximately half of the PCNA-positive cells had differentiated into cells expressing glial fibrillary acidic protein (GFAP) (Curtis et al. 2003) and about 3–5% expressed the neuronal marker βIII-tubulin (Curtis et al. 2003, 2005b). It is not clear why the cells continued to express PCNA, which is considered as indicating ongoing cell division or DNA repair, at a stage when they had matured into βIII-tubulin-positive neurons. Another study reported an increased number of cells with stem cell features (type-B cells) in the subependyma of HD patients, whereas the numbers of progenitors (type-A and -C cells) was unchanged (Curtis et al. 2005a). In summary, the current view is that cell genesis is increased in HD and that some of these cells develop into neurons. Interestingly, as described below, some experiments in animal models support these findings, whereas others suggest that mutant huntingtin may actually impair neurogenesis in the adult brain.

Changes in Neurogenesis in Excitotoxin-induced Models of Huntington's Disease

Changes in neurogenesis have been studied in rats with acute lesions of the neostriatum and in transgenic mouse models of HD. Cell death takes place acutely after injection of the excitotoxin, and therefore, the local environment is likely to differ from that found in the striatum of HD patients where degeneration is slow and gradual. Nonetheless, rats injected with the excitotoxin quinolinic acid into the striatum exhibit a dramatic loss of medium-sized spiny neurons and a marked gliosis at the lesion site and in related brain regions. Therefore, the acute excitotoxic lesion of the striatum mimics certain features of the pathology of advanced HD. Tattersfield et al. (2004) were the first to report that unilateral excitotoxic striatal lesions give rise to a marked increase in the number of BrdU-labeled newborn cells in the ipsilateral SVZ. Some of these cells expressed the neuroblast marker doublecortin (DCX) and migrated into the area of the striatal lesion. After about 3 weeks, approximately 10–20% of these cells had differentiated further and expressed the marker for mature neurons NeuN. A subsequent study showed that some cells in rats with quinolinic acid lesions develop into neurons that express striatal neuronal markers typical for the striatum, such as DARPP-32, parvalbumin, and NPY (Collin et al. 2005). Six weeks after the lesion was induced, however, 87% of the recently generated cells (i.e., BrdU-positive cells) did not express any striatal neuronal marker, and only 6% were DARPP-32-positive. It is not clear what the trigger is for the increased cell proliferation after striatal excitotoxic lesions. We have recently studied cell proliferation in neurospheres generated from cells harvested from the SVZ of animals subjected to unilateral quinolinic acid lesions. As expected, these neurospheres proliferate dramatically faster than spheres generated from cells from intact brains (T. Deierborg et al. in prep.). Interestingly, it appears that microglia-like cells in the neurospheres secrete an unidentified factor that promotes cell proliferation. Thus, microgliosis may be a crucial component of the neuropathology in excitotoxic lesions that stimulates cell proliferation in the adult brain.

Neurogenesis in Transgenic Mouse Models of HD

There are several types of transgenic mouse models of HD that vary regarding, for example, the length of the CAG repeat in the huntingtin transgene, the size of the fragment of huntingtin, the promoter driving the transgene, and consequently the levels of huntingtin expression (Menalled and Chesselet 2002). Neurogenesis has been examined in both the DG and SVZ primarily in two transgenic models of HD. They are the R6/1 and R6/2 mouse lines that both express *exon 1* (representing ~3% of the full protein) of human huntingtin with about 115 and 150 CAG repeats, respectively (Mangiarini et al. 1996). These mice exhibit progressive brain atrophy, but little neuronal death has been detected.

Hippocampal Neurogenesis is Reduced in R6/1 and R6/2 HD Mice

Initially, cell proliferation, as assessed by counts of BrdU-labeled cells, was found to be reduced in the SGZ of R6/1 (Lazic et al. 2004) and R6/2 (Gil et al. 2004) mice at an advanced disease stage. Subsequently, we confirmed the reduced hippocampal cell proliferation with additional cell-cycle-specific markers (Ki-67 and PCNA). We observed that it develops relatively early in R6/2 mice and is indeed coupled to a reduction in neurogenesis, i.e., the number of newborn cells that eventually expressed the neuroblast marker DCX (Fig. 2) and the neuronal marker NeuN were also decreased (Gil et al. 2005; van der Borght and Brundin 2007). Recent

Doublecortin   PSA-NCAM

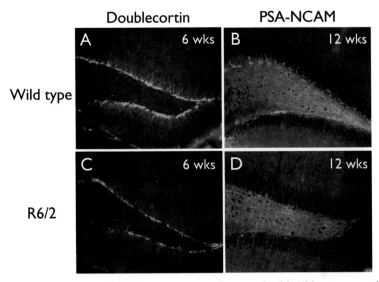

Figure 2. Sections through the hippocampus of a 6-week-old wild-type control (*A*) and an R6/2 (*C*) mouse stained with an antibody against doublecortin (DCX), labeling neuroblasts. There is a clear reduction of DCX-immunoreactive cells in the dentate gyrus (DG) of the R6/2 mouse. Similarly, sections through the brains of 12-week-old mice stained with an antibody against PSA-NCAM demonstrate a reduction in the numbers of presumed newborn cells in R6/2 (*B*) versus wild-type control (*D*) animals. (Data obtained from van der Borght and Brundin 2007.)

studies provide support for these findings and have shown that the age-related decline in PSA-NCAM-positive cells (migrating neuroblasts) is significantly greater in both the hippocampus of R6/1 and R6/2 mice (Fig. 2) (van der Borght and Brundin 2007). Taken together, there is strong evidence that neurogenesis in impaired in the adult hippocampus of R6 mice.

What May be the Functional Consequences of Reduced Hippocampal Neurogenesis in R6/1 and R6/2 HD Mice?

Both R6/1 and R6/2 mice display impairments in spatial learning (Murphy et al. 2000; Smith et al. 2006). R6 mice exhibit multiple defects in neuronal function, e.g., transcriptional dysregulation, receptor changes, defects in synaptic proteins, and alterations in neurotransmitter release, which all might contribute to the cognitive deficits (for discussion, see Smith et al. 2005). It cannot be excluded, however, that the reduction in hippocampal neurogenesis is an additional contributing factor to impaired spatial learning. Indeed, Grote et al. (2005) claim that an improvement in cognitive function in R6/1 mice following treatment with the serotonin reuptake inhibitor fluoxetine may be due to an enhanced hippocampal cell genesis. Interestingly, exposure to an enriched environment can slow progression of symptoms in R6/1 (van Dellen et al. 2000) and R6/2 (Hockly et al. 2002) mice. The enriched environment has multiple effects on brain plasticity in R6 mice, including increasing the level of BDNF (Spires et al. 2004). Environmental enrichment is well known to promote cell genesis in the brain (Kempermann et al. 1997). A recent study showed that environmental enrichment stimulates cell proliferation also in the hippocampus of R6/1 mice (Lazic et al. 2006), suggesting that it could underlie some of the reported symptomatic improvement that the exposure has in R6 mice. It is particularly interesting that the aforementioned treatments actually increase cell genesis in the DG of transgenic mice. It suggests that an inherent capacity to increase their proliferation rate is retained, despite the presence of mutant huntingtin. Not all stimuli, however, known to increase neurogenesis in the DG have the same effect in R6/2 mice. Thus, Phillips et al. (2005) found that systemic administration of kainic acid, which causes seizures and increases hippocampal cell genesis in normal mice, does not enhance neurogenesis in R6/2 mice.

What Mechanisms May Underlie the Reduction in Hippocampal Neurogenesis in R6/1 and R6/2 Mice?

There is currently no single explanation why hippocampal neurogenesis is reduced in the studied transgenic mouse models of HD. One possibility is coupled to endocrine changes in the mice. Due to decreases in dopamine D2 receptor levels in the pituitary and subsequent increases in adrenocorticotrophic hormone release, the R6/2 mice progressively develop increased levels of corticosterone (Bjorkqvist et al. 2006). Because this hormone is known to inhibit hippocampal neurogenesis in other settings (Gould and Tanapat 1999), the Cushing-like syndrome may also have an effect in that direction in R6/2 mice. This hypothesis needs to be tested directly, for example, by examining hippocampal neurogenesis in adrenalectomized R6/2 mice. An additional possibility is that the local trophic environment in the brain is impaired, due to reduced production of growth factors. It is argued that the production of BDNF is reduced in HD and in mouse models of the disease (Altman 1962; Zuccato et al. 2001; Spires et al. 2004; Zuccato and Cattaneo 2007). Because BDNF has been reported to promote neurogenesis in certain brain regions

(Pencea et al. 2001; Mohapel et al. 2005b), it is conceivable that a reduction in the levels of the growth factor impairs hippocampal neurogenesis in R6/1 and R6/2 mice. Neurotransmitters are also known to affect neurogenesis. There is a reduction in serotonergic innervation in R6/2 mice (Reynolds et al. 1999), and, as mentioned earlier, treatment with a serotonin reuptake inhibitor has been suggested to ameliorate a deficit in neurogenesis in R6/1 mice (Grote et al. 2005).

The impairment might also be intrinsic and directly due to alterations in the proliferating cell populations, as opposed to changes in the local trophic environment or endocrine homeostasis. One option is that the mutant huntingtin directly interferes with mitosis in the stem cells and neuronal progenitors. Huntingtin is thought to interact with microtubules, possibly via β-tubulin, and might have a role in the centrosome (Gutekunst et al. 1995; Hoffner et al. 2002). Normal huntingtin contains HEAT-like repeat domains (Andrade and Bork 1995) that also are potential docking sites for proteins that influence chromosomal function (Neuwald and Hirano 2000). Possibly, the mutation interferes with this process. The reduced capacity for cell proliferation may indeed not be limited to the hippocampus in R6/2 mice. We have observed that the generation of pancreatic β cells (assessed by BrdU labeling) is also impaired in adult R6/2 mice (Bjorkqvist et al. 2005). Interestingly, the DG and pancreatic islets are both dependent on the transcription factor NeuroD for their normal development (Liu et al. 2000; Huang et al. 2002). Thus, mice that are null mutant for NeuroD exhibit abnormal cell genesis in both the hippocampal DG and pancreatic islets. Therefore, it is particularly intriguing that huntingtin and some of its interacting proteins normally modulate NeuroD and that mutations in huntingtin can affect this process by causing huntingtin to aggregate (Marcora et al. 2003). This could explain why the hippocampal DG (and pancreas) might be particularly sensitive to mutations in the *huntingtin* gene.

Changes in Cell Proliferation in the SVZ of R6/1 and R6/2 Mice Are Not Consistent

In contrast to the findings in the hippocampus, four separate studies using BrdU injections to label dividing cells report that there are no changes in cell proliferation in the SVZ of R6/1 and R6/2 mice (Gil et al. 2004, 2005; Phillips et al. 2005; Lazic et al. 2006). One recent study suggests specifically that the number of stem cells (disregarding the progenitors) is *increased* in the SVZ of R6/2 mice (Batista et al. 2006). These authors base this conclusion on the three observations. First, 30 days after

BrdU injection, they saw increased numbers of labeled cells in the subependyma of R6/2 mice. These colabled with GFAP and nestin, suggesting that they were stem cells. There was, however, neither a difference in the total number of progenitors in the SVZ nor a statistically significant increase in BrdU/NeuN-colabled cells in the striatum. Second, using electron microscopy, they suggested that the ratio of B cells to A and C cells is increased in R6/2 mice. There was, however, no direct quantification of B cells in this experiment and it is not clear how comparable the sampled regions were between animals. Finally, Batista et al. (2006) reported an increase in clonal neurospheres when SVZ was harvested from symptomatic R6/2 mice compared to wild-type mice. It is not clear if atrophy of the striatum, which would influence the total number of cells harvested during dissection, may have confounded the comparison between wild-type and R6/2 mice. Taken together, the study only suggests that the number of stem cells is increased in the SVZ of R6/2 mice (Batista et al. 2006). Thus, in agreement with an earlier study on R6/2 neurospheres (Phillips et al. 2005), this study provides no evidence for a change in the number of SVZ neuronal precursor cells in R6/2 mice.

Another study examined the effects of subcutaneous injections of fibroblast growth factor-2 (FGF-2) on R6/2 mice and monitored several parameters, including the numbers of BrdU-labeled cells in the SVZ (Jin et al. 2005). In control groups treated with vehicle (which are of interest in this context), the authors reported only a very slight increase in the number of BrdU-positive cells in SVZ of R6/2 mice. As the phenotypic fate of these cells was not examined, it is not possible to state whether neurogenesis per se was affected in R6/2 mice (Jin et al. 2005). Taken together, none of the studies in R6/1 and R6/2 mice provide support for major changes in neurogenesis originating from the SVZ region.

What Is the Relevance of Observations in Transgenic Huntington Mouse Models to Clinical Findings in HD Patients?

It is important to emphasize that no published report exists on the rate of neurogenesis in the *hippocampus* of HD patients, and thus, it remains to be determined whether the mouse studies correlate with clinical findings in this brain area. Regarding the SVZ, the majority of transgenic mouse studies on cell proliferation appear not to agree with the reported changes in neurogenesis in HD patients; i.e., increased neurogenesis has been reported in the SVZ of patients, whereas most mouse studies reveal no change. It is important to stress that the brains of the mouse models and HD patients differ in many respects. The HD patients studied so far regarding neurogenesis have all had much fewer CAG repeats than the R6 mice. The mice do not live as long as the patients and, importantly, display little, if any, cell loss in the striatum. Indeed, the studies on excitotoxic damage to the striatum suggest that neuronal death, and the consequences it has for the surrounding environment, may be a powerful stimulus for cell division. Importantly, the animal studies have typically employed multiple markers for dividing cells, including uptake of injected BrdU. For obvious reasons, the clinical studies have limitations regarding the techniques that can be employed and have primarily relied on immunohistochemical detection of PCNA. The drawback with this marker is that it can also be expressed in cells undergoing DNA repair (Tomasevic et al. 1998), giving rise to false-positive cells. Although this problem also applies to BrdU, it appears to be less critical at the concentrations of BrdU used in most studies. Taken together, there are limitations both to the animal models and to the repertoire of techniques available when studying the human condition that makes comparisons difficult. Future refinements in both arenas may help resolve what, if any, are the true changes in neurogenesis in HD patients.

ALTERATIONS IN NEUROGENESIS IN ALZHEIMER'S DISEASE

Considerable progress has been made in recent years toward better understanding the pathogenesis of AD. It is a common dementing disorder of the aging population that affects more than 10 million individuals in the United States and Europe together. It is characterized by widespread neurodegeneration throughout the association cortex and limbic system, deposition of Aβ in the neuropil and around the blood vessels, and formation of neurofibrillary tangles (Terry et al. 1994). In the initial stages of AD, the neurodegenerative process targets the synaptic terminals (Masliah 1995) and then propagates to axons and dendrites, leading to neuronal dysfunction and eventually to neuronal death (Hyman and Gomez-Isla 1994). Although the mechanisms leading to neurodegeneration in AD are not completely understood, recent studies suggest that alterations in the processing of amyloid precursor protein (APP), resulting in the accumulation of Aβ and APP carboxy-terminal products, have a key role (Sinha et al. 2000; Kamenetz et al. 2003). Several products are derived from APP through alternative proteolytic cleavage pathways, and enormous progress has recently been made in identifying the enzymes involved (Selkoe 1999; Sinha et al. 1999; Vassar et al. 1999; Cai et al. 2001; Luo et al. 2001).

The most significant correlate with the severity of cognitive impairment in AD is the loss of synapses in the frontal cortex and limbic system

(DeKosky and Scheff 1990; Terry et al. 1991; Masliah and Terry 1994; DeKosky et al. 1996). The pathogenic process involves changes in synaptic plasticity that include alterations in formation of synaptic contacts, changes in spine morphology, and abnormal area of synaptic contact (Scheff and Price 2003). However, other cellular mechanisms necessary to maintain synaptic plasticity might also be affected in AD (Cotman et al. 1993; Masliah 2000; Masliah et al. 2001). As indicated earlier, neurogenesis in the mature brain probably has an important role in maintaining synaptic plasticity and memory formation in the hippocampus (van Praag et al. 2002). In the adult nervous system, motor activity and environmental enrichment have been shown to stimulate neurogenesis in the hippocampal DG (Gage et al. 1998; van Praag et al. 2002).

Altered Neurogenesis May Contribute to Cognitive Dysfunction in Alzheimer's Disease

Since neurogenesis in the adult nervous system probably has a role in memory and learning, it is possible that the alterations in synaptic plasticity in AD might involve not only direct damage to the synapses, but also impairment of adult neurogenesis (Fig. 3).

One study reported that hippocampal neurogenesis is increased in patients with AD (Jin et al. 2004b) and suggested that it may be a compensatory mechanism to the neurodegenerative process. However, more

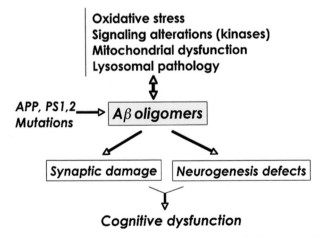

Figure 3. Potential role of oligomers in mechanisms underlying cognitive impairment and neurodegeneration in AD. The toxic effects of the oligomers might target both synaptic terminals and NPCs in the hippocampus.

extensive studies in experimental models of AD-like pathology (see below) and recent studies in human brains (Boekhoorn et al. 2006) suggest that hippocampal neurogenesis is reduced in AD. Instead, the proliferation of nonneuronal cells, i.e., glial and vascular components, is increased. Markers of neurogenesis and cell cycle such as cyclin kinases, DCX, and Musashi-1 (Lovell and Markesbery 2005) are affected in the hippocampus of AD patients. More detailed studies, including analysis of the brains of patients with early AD, are necessary to better understand how neurogenesis changes over time. This may also shed light on the relationship between the changes in neurogenesis and other neuropathological as well as cognitive indicators in AD.

The mechanisms of synaptic pathology and defective neurogenesis in AD are the subject of intense investigation (Fig. 3). Studies in experimental models of AD and in the human brain support the notion that aggregation of Aβ, resulting in the formation of toxic oligomers rather than fibrils, might be ultimately responsible for the synaptic damage that leads to cognitive dysfunction in patients with AD (Walsh and Selkoe 2004; Glabe 2005; Glabe and Kayed 2006). Studies in transgenic mice suggest that Aβ might also interfere with adult neurogenesis by triggering apoptosis and reducing the migration of NPCs.

Changes in Neurogenesis in Animal Models of Alzheimer's Disease

Recent studies have shown that the neurodegenerative process in AD might interfere with neurogenesis in the adult hippocampus (Tatebayashi et al. 2003; Jin et al. 2004b; Lovell and Markesbery 2005). Cholinergic denervation of the hippocampus is one striking feature of AD neuropathology. In experimental animals, lesions of the septal cholinergic neurons that normally innervate the hippocampus result in a small but significant decrease in survival of newborn neurons in the DG (Cooper-Kuhn et al. 2004; Mohapel et al. 2005b; Fontana et al. 2006). Interestingly, treatment with the reversible cholinesterase inhibitor physostigmine can increase hippocampal neurogenesis in normal rats (Mohapel et al. 2005a). Changes in neurogenesis seem to be coupled specifically to the septal input to the hippocampus, since lesions of the glutamatergic input from the entorhinal cortex does not increase hippocampal neurogenesis (Fontana et al. 2006).

In transgenic animal models of AD, there are also significant alterations in adult hippocampal neurogenesis (Dong et al. 2004; Jin et al. 2004a; Wen et al. 2004; Chevallier et al. 2005; Donovan et al. 2006;

Rockenstein et al. 2006). Several transgenic APP models of AD-like pathology have been developed over the past years, most of them expressing mutant forms of APP under the regulatory control of the Thy-1, PrP, or PDGF-β promoters. For example, our Thy1-APP transgenic mouse model develops, beginning at 6 months, progressive synaptic loss in the frontal cortex and hippocampus, memory deficits in the water maze, high levels of Aβ$_{1-42}$ production, and amyloid deposition. In addition, these transgenic mice displayed decreased neurogenesis in the SGZ of the DG (Rockenstein et al. 2006), suggesting that Aβ accumulation contributes to the degenerative process in AD-like models not only by damaging mature neurons, but also by interfering with neurogenesis in the mature CNS (Fig. 4). The alterations in neurogenesis in the DG found in the mThy1-APP transgenic mice (Rockenstein et al. 2006) are consistent with studies in other lines of APP transgenic mice and other models of AD. They exhibit decreased numbers of cells labeled with markers of neurogenesis, such as BrdU and DCX, and display a concomitant increase in the expression of markers of apoptosis (Feng et al. 2001; Haughey et al. 2002; Dong et al. 2004; Wang et al. 2004). Interestingly, a combined AD transgenic

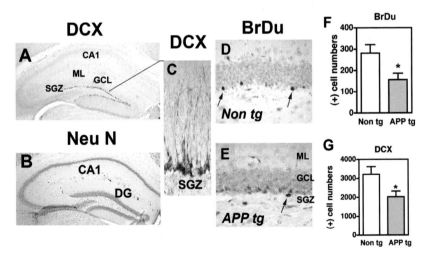

Figure 4. Alterations in neurogenesis in the hippocampus of adult mThy-1 APP transgenic mice. (A) Doublecortin (DCX) expression in the subgranular zone (SGZ) of the mouse hippocampus; (B) comparison with neuronal marker NeuN; (C) detail of NPCs in the SGZ that express DCX; (D) BrdU-positive NPCs in the SGZ of non-transgenic mice; (E) reduced BrdU-positive cells in APP transgenic mice; (F,G) image analysis of sections from wild-type ($n = 8$) and APP transgenic mice ($n = 6$) (12 months old) showed a decrease in the numbers of BrdU- and DCX-positive cells in the SGZ in the APP transgenic mice ($P < 0.01$ by Students t-test, unpaired).

model overexpressing both APP and PS-1 also showed a significant reduction in markers of neurogenesis (Zhang et al. 2007). A different study reported *increased* neurogenesis in a transgenic model in which APP expression is driven by the PDGF promoter (Jin et al. 2004a). However, a more recent and comprehensive analysis showed an increased number of NPCs and decreased in the molecular layer of the DG, markers of neurogenesis in the SGZ. This indicates that in PDGF-promoter-APP mice, there is altered migration, with arrested NPCs accumulating in the DG. These cells undergo apoptosis and thus the net result in these mice is a decreased generation of neuronal cells (Donovan et al. 2006). Furthermore, and consistent with what has been recently reported in AD brains (Boekhoorn et al. 2006), it is possible that increases in cell genesis in transgenic models might reflect glial and vascular changes and not alterations in neurogenesis.

ALTERATIONS IN NEUROGENESIS IN HIV-ASSOCIATED COGNITIVE DISORDERS

The HIV epidemic affects more than 40 million people worldwide and more than 1 million individuals in the United States. Early in the development of AIDS, HIV-infected macrophages traffic in to the CNS (Gendelman et al. 1994), leading to an inflammatory condition with astrogliosis, microglial cell activation (Budka 1991; Wiley and Achim 1994, 1995), myelin loss, and synaptodendritic damage (Ellis et al. 2007). Although the more severe forms of HIV-associated dementia have subsided with the advent of highly active antiretroviral therapy (HAART), chronic forms of HIV encephalitis (HIVE) with moderate neurological and psychiatric disturbances still persist. HIVE represents an important inflammatory condition that leads to cognitive impairment and neurodegeneration (Grant et al. 1995a,b; McArthur and Grant 1997; McArthur et al. 2003). Thus, better understanding the alterations in neurogenesis in patients with HIVE might help to elucidate mechanisms through which inflammatory conditions lead to neurodegeneration and abnormalities in neurogenesis.

HIV May Disrupt Cell Cycle Control in Brain Cells

Recent evidence suggests that the neurodegenerative process in patients with HIVE might target not only mature neurons, but also NPCs in the hippocampus (Krathwohl and Kaiser 2004). HIV proteins such as gp120 inhibit NPC proliferation via chemokine receptors. Furthermore, in HIV patients with cognitive impairment, the numbers of NPCs are reduced in the hippocampus, as evidenced by immunostaining with the proliferation marker Ki-67 (Krathwohl and Kaiser 2004). The mechanisms through which HIV proteins interfere with neurogenesis in the adult brain are unclear, but they might involve signaling alterations (e.g., CDK5/p35/p25) analogous to those involved in synaptic and dendritic damage. Furthermore, previous studies have shown that HIV-Vpr (viral protein R) interferes with the cell cycle in progenitor cells (Amini et al. 2004; Janket et al. 2004), and alterations in CDKs (cyclin-dependent kinases) in patients with HIV have been linked with Kaposi's sarcoma (Murakami-Mori et al. 1998), HPV infection, and HIV-associated nephropathy (Schang 2002). CDK5 has a role in neurogenesis in the developing brain because in mice deficient in this kinase and its activator (p35), neuronal development and migration are arrested (Cicero and Herrup 2005; Gilmore et al. 1998). In the adult nervous system, the role of the p35-CDK5 signaling pathway in neurogenesis is less well understood.

HIV-related Inflammation Might Contribute to Changes in Neurogenesis

In addition to the alterations in neurogenesis caused by HIV proteins, release of proinflammatory factors such as tumor necrosis factor-α (TNF-α), interleukins, and interferons from activated microglia and astrocytes might significantly suppress neurogenesis in the hippocampus of patients with HIV and further contribute to the neurodegenerative pathology in patients with chronic HIV infection. Among the chemokines and chemokine receptors that might have an important role in neurogenesis, recent studies have shown that SDF1 and CXCR4 are very important. Mice deficient in CXCR4 receptors exhibit a malformed DG in which the migration of neuronal progenitors is stalled. Adult neuronal progenitor cells existing in the SGZ, which produce granule neurons, express CXCR4 and other chemokine receptors, and granule neurons express SDF-1, suggesting that SDF1/CXCR4 signaling is also important in adult neurogenesis. Because the cellular receptors for HIV-1 include chemokine receptors, such as CXCR4 and CCR5, it is possible that the virus may interfere with SDF1/CXCR4 signaling in the brain, including disruption of the formation of new granule neurons in the adult brain (Tran and Miller 2005).

CONCLUDING REMARKS

Neurogenesis takes place in all adult mammals that have been studied. The percentage of cells that are replaced, relative to the total number of cells, appears to decrease with evolution. Despite low rates of adult neurogenesis, alterations in adult neurogenesis are a common pathological feature in several human neurodegenerative diseases. We have summarized existing data on changes in adult neurogenesis in PD, HD, and AD. In brief, the proliferation rate is decreased in both the SVZ and hippocampus of PD patients, as indicated by lower numbers of mitotic cells (Hoglinger et al. 2004). In contrast, it is believed that cell genesis is increased in the SVZ of human HD brains and that some of these newborn cells develop into neurons (Curtis et al. 2003, 2005a,b). In AD, hippocampal neurogenesis has been suggested to increase (Jin et al. 2004a). Recent studies, however, suggest that hippocampal neurogenesis is reduced in AD and instead the proliferation of nonneuronal cells, i.e., glial and vascular components, is increased (Boekhoorn et al. 2006).

More detailed studies are needed to achieve a deeper understanding of the temporal and spatial patterns of adult neurogenesis in neurodegenerative diseases. These should include neuropathological studies of patients in early and late stages of each disease, which could help us to better understand how neurogenesis changes over time and to understand its role in the disease processes. Processes controlling adult neurogenesis are complex, and they depend on cellular mechanisms that govern cell proliferation, differentiation, migration, and survival. Therefore, they are vulnerable to perturbations at several levels. Findings of changes in adult neurogenesis in diverse neurodegenerative diseases indicate that diverse and distinct disease mechanisms can modify this complex process. Detailed studies of neurogenesis in neurodegenerative disease may also shed light on the relationship between the changes in neurogenesis and functional changes in, for example, olfaction and cognition in these diseases. In both PD and AD, impairments in olfaction are considered to be early signs of disease (Hawkes 2006). In light of the limited options to treat patients with neurodegenerative diseases, adult neurogenesis may be envisioned as a repair tool to limit neuronal loss, or even enhancing adult neurogenesis might be a novel way to improve the clinical conditions of the affected patients.

ACKNOWLEDGMENTS

The authors thank Beate Winner and Leslie Crews for help in editing and preparing parts of the text, and Karin van der Borght for providing unpublished photographs. This work was supported by National Institutes of Health grants AG18440, AG022074, AG11385, AG10435, and MH62962 and by the Bavarian State Ministry of Sciences, Research and the Arts (ForNeuroCell grant). P.B. is supported by the Swedish Research Council, is coordinator of their strong research environment on brain damage and repair (NeuroFortis), and by the Nordic Center of Excellence on Neurodegeneration.

REFERENCES

Aimone J.B., Wiles J., and Gage F.H. 2006. Potential role for adult neurogenesis in the encoding of time in new memories. *Nat. Neurosci.* **9:** 723–727.

Altman J. 1962. Are new neurons formed in the brains of adult mammals? *Science* **135:** 1127–1128.

Amini S., Khalili K., and Sawaya B.E. 2004. Effect of HIV-1 Vpr on cell cycle regulators. *DNA Cell Biol.* **23:** 249–260.

Andrade M.A. and Bork P. 1995. HEAT repeats in the Huntington's disease protein. *Nat. Genet.* **11:** 115–116.

Armstrong R.J. and Barker R.A. 2001. Neurodegeneration: A failure of neuroregeneration? *Lancet* **358:** 1174–1176.

Baker S.A., Baker K.A., and Hagg T. 2004. Dopaminergic nigrostriatal projections regulate neural precursor proliferation in the adult mouse subventricular zone. *Eur. J. Neurosci.* **20:** 575–579.

Batista C.M., Kippin T.E., Willaime-Morawek S., Shimabukuro M.K., Akamatsu W., and van der Kooy D. 2006. A progressive and cell non-autonomous increase in striatal neural stem cells in the Huntington's disease R6/2 mouse. *J. Neurosci.* **26:** 10452–10460.

Berendse H.W., Booij J., Francot C.M., Bergmans P.L., Hijman R., Stoof J.C., et al. 2001. Subclinical dopaminergic dysfunction in asymptomatic Parkinson's disease patients' relatives with a decreased sense of smell. *Ann. Neurol.* **50:** 34–41.

Bjorklund A., Dunnett S.B., Brundin P., Stoessl A.J., Freed C.R., Breeze R.E., et al. 2003. Neural transplantation for the treatment of Parkinson's disease. *Lancet Neurol.* **2:** 437–445.

Bjorkqvist M., Fex M., Renstrom E., Wierup N., Petersen A., Gil J., et al. 2005. The R6/2 transgenic mouse model of Huntington's disease develops diabetes due to deficient beta-cell mass and exocytosis. *Hum. Mol. Genet.* **14:** 565–574.

Bjorkqvist M., Petersen A., Bacos K., Isaacs J., Norlen P., Gil J., et al. 2006. Progressive alterations in the hypothalamic-pituitary-adrenal axis in the R6/2 transgenic mouse model of Huntington's disease. *Hum. Mol. Genet.* **15:** 1713–1721.

Boekhoorn K., Joels M., and Lucassen P.J. 2006. Increased proliferation reflects glial and vascular-associated changes, but not neurogenesis in the presenile Alzheimer hippocampus. *Neurobiol. Dis.* **24:** 1–14.

Bonfanti L. 2006. PSA-NCAM in mammalian structural plasticity and neurogenesis. *Prog. Neurobiol.* **80:** 129–164.

Borta A. and Hoglinger G.U. 2007. Dopamine and adult neurogenesis. *J. Neurochem.* **100:** 587–595.

Braak H. and Braak E. 2000. Pathoanatomy of Parkinson's disease. *J. Neurol.* (suppl 2) **247:** 110–113.

Braak H., Ghebremedhin E., Rub U., Bratzke H., and Del Tredici K. 2004. Stages in the development of Parkinson's disease-related pathology. *Cell Tissue Res.* **318:** 121–134.

Braak H., Del Tredici K., Rub U., de Vos R.A., Jansen Steur E.N., and Braak E. 2003. Staging of brain pathology related to sporadic Parkinson's disease. *Neurobiol. Aging* **24:** 197–211.

Budka H. 1991. Neuropathology of human immunodeficiency virus infection. *Brain Pathol.* **1:** 163–175.

Cai H., Wang Y., McCarthy D., Wen H., Borchelt D.R., Price D.L., et al. 2001. BACE1 is the major beta-secretase for generation of Abeta peptides by neurons. *Nat. Neurosci.* **4:** 233–234.

Carlen M., Cassidy R.M., Brismar H., Smith G.A., Enquist L.W., and Frisen J. 2002. Functional integration of adult-born neurons. *Curr. Biol.* **12:** 606–608.

Carleton A., Petreanu L.T., Lansford R., Alvarez-Buylla A., and Lledo P.M. 2003. Becoming a new neuron in the adult olfactory bulb. *Nat. Neurosci.* **6:** 507–518.

Chen Y., Ai Y., Slevin J.R., Maley B.E., and Gash D.M. 2005. Progenitor proliferation in the adult hippocampus and substantia nigra induced by glial cell line-derived neurotrophic factor. *Exp. Neurol.* **196:** 87–95.

Chevallier N.L., Soriano S., Kang D.E., Masliah E., Hu G., and Koo E.H. 2005. Perturbed neurogenesis in the adult hippocampus associated with presenilin-1 A246E mutation. *Am. J. Pathol.* **167:** 151–159.

Cicero S. and Herrup K. 2005. Cyclin-dependent kinase 5 is essential for neuronal cell cycle arrest and differentiation. *J. Neurosci.* **25:** 9658–9668.

Collin T., Arvidsson A., Kokaia Z., and Lindvall O. 2005. Quantitative analysis of the generation of different striatal neuronal subtypes in the adult brain following excitotoxic injury. *Exp. Neurol.* **195:** 71–80.

Cooper O. and Isacson O. 2004. Intrastriatal transforming growth factor alpha delivery to a model of Parkinson's disease induces proliferation and migration of endogenous adult neural progenitor cells without differentiation into dopaminergic neurons. *J. Neurosci.* **24:** 8924–8931.

Cooper-Kuhn C.M., Winkler J., and Kuhn H.G. 2004. Decreased neurogenesis after cholinergic forebrain lesion in the adult rat. *J. Neurosci. Res.* **77:** 155–165.

Cotman C., Cummings B., and Pike C. 1993. Molecular cascades in adaptive versus pathological plasticity. In *Neurodegeneration* (ed. A. Gorio), pp. 217–240. Raven Press, New York.

Curtis M.A., Waldvogel H.J., Synek B., and Faull R.L. 2005a. A histochemical and immunohistochemical analysis of the subependymal layer in the normal and Huntington's disease brain. *J. Chem. Neuroanat.* **30:** 55–66.

Curtis M.A., Penney E.B., Pearson J., Dragunow M., Connor B., and Faull R.L. 2005b. The distribution of progenitor cells in the subependymal layer of the lateral ventricle in the normal and Huntington's disease human brain. *Neuroscience* **132:** 777–788.

Curtis M.A., Penney E.B., Pearson A.G., van Roon-Mom W.M., Butterworth N.J., Dragunow M., et al. 2003. Increased cell proliferation and neurogenesis in the adult human Huntington's disease brain. *Proc. Natl. Acad. Sci.* **100:** 9023–9027.

Curtis M.A., Kam M., Nannmark U., Anderson M.F., Axell M.Z., Wikkelso C., et al. 2007. Human neuroblasts migrate to the olfactory bulb via a lateral ventricular extension. *Science* **315:** 1243–1249.

DeKosky S. and Scheff S. 1990. Synapse loss in frontal cortex biopsies in Alzheimer's disease: Correlation with cognitive severity. *Ann. Neurol.* **27:** 457–464.

DeKosky S.T., Scheff S.W., and Styren S.D. 1996. Structural correlates of cognition in dementia: Quantification and assessment of synapse change. *Neurodegeneration* **5:** 417–421.

Dong H., Goico B., Martin M., Csernansky C.A., Bertchume A., and Csernansky J.G. 2004. Modulation of hippocampal cell proliferation, memory, and amyloid plaque deposition in APPsw (Tg2576) mutant mice by isolation stress. *Neuroscience* **127:** 601–609.

Donovan M.H., Yazdani U., Norris R.D., Games D., German D.C., Eisch A.J. 2006. Decreased adult hippocampal neurogenesis in the PDAPP mouse model of Alzheimer's disease. *J. Comp. Neurol.* **495:** 70–83.

Ellis R., Langford D., and Masliah E. 2007. HIV and antiretroviral therapy in the brain: Neuronal injury and repair. *Nat. Rev. Neurosci.* **8:** 33–44.

Emery D.L., Royo N.C., Fischer I., Saatman K.E., and McIntosh T.K. 2003. Plasticity following injury to the adult central nervous system: Is recapitulation of a developmental state worth promoting? *J. Neurotrauma* **20:** 1271–1292.

Eriksson P.S., Perfilieva E., Bjork-Eriksson T., Alborn A.M., Nordborg C., Peterson D.A., et al. 1998. Neurogenesis in the adult human hippocampus. *Nat. Med.* **4:** 1313–1317.

Fallon J., Reid S., Kinyamu R., Opole I., Opole R., Baratta J., et al. 2000. In vivo induction of massive proliferation, directed migration, and differentiation of neural cells in the adult mammalian brain. *Proc. Natl. Acad. Sci.* **97:** 14686–14691.

Feng R., Rampon C., Tang Y.P., Shrom D., Jin J., Kyin M., et al. 2001. Deficient neurogenesis in forebrain-specific presenilin-1 knockout mice is associated with reduced clearance of hippocampal memory traces. *Neuron* **32:** 911–926.

Fontana X., Nacher J., Soriano E., and del Rio J.A. 2006. Cell proliferation in the adult hippocampal formation of rodents and its modulation by entorhinal and fimbria-fornix afferents. *Cereb. Cortex* **16:** 301–312.

Frielingsdorf H., Schwarz K., Brundin P., and Mohapel P. 2004. No evidence for new dopaminergic neurons in the adult mammalian substantia nigra. *Proc. Natl. Acad. Sci.* **101:** 10177–10182.

Gage F.H., Kempermann G., Palmer T.D., Peterson D.A., and Ray J. 1998. Multipotent progenitor cells in the adult dentate gyrus. *J. Neurobiol.* **36:** 249–266.

Galvin J.E., Uryu K., Lee V.M.-Y., and Trojanowski J.Q. 1999. Axon pathology in Parkinson's disease and Lewy body dementia hippocampus contains α-, β- and γ-synuclein. *Proc. Natl. Acad. Sci.* **96:** 13450–13455.

Gendelman H., Lipton S., Tardieu M., Bukrinsky M., and Nottet H. 1994. The neuropathogenesis of HIV-1 infection. *J. Leukocyte Biol.* **56:** 389–398.

Gil J.M., Leist M., Popovic N., Brundin P., and Petersen A. 2004. Asialoerythropoietin is not effective in the R6/2 line of Huntington's disease mice. *BMC Neurosci.* **5:** 17.

Gil J.M., Mohapel P., Araujo I.M., Popovic N., Li J.Y., Brundin P., et al. 2005. Reduced hippocampal neurogenesis in R6/2 transgenic Huntington's disease mice. *Neurobiol. Dis.* **20:** 744–751.

Gilmore E.C., Ohshima T., Goffinet A.M., Kulkarni A.B., and Herrup K. 1998. Cyclin-dependent kinase 5-deficient mice demonstrate novel developmental arrest in cerebral cortex. *J. Neurosci.* **18:** 6370–6377.

Glabe C.C. 2005. Amyloid accumulation and pathogenesis of Alzheimer's disease: Significance of monomeric, oligomeric and fibrillar Abeta. *Subcell. Biochem.* **38:** 167–177.

Glabe C.G. and Kayed R. 2006. Common structure and toxic function of amyloid oligomers implies a common mechanism of pathogenesis. *Neurology* **66:** S74–S78.

Gould E. and Tanapat P. 1999. Stress and hippocampal neurogenesis. *Biol. Psychiatry* **46:** 1472–1479.

Gould E., Beylin A., Tanapat P., Reeves A., and Shors T.J. 1999. Learning enhances adult neurogenesis in the hippocampal formation. *Nat. Neurosci.* **2:** 260–265.

Grant I., Heaton R.K., and Atkinson J.H. 1995a. Neurocognitive disorders in HIV-1 infection. HNRC Group. HIV Neurobehavioral Research Center. *Curr. Top. Microbiol. Immunol.* **202:** 11–32.

Grant I., Heaton R., Atkinson J., and Group H. 1995b. Neurocognitive disorder in HIV-1 infection. In *Current topics in microbiology and immunology. HIV and dementia* (ed. M. Oldstone and L. Vitovic), pp. 9–30. Springer-Verlag, Heidelberg.

Grote H.E., Bull N.D., Howard M.L., van Dellen A., Blakemore C., Bartlett P.F., et al. 2005. Cognitive disorders and neurogenesis deficits in Huntington's disease mice are rescued by fluoxetine. *Eur. J. Neurosci.* **22:** 2081–2088.

Gutekunst C.A., Levey A.I., Heilman C.J., Whaley W.L., Yi H., Nash N.R., et al. 1995. Identification and localization of huntingtin in brain and human lymphoblastoid cell lines with anti-fusion protein antibodies. *Proc. Natl. Acad. Sci.* **92:** 8710–8714.

Hagell P. and Brundin P. 2001. Cell survival and clinical outcome following intrastriatal transplantation in Parkinson disease. *J. Neuropathol. Exp. Neurol.* **60:** 741–752.

Harding A.J. and Halliday G.M. 2001. Cortical Lewy body pathology in the diagnosis of dementia. *Acta Neuropathol.* **102:** 355–363.

Hashimoto M., Rockenstein E., and Masliah E. 2003. Transgenic models of alpha-synuclein pathology: Past, present, and future. *Ann. N. Y. Acad. Sci.* **991:** 171–188.

Hashimoto M., Kawahara K., Bar-On P., Rockenstein E., Crews L., and Masliah E. 2004. The Role of alpha-synuclein assembly and metabolism in the pathogenesis of Lewy body disease. *J. Mol. Neurosci.* **24:** 343–352.

Haughey N.J., Nath A., Chan S.L., Borchard A.C., Rao M.S., and Mattson M.P. 2002. Disruption of neurogenesis by amyloid beta-peptide, and perturbed neural progenitor cell homeostasis, in models of Alzheimer's disease. *J. Neurochem.* **83:** 1509–1524.

Hawkes C. 2006. Olfaction in neurodegenerative disorder. *Adv. Otorhinolaryngol.* **63:** 133–151.

Hockly E., Cordery P.M., Woodman B., Mahal A., van Dellen A., Blakemore C., et al. 2002. Environmental enrichment slows disease progression in R6/2 Huntington's disease mice. *Ann. Neurol.* **51:** 235–242.

Hoffner G., Kahlem P., and Djian P. 2002. Perinuclear localization of huntingtin as a consequence of its binding to microtubules through an interaction with beta-tubulin: Relevance to Huntington's disease. *J. Cell. Sci.* **115:** 941–948.

Hoglinger G.U., Rizk P., Muriel M.P., Duyckaerts C., Oertel W.H., Caille I., et al. 2004. Dopamine depletion impairs precursor cell proliferation in Parkinson disease. *Nat. Neurosci.* **7:** 726–735.

Huang H.P., Chu K., Nemoz-Gaillard E., Elberg D., and Tsai M.J. 2002. Neogenesis of beta-cells in adult BETA2/NeuroD-deficient mice. *Mol. Endocrinol.* **16:** 541–551.

Huisman E., Uylings H.B., and Hoogland P.V. 2004. A 100% increase of dopaminergic cells in the olfactory bulb may explain hyposmia in Parkinson's disease. *Mov. Disord.* **19:** 687–692.

Hyman B. and Gomez-Isla T. 1994. Alzheimer's disease is a laminar regional and neural system specific disease, not a global brain disease. *Neurobiol. Aging* **15:** 353–354.

Jakubs K., Nanobashvili A., Bonde S., Ekdahl C.T., Kokaia Z., Kokaia M., et al. 2006. Environment matters: Synaptic properties of neurons born in the epileptic adult brain develop to reduce excitability. *Neuron* **52:** 1047–1059.

Janket M.L., Manickam P., Majumder B., Thotala D., Wagner M., Schafer E.A., et al. 2004. Differential regulation of host cellular genes by HIV-1 viral protein R (Vpr): cDNA microarray analysis using isogenic virus. *Biochem. Biophys. Res. Commun.* **314:** 1126–1132.

Jellinger K.A. 1991. Pathology of Parkinson's disease. Changes other than the nigrostriatal pathway. *Mol. Chem. Neuropathol.* **14:** 153–197.

Jellinger K.A. and Attems J. 2006. Does striatal pathology distinguish Parkinson disease with dementia and dementia with Lewy bodies? *Acta Neuropathol.* **112:** 253–260.

Jin K., Galvan V., Xie L., Mao X.O., Gorostiza O.F., Bredesen D.E., et al. 2004a. Enhanced neurogenesis in Alzheimer's disease transgenic (PDGF-APPSw,Ind) mice. *Proc. Natl. Acad. Sci.* **101:** 13363–13367.

Jin K., LaFevre-Bernt M., Sun Y., Chen S., Gafni J., Crippen D., et al. 2005. FGF-2 promotes neurogenesis and neuroprotection and prolongs survival in a transgenic mouse model of Huntington's disease. *Proc. Natl. Acad. Sci.* **102:** 18189–18194.

Jin K., Peel A.L., Mao X.O., Xie L., Cottrell B.A., Henshall D.C., et al. 2004b. Increased hippocampal neurogenesis in Alzheimer's disease. *Proc. Natl. Acad. Sci.* **101:** 343–347.

Johnson J., Hague S.M., Hanson M., Gibson A., Wilson K.E., Evans E.W., et al. 2004. SNCA multiplication is not a common cause of Parkinson disease or dementia with Lewy bodies. *Neurology* **63:** 554–556.

Jordan J.D., Ming G.L., and Song H. 2006. Adult neurogenesis as a potential therapy for neurodegenerative diseases. *Discov. Med.* **6:** 144–147.

Kamenetz F., Tomita T., Hsieh H., Seabrook G., Borchelt D., Iwatsubo T., et al. 2003. APP processing and synaptic function. *Neuron* **37:** 925–937.

Kay J.N. and Blum M. 2000. Differential response of ventral midbrain and striatal progenitor cells to lesions of the nigrostriatal dopaminergic projection. *Dev. Neurosci.* **22:** 56–67.

Kempermann G. 2006. *Adult neurogenesis, stem cells and neuronal development in the adult brain.* Oxford University Press, New York.

Kempermann G., Brandon E.P., and Gage F.H. 1998. Environmental stimulation of 129/SvJ mice causes increased cell proliferation and neurogenesis in the adult dentate gyrus. *Curr. Biol.* **8:** 939–942.

Kempermann G., Kuhn H.G., and Gage F.H. 1997. More hippocampal neurons in adult mice living in an enriched environment. *Nature* **386:** 493–495.

Kobayashi T., Ahlenius H., Thored P., Kobayashi R., Kokaia Z., and Lindvall O. 2006. Intracerebral infusion of glial cell line-derived neurotrophic factor promotes striatal neurogenesis after stroke in adult rats. *Stroke* **37:** 2361–2367.

Kokaia Z., Thored P., Arvidsson A., and Lindvall O. 2006. Regulation of stroke-induced neurogenesis in adult brain—recent scientific progress. *Cereb. Cortex* (suppl 1) **16:** i162–i167.

Krathwohl M.D. and Kaiser J.L. 2004. HIV-1 promotes quiescence in human neural progenitor cells. *J. Infect. Dis.* **190:** 216–226.

Kruger R., Kuhn W., Muller T., Woitalla D., Graeber M., Kosel S., et al. 1998. Ala30Pro mutation in the gene encoding α-synuclein in Parkinson's disease. *Nat. Genet.* **18:** 106–108.

Laplagne D.A., Esposito M.S., Piatti V.C., Morgenstern N.A., Zhao C., van Praag H., et al. 2006. Functional convergence of neurons generated in the developing and adult hippocampus. *PLoS Biol.* **4:** e409.

Lazic S.E., Grote H., Armstrong R.J., Blakemore C., Hannan A.J., van Dellen A., et al. 2004. Decreased hippocampal cell proliferation in R6/1 Huntington's mice. *Neuroreport* **15:** 811–813.

Lazic S.E., Grote H.E., Blakemore C., Hannan A.J., van Dellen A., Phillips W., et al. 2006. Neurogenesis in the R6/1 transgenic mouse model of Huntington's disease: Effects of environmental enrichment. *Eur. J. Neurosci.* **23:** 1829–1838.

Lichtenwalner R.J. and Parent J.M. 2006. Adult neurogenesis and the ischemic forebrain. *J. Cereb. Blood Flow. Metab.* **26:** 1–20.

Lie D.C., Dziewczapolski G., Willhoite A.R., Kaspar B.K., Shults C.W., and Gage F.H. 2002. The adult substantia nigra contains progenitor cells with neurogenic potential. *J. Neurosci.* **22:** 6639–6649.

Lippa C.F., Duda J.E., Grossman M., Hurtig H.I., Aarsland D., Boeve B.F., et al. 2007. DLB and PDD boundary issues: Diagnosis, treatment, molecular pathology, and biomarkers. *Neurology* **68:** 812–819.

Liu M., Pleasure S.J., Collins A.E., Noebels J.L., Naya F.J., Tsai M.J., et al. 2000. Loss of BETA2/NeuroD leads to malformation of the dentate gyrus and epilepsy. *Proc. Natl. Acad. Sci.* **97:** 865–870.

Longo F.M., Yang T., Xie Y., and Massa S.M. 2006. Small molecule approaches for promoting neurogenesis. *Curr. Alzheimer Res.* **3:** 5–10.

Lovell M.A. and Markesbery W.R. 2005. Ectopic expression of Musashi-1 in Alzheimer disease and Pick disease. *J. Neuropathol. Exp. Neurol.* **64:** 675–680.

Luo Y., Bolon B., Kahn S., Bennett B.D., Babu-Khan S., Denis P., et al. 2001. Mice deficient in BACE1, the Alzheimer's beta-secretase, have normal phenotype and abolished beta-amyloid generation. *Nat. Neurosci.* **4:** 231–232.

Maguire-Zeiss K.A. and Federoff H.J. 2003. Convergent pathobiologic model of Parkinson's disease. *Ann. N. Y. Acad. Sci.* **991:** 152–166.

Mangiarini L., Sathasivam K., Seller M., Cozens B., Harper A., Hetherington C., et al. 1996. Exon 1 of the HD gene with an expanded CAG repeat is sufficient to cause a progressive neurological phenotype in transgenic mice. *Cell* **87:** 493–506.

Mao L., Lau Y.S., Petroske E., and Wang J.Q. 2001. Profound astrogenesis in the striatum of adult mice following nigrostriatal dopaminergic lesion by repeated MPTP administration. *Brain Res. Dev. Brain Res.* **131:** 57–65.

Marcora E., Gowan K., and Lee J.E. 2003. Stimulation of NeuroD activity by huntingtin and huntingtin-associated proteins HAP1 and MLK2. *Proc. Natl. Acad. Sci.* **100:** 9578–9583.

Masliah E. 1995. Mechanisms of synaptic dysfunction in Alzheimer's disease. *Histol. Histopathol.* **10:** 509–519.

———. 2000. The role of synaptic proteins in Alzheimer's disease. *Ann. N.Y. Acad. Sci.* **924:** 68–75.

Masliah E. and Terry R. 1994. The role of synaptic pathology in the mechanisms of dementia in Alzheimer's disease. *Clin. Neurosci.* **1:** 192–198.

Masliah E., Mallory M., Alford M., DeTeresa R., Hansen L.A., McKeel D.W., Jr., et al. 2001. Altered expression of synaptic proteins occurs early during progression of Alzheimer's disease. *Neurology* **56:** 127–129.

McArthur J. and Grant I. 1997. HIV neurocognitive disorders. In *Neurological and neuropsychiatric manifestations of HIV-1 infection* (ed. H. Gendelman et al.), pp. 499–523. Chapman and Hall, New York.

McArthur J.C., Haughey N., Gartner S., Conant K., Pardo C., Nath A., et al. 2003. Human immunodeficiency virus-associated dementia: an evolving disease. *J. Neurovirol.* **9:** 205–221.

McKeith I., Galasko D., Kosaka K., Perry E., Dickson D., Hansen L., et al. 1996. Clinical and pathological diagnosis of dementia with Lewy bodies (DLB): Report of the CDLB International Workshop. *Neurology* **47:** 1113–1124.

McKeith I., Mintzer J., Aarsland D., Burn D., Chiu H., Cohen-Mansfield J., et al. 2004. Dementia with Lewy bodies. *Lancet Neurol.* **3:** 19–28.

Menalled L.B. and Chesselet M.F. 2002. Mouse models of Huntington's disease. *Trends. Pharmacol. Sci.* **23:** 32–39.

Miles D.K. and Kernie S.G. 2006. Activation of neural stem and progenitor cells after brain injury. *Prog. Brain. Res.* **157:** 187–197.

Ming G.L. and Song H. 2005. Adult neurogenesis in the mammalian central nervous system. *Annu. Rev. Neurosci.* **28:** 223–250.

Mohapel P., Leanza G., Kokaia M., and Lindvall O. 2005a. Forebrain acetylcholine regulates adult hippocampal neurogenesis and learning. *Neurobiol. Aging* **26:** 939–946.

Mohapel P., Mundt-Petersen K., Brundin P., and Frielingsdorf H. 2006. Working memory training decreases hippocampal neurogenesis. *Neuroscience* **142:** 609–613.

Mohapel P., Frielingsdorf H., Haggblad J., Zachrisson O., and Brundin P. 2005b. Platelet-derived growth factor (PDGF-BB) and brain-derived neurotrophic factor (BDNF) induce striatal neurogenesis in adult rats with 6-hydroxydopamine lesions. *Neuroscience* **132:** 767–776.

Murakami-Mori K., Mori S., and Nakamura S. 1998. Endogenous basic fibroblast growth factor is essential for cyclin E-CDK2 activity in multiple external cytokine-induced proliferation of AIDS-associated Kaposi's sarcoma cells: Dual control of AIDS-associated Kaposi's sarcoma cell growth and cyclin E-CDK2 activity by endogenous and external signals. *J. Immunol.* **161:** 1694–1704.

Murphy K.P., Carter R.J., Lione L.A., Mangiarini L., Mahal A., Bates G.P., et al. 2000. Abnormal synaptic plasticity and impaired spatial cognition in mice transgenic for exon 1 of the human Huntington's disease mutation. *J. Neurosci.* **20:** 5115–5123.

Nakatomi H., Kuriu T., Okabe S., Yamamoto S., Hatano O., Kawahara N., et al. 2002. Regeneration of hippocampal pyramidal neurons after ischemic brain injury by recruitment of endogenous neural progenitors. *Cell* **110:** 429–441.

Neuwald A.F. and Hirano T. 2000. HEAT repeats associated with condensins, cohesins, and other complexes involved in chromosome-related functions. *Genome Res.* **10:** 1445–1452.

Oumesmar B.N., Vignais L., Duhamel-Clerin E., Avellana-Adalid V., Rougon G., and Baron-Van Evercooren A. 1995. Expression of the highly polysialylated neural cell adhesion molecule during postnatal myelination and following chemically induced demyelination of the adult mouse spinal cord. *Eur. J. Neurosci.* **7:** 480–491.

Parent J.M. 2003. Injury-induced neurogenesis in the adult mammalian brain. *Neuroscientist* **9:** 261–272.

Pencea V., Bingaman K.D., Wiegand S.J., and Luskin M.B. 2001. Infusion of brain-derived neurotrophic factor into the lateral ventricle of the adult rat leads to new neurons in the parenchyma of the striatum, septum, thalamus, and hypothalamus. *J. Neurosci.* **21:** 6706–6717.

Phillips W., Morton A.J., and Barker R.A. 2005. Abnormalities of neurogenesis in the R6/2 mouse model of Huntington's disease are attributable to the in vivo microenvironment. *J. Neurosci.* **25:** 11564–11576.

Polymeropoulos M., Lavedan C., Leroy E., Ide S., Dehejia A., Dutra A., et al. 1997. Mutation in the α-synuclein gene identified in families with Parkinson's disease. *Science* **276:** 2045–2047.

Reimers D., Herranz A.S., Diaz-Gil J.J., Lobo M.V., Paino C.L., Alonso R., et al. 2006. Intrastriatal infusion of liver growth factor stimulates dopamine terminal sprouting and partially restores motor function in 6-hydroxydopamine-lesioned rats. *J. Histochem. Cytochem.* **54:** 457–465.

Reynolds G.P., Dalton C.F., Tillery C.L., Mangiarini L., Davies S.W., and Bates G.P. 1999. Brain neurotransmitter deficits in mice transgenic for the Huntington's disease mutation. *J. Neurochem.* **72:** 1773–1776.

Rockenstein E., Mante M., Adame A., Crews L., Moessler H., and Masliah E. 2006. Effects of cerebrolysin trade mark on neurogenesis in an APP transgenic model of Alzheimer's disease. *Acta Neuropathol.* **113:** 265–275.

Romanko M.J., Rola R., Fike J.R., Szele F.G., Dizon M.L., Felling R.J., et al. 2004. Roles of the mammalian subventricular zone in cell replacement after brain injury. *Prog. Neurobiol.* **74:** 77–99.

Sanai N., Tramontin A.D., Quinones-Hinojosa A., Barbaro N.M., Gupta N., Kunwar S., et al. 2004. Unique astrocyte ribbon in adult human brain contains neural stem cells but lacks chain migration. *Nature* **427:** 740–744.

Santarelli L., Saxe M., Gross C., Surget A., Battaglia F., Dulawa S., et al. 2003. Requirement of hippocampal neurogenesis for the behavioral effects of antidepressants. *Science* **301:** 805–809.

Schang L.M. 2002. Cyclin-dependent kinases as cellular targets for antiviral drugs. *J. Antimicrob. Chemother* **50:** 779–792.

Scharfman H.E. and Hen R. 2004. Neuroscience. Is more neurogenesis always better? *Science* **315:** 336–338.

Scheff S.W. and Price D.A. 2003. Synaptic pathology in Alzheimer's disease: A review of ultrastructural studies. *Neurobiol. Aging* **24:** 1029–1046.

Selkoe D.J. 1999. Translating cell biology into therapeutic advances in Alzheimer's disease. *Nature* (suppl.) **399:** A23–A31.

Shan X., Chi L., Bishop M., Luo C., Lien L., Zhang Z., et al. 2006. Enhanced de novo neurogenesis and dopaminergic neurogenesis in the substantia nigra of MPTP-induced Parkinson's disease-like mice. *Stem Cells* **24:** 1280–1287.

Shults C.W. 2006. Lewy bodies. *Proc. Natl. Acad. Sci.* **103:** 1661–1668.

Singleton A.B., Farrer M., Johnson J., Singleton A., Hague S., Kachergus J., et al. 2003. Alpha-Synuclein locus triplication causes Parkinson's disease. *Science* **302:** 841.

Sinha S., Anderson J.P., Barbour R., Basi G.S., Caccavello R., Davis D., et al. 1999. Purification and cloning of amyloid precursor protein beta-secretase from human brain. *Nature* **402:** 537–540.

Sinha S., Anderson J., John V., McConlogue L., Basi G., Thorsett E., et al. 2000. Recent advances in the understanding of the processing of APP to beta amyloid peptide. *Ann. N.Y. Acad. Sci.* **920:** 206–208.

Smith R., Brundin P., and Li J.Y. 2005. Synaptic dysfunction in Huntington's disease: A new perspective. *Cell Mol. Life Sci.* **62:** 1901–1912.

Smith R., Chung H., Rundquist S., Maat-Schieman M.L., Colgan L., Englund E., et al. 2006. Cholinergic neuronal defect without cell loss in Huntington's disease. *Hum. Mol. Genet.* **15:** 3119–3131.

Sobel N., Thomason M.E., Stappen I., Tanner C.M., Tetrud J.W., Bower J.M., et al. 2001. An impairment in sniffing contributes to the olfactory impairment in Parkinson's disease. *Proc. Natl. Acad. Sci.* **98:** 4154–4159.

Spillantini M., Schmidt M., Lee V.-Y., Trojanowski J., Jakes R., and Goedert M. 1997. α-Synuclein in Lewy bodies. *Nature* **388:** 839–840.

Spires T.L., Grote H.E., Varshney N.K., Cordery P.M., van Dellen A., Blakemore C., et al. 2004. Environmental enrichment rescues protein deficits in a mouse model of Huntington's disease, indicating a possible disease mechanism. *J. Neurosci.* **24:** 2270–2276.

Steiner B., Winter C., Hosman K., Siebert E., Kempermann G., Petrus D.S., et al. 2006. Enriched environment induces cellular plasticity in the adult substantia nigra and improves motor behavior function in the 6-OHDA rat model of Parkinson's disease. *Exp. Neurol.* **199:** 291–300.

Takeda A., Mallory M., Sundsmo M., Honer W., Hansen L., and Masliah E. 1998. Abnormal accumulation of NACP/α-synuclein in neurodegenerative disorders. *Am. J. Pathol.* **152:** 367–372.

Tatebayashi Y., Lee M.H., Li L., Iqbal K., and Grundke-Iqbal I. 2003. The dentate gyrus neurogenesis: A therapeutic target for Alzheimer's disease. *Acta Neuropathol.* **105:** 225–232.

Tattersfield A.S., Croon R.J., Liu Y.W., Kells A.P., Faull R.L., and Connor B. 2004. Neurogenesis in the striatum of the quinolinic acid lesion model of Huntington's disease. *Neuroscience* **127:** 319–332.

Teramoto T., Qiu J., Plumier J.C., and Moskowitz M.A. 2003. EGF amplifies the replacement of parvalbumin-expressing striatal interneurons after ischemia. *J. Clin. Invest.* **111:** 1125–1132.

Terry R., Hansen L., and Masliah E. 1994. Structural basis of the cognitive alterations in Alzheimer disease. In *Alzheimer disease* (ed. R. Terry and R. Katzman), pp. 179–196. Raven Press, New York.

Terry R., Masliah E., Salmon D., Butters N., DeTeresa R., Hill R., et al. 1991. Physical basis of cognitive alterations in Alzheimer disease: Synapse loss is the major correlate of cognitive impairment. *Ann. Neurol.* **30:** 572–580.

Tomasevic G., Kamme F., and Wieloch T. 1998. Changes in proliferating cell nuclear antigen, a protein involved in DNA repair, in vulnerable hippocampal neurons following global cerebral ischemia. *Brain. Res. Mol. Brain Res.* **60:** 168–176.

Tran P.B. and Miller R.J. 2005. HIV-1, chemokines and neurogenesis. *Neurotox. Res.* **8:** 149–158.

Trojanowski J.Q. and Lee V.M. 1998. Aggregation of neurofilament and alpha-synuclein proteins in Lewy bodies: Implications for the pathogenesis of Parkinson disease and Lewy body dementia. *Arch. Neurol.* **55:** 151–152.

Wakabayashi K., Matsumoto K., Takayama K., Yoshimoto M., and Takahashi H. 1997. NACP, a presynaptic protein, immunoreactivity in Lewy bodies in Parkinson's disease. *Neurosci. Lett.* **239:** 45–48.

Walsh D.M. and Selkoe D.J. 2004. Oligomers on the brain: The emerging role of soluble protein aggregates in neurodegeneration. *Protein Pept. Lett.* **11:** 213–228.

van Dellen A., Blakemore C., Deacon R., York D., and Hannan A.J. 2000. Delaying the onset of Huntington's in mice. *Nature* **404:** 721–722.

van der Borght K. and Brundin P. 2004. Reduced expression of PSA-NCAM in the hippocampus and piriform cortex of the R6/1 and R6/2 mouse models of Huntington's disease. *Exp. Neurol.* **204:** 473–478.

Van Kampen J.M. and Eckman C.B. 2006. Dopamine D3 receptor agonist delivery to a model of Parkinson's disease restores the nigrostriatal pathway and improves locomotor behavior. *J. Neurosci.* **26:** 7272–7280.

Van Kampen J.M. and Robertson H.A. 2005. A possible role for dopamine D3 receptor stimulation in the induction of neurogenesis in the adult rat substantia nigra. *Neuroscience* **136:** 381–386.

van Praag H., Kempermann G., and Gage F.H. 1999. Running increases cell proliferation and neurogenesis in the adult mouse dentate gyrus. *Nat. Neurosci.* **2:** 266–270.

van Praag H., Schinder A.F., Christie B.R., Toni N., Palmer T.D., and Gage F.H. 2002. Functional neurogenesis in the adult hippocampus. *Nature* **415:** 1030–1034.

Vassar R., Bennett B.D., Babu-Khan S., Kahn S., Mendiaz E.A., Denis P., et al. 1999. Beta-secretase cleavage of Alzheimer's amyloid precursor protein by the transmembrane aspartic protease BACE. *Science* **286:** 735–741.

Wang R., Dineley K.T., Sweatt J.D., and Zheng H. 2004. Presenilin 1 familial Alzheimer's disease mutation leads to defective associative learning and impaired adult neurogenesis. *Neuroscience* **126:** 305–312.

Wen P.H., Hof P.R., Chen X., Gluck K., Austin G., Younkin S.G., et al. 2004. The presenilin-1 familial Alzheimer disease mutant P117L impairs neurogenesis in the hippocampus of adult mice. *Exp. Neurol.* **188:** 224–237.

Wiley C. and Achim C. 1994. HIV encephalitis is the pathologic correlate of dementia in AIDS. *Ann. Neurol.* **36:** 673–676.

―――. 1995. Human immunodeficiency virus encephalitis and dementia. *Ann. Neurol.* **38:** 559–560.

Winner B., Geyer M., Couillard-Despres S., Aigner R., Bogdahn U., Aigner L., et al. 2006. Striatal deafferentation increases dopaminergic neurogenesis in the adult olfactory bulb. *Exp. Neurol.* **197:** 113–121.

Winner B., Lie D.C., Rockenstein E., Aigner R., Aigner L., Masliah E., et al. 2004. Human wild-type alpha-synuclein impairs neurogenesis. *J. Neuropathol. Exp. Neurol.* **63:** 1155–1166.

Winner B., Rockenstein E., Lie D.C., Aigner R., Mante M., Bogdahn U., et al. 2007. Mutant alpha-synuclein exacerbates age-related decrease of neurogenesis. *Neurobiol. Aging* (in press).

Yamada M., Onodera M., Mizuno Y., and Mochizuki H. 2004. Neurogenesis in olfactory bulb identified by retroviral labeling in normal and 1-methyl-4-phenyl-1,2,3, 6-tetrahydropyridine-treated adult mice. *Neuroscience* **124:** 173–181.

Yoshimi K., Ren Y.R., Seki T., Yamada M., Ooizumi H., Onodera M., et al. 2005. Possibility for neurogenesis in substantia nigra of parkinsonian brain. *Ann. Neurol.* **58:** 31–40.

Zhang C., McNeil E., Dressler L., and Siman R. 2007. Long-lasting impairment in hippocampal neurogenesis associated with amyloid deposition in a knock-in mouse model of familial Alzheimer's disease. *Exp. Neurol.* **204:** 77–87.

Zhao M., Momma S., Delfani K., Carlen M., Cassidy R.M., Johansson C.B., et al. 2003. Evidence for neurogenesis in the adult mammalian substantia nigra. *Proc. Natl. Acad. Sci.* **100:** 7925–7930.

Zuccato C. and Cattaneo E. 2007. Role of brain-derived neurotrophic factor in Huntington's disease. *Prog. Neurobiol.* **81:** 294–330.

Zuccato C., Ciammola A., Rigamonti D., Leavitt B.R., Goffredo D., Conti L., et al. 2001. Loss of huntingtin-mediated BDNF gene transcription in Huntington's disease. *Science* **293:** 493–498.

25

Epilepsy and Adult Neurogenesis

Sebastian Jessberger
Laboratory of Genetics, Salk Institute for Biological Studies
La Jolla, California 920937

Jack M. Parent
Department of Neurology, University of Michigan
Ann Arbor, Michigan 48109

THE EPILEPSIES ARE A DIVERSE GROUP of neurological disorders that share the central feature of spontaneous recurrent seizures. Some epilepsies result from inherited mutations in single or multiple genes, termed idiopathic or primary epilepsies, whereas symptomatic or secondary epilepsies develop as a consequence of acquired brain abnormalities such as from tumor, trauma, stroke, infection, or developmental malformation. Of acquired epilepsies, mesial temporal lobe epilepsy (mTLE) is a particularly common and often intractable form. In addition to pharmacoresistant seizures, the syndrome of mTLE almost always involves impairments in cognitive function (Helmstaedter 2002; Elger et al. 2004; von Lehe et al. 2006) that may progress even with adequate seizure control (Blume 2006).

Seizure activity from mTLE typically arises from the hippocampus or other mesial temporal lobe structures. Simple and complex partial seizures, the most common seizure types in this epilepsy syndrome, often become medically refractory and may respond only to surgical resection of the epileptogenic tissue. Hippocampi in these cases usually show substantial structural abnormalities that include pyramidal cell loss, astrogliosis, dentate granule cell axonal reorganization (mossy fiber sprouting), and dispersion of the granule cell layer (Blumcke et al. 1999).

Humans with mTLE often have a history of an early "precipitating" insult, such as a prolonged or complicated febrile seizure, followed by a latent period and then the development of epilepsy in later childhood or adolescence. These historical findings have led to the development of what are currently the most common animal models, the status epilepticus (SE) models, used to study epileptogenic mechanisms in mTLE. In these models, a prolonged seizure induced by chemoconvulsant (typically kainic acid or pilocarpine) treatment or electrical stimulation leads to an initial brain injury followed, after a latent period of days to weeks, by spontaneous recurrent seizures. These models recapitulate much of the pathology of human mTLE (for review, see Buckmaster 2004). Experimental paradigms are necessary to investigate mechanisms underlying mTLE as surgical specimens from mTLE cases are collected at late stages of the disease and thus are unlikely to reveal early features critical for the disease process. Studies of experimental mTLE indicate that excess neural activity in the course of seizures not only damages existing mature structures of the hippocampal formation, but also dramatically affects endogenous neural stem cells (NSCs) within the adult rodent dentate gyrus (DG) (Bengzon et al. 1997; Parent et al. 1997; Scott et al. 1998). We discuss here the consequences of seizure activity on proliferation of NSCs, maturation and integration of newborn neurons, and the functional relevance of seizure-induced neurogenesis.

SEIZURE-INDUCED CELL PROLIFERATION

Prolonged seizure activity leads after a latent period of several days to a dramatic increase in cell proliferation judged by Ki-67 expression or short-pulse bromodeoxyuridine (BrdU) labeling in the DG (Parent et al. 1997; Gray and Sundstrom 1998; Jessberger et al. 2005) and rostral subventricular zone (SVZ) (Parent et al. 2002). In the dentate, the immediate proliferative response appears to be mediated by radial glia-like type-1 cells, whereas at the peak of cell proliferation, abnormal activation of doublecortin (DCX)-expressing neuroblasts occurs (Huttmann et al. 2003; Jessberger et al. 2005). BrdU labeling prior to SE has shown that most of the proliferating cells that respond to seizure activity are mitotically active even prior to the insult (Parent et al. 1999). The severity and duration of seizure activity do not seem to be a major factor as even single seizure-like discharges induce cell proliferation (Bengzon et al. 1997). However, the survival of seizure-generated granule cells, at least in certain SE models, appears to decrease with increased seizure severity (Mohapel et al. 2004), an effect that is potentially mediated by subsequent inflammation (Ekdahl et al. 2003). Cell proliferation returns to baseline levels approximately 3–4 weeks following the initial SE (Parent et al.

1997). Recent evidence suggests that at later stages following SE, the potential for adult neurogenesis might even be reduced (Hattiangady et al. 2004). The reasons for reduced neurogenesis late after seizures might be either "exhaustion" of the NSC pool or alterations in the neurogenic niche preventing support and proper function of NSCs.

A critical question that remains to be answered is how seizure activity translates into increased cell proliferation. It seems that NSCs are capable of "sensing" electrical activity (Deisseroth et al. 2004). Direct mechanisms may involve activation of glutamate and γ-amino-n-butyric acid (GABA) receptors expressed by dentate NSCs (Gould et al. 1994; Tozuka et al. 2005) or changes in histone acetylation induced by seizures (Huang et al. 2002; Jessberger et al. 2007b). Alternatively, seizure-induced expression of trophic factors such as brain-derived neurotrophic factor (BDNF), vascular endothelial growth factor (VEGF), and others by surrounding tissue could indirectly induce NSC proliferation (Isackson et al. 1991; Gall 1993; Newton et al. 2003). A combination of mechanisms seems to be most probable, although defining them precisely is very challenging given the substantial alteration in gene expression that occurs following SE (Elliott and Lowenstein 2004).

MATURATION AND INTEGRATION OF SEIZURE-GENERATED GRANULE CELLS

The accelerated NSC proliferation likewise is reflected by a marked increase in net neurogenesis (which is the number of new neurons actually generated) (Fig. 1). Similar to normal conditions, approximately

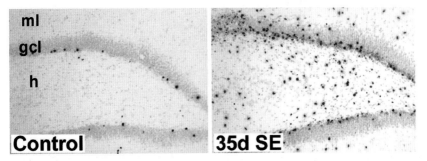

Figure 1. Pilocarpine-induced SE increases dentate gyrus (DG) cell proliferation. DG BrdU labeling in adult rats 35 days after pilocarpine-induced SE (*right*) or saline treatment in a control (*left*). BrdU immunoreactivity is increased markedly in the inner granule cell layer (gcl), hilus (h), and molecular layer (ml) of the animal that experienced 2 hours of continuous seizure activity (*right*). BrdU was given on days 7–21 after pilocarpine or saline treatment.

75–90% of all newly generated cells express markers characteristic of dentate granule cells 4 weeks after labeling dividing cells with BrdU or retroviral reporter vectors (Parent et al. 1997; Jessberger et al. 2005 and 2007a). Interestingly, seizure activity appears to accelerate the functional maturation and integration of adult-born granule cells, although the consequence of these effects on hippocampal network function are unknown (Overstreet-Wadiche et al. 2006). Other populations of seizure-generated granule cells in most TLE models studied, however, show severe morphological abnormalities that might critically affect dentate connectivity. Basically, two features are altered in the course of seizure-induced neurogenesis: the formation of hilar basal dendrites and the ectopic migration of newborn granule cells into the polymorphic cell layer.

Granule cells that are born under normal conditions in the adult DG are highly polarized neurons. They have a single dendrite arising from the apical portion of the cell body branching in the distal granule cell layer (GCL) or proximal molecular layer (ML). Within the ML, excitatory synapses are formed onto granule cell dendrites by perforant path axons connecting dentate granule cells to the entorhinal cortex (van Praag et al. 2002; Schmidt-Hieber et al. 2004; Ge et al. 2006; Zhao et al. 2006). In striking contrast to cells born under normal conditions, a portion of seizure-generated granule cells extend an additional basal dendrite toward the hilus (Fig. 2) (Ribak et al. 2000; Dashtipour et al. 2003; Shapiro et al. 2005). During the immature DCX-expressing stage of neuronal development, hilar basal dendrites form nonspiny immature synapses onto mossy fiber axons (Shapiro and Ribak 2006). At later stages of granule cell development (>4 weeks postmitotic), hilar basal dendrites are covered with numerous spiny processes, among them mature mushroom spines (Jessberger et al. 2007a). Despite the abnormal integration that might result in the establishment of recurrent excitatory networks (Overstreet-Wadiche et al. 2006), granule cells that extend hilar basal dendrites become stably integrated into the dentate circuitry (Jessberger et al. 2007a), thus leading to lasting changes in dentate connectivity. The molecular mechanisms responsible for the extension of hilar basal dendrites are unknown but may involve an alteration in the glial scaffold (Shapiro et al. 2005). In contrast to hilar basal dendrites, sprouting of mossy fibers following seizures does not depend on newborn granule cells (Parent et al. 1999). Granule cells located in the GCL that were born following seizures do not send axonal sprouts to the ML; recent work suggests that adult-generated neurons that were born at least 4 weeks prior to SE induction do so (Jessberger et al. 2007a).

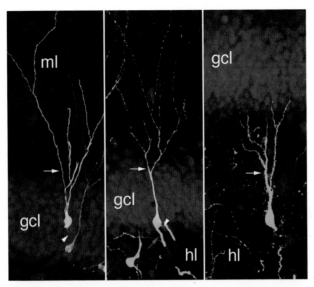

Figure 2. Neurogenesis following kainic-acid-induced seizures is largely aberrant. Retroviral labeling reveals the highly polarized morphology with a single apical dendrite (*arrow*) and an axon (*arrowhead*) extending from newborn granule cells that were born under normal conditions (*left panel*). Kainic-acid-induced seizures lead to the formation of a basal dendrite (*blue arrow*) in addition to the apical dendritic (*arrow*) and axonal processes (*arrowhead, middle panel*). Another portion of seizure-generated granule cells ectopically migrates into the hilus (*right panel*). Surprisingly, some ectopic granule cells morphologically appear to be very normal despite the aberrant localization. (gcl) Granule cell layer; (hl) hilus; (ml) molecular layer.

In addition to aberrant dendritic growth, SE alters the migration of adult-born neurons. Neuroblasts generated in the SVZ migrate more rapidly to the olfactory bulb, and a portion exits the migratory stream prematurely to enter nonolfactory forebrain regions (Parent et al. 2002). Few, if any, of the neuroblasts that reach the cortex appear to survive. In the DG of adult rats, a substantial fraction of seizure-generated granule cells ectopically migrates toward the hilar/CA3 border (Figs. 2 and 3) (Parent et al. 1997; Scharfman et al. 2000). Despite their abnormal localization, the intrinsic electrophysiological features of ectopic granule cells are remarkably similar to those of cells born under control conditions. However, ectopic granule cells burst in synchrony with CA3 pyramidal cells, indicating aberrant integration into the dentate circuitry after SE (Scharfman et al. 2002). Remarkably, ectopic cells still receive normal synaptic input from the perforant path (Scharfman et al. 2002, 2003).

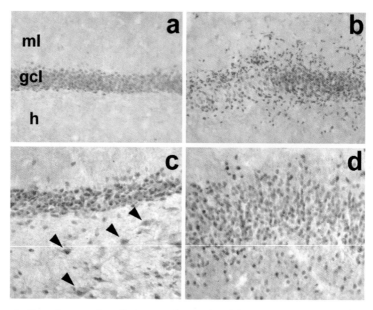

Figure 3. Ectopic granule cells in experimental and human TLE. (*Top panels*) Prox1 immunoreactivity in adult rat dentate gyrus (DG) 35 days after saline treatment (*a*) or pilocarpine-induced SE (*b*) shows many ectopic granule neurons in the epileptic rat (*b*) but not in the control (*a*). (*Bottom panels*) NeuN immunoreactivity in control human (*c*; temporal lobe tumor) and human TLE (*d*) DG shows granule cell layer (gcl) dispersion and ectopic granule-like neurons in the hilus (h) and molecular layer (ml) only in the patient with TLE who had mesial temporal sclerosis (*d*). Arrowheads (*c*) point to larger, NeuN-immunoreactive hilar neurons in the control tissue that are not seen in the TLE subject (*d*) due to hilar cell loss, or in either rat (*a,b*) because Prox1 is expressed specifically in dentate granule cells.

Why some seizure-generated granule cells migrate into the hilus remains unclear. Recent work suggests that SE induces abnormal chain migration of granule cell progenitors toward the hilus (Parent et al. 2006), and this aberrant migratory behavior may be initiated by loss of DG reelin expression (Gong et al. 2007). Whereas abnormal features such as basal dendrites and ectopic migration led to the hypothesis that seizure-induced neurogenesis contributes to the epileptic disease process (Parent 2002; Parent and Lowenstein 2002), recent evidence sheds new light on a potential compensatory role seizure-generated neurons might have. In contrast to SE induced by kainic acid or pilocarpine, electrical induction of SE leaves the morphology of seizure-generated granule cells relatively normal (Jakubs et al. 2006). In this model, seizure-generated granule cells have less excitatory but increased inhibitory input resulting in overall decreased excitability compared to newborn cells generated in running rats (Jakubs et al. 2006). Given this finding, the heightened excitability within the epileptic hippocampal circuitry might be compensated through more inhibited newborn granule cells. Without a doubt, the question of whether seizure-generated granule cells are part of the disease or an attempt of the injured brain to repair itself is far from being conclusively answered. However, future studies that reproduce consequences of seizure activity on adult neurogenesis in otherwise intact animals likely will help answer this critical question.

FUNCTIONAL RELEVANCE OF SEIZURE-INDUCED NEUROGENESIS

Seizure-associated neurogenesis may have a role in two features of TLE that are poorly understood from a mechanistic point of view. The first one is the progression from a brain insult to the clinical syndrome of epilepsy, a process called epileptogenesis (Dalby and Mody 2001; Magloczky and Freund 2005). What is the evidence that altered neurogenesis might be involved in epileptogenesis? Unquestionably, SE leads to heightened excitability and recurrent excitatory networks within the hippocampal circuitry. One reason for this might be the excitotoxic loss of inhibitory, GABAergic neurons, although debate is ongoing regarding the relevance of cell death or interneuronal disconnection in the establishment of an epileptic circuitry (Bernard et al. 1998; Dalby and Mody 2001; Sloviter et al. 2003; Ratzliff et al. 2004). Seizure-generated granule cells show two features that might indicate an epileptogenic role. Seizure-generated granule cells with hilar basal dendrites receive excitatory input from mossy fibers and could thus form a recurrent excitatory circuit (Fig. 4) (Shapiro and Ribak 2006). Similarly, ectopic granule cells appear to be abnormally synchronized with spontaneous, rhythmic bursts of CA3 pyramidal neurons (Scharfman et al. 2000). Compatible with these data is the finding that the reduction of seizure-generated neurons impairs epileptogenesis and reduces the frequency of spontaneous recurrent seizures (Jung et al. 2004, 2006). Abnormal network formation that results from altered neurogenesis after brain injury may also support seizure propagation or adversely influence seizure termination mechanisms. However, the image is not as clear-cut as other evidence speaks against an epileptogenic role of seizure-generated neurons and rather favors an antiepileptogenic effect (Jakubs et al. 2006).

Another consequence of TLE is the common occurrence of cognitive dysfunction that often leads to substantial morbidity (Helmstaedter et al. 2003; Elger et al. 2004). There is growing evidence that adult neurogenesis

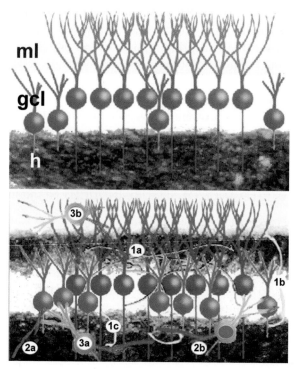

Figure 4. Aberrant integration of adult-born neurons after SE alters dentate gyrus (DG) circuitry. (*Top panel*) Schematic of the normal DG with mature granule neurons in the granule cell layer (gcl) closer to the molecular layer (ml) where they send their dendrites, and immature granule neurons closer to the hilus (h) into which their mossy fiber axons (*dark brown*) enter. In experimental TLE (*bottom panel*), SE increases neurogenesis and also alters the integration of differentiating neurons. Mossy fiber sprouting onto preexisting dentate granule cell dendrites (*dark brown* in ml) may result from plasticity of mature (*1a*) as well as immature (*1b*) granule cells. Mossy fibers also sprout aberrantly onto adult-born granule neurons with hilar basal dendrites (*1c*). Persistent hilar basal dendrites appear on granule neurons in both the gcl (*2b*) and h (*2c*). Some seizure-generated neurons migrate ectopically to reside in the h (*3a*) or ml (*3c*).

is involved in certain forms of hippocampus-dependent learning and memory under normal conditions (Shors et al. 2001, 2002; Santarelli et al. 2003; Snyder et al. 2005; Meshi et al. 2006; Saxe et al. 2006; Winocur et al. 2006). Seizure-generated neurons might contribute to cognitive impairment associated with TLE in three respects. First, the "normal" function of adult-generated neurons that might depend on specific plasticity of immature neurons (Schmidt-Hieber et al. 2004) might be disrupted due to an altered integration pattern (Overstreet-Wadiche et al. 2006, Jessberger et al. 2007a) or due to decreased levels of adult neurogenesis at late stages following prolonged seizure acitivity (Hattiangady et al. 2004). Second, the well-documented aberrant integration (hilar basal dendrites and ectopic granule cells) in most rodent models of TLE could critically interfere with synaptic transmission and information processing (Fig. 4). Supporting this hypothesis is the finding that inhibition of seizure-induced neurogenesis with the histone deacetylase (HDAC) inhibitor and antiepileptic drug valproic acid (VPA) protects kainic-acid-treated animals from impairment in a hippocampus-dependent object recognition task (Jessberger et al. 2007b). A third potential influence is that chronic suppression of neurogenesis in the epileptic brain (Hattiangady et al. 2004) could interfere with hippocampus-dependent learning and memory. As is the case for adult neurogenesis under normal conditions, however, the functional role of seizure-generated neurons in epileptogenesis or cognitive impairment will only be satisfyingly answered if ablation strategies with a higher specificity and selectivity than those that are currently available are developed.

SUMMARY

The finding that adult neurogenesis is altered dramatically during epileptogenesis challenges the conceptual understanding of the cause and consequences of TLE. Most rodent models of TLE are associated with severe alterations in morphology and connectivity of adult-born neurons such as the extension of hilar basal dendrites and the ectopic migration into the hilus of granule cell progenitors. To date, however, the role that new neurons actually have in the epileptic disease process remains unclear. Simply put, are they good or bad? In the end, both descriptions may prove valid in that some aspects of seizure-induced neurogenesis might be beneficial and others harmful for the epileptic brain. In any case, the characterization and identification of cellular and molecular mechanisms underlying seizure-induced neurogenesis will bring us one step further toward understanding ongoing plasticity in the adult brain. Without a doubt, seizure-induced neurogenesis represents a promising target for intervention in the treatment of human epilepsy and its comorbidities.

REFERENCES

Bengzon J., Kokaia Z., Elmér E., Nanobashvili A., Kokaia M., and Lindvall O. 1997. Apoptosis and proliferation of dentate gyrus neurons after single and intermittent limbic seizures. *Proc. Natl. Acad. Sci.* **94:** 10432–10437.

Bernard C., Esclapez M., Hirsch J.C., and Ben-Ari Y. 1998. Interneurones are not so dormant in temporal lobe epilepsy: A critical reappraisal of the dormant basket cell hypothesis. *Epilepsy Res.* **32:** 93–103.

Blumcke I., Beck H., Lie A.A., and Wiestler O.D. 1999. Molecular neuropathology of human mesial temporal lobe epilepsy. *Epilepsy Res.* **36:** 205–223.

Blume W.T. 2006. The progression of epilepsy. *Epilepsia* (suppl. 1) **47:** 71–78.

Buckmaster P.S. 2004. Laboratory animal models of temporal lobe epilepsy. *Comp. Med.* **54:** 473–485.

Dalby N.O. and Mody I. 2001. The process of epileptogenesis: A pathophysiological approach. *Curr. Opin. Neurol.* **14:** 187–192.

Dashtipour K., Wong A.M., Obenaus A., Spigelman I., and Ribak C.E. 2003. Temporal profile of hilar basal dendrite formation on dentate granule cells after status epilepticus. *Epilepsy Res.* **54:** 141–151.

Deisseroth K., Singla S., Toda H., Monje M., Palmer T.D., and Malenka R.C. 2004. Excitation-neurogenesis coupling in adult neural stem/progenitor cells. *Neuron* **42:** 535–552.

Ekdahl C.T., Claasen J.H., Bonde S., Kokaia Z., and Lindvall O. 2003. Inflammation is detrimental for neurogenesis in adult brain. *Proc. Natl. Acad. Sci.* **100:** 13632–13637.

Elger C.E., Helmstaedter C., and Kurthen M. 2004. Chronic epilepsy and cognition. *Lancet Neurol.* **3:** 663–672.

Elliott R.C. and Lowenstein D.H. 2004. Gene expression profiling of seizure disorders. *Neurochem. Res.* **29:** 1083–1092.

Gall C.M. 1993. Seizure-induced changes in neurotrophin expression: Implications for epilepsy. *Exp. Neurol.* **124:** 150–166.

Ge S., Goh E.L., Sailor K.A., Kitabatake Y., Ming G.L., and Song H. 2006. GABA regulates synaptic integration of newly generated neurons in the adult brain. *Nature* **439:** 589–593.

Gong C., Wang T.W., Huang H.S., and Parent J.M. 2007. Reelin regulates neuronal progenitor migration in intact and epileptic hippocampus. *J. Neurosci.* **27:** 1803–1811.

Gould E., Cameron H.A., and McEwen B.S. 1994. Blockade of NMDA receptors increases cell death and birth in the developing rat dentate gyrus. *J. Comp. Neurol.* **340:** 551–565.

Gray W.P. and Sundstrom L.E. 1998. Kainic acid increases the proliferation of granule cell progenitors in the dentate gyrus of the adult rat. *Brain Res.* **790:** 52–59.

Hattiangady B., Rao M.S., and Shetty A.K. 2004. Chronic temporal lobe epilepsy is associated with severely declined dentate neurogenesis in the adult hippocampus. *Neurobiol. Dis.* **17:** 473–490.

Helmstaedter C. 2002. Effects of chronic epilepsy on declarative memory systems. *Prog. Brain Res.* **135:** 439–453.

Helmstaedter C., Kurthen M., Lux S., Reuber M., and Elger C.E. 2003. Chronic epilepsy and cognition: A longitudinal study in temporal lobe epilepsy. *Ann. Neurol.* **54:** 425–432.

Huang Y., Doherty J.J., and Dingledine R. 2002. Altered histone acetylation at glutamate receptor 2 and brain-derived neurotrophic factor genes is an early event triggered by status epilepticus. *J. Neurosci.* **22:** 8422–8428.

Huttmann K., Sadgrove M., Wallraff A., Hinterkeuser S., Kirchhoff F., Steinhauser C., and Gray W.P. 2003. Seizures preferentially stimulate proliferation of radial glia-like astrocytes in the adult dentate gyrus: Functional and immunocytochemical analysis. *Eur. J. Neurosci.* **18:** 2769–2778.

Isackson P.J., Huntsman M.M., Murray K.D., and Gall C.M. 1991. BDNF mRNA expression is increased in adult rat forebrain after limbic seizures: Temporal patterns of induction distinct from NGF. *Neuron* **6:** 937–948.

Jakubs K., Nanobashvili A., Bonde S., Ekdahl C.T., Kokaia Z., Kokaia M., and Lindvall O. 2006. Environment matters: Synaptic properties of neurons born in the epileptic adult brain develop to reduce excitability. *Neuron* **52**: 1047–1059.

Jessberger S., Romer B., Babu H., and Kempermann G. 2005. Seizures induce proliferation and dispersion of doublecortin-positive hippocampal progenitor cells. *Exp. Neurol.* **196**: 342–351.

Jessberger S., Zhao C., Toni N., Clemenson G.D., Li Y., and Gage F.H. 2007a. Aberrant integration of seizure-generated granule cells. *J. Neurosci.* (in press).

Jessberger S., Nakashima K., Clemenson G.D., Mejia E., Mathews E., Ure K., Ogawa S., Sinton C., Gage F.H., and Hsieh J. 2007b. Epigenetic modulation of seizure-induced neurogenesis and cognitive decline. *J. Neurosci.* (in press).

Jung K.H., Chu K., Kim M., Jeong S.W., Song Y.M., Lee S.T., Kim J.Y., Lee S.K., and Roh J.K. 2004. Continuous cytosine-*b*-D-arabinofuranoside infusion reduces ectopic granule cells in adult rat hippocampus with attenuation of spontaneous recurrent seizures following pilocarpine-induced status epilepticus. *Eur. J. Neurosci.* **19**: 3219–3226.

Jung K.H., Chu K., Lee S.T., Kim J., Sinn D.I., Kim J.M., Park D.K., Lee J.J., Kim S.U., Kim M., Lee S.K., and Roh J.K. 2006. Cyclooxygenase-2 inhibitor, celecoxib, inhibits the altered hippocampal neurogenesis with attenuation of spontaneous recurrent seizures following pilocarpine-induced status epilepticus. *Neurobiol. Dis.* **23**: 237–246.

Magloczky Z. and Freund T.F. 2005. Impaired and repaired inhibitory circuits in the epileptic human hippocampus. *Trends Neurosci.* **28**: 334–340.

Meshi D., Drew M.R., Saxe M., Ansorge M.S., David D., Santarelli L., Malapani C., Moore H., and Hen R. 2006. Hippocampal neurogenesis is not required for behavioral effects of environmental enrichment. *Nat. Neurosci.* **9**: 729–731.

Mohapel P., Ekdahl C.T., and Lindvall O. 2004. Status epilepticus severity influences the long-term outcome of neurogenesis in the adult dentate gyrus. *Neurobiol. Dis.* **15**: 196–205.

Newton S.S., Collier E.F., Hunsberger J., Adams D., Terwilliger R., Selvanayagam E., Duman R.S. 2003. Gene profile of electroconvulsive seizures: Induction of neurotrophic and angiogenic factors. *J. Neurosci.* **23**: 10841–10851.

Overstreet-Wadiche L.S., Bromberg D.A., Bensen A.L., and Westbrook G.L. 2006. Seizures accelerate functional integration of adult-generated granule cells. *J. Neurosci.* **26**: 4095–4103.

Parent J.M. 2002. The role of seizure-induced neurogenesis in epileptogenesis and brain repair. *Epilepsy Res.* **50**: 179–189.

Parent J.M. and Lowenstein D.H. 2002. Seizure-induced neurogenesis: Are more new neurons good for an adult brain? *Prog. Brain Res.* **135**: 121–131.

Parent J.M., Valentin V.V., and Lowenstein D.H. 2002. Prolonged seizures increase proliferating neuroblasts in the adult rat subventricular zone-olfactory bulb pathway. *J. Neurosci.* **22**: 3174–3188.

Parent J.M., Tada E., Fike J.R., and Lowenstein D.H. 1999. Inhibition of dentate granule cell neurogenesis with brain irradiation does not prevent seizure-induced mossy fiber synaptic reorganization in the rat. *J. Neurosci.* **19**: 4508–4519.

Parent J.M., Elliott R.C., Pleasure S.J., Barbaro N.M., and Lowenstein D.H. 2006. Aberrant seizure-induced neurogenesis in experimental temporal lobe epilepsy. *Ann. Neurol.* **59**: 81–91.

Parent J.M., Yu T.W., Leibowitz R.T., Geschwind D.H., Sloviter R.S., and Lowenstein D.H. 1997. Dentate granule cell neurogenesis is increased by seizures and contributes to aberrant network reorganization in the adult rat hippocampus. *J. Neurosci.* **17**: 3727–3738.

Ratzliff A.H., Howard A.L., Santhakumar V., Osapay I., and Soltesz I. 2004. Rapid deletion of mossy cells does not result in a hyperexcitable dentate gyrus: Implications for epileptogenesis. *J. Neurosci.* **24:** 2259–2269.

Ribak C.E., Tran P.H., Spigelman I., Okazaki M.M., and Nadler J.V. 2000. Status epilepticus-induced hilar basal dendrites on rodent granule cells contribute to recurrent excitatory circuitry. *J. Comp. Neurol.* **428:** 240–253.

Santarelli L., Saxe M., Gross C., Surget A., Battaglia F., Dulawa S., Weisstaub N., Lee J., Duman R., Arancio O., Belzung C., and Hen R. 2003. Requirement of hippocampal neurogenesis for the behavioral effects of antidepressants. *Science* **301:** 805–809.

Saxe M.D., Battaglia F., Wang J.W., Malleret G., David D.J., Monckton J.E., Garcia A.D., Sofroniew M.V., Kandel E.R., Santarelli L., Hen R., and Drew M.R. 2006. Ablation of hippocampal neurogenesis impairs contextual fear conditioning and synaptic plasticity in the dentate gyrus. *Proc. Natl. Acad. Sci.* **103:** 17501–17506.

Scharfman H.E., Goodman J.H., and Sollas A.L. 2000. Granule-like neurons at the hilar/CA3 border after status epilepticus and their synchrony with area CA3 pyramidal cells: Functional implications of seizure-induced neurogenesis. *J. Neurosci.* **20:** 6144–6158.

Scharfman H.E., Sollas A.L., and Goodman J.H. 2002. Spontaneous recurrent seizures after pilocarpine-induced status epilepticus activate calbindin-immunoreactive hilar cells of the rat dentate gyrus. *Neuroscience* **111:** 71–81.

Scharfman H.E., Sollas A.E., Berger R.E., Goodman J.H., and Pierce J.P. 2003. Perforant path activation of ectopic granule cells that are born after pilocarpine-induced seizures. *Neuroscience* **121:** 1017–1029.

Schmidt-Hieber C., Jonas P., and Bischofberger J. 2004. Enhanced synaptic plasticity in newly generated granule cells of the adult hippocampus. *Nature* **429:** 184–187.

Scott B.W., Wang S., Burnham W.M., De Boni U., and Wojtowicz J.M. 1998. Kindling-induced neurogenesis in the dentate gyrus of the rat. *Neurosci. Lett.* **248:** 73–76.

Shapiro L.A. and Ribak C.E. 2006. Newly born dentate granule neurons after pilocarpine-induced epilepsy have hilar basal dendrites with immature synapses. *Epilepsy Res.* **69:** 53–66.

Shapiro L.A., Korn M.J., and Ribak C.E. 2005. Newly generated dentate granule cells from epileptic rats exhibit elongated hilar basal dendrites that align along GFAP-immuno-labeled processes. *Neuroscience* **136:** 823–831.

Shors T.J., Townsend D.A., Zhao M., Kozorovitskiy Y., and Gould E. 2002. Neurogenesis may relate to some but not all types of hippocampal-dependent learning. *Hippocampus* **12:** 578–584.

Shors T.J., Miesegaes G., Beylin A., Zhao M., Rydel T., and Gould E. 2001. Neurogenesis in the adult is involved in the formation of trace memories. *Nature* **410:** 372–376.

Sloviter R.S., Zappone C.A., Harvey B.D., Bumanglag A.V., Bender R.A., and Frotscher M. 2003. "Dormant basket cell" hypothesis revisited: Relative vulnerabilities of dentate gyrus mossy cells and inhibitory interneurons after hippocampal status epilepticus in the rat. *J. Comp. Neurol.* **459:** 44–76.

Snyder J.S., Hong N.S., McDonald R.J., and Wojtowicz J.M. 2005. A role for adult neurogenesis in spatial long-term memory. *Neuroscience* **130:** 843–852.

Tozuka Y., Fukuda S., Namba T., Seki T., and Hisatsune T. 2005. GABAergic excitation promotes neuronal differentiation in adult hippocampal progenitor cells. *Neuron* **47:** 803–815.

van Praag H., Schinder A.F., Christie B.R., Toni N., Palmer T.D., and Gage F.H. 2002. Functional neurogenesis in the adult hippocampus. *Nature* **415:** 1030–1034.

von Lehe M., Lutz M., Kral T., Schramm J., Elger C.E., and Clusmann H. 2006. Correlation of health-related quality of life after surgery for mesial temporal lobe epilepsy with two seizure outcome scales. *Epilepsy Behav.* **9:** 73–82.

Winocur G., Wojtowicz J.M., Sekeres M., Snyder J.S., and Wang S. 2006. Inhibition of neurogenesis interferes with hippocampus-dependent memory function. *Hippocampus* **16:** 296–304.

Zhao C., Teng E.M., Summers R.G., Jr., Ming G.L., and Gage F.H. 2006. Distinct morphological stages of dentate granule neuron maturation in the adult mouse hippocampus. *J. Neurosci.* **26:** 3–11.

26

Neurogenesis following Stroke Affecting the Adult Brain

Olle Lindvall[1,2] and Zaal Kokaia[2,3]
[1]Laboratory of Neurogenesis and Cell Therapy
Section of Restorative Neurology
Wallenberg Neuroscience Center
[2]Laboratory of Neural Stem Cell Biology
Section of Restorative Neurology
University Hospital, SE-221 84 Lund, Sweden
[3]Lund Strategic Research Center for Stem Cell Biology
and Cell Therapy, Lund, Sweden

STROKE IS CAUSED BY OCCLUSION OF A CEREBRAL ARTERY, which gives rise to focal ischemia with irreversible injury in a core region and partially reversible damage in the surrounding penumbra zone. In another type of insult, abrupt and near-total interruption of cerebral blood flow as a consequence of cardiac arrest or coronary artery occlusion leads to global ischemia and selective death of certain vulnerable neuronal populations such as the pyramidal neurons of hippocampal CA1. During the last decade, these ischemic insults have been reported to induce the formation of new neurons in the adult rodent brain from neural stem cells (NSCs) located in two regions: the subventricular zone (SVZ), lining the lateral ventricle, and the subgranular zone (SGZ) in the dentate gyrus (DG). Ischemia-induced neurogenesis is triggered both in areas where new neurons are normally formed, such as the DG, and in areas that are nonneurogenic in the intact brain, e.g., the striatum. These findings have raised several important issues: (1) Is the evidence for the formation of new neurons really solid or could there be other interpretations such as aberrant DNA synthesis caused by the ischemic insult in already existing, mature neurons? (2) What are the functional consequences of ischemia-induced neurogenesis? (3) Because the neurogenic response is minor and recovery after stroke incomplete, how can this presumed self-repair mechanism be boosted?

In this chapter, we summarize the current status of research on neurogenesis after stroke. We also discuss the basic scientific problems that need to be addressed before this potential self-repair mechanism should be considered in a clinical-therapeutic perspective. Stroke is a leading cause of chronic disability in humans. No effective treatment to promote recovery in patients exists. Many different types of neurons and glial cells die in stroke. To repair the stroke-damaged brain may therefore seem unrealistic. However, even reestablishment of only a fraction of damaged neuronal circuitries could have important clinical implications. We focus here on the stroke-induced formation of new neurons in damaged areas where neurogenic mechanisms do not normally operate. The modulation of neurogenesis in the DG by focal ischemic stroke, as well as by many other physiological and pathological conditions (see, e.g., Abrous et al. 2005), is not covered here.

ANIMAL MODELS OF STROKE

Experimental models, which mimic the conditions during ischemic stroke in humans, have been developed in animals. These models cause motor, sensory, and cognitive deficits similar to what is observed in stroke patients, and studies in postmortem specimens confirm the relevance of the animal models for the human condition (Leifer and Kowall 1993). In the most common model for neurogenesis research, stroke is induced by transient middle cerebral artery occlusion (MCAO) in rats and mice, which is accomplished by insertion of a filament through the internal carotid artery to the origin of the middle cerebral artery (MCA). Recirculation is restored by withdrawal of the filament. Depending on the duration of the occlusion, either only the dorsolateral part of the rat striatum will be damaged (30 minutes MCAO) or the lesion will extend into the overlying parietal cortex (2 hours MCAO). In another model of ischemic stroke, the MCA of spontaneously hypertensive rats is ligated with a thread distal to the origin of the striatal branches. In normal rats, the same procedure is combined with bilateral occlusion of the carotid arteries during about 1 hour. Both methods lead to selective ischemic lesions of the cerebral cortex without damage to the striatum.

Other animal models of stroke are also being used to study neurogenesis. Subjecting exposed crania of rats to a ring-shaped laser-irradiation beam with simultaneous systemic infusion of photosensitizer induces photothrombotic stroke with region-at-risk cortical tissue located within the ischemic ring locus (Gu et al. 1999). Wiping the pia and attached blood vessels from the cortical surface causes permanent devascularization and damage to the cerebral cortex (Gonzalez and Kolb 2003). Embolic ischemic lesions to the striatum and cerebral cortex are induced by placing a blood clot at the origin of the MCA (Zhang et al. 1997).

EVIDENCE FOR STRIATAL NEUROGENESIS AFTER STROKE IN ANIMALS

The observation that neurogenesis after stroke occurs in areas of the adult brain where new neurons are not normally formed has raised a lot of questions about the solidity of the data and alternative explanations. The main evidence for striatal neurogenesis (Fig. 1A) following stroke in animal models can be summarized as follows: (1) Cells coexpressing the thymidine analog bromodeoxyuridine (BrdU), given intraperitoneally after stroke, and markers of immature (e.g., doublecortin [DCX], PSA-NCAM [polysialic-acid-neural cell adhesion molecule], Hu) and mature neurons (e.g., NeuN and DARPP-32) are detected in the damaged striatum (Arvidsson et al. 2002; Parent et al. 2002). After BrdU injections, given early or late after the insult, the colabeled cells are first DCX-immunoreactive over 2–3 weeks but then gradually lose this expression and become labeled with NeuN, consistent with a maturation process (Thored et al. 2006). (2) Cells expressing neuroblast markers such as DCX are detected in the stroke-damaged striatum and coexpress transcription factors specific for developing striatal projection neurons, i.e., Pbx and Meis2 (Arvidsson et al. 2002). (3) Injection of Cre-encoding plasmid into the lateral ventricles of transgenic mice carrying a floxed green fluorescent protein (GFP) gene, specifically labeling SVZ cells and their progeny, gives rise to GFP-labeled cells in the striatum coexpressing NeuN and having a mature neuronal morphology at 90 days after stroke (Yamashita et al. 2006).

A single study has so far used transgenic Nestin-CreER mice to trace Nestin[+] progenitors and their progeny after tamoxifen induction in a stroke model (Burns et al. 2007). At 1 week after MCAO, only very few BrdU[+] cells also expressed GFP, and no GFP[+] cells coexpressed the neuroblast marker DCX. These results indicated very little contribution of cell replacement from SVZ progenitors. However, neither the extent of the damage (which markedly influences the magnitude of neurogenesis after stroke) nor the number of DCX[+] neuroblasts recruited to striatum was reported. The contribution of SVZ progenitors to striatal neurogenesis can only be conclusively assessed when a substantial number of

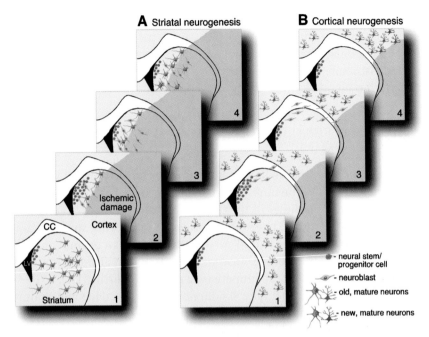

Figure 1. Schematic representation of stroke-induced neurogenesis in the striatum (*A*) and cerebral cortex (*B*). (*A*) Neural stem or progenitor cells reside in the subventricular zone (SVZ) adjacent to the lateral ventricle (*1*). Focal ischemic insults, which lead to pronounced loss of striatal and cortical neurons, give rise to increased proliferation of progenitors (*2*). Neuroblasts formed after, and to some extent before, the stroke migrate to the damaged part of the striatum (*3*), where they differentiate and express markers specific for mature striatal projection neurons (*4*). The migration of neuroblasts and formation of new neurons continue during 1 year after the insult. (*B*) Focal ischemia-induced damage to the cortex triggers proliferation of SVZ progenitors (*2*). Neuroblasts migrate toward the cortical damage (*3*) and, in the peri-infarct area, start to express markers of mature neurons (*4*). In some studies, evidence for cortical neurogenesis has been obtained in animals with ischemic damage in both striatum and cortex.

neuroblasts are detected in the striatum. Moreover, the Nestin-CreER system does not label Nestin$^+$ SGZ progenitors, and it cannot be excluded that the population of SVZ progenitors activated after stroke is also undetectable with this system.

EVIDENCE FOR CORTICAL NEUROGENESIS AFTER STROKE IN ANIMALS

Selective apoptotic degeneration of neurons in cerebral cortex leads to the formation of new corticothalamic neurons (Magavi et al. 2000).

However, in the initial studies describing stroke-induced striatal neuro-genesis, significant numbers of new neurons in the ischemic cerebral cortex were not detected (R.L. Zhang et al. 2001; Arvidsson et al. 2002; Parent et al. 2002). Evidence is now emerging that under certain circumstances, limited accumulation of new neurons may occur in the cerebral cortex after stroke (Fig. 1B). Jin et al. (2003) reported migration of DCX^+ neuroblasts from SVZ to the boundaries of the cortical lesion. However, whether these neuroblasts were formed in response to the ischemic insult was unclear (Jin et al. 2003). Several other studies have described the presence of new cells expressing markers of neuroblasts or mature neurons in the ischemia-damaged cerebral cortex (Jiang et al. 2001; R. Zhang et al. 2006; Ziv et al. 2007). It has also been shown that cortical neurogenesis after stroke can be triggered (Kolb et al. 2007) or enhanced (Taguchi et al. 2004; R. Zhang et al. 2006; Wang et al. 2007; Ziv et al. 2007) by additional manipulation, e.g., growth factor infusion. The most convincing evidence for cortical neurogenesis after stroke was recently presented by Leker et al. (2007) in the distal MCAO model, which causes selective cortical damage. When BrdU was injected during days 1–5 after the insult, the majority of $BrdU^+$ cells in the peri-infarct cortex at 1 week were expressing markers for NSCs such as Sox2 and Nestin. However, at 3 months after the stroke, about 12% of $BrdU^+$ cells were also $NeuN^+$, indicating maturation, at least partly, of NSCs into neurons. The larger variability in the magnitude of cortical neurogenesis as compared to striatal neurogenesis in different studies could be related to the stroke model used, the pattern, location, extent, and dynamics of the ischemic lesion, and the resulting differences in the cues responsible for migration and neuronal differentiation of the new cells.

EVIDENCE FOR NEUROGENESIS AFTER STROKE IN HUMANS

The SVZ has been identified as the main source of neuroblasts that are generated after stroke and migrate toward the damage in rodents. In addition, the SVZ in the adult human brain contains NSCs (Eriksson et al. 1998; Sanai et al. 2004; Quinones-Hinojosa et al. 2006), and, similar to rodents, progenitor cells and neuroblasts reach the olfactory bulb via the rostral migratory stream to become mature neurons (Curtis et al. 2007). Thus, also in humans, the SVZ could have a latent potential for producing new neurons for repair of the stroke-damaged brain. However, the evidence for stroke-induced neurogenesis in the human brain after stroke is so far very limited. Suggesting a neurogenic response, Macas et al. (2006) found an increased number of cells expressing the proliferation marker Ki-67 in the ipsilateral SVZ and of neuronal progenitor cells in the parenchyma close to the ventricular wall. Jin et al. (2006) reported that cells expressing markers of cell proliferation and immature neurons, which were preferentially located close to blood vessels, were found in the area surrounding cortical infarcts. One attractive possibility to reveal whether stroke-induced neurogenesis is significant in the human brain could be to measure the levels of ^{14}C for birth dating of neurons in patients, as has been done in healthy individuals and shown absence of cortical neurogenesis (Bhardwaj et al. 2006).

MECHANISMS INVOLVED IN STROKE-INDUCED NEUROGENESIS

Proliferation

Overall cell proliferation in the ipsilateral rat SVZ is increased 1–2 weeks after MCAO (Jin et al. 2001; Arvidsson et al. 2002; Zhang et al. 2004a,b) but returned to baseline at 6 weeks (Thored et al. 2006). However, the ipsilateral SVZ is expanded at 2, 6, and 16 weeks after MCAO (Thored et al. 2006), and the number and size of neurospheres isolated from the ipsilateral SVZ are increased at 6 weeks. Taken together, these findings provide evidence that stroke leads to long-term alterations in the stem cell niche in the SVZ.

During embryonic development, the orientation of mitotic cleavage regulates neurogenesis. Zhang et al. (2004a) have shown that not only the number of mitotic cells increases, but also the cleavage orientation in the SVZ changes from horizontal to vertical during the first week after stroke in the adult brain. Thus, stroke leads to a transient alteration in SVZ cell division from asymmetric to symmetric, thereby expanding the pool of progenitor cells. Stroke was also found to increase the proportion of dividing cells in the SVZ and reduce their cell cycle length, largely due to a shortening of the G_1 phase (R.L. Zhang et al. 2006). These actively proliferating cells were most likely progenitor cells. Whether stroke influences proliferation and cell cycle length of SVZ NSCs, which constitute about 2% of the SVZ cell population in adult rodents (Morshead et al. 1994; Doetsch et al. 1997), is not known.

Interestingly, Zhang et al. (2007b) have recently demonstrated that the ependymal cells lining the lateral ventricular wall proliferate acutely and transiently after stroke. These authors also provide evidence that the ependymal cells are transformed into radial glia cells and that these cells are surrounded by type-C (progenitors) and type-A (neuroblasts) cells in the SVZ. Whether the ependymal cell-derived radial glial cells really act as neuronal progenitors following stroke and give rise to striatal neurons is not known.

Both Notch1 and the naturally occurring activator, Jagged1, are expressed in the adult SVZ (Stump et al. 2002). Reduced Jagged1/Notch1 signaling decreased SVZ cell proliferation in mice (Nyfeler et al. 2005). Conversely, administration of Notch ligands increased the number of proliferating cells in SVZ and new cells expressing the immature neuronal marker Hu in the cerebral cortex of intact rats (Androutsellis-Theotokis et al. 2006). Also following cortical stroke, supplying Notch ligands stimulated cell proliferation in SVZ (Androutsellis-Theotokis et al. 2006). Taken together, these findings indicate that Notch signaling is important for NSC maintenance in the SVZ and suggest that Notch ligands may be useful to enhance stem cell expansion during the recovery phase after stroke.

Survival

About 80% of the stroke-generated striatal neurons die during the first 2 weeks after their formation (Arvidsson et al. 2002). Caspases are probably involved in this neuronal loss, and administration of caspase inhibitors markedly promotes neuroblast survival in the damaged striatum (Thored et al. 2006). It is conceivable that the inflammatory changes accompanying the ischemic damage contribute to the poor survival of the new striatal neurons similar to what has been described for new hippocampal neurons in an inflammatory environment (Ekdahl et al. 2003; Monje et al. 2003). In agreement, Hoehn et al. (2005) found that administration of the nonsteroidal anti-inflammatory drug indomethacin increased the number of neuroblasts in the striatum after stroke. In addition, Liu et al. (2007) recently reported that minocycline, administered during 4 weeks after transient MCAO in rats, reduced the number of activated microglia and increased the number of new, mature neurons in the DG.

The role of inflammation for stroke-induced neurogenesis is, however, complex, and microglia can most likely also be beneficial, e.g., by producing factors that are important for migration of newly generated neuroblasts (see discussion below). Interestingly, we have recently demonstrated the presence of an increased number of activated microglia in the SVZ during several months concomitantly with long-term neurogenesis (W. Gomes Leal et al., unpubl.). It is tempting to speculate that factors released from these microglia could contribute to the maintenance of the neurogenic response.

It has been proposed that autoimmune, central nervous system (CNS)-specific T cells, by interacting with resident microglia, can promote progenitor proliferation in the SVZ and possibly also neuronal survival and differentiation (Ziv et al. 2006). Modulation of the immune system by down-regulation of the activity of regulatory T cells, using sub-cutaneous injection of the copolymer poly-YE directly after induction of permanent MCAO in rats, increased hippocampal and induced cortical neurogenesis (Ziv et al. 2007). This procedure probably enhanced the rapid recruitment of the relevant T cells recognizing CNS antigens.

Recruitment to Damaged Area

The new neuroblasts exhibit distinct migratory behaviors and can divide on their way through the striatum toward the damage (Zhang et al. 2007a). Radial glia, which are distributed in both the SVZ and the ischemic stria-tum after stroke, are probably involved in guiding the migrating neuro-blasts, as evidenced by both in vivo and in vitro data (Zhang et al. 2007b). There is also a close association between the newly formed neurons and the vasculature when they migrate toward the damaged area. At 18 days following 30 minutes MCAO in mice, Yamashita et al. (2006) found chains of neuroblasts wound around endothelial cells, the neuroblasts being elon-gated parallel to the blood vessels and the chains oriented in the mediolat-eral direction. In another model, at 7 days after focal cortical stroke in mice (Ohab et al. 2006), neuroblasts were located in large numbers in physical proximity to endothelial cells in the peri-infarct cortex, where active vas-cular remodeling occurred. We have found (P. Thored et al., unpubl.) that the newly formed neuroblasts are located in the vicinity of blood vessels throughout long-term neurogenesis (at least up to 16 weeks after the insult). The majority of the neuroblasts migrated toward the ischemic dam-age through the striatal area, which selectively, as compared to other stri-atal areas adjacent to SVZ, exhibited both long-lasting increase of vessel density and, during the first 2 weeks after the insult, endothelial cell pro-liferation, consistent with angiogenesis. The blood vessels probably have a role both for migration and survival and differentiation of the closely located neuroblasts by endothelial release of factors such as stromal-cell-derived factor-1α (SDF-1α) and brain-derived neurotrophic factor (BDNF) (Louissaint et al. 2002; Ohab et al. 2006; Wang et al. 2006).

The molecular mechanisms involved in directing the new neurons to the damaged area are only partly known. The most well-established mechanism regulating migration is the inflammatory chemoattractant SDF-1α and its cognate receptor CXCR4. Following stroke, SDF-1α is expressed in reactive astrocytes and activated microglia in the damaged area (Thored et al. 2006), and SDF-1α levels are increased in the stroke hemisphere (Robin et al. 2006). CXCR4 is expressed on the neuronal pro-genitors and stroke-generated neuroblasts (Robin et al. 2006). When blocking CXCR4, the migration of neuroblasts was markedly attenuated (Robin et al. 2006; Thored et al. 2006). Taken together, these data indicate that SDF-1α/CXCR4 signaling regulates the directed migration of new striatal neurons toward the ischemic damage.

Another factor of importance for neuroblast migration after stroke is monocyte chemoattractant protein-1 (MCP-1) (Yan et al. 2006). MCP-1 is up-regulated in activated microglia and reactive astrocytes in the cortex and striatum, and the new neuroblasts express the MCP-1 receptor CCR2 after stroke. Arguing for involvement of this factor in the migration, transgenic mice lacking MCP-1 or CCR2 showed less recruitment of neuroblasts into the striatum after stroke (Yan et al. 2006).

An important question is how the newly formed neuroblasts can migrate through the ischemic and partially damaged striatum. Lee et al. (2006) have recently reported that the extracellular protease matrix metalloproteinase-9 (MMP-9) is colocalized with the neuroblast marker DCX in the SVZ and striatum following stroke in mice. Administration of an MMP inhibitor markedly reduced neuroblast migration. Taken together, these data indicate that extracellular proteolysis through the action of MMP is involved in neuroblast migration after stroke (Lee et al. 2006).

Erythropoietin (EPO) administered to the stroke-damaged brain promotes neurogenesis (Table 1). However, endogenous EPO also seems to be involved in the regulation of neurogenesis after stroke. Conditional knockdown of the EPO receptor in mice subjected to cortical stroke led to reduced SVZ cell proliferation and impaired continued migration of neuroblasts to peri-infarct cortex (Tsai et al. 2006).

The extension of the damage has an important role for the degree of recruitment of new neuroblasts to the striatum after stroke. Following 30 minutes MCAO, which gave rise to a lesion restricted to the striatum, the number of new striatal neuroblasts was markedly fewer as compared to a 2-hour insult, which caused more extensive striatal damage and, in addition, injury to the overlying parietal cortex (Thored et al. 2006). The number of neuroblasts correlated significantly with the volume of striatal injury. Thus, the extent of the ischemic lesion influences the degree of activation of the molecular and cellular mechanisms that promote recruitment of new striatal neurons after stroke.

Stroke mainly affects older people; it is therefore of major clinical importance whether new neurons can also be recruited to the ischemically damaged area in the aged brain. No differences between young (3 months) and old (15 months) rats were observed in number and distributional pattern of new neuroblasts and mature neurons in the damaged striatum after stroke (Darsalia et al. 2005). Thus, the stroke-damaged aged brain

Table 1. Experimental studies indicating strategies to promote neurogenesis in nonneurogenic areas of the adult brain following focal ischemic stroke

Model/duration	Species/strain	Compound/treatment	Effect on neurogenesis	Effects on functional recovery	Reference
Embolic MCAO/permanent	Rat/Wistar	nitric oxide donor/intravenous	proliferation (BrdU$^+$) ↑ in SVZ	improved rotorod and adhesive tape removal tests	R. Zhang et al. (2001)
Embolic MCAO/permanent	Rat/Wistar	Sildenafil/orally	proliferation (BrdU$^+$) ↑ in SVZ, neuroblasts (Tuj1+) ↑ in SVZ and striatum	improved foot-fault and adhesive tape removal tests	Zhang et al. (2002)
MCAO/2 hr	Rat/Wistar	statins/orally	BrdU$^+$ cells, BrdU$^+$/Tuj1+ neuroblasts ↑ in SVZ	improved adhesive tape removal tests, Neurological Severity Score	Chen et al. (2003)
MCAO/30 min	Rat/Wistar	AAV-BDNF/intranigral	neuroblasts (DCX$^+$) ↑ in striatum	n.d.	Gustafsson et al. (2003)
MCAO/20 min	Mouse/129S6/SvEvTac	EGF	neuroblasts (DCX$^+$), new mature neurons (BrdU$^+$/NeuN$^+$) ↑ in striatum	n.d.	Teramoto et al. (2003)
MCAO/90 min	Rat/Sprague-Dawley	VEGF/intraventricular	progenitor survival (BrdU$^+$) ↑ in SVZ	improved Neurological Severity Score	Sun et al. (2003)
MCAO/90 min	Rat/Sprague-Dawley	HB-EGF/intraventricular	proliferation (BrdU$^+$) ↑ in SVZ; neuroblast migration (DCX$^+$) ↓ in striatum	improved Neurological Severity Score	Jin et al. (2004)

Model	Species/Strain	Treatment	Findings	Functional outcome	Reference
Distal MCAO/permanent	Mouse/SCID	CD34+ cord blood cells/intravenous	neuroblasts (PSA-NCAM+) and BrdU+/NeuN+ mature neurons ↑ in cortex	decreased and increased rearing and locomotion in light and dark, respectively; normalized hyperactivity and auditory stimulation-induced excessive startle	Taguchi et al. (2004)
Embolic MCAO/permanent	Rat/Wistar	EPO/intraperitonial	BrdU+, Nestin+, DCX+ cells ↑ in SVZ	improved foot-fault and corner tests	Wang et al. (2004)
MCAO/2 hr	Rat/Sprague-Dawley	indomethacin/orally	new (BrdU+) cells coexpressing DCX, Nestin, GFAP, or NG2 ↑ in striatum and cortex	n.d.	Hoehn et al. (2005)
Distal MCAO/permanent	Rat/hypertensive	enriched environment	proliferation (BrdU+) and neuroblasts (DCX+) ↑ in SVZ	improved rotorod and foot placement test	Komitova et al. (2005)
MCAO/90 min	Rat/Wistar	G-CSF/intravenous	DCX+ neuroblasts ↑ in cortex	improved rotorod, adhesive tape removal test, Neurological Severity Score	Schneider et al. (2005)
Distal MCAO/permanent	Rat/hypertensive	1. Notch ligand Dll4/intraventricular 2. Dll4+FGF2/intraventricular	1. BrdU+ cells ↑ in SVZ bilaterally 2. BrdU+ cells ↑ in SVZ bilaterally	1. no effect 2. improved motor skills	Androutsellis-Theotokis et al. (2006)

(continued)

Table 1. (*continued*)

Model/duration	Species/strain	Compound/treatment	Effect on neurogenesis	Effects on functional recovery	Reference
Distal MCAO/permanent	Mouse	G-CSF and stem cell factor/subcutaneous	SVZ progenitors (BrdU$^+$/Musashi-1$^+$) ↑	improved rotorod test and Morris water maze	Kawada et al. (2006)
MCAO/2 hr	Rat/Wistar	GDNF/intrastriatal	progenitor proliferation (BrdU$^+$, Ki-67$^+$) ↑ in SVZ, neuroblasts (DCX$^+$) ↑, mature neurons (BrdU$^+$/NeuN$^+$) ↑ in striatum	n.d.	Kobayashi et al. (2006)
Cortical devascularization stroke	Rat/Long-Evans	EGF+EPO/intraventricular	cell migration (β-gal-labeled) ↑ from SVZ to cortex	improved forelimb asymmetry and inhibition; skilled reaching	Kolb et al. (2006)
Distal MCAO/permanent	Mouse/C57BL	1.SDF-1β/subcutaneous 2. angiopoietin/subcutaneous 3. CXCR4 antagonist/subcutaneous	1. neuroblasts (DCX$^+$) ↑ in cortex 2. neuroblasts (DCX$^+$) ↑ in cortex 3. dispersal of neuroblasts (DCX$^+$) in cortex	1. improved whisker-guided forelimb reaching 2. improved whisker-guided forelimb reaching 3. n.d.	Ohab et al. (2006)
MCAO/2 hr	Rat/Wistar	1. caspase inhibitor/intraventricular 2. CXCR4 antagonist/intraventricular	1. neuroblast (DCX$^+$, DCX$^+$/BrdU$^+$) survival ↑ in striatum 2. neuroblast migration (DCX$^+$) ↓ in striatum	n.d.	Thored et al. (2006)

Model/occlusion	Species/strain	Treatment/route	Cellular findings	Behavioral outcome	Reference
Embolic MCAO/permanent	Rat/Wistar	Tadalafil/orally	BrdU$^+$ cells ↑ in SVZ	improved foot-fault and adhesive tape removal tests, Neurological Severity Score	L. Zhang et al. (2006)
MCAO/30 min	Rat/Sprague-Dawley	Bcl-2 plasmid/intraventricular	BrdU$^+$ cells coexpressing DCX and ChAT ↑ in striatum, and Tuj-1, MAP2, GAD67 ↑ in striatum and cortex	n.d.	R. Zhang et al. (2006)
Distal MCAO/permanent	Rat/hypertensive	AAV-FGF/intracerebral	BrdU$^+$ cells ↑ in SVZ, BrdU$^+$ cells coexpressing Sox2, Pax6, Hu, and NeuN ↑ in peri-infarct cortex	improved standardized disability scale	Leker et al. (2007)
MCAO/permanent	Mouse/C57BL	estradiol/subcutaneous silastic capsule	proliferation (BrdU$^+$) ↑ in dorsal SVZ	n.d.	Suzuki et al. (2007)
MCAO/permanent	Transgenic mice/C57BL/6	VEGF overexpression	BrdU$^+$ cells, BrdU$^+$/DCX$^+$ neuroblasts ↑ in SVZ, BrdU$^+$/NeuN$^+$ mature neurons ↑ in cortex	improved rotorod test	Wang et al. (2007)
MCAO/permanent	Rat/Sprague-Dawley	down-regulator of T-cell activity (poly-YE)/subcutaneous	new mature (BrdU$^+$/MAP2$^+$) neurons ↑ in cortex	improved Neurological Severity Score, Morris water maze	Ziv et al. (2007)

(AAV) Adeno-associated virus; (BDNF) brain-derived neurotrophic factor; (BrdU) bromodeoxyuridine; (DCX) doublecortin; (EPO) erythropoietin; (G-CSF) granulocyte colony-stimulating factor; (GDNF) glial cell line-derived neurotrophic factor; (MCAO) middle cerebral artery occlusion; (n.d.) not demonstrated; (SVZ) subventricular zone; (VEGF) vascular endothelial growth factor.

is also permissive for neuroblast migration, and this potential self-repair mechanism seems to operate similarly in young and old brains.

Differentiation

In most studies, NeuN or MAP-2 coexpression in BrdU-labeled cells has been taken as evidence for the formation of new mature neurons after stroke. Little is known about the specific phenotype of these neurons. Arvidsson et al. (2002) estimated that 42% of the mature, new BrdU$^+$/NeuN$^+$ neurons formed after stroke in the striatum expressed DARPP-32, which is a specific marker for striatal medium-sized spiny neurons. This is the phenotype of most of the neurons destroyed by the ischemic lesion. Similarly, Parent et al. (2002) reported that a majority of new neurons in the striatal peri-infarct area after stroke were DARPP-32$^+$. In another model of striatal injury (excitotoxic lesion), Collin et al. (2005) described the generation of cells expressing markers of interneurons (parvalbumin and neuropeptide Y). Such cells were not detected following MCAO in rats (Parent et al. 2002). In contrast, Teramoto et al. (2003) did not observe BrdU$^+$ cells expressing DARPP-32; however, they did observe a substantial number of BrdU$^+$ cells colabeled with parvalbumin after stroke in mice, which had been infused with epidermal growth factor (EGF).

Morphological and Functional Integration

When new striatal neurons are generated by adenoviral overexpression of BDNF and Noggin in the ventricular wall of intact animals, the new neurons establish efferent connections with neurons in the globus pallidus (Chmielnicki et al. 2004). Whether the stroke-generated neurons form afferent and efferent connections is not known. The situation may be very different from that in the intact brain with stroke causing damage also in target areas such as globus pallidus. Yamashita et al. (2006) have reported that the axons of the new striatal neurons contain abundant presynaptic vesicles and form synapses with neighboring cells, indicating some level of integration into existing neural circuitries.

A crucial question is whether the new cells develop into neurons with functional properties, including synaptic connectivity, characteristic of those neurons that they should replace. Due to technical difficulties, it has not yet been possible to perform electrophysiological recordings on stroke-generated cells expressing neuronal markers, and it is therefore not known if they become functional neurons and are synaptically integrated into existing neural circuitries.

Another important issue is whether the new neurons develop their functional synaptic connectivity to improve or worsen function in the diseased brain. For stroke-generated neurons, this is unknown. We have recently reported that a pathological environment encountering new granule cells, formed in the dentate SGZ, influences the development of their synaptic connectivity (Jakubs et al. 2006). New cells generated after status epilepticus, which gave rise to chronic seizures, neuronal death, and inflammation, exhibited functional connectivity consistent with decreased excitability, i.e., reduced excitatory and increased inhibitory drive. These data suggest that new neurons born in the adult brain exhibit a high degree of plasticity in their afferent synaptic inputs, which can act to mitigate pathological brain function.

STRATEGIES TO STIMULATE ISCHEMIA-INDUCED NEUROGENESIS

It is important to emphasize that neurogenesis after stroke is not just one process but comprises several steps: (1) proliferation of stem/progenitor cells, (2) survival of immature or mature neurons, (3) migration of new neuroblasts to appropriate location, (4) differentiation of new neuroblasts to phenotype of neurons that need to be replaced, and (5) development of functional synaptic connectivity counteracting disease symptoms. Considering our own findings of long-term neurogenesis after stroke, one could argue that the main problem for effectiveness is not insufficient production of new neuroblasts. Therefore, the proliferation step may not be the most suitable target, but the major emphasis should probably be put on approaches improving the survival and differentiation of the new neurons.

Many approaches have been reported to promote neurogenesis after stroke (Table 1), and some of them have been associated with improved functional recovery. Of particular interest in a therapeutic perspective are those strategies that could be applicable also in a clinical setting. For example, intrastriatal infusion of glial cell-line-derived neurotrophic factor (GDNF) increased cell proliferation in ipsilateral SVZ and recruitment of new neuroblasts into the striatum after MCAO and improved survival of new mature neurons (Kobayashi et al. 2006). Similar intrastriatal infusion of GDNF protein has been shown to induce sprouting of dopaminergic neurons in a Parkinson's patient (Love et al. 2005). Thus, administration of this factor may become of therapeutic value to promote neuroregenerative responses in stroke patients.

FUNCTIONAL CONSEQUENCES OF ISCHEMIA-INDUCED NEUROGENESIS

A lot of circumstantial evidence has been presented suggesting that endogenous neurogenesis can contribute to functional recovery after stroke. Spontaneous motor recovery after stroke in rats affecting only the striatum occurred over several months concomitantly with striatal neurogenesis (Thored et al. 2006). Administration of molecules that promote neural proliferation in the SVZ and striatal neurogenesis after stroke (see Table 1) has been reported to be associated with improved functional outcome. However, with available methods, it is not possible to prove a definite causal relationship between neurogenesis and behavioral improvement after stroke. Suppression or enhancement of stroke-induced neurogenesis through delivery of mitosis inhibitors and irradiation or administration of trophic factors, respectively, will also affect other processes. Generation of conditional transgenic mice, in which the newly formed neuroblasts carry a gene for selective ablation, could solve this problem.

PERSPECTIVES

The demonstration of a mechanism by which endogenous NSCs generate new neurons migrating toward the ischemically damaged area after a stroke affecting the adult brain has raised a lot of interest in both the scientific and clinical communities. It is now a major challenge to determine how the reservoirs of NSCs located in the SVZ and SGZ could be efficiently recruited for repair of the stroke-damaged brain. However, many fundamental issues remain to be addressed regarding the regulation and functional relevance of this potential self-repair mechanism. Are the new neurons formed after stroke functional neurons, and are they good or bad? Are more neurons really better for functional recovery? How can we optimize neurogenesis to generate sufficient numbers of new, functionally integrated neurons with correct phenotype? Neurogenesis could have a role in animal models of stroke, but what occurs in human disease?

Obviously, much more knowledge about the mechanisms of cell proliferation, differentiation, migration, survival, and functional integration in stroke-induced neurogenesis is needed. Also, what is the maximum level of improvement after stroke that may be achieved by neurogenesis? Because stroke-induced neurogenesis can continue for many months

(Thored et al. 2006), various interventions to promote neurogenesis are not restricted to the acute postischemic period but can be applied over an extended time period. It seems unlikely that stimulation of neurogenesis from endogenous stem cells per se could be developed into a novel therapeutic approach and provide optimum repair of the stroke-damaged brain. Such a strategy probably needs to be combined with transplantation of NSCs or their progeny in the vicinity of the damaged area. For efficient repair, it may also be necessary to provide the new cells, generated from endogenous and grafted stem cells, with a platform in the form of a synthetic extracellular matrix so that they can reform appropriate brain structure.

SUMMARY

During the past few years, a bulk of experimental evidence has been presented supporting the idea that the stroke-damaged adult brain makes an attempt to repair itself by producing new neurons also in areas where neurogenesis does not normally occur, e.g., the striatum and cerebral cortex. We summarize here the current status of this research. Knowledge about mechanisms regulating the different steps of neurogenesis after stroke is rapidly increasing but still incomplete. The functional consequences of stroke-induced neurogenesis and the level of integration of the new neurons into existing neural circuitries are not known. To have a substantial impact on the recovery after stroke, this potential mechanism for self-repair needs to be markedly enhanced, primarily by increasing the survival and differentiation of the generated neuroblasts. Moreover, for efficient repair, optimization of neurogenesis most likely needs to be combined with promotion of other endogenous neuroregenerative responses, e.g., protection and sprouting of remaining mature neurons and transplantation of stem-cell-derived neurons and glia cells.

ACKNOWLEDGMENTS

Our own research was supported by the Swedish Research Council, Juvenile Diabetes Research Foundation, EU project LSHB-2006-037526 (StemStroke), and the Söderberg, Crafoord, Segerfalk, and Kock Foundations. The Lund Stem Cell Center is supported by a Center of Excellence grant in Life Sciences from the Swedish Foundation for Strategic Research.

REFERENCES

Abrous D.N., Koehl M., and Le Moal M. 2005. Adult neurogenesis: From precursors to network and physiology. *Physiol. Rev.* **85:** 523–569.

Androutsellis-Theotokis A., Leker R.R., Soldner F., Hoeppner D.J., Ravin R., Poser S.W., Rueger M.A., Bae S.K., Kittappa R., and Mckay R.D. 2006. Notch signalling regulates stem cell numbers in vitro and in vivo. *Nature* **442:** 823–826.

Arvidsson A., Collin T., Kirik D., Kokaia Z., and Lindvall O. 2002. Neuronal replacement from endogenous precursors in the adult brain after stroke. *Nat. Med.* **8:** 963–970.

Bhardwaj R.D., Curtis M.A., Spalding K.L., Buchholz B.A., Fink D., Bjork-Eriksson T., Nordborg C., Gage F.H., Druid H., Eriksson P.S., and Frisen J. 2006. Neocortical neurogenesis in humans is restricted to development. *Proc. Natl. Acad. Sci.* **103:** 12564–12568.

Burns K.A., Ayoub A.E., Breunig J.J., Adhami F., Weng W.L., Colbert M.C., Rakic P., and Kuan C.Y. 2007. Nestin-creer mice reveal DNA synthesis by nonapoptotic neurons following cerebral ischemia-hypoxia. *Cereb. Cortex* (in press).

Chen J., Zhang Z.G., Li Y., Wang Y., Wang L., Jiang H., Zhang C., Lu M., Katakowski M., Feldkamp C.S., and Chopp M. 2003. Statins induce angiogenesis, neurogenesis, and synaptogenesis after stroke. *Ann. Neurol.* **53:** 743–751.

Chmielnicki E., Benraiss A., Economides A.N., and Goldman S.A. 2004. Adenovirally expressed noggin and brain-derived neurotrophic factor cooperate to induce new medium spiny neurons from resident progenitor cells in the adult striatal ventricular zone. *J. Neurosci.* **24:** 2133–2142.

Collin T., Arvidsson A., Kokaia Z., and Lindvall O. 2005. Quantitative analysis of the generation of different striatal neuronal subtypes in the adult brain following excitotoxic injury. *Exp. Neurol.* **195:** 71–80.

Curtis M.A., Kam M., Nannmark U., Anderson M.F., Axell M.Z., Wikkelso C., Holtas S., Van Roon-Mom W.M., Bjork-Eriksson T., Nordborg C., et al. 2007. Human neuroblasts migrate to the olfactory bulb via a lateral ventricular extension. *Science* **315:** 1243–1249.

Darsalia V., Heldmann U., Lindvall O., and Kokaia Z. 2005. Stroke-induced neurogenesis in aged brain. *Stroke* **36:** 1790–1795.

Doetsch F., Garcia-Verdugo J.M., and Alvarez-Buylla A. 1997. Cellular composition and three-dimensional organization of the subventricular germinal zone in the adult mammalian brain. *J. Neurosci.* **17:** 5046–5061.

Ekdahl C.T., Claasen J.H., Bonde S., Kokaia Z., and Lindvall O. 2003. Inflammation is detrimental for neurogenesis in adult brain. *Proc. Natl. Acad. Sci.* **100:** 13632–13637.

Eriksson P.S., Perfilieva E., Björk-Eriksson T., Alborn A.M., Nordborg C., Peterson D.A., and Gage F.H. 1998. Neurogenesis in the adult human hippocampus. *Nat. Med.* **4:** 1313–1317.

Gonzalez C.L. and Kolb B. 2003. A comparison of different models of stroke on behaviour and brain morphology. *Eur. J. Neurosci.* **18:** 1950–1962.

Gu W., Jiang W., and Wester P. 1999. A photothrombotic ring stroke model in rats with sustained hypoperfusion followed by late spontaneous reperfusion in the region at risk. *Exp. Brain Res.* **125:** 163–170.

Gustafsson E., Andsberg G., Darsalia V., Mohapel P., Mandel R.J., Kirik D., Lindvall O., and Kokaia Z. 2003. Anterograde delivery of brain-derived neurotrophic factor to striatum via nigral transduction of recombinant adeno-associated virus increases neuronal death but promotes neurogenic response following stroke. *Eur. J. Neurosci.* **17:** 2667–2678.

Hoehn B.D., Palmer T.D., and Steinberg G.K. 2005. Neurogenesis in rats after focal cerebral ischemia is enhanced by indomethacin. *Stroke* 36: 2718–2724.

Jakubs K., Nanobashvili A., Bonde S., Ekdahl C.T., Kokaia Z., Kokaia M., and Lindvall O. 2006. Environment matters: Synaptic properties of neurons born in the epileptic adult brain develop to reduce excitability. *Neuron* 52: 1047–1059.

Jiang W., Gu W., Brannstrom T., Rosqvist R., and Wester P. 2001. Cortical neurogenesis in adult rats after transient middle cerebral artery occlusion. *Stroke* 32: 1201–1207.

Jin K., Sun Y., Xie L., Childs J., Mao X.O., and Greenberg D.A. 2004. Post-ischemic administration of heparin-binding epidermal growth factor-like growth factor (HB-EGF) reduces infarct size and modifies neurogenesis after focal cerebral ischemia in the rat. *J. Cereb. Blood Flow Metab.* 24: 399–408.

Jin K., Minami M., Lan J.Q., Mao X.O., Batteur S., Simon R.P., and Greenberg D.A. 2001. Neurogenesis in dentate subgranular zone and rostral subventricular zone after focal cerebral ischemia in the rat. *Proc. Natl. Acad. Sci.* 98: 4710–4715.

Jin K., Sun Y., Xie L., Peel A., Mao X.O., Batteur S., and Greenberg D.A. 2003. Directed migration of neuronal precursors into the ischemic cerebral cortex and striatum. *Mol. Cell. Neurosci.* 24: 171–189.

Jin K., Wang X., Xie L., Mao X.O., Zhu W., Wang Y., Shen J., Mao Y., Banwait S., and Greenberg D.A. 2006. Evidence for stroke-induced neurogenesis in the human brain. *Proc. Natl. Acad. Sci.* 103: 13198–13202.

Kawada H., Takizawa S., Takanashi T., Morita Y., Fujita J., Fukuda K., Takagi S., Okano H., Ando K., and Hotta T. 2006. Administration of hematopoietic cytokines in the subacute phase after cerebral infarction is effective for functional recovery facilitating proliferation of intrinsic neural stem/progenitor cells and transition of bone marrow-derived neuronal cells. *Circulation* 113: 701–710.

Kobayashi T., Ahlenius H., Thored P., Kobayashi R., Kokaia Z., and Lindvall O. 2006. Intracerebral infusion of glial cell line-derived neurotrophic factor promotes striatal neurogenesis after stroke in adult rats. *Stroke* 37: 2361–2367.

Kolb B., Morshead C., Gonzalez C., Kim M., Gregg C., Shingo T., and Weiss S. 2007. Growth factor-stimulated generation of new cortical tissue and functional recovery after stroke damage to the motor cortex of rats. *J. Cereb. Blood Flow Metab.* 27: 983–997.

Komitova M., Mattsson B., Johansson B.B., and Eriksson P.S. 2005. Enriched environment increases neural stem/progenitor cell proliferation and neurogenesis in the subventricular zone of stroke-lesioned adult rats. *Stroke* 36: 1278–1282.

Lee S.R., Kim H.Y., Rogowska J., Zhao B.Q., Bhide P., Parent J.M., and Lo E.H. 2006. Involvement of matrix metalloproteinase in neuroblast cell migration from the subventricular zone after stroke. *J. Neurosci.* 26: 3491–3495.

Leifer D. and Kowall N.W. 1993. Immunohistochemical patterns of selective cellular vulnerability in human cerebral ischemia. *J. Neurol. Sci.* 119: 217–228.

Leker R.R., Soldner F., Velasco I., Gavin D.K., Androutsellis-Theotokis A., and Mckay R.D. 2007. Long-lasting regeneration after ischemia in the cerebral cortex. *Stroke* 38: 153–161.

Liu Z., Fan Y., Won S.J., Neumann M., Hu D., Zhou L., Weinstein P.R., and Liu J. 2007. Chronic treatment with minocycline preserves adult new neurons and reduces functional impairment after focal cerebral ischemia. *Stroke* 38: 146–152.

Louissaint A., Jr., Rao S., Leventhal C., and Goldman S.A. 2002. Coordinated interaction of neurogenesis and angiogenesis in the adult songbird brain. *Neuron* 34: 945–960.

Love S., Plaha P., Patel N.K., Hotton G.R., Brooks D.J., and Gill S.S. 2005. Glial cell line-derived neurotrophic factor induces neuronal sprouting in human brain. *Nat. Med.* **11:** 703–704.

Macas J., Nern C., Plate K.H., and Momma S. 2006. Increased generation of neuronal progenitors after ischemic injury in the aged adult human forebrain. *J. Neurosci.* **26:** 13114–13119.

Magavi S.S., Leavitt B.R., and Macklis J.D. 2000. Induction of neurogenesis in the neo-cortex of adult mice. *Nature* **405:** 951–955.

Monje M.L., Toda H., and Palmer T.D. 2003. Inflammatory blockade restores adult hip-pocampal neurogenesis. *Science* **302:** 1760–1765.

Morshead C.M., Reynolds B.A., Craig C.G., Mcburney M.W., Staines W.A., Morassutti D., Weiss S., and Van Der Kooy D. 1994. Neural stem cells in the adult mammalian fore-brain: A relatively quiescent subpopulation of subependymal cells. *Neuron* **13:** 1071–1082.

Nyfeler Y., Kirch R.D., Mantei N., Leone D.P., Radtke F., Suter U., and Taylor V. 2005. Jagged1 signals in the postnatal subventricular zone are required for neural stem cell self-renewal. *EMBO J.* **24:** 3504–3515.

Ohab J.J., Fleming S., Blesch A., and Carmichael S.T. 2006. A neurovascular niche for neu-rogenesis after stroke. *J. Neurosci.* **26:** 13007–13016.

Parent J.M., Vexler Z.S., Gong C., Derugin N., and Ferriero D.M. 2002. Rat forebrain neu-rogenesis and striatal neuron replacement after focal stroke. *Ann. Neurol.* **52:** 802–813.

Quinones-Hinojosa A., Sanai N., Soriano-Navarro M., Gonzalez-Perez O., Mirzadeh Z., Gil-Perotin S., Romero-Rodriguez R., Berger M.S., Garcia-Verdugo J.M., and Alvarez-Buylla A. 2006. Cellular composition and cytoarchitecture of the adult human sub-ventricular zone: A niche of neural stem cells. *J. Comp. Neurol.* **494:** 415–434.

Robin A.M., Zhang Z.G., Wang L., Zhang R.L., Katakowski M., Zhang L., Wang Y., Zhang C., and Chopp M. 2006. Stromal cell-derived factor 1alpha mediates neural pro-genitor cell motility after focal cerebral ischemia. *J. Cereb. Blood Flow Metab.* **26:** 125–134.

Sanai N., Tramontin A.D., Quinones-Hinojosa A., Barbaro N.M., Gupta N., Kunwar S., Lawton M.T., Mcdermott M.W., Parsa A.T., Manuel-Garcia Verdugo J., et al. 2004. Unique astrocyte ribbon in adult human brain contains neural stem cells but lacks chain migration. *Nature* **427:** 740–744.

Schneider A., Kruger C., Steigleder T., Weber D., Pitzer C., Laage R., Aronowski J., Maurer M.H., Gassler N., Mier W., et al. 2005. The hematopoietic factor G-CSF is a neuronal ligand that counteracts programmed cell death and drives neurogenesis. *J. Clin. Invest.* **115:** 2083–2098.

Stump G., Durrer A., Klein A.L., Lutolf S., Suter U., and Taylor V. 2002. Notch1 and its ligands delta-like and jagged are expressed and active in distinct cell populations in the postnatal mouse brain. *Mech. Dev.* **114:** 153–159.

Sun Y., Jin K., Xie L., Childs J., Mao X.O., Logvinova A., and Greenberg D.A. 2003. VEGF-induced neuroprotection, neurogenesis, and angiogenesis after focal cerebral ischemia. *J. Clin. Invest.* **111:** 1843–1851.

Suzuki S., Gerhold L.M., Bottner M., Rau S.W., Dela Cruz C., Yang E., Zhu H., Yu J., Cashion A.B., Kindy M.S., et al. 2007. Estradiol enhances neurogenesis following ischemic stroke through estrogen receptors alpha and beta. *J. Comp. Neurol.* **500:** 1064–1075.

Taguchi A., Soma T., Tanaka H., Kanda T., Nishimura H., Yoshikawa H., Tsukamoto Y., Iso H., Fujimori Y., Stern D.M., et al. 2004. Administration of cd34+ cells after stroke enhances neurogenesis via angiogenesis in a mouse model. *J. Clin. Invest.* **114:** 330–338.

Teramoto T., Qiu J., Plumier J.C., and Moskowitz M.A. 2003. Egf amplifies the replacement of parvalbumin-expressing striatal interneurons after ischemia. *J. Clin. Invest.* **111:** 1125–1132.

Thored P., Arvidsson A., Cacci E., Ahlenius H., Kallur T., Darsalia V., Ekdahl C.T., Kokaia Z., and Lindvall O. 2006. Persistent production of neurons from adult brain stem cells during recovery after stroke. *Stem Cells* **24:** 739–747.

Tsai P.T., Ohab J.J., Kertesz N., Groszer M., Matter C., Gao J., Liu X., Wu H., and Carmichael S.T. 2006. A critical role of erythropoietin receptor in neurogenesis and post-stroke recovery. *J. Neurosci.* **26:** 1269–1274.

Wang H., Ward N., Boswell M., and Katz D.M. 2006. Secretion of brain-derived neurotrophic factor from brain microvascular endothelial cells. *Eur. J. Neurosci.* **23:** 1665–1670.

Wang L., Zhang Z., Wang Y., Zhang R., and Chopp M. 2004. Treatment of stroke with erythropoietin enhances neurogenesis and angiogenesis and improves neurological function in rats. *Stroke* **35:** 1732–1737.

Wang Y., Jin K., Mao X.O., Xie L., Banwait S., Marti H.H., and Greenberg D.A. 2007. VEGF-overexpressing transgenic mice show enhanced post-ischemic neurogenesis and neuromigration. *J. Neurosci. Res.* **85:** 740–747.

Yamashita T., Ninomiya M., Hernandez Acosta P., Garcia-Verdugo J.M., Sunabori T., Sakaguchi M., Adachi K., Kojima T., Hirota Y., Kawase T., et al. 2006. Subventricular zone-derived neuroblasts migrate and differentiate into mature neurons in the post-stroke adult striatum. *J. Neurosci.* **26:** 6627–6636.

Yan Y.P., Sailor K.A., Lang B.T., Park S.W., Vemuganti R., and Dempsey R.J. 2007. Monocyte chemoattractant protein-1 plays a critical role in neuroblast migration after focal cerebral ischemia. *J. Cereb. Blood Flow Metab.* **27:** 1213–1224.

Zhang L., Zhang Z., Zhang R.L., Cui Y., LaPointe M.C., Silver B., and Chopp M. 2006. Tadalafil, a long-acting type 5 phosphodiesterase isoenzyme inhibitor, improves neurological functional recovery in a rat model of embolic stroke. *Brain Res.* **1118:** 192–198.

Zhang R., Zhang L., Zhang Z., Wang Y., Lu M., Lapointe M., and Chopp M. 2001. A nitric oxide donor induces neurogenesis and reduces functional deficits after stroke in rats. *Ann. Neurol.* **50:** 602–611.

Zhang R., Wang Y., Zhang L., Zhang Z., Tsang W., Lu M., Zhang L., and Chopp M. 2002. Sildenafil (Viagra) induces neurogenesis and promotes functional recovery after stroke in rats. *Stroke* **33:** 2675–2680.

Zhang R., Xue Y.Y., Lu S.D., Wang Y., Zhang L.M., Huang Y.L., Signore A.P., Chen J., and Sun F.Y. 2006. Bcl-2 enhances neurogenesis and inhibits apoptosis of newborn neurons in adult rat brain following a transient middle cerebral artery occlusion. *Neurobiol. Dis.* **24:** 345–356.

Zhang R.L., Zhang Z.G., Zhang L., and Chopp M. 2001. Proliferation and differentiation of progenitor cells in the cortex and the subventricular zone in the adult rat after focal cerebral ischemia. *Neuroscience* **105:** 33–41.

Zhang R.L., Chopp M., Zhang Z.G., Jiang Q., and Ewing J.R. 1997. A rat model of focal embolic cerebral ischemia. *Brain Res.* **766:** 83–92.

Zhang R.L., Zhang Z.G., Lu M., Wang Y., Yang J.J., and Chopp M. 2006. Reduction of the cell cycle length by decreasing g1 phase and cell cycle reentry expand neuronal progenitor cells in the subventricular zone of adult rat after stroke. *J. Cereb. Blood Flow Metab.* **26:** 857–863.

Zhang R.L., Zhang Z., Zhang C., Zhang L., Robin A., Wang Y., Lu M., and Chopp M. 2004a. Stroke transiently increases subventricular zone cell division from asymmetric to symmetric and increases neuronal differentiation in the adult rat. *J. Neurosci.* **24:** 5810–5815.

Zhang R.L., Letourneau Y., Gregg S.R., Wang Y., Toh Y., Robin A.M., Zhang Z.G., Zhang Z., and Chopp M. 2007a. Neuroblast division during migration toward the ischemic striatum: A study of dynamic migratory and proliferative characteristics of neuroblasts from the subventricular zone. *J. Neurosci.* **27:** 3157–3162.

Zhang R.L., Zhang Z.G., Wang Y., Letourneau Y., Liu X.S., Zhang X., Gregg S.R., Wang L., and Chopp M. 2007b. Stroke induces ependymal cell transformation into radial glia in the subventricular zone of the adult rodent brain. *J. Cereb. Blood Flow Metab.* **27:** 1201–1212.

Zhang R.L., Zhang Z., Wang L., Wang Y., Gousev A., Zhang L., Ho K.L., Morshead C., and Chopp M. 2004b. Activated neural stem cells contribute to stroke-induced neurogenesis and neuroblast migration toward the infarct boundary in adult rats. *J. Cereb. Blood Flow Metab.* **24:** 441–448.

Ziv Y., Ron N., Butovsky O., Landa G., Sudai E., Greenberg N., Cohen H., Kipnis J., and Schwartz M. 2006. Immune cells contribute to the maintenance of neurogenesis and spatial learning abilities in adulthood. *Nat. Neurosci.* **9:** 268–275.

Ziv Y., Finkelstein A., Geffen Y., Kipnis J., Smirnov I., Shpilman S., Vertkin I., Kimron M., Lange A., Hecht T., et al. 2007. A novel immune-based therapy for stroke induces neuroprotection and supports neurogenesis. *Stroke* **38:** 774–782.

27

Adult Neurogenesis in Teleost Fish

Günther K.H. Zupanc
School of Engineering and Science
Jacobs University Bremen
Bremen, Germany

IN CONTRAST TO MAMMALS, ADULT NEUROGENESIS is extremely pronounced in teleost fish. As shown in this chapter, this includes an unsurpassed potential to replace neurons lost to injury by newly generated ones, leading to both structural and functional recovery after CNS trauma. These features make it particularly interesting to study the generation of new neurons in adult fish—not only to explore a fascinating biological phenomenon, but also to gain a better understanding of the evolutionary constraints that limit adult neurogenesis in the mammalian brain.

RATE OF CELL PROLIFERATION IN THE ADULT BRAIN

Quantitative analysis has suggested that the rate of cell proliferation is at least one, if not two, orders of magnitude higher in the adult fish brain than in the adult mammalian brain. In the most intensively studied model system, the brown ghost knifefish (*Apteronotus leptorhynchus*), approximately 100,000 cells are generated in the whole brain within any 2-hour period (Zupanc and Horschke 1995). In zebrafish, approximately 6000 cells are born within any 30-minute period (Hinsch and Zupanc 2007). These cell numbers correspond to approximately 0.2% and 0.06% of the total population of brain cells in each of the two species, respectively. Thus, the number of cells that enter the S phase of mitosis per time unit, relative to the total number of cells in the adult brain, appears to be similar in *A. leptorhynchus* and zebrafish.

In contrast, in the subventricular zone (SVZ) of adult mice, approximately 30,000 cells are formed every day (Lois and Alvarez-Buylla 1994), corresponding to 0.03% of the estimated 110 million cells in the whole brain of adult mice (Williams 2000). In the dentate gyrus of the hippocampus of adult rats, 9000 cells are produced every day (Cameron and McKay 2001), which corresponds to 0.003% of the estimated 330 million cells in the whole brain of the adult rat (Herculano-Houzel and Lent 2005).

PROLIFERATION ZONES IN THE ADULT TELEOSTEAN BRAIN

As in the adult mammalian brain, in the adult teleostean brain, the vast majority of the mitotic cells are found at high concentrations in small well-defined areas of the brain ("proliferation zones") (Fig. 1). A large number of these zones are situated at or near the surfaces of ventricles or related systems. Although other zones, particularly in the cerebellum, are located in regions distant from any ventricle, many are derived from areas located at ventricular surfaces during embryonic stages of development. Then, as a result of the everted development of the fish brain during embryogenesis, the associated ventricular lumina are obliterated and translocated (Powels 1978a,b).

Whereas in the adult mammalian brain, the generation of new neurons is restricted to the olfactory bulb (OB) and the hippocampus, mapping studies have revealed dozens of such proliferation zones in the adult teleost

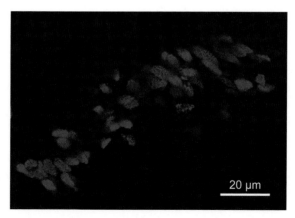

Figure 1. Confocal image of the proliferation zone in the periventricular gray zone of the optic tectum, revealed by labeling of mitotic cells with BrdU in the adult zebrafish brain. A post-BrdU-administration survival time of 2 days was used. (After Zupanc et al. 2005.)

fish brain (Kranz and Richter 1970a,b; Richter and Kranz 1970a,b; Zupanc and Horschke 1995; Zikopoulos et al. 2000; Ekström et al. 2001; Zupanc et al. 2005; Grandel et al. 2006). Quantitative analysis has shown that the cerebellum is the site of origin of approximately 75% of all new brain cells in *A. leptorhynchus* (Zupanc and Horschke 1995) and of approximately 60% of all new cells in zebrafish (Hinsch and Zupanc 2007). The remaining 25% and 40%, respectively, of all cells are produced in various regions of the telencephalon, diencephalon, mesencephalon, and rhombencephalon. Among these regions, two are of particular interest from a comparative point of view—the OB and the lateral and posterior zones of the dorsal telencephalon. On the basis of neuroanatomical evidence (Northcutt and Braford 1980; Nieuwenhuys and Meek 1990; Braford 1995; Northcutt 1995; Butler 2000; Vargas et al. 2000) and results of functional studies (Rodrígues et al. 2002; Portavella et al. 2004), part of the latter zones is thought to be homologous to the mammalian hippocampus. Thus, these results are in agreement with the notion that adult neurogenesis in the OB and the hippocampus is a conserved vertebrate trait.

ADULT STEM CELLS IN THE TELECOST BRAIN

Isolation of cells from the proliferation zones in the dorsal telencephalon and the cerebellum of the adult teleost brain has revealed a cellular population with stem-cell-like properties (Hinsch and Zupanc 2006). After 3–4 days in culture, neurospheres originating from these cells develop. They grow through cell proliferation and reach diameters of up to 140 μm within 3 weeks. Growth of such neurospheres can be promoted by epidermal growth factor and basic fibroblast growth factor. Differentiation is induced by exposing the neurospheres to fetal bovine serum and laminin as coating substrate. Under the latter conditions, neurospheres give rise to both neurons and glial cells (Fig. 2). Because, in addition to this multipotency, the cells isolated from the proliferation zones in the adult brain show the ability for self-renewal, they can be regarded as true stem cells.

MIGRATION AND GUIDANCE OF YOUNG CELLS

Depending on the brain region, the new cells either remain near the areas where they are born or migrate within the first 1–2 weeks following their generation from the proliferation zone to specific target areas. In the retina, an annulus of progenitor cells exists at the junction of the retina and the iris. These progenitors generate neurons that are continuously

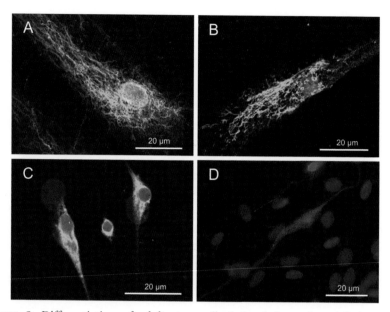

Figure 2. Differentiation of adult stem cells isolated from the adult brain of *Apteronotus leptorhynchus*. The process of differentiation was induced by plating neurospheres onto a laminin-coated surface and cultivating them in medium containing fetal bovine serum. Approximately 24 hours after plating, the cells started to differentiate into various cell types. (*A*) GFAP-immunoreactive astrocyte. (*B*) Vimentin-positive cell. (*C*) Hu-expressing neurons. (*D*) MAP2 (2a + 2b)-immunopositive neuron. Nuclei were stained with either propidium iodide (*red*) or DAPI (*blue*). Note absence of immunolabeling in some cells in *C* and *D*. (After Hinsch and Zupanc 2006.)

added appositionally to the margin at the extant retina (Johns 1977; Meyer 1978; Hagedorn and Fernald 1992; Marcus et al. 1999). Similarly, in the optic tectum—which is homologous to the superior colliculus of mammals and to which retinal ganglion cells project—the majority of the new cells are generated at the caudal pole, where they remain during their subsequent development (Raymond and Easter 1983; Mansour-Robaey and Pinganaud 1990; Nguyen et al. 1999; Wullimann and Puelles 1999; Ekström et al. 2001; Candal et al. 2005; Zupanc et al. 2005). As a result, the optic tectum grows asymmetrically by expanding primarily from its caudal end.

In contrast, in the corpus cerebelli (one of the major subdivisions of the teleostean cerebellum), the new cells are generated in specific proliferation zones near the midline in the dorsal molecular layer and, in some species, also in the ventral molecular layer. Subsequently, the young cells migrate into the granular layer where they evenly spread (Fig. 3) (Zupanc

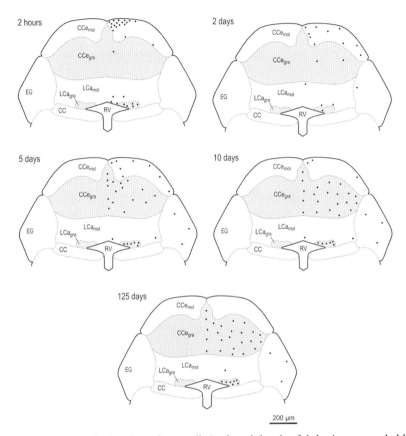

Figure 3. Pattern of migration of new cells in the adult zebrafish brain, as revealed by analysis of the distribution of BrdU-labeled cells in the corpus cerebelli, eminentia granularis, and rostral part of lobus caudalis cerebelli after post-BrdU administration survival times ranging from 2 hours to 125 days. The locations of labeled cells are indicated by black dots. The number of dots roughly represents the density of labeled cells in this brain region. Note migration of BrdU-labeled cells from the dorsal molecular layer of the corpus cerebelli into the associated granular layer, but absence of apparent migrational activity of the new cells in the eminentia granularis and the lobus caudalis cerebelli. (CC) Crista cerebellaris; (CCe$_{gra}$) granule cell layer of corpus cerebelli; (CCe$_{mol}$) molecular layer of corpus cerebelli; (EG) eminentia granularis; (LCa$_{gra}$) granule cell layer of lobus caudalis cerebelli; (LCa$_{mol}$) molecular layer of lobus caudalis cerebelli; (RV) rhombencephalic ventricle. (After Zupanc et al. 2005.)

et al. 1996, 2005). Thus, the corpus cerebelli grows in a rather symmetric fashion.

Several lines of evidence, obtained in *A. leptorhynchus,* suggest that during their migration, the young cells are guided by radial glial fibers that delineate the migratory path (Zupanc and Clint 2003). The population of these fibers is formed by two subpopulations, one expressing vimentin and the other glial fibrillary acidic protein (GFAP). The morphology and distribution of these two fiber populations are similar but not identical, thus indicating a partial overlap. It is possible that in the course of their development, similar to the situation in mammals (Voigt 1989), the fish radial glia first express vimentin, and then, while the vimentin expression is gradually reduced, an increasing amount of GFAP is produced.

The hypothesis that radial glial fibers in the cerebellum provide a scaffolding for the migrating young cells receives further support by the results of double-labeling experiments. Labeling of S-phase cells with 5-bromo-2'-deoxyuridine (BrdU), followed by sacrifice of the fish 2 days postadministration of this thymidine analog (i.e., at a time when the young cells show maximum migratory activity), reveals elongated BrdU-labeled cells in close apposition to GFAP-labeled radial glial fibers (Zupanc and Clint 2003).

REGULATION OF CELL NUMBER BY APOPTOTIC CELL DEATH

After arrival at their target sites within the cerebellum of A. leptorhynchus, the areal density of BrdU-labeled cells drops by about 50% in the period 4–7 weeks after their generation (Zupanc et al. 1996; Ott et al. 1997). This decrease is thought to be caused by apoptotic cell death. Experimental evidence in favor of this notion has been suggested by using terminal deoxynucleotidyl-transferase-mediated dUTP-biotin nick end-labeling (TUNEL) of 3'-OH ends of DNA (Soutschek and Zupanc 1996). Studies applying TUNEL to brain sections of A. leptorhynchus have shown that a large number of cells continuously undergo apoptosis in the granular layers of the three subdivisions of the cerebellum (Soutschek and Zupanc 1996). In contrast, the number of apoptotic cells is low in the molecular layers. This suggests that apoptosis is used as a mechanism to regulate the numbers of young cells after they have reached their target areas. A similar observation has been made in the teleostean retina during postembryonic development. Apoptotic cell death occurs predominantly in areas where the new cells differentiate and become integrated into visual circuits (Candal et al. 2005).

Studies in mammals have shown that during embryonic development, apoptosis leads to the elimination of young cells that, after arrival at the target site, have failed to make proper connections with other neurons and to receive adequate amounts of specific survival factors produced by cells in the target area (Raff 1992; Raff et al. 1993). Although not yet examined in the context of adult neurogenesis, it is possible that a similar mechanism regulates the number of cells born in the adult brain.

LONG-TERM PERSISTENCE OF THE CELLS GENERATED IN THE ADULT BRAIN

Long-term persistence of the cells generated in the adult fish brain has been shown in both A. *leptorhynchus* and zebrafish. In the former species, it has been shown that virtually all of the cells that survive the massive wave of apoptotic cell death occurring 4–7 weeks after their generation persist for at least 440 days in total (Ott et al. 1997). Because the latter period corresponds to approximately half the adult life span of this species in captivity, it is likely that a large portion of the adult-born cells survive for the rest of the fish's life.

In zebrafish, long-term persistence of cells generated in the adult brain up to 1010 days has been demonstrated (Zupanc et al. 2005; Hinsch and Zupanc 2007). Quantitative analysis has shown that at 446–656 days following the administration of BrdU, about half of the cells labeled at 10 days are still present (Hinsch and Zupanc 2007). Because the entire life span of zebrafish has been estimated to be approximately 3.5 years (Gerhard et al. 2002), and the fish received the single dose of BrdU at an age of approximately 1 year, these experiments show persistence of the new cells generated in the adult brain for the rest of the fish's life expectancy.

In mammals, persistence of brain cells generated during adulthood has been shown in both the hippocampus and the OB. In the hippocampus of mice, granular neurons survive for at least 11 months (Kempermann et al. 2003), which is equivalent to approximately half the life span of mice. In the OB of mice and rats, new cells survive for at least 112 days and 550 days, respectively (Kaplan et al. 1985; Corotto et al. 1993; Winner et al. 2002).

In A. *leptorhynchus,* the long-term survival, together with the continuous production of new cells, leads to a permanent growth of the entire brain, except for very old fish in which the growth rate appears to reach a plateau. This parallels the continuous growth of the body in this species. Quantitative analysis has shown that, although the body weight of the fish increases from 1 to 16 g, the total number of brain cells doubles from 5×10^7 to 1×10^8 (Zupanc and Horschke 1995).

NEURONAL DIFFERENTIATION

A central question arising in the context of the generation of new cells in the adult fish brain concerns their cellular identity. Double-labeling experiments using immunohistochemistry against BrdU and the neuron-specific protein Hu have shown that approximately 50% of the long-term surviving cells born in the adult zebrafish brain develop into neurons (Zupanc et al. 2005; Hinsch and Zupanc 2007). The BrdU/Hu-positive cells are particularly abundant in the dorsal telencephalon, including the region presumably homologous to the mammalian hippocampus, but are also found in other areas of the adult brain.

Experiments in which BrdU immunohistochemistry was combined with retrograde tract tracing have shown that in the cerebellum of A. leptorhynchus and zebrafish, at least part of the young cells develop into granule cell neurons (Zupanc et al. 1996, 2005). Because in these studies the BrdU-labeled granule cells were retrogradely traced by application of tracer substance to the molecular layers, the experiments also indicate that these cells have developed axons traveling from their somata in the granule cell layers to the corresponding molecular layers to form the parallel fiber network. Thus, it appears that they have integrated into the existing neural network of cerebellar neurons. It is unknown whether the axons of the newly generated granule cells make proper synaptic connections with other cells, such as Purkinje cells. It also remains to be examined whether the new granule cells are functional, and whether their physiological properties are similar to those of the "older" granular neurons.

GENOMIC REGULATION OF THE DEVELOPMENT OF NEW CELLS

How are cell proliferation and subsequent differentiation regulated at the genomic level? Traditionally, it was thought that the regulation of these developmental processes operates on a constant genome and is restricted to alterations at transcriptional and posttranslational levels. However, chromosomal analysis of the newly generated cells in the adult brain of A. leptorhynchus has provided evidence for somatic genomic alterations, resulting in a genetic mosaic of these cells (Rajendran et al. 2007).

Metaphase chromosome spreads of brain cells in this species have revealed a euploid complement of chromosomes, assumed to be 22, in only 22% of all cells (Fig. 4A–D). The remaining cells show numerical chromosomal abnormalities. Chromosome loss was observed in approximately 84% of the aneuploid cells, whereas chromosome gain occurred in the remaining 16% of the cells. Surprisingly, not only single chromosomes,

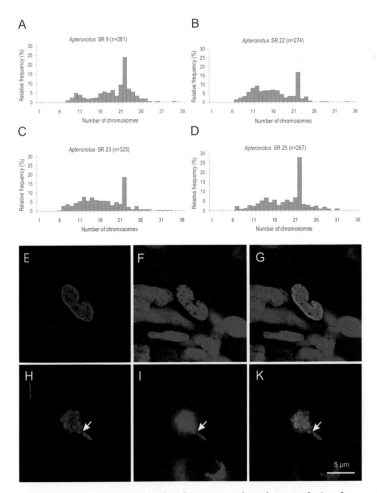

Figure 4. (*A–D*) Chromosome number histograms based on analysis of metaphase chromosome spreads of cells from the adult brain of four individual fish of *Apteronotus leptorhynchus*. Between 17% and 28% of the cells are euploid, having 22 chromosomes. Among the aneuploid cells, the majority lost chromosomes, whereas a rather minor portion gained chromosomes. (n) Number of cells analyzed. (*E–K*) High-power images obtained through deconvolution microscopy of phosphorylated histone H3-labeled cells (*red*) in the molecular layer of the corpus cerebelli. The nuclei were counterstained with DAPI (*blue*) (*E–G*) Morphologically normal metaphase profile. (*H–K*) Prometaphase/metaphase profile with single laggard. Note the clear gap between the laggard and the daughter nucleus (*arrow*). In addition to the phosphorylated histone H3 label (*E, H*) and the DAPI counterstain (*F, I*), the overlay (*G, K*) is shown. (After Rajendran et al. 2007.)

but also multiple chromosomes might be lost or gained. In the most extreme cases found in this study, up to 16 chromosomes were lost and up to 12 chromosomes gained, relative to the euploid number of 22 chromosomes.

Aneuploidy associated with newborn cells has also been found in mammals both during embryonic development and during postnatal stages of development. Spectral karyotype analysis of metaphase spreads of mouse embryonic cerebral cortical neuroblasts revealed aneuploidy in approximately 33% of the cells, with the vast majority of them being hypoploid (Rehen et al. 2001). Similar results have been obtained by karyotyping of mitotic cells in the murine SVZ (Kaushal et al. 2003), as well as by examination of cerebral cortical neurons and cerebellar Purkinje cells through use of a nuclear transfer technique (Osada et al. 2002). Significant levels of aneuploidy have also been reported for individual chromosomes of cells in the normal human brain during embryonic stages of development and during adulthood (Rehen et al. 2005; Yurov et al. 2005). These findings suggest that aneuploidy is a rather universal phenomenon among newborn cells, possibly occurring in any vertebrate species.

As shown in both normal brain cells and peripheral tumor cells, aneuploidy can be accounted for by segregation defects during mitotic division (Yang et al. 2003; Stewénius et al. 2005; for review, see Gisselsson and Höglund 2005). In the adult brain of A. leptorhynchus, labeling of condensed chromosomes of M-phase cells by phosphorylated histone H3 has revealed three types of such defects—laggards, anaphase bridges, and micronuclei (Fig. 4E–K) (Rajendran et al. 2007). Quantitative analysis has shown that such chromosome segregation defects are quite common— approximately 14% of all phosphorylated histone H3 cells display one or several of these defects. Examination of the mechanism mediating anaphase bridging has shown that this defect is caused by abnormally short telomeres during mitosis and typically results in either intercentromeric chromatin fragmentation or centromere detachment. The latter, in turn, lead to pericentromeric chromosome rearrangements and loss of whole chromosomes, respectively (Stewénius et al. 2005).

Although less well examined, a potential consequence of acquired chromosomal aneuploidy is a modification of expression levels of genes residing on the affected chromosomes. This is particularly important because several investigations in mammalian systems have suggested that a significant number of the aneuploid cells show long-term persistence. In adult mice, at least some of the aneuploid cells born in the SVZ are competent to migrate into their target region, the OB, and survive after differentiation into neurons (Kaushal et al. 2003). Within the OB, a combination of retrograde tracing with fluorescent in situ hybridization for chromosome-specific loci and with immediate-early gene immunolabeling indicated that these aneuploid neurons are functionally active and form proper anatomical connections (Kingsbury et al. 2005). Along similar lines, hybrid embryonic-stem-cell-derived tetraploid mice have been shown to exhibit a normal phenotype (with the exception of elevated body weight and hematocrit) in 90 morphological, physiological, and behavioral parameters, compared to euploid controls raised from normal matings (Schwenk et al. 2003). These results show that aneuploidy does not necessarily result in death of the individual aneuploid cells or in dysfunction of the whole animal. Rather, the genetic mosaicism caused by aneuploidy could provide a molecular basis for the phenotypic plasticity commonly shown by cells during embryonic and postnatal development.

TOWARD A FUNCTIONAL UNDERSTANDING OF ADULT NEUROGENESIS IN FISH: THE NUMERICAL MATCHING HYPOTHESIS

Why does neurogenesis cease in all but two regions of the mammalian brain at the time of, or shortly after, birth, but persist throughout life in numerous regions of the teleost fish brain? A possible answer is given by comparative analysis of the differences in the development of motor structures and sensory organs between teleosts and mammals (Zupanc 1999, 2001, 2006a,b). Although many species of teleosts and mammals grow during postembryonic stages of development and even throughout life, there is a distinct difference in the growth pattern between these two taxonomic classes. Whereas in mammals, postembryonic growth is the result of an increase in size, but not in number of individual muscle fibers (Rowe and Goldspink 1969), in fish, both the number of muscle fibers and the volume of individual fibers increase (Weatherley and Gill 1985; Koumans and Akster 1995; Zimmerman and Lowery 1999; for review, see Rowlerson and Veggetti 2001). It is therefore possible that the hyperplasia of peripheral motor elements prompts a concomitant increase in the number of central neurons involved in neuronal control of associated muscle activity.

Along similar lines, the number of sensory receptor cells, receptor organs, or receptor units in the periphery increases with age, as has been shown in several species of fish. Such a formation of new sensory elements has been shown for sensory hair cells in the inner ear of sharks (Corwin 1981), for retinal cells in the eyes of goldfish (Johns and Easter 1977), and for electrosensory receptor organs in the gymnotiform fish

Sternopygus darienses (Zakon 1984). Thus, like the formation of new motor elements, the continuous increase in the number of sensory elements in fish might lead to the generation of new neurons involved in the processing of sensory information, ensuring numerical matching of central neurons and peripheral sensory elements.

In fish, the matching hypothesis can explain the enormous mitotic activity in the cerebellum. On the basis of anatomical, physiological, and behavioral experiments, the cerebellum, notably of weakly electric fish, has been proposed to have an important role not only in the well-established function of control and coordination of movements, but also in sensory processing. The latter has been shown particularly in the context of the involvement of the cerebellum in tracking movements of objects around the animal, and in the generation and subtraction of sensory expectations (for reviews, see Paulin 1993; Bell et al. 1997). On the other hand, the lack of adult neurogenesis in the mammalian cerebellum could be causally linked to the absence of changes in the number of muscle fibers and sensory elements during periods of growth in the periphery.

At the central level, numerical matching has been shown between the presynaptic granule cell population and its postsynaptic target neurons, the Purkinje cells, by making use of the neurologically mutant mouse strain, lurcher. The cerebella of lurcher heterozygotes are characterized by early postnatal degeneration of several cell types, beginning with the loss of Purkinje cells during the second week of life. This stage is followed by degeneration of granule and olivary neurons and Bergmann glia (Caddy and Biscoe 1976; Swisher and Wilson 1977). In wild-type mice, there is a constant ratio of about 175 granule cells for each Purkinje cell. In lurcher chimeric mice, variation in the number of Purkinje cells is genetically caused, whereas the loss of granule cells appears to be a secondary, phenotypic consequence caused by Purkinje-cell loss (Wetts and Herrup 1983).

In fish, there is some experimental evidence supporting the hypothesis that changes in the number of peripheral sensory elements leads to changes in the production of corresponding central elements. Raymond et al. (1983) found that, in goldfish, permanent removal of the optic input by enucleation of the eye results in sustained depression of mitotic activity in the tectal proliferation zone on the denervated side compared to the intact side. Temporary denervation by optic-nerve crush initially has a similar effect, but on reinnervation of the tectum by the regenerating optic fibers, proliferation is enhanced on the experimental side compared to the control side.

NEURONAL REGENERATION

The phenomenon of adult neurogenesis in teleost fish is closely linked to their enormous ability to replace neurons lost to injury by newly generated neurons. This so-called neuronal regeneration has been particularly well studied in the following three neuronal systems: the retina and its primary projection target, the optic tectum (for review, see Otteson and Hitchcock 2003); the spinal cord (for review, see Waxman and Anderson 1986); and the cerebellum (for reviews, see Zupanc 1999, 2001, 2006a,b; Zupanc and Clint 2003; Zupanc and Zupanc 2006). In the following, I focus on the latter neuronal system.

Investigations on neuronal regeneration in the teleostean cerebellum are based on a well-defined lesion paradigm in *A. leptorhynchus* (Fig. 5A) (Zupanc et al. 1998). This paradigm takes advantage of the prominent position of one subdivision of the cerebellum, the corpus cerebelli, which forms a roof on top of the brain, except over the telencephalon. Cerebellar lesions can therefore be made in a defined way by puncturing the skull of the anesthetized fish with a sterile surgical blade, without damaging other parts of the brain.

Nissl stains of sections taken through the area of such lesions after various postlesion survival times have indicated that neuronal tissue is restored within a few weeks following the injury. This enormous regenerative potential appears to be mediated by two major processes: elimination of damaged cells through apoptosis, and replacement of these cells by newly generated ones.

The first apoptotic cells are detectable at the lesion site as early as 5 minutes after application of the stab wound (Zupanc et al. 1998). Thirty minutes after the lesion, the number of cells undergoing apoptosis reaches maximum levels. At 2 days postlesion, the number of apoptotic cells starts to gradually decline until background levels are reached approximately 20 days after the lesion. Over this time period, only a few cells appear to undergo necrotic cell death.

The elimination of injured cells almost exclusively by apoptosis is remarkable because this contrasts with the situation in mammals. In the latter taxon, necrotic cell death dominates within the core of the injury. Apoptosis occurs in a significant number of cells only in surrounding areas (for reviews, see Beattie et al. 2000; Vajda 2002; Liou et al. 2003). Necrosis often leads to inflammation at the site of the injury (for review, see Kerr et al. 1995). This inflammatory response triggers further necrotic events, thus gradually transforming the site of injury into large cavities devoid of cells (Zhang et al. 1997). These cavities are typically bordered

by astrocytic scars that act as mechanical and biochemical barriers preventing the ingrowth of nerve fibers and the migration of cells into the
lesion site (for review, see Reier et al. 1983).

In contrast to necrosis, apoptosis is characterized by cell shrinkage,
nuclear condensation, and production of membrane-enclosed particles
that are digested by other cells. Most significantly, the side effects that
accompany necrosis, such as inflamation of the surrounding tissue, are
typically absent in apoptosis. Thus, use of this "clean" type of cell death
for the elimination of damaged cells appears to be an essential component of the enormous regenerative capability of the adult fish brain.

The cellular debris caused by apoptosis is thought to be removed by
the action of microglia/macrophages. Their number is very low in the

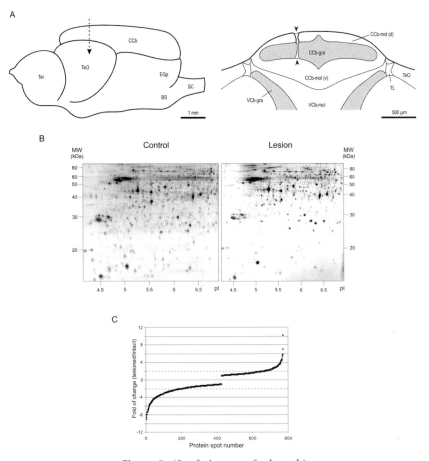

Figure 5. *(See facing page for legend.)*

intact brain, but starts to increase at and near the lesion site approximately 3 days after application of the lesion, and reaches maximum levels at 10 days (Zupanc et al. 2003). One month after the lesion, the number of microglia/macrophages has returned to background levels. A similar increase in the number of microglia/macrophages in response to CNS lesions, paralleled by active phagocytic activity, has been found in other systems and species (for reviews, see Kreutzberg 1996; Moore and Thanos 1996; Streit et al. 1999).

In the cerebellum of A. *leptorhynchus*, the cells eliminated through apoptosis are replaced by newly generated cells within a few weeks following the lesion. Most of these cells are produced 1–10 days after the injury. During this time period, the rate of cell proliferation is increased severalfold at the site of the lesion, compared to unlesioned sites of the corpus cerebelli (Zupanc and Ott 1999). As in the intact cerebellum, the cells destined to replace the ones lost to the injury are guided to the lesion site by radial glial fibers (Clint and Zupanc 2001). The density of these fibers starts to increase approximately 8 days postlesion. This effect is restricted to the sites at and near the lesion.

At least some, but presumably most, of the new cells recruited for tissue repair develop into granular neurons—the phenotype of most of

Figure 5. Proteome analysis of regeneration-associated proteins in the adult brain of teleost fish. (A) The cerebellar lesion paradigm. (*Left*) Side view of the brain of *Apteronotus leptorhynchus*. One subdivision of the cerebellum, the corpus cerebelli (CCb), forms a roof on top of the brain, except over the telencephalon (Tel). Cerebellar lesions can therefore be made in a defined way by puncturing the skull of the fish with a surgical blade, without damaging other parts of the brain. (*Right*) Transverse section through the corpus cerebelli taken at the level indicated by the arrow in the left drawing. The lesion path (*arrowheads*) travels through the dorsal molecular layer (CCb-mol[d]) and the granular layer (CCb-gra) about halfway between the midline of the brain and the lateral edge of the granular layer. (BS) Brain stem; (CCb-mol[v]), ventral molecular layer of corpus cerebelli; (EGp) eminentia granularis pars posterior; (SC) spinal cord; (TeO) optic tectum; (TL) torus longitudinalis; (VCb-gra) granule cell layer of valvula cerebelli; (VCb-mol) molecular layer of valvula cerebelli. (B) Images of 2D gels of proteins from the intact corpus cerebelli ("Control") and lesioned corpus cerebelli ("Lesion"). (C) Scatter plot of the fold changes of 772 protein spots in the corpus cerebelli of the brown ghost knifefish 3 days after a lesion, compared to unlesioned controls. Decrease in protein abundance is indicated by negative changes, increases by positive changes. Proteins are sorted in ascending order of the spot intensity change. The analysis is based on a comparison of the averaged spot intensities of four 2D gels from lesioned brains and three 2D gels from intact brains. The dotted lines indicate the +2-fold and −2-fold threshold. (After Zupanc and Zupanc 2006; Zupanc et al. 2006.)

the neurons that are lost to the injury. Retrograde tracing experiments combined with labeling of S-phase cells with BrdU have furthermore shown that these new granule cell neurons send axons into the associated molecular layer (Zupanc and Ott 1999). This suggests that the new granular neurons integrate into the existing neural network of the cerebellum.

Differential proteome analysis has suggested that a large number of proteins are involved in the individual steps of the regenerative process. Examination of cerebellar tissue collected 3 days after application of lesions (when a marked increase in the generation of new cells recruited for the replacement of the cells lost to injury occurs [Zupanc and Ott 1999]) has revealed alterations in protein-spot intensity on 2D gels in 53 out of nearly 800 protein spots (Fig. 5B) (Zupanc et al. 2006). Twenty-four of these differentially regulated proteins could be identified by peptide mass fingerprinting and mass spectrometry/mass spectrometry fragmentation. They include cytoskeletal proteins, proteins potentially involved in cell proliferation, cellular motility, neuroprotection, and energy metabolism, as well as one transcriptional regulator. Thus, this study has shown that identification of the multitude of proteins involved in the individual steps of the process of neuronal regeneration is feasible in the foreseeable future.

SUMMARY

Fish are distinctive in their enormous potential to continuously produce new neurons in the adult brain. Whereas in mammals, adult neurogenesis is restricted to the OB and the hippocampus, in fish, new neurons are generated not only in structures homologous to these two regions, but also in dozens of other brain areas. These neurons arise from adult stem cells capable of multipotency and self-renewal. In some regions of the fish brain, the new cells remain near the proliferation zones in the course of their further development. In others, they migrate, often guided by radial glial fibers, to specific target areas. Approximately 50% of the young cells undergo apoptotic cell death, whereas the others survive for the rest of the fish's life. A large number of the surviving cells differentiate into neurons. One mechanism mediating gene expression during development of the new neurons appears to be the loss and gain of chromosomes by somatic-genomic alteration. The continued generation of new neurons in the adult brain is closely related to the ability of fish to replace neurons lost to injury by newly generated neurons. This so-called neuronal regeneration involves a series of highly orchestrated processes. A combination of a well-defined cerebellar lesion paradigm with differential proteome analysis has shown that identification of the multitude of proteins mediating the regenerative potential of the adult fish brain is feasible in the foreseeable future.

ACKNOWLEDGMENTS

Investigations of the author have been supported by funds of the Max-Planck Gesellschaft, Bundesministerium für Forschung und Technologie, Royal Society, Leverhulme Trust, Biotechnology and Biological Sciences Research Council, Wellcome Trust, Deutsche Forschungsgemeinschaft, International University Bremen, Wilhelm Herbst Stiftung, and Tönjes Vagt Stiftung. I thank Marianne M. Zupanc for critically reading various drafts of the manuscript.

REFERENCES

Beattie M.S., Farooqui A.A., and Bresnahan J.C. 2000. Review of current evidence for apoptosis after spinal cord injury. *J. Neurotrauma* **17:** 915–925.

Bell C., Bodznick D., Montgomery J., and Bastian J. 1997. The generation and subtraction of sensory expectations within cerebellum-like structures. *Brain Behav. Evol.* (suppl. 1): **50:** 17–31.

Braford M.R. 1995. Comparative aspects of forebrain organization in the ray-finned fishes: Touchstones or nor? *Brain Behav. Evol.* **46:** 259–274.

Butler A.B. 2000. Topography and topology of the teleost telencephalon: A paradoxon resolved. *Neurosci. Lett.* **293:** 95–98.

Caddy K.W.T. and Biscoe T.J. 1976. The number of Purkinje cells and olive neurones in the normal and lurcher mutant mouse. *Brain Res.* **111:** 396–398.

Cameron H.A. and McKay R.D. 2001. Adult neurogenesis produces a large pool of new granule cells in the dentate gyrus. *J. Comp. Neurol.* **435:** 406–417.

Candal E., Anadón R., DeGrip W.J., and Rodríguez-Moldes I. 2005. Patterns of cell proliferation and cell death in the developing retina and optic tectum of the brown trout. *Dev. Brain Res.* **154:** 101–119.

Clint S.C. and Zupanc G.K.H. 2001. Neuronal regeneration in the cerebellum of adult teleost fish, *Apteronotus leptorhynchus:* Guidance of migrating young cells by radial glia. *Dev. Brain Res.* **130:** 15–23.

Corotto F.S., Henegar J.A., and Maruniak J.A. 1993. Neurogenesis persists in the subependymal layer of the adult mouse brain. *Neurosci. Lett.* **149:** 111–114.

Corwin J.T. 1981. Postembryonic production and aging of inner ear hair cells in sharks. *J. Comp. Neurol.* **201:** 541–553.

Ekström P., Johnsson C.-M., and Ohlin L.-M. 2001. Ventricular proliferation zones in the brain of an adult teleost fish and their relation to neuromeres and migration (secondary matrix) zones. *J. Comp. Neurol.* **436:** 92–110.

Gerhard G.S., Kauffman E.J., Wang X., Stewart R., Moore J.L., Kasales C.J., Demidenko E., and Cheng K.C. 2002. Life spans and senescent phenotypes in two strains of zebrafish (*Danio rerio*). *Exp. Gerontol.* **37:** 1055–1068.

Gisselsson D. and Höglund M. 2005. Connecting mitotic instability and chromosome aberrations in cancer: Can telomeres bridge the gap? *Semin. Cancer Biol.* **15:** 13–23.

Grandel H., Kaslin J., Ganz J., Wenzel I., and Brand M. 2006. Neural stem cells and neurogenesis in the adult zebrafish brain: Origin, proliferation dynamics, migration and cell fate. *Dev. Biol.* **295:** 263–277.

Hagedorn M. and Fernald R.D. 1992. Retinal growth and cell addition during embryogenesis in the teleost, *Haplochromis burtoni*. *J. Comp. Neurol.* **321:** 193–208.

Herculano-Houzel S. and Lent R. 2005. Isotropic fractionator: A simple, rapid method for the quantification of total cell and neuron numbers in the brain. *J. Neurosci.* **25:** 2518–2521.

Hinsch K. and Zupanc G.K.H. 2006. Isolation, cultivation, and differentiation of neural stem cells from adult fish brain. *J. Neurosci. Methods* **158:** 75–88.

———. 2007. Generation and long-term persistence of new neurons in the adult zebrafish brain: A quantitative analysis. *Neuroscience* **146:** 679–696.

Johns P.R. 1977. Growth of the adult goldfish eye. III. Source of the new retinal cells. *J. Comp. Neurol.* **176:** 343–358.

Johns P.R. and Easter S.S.J. 1977. Growth of the adult goldfish eye: II. Increase in retinal cell number. *J. Comp. Neurol.* **176:** 331–342.

Kaplan M.S., McNelly N.A., and Hinds J.W. 1985. Population dynamics of adult-formed granule neurons of the rat olfactory bulb. *J. Comp. Neurol.* **239:** 117–125.

Kaushal D., Contos J.J.A., Treuner K., Yang A.H., Kingsbury M.A., Rehen S.K., McConnell M.J., Okabe M., Barlow C., and Chun J. 2003. Alteration of gene expression by chromosome loss in the postnatal mouse brain. *J. Neurosci.* **23:** 5599–5606.

Kempermann G., Gast D., Kronenberg G., Yamaguchi M., and Gage F.H. 2003. Early determination and long-term persistence of adult-generated new neurons in the hippocampus of mice. *Development* **130:** 391–399.

Kerr J.F.R., Gobé G.C., Winterford C.M., and Harmon B.V. 1995. Anatomical methods in cell death. In *Cell death* (ed. L.M. Schwartz and B.A. Osborne), pp. 1–27. Academic Press, San Diego.

Kingsbury M.A., Friedman B., McConnell M.J., Rehen S.K., Yang A.H., Kaushal D., and Chun J. 2005. Aneuploid neurons are functionally active and integrated into brain circuitry. *Proc. Natl. Acad. Sci.* **102:** 6143–6147.

Koumans J.T.M. and Akster H.A. 1995. Myogenic cells in development and growth of fish. *Comp. Biochem. Physiol.* **110A:** 3–20.

Kranz D. and Richter W. 1970a. Autoradiographische Untersuchungen über die Lokalisation der Matrixzonen des Diencephalons von juvenilen und adulten *Lebistes reticulatus* (Teleostei). *Z. mikrosk.-anat. Forsch.* **82:** 42–66.

———. 1970b. Autoradiographische Untersuchungen zur DNS-Synthese im Cerebellum und in der Medulla oblongata von Teleostiern verschiedenen Lebensalters. *Z. mikrosk.-anat. Forsch.* **82:** 264–292.

Kreutzberg G.W. 1996. Microglia: A sensor for pathological events in the CNS. *Trends Neurosci.* **19:** 312–318.

Liou A.K.F., Clark R.S., Henshall D.C., Yin X.-M., and Chen J. 2003. To die or not to die for neurons in ischemia, traumatic brain injury and epilepsy: A review on the stress-activated signaling pathways and apoptotic pathways. *Prog. Neurobiol.* **69:** 103–142.

Lois C. and Alvarez-Buylla A. 1994. Long-distance neuronal migration in the adult mammalian brain. *Science* **264:** 1145–1148.

Mansour-Robaey S. and Pinganaud G. 1990. Quantitative and morphological study of cell proliferation during morphogenesis in the trout visual system. *J. Hirnforsch.* **31:** 495–504.

Marcus R.C., Delaney C.L., and Easter S.S. 1999. Neurogenesis in the visual system of embryonic and adult zebrafish (*Danio rerio*). *Vis. Neurosci.* **16:** 417–424.

Meyer R.L. 1978. Evidence from thymidine labelling for continuing growth of retina and tectum in juvenile goldfish. *Exp. Neurol.* **59:** 99–111.

Moore S. and Thanos S. 1996. The concept of microglia in relation to central nervous system disease and regeneration. *Prog. Neurobiol.* **48:** 441–460.

Nguyen V., Deschet K., Henrich T., Godet E., Joly J.S., Wittbrodt J., Chourrout D., and Bourrat F. 1999. Morphogenesis of the optic tectum in the medaka (*Oryzias latipes*): A morphological and molecular study, with special emphasis on cell proliferation. *J. Comp. Neurol.* **413:** 385–404.

Nieuwenhuys R. and Meek J. 1990. The telencephalon of actinopterygian fishes. In *Comparative structure and evolution of the cerebral cortex* (ed. E.G. Jones and A. Peters), pp. 31–73. Plenum, New York.

Northcutt R.G. 1995. The forebrain of gnathostomes: In search of a morphotype. *Brain Behav. Evol.* **46:** 275–318.

Northcutt R.G. and Braford M.R. 1980. New observations on the organization and evolution of the telencephalon of actinopterygian fishes. In *Comparative neurology of the telencephalon* (ed. S.O.E. Ebbesson), pp. 41–98. Plenum, New York.

Osada T., Kusakabe H., Akutsu H., Yagi T., and Yanagimachi R. 2002. Adult murine neurons: Their chromatin and chromosome changes and failure to support embryonic development as revealed by nuclear transfer. *Cytogenet. Genome Res.* **97:** 7–12.

Ott R., Zupanc G.K.H., and Horschke I. 1997. Long-term survival of postembryonically born cells in the cerebellum of gymnotiform fish, *Apteronotus leptorhynchus. Neurosci. Lett.* **221:** 185–188.

Otteson D.C. and Hitchcock P.F. 2003. Stem cells in the teleost retina: Persistent neurogenesis and injury-induced regeneration. *Vision Res.* **43:** 927–936.

Paulin M.G. 1993. The role of the cerebellum in motor control and perception. *Brain Behav. Evol.* **41:** 39–50.

Portavella M., Torres B., and Salas C. 2004. Avoidance response in goldfish: Emotional and temporal involvement of medial and lateral telencephalic pallium. *J. Neurosci.* **24:** 2335–2342.

Pouwels E. 1978a. On the development of the cerebellum of the trout, *Salmo gairdneri:* I. Patterns of cell migration. *Anat. Embryol.* **152:** 291–308.

———. 1978b. On the development of the cerebellum of the trout, *Salmo gairdneri:* III. Development of neuronal elements. *Anat. Embryol.* **153:** 37–54.

Raff M.C. 1992. Social controls on cell survival and cell death. *Nature* **356:** 397–400.

Raff M.C., Barres B.A., Burne J.F., Coles H.S., Ishizaki Y., and Jacobson M.D. 1993. Programmed cell death and the control of cell survival: Lessons from the nervous system. *Science* **262:** 695–700.

Rajendran R.S., Zupanc M.M., Lösche A., Westra J., Chun J., and Zupanc G.K.H. 2007. Numerical chromosome variation and mitotic segregation defect in the adult brain of teleost fish. *Dev. Neurobiol.* **67:** 1334–1347.

Raymond P.A. and Easter S.S. 1983. Postembryonic growth of the optic tectum in goldfish. I. Location of germinal cells and numbers of neurons produced. *J. Neurosci.* **3:** 1077–1091.

Raymond P., Easter S., Burnham J., and Powers M. 1983. Postembryonic growth of the optic tectum in goldfish. II. Modulation of cell proliferation by retinal fiber input. *J. Neurosci.* **3:** 1092–1099.

Rehen S.K., McConnell M.J., Kaushal D., Kingsbury M.A., Yang A.H., and Chun J. 2001. Chromosomal variation in neurons of the developing and adult mammalian nervous system. *Proc. Natl. Acad. Sci.* **98:** 13361–13366.

Rehen S.K., Yung Y.C., McCreight M.P., Kaushal D., Yang A.H., Almeida B.S.V., Kingsbury M.A., Cabral K.M.S., McConnell M.J., Anliker B., et al. 2005. Constitutional aneuploidy in the normal human brain. *J. Neurosci.* **25:** 2176–2180.

Reier P.J., Stensaas L.J., and Guth L. 1983. The astrocytic scar as an impediment to regeneration in the central nervous system. In *Spinal cord reconstruction* (ed. C.C. Kao et al.), pp. 163–195. Raven Press, New York.

Richter W. and Kranz D. 1970a. Autoradiographische Untersuchungen über die Abhängigkeit des ^{3}H-Thymidin-Index vom Lebensalter in den Matrixzonen des Telencephalons von *Lebistes reticulatus* (Teleostei). *Z. mikrosk.-anat. Forsch.* **81:** 530–554.

———. 1970b. Die Abhängigkeit der DNS-Synthese in den Matrixzonen des Mesencephalons vom Lebensalter der Versuchstiere (*Lebistes reticulatus* - Teleostei): Autoradiographische Untersuchungen. *Z. mikrosk.-anat. Forsch.* **82:** 76–92.

Rodríguez F., López J.C., Vargas J.P., Gómez Y., Broglio C., and Salas C. 2002. Conservation of spatial memory function in the pallial forebrain of reptiles and ray-finned fishes. *J. Neurosci.* **22:** 2894–2903.

Rowe R.W.D. and Goldspink G. 1969. Muscle fibre growth in five different muscles in both sexes of mice. *J. Anat.* **104:** 519–530.

Rowlerson A. and Veggetti A. 2001. Cellular mechanisms of post-embryonic muscle growth in aquaculture species. In *Muscle development and growth* (ed. I.A. Johnston), pp. 103–140. Academic Press, San Diego.

Schwenk F., Zevnik B., Brüning J., Röhl M., Willuweit A., Rode A., Hennek T., Kauselmann G., Jaenisch R., and Kühn R. 2003. Hybrid embryonic stem cell-derived tetraploid mice show apparently normal morphological, physiological, and neurological characteristics. *Mol. Cell. Biol.* **23:** 3982–3989.

Soutschek J. and Zupanc G.K.H. 1996. Apoptosis in the cerebellum of adult teleost fish, *Apteronotus leptorhynchus*. *Dev. Brain Res.* **97:** 279–286.

Stewénius Y., Gorunova L., Jonson T., Larsson N., Höglund M., Mandahl N., Mertens F., Mitelman F., and Gisselsson D. 2005. Structural and numerical chromosome changes in colon cancer develop through telomere-mediated anaphase bridges, not through mitotic multipolarity. *Proc. Natl. Acad. Sci.* **102:** 5541–5546.

Streit W.J., Walter S.A., and Pennell N.A. 1999. Reactive microgliosis. *Prog. Neurobiol.* **57:** 563–581.

Swisher D.A. and Wilson D.B. 1977. Cerebellar histogenesis in the lurcher *(Lc)* mutant mouse. *J. Comp. Neurol.* **173:** 1038–1041.

Vajda F.J. 2002. Neuroprotection and neurodegenerative disease. *J. Clin. Neurosci.* **9:** 4–8.

Vargas J.P., Rodríguez F., López J.C., Arias J.L., and Salas C. 2000. Spatial learning-induced increase in the argyophilic nucleolar organizer region of dorsolateral telencephalic neurons in goldfish. *Brain Res.* **865:** 77–84.

Voigt T. 1989. Development of glial cells in the cerebral wall of ferrets: Direct tracing of their transformation from radial glia into astrocytes. *J. Comp. Neurol.* **289:** 74–88.

Waxman S.G. and Anderson M.J. 1986. Regeneration of central nervous structures: *Apteronotus* spinal cord as a model system. In *Electroreception* (ed. T.H. Bullock and W. Heiligenberg), pp. 183–208. John Wiley & Sons, New York.

Weatherley A.H. and Gill H.S. 1985. Dynamics of increase in muscle fibres in fishes in relation to size and growth. *Experientia* **41:** 353–354.

Wetts R. and Herrup K. 1983. Direct correlation between Purkinje and granule cell number in the cerebella of lurcher chimeras and wild-type mice. *Brain Res.* **312:** 41–47.

Williams R.W. 2000. Mapping genes that modulate brain development: A quantitative genetic approach. In *Mouse brain development* (ed. A.F. Goffinet and P. Rakic), pp. 21–49. Springer-Verlag, New York.

Winner B., Cooper-Kuhn C.M., Aigner R., Winkler J., and Kuhn H.G. 2002. Long-term survival and cell death of newly generated neurons in the adult rat olfactory bulb. *Eur. J. Neurosci.* **16:** 1681–1689.

Wullimann M.F. and Puelles L. 1999. Postembryonic neural proliferation in the zebrafish forebrain and its relationship to prosomeric domains. *Anat. Embryol.* **199:** 329–348.

Yang A.H., Kaushal D., Rehen S.K., Kriedt K., Kingsbury M.A., McConnell M.J., and Chun J. 2003. Chromosome segregation defects contribute to aneuploidy in normal neural progenitor cells. *J. Neurosci.* **23:** 10454–10462.

Yurov Y.B., Iourov I.Y., Monakhov V.V., Soloviev I.V., Vostrikov V.M., and Vorsanova S.G. 2005. The variation of aneuploidy frequency in the developing and adult human brain revealed by an interphase FISH study. *J. Histochem. Cytochem.* **53:** 385–390.

Zakon H.H. 1984. Postembryonic changes in the peripheral electrosensory system of a weakly electric fish: Addition of receptor organs with age. *J. Comp. Neurol.* **228:** 557–570.

Zhang Z., Krebs C.J., and Guth L. 1997. Experimental analysis of progressive necrosis after spinal cord trauma in the rat: Etiological role of the inflammatory response. *Exp. Neurol.* **143:** 141–152.

Zikopoulos B., Kentouri M., and Dermon C.R. 2000. Proliferation zones in the adult brain of a sequential hermaphrodite teleost species *(Sparus aurata). Brain Behav. Evol.* **56:** 310–322.

Zimmerman A.M. and Lowery M.S. 1999. Hyperplastic development and hypertrophic growth of muscle fibers in the white seabass (*Atractoscion nobilis*). *J. Exp. Zool.* **284:** 299–308.

Zupanc G.K.H. 1999. Neurogenesis, cell death and regeneration in the adult gymnotiform brain. *J. Exp. Biol.* **202:** 1435–1446.

———. 2001. Adult neurogenesis and neuronal regeneration in the central nervous system of teleost fish. *Brain Behav. Evol.* **58:** 250–275.

———. 2006a. Neurogenesis and neuronal regeneration in the adult fish brain. *J. Comp. Physiol. A* **192:** 649–670.

———. 2006b. Adult neurogenesis and neuronal regeneration in the teleost fish brain: implications for the evolution of a primitive vertebrate trait. In *The evolution of nervous systems in non-mammalian vertebrates* (ed. T.H. Bullock and L.R. Rubenstein), pp. 485–520. Academic Press, Oxford.

Zupanc G.K.H. and Clint S.C. 2003. Potential role of radial glia in adult neurogenesis of teleost fish. *Glia* **43:** 77–86.

Zupanc G.K.H. and Horschke I. 1995. Proliferation zones in the brain of adult gymnotiform fish: A quantitative mapping study. *J. Comp. Neurol.* **353:** 213–233.

Zupanc G.K.H. and Ott R. 1999. Cell proliferation after lesions in the cerebellum of adult teleost fish: time course, origin, and type of new cells produced. *Exp. Neurol.* **160:** 78–87.

Zupanc G.K.H. and Zupanc M.M. 2006. New neurons for the injured brain: Mechanisms of neuronal regeneration in adult teleost fish. *Regenerative Med.* **1:** 207–216.

Zupanc G.K.H., Hinsch K., and Gage F.H. 2005. Proliferation, migration, neuronal differentiation, and long-term survival of new cells in the adult zebrafish brain. *J. Comp. Neurol.* **488:** 290–319.

Zupanc G.K.H., Horschke I., Ott R., and Rascher G.B. 1996. Postembryonic development of the cerebellum in gymnotiform fish. *J. Comp. Neurol.* **370:** 443–464.

Zupanc G.K.H., Kompass K.S., Horschke I., Ott R., and Schwarz H. 1998. Apoptosis after injuries in the cerebellum of adult teleost fish. *Exp. Neurol.* **152:** 221–230.

Zupanc G.K.H., Clint S.C., Takimoto N., Hughes A.T.L., Wellbrock U.M., and Meissner D. 2003. Spatio-temporal distribution of microglia/macrophages during regeneration in the cerebellum of adult teleost fish, *Apteronotus leptorhynchus:* A quantitative analysis. *Brain Behav. Evol.* **62:** 31–42.

Zupanc M.M., Wellbrock U.M., and Zupanc G.K.H. 2006. Proteome analysis identifies novel protein candidates involved in regeneration of the cerebellum of teleost fish. *Proteomics* **6:** 577–696.

28

Neurogenesis in the Adult Songbird: A Model for Inducible Striatal Neuronal Addition

Steve A. Goldman
Division of Cell and Gene Therapy
Departments of Neurology and Neurosurgery
The University of Rochester Medical Center
Rochester, New York 14642

THE DAMAGED BRAIN HAS TRADITIONALLY BEEN THOUGHT to exhibit little significant structural repair after injury. In part, this appears to reflect the failure of the mature forebrain to generate new neurons, except for a few discrete, relatively archaic regions of the brain, the hippocampus and olfactory bulb (OB) (Altman and Das 1966; Bayer et al. 1982; for review, see Goldman 1998; Alvarez-Buylla and Garcia-Verdugo 2002; Gage 2002). The limitation on neuronal addition to the adult brain has clearly been selected, and thus comprises an adaptation of likely, if unclear, evolutionary benefit. Among other possibilities, the lack of persistent neurogenesis in most regions of the adult mammalian brain may be associated with the need to stabilize the retention of long-term memories and entrained behaviors (Rakic 2002). Perhaps as a result, the adult mammalian neocortex exhibits no constitutive neuronal addition, and little or none after injury, except for discrete experimental lesions of defined neuronal populations (Magavi et al. 2000). In contrast, the subcortical neostriatum retains the capacity to regenerate neurons after stroke and major traumatic injury (Arvidsson et al. 2002; Parent et al. 2002; Jin et al. 2003a; Parent 2003). However, the numbers of striatal neurons generated in response to stroke have thus far been described as comprising only a small fraction of the population lost to the ischemic insult and have not yet been demonstrated to contribute to functional recovery.

In general terms, the lack of compensatory neuronal replacement in most adult brain regions has impeded not only the recovery of function after brain injury, but also the development of cell-targeted treatments for those injuries. Yet, the overall lack of neuronal production in the adult vertebrate forebrain, whether compensatory or constitutive, appears to reflect not a lack of appropriate precursor cells, but rather their tonic inhibition from generating new neurons able to survive and migrate (for review, see Goldman 2003). In the 1980s, a series of studies indicated that neurogenesis persists in a number of gonadal steroid-responsive regions of the adult avian brain and that neuronal progenitor cells are distributed widely throughout the ventricular subependyma of the adult avian forebrain (Goldman and Nottebohm 1983; Alvarez-Buylla and Nottebohm 1988). In the 1990s, studies arising from a rapidly expanding spectrum of laboratories revealed that more broadly defined neural stem cells, as well as their derived neuronally restricted transit-amplifying progenitor cells, were similarly dispersed throughout the mammalian ventricular subependyma (Lois and Alvarez-Buylla 1993; Morshead et al. 1994; Kirschenbaum and Goldman 1995; Doetsch and Alvarez-Buylla 1996; Weiss et al. 1996; Pincus et al. 1998). These subependymal neural stem and progenitor cells continue to produce neurons that are recruited to a few discrete regions: The olfactory subependyma and bulb and the dentate gyrus (DG) of the hippocampus. More generally, though, neural stem cells in adult mammals appear to either generate new astrocytes and oligodendrocytes or rather become vestigial (Morshead and van der Kooy 1992). Together, these observations led to the concept that persistent neuroepithelial stem cells and their more restricted daughter cells, comprising distinct populations of transit-amplifying neuronal and glial progenitors, are maintained in the ventricular lining of all adult vertebrates (for review, see Goldman and Luskin 1998; Gage 2000; Alvarez-Buylla et al. 2001), including humans (Kirschenbaum et al. 1994; Pincus et al. 1998; Sanai et al. 2004). The differentiated fate of these cells varies across species and regions but is invariably dependent on regional cues that serve to establish local niches for neurogenesis and gliogenesis (for review, see Alvarez-Buylla and Lim 2004).

Whereas mammals appear to have widespread populations of potentially neurogenic stem cells, but actually exhibit neuronal addition to only a few discrete regions, birds recruit new neurons far more broadly, with ongoing neuronal addition to a number of regions in the neostriatum, as well as in the more phylogenetically primitive paraolfactory and parahippocampal regions (Goldman and Nottebohm 1983; Nottebohm 1985; Alvarez-Buylla and Nottebohm 1988; Alvarez-Buylla et al. 1990; Barnea and Nottebohm 1994). The addition of new neurons to the neostriatum is of particular interest in that the regions involved are responsive to gonadal hormones and exhibit androgen- and estrogen-dependent modulation of neuronal addition and survival. These regions include the higher vocal center (HVC), as well as adjacent areas of the mediocaudal neostriatum involved in auditory perception. In songbirds, these regions of the brain are robustly neurogenic, exhibit well-described hormonal responsiveness with discrete anatomic concomitants, and are associated with a well-described and readily quantifiable behavioral output, song. As a result, the songbird HVC and its associated mediocaudal neostriatum have evolved as uniquely valuable model systems within which to identify permissive conditions for the migration and integration of new neurons in the adult brain (for review, see Nottebohm 1985, 2002; Goldman 1998). This chapter focuses on the molecular mechanisms and cell–cell interactions that permit neuronal production, migration, and integration in the adult songbird brain. It then considers how our nascent understanding of neurogenesis in the adult avian brain may inform new approaches to inducing neuronal replacement in the diseased adult neostriatum and, in particular, in its human homolog, the caudate-putamen. (For more detailed consideration of induced neurogenesis as a repair strategy, see Goldman 2004, 2005.)

NEUROGENESIS IN THE ADULT SONGBIRD FOREBRAIN

Reports of ependymal neurogenesis in the spinal cord of both bony and cartilaginous adult fish (Leonard et al. 1978; Birse et al. 1980; Waxman and Anderson 1985) led to the search for adult neurogenesis among higher species. Perhaps the most robust and best-understood example of adult neurogenesis was among the first discovered, that of the adult songbird forebrain. The songbird exhibits topographically widespread neurogenesis throughout adulthood, most remarkably so in an area of the dorsomediocaudal neostriatum concerned with song learning, the HVC (Goldman and Nottebohm 1983). This forebrain region generates new neurons within the ventricular zone or, as classically designated in adults, the subependymal zone (SZ) (Boulder Committee 1970). Both humoral and contact-mediated signals modulate the migration and survival of new neurons in the HVC through an orchestrated set of hormonally modulated paracrine interactions (Louissaint et al. 2002). New neurons of the songbird brain depart the SZ to enter the brain parenchyma by migrating upon radial guide fibers, which emanate from cell bodies in the ventricular epithelium (Alvarez-Buylla and Nottebohm 1988). The radial guide cells coderive with new neurons from a common progenitor, which is widespread throughout the songbird SZ (Goldman et al. 1996b). Neuronal production in the SZ is thus followed by the migration of the new daughter cells into the subjacent parenchyma along these radial fibers; the neuroblasts thereby infiltrate the HVC, where they cease migration and differentiate into physiologically functional, synaptically integrated members of the local neuronal network (Fig. 1) (Paton and Nottebohm 1984). The extent of this ongoing neuronal recruitment to the HVC is notable: The adult canary HVC alone recruits over 1.4% of its neurons daily, with a gonadal hormone-dependent regulation of post-mitotic neuronal survival (Rasika et al. 1994; Hidalgo et al. 1995). Yet, neurogenesis in the adult songbird brain is hardly limited to the HVC: Similarly robust neuronal production has been noted in the adjacent mediocaudal striatum, as well as in the lobus paraolfactorius and parahippocampus (Alvarez-Buylla et al. 1990, 1994).

Many of the newly generated neurons in the HVC go on to establish long-distance projections to one of the HVC's distant target nuclei, robustus archistriatalis (RA) (Alvarez-Buylla and Kirn 1997), indicating that projection neurons, as well as local interneurons (Paton et al. 1985), can derive from adult progenitors. Interestingly, the successful long-distance extension of new axons through adult brain parenchyma to distant forebrain centers, as exhibited by HVC neurons projecting to the RA, has also been noted by newly induced medium spiny neurons of the adult rodent neostriatum (Chmielnicki et al. 2004). Together, these observations suggest that in the otherwise relatively nonpermissive environment of the adult brain, newly generated neurons may be uniquely competent to exhibit long-distance axogenesis; as such, they may not be subject to the same repulsive constraints upon axonal extension as their already resident, more stable neuronal counterparts.

It is important to note that neurogenesis by the adult songbird subependymal zone may be assessed in vitro as well as in vivo. Indeed, the establishment of long-term explant cultures of the songbird's neostriatal SZ first made it possible to examine both the contact-mediated and humoral regulation of neuronal production and migration in the adult brain (Goldman 1990; Goldman et al. 1992). Cultured explants of the HVC displayed substantial neuronal outgrowth, which was observed only from the SZ, and was composed entirely of newly generated neurons, as defined by their in vitro incorporation of [^{3}H]thymidine (Goldman 1990; Goldman and Nedergaard 1992). Neurogenesis could be supported for extended periods of time in these cultures and was inversely proportional to the ambient serum concentration (Goldman et al. 1992). These

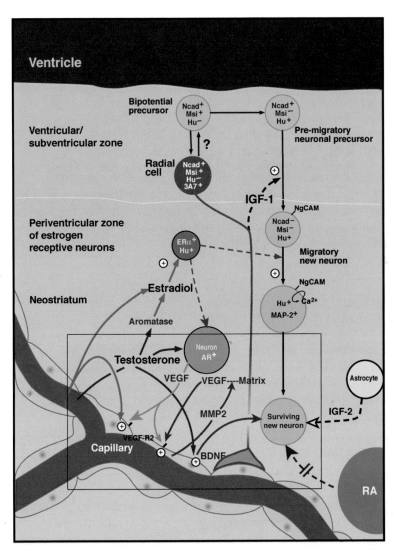

Figure 1. Neuronal migration into the adult songbird neostriatum. This schematic summarizes much of what is known about neuronal recruitment into the avian brain, particularly as it proceeds in the mediocaudal neostriatal region including the song control nucleus HVC (see text). (3A7) Vimentin-associated filament recognized by Mab 3A7 (Goldman et al. 1993); (ER) estrogen receptor; (musashi) an RNA-binding protein expressed by neural progenitors (Sakakibara and Okano 1997; Keyoung et al. 2001); (IGF-1 and IGF-2) insulin-like growth factors 1 and 2; (MAP-2) microtubule-associated protein-2; (Ncad) N-cadherin; (NgCAM) neuro-glial cell adhesion molecule. (Modified from Louissaint et al. 2002.)

findings suggested that HVC neurogenesis in vivo might be modulated or frankly suppressed by serum-borne or induced agents, an observation akin to the tonic inhibition of neurogenesis from resident neural stem cells noted in most regions of the adult rodent brain.

RADIAL CELL SUPPORT OF NEURONAL MIGRATION INTO THE ADULT SONGBIRD BRAIN

Neuronal migration into the adult songbird brain occurs along a system of guide fibers that emanate from radial guide cells of the ventricular epithelium (Goldman and Nottebohm 1983; Alvarez-Buylla and Nottebohm 1988; Alvarez-Buylla et al. 1990). In cultures of the adult HVC SZ, radial fibers emanate from the ciliated ependymal cells that initially depart the SZ explants (Goldman et al. 1993). These adult radial cells can generate fibers that span >2 cm in length in culture and support the in vitro migration of SZ-derived neurons. In vivo, these radial cells appear to comprise a unique subpopulation of pseudostratified ependymal cells, characterized by a single apical cilium, that abut the ventricular lumen and constitute the major dividing cell type of the adult avian SZ (Alvarez-Buylla et al. 1998). The fibers emanating from these ependymal cell bodies extend widely throughout the avian telencephalon, providing an extensive scaffolding by which new neurons migrate and integrate into distant parenchymal destinations.

An early intriguing observation in cultures of the adult HVC SZ was the almost invariate association of new neurons, ciliated ependymal cells, and ependymally derived radial cells, suggesting their coderivation (Goldman et al. 1993). When a lacZ-bearing retroviral vector was introduced into the neostriatal SZ of adult finches, coderived clones of neurons and radial cells were typically observed (Goldman et al. 1996b). These results indicated that neurons and their radial guide cells were cogenerated, suggesting in turn that the common progenitors of radial cells and neurons are the radial cells themselves (Alvarez-Buylla et al. 1990; Gray and Sanes 1992). Indeed, the demonstration that radial cells comprise most, if not all, dividing cells in the adult canary SZ (Alvarez-Buylla et al. 1998) further supported the notion that radial cells are the parents as well as the siblings of new neurons. This model of radial cell neurogenesis in the adult avian brain is now recognized as essentially an adult manifestation of a more ubiquitous process of neuronal production by radial glia in development.

REGULATION OF NEURONAL DEPARTURE FROM THE SZ

The discovery of the Hu proteins, a family of neuron-specific RNA-binding proteins homologous to *Drosophila* elav (Robinow and White 1991), provided a reagent with which to identify newly generated neurons almost immediately upon their lineage commitment to neuronal fate (Marusich and Weston 1992; Marusich et al. 1994; Barami et al. 1995). In the adult HVC, Hu expression appears in neuronal daughter cells within hours of their parental cell division. As a result, the appearance of Hu protein expression has been used to follow the early development of newly generated neurons still within the adult songbird SZ. Using immunolocalization of Hu, preceded at various intervals by injection of [^{3}H]thymidine as an autoradiographic marker of cell division, neuronal daughter cells, defined as [^{3}H]thymidine/Hu$^+$, were found to persist in the SZ for at least 4 days before initiating migration. This delay in parenchymal penetration is not observed in embryonic neurogenesis, and its existence suggested a time window during which departure from the SZ might be dynamically regulated in the adult (Barami et al. 1995). In effect, these data suggested that new neurons may await permissive signals before migrating from the adult subependyma.

A number of moieties appear to potentially constrain neuronal daughter cells to the adult SZ. Perhaps foremost among these is N-cadherin, which is heavily expressed by adult ependymal and subependymal cells and which appears to restrict cells to these layers. Barami et al. (1994) noted that although Hu$^+$/cadherin$^+$ cells were abundant within the adult canary SZ, N-cadherin expression fell sharply before a new Hu$^+$ migrant exited the SZ. In addition, when explants of the adult songbird SZ were exposed to anti-N-cadherin Fabs, neuronal migration sharply accelerated (Barami et al. 1994). These observations suggested that N-cadherin restricts neuronal departure from the adult SZ; as such, the signals that initiate neuronal migration may act in part by down-regulating N-cadherin expression or functional competence.

IGF-1 IS EXPRESSED BY ADULT RADIAL CELLS AND PROMOTES NEURONAL RECRUITMENT

Because new neurons and their radial cell partners appear to coderive from a common SZ progenitor, Jiang et al. (1998) asked whether the radial cells might provide trophic support for their newly generated neuronal siblings.

They focused on the insulin-like growth factor-1 (IGF-1), which had previously been shown to support the survival and differentiation of neuronal progenitor cells in rodents (DiCicco-Bloom and Black 1988). IGF-1 was indeed expressed heavily by adult finch radial cells and their fibers, with little expression elsewhere (Jiang et al. 1998). IGF-2, in contrast, was expressed predominantly by parenchymal astrocytes. Scharff and colleagues (Holzenberger et al. 1997b) reported that adult canary radial cells expressed IGF-2 immunoreactivity, but not mRNA, suggesting that radial cells might generate IGF-1 while receiving and potentially sequestering IGF-2. However, because IGF-1 and IGF-2 appear to act at the same receptor in birds (de la Rosa et al. 1994), the significance of their differential expression in this system remains unclear. Indeed, both IGF-1 and IGF-2 were found to support the generation and departure, but not the survival, of new neurons arising from cultured explants of the adult SZ (Jiang et al. 1998). Together, these results suggested that in neurogenic regions of the adult avian forebrain, IGF-1 might act as a radial-cell-associated neuronal specification and/or departure factor, serving to regulate neuronal recruitment and migration into the adult brain.

GONADAL STEROID SUPPORT OF NEURONAL RECRUITMENT IN THE ADULT SONGBIRD BRAIN

The integration and survival of new neurons in the HVC are modulated by the gonadal steroids testosterone and estradiol (Rasika et al. 1994; Hidalgo et al. 1995), whose actions mediate the seasonal hypertrophy of the HVC in adult canaries (Nottebohm 1981). Neurogenic regions of the adult songbird telencephalon, in particular the mediocaudal neostriatum, are associated with a layer of subventricular cells that express estrogen receptor (ER) protein (Gahr 1990). In the HVC, this layer is expanded to include the entire volume of the nucleus. This is intriguing because testosterone both supports the survival of new neurons in the adult avian HVC (Rasika et al. 1994) and is readily converted to estradiol in the neostriatal parenchyma by aromatase (Schlinger and Arnold 1991). Thus, we postulated that estrogen might be an active neurotrophic steroid in the adult songbird brain. This was indeed the case: In a cohort of ovariectomized female canaries, both the number and proportion of new neurons that survived their first month were substantially higher in estrogen-treated birds than in their untreated counterparts (Hidalgo et al. 1995). However, estrogen did not appear to act directly as a neurotrophin in this system, in that the new neurons did not express the

ER at any time during their first postmitotic month. In addition, the rate of division of mitotic neuronal progenitors was not influenced by estrogen (Brown et al. 1993). Rather, the estrogen-receptive cells appeared to constitute a layer of mitotically quiescent subventricular neurons (Hidalgo et al. 1995) that were traversed by the new neurons during their first few days of migration (see Fig. 1).

Both within and medial to HVC, these ER^+ neurons included cells that projected to area X, an anterior striatal region that regulates song acquisition and learning (Fig. 2) (Bottjer and Johnson 1997). In light of

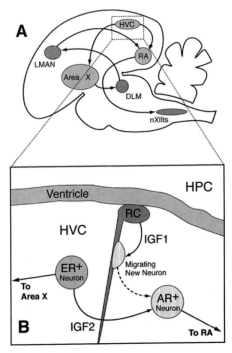

Figure 2. Paracrine interactions among neurons of the adult vocal control nucleus HVC. (A) Anatomy of the vocal control system in the adult canary brain, for simplicity's sake omitting the auditory integration centers. (Modified, with permission, from Bottjer and Johnson 1997 © Elsevier.) (B) Known cooperative paracrine interactions among both androgen- and estrogen-receptive neurons and between new and resident neurons in the adult canary. (Modified from Goldman 1998.) (DLM) Dorsolateral medial nucleus of the thalamus; (HVC) higher vocal center; (LMN) lateral magnocellular nucleus of the neostriatum; (nXIIts) tracheosyringeal motor neurons of the XIIth nerve nucleus; (RA) nucleus robustus of the archistriatum; (AR) androgen (testosterone) receptor; (ER) estrogen receptor; (HPC) parahippocampus; (IGF-1 and IGF-2) insulin-like growth factors 1 and 2; (RC) radial cell.

estrogen's neurotrophic effect in this system, the estrogen-responsive subventricular layer of the HVC and adjacent neostriatum may serve to act as a cellular "gatekeeper"; it is ideally situated to modulate the survival of those new neurons traversing it in an estrogen-dependent fashion. Yet, since the migrants do not express ER themselves, any estrogenic rescue of new neurons must be indirect, presumably through paracrine agents released in a gonadal steroid-dependent fashion. Although the specific agents released by estrogen-responsive subventricular zone cells in the songbird neostriatum have yet to be identified, examples of steroid-modulated neurotrophin release abound in other systems such as the rodent hypothalamus, in which brain-derived neurotrophic factor (BDNF) may be released from ER^+ cells in response to estrogen (Sohrabji et al. 1994).

MODULATION OF MIGRATION AND SURVIVAL DURING TRANSIT

Upon departing the SZ, new neurons of the adult songbird brain express a number of Ig-class adhesion molecules. NgCAM is of particular interest because the migration of new neurons in cultures of the adult finch SZ is disrupted by the addition of anti-NgCAM Fab (Barami et al. 1994). To determine whether NgCAM's role was limited to the adhesive support of migration, or whether early neuronal survival was dependent on NgCAM signaling, we used confocal calcium imaging to examine the effects of NgCAM protein and anti-NgCAM Ig on calcium-mediated signal transduction in newly generated HVC neurons in vitro. Newly generated bipolar neuronal migrants responded to NgCAM with reversible increments in cytosolic calcium, but only over the period spanning 6–9 days in vitro (Goldman et al. 1996a). This restricted period corresponds to the postmitotic age at which new neurons leave the adult SZ to enter the brain parenchyma (Barami et al. 1995). The calcium response to NgCAM was G-protein-dependent, mediated through voltage-gated calcium channels, and specific to NgCAM. Notably, surface expression of NgCAM was maintained long after the calcium response to its addition ceased (Barami et al. 1995). These findings suggested that newly generated, migratory neuroblasts in the HVC exhibit a dynamic coupling and uncoupling of calcium-signaling pathways to the stably expressed NgCAM molecule.

Because neuronal passage through the estrogen-receptive subventricular layer coincides with the initiation of NgCAM-dependent calcium signaling by the migrants, Williams et al. (1999) then asked whether migration-associated, NgCAM-dependent calcium signaling might require estrogen exposure. Using cultured explants of the adult HVC, they found that indeed only those migrating neuroblasts raised in the presence of estrogen developed calcium responses to NgCAM; their counterparts raised in estrogen-free media never did so. Moreover, those SZ-derived neurons not exposed to estrogen underwent apoptotic death within the few days thereafter, in contrast to their counterparts maintained in estrogen. These findings indicated that the coupling of NgCAM to calcium signaling pathways requires estrogen and that this process contributes to neuronal survival during parenchymal migration (Williams et al. 1999).

NEUROTROPHIC MODULATION OF POSTMIGRATORY NEURONAL MATURATION AND PHENOTYPE

IGF-1, as noted, is expressed by radial cells and supports both the initial departure and survival of new neurons in the adult HVC (Jiang et al. 1998). Yet the role of the insulin-like growth factors in this system may extend to mature HVC neurons as well. Scharff, Nottebohm, and colleagues (Holzenberger et al. 1997a) reported that IGF-2 may act as an interneuronal paracrine agent within HVC. They noted that IGF-2 mRNA was selectively expressed by X-projecting neurons in HVC, and both bound and sequestered by neighboring RA-projecting neurons, which include the newly generated pool (Alvarez-Buylla and Kirn 1997). Johnson and Bottjer (1995) had previously reported that these RA-projecting neurons were largely sensitive to androgens, as were X-projecting neurons to estrogen. Together, these data suggested that androgen-receptive RA-projecting neurons, which include the newly generated cells, might be acted upon by IGF-2 secreted, in an estrogen-dependent fashion, by their area X-projecting neighbors. This in turn suggested a gonadal steroid-modulated, IGF2-mediated paracrine interaction between the two HVC neuronal phenotypes, one stable and the other seasonally replaced. Yet the effect of this interaction, if any, on the recruitment of new neurons to HVC, or their instruction as RA-projecting cells, remains unclear. Among other possibilities, radial cell-associated IGF-1, and possibly glial IGF-2, might be involved in the recruitment of new neurons to HVC (Jiang et al. 1998), whereas steroid-modulated IGF-2 derived from X-projecting neurons might promote the differentiation and survival of the newly recruited neurons as RA-projection neurons.

Like IGF-1 and IGF-2, the neurotrophin family member BDNF can potentiate neuronal differentiation and survival in the adult HVC. The neurotrophin family members were first noted to be critical determinants of neuronal recruitment in the adult mammalian brain (Kirschenbaum and Goldman 1995). In adult canaries, BDNF contributes to both the androgen- and activity-dependent regulation of survival of new neurons in HVC (Rasika et al. 1999; Alvarez-Borda and Nottebohm 2002). BDNF appears to be regulated in HVC not only within endogenous neuronal stores, but also by activity-dependent neuronal inputs into HVC; in particular, activity-dependent increases in BDNF release have been implicated in neuronal addition to both the mediocaudal neostriatum (MCN) and HVC (Li et al. 2000). In addition to these neuronal stores, BDNF is made available to HVC by gonadal steroid-activated endothelial cells (Louissaint et al. 2002). The latter appear to provide a significant local source of BDNF to new neuronal migrants infiltrating HVC from the adjacent subependyma. Thus, since gonadal steroids regulate both afferent activity into HVC (by influencing androgen-dependent inputs from area X and MAN) and endothelial production of BDNF (by triggering angiogenesis), they would appear to comprise master regulators of BDNF availability within the adult HVC.

Activity-dependent modulation of neurogenesis in the adult HVC is likely based on a number of afferent effectors. Besides the neurotrophins, a variety of neurotransmitter inputs may also modulate neuronal addition to HVC. In particular, HVC enjoys an abundant dopaminergic input, and testosterone treatment expands both the density and distribution of local dopaminergic innervation to HVC (Appeltants et al. 2003). Importantly in this respect, dopamine has been implicated in the modulation of neurogenesis in several mammalian systems, as the dopamine D2 receptor is expressed by mitotically competent subependymal progenitor cells, and dopamine appears to act on it to suppress progenitor proliferation (Kippin et al. 2005), potentially biasing these cells to terminal neuronal differentiation. In the canary, the A10 and A11 groups of dopaminergic neurons, arising in the ventral tegmentum and the mesencephalic central gray, respectively, provide abundant dopaminergic innervation of HVC (Appeltants et al. 2000). The ventral tegmental A10 dopaminergic neurons, as well as those of the adjacent substantia nigra pars compacta (A9), also innervate area X, which has reciprocal connections with HVC. It is intriguing to note that these A9/A10 neurons serve to regulate context-dependent song: Finches singing in a mate-directed fashion release more dopamine into area X than do their controls, an event that correlates with increased song complexity (Sasaki et al. 2006). Thus, these mesencephalic dopaminergic centers can influence song behavior in the acute setting by modulating dopamine release in a primary afferent nucleus to HVC, whereas they are simultaneously well-positioned to influence song learning in a more sustained fashion by directly regulating neurogenesis and neuronal recruitment in HVC.

ANGIOGENESIS IS A NECESSARY ANTECEDENT
TO NEURONAL RECRUITMENT

As gonadal steroid-associated cell genesis in the adult canary brain was explored, it quickly became clear that testosterone and estrogen were both powerful triggers of HVC endothelial cell division (Goldman and Nottebohm 1983; Hidalgo et al. 1995). Testosterone treatment in particular yielded a >25-fold increase in the mitotic index of HVC endothelial cells (Goldman and Nottebohm 1983). This burst of endothelial cell division was transient and self-limited; endothelial mitotic indices fell to baseline within 2 weeks after testosterone implantation, despite persistently elevated androgen levels during that period. Almost two decades later, Louissaint et al. (2002) followed up this observation by studying the concurrence of HVC angiogenesis and neuronal recruitment in response to testosterone. Specifically, they asked whether androgen-initiated angiogenesis might directly incite neuronal recruitment. They found that androgen-associated endothelial proliferation not only contributed to, but was required for the acceptance of new neurons by adult brain parenchyma (Louissaint et al. 2002). When adult female canaries were treated with testosterone, the birds rapidly up-regulated both vascular endothelial growth factor (VEGF) and its endothelial receptor VEGF-R2/KDR in HVC. Importantly, whereas VEGF was induced by androgen, VEGF-R2 was induced by estradiol. Thus, both androgen and estrogen were required for the induction of angiogenesis, indicating the importance of cooperative interactions among the gonadal steroids to HVC neurogenesis.

The need for both androgen- and estrogen-triggered signals in driving endothelial cell division highlights the potential importance of aromatase, the enzyme that converts testosterone to estradiol in the adult HVC (Schlinger and Arnold 1991; Shen et al. 1995). By regulating the relative proportions of androgen and estradiol generated from locally available testosterone, aromatase occupies a potentially key position from which it may orchestrate androgen-induced angiogenesis and neurogenesis. Louissaint et al. (2002) noted that within the mediocaudal neostriatum, VEGF induced in response to testosterone was restricted to the HVC, within which its rise was associated with a burst in mitotic angiogenesis. They also noted that fluorescence-activated cell sorting (FACS)-isolated HVC endothelial cells produced BDNF and substantially increased their secretion of BDNF in response to testosterone. In parallel studies, they noted that BDNF addition promoted the departure and migration of neurons arising from explants of adult HVC. In vivo, HVC BDNF rose by the third week after testosterone; in situ hybridization revealed that much of this induced BDNF mRNA was endothelial. Importantly, the androgen-induced rise in BDNF lagged by at least 2 weeks the local rise in VEGF and VEGF-R2, suggesting that endothelial activation was a necessary antecedent to the rise in HVC BDNF (Louissaint et al. 2002).

On this basis, Louissaint et al. (2002) postulated that testosterone-induced VEGF was the trigger for HVC angiogenesis and that the resulting expansion of the HVC capillary bed would present a significant source of BDNF, which might serve both to attract new neuronal migrants to HVC and to support their survival once there. To test this hypothesis, they treated testosterone-treated females with an inhibitor of the VEGF-R2 tyrosine kinase and found that neuronal addition to HVC was indeed suppressed by inhibition of VEGF-R2 tyrosine kinase activity (Louissaint et al. 2002). These observations served to establish androgen-induced VEGF and the VEGF-triggered expansion of the HVC vascular bed as necessary antecedents to the local rise in HVC BDNF, which was in turn required for the recruitment of new neurons to the adult HVC and their survival therein.

This link between endothelial mobilization and neuronal production has now found parallels in both the normal and disturbed mammalian brain. In the normal hippocampus, angiogenesis and neurogenesis are spatially colocalized (Palmer et al. 2000), to the extent that angioblasts and neural stem cells may both respond to VEGF as a common mitogen (Cao et al. 2004). In the mammalian ventricular wall, endothelial cells may regulate neural stem cell turnover (Shen et al. 2004), whereas the compensatory neurogenesis that is noted after striatal ischemia may similarly require signals from mobilized endothelial cells (Ohab et al. 2006). Together, these findings suggest that the causal relationship between angiogenesis and neurogenesis revealed in the songbird brain may be prototypic of the paracrine interactions among endothelial and neural cells required for neuronal addition to the mature brain.

MATRIX METALLOPROTEINASES COMPRISE REGULATORY CHECKPOINTS FOR NEUROGENESIS

Testosterone-induced neuronal addition to the adult HVC thus requires the androgenic induction of VEGF, followed by VEGF-stimulated angiogenesis. The expanded vasculature acts as a source of BDNF, which supports the immigration of new neurons from the overlying ventricular zone. In tumorigenesis, a similar process of adult angiogenesis is regulated by matrix metalloproteinase (MMP) activity, in particular that of the gelatinases, which may act to release VEGF from matrix-bound stores, thereby initiating endothelial cell proliferation (Bergers et al. 2000; Egeblad and Werb 2002). To assess whether the MMPs might subserve a similar role in initiating HVC angiogenesis, Kim et al. (2006) investigated the role of the gelatinases in neuronal addition to the HVC of adult female canaries. Using in situ zymography of the neostriatum, they noted that testosterone-induced perivascular gelatinase activity was most prominent in the HVC, high-resolution gels of which revealed distinct MMP activities that comigrated with MMP2 and MMP9. PCR cloning yielded canary MMP2 and MMP9 orthologs, and quantitative PCR (qPCR) of these revealed that HVC MMP2 mRNA levels doubled within 8 days of testosterone, whereas MMP9 transcript levels were stable. These data indicated that MMP2 was the androgen-regulated gelatinase in HVC. Accordingly, isolated adult canary forebrain endothelial cells were found to secrete MMP2, and VEGF substantially increased endothelial MMP2 gelatinase activity.

To assess the importance of androgen-regulated, VEGF-induced MMP2 to adult angiogenesis and neurogenesis, Kim et al. (2006) then treated testosterone-implanted females with the MMP2/9 inhibitor SB-3CT, which substantially decreased HVC gelatinase activity in response to testosterone. The SB-3CT-treated birds exhibited a decreased endothelial mitotic index and manifested substantially diminished neuronal recruitment to the HVC (Kim et al. 2006). Thus, the androgenic induction of endothelial MMP2 appears to be an important regulator of neuronal addition to the adult HVC. In broader terms, these observations serve to again highlight the critical role of the gonadal steroids in orchestrating a complex yet coherent set of cellular and molecular interactions, whose net effect is the establishment of local permissiveness to neuronal migration and survival.

INITIATION AND REGULATION OF HVC INVOLUTION AFTER ANDROGEN WITHDRAWAL

The seasonal acquisition and extinction of song in canaries is accompanied by a seasonal hypertrophy and involution of the HVC. Despite two decades of studies of the mechanisms subserving androgen-associated hypertrophy of the HVC, little work has been done exploring the processes involved in the involution of the HVC in the setting of steroid withdrawal. Kirn et al. reported the death of a subpopulation of newly added HVC neurons following each season, but the molecular basis and concomitants of neuronal death have not been hitherto explored. Among a wide range of unstudied issues is the extent to which vascular involution may precede, or be required for, neuronal loss. Although the antecedent loss of vascular trophic support may be a concomitant or even a prerequisite for neuronal loss in HVC, this scenario remains entirely hypothetical; no published studies have yet addressed the role of vascular stability in the maintenance of HVC neurons or, in particular, in the retention of the nominally renewable pool of HVC to RA projection neurons.

Despite the dearth of work focused on the molecular mechanisms of seasonal neuronal loss in HVC, several studies have provided adventitious insight into this process. Lombardino and Nottebohm and their colleagues used expression array analysis to determine that the ubiquitination gene UCHL1, which negatively regulates proteosomal processing in adult neurons, is relatively deficient in new neurons in the adult HVC (Lombardino et al. 2005). In particular, they reported that UCHL1 is relatively underexpressed in cells of the HVC to RA projection, which comprise the "replaceable" pool of neurons in this region. Similarly, they determined that both olfactory and dentate granule neurons, analogous mammalian replaceable pools, are also deficient in this gene. In contrast, more long-lasting neuronal pools, including those of the stable HVC to area X projection from HVC, express constitutively higher levels of UCHL1. On this basis, Lombardino et al. (2005) concluded that neuronal turnover is associated with low expression levels of UCHL1 and postulated that UCHL1 might serve a protective stabilization function in normal adult neurons and that its loss or dysfunction might predispose an otherwise healthy neuron to premature demise. This observation finds resonance in the subsequent demonstration that mutations in UCHL1 may be linked to a rare autosomal dominant form of Parkinson's disease, in which dopaminergic neurons of the substantia nigra are lost, antecedent to a more widespread loss of neurons throughout the major subcortical structures (for review, see Chung et al. 2003; Schapira 2006).

MAINTENANCE AND CHARACTERIZATION OF A NEUROGENIC NICHE IN THE ADULT SONGBIRD HVC

Members of the transforming growth factor-β (TGF-β) family, in particular activin and follistatin, have been implicated as possible gonadal modulators of neuronal recruitment in the adult HVC (Hidalgo et al. 1995). Interestingly, activin and follistatin have each been identified as antagonizing the reception and actions of the bone morphogenetic proteins (BMPs), which have otherwise been identified as potentiating glial dedifferentiation in both development and adulthood. One might thus predict that neurogenic regions of the adult canary ventricular zone might be distinguished from nonneurogenic regions by sustained expression of BMP inhibitors. One of the broadest spectrum and most potent of these is the soluble BMP antagonist noggin, the expression of which has already been identified in persistently neurogenic regions of the adult rodent brain (Lim et al. 2000b; Chmielnicki et al. 2004). Surprisingly, however, the ontogeny and distribution of BMP inhibition in regions of sustained neurogenesis have never been explored in the adult songbird brain. More refined molecular identification of both neurogenic and glial suppressive agents available within the HVC should permit us to better define those features distinguishing neurogenic and nonneurogenic regions of the adult canary ventricular zone. Such studies should permit us to define the cell–cell interactions, and the molecular modulators thereof, that provide a permissive environment for adding neurons to the adult canary brain. In broader terms, by using this information to faithfully recapitulate the molecular environment of the neurogenic songbird HVC, one may hope to induce and support neurogenesis in otherwise nonneurogenic regions of the adult mammalian brain.

MAMMALIAN STRIATAL NEUROGENESIS

As in birds, both neural stem cells and their committed neuronal progeny persist in the subependymal layer of the mammalian striatal wall. Yet, in contrast to birds, who enjoy widespread parenchymal neuronal migration, in mammals, the neuronal progeny of these cells are typically restricted to the olfactory subependyma and its rostral extension to the OB, such that the normal adult mammalian neostriatum appears to exhibit little or no neuronal addition in adulthood. Nonetheless, some compensatory neuronal recruitment from resident subependymal progenitor cells has been noted in response to both stroke and degenerative striatal neuronal loss (Arvidsson et al. 2002; Parent et al. 2002; Jin et al. 2003a). Furthermore, a number of groups have now successfully targeted endogenous progenitor cells for directed mobilization and neuronal induction, using both neurotrophic and chemotaxic cytokines delivered to the brain by both protein infusion and viral vectors (Benraiss et al. 2001; Pencea et al. 2001; Jin et al. 2002a,b, 2003b; Chmielnicki et al. 2004).

Experiments a decade ago in both the rodent neostriatum and the canary HVC indicated that BDNF can signal the differentiation and survival of new neurons from ventricular zone neural stem cells (Kirschenbaum and Goldman 1995; Rasika et al. 1999). On that basis, Benraiss et al. (2001) then noted that a single intraventricular injection of adenoviral BDNF (AdBDNF), which resulted in widespread ependymal production of BDNF, induced the production of new neurons from subependymal progenitor cells. A parallel study using BDNF protein infusion exhibited similar results (Pencea et al. 2001). In both cases, a large number of the new neurons invaded the neostriatum, an otherwise nonneurogenic region in mammals that is critically important to the direction and coordination of movement. The new neurons largely integrated as medium spiny neurons that extended projections to the globus pallidus. These are the cells that are typically lost in Huntington's disease, which suggested that AdBDNF-induced cells may be able to directly replace the very phenotype lost in the course of Huntington's (Ivkovic and Ehrlich 1999). Importantly, the new striatal neurons stably persisted for months after their production, even though viral BDNF expression lasted only 1 month, suggesting that once integrated, the newly generated cells could survive independently of exogenous BDNF overexpression (Benraiss et al. 2001). Following up on these studies with the intent of increasing the efficiency of induced striatal neurogenesis, Chmielnicki et al. (2004) next determined that BDNF-induced striatal neuronal addition could be enhanced by simultaneously using noggin to block glial differentiation by the targeted neural stem cells. Noggin is a potent inhibitor of the progliogenic BMPs (Zimmerman et al. 1996; Lim et al. 2000a), and its viral overexpression may be used to suppress the otherwise tonically high incidence of astrocyte production by these cells. Chmielnicki et al. noted that AdNoggin strongly potentiated AdBDNF-induced striatal neurogenesis, roughly tripling the rate of striatal neuronal addition to >400/mm^3/month (Chmielnicki and Goldman 2002; Chmielnicki et al. 2004), presumably by expanding the pool of progenitors potentially responsive to BDNF. Thus, the experimental provision of an effectively heterotopic niche for neurogenesis, by providing instructional cues discerned in part by studying the normally neurogenic canary HVC, proved to be a feasible means of recruiting large numbers of new neurons to otherwise nonneurogenic regions of the adult mammalian brain.

Extending this logic, Cho et al. (2004, 2007) then asked whether this approach of inducing compensatory neurogenesis in otherwise nonneurogenic regions might prove a means of treating neurodegenerative diseases, Huntington's disease in particular. To that end, they assessed the effect of BDNF and noggin overexpression in R6-2 mice, animals that are transgenic for an approximately 150-polyglutamine repeat in the huntingtin gene, and which exhibit a relatively severe Huntington's phenotype (Mangiarini et al. 1996). They found that R6-2 mice treated with AdBDNF and AdNoggin generated new medium spiny neurons that projected to the globus pallidus, in numbers no different from those of identically treated normal rats and mice. Importantly, the animals so treated survived significantly longer than their untreated counterparts, and the effect of BDNF and noggin treatment was entirely dependent on the new neurons thereby recruited (Cho et al. 2004, 2007).

At the very least, these findings offer promise for using neuronal induction from endogenous progenitor cells as a means of treating Huntington's disease (Goldman 2005; Goldman and Benraiss 2006), a disorder that is otherwise inevitably and rapidly fatal. More generally, activating endogenous stem cells in this manner may prove to be an option not only for Huntington's disease, but also for such other causes of striatal neuronal loss as striatonigral degeneration and lenticulostriate stroke. In yet broader terms, this set of studies illustrates how studying the mechanistic underpinnings of neurogenesis in the songbird brain has not only informed our understanding of neurogenesis in adult mammals, but also provided us the ability to induce neuronal addition in otherwise nonneurogenic regions of the brain, and to do so in the service of medicine.

ACKNOWLEDGMENTS

S.A.G. is supported by the National Institute of Neurological Disorders and Stroke, the Mathers Charitable Foundation, the Adelson Program in Neural Repair, the James S. McDonnell Foundation, the Children's Neurobiological Solutions (CNS) Foundation, the Ataxia-Telangiectasia Children's Program, and the National Multiple Sclerosis Society.

REFERENCES

Altman J. and Das G.D. 1966. Autoradiographic and histological studies of postnatal neurogenesis. I. A longitudinal investigation of the kinetics, migration and transformation of cells incorporating tritiated thymidine in neonate rats, with special reference to postnatal neurogenesis in some brain regions. *J. Comp. Neurol.* **126:** 337–389.

Alvarez-Buylla A. and Garcia-Verdugo J.M. 2002. Neurogenesis in adult subventricular zone. *J. Neurosci.* **22:** 629–634.

Alvarez-Buylla A. and Kirn J. 1997. Birth, migration, incorporation and death of vocal control neurons in adult songbirds. *J. Neurosci.* **33:** 585–601.

Alvarez-Buylla A. and Lim D.A. 2004. For the long run: Maintaining germinal niches in the adult brain. *Neuron* **41:** 683–686.

Alvarez-Buylla A. and Nottebohm F. 1988. Migration of young neurons in adult avian brain. *Nature* **335:** 353–354.

———. 2002. Gonads and singing play separate, additive roles in new neuron recruitment in adult canary brain. *J. Neurosci.* **22:** 8684–8690.

Alvarez-Buylla A., Garcia-Verdugo J.M., and Tramontin A. 2001. A unified hypothesis on the lineage of neural stem cells. *Nat. Rev. Neurosci.* **2:** 287–293.

Alvarez-Buylla A., Ling C.Y., and Yu W.S. 1994. Contribution of neurons born during embryonic, juvenile, and adult life to the brain of adult canaries: Regional specificity and delayed birth of neurons in the song-control nuclei. *J. Comp. Neurol.* **347:** 233–248.

Alvarez-Buylla A., Theelen M., and Nottebohm F. 1990. Proliferation "hot spots" in adult avian ventricular zone reveal radial cell division. *Neuron* **5:** 101–109.

Alvarez-Buylla A., Garcia-Verdugo J., Mateo A., and Merchant-Larios H. 1998. Primary neural precursors and intermitotic nuclear migration in the ventricular zone of adult canaries. *J. Neurosci.* **18:** 1020–1037.

Appeltants D., Ball G., and Balthazart J. 2003. Song activation by testosterone is associated with an increased catecholaminergic innervation of the song control system in female canaries. *Neuroscience* **121:** 801–814.

Appeltants D., Absil P., Balthazart J., and Ball G. 2000. Identification of the origin of catecholaminergic inputs to HVC in canaries by retrograde tracing combined with tyrosine hydroxylase immunocytochemistry. *J. Chem. Neuroanat.* **18:** 117–133.

Arvidsson A., Collin T., Kirik D., Kokaia Z., and Lindvall O. 2002. Neuronal replacement from endogenous precursors in the adult brain after stroke. *Nat. Med.* **8:** 963–970.

Barami K., Iversen K., Furneaux H., and Goldman S.A. 1995. Hu protein as an early marker of neuronal phenotypic differentiation by subependymal zone cells of the adult songbird forebrain. *J. Neurobiol.* **28:** 82–101.

Barami K., Kirschenbaum B., Lemmon V., and Goldman S.A. 1994. N-cadherin and Ng-CAM/8D9 are involved serially in the migration of newly generated neurons into the adult songbird brain. *Neuron* **13:** 567–582.

Barnea A. and Nottebohm F. 1994. Seasonal recruitment of hippocampal neurons in adult free-ranging black-capped chickadees. *Proc. Natl. Acad. Sci.* **91:** 11217–11221.

Bayer S., Yackel J., and Puri P. 1982. Neurons in the rat dentate gyrus granular layer substantially increase during juvenile and adult life. *Science* **216:** 890–892.

Benraiss A., Chmielnicki E., Lerner K., Roh D., and Goldman S.A. 2001. Adenoviral brain-derived neurotrophic factor induces both neostriatal and olfactory neuronal recruitment from endogenous progenitor cells in the adult forebrain. *J. Neurosci.* **21:** 6718–6731.

Bergers G., Brekken R., McMahon G., Vu T., Itoh T., Tmaki K., Tanzawa K., Thorpe P., Itohara S., Werb Z., and Hanahan D. 2000. Matrix metalloproteinase-9 triggers the angiogenic switch during carcinogenesis. *Nat. Cell Biol.* **2:** 737–744.

Birse S.C. Leonard R.B., and Coggeshall R.E. 1980. Neuronal increase in various areas of the nervous system of the guppy, *Lebistes*. *J. Comp. Neurol.* **194:** 291–301.

Bottjer S. and Johnson F. 1997. Circuits, hormones, and learning: Vocal behavior in songbirds. *J. Neurobiol.* **33:** 602–618.

BoulderCommittee. 1970. Embryonic vertebrate central nervous system: Revised terminology. *Anat. Rec.* **166:** 257–261.

Brown S., Johnson F., and Bottjer S. 1993. Neurogenesis in adult canary telencephalon is independent of gonadal hormone levels. *J. Neurosci.* **13:** 2024–2032.

Cao L., Jiao X., Zuzga D.S., Liu Y., Fong D.M., Young D., and During M.J. 2004. VEGF links hippocampal activity with neurogenesis, learning and memory. *Nat. Genet.* **36:** 827–835.

Chmielnicki E. and Goldman S.A. 2002. Induced neurogenesis by endogenous progenitor cells in the adult mammalian brain. *Prog. Brain Res.* **138:** 451–464.

Chmielnicki E., Benraiss A., Economides A.N., and Goldman S.A. 2004. Adenovirally expressed noggin and brain-derived neurotrophic factor cooperate to induce new medium spiny neurons from resident progenitor cells in adult striatal ventricular zone. *J. Neurosci.* **24:** 2133–2142.

Cho S.-R., Benraiss A., Chmielnicki E., Samdani A., Economides A., and Goldman S.A. 2007. Mobilization of endogenous progenitor cells regenerates striatal neurons and slows disease progression in a transgenic model of Huntington's Disease. *J. Clin. Invest.* (in press).

Cho S.-R., Chmielnicki E., and Goldman S.A. 2004. Adenoviral co-delivery of BDNF and noggin induces striatal neuronal replacement and delays motor impairment in a transgenic model of Huntington's Disease. *Molec. Ther.* **9:** S86–S87.

Chung K., Dawson V., and Dawson T. 2003. New insights into Parkinson's Disease. *J. Neurol.* **250:** S3:15–24.

de la Rosa E., Bondy C., Hernandez-Sanchez C., Wu X., Zhou J., Lopez-Carranza A., Scavo L., and de Pablo F. 1994. Insulin and insulin-like growth factor system component gene expression in the chicken retina from early neurogenesis until late development and their effect on neuroepithelial cells. *Eur. J. Neurosci.* **6:** 1801–1810.

DiCicco-Bloom E. and Black I.B. 1988. Insulin growth factors regulate the mitotic cycle in cultured rat sympathetic neuroblasts. *Proc. Natl. Acad. Sci.* **85:** 4066–4070.

Doetsch F. and Alvarez-Buylla A. 1996. Network of tangential pathways for neuronal migration in adult mammalian brain. *Proc. Natl. Acad. Sci.* **93:** 14895–14900.

Egeblad M. and Werb Z. 2002. New functions for the matrix metalloproteinases in cancer progression. *Nat. Rev. Cancer* **2:** 163–175.

Gage F.H. 2000. Mammalian neural stem cells. *Science* **287:** 1433–1438.

———. 2002. Neurogenesis in the adult brain. *J. Neurosci.* **22:** 612–613.

Gahr M. 1990. Delineation of a brain nucleus: Comparisons of cytochemical, hodological, and cytoarchitectural views of the song control nucleus HVC of the adult canary. *J. Comp. Neurol.* **294:** 30–36.

Goldman S. 1998. Adult neurogenesis: From canaries to the clinic. *J. Neurobiol.* **36:** 267–286.

Goldman S.A. 1990. Neuronal development and migration in explant cultures of the adult canary forebrain. *J. Neurosci.* **10:** 2931–2939.

———. 2003. Glia as neural progenitor cells. *Trends Neurosci.* **26:** 590–596.

———. 2004. Directed mobilization of endogenous neural progenitor cells: The intersection of stem cell biology and gene therapy. *Curr. Opin. Molec. Ther.* **6:** 466–472.

———. 2005. Stem and progenitor cell-based therapy of the human central nervous sytem. *Nat. Biotech.* **23:** 862–871.

Goldman S.A. and Benraiss A. 2006. Method of inducing neuronal production in the brain and spinal cord. US Patent no. 7,037,493.

Goldman S.A. and Luskin M.B. 1998. Strategies utilized by migrating neurons of the postnatal vertebrate forebrain. *Trends Neurosci.* **21:** 107–114.

Goldman S.A. and Nedergaard M. 1992. Newly generated neurons of the adult songbird brain become functionally active in long-term culture. *Dev. Brain Res.* **68:** 217–223.

Goldman S.A. and Nottebohm F. 1983. Neuronal production, migration, and differentiation in a vocal control nucleus of the adult female canary brain. *Proc. Natl. Acad. Sci.* **80:** 2390–2394.

Goldman S.A., Lemmon V., and Chin S.S. 1993. Migration of newly generated neurons upon ependymally derived radial guide cells in explant cultures of the adult songbird forebrain. *Glia* **8:** 150–160.

Goldman S.A., Williams S., Barami K., Lemmon V., and Nedergaard M. 1996a. Transient coupling of Ng-CAM expression to NgCAM-dependent calcium signaling during migration of new neurons in the adult songbird brain. *Mol. Cell. Neurosci.* **7:** 29–45.

Goldman S.A., Zaremba A., and Niedzwiecki D. 1992. In vitro neurogenesis by neuronal precursor cells derived from the adult songbird brain. *Neurosci.* **12:** 2532–2541.

Goldman S.A., Zukhar A., Barami K., Mikawa T., and Niedzwiecki D. 1996b. Ependymal/subependymal zone cells of postnatal and adult songbird brain generate both neurons and nonneuronal siblings in vitro and in vivo. *J. Neurobiol.* **30:** 505–520.

Gray G. and Sanes J. 1992. Lineage of radial glia in the chicken optic tectum. *Development* **114:** 271–283.

Hidalgo A., Barami K., Iversen K., and Goldman S.A. 1995. Estrogens and non-estrogenic ovarian influences combine to promote the recruitment and decrease the turnover of new neurons in the adult female canary brain. *J. Neurobiol.* **27:** 470–487.

Holzenberger M., Jarvis E., Chong C., Grossman M., Nottebohm F., and Scharff C. 1997a. Selective expression of insulin-like growth factor II in the songbird brain. *J. Neurosci.* **17:** 6974–6987.

———. 1997b. Selective expression of insulin-like growth factor II in the songbird brain. *J. Neurosci.* **17:** 6974–6987.

Ivkovic S. and Ehrlich M. 1999. Expression of the striatal DARPP-32/ARPP-21 phenotype in GABAergic neurons requires neurotrophins in vivo and in vitro. *J. Neurosci.* **19:** 5409–5419.

Jiang J., McMurtry J., Niedzwiecki D., and Goldman S.A. 1998. Insulin-like growth factor-1 is a radial cell-associated neurotrophin that promotes neuronal recruitment from the adult songbird edpendyma/subependyma. *J. Neurobiol.* **36:** 1–15.

Jin K., Mao X.O., Sun Y., Xie L., and Greenberg D.A. 2002a. Stem cell factor stimulates neurogenesis in vitro and in vivo. *J. Clin. Invest.* **110:** 311–319.

Jin K., Zhu Y., Sun Y., Mao X.O., Xie L., and Greenberg D.A. 2002b. Vascular endothelial growth factor (VEGF) stimulates neurogenesis in vitro and in vivo. *Proc. Natl. Acad. Sci.* **99:** 11946–11950.

Jin K., Sun Y., Xie L., Peel A., Mao X.O., Batteur S., and Greenberg D.A. 2003a. Directed migration of neuronal precursors into the ischemic cerebral cortex and striatum. *Mol. Cell. Neurosci.* **24:** 171–189.

Jin K., Sun Y., Xie L., Batteur S., Mao X.O., Smelick C., Logvinova A., and Greenberg D.A. 2003b. Neurogenesis and aging: FGF-2 and HB-EGF restore neurogenesis in hippocampus and subventricular zone of aged mice. *Aging Cell* **2:** 175–183.

Johnson F. and Bottjer S. 1995. Differential estrogen accumulation among populations of projection neurons in the higher vocal center of male canaries. *J. Neurobiol.* **26:** 87–108.

Keyoung H.M., Roy N.S., Benraiss A., Louissaint A., Jr., Suzuki A., Hashimoto M., Rashbaum W.K., Okano H., and Goldman S.A. 2001. High-yield selection and extraction of two promoter-defined phenotypes of neural stem cells from the fetal human brain. *Nat. Biotechnol.* **19:** 843–850.

Kim D.-H., Lilliehook C., Roides B., Chen Z., Chang M., Mobashery S., and Goldman S. 2006. Testosterone-induced MMP activation is a regulatory checkpoint for neuronal addition to the adult songbird brain. *Soc. Neurosci.* **36:** 418.21.

Kippin T., Kapur S., and van der Kooy D. 2005. Dopamine specifically inhibits forebrain neural stem cell proliferation, suggesting a novel effect of antipsychotic drugs. *J. Neurosci.* **25:** 5814–5823.

Kirschenbaum B. and Goldman S.A. 1995. Brain-derived neurotrophic factor promotes the survival of neurons arising from the adult rat forebrain subependymal zone. *Proc. Nat. Acad. Sci.* **92:** 210–214.

Kirschenbaum B., Nedergaard M., Preuss A., Barami K., Fraser R.A., and Goldman S.A. 1994. In vitro neuronal production and differentiation by precursor cells derived from the adult human forebrain. *Cerebral Cortex* **4:** 576–589.

Leonard R., Coggeshall R., and Willis W. 1978. A documentation of an age-related increase in neuronal and axonal numbers in the stingray. *J. Comp. Neurol.* **179:** 13–22.

Li X.-C., Jarvis E.D., Alvarez-Borda B., Lim D.A., and Nottebohm F. 2000. A reationship between behavior, neurotrophin expression, and new neuron survival. *Proc. Natl. Acad. Sci.* **97:** 8584–8589.

Lim D., Tramontin A., Trevejo J., Herrera D., Garcia-Verdugo J., and Alvarez-Buylla A. 2000a. Noggin antagonizes BMP signaling to create a niche for adult neurogenesis. *Neuron* **28:** 713–726.

———. 2000b. Noggin antagonizes BMP signaling to create a niche for adult neurogenesis. *Neuron* **28:** 713–726.

Lois C. and Alvarez-Buylla A. 1993. Proliferating subventricular zone cells in the adult mammalian forebrain can differentiate into neurons and glia. *Proc. Natl. Acad. Sci.* **90:** 2074–2077.

Lombardino A.J., Li X.C., Hertel M., and Nottebohm F. 2005. Replaceable neurons and neurodegenerative disease share depressed UCHL1 levels. *Proc. Natl. Acad. Sci.* **102:** 8036–8041.

Louissaint A., Jr., Rao S., Leventhal C., and Goldman S.A. 2002. Coordinated interaction of neurogenesis and angiogenesis in the adult songbird brain. *Neuron* **34:** 945–960.

Magavi S., Leavitt B., and Macklis J. 2000. Induction of neurogenesis in the neocortex of adult mice. *Nature* **405:** 951–955.

Mangiarini L., Sathasivam K., Seller M., Cozens B., Harper A., Hetherington C., Lawton M., Trottier Y., Lehrach H., Davies S.W., and Bates G.P. 1996. Exon 1 of the HD gene with an expanded CAG repeat is sufficient to cause a progressive neurological phenotype in transgenic mice. *Cell* **87:** 493–506.

Marusich M. and Weston J. 1992. Identification of early neurogenic cells in the neural crest lineage. *Dev. Biol.* **149:** 295–306.

Marusich M., Furneaux H., Henion P., and Weston J. 1994. Hu neuronal proteins are expressed in proliferating neurogenic cells. *J. Neurobiol.* **25:** 143–155.

Morshead C. and van der Kooy D. 1992. Postmitotic death is the fate of constitutively proliferating cells in the subependymal layer of the adult mouse brain. *J. Neurosci.* **12:** 249–256.

Morshead C.M., Reynolds B.A., Craig C.G., McBurney M.W., Staines W.A., Morassutti D., Weiss S., and van der K.D. 1994. Neural stem cells in the adult mammalian forebrain: A relatively quiescent subpopulation of subependymal cells. *Neuron* **13:** 1071–1082.

Nottebohm F. 1981. A brain for all seasons: Cyclical anatomical changes in song control nuclei of the canary brain. *Science* **214:** 1368–1370.

———. 1985. Neuronal replacement in adulthood. *Ann. N.Y. Acad. Sci.* **457:** 143–161.

———. 2002. Why are some neurons replaced in adult brain? *J. Neurosci.* **22:** 624–628.

Ohab J., Fleming S., Blesch A., and Carmichael S.T. 2006. A neurovascular niche for neurogenesis after stroke. *J. Neurosci.* **26:** 13007–13016.

Palmer T.D., Willhoite A.R., and Gage F.H. 2000. Vascular niche for adult hippocampal neurogenesis. *J. Comp. Neurol.* **425:** 479–494.

Parent J.M. 2003. Injury-induced neurogenesis in the adult mammalian brain. *Neuroscientist* **9:** 261–272.

Parent J.M., Vexler Z.S., Gong C., Derugin N., and Ferriero D.M. 2002. Rat forebrain neurogenesis and striatal neuron replacement after focal stroke. *Ann. Neurol.* **52:** 802–813.

Paton J.A., and Nottebohm F.N. 1984. Neurons generated in the adult brain are recruited into functional circuits. *Science.* **225:** 1046–1048.

Paton J.A., O'Loughlin B.E., and Nottebohm F. 1985. Cells born in adult canary forebrain are local interneurons. *J. Neurosci.* **5:** 3088–3093.

Pencea V., Bingaman K.D., Wiegand S.J., and Luskin M.B. 2001. Infusion of brain-derived neurotrophic factor into the lateral ventricle of the adult rat leads to new neurons in the parenchyma of the striatum, septum, thalamus, and hypothalamus. *J. Neurosci.* **21:** 6706–6717.

Pincus D.W., Keyoung H.M., Harrison-Restelli C., Goodman R.R., Fraser R.A., Edgar M., Sakakibara S., Okano H., Nedergaard M., and Goldman S.A. 1998. Fibroblast growth factor-2/brain-derived neurotrophic factor-associated maturation of new neurons generated from adult human subependymal cells. *Ann. Neurol.* **43:** 576–585.

Rakic P. 2002. Adult neurogenesis in mammals: An identity crisis. *J. Neurosci.* **22:** 614–618.

Rasika S., Alvarez-Buylla A., and Nottebohm F. 1999. BDNF mediates the effects of testosterone on the survival of new neurons in an adult brain. *Neuron* **22:** 53–62.

Rasika S., Nottebohm F., and Alvarez-Buylla A. 1994. Testosterone increases the recruitment and/or survival of new high vocal center neurons in adult female canaries. *Proc. Nat. Acad. Sci.* **89:** 8591–8595.

Robinow S. and White K. 1991. Characterization and spatial distribution of the elav protein during *Drosophila melanogaster* development. *J. Neurobiol.* **22:** 443–461.

Sakakibara S. and Okano H. 1997. Expression of neural RNA-binding proteins in the postnatal CNS: Implications of their roles in neuronal and glial cell development. *J. Neurosci.* **17:** 8300.

Sanai N., Tramontin A.D., Quinones-Hinojosa A., Barbaro N.M., Gupta N., Kunwar S., Lawton M.T., McDermott M.W., Parsa A.T., Manuel-Garcia V.J., Berger M.S., and Alvarez-Buylla A. 2004. Unique astrocyte ribbon in adult human brain contains neural stem cells but lacks chain migration. *Nature* **427:** 740–744.

Sasaki A., Sotnikova T.D., Gainetdinov R.R., and Jarvis E.D. 2006. Social context-dependent singing-regulated dopamine. *J. Neurosci.* **26:** 9010–9014.

Schapira A. 2006. Etiology of Parkinson's Disease. *Neurology* **66:** S10–23.

Schlinger B. and Arnold A. 1991. Brain is the major site of estrogen synthesis in a male songbird. *Proc. Nat. Acad. Sci.* **88:** 4191–4194.

Shen P., Schlinger B., Campagnoni A., and Arnold A. 1995. An atlas of aromatase mRNA expression in the zebra finch brain. *J. Comp. Neurol.* **360:** 172–184.

Shen Q., Goderie S.K., Jin L., Karanth N., Sun Y., Abramova N., Vincent P., Pumiglia K., and Temple S. 2004. Endothelial cells stimulate self-renewal and expand neurogenesis of neural stem cells. *Science* **304:** 1338–1340.

Sohrabji F., Miranda R., and Toran-Allerand C.D. 1994. Estrogen differentially regulates estrogen and nerve growth factor mRNAs in adult sensory neurons. *J. Neurosci.* **14:** 459–471.

Waxman S. and Anderson M. 1985. Generation of electromotor neurons in *Stenarchus albifrons:* Differences between normally growing and regenerating spinal cord. *Dev. Biol.* **112:** 338–344.

Weiss S., Dunne C., Hewson J., Wohl C., Wheatley M., Peterson A.C., and Reynolds B.A. 1996. Multipotent CNS stem cells are present in the adult mammalian spinal cord and ventricular neuroaxis. *J. Neurosci.* **16:** 7599–7609.

Williams S., Leventhal C., Lemmon V., Nedergaard M., and Goldman S.A. 1999. Estrogen promotes the initial migration and inception of NgCAM-dependent calcium-signaling by new neurons of the adult songbird brain. *Mol. Cell. Neurosci.* **13:** 41–55.

Zimmerman L.B., De Jesus-Escobar J.M., and Harland R.M. 1996. The Spemann organizer signal noggin binds and inactivates bone morphogenetic protein 4. *Cell* **86:** 599–606.

29

Adult Human Neurogenesis: A Response to Cell Loss and New Circuitry Requirements?

Maurice A. Curtis and Peter S. Eriksson
Center for Brain Repair and Rehabilitation
Institute of Neuroscience and Physiology
The Sahlgrenska Academy at Göteborg University
Medicinaregatan 11, SE 413 19
Göteborg, Sweden

Richard L.M. Faull
Department of Anatomy with Radiology
The University of Auckland
Auckland, New Zealand

> *... the functional specialization of the brain imposes on the neurones two great lacunae; proliferation inability and irreversability of intraprotoplasmic differentiation. It is for this reason that, once the development was ended, the founts of growth and regeneration of axons and dendrites dried up irrevocably. In adult centers the nerve paths are something fixed, ended, immutable. Everything may die, nothing may be regenerated. It is for the science of the future to change, if possible, this harsh decree.*

ALMOST 100 YEARS HAVE PAST SINCE SANTIAGO RAMON Y CAJAL wrote those words. In comparison, today there has been a giant paradigm shift from believing in a fixed and immutable brain structure to one of plasticity and one with neurogenic potential, even in the adult human brain. Although the exact functional significance remains to be fully appreciated, it is clear that the integration of new cells from the subventricular zone (SVZ) into the olfactory bulb (OB) or from the subgranule layer into the dentate gyrus (DG) allows mammals to engage in olfaction and in memory and learning. Despite these being vital for normal brain function, the most robust up-regulator of neurogenesis in humans is a response to neurodegeneration in specific brain regions. The specific and dramatic loss of striatal projection neurons in Huntington's disease (HD), the specific loss of hippocampal cholinergic neurons in Alzheimer's disease (AD), and the necrotic death of cells in the core and penumbral region of tissue beyond the point of vessel occlusion in stroke all lead to exacerbation of the neurogenic response. The important driver of this process is therefore neurodegeneration and the requirement for the production of new neurons. In this chapter, we discuss each disease as it stands from the perspective of the human data available and use animal disease model data to illuminate where human data is unavailable. Furthermore, we would like to hypothesize a novel idea as to the importance of neurogenesis in the adolescent and adult brain under certain pathological conditions. Here, we argue that disorders such as schizophrenia may develop as a result of inadequate adult neurogenesis and thus may be a primary progenitor cell disorder rather than a postdisease response.

METHODS FOR DETECTION OF NEUROGENESIS IN VIVO

Tritiated Thymidine

The first demonstration of neurogenesis in vivo in the rodent brain occurred with the introduction of a radioactive thymidine analog (tritiated thymidine, [^{3}H]thymidine) into the animal. Cells that have undergone replication of their DNA in the normal course of cell division or DNA repair had the possibility of incorporating [^{3}H]thymidine and not the biological thymidine, but cells that were not undergoing DNA replication or repair were unaffected by the analog. Subsequently, the animal was sacrificed and the brain was sectioned and placed on [^{3}H]thymidine-sensitive film; the exposure of the film would indicate where in the tissue the DNA had been replicated. When this technique is combined with a histological staining, it is possible to detect from which cell the radioactive signal had come. The early brain studies performed by Altman and Das revealed that many of the cells that had incorporated [^{3}H]thymidine had the morphology of mature neurons (Altman 1962, 1969; Altman and Das 1965a,b, 1966, 1967; Das and Altman 1970, 1971).

Today, this technique is not commonly used because of the lengthy autoradiographic incubations and the potential for uncertainty as to the morphology of a cell believed to be a neuron. However, this method is still valid and has paved the way for innumerable other studies on neurogenesis.

Bromodeoxyuridine Labeling

The most common method currently used for the detection of neurogenesis in the brain is to inject the animal with bromodeoxyuridine (BrdU), which is also a thymidine analog; however, rather than being radioactive, it can be detected by an antibody using a pretreatment to denature the DNA but with otherwise standard immunohistochemical techniques. Because BrdU does weaken the DNA strand, high doses of BrdU do make the DNA more susceptible to strand breakage. BrdU was, in fact, initially used for radiosensitization. However, at the doses required to detect cell proliferation, the compound is stable in the cell and appears to cause minimal damage. The main advantage of using BrdU for detecting cell proliferation is that it can be used in conjunction with a plethora of other antibodies raised against other cell proteins. Hence, double, triple, and even quadruple labeling can be readily achieved using appropriate fluorochromes. The detection of the localization of the different markers is performed using a laser-scanning or two-photon confocal microscope. This technique has also been applied to humans and has allowed the first demonstration of human neurogenesis in the hippocampus. As a method of determining the spread of oropharyngeal carcinoma in a cohort of patients, 250 mg of BrdU was administered in a single injection. Subsequently, it was recognized that the BrdU might also have been incorporated into dividing cells in the brain and that if they were then double-labeled with mature neuronal markers, this would reveal whether new neurons were being born in the adult brain (Eriksson et al. 1998). One weakness in the BrdU method of determining neurogenesis, particularly in the human brain, is that the lipofuschin content, especially in sections derived from older subjects, contributes significant fluorescent background which hampers the detection of the BrdU labeling (Cooper-Kuhn and Kuhn 2002). Furthermore, the cells that divide rapidly dilute out the BrdU, thus underestimating the real number of dividing cells (Karpowicz et al. 2005). However, today, BrdU labeling and colabeling with a neuronal marker remain the most widely used technique for detecting neurogenesis.

Radiocarbon Birth Dating Neurons

The most recent method that uniquely demonstrates neurogenesis and even the age of neurons in the brain capitalizes on the variation in the atmosphere's radioactive ^{14}C content over time. In the 1950s and 1960s, when above-ground nuclear testing was being carried out, the atmospheric levels of ^{14}C rose dramatically. However, after the test ban treaty in 1963, the levels of ^{14}C steadily and rapidly decreased due to absorption of the ^{14}C into the biosphere. Since carbon is a major constituent of the body, higher levels of ^{14}C could also be detected in the tissues of those born during the periods of high atmospheric ^{14}C (Spalding et al. 2005). ^{14}C is stable in the DNA of tissues not undergoing cell division, and thus by detecting the levels of ^{14}C in neurons postmortem, it is possible to determine the age of the neurons. This technique has been employed predominantly by a group from Stockholm led by Professor Frisen. These investigators have been interested in the age of the neurons in the human cortex and the ^{14}C method is one suitable way to examine large populations of neurons (Spalding et al. 2005; Bhardwaj et al. 2006). The technique used for sorting the cells into neurons or other cell types in the brain is fluorescence-activated cell sorting (FACS), and Frisen and colleagues did this by antibody labeling the nuclei of neurons and detecting immunoreactivity for neuronal nuclei (NeuN). Subsequently, using a radiomass spectrometer, the ^{14}C levels were measured in the NeuN-positive cells (Spalding et al. 2005). One of the big questions in neuroscience in the past few years has been is there any neurogenesis in the adult human neocortex? Despite conflicting stories in the monkey brain, the results from a combined study using the ^{14}C method together with BrdU labeling in the human cerebral cortex failed to reveal any neurogenesis in the human cortex (Bhardwaj et al. 2006). The downside of the ^{14}C technique is that very small numbers of newly born neurons could potentially remain undetected, and if the newly born neurons have a high turnover, they might also go undetected. However, for the detection of stably integrating cells of a certain kind, this method remains a very sound technique.

HIPPOCAMPAL NEUROGENESIS OCCURS IN THE NORMAL HUMAN BRAIN

Although adult hippocampal neurogenesis was first suggested in the 1960s, the demonstration of its existence in the human brain was not confirmed until 1998 (Eriksson et al. 1998). The study was the result of an extensive and long-lasting hunt for brain tissue from patients treated with BrdU either for therapeutic purposes (the drug was originally developed as an anticancer, radiosensitizing drug) or for "tumor staging," e.g., determining the degree of malignancy of a tumor. The idea underpinning the discovery of such a tissue source was that if the BrdU could be used to estimate proliferation among tumor cells, it could then also be used to birth date neuronal cells. Therefore, identifying cells that had a BrdU-established birth date as well as neuronal characteristics would demonstrate adult human neurogenesis. Identifying hospitals or clinics that had access to such materials was difficult, and numerous unfruitful inquires were made until one night while on-call at the emergency room, a serendipitous meeting took place between me (P. Eriksson), a neurologist under training at the time, and Thomas Björk-Eriksson, an oncologist also under training at the time. Dr. Björk-Eriksson was conducting a small study on the value of BrdU staging of tumors in the head and neck region. The late-night coffee break and the resulting discussions on BrdU treatment inspired our collaboration, and in effect the so-sought-after tissue was literally right in our "backyard" in Gothenburg.

To date, the precise functional impact of adult hippocampal neurogenesis strictly speaking remains elusive, although several animal studies suggest that hippocampal neurogenesis contributes to spatial memory. Furthermore, the importance of adult neurogenesis can be teased out when looking at conditions in which neurogenesis is reduced in the brain such as in depression or after radiotherapy. In children who receive radiotherapy to the brain, there is a characteristic cognitive deficit associated with it that follows treatment. The condition improves after treatment with a time course similar to that required to restore normal levels of progenitor cell numbers in the brain (i.e., 3–6 weeks). Furthermore, in depression, the effect of serotonin reuptake inhibition by compounds such as fluoxetine does not account for the improvement in symptomatology given that the best results are seen 3–6 weeks after fluoxetine administration. A study performed by Chen et al. (2000)—in which mice were chronically administered the antidepressant compound lithium and the cell cycle marker BrdU—demonstrated a 25% increase in the number of proliferating cells in the DG of the hippocampus compared to controls. Lithium has been shown to increase levels of B-cell lymphoma protein-2 (bcl-2) that exerts a major antiapoptotic neuroprotective effect in vitro and in vivo and may exert a trophic effect as well (Nonaka et al. 1998; Manji et al. 1999, 2000a,b; Chen et al. 2000; Manev et al. 2001; Willing et al. 2002; Malberg and Duman 2003; Senatorov et al. 2004). Although little is known about the intracellular signaling involved in adult neurogenesis, it has been demonstrated that long-term treatment with antidepressants leads to an up-regulation of brain-derived neurotrophic factor (BDNF) as a result of activation of a cAMP-cAMP response element-binding (CREB) protein cascade (Malberg et al. 2000; Duman et al. 2001). Thus, in an indirect way, the importance of hippocampal neurogenesis is being established.

Previous studies have established so-called place cells originally described by O'Keefe and Dostrovsky (1971) that have the ability to fire when a rat reaches a specific spatial location. Later, additional experimental studies using various paradigms altering hippocampal function (lesions, drugs, environmental stimulations, stress, and many mutations) could verify altered performance in spatial memory paradigms such as the Morris water maze test, radial arm maze, circular maze, and contextual conditioning (Morris et al. 1982; Morris et al. 1986; Tsien et al. 1996; Kim and Diamond 2002; Maekawa et al. 2005). In addition, the hippocampus is thought to have access to cortically processed sensory information and to memory storage units in the neocortex through afferent and efferent connections (Lavenex and Amaral 2000). Furthermore, this circuit has been viewed as one of the sites in which association of multimodal sensory input and formation of associated memory occurs (Black et al. 1977; McNaughton and Morris 1987). The human hippocampus is also thought to be highly important for the function of declarative memory (memory of facts and events) (Squire and Cave 1991; Eichenbaum 2000a,b; Eichenbaum and Harris 2000). Recently, the notion that adult neurogenesis is important for spatial memory has been questioned in studies where no effects on special learning could be documented. A novel hypothesis was proposed where newborn neurons within the DG were proposed to contribute temporal information on newly formed declarative memories (Aimone et al. 2006). The remaining debate and lack of definitive confirmation of the role of adult hippocampal neurogenesis can be attributed to the lack of models suitable for testing hypotheses on this phenomenon. Development of molecular models where neurogenesis can be conditionally shut down at a certain time point without having the caveats of secondary effects is essential for understanding the role of this phenomenon. What is clear is that the production of new neurons appears to be important for the brain to cope with a changing external environment and to produce new memories in a temporally relevant manner. In the hippocampus, it may not be necessary to have long-term integration of the new cells in order for them to have a temporally important effect on memory formation; thus, the ongoing cell loss of granule cells may drive new production and short-term integration of precursor cells into existing host circuitry.

THE HUMAN SVZ HARBORS PROGENITORS
THAT BECOME ADULT-BORN NEURONS

During development, neurons migrate from their place of birth in the SVZ differentiating en route and at their final destination where they become incorporated in the functional circuitry of the forebrain. In the adult brain, rodent studies reveal that neurogenesis continues in the SVZ throughout adult life from which cell replacement occurs, particularly in the OB (Luskin and Boone 1994; Altman 1969; Lois et al. 1996; Benraiss et al. 2001). The fate of the replacement cells is at least in part determined in the SVZ. Our studies (Curtis et al. 2003; Curtis et al. 2005a,b) and others (Quinones-Hinojosa et al. 2006) on the biology of the SVZ in the normal and diseased human brain demonstrate that as in the rodent, the SVZ retains a neurogenic potential in the adult human brain.

In the adult human brain SVZ, our studies (Curtis et al. 2003; Curtis et al. 2005a,b) and others (Doetsch et al. 1997; Quinones-Hinojosa et al. 2006) have demonstrated the presence of three major cell types (A, B, and C) that are differentially distributed in three layers in the SVZ immediately below the ependymal layer. Type-A cells are found in small numbers scattered in the cell-poor layer (layer 2) immediately beneath the ependymal cells; rodent studies have identified type-A cells as migrating neuroblasts/immature neurons destined for the OB (Lois et al. 1996; Doetsch et al. 1997). Layer 3 comprises Type-B cells (B1 and B2 have different nuclear morphology) that are glial fibrillary acidic protein (GFAP)-positive and are the most numerous cell type in the SVZ; they form a well-delineated and conspicuous layer 3 in the SVZ interposed between the cell-poor layer 2 superficially and the deeper layer 4 consisting of myelinated fibres. In the adult rodent brain, the type-B astrocytes in the SVZ have been identified as the primary progenitors of new neurons (Doetsch et al. 1999; Imura et al. 2003). Type-C cells are less numerous and are located deep in layer 3 very close to the myelin in layer 4. In our human studies, type-C cells express βIII-tubulin and sometimes coexpress the endogenous proliferating cell nuclear antigen (PCNA) protein, a combination indicating neurogenesis in the human SVZ (Curtis et al. 2005b). In the mouse brain, there is a ratio of type-A:B:C cells of 3:2:1 in contrast to the normal human SVZ that has an A:B:C ratio of 1:3:1; this variation may reflect the high demand in the mouse of type-A cells to migrate to the OB as replacement interneurons (Lois et al. 1996;

Doetsch et al. 1997; Curtis et al. 2005a). Thus, in general, the SVZ in humans contains progenitor/glial cells, migratory cells, and neuroblasts, demonstrating that the human SVZ retains a neurogenic (germinal) potential.

OLFACTORY NEUROGENESIS OCCURS, TO REPLACE DYING OLFACTORY NEURONS, AFTER TARGETED LONG-DISTANCE MIGRATION OF PROGENITOR CELLS FROM THE SVZ

For olfactory neurogenesis to take place in the rodent and subhuman primate brain, neuroblast/progenitor cells migrate the long distance from the SVZ surrounding the lateral ventricles to the OB via the rostral migratory stream (RMS). However, progenitor cells that differentiate to form neurons and replace granule cells in the OB have been thought to arise locally within the OB in the human brain (Bedard and Parent 2004). Many species possess an open tube between the lateral ventricle and olfactory ventricle, and this allows the free flow of cerebrospinal fluid (CSF) as well as a continuous SVZ between the two regions (Michigan State University Brain Diversity Bank 2007; Rae 1994). Recent studies in our laboratory have revealed that the human brain also has an RMS similar to that of the rodent and in particular (1) contains numerous migratory progenitor cells, (2) takes a similar route but with adjustment for variation in species-specific forebrain differences (Fig. 1), and (3) allows migratory progenitor cells to differentiate in the OB to form neurons, as evidenced by BrdU labeling colocalized with the neuronal marker NeuN (Curtis et al. 2007).

In the human brain, the RMS takes a course rostroventrally to the striatum and then the cells migrate forward in the olfactory tract to the OB (Fig. 2a). The human brain follows the basic structural organization of the mammalian brain; however, compared to the rodent, the human forebrain is extensively developed. Furthermore, the human OB, and hence the olfactory interneuron replacement system, is comparatively smaller than in rodents, and thus, the RMS has until recently remained illusive in the human brain.

The RMS begins as a cleft in the floor of the anterior horn of the lateral ventricle where large numbers of PCNA-positive (proliferating cells) cells are continuous with the SVZ that overlies the caudate nucleus (CN; Fig. 2b). The stream of cells first takes a caudal and ventral track along the undersurface of the CN, caudal to the genu of the corpus callosum and the frontal cortical white matter of the gyrus rectus (see Fig. 1). When the descending limb reaches the ventral CN, the stream takes a

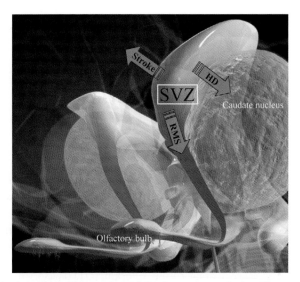

Figure 1. Diagramatic representation of the human ventriculo-olfactory neurogenic system (VONS). From the subventricular zone (SVZ) near the striatum, there is a massive up-regulation of progenitor cells and neurogenesis in response to the degenerating striatum. The arrows indicate the direction of migration that would be required for the progenitor cells to migrate to bring about cell replacement in stroke or Huntington's disease (HD). Furthermore, the rostral migratory stream (RMS) also begins in the SVZ and descends rostral to the striatum before taking a rostral and ventral course into the olfactory bulb (OB) where interneuron replacement occurs. Both for the OB and for the striatum, cell death drives progenitor proliferation in the SVZ.

rostral turn to form the rostral limb of the RMS; the rostral limb passes ventrally and rostrally to enter the anterior olfactory cortex (AOC) and traverses the olfactory tract to enter the OB (Fig. 1). Near the SVZ, the mediolateral extent that contained proliferating cells covered 2.1 mm and began the "descending limb"; at the level of the rostral limb, proliferating cells only extend mediolaterally over approximately 0.6 mm. Thus, the overall mediolateral extent of the RMS is 2.7 mm from the SVZ to the olfactory tract, and the total length of the RMS pathway from the SVZ to the start of the olfactory tract is approximately 17 mm.

The RMS takes a path from medial, by the SVZ, to lateral, beneath the CN; close to the SVZ, the RMS is very wide. In the olfactory tract, the proliferating cells are present in a flat band in the core of the olfactory tract from the caudal region near the AOC to the rostral extent where the olfactory tract enters the OB. The RMS contains large numbers of neuroblasts as evidenced by βIII-tubulin-positive staining of cell

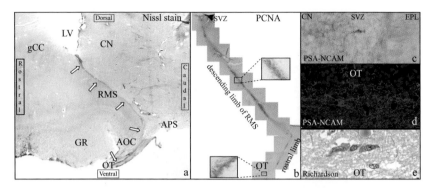

Figure 2. The human RMS from the SVZ to the anterior olfactory cortex. (*a*) Nissl staining of the sagittal human forebrain sections reveals intense stained cells that delineate the location of the RMS behind the frontal cortex white matter of the gyrus rectus (GR) and in front of the caudate nucleus (CN) and anterior perforated substance (APS); arrows indicate the pathway. (AOC) Anterior olfactory cortex; (LV) lateral ventricle; (gCC) genu of the corpus callosum; (OT) olfactory tract. (*b*) A montage of PCNA labeling of the human RMS shows the presence of numerous proliferating cells in both the descending and rostral limbs of the RMS. Bar, 5 mm. (*c*) PSA-NCAM cells are mostly located beneath the gap region of the SVZ and are oriented toward the CN. (*d*) PSA-NCAM with a Hoechst nuclear counterstain in the olfactory tract indicates the direction of the migrating cells in the long axis of the tract and reveals much labeling in the periphery of the tract. (*e*) Progenitor cells with migratory features: elongated cell bodies and nuclei close to each other with the leading process of one progenitor overlapping another progenitor.

processes and occasionally in nuclei. Thus, overall, the RMS is a funnel-shaped structure that is widest in the mediolateral extent at the SVZ and becomes narrower at the rostral limb; this is illustrated diagramatically in Figure 1. Immunostaining for polysialic-acid-neural cell adhesion molecule (PSA-NCAM), which is expressed by migrating cells in the rodent RMS (Bedard et al. 2002; Bonfanti 2006), is present in abundance in the SVZ (Fig. 2c), descending and rostral limbs of the RMS, the olfactory tract core (Fig. 2d) and periphery, and the OB. In the descending limb and rostral limb of the RMS, there are numerous bipolar and elongated PSA-NCAM-positive cells. The olfactory tract core and periphery also have an abundance of PSA-NCAM-positive cells that have a morphology and staining pattern similar to that seen in the descending limb of the RMS. PSA-NCAM-positive cells in the OB are more randomly oriented, have fewer processes, and are larger than the cells in the olfactory tract core, indicating their possible early stage of maturation. In the rostral limb of the RMS, the morphology of the PSA-NCAM cells was similar to those in the descending limb. Double-labeling and laser-scanning confocal microscopy for PSA-NCAM and βIII-tubulin revealed that in the SVZ, PSA-NCAM and βIII-tubulin are colocalized on the same bipolar cells. Double-labeled PSA-NCAM- and βIII-tubulin-positive cells are present in the olfactory tract as well as in the OB. Electron microscopy and light microscopy of Richardson-stained sections (Fig. 2e) of the longitudinally cut tract revealed that progenitor cells had a pointed narrow leading process that contained a few mitochondria, whereas the trailing area of the cell was densely packed with cell organelles and was extensively wider than the leading process.

Again, in this study on the olfactory system in humans, the usefulness of birth-dating cells was invaluable, and BrdU-labeling paradigm was the same as that used for the 1998 hippocampal study. The OB from patients that had been administered BrdU revealed double labeling in the OB with NeuN, a mature neuronal marker, and an antibody against BrdU. Laser-scanning confocal z-series analysis revealed the colocalization of BrdU and NeuN in the same cells in the periglomerular layer.

For the olfactory system, the underpinning driver of new cell production in the SVZ, their long-distance migration, and eventual maturation into granule cells is the loss of the granule cell population in the OB. However, in general, the OB must be intact for replacement to take place. In experiments where the OB was removed from the rodent brain, the SVZ reduced its production of progenitor cells over time. When the OB tissue was replaced with "like" tissue, the SVZ up-regulated progenitor cell production, whereas when the OB region was replaced with cortical tissue, the SVZ remained inactive. The replacement of olfactory interneurons in a lamina-specific way in the human brain after long-distance migration indicates the precise signaling in the ventriculo-olfactory neurogenic system (VONS) in the human brain.

NEUROGENESIS IS UP-REGULATED IN HUNTINGTON'S DISEASE

Huntington's disease (HD) is an autosomal dominant neurodegenerative disorder that causes progressive cell death in the striatum and is caused by an expanded CAG repeat in the huntingtin gene IT15 on the short arm of chromosome 4 (Gusella et al. 1983; Vonsattel et al. 1985; Snell et al. 1993; Huntington's Disease Collaborative Research Group 1993). In our studies, we have shown that in response to the progressive and massive cell death in the striatum in HD, the SVZ is significantly altered with respect to thickness, cellular composition, numbers of progenitor cells, and mature cell types present (Curtis et al. 2003, 2005a). In particular, there is an approximate overall 2.8-fold increase in SVZ thickness (mostly in layer 3) compared with the normal SVZ that results from a large increase in the number of proliferating cells in the HD SVZ in HD grades 1–3. The greatest number of proliferating cells was in the central and ventral regions of the SVZ in HD brains, whereas the lowest number of proliferating cells was in the rostral locations where there was no significant difference in the SVZ cell turnover between the HD and normal brains (Curtis et al. 2005b). Of particular interest in our studies was that despite an approximate 50% increase in the number of type-A and type-C cells in the HD SVZ, most of the increase in the number of cells was due to an increase in the number of type-B (glial) cells comprising layer 3, which resulted in a shift in the ratio of type-A:B:C from 1:3:1 in normals to 1:7.5:1 in HD grade 3 (Curtis et al. 2005a). As expected, the increase in the number of type-B/glial progenitor cells corresponded to the increased numbers of GFAP-positive cells in the HD SVZ. In addition, the number of newly produced neurons was increased in the HD SVZ in proportion to the increase in PCNA-positive dividing cells. Thus, although there was a 2.6-fold increase in the numbers of new neurons, the proportion of new neurons to astrocytes remained unaltered. Furthermore, our HD studies showed that the degree of cell proliferation in the SVZ increased with pathological severity (i.e., with increasing cell death in the striatum) and with increasing CAG repeats in the HD gene. Interestingly, in animal models of HD, the basic requirement for increased progenitor cell production in the SVZ seems also to depend on the presence of cell death in the forebrain. In transgenic mouse models of HD where there is minimal cell loss in the striatum, the SVZ is basically unaltered (Sathasivam et al. 1999; Slow et al. 2005), but in rat striatal lesion models, there is a marked increase in SVZ progenitor cell proliferation and neurogenesis (Tattersfield et al. 2004; Phillips et al. 2005). Furthermore, quinolinic acid striatal lesions in the adult rat brain cause progenitor cells to migrate from the SVZ toward the site of the lesion where they differentiate into neurons (Tattersfield et al. 2004). These collective findings from human and rodent studies suggest that cell death in the striatum is a powerful stimulant of progenitor cell proliferation and neurogenesis in the SVZ (see Fig. 1).

Other very interesting recent developments on neurogenesis have been obtained from HD transgenic mice studies. The most commonly reported transgenic mice are R6/1 and R6/2; both contain exon 1 of the human HD gene with 115 and 150 CAG repeats, respectively. In comparison, in human HD, CAG repeat lengths in this range lead to severe symptomatology and death in early infancy, whereas the more common HD CAG repeat range is 40–55 and usually results in a disease onset at

40 or more years. Of particular and special interest from the transgenic mice studies is the finding that R6/1 and R6/2 mice housed in an enriched environment (objects of novelty in the home cage that mice can explore and play with) demonstrate a delayed onset of HD-like symptoms and a delayed loss of cannabinoid CB1 receptors compared with those housed normally (van Dellen et al. 2000; Hockly et al. 2002; Glass et al. 2004). In particular, mice housed in an enriched environment were behaviorally better in motor ability such that they were more able to perform simple tasks such as turning on a suspended horizontal rod, a task that HD mice from normal cages struggled with. In addition, grip-strength and rota-rod performance were better in those mice housed in an enriched environment. Furthermore, the mice housed in enriched environments took longer to develop the seizures that are characteristic of their genotype and, upon examination of the brains, appeared to have less volume loss in the peristriatal region compared with normally housed controls. Overall, these studies indicate that the enriched environment delayed the onset of the symptoms and disease progression caused by the HD gene in transgenic mice. However, although neither of these studies examined the SVZ or the effect that enrichment had on the germinal zones (van Dellen et al. 2000; Hockly et al. 2002), further very interesting studies by Lazic et al. (2004, 2006) on the R6/1 mice examined the affect that environmental enrichment has on cell proliferation and neurogenesis in both the hippocampus and the SVZ; they showed a reduction in the HD trans-genic mice compared with wild-type controls and that this reduction was, at least in part, ameliorated by environmental enrichment (Lazic et al. 2004, 2006). These studies raise the very interesting possibility that environmental factors may have a role in cell proliferation and neurogenesis in neurodegenerative diseases.

Of interest, the SVZ environment is enriched in cell types, endogenous compounds, and receptors that are known to actively regulate the cell cycle and/or the differentiation of precursors. The human SVZ is enriched in neuropeptide Y (NPY)-positive cells (Curtis et al. 2005a). In rodents, NPY powerfully promotes the production of progenitor cells as evidenced by a 50% reduction in progenitor cells in NPY-deficient knock-out mice (Hansel et al. 2001). In the human brain, NPY cells express other transmitters such as nitric oxide synthase (NOS) which influences progenitor cell proliferation, migration, and neurite outgrowth (Reif et al. 2004; Chen et al. 2005). Furthermore, certain γ-amino-n-butyric acid (GABA) receptors are enriched in the SVZ of the normal brain compared with other brain regions, and in HD, these $GABA_A$ receptors are significantly increased in the SVZ compared with normal brains (Curtis et al. 2005a).

One such example is the GABA$_A$ receptor subunit γ2 that is located on cell bodies and on the neuropil in the SVZ. The γ2 subunit of the GABA$_A$ receptor is involved with desensitization of the receptor complex to GABA in the developing SVZ/ventricular zone (Saxena and Macdonald 1994; LoTurco et al. 1995). In the rat brain, the matured progeny of the SVZ express only low levels of the γ2 subunit compared with the very high levels in the SVZ during development. The neurotransmitter GABA is also an important trophic factor for the maturation of neurons during development (Saxena and Macdonald 1994; LoTurco et al. 1995). It is clear from these studies that the adult human SVZ maintains a germinal capacity for the ongoing development and possible repair of the nervous system and has the capacity to increase cell proliferation and neurogenesis in response to neurodegenerative cell death in adult life.

NEUROGENESIS IS UP-REGULATED IN OTHER NEURODEGENERATIVE DISEASES AND BRAIN INJURIES

To date, only one study has reported results from the study of neurogenesis in human brains affected by stroke: Jin et al. (2006) made extensive use of endogenously expressed cell proliferation, cell stage, and lineage markers in various combinations to identify neurogenic cells in the peri-infarct (penumbral) region of cortical strokes. The plethora of protein markers they detected included Ki-67, *Saccharomyces cereviseae* minichromosome maintenance-2 homolog (MCM2), PCNA, doublecortin (DCX), turned on after division (TOAD), βIII-tubulin, and embryonic cell adhesion molecule (ENCAM). The combination of proliferation and neuronal lineage markers revealed the presence of Ki-67 and TUC-4 in the same cells that expressed DCX and βIII-tubulin, cell division, and neuronal lineage markers, respectively (Jin et al. 2006).

The effect of stroke on the SVZ has been extensively examined in the rodent brain. Commonly, unilateral middle cerebral artery occlusion (MCAO) is used to induce a stroke-like lesion in the ipsilateral cerebral hemisphere. Using the proliferative marker BrdU, two groups have reported that endogenous precursors that reside in the SVZ migrate from the SVZ toward the infarcted striatal region in an attempt to replace neurons lost following MCAO (Arvidsson et al. 2002; Parent et al. 2002). The study by Arvidsson et al. (2002) revealed a 31-fold increase in the number of newly produced neurons in the striatum on the lesioned side compared to the control side of the brain and most of the new neurons were located within the damaged area. Despite these remarkable findings, 6 weeks after

MCAO, the estimated fraction of newborn neurons was only about 0.2% of those that were lost due to infarction (Arvidsson et al. 2002). Thus, there is a need for a substantial up-regulation of this repair process before recovery of a full complement of striatal neurons might be achieved. Experimentally, pharmacological treatment with cytokines and chemokines can induce cell proliferation, migration, and differentiation. The factors with a known activity in producing cell proliferation include insulin-like growth factor-1 (Arsenijevic and Weiss 1998; Arsenijevic et al. 2001; Aberg et al. 2003), fibroblast growth factor-2 (Craig et al. 1996; Ciccolini and Svendsen 1998; Martens et al. 2000), epidermal growth factor (Kuhn et al. 1997), and brain derived neurotrophic factor (Zigova et al. 1998; Pencea et al. 2001; Tirassa et al. 2003). In addition, antidepressant drugs such as lithium and fluoxetine can increase the number of endogenous progenitor cells in the niche regions of the brain (Chen et al. 2000; Malberg and Duman 2003). However, of very recent interest is that environmental enrichment of caged laboratory animals, by the repeated introduction of novel objects, increases the number of progenitor cells in the SVZ. In one study reported by Komitova et al. (2005), MCAO-treated rats were housed in an enriched environment that began either 24 hours or 7 days post-MCAO. Five weeks after MCAO, both groups of animals that were housed in an enriched environment had similar increased numbers of progenitor cells in the SVZ compared to normally caged lesioned controls. These results indicate that environmental enrichment not only increases the number of progenitors in the SVZ, but also lengthens the survival of progenitors in the SVZ (Komitova et al. 2005). More extensive research is required to identify and understand the molecular mechanisms by which environmental enrichment enhances progenitor cell proliferation and neurogenesis in the SVZ. Interestingly, despite the large number of exogenous compounds that up-regulate the neurogenic response in the brain, the important endogenous stimulant of neurogenesis is in fact the requirement for replacement of neurons because of neurodegenerative disease.

Recent studies in Alzheimer's disease (AD)-affected brains also reveal an apparent up-regulation of neurogenic proteins in the hippocampus, the region preferentially affected by the pathological loss of cholinergic neurons. More specifically, in studies also performed by Jin et al. (2004) on postmortem human Alzheimer brains, DCX, TUC-4, PSA-NCAM, and NeuroD proteins were up-regulated. However, most of the findings in AD come from western blot experiments that do not allow morphological information to be gathered; the important neurogenic proteins are expressed in the diseased hippocampus. Of interest in these studies is the differential high expression of DCX and TUC-4 in the dentate granule cell layer in AD, a region severely affected, compared with the low levels of these proteins in Ammon's horns, a relatively unaffected region of the hippocampal formation (Jin et al. 2004). However, in a younger cohort of presenile patients, these results were not repeated with a more aggressive disease (Boekhoorn et al. 2006). Although the number of cells expressing the cell cycle marker Ki-67 was significantly increased, this was largely due to the proliferation of nonneuronal cells. Furthermore, animal studies using other AD models based on presenelin 1 (PS1) transgenes or intraventricular infusion of β-amyloid reported an impairment of adult neurogenesis (Haughey et al. 2002a,b; Wang et al. 2004; Wen et al. 2004; Donovan et al. 2006).

PRIMARY PROGENITOR DISORDERS IN THE NERVOUS SYSTEM

So far, no primary progenitor disorder of the brain has been firmly established. The lack of such disorders hampers our understanding of the role of, for example, adult hippocampal neurogenesis. One simple attempt to define an adult primary progenitor disorder could be made using the following criteria: (1) The disorder should be caused by a primary defect in the function of progenitor/stem cells. (2) The disorder should generate delayed (adult or adolescence) onset symptoms.

There are several disorders that potentially could fit these two very simple criteria. Schizophrenia is one condition often referred to as a developmental disorder that could be a primary progenitor cell disorder. The characteristic age at onset of this disorder—late adolescence and young adulthood—suggest that schizophrenia could be the result of dysregulation of adult neurogenesis. Schizophrenia is in general viewed as a disorder resulting from a developmental defect. Evidence put forward in favor of this hypothesis are, for example, reduced thickness of the cerebral cortex and cellular disarray of the hippocampus in tissue from schizophrenic subjects (Harrison 1999a,b,c). These findings could also be interpreted as support for a dysregulation of adult neural progenitors.

The importance of the hippocampal system in schizophrenia is demonstrated mainly by a subtle but significant volume difference in schizophrenia; this is a consistent neuroanatonical finding in schizophrenia research (Bogerts 1997; Heckers and Konradi 2002). This in turn could indicate that hippocampal neurogenesis may be disrupted in schizophrenia (Eriksson 2006).

A recent study has provided evidence that disrupted-in-schizophrenia 1 (DISC 1)—a gene associated with schizophrenia in multiple genetic studies in patients—is very highly expressed in the DG of the hip-pocampus (Austin et al. 2004). DISC 1 is a multifunctional protein involved in extracellular signaling, neurite outgrowth, and neuronal migration (Morris et al. 2003). These findings are interesting considering the evidence that neurite and synapse architecture is abnormal in schizo-phrenia (McGlashan and Hoffman 2000). Although no studies have demonstrated a function for this gene in neurogenesis, these findings may suggest that a disruption of *Disc1* could result in aberrant adult dentate neurogenesis.

Recently, a genome-wide scan of schizophrenia families in Iceland identified neuregulin 1 (*NRG1*) as a gene associated with susceptibility to schizophrenia (Stefansson et al. 2002). This was recently followed up by Anton et al. (2004) demonstrating a direct link between neuronal stem/progenitor cells and schizophrenia. These authors were able to show that the receptor tyrosine kinase ErbB4 is expressed prominently by the neu-roblasts present in the SVZ and the RMS. The neuregulins (NRG1–NRG3), previously identified as ErbB4 ligands, were detected either in the RMS or in adjacent regions in the adult rodent brain. Mice deficient in ErbB4 had altered neuroblast chain organization and migration and deficits in the placement and differentiation of newly formed olfactory inter-neurons. These findings suggest that ErbB4 activation helps to regulate the organization of neural migration through the RMS and influences the differentiation of olfactory interneurons. In the human brain, alterations in the ErbB4 system may in turn result in impaired NRG-ErbB4-mediated stem/progenitor cell progeny differentiation and migration in VONS, and resultant changes in neural circuitry in the forebrain through aberrant cell genesis, that may enhance the risk for developing schizophrenia. Interestingly, mice heterozygous with respect to either *Nrg1* or *Erbb4* have behavioral phenotypes that resemble those associated with schizophrenia. Furthermore, these mice show behavioral phenotypes that are partially reversible with clozapine, an atypical antipsychotic drug used to treat schizophrenia (Stefansson et al. 2002).

Interestingly, Wakade et al. (2002) recently showed that atypical neuroleptics stimulate a two- to threefold increase in newly divided cells in the SVZ (Wakade et al. 2002). This in turn may indicate novel mech-anisms of action for neuroleptics involving modulation of cell genesis in the adult forebrain. With respect to SVZ-derived new olfactory neurons, it is interesting to note that testing of olfactory functioning in patients with schizophrenia has gained increasing attention since deficits in odor identification and detection threshold sensitivity, discrimination, and memory have been associated with this disorder. The observed olfactory deficits are not explained by medication use, cognitive impairment, or smoking status; instead, they support the hypothesis of primary dysfunction in the olfactory system. Furthermore, structural abnormalities in the olfactory system, as well as disruptions of the basic physiology of this system, have been described recently (Moberg et al. 1999, 2006). Because the olfactory system continuously regenerates throughout life, it provides another link between an altered SVZ cell genesis from adult neuronal stem/progenitor cells and schizophrenia.

Newly generated cells that deviate from the rostral migration toward the OB instead migrate tangentially and generate glial cells in the striatum and cerebral cortex. Interestingly a reduction in the GFAP-immunoreactive area fraction and significant narrowing of layer V in schizophrenia suggest that atrophy of astroglial processes may contribute to the reduction in intercellular neuropil, proposed as one example of primary structural abnormality of the prefrontal cortex in schizophrenia (Rajkowska et al. 2002). Thus, changes or even reduced numbers in the glial population might be related to the neuropil changes manifested by the reduction in the width of layer V found in patients with schizophrenia (Rajkowska et al. 2002). Since prefrontal neurons of layer V send glutamatergic axons to the striatum, the astroglial pathology found in layer V in the prefrontal cortex may result in aberrant synaptic architecture and thus be involved in pathological interactions between the glutamatergic and dopaminergic systems suggested in schizophrenia (Rajkowska et al. 2002). Future studies evaluating the possibility of disturbances in the generation or migration of newly formed cells (e.g., the prefrontal cortex) in the adult human brain are, however, needed to verify the involvement of aberrant neurogenesis in the pathology of schizophrenia.

Another indirect line of evidence can be derived from animal studies on the effect of enriched environment showing pronounced effects on the stem cell pool in the adult brain. Studies on the effect of the environment in schizophrenic subjects have similarly shown pronounced effects on the development and progression of this disorder (Raine et al. 2003).

The role of stem/progenitor cells in the adult brain is still uncertain, but evidence is now pointing toward roles in normal brain function such as memory formation and olfactory system function, as well as in various diseases like Huntington's, Alzheimer's, stroke, and the development of schizophrenia perhaps through perturbed adult neuro- and gliogenesis. Detailed analyses of human tissue samples obtained postmortem may provide further important knowledge on the role of neurogenesis that cannot be generated through studies in animals.

ACKNOWLEDGMENTS

This work was supported by grants from the Swedish Medical Research Council project (12X-1253), the Sahlgrenska Academy, Göteborg University (PSE), and the Health Research Council of New Zealand and the Neurological Foundation of New Zealand (R.L.M.F.). M.A.C. was funded as a Wrightson postdoctoral fellow by the Neurological Foundation of New Zealand.

REFERENCES

Aberg M.A., Aberg N.D., Palmer T.D., Alborn A.M., Carlsson-Skwirut C., Bang P., Rosengren L.E., Olsson T., Gage F.H., and Eriksson P.S. 2003. IGF-I has a direct proliferative effect in adult hippocampal progenitor cells. *Mol. Cell. Neurosci.* **24:** 23–40.

Aimone J.B., Wiles J., and Gage F.H. 2006. Potential role for adult neurogenesis in the encoding of time in new memories. *Nat. Neurosci.* **9:** 723–727.

Altman J. 1962. Are new neurons formed in the brains of adult mammals? *Science* **135:** 1127–1128.

———. 1969. Autoradiographic and histological studies of postnatal neurogenesis. IV. Cell proliferation and migration in the anterior forebrain, with special reference to persisting neurogenesis in the olfactory bulb. *J. Comp. Neurol.* **137:** 433–457.

Altman J. and Das G.D. 1965a. Autoradiographic and histological evidence of postnatal hippocampal neurogenesis in rats. *J. Comp. Neurol.* **124:** 319–335.

———. 1965b. Post-natal origin of microneurones in the rat brain. *Nature* **207:** 953–956.

———. 1966. Autoradiographic and histological studies of postnatal neurogenesis. I. A longitudinal investigation of the kinetics, migration and transformation of cells incorporating tritiated thymidine in neonate rats, with special reference to postnatal neurogenesis in some brain regions. *J. Comp. Neurol.* **126:** 337–389.

———. 1967. Postnatal neurogenesis in the guinea-pig. *Nature* **214:** 1098–1101.

Anton E.S., Ghashghaei H.T., Weber J.L., McCann C., Fischer T.M., Cheung I.D., Gassmann M., Messing A., Klein R., Schwab M.H., Lloyd K.C., and Lai C. 2004. Receptor tyrosine kinase ErbB4 modulates neuroblast migration and placement in the adult forebrain. *Nat. Neurosci.* **7:** 1319–1328.

Arsenijevic Y. and Weiss S. 1998. Insulin-like growth factor-I is a differentiation factor for postmitotic CNS stem cell-derived neuronal precursors: Distinct actions from those of brain-derived neurotrophic factor. *J. Neurosci.* **18:** 2118–2128.

Arsenijevic Y., Weiss S., Schneider B., and Aebischer P. 2001. Insulin-like growth factor-I is necessary for neural stem cell proliferation and demonstrates distinct actions of epidermal growth factor and fibroblast growth factor-2. *J. Neurosci.* **21:** 7194–7202.

Arvidsson A., Collin T., Kirik D., Kokaia Z., and Lindvall O. 2002. Neuronal replacement from endogenous precursors in the adult brain after stroke. *Nat. Med.* **8:** 963–970.

Austin C.P., Ky B., Ma L., Morris J.A., and Shughrue P.J. 2004. Expression of Disrupted-In-Schizophrenia-1, a schizophrenia-associated gene, is prominent in the mouse hippocampus throughout brain development. *Neuroscience* **124:** 3–10.

Bedard A. and Parent A. 2004. Evidence of newly generated neurons in the human olfactory bulb. *Brain Res. Dev. Brain Res.* **151:** 159–168.

Bedard A., Levesque M., Bernier P.J., and Parent A. 2002. The rostral migratory stream in adult squirrel monkeys: Contribution of new neurons to the olfactory tubercle and involvement of the antiapoptotic protein Bcl-2. *Eur. J. Neurosci.* **16:** 1917–1924.

Benraiss A., Chmielnicki E., Lerner K., Roh D., and Goldman S.A. 2001. Adenoviral brain-derived neurotrophic factor induces both neostriatal and olfactory neuronal recruitment from endogenous progenitor cells in the adult forebrain. *J. Neurosci.* **21:** 6718–6731.

Bhardwaj R.D., Curtis M.A., Spalding K.L., Buchholz B.A., Fink D., Bjork-Eriksson T., Nordborg C., Gage F.H., Druid H., Eriksson P.S., and Frisen J. 2006. Neocortical neurogenesis in humans is restricted to development. *Proc. Natl. Acad. Sci.* **103:** 12564–12568.

Black A.H., Nadel L., and O'Keefe J. 1977. Hippocampal function in avoidance learning and punishment. *Psychol. Bull.* **84:** 1107–1129.

Boekhoorn K., Joels M., and Lucassen P.J. 2006. Increased proliferation reflects glial and vascular-associated changes, but not neurogenesis in the presenile Alzheimer hippocampus. *Neurobiol. Dis.* **24:** 1–14.

Bogerts B. 1997. The temporolimbic system theory of positive schizophrenic symptoms. *Schizophr. Bull.* **23:** 423–435.

Bonfanti L. 2006. PSA-NCAM in mammalian structural plasticity and neurogenesis. *Prog. Neurobiol.* **80:** 129–164.

Chen G., Rajkowska G., Du F., Seraji-Bozorgzad N., and Manji H.K. 2000. Enhancement of hippocampal neurogenesis by lithium. *J. Neurochem.* **75:** 1729–1734.

Chen J., Zacharek A., Zhang C., Jiang H., Li Y., Roberts C., Lu M., Kapke A., and Chopp M. 2005. Endothelial nitric oxide synthase regulates brain-derived neurotrophic factor expression and neurogenesis after stroke in mice. *J. Neurosci.* **25:** 2366–2375.

Ciccolini F. and Svendsen C.N. 1998. Fibroblast growth factor 2 (FGF-2) promotes acquisition of epidermal growth factor (EGF) responsiveness in mouse striatal precursor cells: Identification of neural precursors responding to both EGF and FGF-2. *J. Neurosci.* **18:** 7869–7880.

Cooper-Kuhn C.M. and Kuhn H.G. 2002. Is it all DNA repair? Methodological considerations for detecting neurogenesis in the adult brain. *Brain Res. Dev. Brain Res.* **134:** 13–21.

Craig C.G., Tropepe V., Morshead C.M., Reynolds B.A., Weiss S., and van der Kooy D. 1996. *In vivo* growth factor expansion of endogenous subependymal neural precursor cell populations in the adult mouse brain. *J. Neurosci.* **16:** 2649–2658.

Curtis M.A., Penney E.B., Pearson A.G., van Roon-Mom W.M.C., Butterworth N.J., Dragunow M., Connor B., and Faull R.L.M. 2003. Increased cell proliferation and neurogenesis in the adult human Huntington's disease brain. *Proc. Natl. Acad. Sci.* **100:** 9023–9027.

Curtis M.A., Waldvogel H.J., Synek B., and Faull R.L.M. 2005a. A histochemical and immunohistochemical analysis of the subependymal layer in the normal and Huntington's disease brain. *J. Chem. Neuroanat.* **30:** 55–66.

Curtis M.A., Penney E.B., Pearson J., Dragunow M., Connor B., and Faull R.L.M. 2005b. The distribution of progenitor cells in the subependymal layer of the lateral ventricle in the normal and Huntington's disease human brain. *Neuroscience* **132:** 777–788.

Curtis M.A., Kam M., Nannmark U., Anderson M.F., Axell M.Z., Wikkelso C., Holtas S., van Roon-Mom W.M., Bjork-Eriksson T., Nordborg C., et al. 2007. Human neuroblasts migrate to the olfactory bulb via a lateral ventricular extension. *Science* **315:** 1243–1249.

Das G.D. and Altman J. 1970. Postnatal neurogenesis in the caudate nucleus and nucleus accumbens septi in the rat. *Brain Res.* **21:** 122–127.

———. 1971. Postnatal neurogenesis in the cerebellum of the cat and tritiated thymidine autoradiography. *Brain Res.* **30:** 323–330.

Doetsch F., Garcia-Verdugo J.M., and Alvarez-Buylla A. 1997. Cellular composition and three-dimensional organization of the subventricular germinal zone in the adult mammalian brain. *J. Neurosci.* **17:** 5046–5061.

Doetsch F., Caille I., Lim D.A., Garcia-Verdugo J.M., and Alvarez-Buylla A. 1999. Subventricular zone astrocytes are neural stem cells in the adult mammalian brain. *Cell* **97:** 703–716.

Donovan M.H., Yazdani U., Norris R.D., Games D., German D.C., and Eisch A.J. 2006. Decreased adult hippocampal neurogenesis in the PDAPP mouse model of Alzheimer's disease. *J. Comp. Neurol.* **495:** 70–83.

Duman R.S., Nakagawa S., and Malberg J. 2001. Regulation of adult neurogenesis by antidepressant treatment. *Neuropsychopharmacology* **25:** 836–844.

Eichenbaum H. 2000a. A cortical-hippocampal system for declarative memory. *Nat. Rev. Neurosci.* **1:** 41–50.

———. 2000b. Hippocampus: Mapping or memory? *Curr. Biol.* **10:** R785–787.

Eichenbaum H. and Harris K. 2000. Toying with memory in the hippocampus. *Nat. Neurosci.* **3:** 205–206.

Eriksson P.S. 2006. Schizophrenia—A stem cell disorder. *Exp. Neurol.* **199:** 26–27.

Eriksson P.S., Perfilieva E., Bjork-Eriksson T., Alborn A.M., Nordborg C., Peterson D.A., and Gage F.H. 1998. Neurogenesis in the adult human hippocampus. *Nat. Med.* **4:** 1313–1317.

Glass M., van Dellen A., Blakemore C., Hannan A.J., and Faull R.L. 2004. Delayed onset of Huntington's disease in mice in an enriched environment correlates with delayed loss of cannabinoid CB1 receptors. *Neuroscience* **123:** 207–212.

Gusella J.F., Wexler N.S., Conneally P.M., Naylor S.L., Anderson M.A., Tanzi R.E., Watkins P.C., Ottina K., Wallace M.R., Sakaguchi A.Y. et al. 1983. A polymorphic DNA marker genetically linked to Huntington's disease. *Nature* **306:** 234–238.

Hansel D.E., Eipper B.A., and Ronnett G.V. 2001. Neuropeptide Y functions as a neuroproliferative factor. *Nature* **410:** 940–944.

Harrison P.J. 1999a. Neurochemical alterations in schizophrenia affecting the putative receptor targets of atypical antipsychotics. Focus on dopamine (D1, D3, D4) and 5-HT2a receptors. *Br. J. Psychiatry Suppl.* **38:** 12–22.

———. 1999b. The neuropathological effects of antipsychotic drugs. *Schizophr. Res.* **40:** 87–99.

———. 1999c. The neuropathology of schizophrenia. A critical review of the data and their interpretation. *Brain* **122:** 593–624.

Haughey N.J., Liu D., Nath A., Borchard A.C., and Mattson M.P. 2002a. Disruption of neurogenesis in the subventricular zone of adult mice, and in human cortical neuronal precursor cells in culture, by amyloid beta-peptide: Implications for the pathogenesis of Alzheimer's disease. *Neuromolecular Med.* **1:** 125–135.

Haughey N.J., Nath A., Chan S.L., Borchard A.C., Rao M.S., and Mattson M.P. 2002b. Disruption of neurogenesis by amyloid beta-peptide, and perturbed neural progenitor cell homeostasis, in models of Alzheimer's disease. *J. Neurochem.* **83:** 1509–1524.

Heckers S. and Konradi C. 2002. Hippocampal neurons in schizophrenia. *J. Neural Transm.* **109:** 891–905.

Hockly E., Cordery P.M., Woodman B., Mahal A., van Dellen A., Blakemore C., Lewis C.M., Hannan A.J., and Bates G.P. 2002. Environmental enrichment slows disease progression in R6/2 Huntington's disease mice. *Ann. Neurol.* **51:** 235–242.

The Huntington's Disease Collaborative Research Group. 1993. A novel gene containing a trinucleotide repeat that is expanded and unstable on Huntington's disease chromosomes. *Cell* **72:** 971–983.

Imura T., Kornblum H.I., and Sofroniew M.V. 2003. The predominant neural stem cell isolated from postnatal and adult forebrain but not early embryonic forebrain expresses GFAP. *J. Neurosci.* **23:** 2824–2832.

Jin K., Peel A.L., Mao X.O., Xie L., Cottrell B.A., Henshall D.C., and Greenberg D.A. 2004. Increased hippocampal neurogenesis in Alzheimer's disease. *Proc. Natl. Acad. Sci.* **101:** 343–347.

Jin K., Wang X., Xie L., Mao X.O., Zhu W., Wang Y., Shen J., Mao Y., Banwait S., and Greenberg D.A. 2006. Evidence for stroke-induced neurogenesis in the human brain. *Proc. Natl. Acad. Sci.* **103:** 13198–13202.

Karpowicz P., Morshead C., Kam A., Jervis E., Ramunas J., Cheng V., and van der Kooy D. 2005. Support for the immortal strand hypothesis: Neural stem cells partition DNA asymmetrically in vitro. *J. Cell Biol.* **170:** 721–732.

Kim J.J. and Diamond D.M. 2002. The stressed hippocampus, synaptic plasticity and lost memories. *Nat. Rev. Neurosci.* **3:** 453–462.

Komitova M., Mattsson B., Johansson B.B., and Eriksson P.S. 2005. Enriched environment increases neural stem/progenitor cell proliferation and neurogenesis in the subventricular zone of stroke-lesioned adult rats. *Stroke* **36:** 1278–1282.

Kuhn H.G., Winkler J., Kempermann G., Thal L.J., and Gage F.H. 1997. Epidermal growth factor and fibroblast growth factor-2 have different effects on neural progenitors in the adult rat brain. *J. Neurosci.* **17:** 5820–5829.

Lavenex P. and Amaral D.G. 2000. Hippocampal-neocortical interaction: A hierarchy of associativity. *Hippocampus* **10:** 420–430.

Lazic S.E., Grote H., Armstrong R.J., Blakemore C., Hannan A.J., van Dellen A., and Barker R.A. 2004. Decreased hippocampal cell proliferation in R6/1 Huntington's mice. *Neuroreport* **15:** 811–813.

Lazic S.E., Grote H.E., Blakemore C., Hannan A.J., van Dellen A., Phillips W., and Barker R.A. 2006. Neurogenesis in the R6/1 transgenic mouse model of Huntington's disease: Effects of environmental enrichment. *Eur. J. Neurosci.* **23:** 1829–1838.

Lois C., Garcia-Verdugo J.M., and Alvarez-Buylla A. 1996. Chain migration of neuronal precursors. *Science* **271:** 978–981.

LoTurco J.J., Owens D.F., Heath M.J., Davis M.B., and Kriegstein A.R. 1995. GABA and glutamate depolarize cortical progenitor cells and inhibit DNA synthesis. *Neuron* **15:** 1287–1298.

Luskin M.B. and Boone M.S. 1994. Rate and pattern of migration of lineally-related olfactory bulb interneurons generated postnatally in the subventricular zone of the rat. *Chem. Senses* **19:** 695–714.

Maekawa M., Takashima N., Arai Y., Nomura T., Inokuchi K., Yuasa S., and Osumi N. 2005. Pax6 is required for production and maintenance of progenitor cells in postnatal hippocampal neurogenesis. *Genes Cells* **10:** 1001–1014.

Malberg J.E. and Duman R.S. 2003. Cell proliferation in adult hippocampus is decreased by inescapable stress: Reversal by fluoxetine treatment. *Neuropsychopharmacology* **28:** 1562–1571.

Malberg J.E., Eisch A.J., Nestler E.J., and Duman R.S. 2000. Chronic antidepressant treatment increases neurogenesis in adult rat hippocampus. *J. Neurosci.* **20:** 9104–9110.

Manev R., Uz T., and Manev H. 2001. Fluoxetine increases the content of neurotrophic protein S100beta in the rat hippocampus. *Eur. J. Pharmacol.* **420:** R1–2.

Manji H.K., Moore G.J., and Chen G. 1999. Lithium at 50: Have the neuroprotective effects of this unique cation been overlooked? *Biol. Psychiatry* **46:** 929–940.

———. 2000a. Clinical and preclinical evidence for the neurotrophic effects of mood stabilizers: Implications for the pathophysiology and treatment of manic-depressive illness. *Biol. Psychiatry* **48:** 740–754.

———. 2000b. Lithium up-regulates the cytoprotective protein Bcl-2 in the CNS *in vivo*: A role for neurotrophic and neuroprotective effects in manic depressive illness. *J. Clin. Psychiatry* **61:** 82–96.

Martens D.J., Tropepe V., and van Der Kooy D. 2000. Separate proliferation kinetics of fibroblast growth factor-responsive and epidermal growth factor-responsive neural stem cells within the embryonic forebrain germinal zone. *J. Neurosci.* **20:** 1085–1095.

McGlashan T.H. and Hoffman R.E. 2000. Schizophrenia as a disorder of developmentally reduced synaptic connectivity. *Arch. Gen. Psychiatry* **57:** 637–648.

McNaughton N. and Morris R.G. 1987. Chlordiazepoxide, an anxiolytic benzodiazepine, impairs place navigation in rats. *Behav. Brain Res.* **24:** 39–46.

Michigan State University Brain Diversity Bank. 2007. Comparative mammalian brain collections at: http://www.brainmuseum.org/Specimens/lagomorpha/domesticrabbit/sections/thumbnail.html

Moberg P.J., Agrin R., Gur R.E., Gur R.C., Turetsky B.I., and Doty R.L. 1999. Olfactory dysfunction in schizophrenia: A qualitative and quantitative review. *Neuropsychopharmacology* **21:** 325–340.

Moberg P.J., Arnold S.E., Doty R.L., Gur R.E., Balderston C.C., Roalf D.R., Gur R.C., Kohler C.G., Kanes S.J., Siegel S.J., and Turetsky B.I. 2006. Olfactory functioning in schizophrenia: Relationship to clinical, neuropsychological, and volumetric MRI measures. *J. Clin. Exp. Neuropsychol.* **28:** 1444–1461.

Morris J.A., Kandpal G., Ma L., and Austin C.P. 2003. DISC1 (Disrupted-In-Schizophrenia 1) is a centrosome-associated protein that interacts with MAP1A, MIPT3, ATF4/5 and NUDEL: Regulation and loss of interaction with mutation. *Hum. Mol. Genet.* **12:** 1591–1608.

Morris R., Cahusac P.M., Salt T.E., Morris R.G., and Hill R.G. 1982. A behavioural model for the study of facial nociception and the effects of descending modulatory systems in the rat. *J. Neurosci. Methods* **6:** 245–252.

Morris R.G., Hagan J.J., and Rawlins J.N. 1986. Allocentric spatial learning by hippocampectomised rats: A further test of the "spatial mapping" and "working memory" theories of hippocampal function. *Q. J. Exp. Psychol. B.* **38:** 365–395.

Nonaka S., Katsube N., and Chuang D.M. 1998. Lithium protects rat cerebellar granule cells against apoptosis induced by anticonvulsants, phenytoin and carbamazepine. *J. Pharmacol. Exp. Therapeut.* **286:** 539–547.

O'Keefe J. and Dostrovsky J. 1971. The hippocampus as a spatial map. Preliminary evidence from unit activity in the freely-moving rat. *Brain Res.* **34:** 171–175.

Parent J.M., Vexler Z.S., Gong C., Derugin N., and Ferriero D.M. 2002. Rat forebrain neurogenesis and striatal neuron replacement after focal stroke. *Ann. Neurol.* **52:** 802–813.

Pencea V., Bingaman K.D., Wiegand S.J., and Luskin M.B. 2001. Infusion of brain-derived neurotrophic factor into the lateral ventricle of the adult rat leads to new neurons in the parenchyma of the striatum, septum, thalamus, and hypothalamus. *J. Neurosci.* **21:** 6706–6717.

Phillips W., Morton A.J., and Barker R.A. 2005. Abnormalities of neurogenesis in the R6/2 mouse model of Huntington's disease are attributable to the in vivo microenvironment. *J. Neurosci.* **25:** 11564–11576.

Quinones-Hinojosa A., Sanai N., Soriano-Navarro M., Gonzalez-Perez O., Mirzadeh Z., Gil-Perotin S., Romero-Rodriguez R., Berger M.S., Garcia-Verdugo J.M., and Alvarez-Buylla A. 2006. Cellular composition and cytoarchitecture of the adult human subventricular zone: A niche of neural stem cells. *J. Comp. Neurol.* **494:** 415–434.

Rae A.S. 1994. Nodules of cellular proliferation in sheep olfactory ventricle. *Clin. Neuropathol.* **13:** 17–18.

Raine A., Mellingen K., Liu J., Venables P., and Mednick S.A. 2003. Effects of environmental enrichment at ages 3–5 years on schizotypal personality and antisocial behavior at ages 17 and 23 years. *Am. J. Psychiatry* **160:** 1627–1635.

Rajkowska G., Miguel-Hidalgo J.J., Makkos Z., Meltzer H., Overholser J., and Stockmeier C. 2002. Layer-specific reductions in GFAP-reactive astroglia in the dorsolateral prefrontal cortex in schizophrenia. *Schizophr. Res.* **57:** 127–138.

Reif A., Schmitt A., Fritzen S., Chourbaji S., Bartsch C., Urani A., Wycislo M., Mossner R., Sommer C., Gass P., and Lesch K.P. 2004. Differential effect of endothelial nitric oxide synthase (NOS-III) on the regulation of adult neurogenesis and behaviour. *Eur. J. Neurosci.* **20:** 885–895.

Sathasivam K., Hobbs C., Mangiarini L., Mahal A., Turmaine M., Doherty P., Davies S.W., and Bates G.P. 1999. Transgenic models of Huntington's disease. *Philos. Trans. R. Soc. Lond. B. Biol. Sci.* **354:** 963–969.

Saxena N.C. and Macdonald R.L. 1994. Assembly of GABA$_A$ receptor subunits: Role of the delta subunit. *J. Neurosci.* **14:** 7077–7086.

Senatorov V.V., Ren M., Kanai H., Wei H., and Chuang D.M. 2004. Short-term lithium treatment promotes neuronal survival and proliferation in rat striatum infused with quinolinic acid, an excitotoxic model of Huntington's disease. *Mol. Psychiatry* **9:** 371–385.

Slow E.J., Graham R.K., Osmand A.P., Devon R.S., Lu G., Deng Y., Pearson J., Vaid K., Bissada N., Wetzel R., Leavitt B.R., and Hayden M.R. 2005. Absence of behavioral abnormalities and neurodegeneration in vivo despite widespread neuronal huntingtin inclusions. *Proc. Natl. Acad. Sci.* **102:** 11402–11407.

Snell R.G., MacMillan J.C., Cheadle J.P., Fenton I., Lazarou L.P., Davies P., MacDonald M.E., Gusella J.F., Harper P.S., and Shaw D.J. 1993. Relationship between trinucleotide repeat expansion and phenotypic variation in Huntington's disease.[comment]. *Nat. Genet.* **4:** 393–397.

Spalding K.L., Bhardwaj R.D., Bucholz B.A., Druid H., and Frisen J. 2005. Retrospective birth dating of cells in humans. *Cell* **122:** 133–143.

Squire L.R. and Cave C.B. 1991. The hippocampus, memory, and space. *Hippocampus* **1:** 269–271.

Stefansson H., Sigurdsson E., Steinthorsdottir V., Bjornsdottir S., Sigmundsson T., Ghosh S., Brynjolfsson J., Gunnarsdottir S., Ivarsson O., Chou T.T., Hjaltason O., Birgisdottir B., Jonsson H., Gudnadottir V.G., Gudmundsdottir E., Bjornsson A., Ingvarsson B., Ingason A., Sigfusson S., Hardardottir H., Harvey R.P., Lai D., Zhou M., Brunner D., Mutel V., Gonzalo A., Lemke G., Sainz J., Johannesson G., Andresson T., Gudbjartsson D., Manolescu A., Frigge M.L., Gurney M.E., Kong A., Gulcher J.R., Petursson H., and

Stefansson K. 2002. Neuregulin 1 and susceptibility to schizophrenia. *Am. J. Hum. Genet.* **71:** 877–892.

Tattersfield A.S., Croon R.J., Liu Y.W., Kells A., Faull R.L.M., and Connor B. 2004. Neurogenesis in the striatum of the quinolinic acid lesion model of Huntington's disease. *Neuroscience* **127:** 319–332.

Tirassa P., Triaca V., Amendola T., Fiore M., and Aloe L. 2003. EGF and NGF injected into the brain of old mice enhance BDNF and ChAT in proliferating subventricular zone. *J. Neurosci. Res.* **72:** 557–564.

Tsien J.Z., Huerta P.T., and Tonegawa S. 1996. The essential role of hippocampal CA1 NMDA receptor-dependent synaptic plasticity in spatial memory. *Cell* **87:** 1327–1338.

van Dellen A., Blakemore C., Deacon R., York D., and Hannan A.J. 2000. Delaying the onset of Huntington's in mice. *Nature* **404:** 721–722.

Vonsattel J.P., Myers R.H., Stevens T.J., Ferrante R.J., Bird E.D., and Richardson E.P., Jr. 1985. Neuropathological classification of Huntington's disease. *J. Neuropathol. Exp. Neurol.* **44:** 559–577.

Wakade C.G., Mahadik S.P., Waller J.L., and Chiu F.C. 2002. Atypical neuroleptics stimulate neurogenesis in adult rat brain. *J. Neurosci. Res.* **69:** 72–79.

Wang R., Dineley K.T., Sweatt J.D., and Zheng H. 2004. Presenilin 1 familial Alzheimer's disease mutation leads to defective associative learning and impaired adult neurogenesis. *Neuroscience* **126:** 305–312.

Wen P.H., Hof P.R., Chen X., Gluck K., Austin G., Younkin S.G., Younkin L.H., DeGasperi R., Gama Sosa M.A., Robakis N.K., Haroutunian V., and Elder G.A. 2004. The presenilin-1 familial Alzheimer disease mutant P117L impairs neurogenesis in the hippocampus of adult mice. *Exp. Neurol.* **188:** 224–237.

Willing A.E., Zigova T., Milliken M., Poulos S., Saporta S., McGrogan M., Snable G., and Sanberg P.R. 2002. Lithium exposure enhances survival of NT2N cells (hNT neurons) in the hemiparkinsonian rat. *Eur. J. Neurosci.* **16:** 2271–2278.

Zigova T., Pencea V., Wiegand S.J., and Luskin M.B. 1998. Intraventricular administration of BDNF increases the number of newly generated neurons in the adult olfactory bulb. *Mol. Cell. Neurosci.* **11:** 234–245.

30

Adult Hippocampal Neurogenesis in Natural Populations of Mammals*

Irmgard Amrein and Hans-Peter Lipp
Institute of Anatomy, University of Zurich-Irchel
CH-8057 Zurich, Switzerland

Rudy Boonstra
Centre for the Neurobiology of Stress
University of Toronto at Scarborough
Scarborough, Ontario, M1C 1A4, Canada

J. Martin Wojtowicz
Department of Physiology, University of Toronto
Toronto, Ontario, M5S 1A8, Canada

THIS CHAPTER IS BASED ON THE PREMISE that if we are to acquire a deep understanding of adult neurogenesis—what it is selected for (i.e., its functional and adaptive significance), what causes it to go up or down (e.g., species constraints, reproductive hormones, seasonality, stress, and environmental conditions), and why it declines with age—the research must ultimately be grounded on an evolutionary and ecological foundation. The aphorism of Dobzansky (1973) is particularly apropos: "Nothing in biology makes sense, except in the light of evolution." Thus, simply focusing on humans and those laboratory species we select for will not be sufficient to crack this enigma. Such a deep understanding may also aid in ameliorating debilitating aspects of the human condition after injury or in disease. This chapter advocates for studies that deal with animals that live out their lives in the context of what they were actually selected to do. Given the paucity of studies from nature, it raises more questions than it answers. It focuses largely on mammals.

*The authors all contributed equally to the article.
Adult Neurogenesis ©2008 Cold Spring Harbor Laboratory Press 978-087969-784-6

The formation of new neurons in adult animals is a highly conserved trait in vertebrates, occurring in all groups, from fish to mammals in various brain regions. It is linked to a diversity of life history traits such as lifelong body growth in fishes and rats and seasonal variation in song control nuclei in birds (Lindsey and Tropepe 2006). In mammals, adult neurogenesis occurs physiologically in two germinal areas: the subventricular zone (SVZ), which lies adjacent to the lateral wall of the lateral ventricle and generates GABAergic olfactory bulb neurons, and the subgranular zone (SGV) of the dentate gyrus (DG) of the hippocampal formation. The critical question in understanding adult hippocampal neurogenesis is why the mature neural synaptic connections may not be sufficient to allow animals to cope with their environment. What memory or cognitive need(s) dictates the continual generation of new neurons in adults that simple rewiring or strengthening of the connections cannot?

The study of adult neurogenesis in natural populations has been essentially ignored. Birds (~9900 extant species) originated from reptiles and have been separated for approximately 300 million years from the line that eventually led to mammals (~5400 extant species). Thus far, our understanding of how neurogenesis functions in free-living animals in the natural world has been restricted to only one bird species (black-capped chickadees; Barnea and Nottebohm 1994) and a few mammal species as described in detail below. For the rest, all adult neurogenesis studies target either species bred and raised specifically for the laboratory (rodents) or in the laboratory (various primates) or domesticated animals. However, even these latter studies are not frequent.

Mice and rats have been enormously useful models to understand the molecular and biochemical basis of the regulation of adult neurogenesis and their implications for behavior (Lledo et al. 2006). They offer the degree of environmental control and genetic homogeneity that may be needed to limit experimental variation. However, findings from the laboratory may be of limited applicability as to why adult neurogenesis occurs in the first place, and thus what it is selected for, both in wild species functioning in their natural environment and perhaps in humans.

First, laboratory rodents may be suboptimal models to understand what happens in the natural world, as the selective regimes laboratory animals experience are significantly different from those experienced by animals in nature (Künzl et al. 2001; Wolff 2003). Laboratory rodents are often less aggressive, less aware of their environment, explore less, are more social, and respond more to stressors than their natural counterparts. Indeed, based on an analysis of aging in laboratory and natural populations of the house mouse, it has been concluded that laboratory-adapted stocks of rodents may be particularly inappropriate for the analysis of the genetic and physiological factors that regulate aging in mammals. Findings derived from them may be unable to provide much information about the mechanisms of aging in mammals, except under the highly artificial conditions of the laboratory (Miller et al. 2002). The same may apply to the study of adult neurogenesis.

Second, at the species level, adult neurogenesis may have a central role in evolutionary adaptations to dealing with particular ecological pressures, and differences among species may be essential to life history adaptations (Wingfield and Sapolsky 2003; Boonstra 2005; Smulders 2006). Ecological and evolutionary complexity in the natural world may result in a diverse array of physiological solutions to ecological problems. Thus, one set of guiding generalities to explain the role of neurogenesis and how it functions may not fit all avian and mammalian life histories. In particular, mammals span the gamut of life history variation, with species at one end of the spectrum having high reproductive rates, rapid development, and short life spans (e.g., rats and mice) and those at the other end having the opposite traits (e.g., bats, marmots, elephants, deer, some carnivores, and long-lived primates such as humans). Species with the former traits have been studied heavily to gain insights into the functioning of neurogenesis. In contrast, species with the opposite suite of life history traits have not been well-studied, and thus neurogenesis may operate in a different mode. Therefore, it is critical to also study species that span both the phylogenetic spectrum and the gamut of longevity from short-lived to long-lived species.

Third, at the individual level, adult neurogenesis may be the template that allows animals to cope with variation in their environment—the environmental certainty and uncertainty that are the daily fare of existence of all organisms (e.g., winter vs. summer, times of low vs. high social pressure, times of low and high predation pressure, and times of severe vs. benign environmental conditions).

The key to understanding the role that adult hippocampal neurogenesis may have in memory, and thus in predicting the future, is the inherent time lag between the production of an incipient neuron and its integration as a fully functional neuron with dendrites and axons. Such a lag typically takes 2–4 weeks in rodents (Piatti et al. 2006) and approximately 5 weeks in monkeys (Ngwenya et al. 2006). Thus, if adult neurogenesis has fitness consequences, birth of new neurons must be anticipatory, i.e., related to future, not present, needs. These needs may give insight into the adaptive purpose of neurogenesis. Thus, do rates of neurogenesis vary over the biological year (i.e., most animals have distinct breeding and nonbreeding seasons)? Is it related to particular behavioral and cognitive demands?

IS ADULT HIPPOCAMPAL NEUROGENESIS A FEATURE COMMON TO MAMMALS?

Presence in the Order Rodentia

Reports on adult neurogenesis in wild-living mammals are scarce, and the methods are not always comparable. In many studies, performed prior to widespread application of the bromodeoxyuridine (BrdU) method in the early 1990s, the evidence is only suggestive and would not stand the scrutiny of today's requirements. More recent reports are generally more reliable, but they need to be interpreted with caution, because the range of phenotypic markers is limited. Neurogenesis in the hippocampus of wild yellow-necked wood mice (*Apodemus flavicollis*), wood mice (*Apodemus sylvaticus*), bank voles (*Clethrionomys glareolus*), and European pine voles (*Microtus subterraneus*) has been visualized immunohistochemically using markers against Ki-67 (a protein active during mitosis), as well as doublecortin and NeuroD (both found in neuronal lineage precursor cells and/or young neurons) (Amrein et al. 2004b). In wild American meadow voles (*Microtus pennsylvanicus*), proliferating cells were labeled by incorporation of injected [^{3}H]thymidine (Galea and McEwen 1999). In Eastern gray squirrels (*Sciurus carolinensis*) from the United States, proliferating cells were found after injection of BrdU, subsequently incorporated in the DNA of dividing cells (Lavenex et al. 2000). In a study comparing wild yellow-pine chipmunks (*Tamias amoenus*) and Eastern grey squirrels from Canada, cell proliferation and neuronal fates were visualized immunohistochemically using Ki-67 and doublecortin (Barker et al. 2005). Preliminary data in tropical bats (Chiroptera) from South America and Africa obtained by immunohistochemical analysis with markers against Ki-67, MCM2 (a marker for nonactive precursor cells), doublecortin, and NeuroD showed sparse or no adult hippocampal neurogenesis (Amrein 2005; Amrein et al. 2006).

Investigations in domesticated and laboratory-bred mammals other than mice and rats reported adult neurogenesis in the hippocampus of several rodent species. In guinea pigs (*Cavia cavia*), Altman and Das (1967) were the first to show postnatal cell proliferation using [^{3}H]thymidine for visualization, whereas later studies also included adults using BrdU in combination with a double-labeling technique discriminating neurons from glia by means of NeuN (a marker for neurons) and GFAP (glial fibrillary acid protein) (Guidi et al. 2005). In gerbils (*Meriones unguiculatus*), adult neurogenesis was documented with BrdU (Dawirs et al. 2000), and in prairie voles (*Microtus ochrogaster*) by using BrdU-combined double and triple labeling with GFAP, Map-2, NeuN, and TuJ1 (Fowler et al. 2002).

Photoperiod-dependent adult neurogenesis in various brain regions of golden hamsters (*Mesocricetus auratus*) was described by Huang et al. (1998) using markers for BrdU in combination with NeuN and GFAP.

Presence in the Other Mammalian Orders

The presence of adult hippocampal proliferation and neurogenesis has been indicated in rabbits (*Oryctolagus cuniculus*, order Lagomorpha), using [³H]thymidine (Gueneau et al. 1982) or immunohistochemical markers against the M1 subunit ribonucleotide reductase (RNR, a rarely used proliferation marker), GFAP, calbindin, neurofilament, and Nestin (Zhu et al. 2003). With the same protocol, Zhu et al. (2003) found adult neurogenesis in pigs (*Sus domesticus*, order Artiodactyla) and sheep (*Ovis aries*, order Artiodactyla). Tree shrews (*Tupaia glis*, order Scandentia) were investigated using BrdU and neuron-specific enolase (NSE) (Gould et al. 1997). In marsupials, there is a report for one species, the small, mouse-like fat-tailed dunnart (*Sminthopsis crassicaudata*, order Dasyuromorphia), based on [³H]thymidine incorporation and immunohistochemistry for GFAP, PSA-NCAM (polysialic-acid-neural cell adhesion molecule), and calbindin (Harman et al. 2003).

In nonhuman primates, adult hippocampal neurogenesis was observed in rhesus and cynomolgus monkeys (*Macaca mulatta* and *Macaca fascicularis*, order Primates) by administration of BrdU in combination with extensive immunohistochemistry (TOAD-64, calbindin, NSE, GFAP, PCNA, O4, CNP, NeuN, and TuJ) (Gould et al. 1999; Kornack and Rakic 1999). In Japanese macaques (*Macaca fuscata*), BrdU was used in combination with markers for Musashi1, Nestin, NeuN, β-tubulin class III, GFAP, S100β, CNP, and GAD (Tonchev et al. 2003). Adult neurogenesis has also been reported for the New World marmoset monkey (*Callithrix jacchus*), using BrdU combined with immunohistochemistry against the neuron-specific enolase NSE (Gould et al. 1998). In comparison to rodents, monkeys show markedly lower basal proliferation rates and a lower number of cells taking neuronal fate (Tonchev and Yamashima 2006), as well as much longer maturation times of the newly born cells (Ngwenya et al. 2006).

To summarize, of 25 mammalian orders, adult hippocampal neurogenesis of variable degrees has been investigated and reported in only six orders (except for rodents, all orders with few or only one species), whereas one order (order Chiroptera) revealed thus far no adult neurogenesis for the majority of species investigated (only 3 species of 16 showing sparse neurogenesis and the others none) (Amrein 2005; Amrein et al. 2006 and unpubl.). The remaining 18 mammalian orders have not been investigated. Moreover, in many studies, quantification is missing, incomplete, or anecdotal. Obviously, one may assume, but cannot ascertain, that adult hippocampal neurogenesis is a widespread mammalian trait. In addition, the diversity of techniques has thus far hampered a systematic comparison of neurogenesis rates in the species investigated. Nevertheless, some important findings have emerged.

AGE DEPENDENCE OF NEUROGENESIS

The most common finding thus far, in all species investigated for this trait is an age-dependent decline in adult hippocampal neurogenesis. Protracted neurogenesis of granule cells peaks at puberty and declines steadily thereafter, as documented in mice, rats, and monkeys (Gould et al. 1999; Kempermann et al. 1998; Kuhn et al. 1996; Seki and Arai 1995), albeit with considerable species differences. In wild species, age determination is never as precise as in laboratory species and is generally restricted to age classes, such as juvenile, adult, or old. Nonetheless, it has been clearly shown that older wood mice, voles, chipmunks, and squirrels show a decline in ongoing proliferation compared to that of young and adults (Amrein et al. 2004b; Barker et al. 2005). Thus, the decline in ongoing proliferation activity in elderly laboratory-bred animals is not a domestication effect but appears to be a truly general phenomenon, probably occurring in humans as well (Fahrner et al. 2007).

INDIVIDUAL AND SEX DIFFERENCES OF ADULT NEUROGENESIS IN WILD POPULATIONS

Seasonal reproduction is the norm for most species in the wild (Bronson and Heideman 1994). Wild rodents show a recognizable effect of hormonal fluctuations on adult neurogenesis. Seasonal fluctuation in proliferation activity due to reproductive state has been reported in wild-trapped female meadow voles, whereas nonproductive females with low levels of estradiol showed a higher number of proliferating cells (Galea and McEwen 1999; Ormerod and Galea 2003). These findings were replicated in the laboratory, where manipulations of hormone levels in male and female voles revealed different effects on proliferation and survival of newborn cells (Ormerod and Galea 2001; Ormerod et al. 2004), an effect that appeared also to be moderately correlated with spatial memory in water-maze tasks (Ormerod et al. 2004). Furthermore, reduction of daylight length doubled the rate of adult neurogenesis in hamsters (Huang et al. 1998). Whereas wild house mice breed seasonally (Berry 1981), laboratory mice and rats reproduce continually due to defective melatonin processing (Kennaway et al. 2002); thus, evidence for a seasonally cycling hormone level and corresponding proliferation regulation is unavailable. However, the effect of manipulated hormone levels on adult neurogenesis of laboratory rodents is well-documented (for a review, see Galea et al. 2006).

Whether cell proliferation, survival, and fate of the new cells in wild rodents can be altered with experimental factors other than hormones (i.e., activity, diet, and learning tasks) is still unknown. Lavenex et al. (2000) investigated whether the basal proliferation activity in wild squirrels shows seasonal variations correlated with their caching activities, but they failed to find such fluctuations.

ADAPTIVE SIGNIFICANCE AND FUNCTION OF ADULT NEUROGENESIS ACROSS WILD LIVING SPECIES

The central motivating factor for studying adult hippocampal neurogenesis in natural populations has been the putative (and still debated) beneficial role of newly born granule cells on spatial learning and memory performance of rodents (Abrous and Wojtowicz, Chapter 21 in this volume). If the observation that adult rats generate 10,000 new neurons daily (Cameron and McKay 2001; McDonald and Wojtowicz 2005) can be extended to most other mammal species, it is highly unlikely that this trait is some vestigial holdover from the past, as all animals are living in an energy-limited world. Thus, one would expect to find regulated levels of neurogenesis in species with differential demands for memory abilities. To some extent, these expectations have been met.

Barker et al. (2005) studied chipmunks and squirrels with differential food-caching strategies that prepare them for winter. Chipmunks have small territories with a single food cache, whereas squirrels use multiple storage places located in larger territories. Indeed, chipmunks showed lower basal proliferation rate than did squirrels but, interestingly, not a lower number of immature neurons. A possible explanation of these findings is the enhanced turnover rate of the newly generated cells in the squirrels in comparison to chipmunks.

Likewise, comparison of Russian rodents having different home ranges indicated higher proliferation activity in wood mice, known to patrol large territories, in comparison to bank voles, which patrol smaller ones (Kikkawa 1964; Niethammer and Krapp 1978, 1982; Dell'Omo and Shore 1996; Amrein et al. 2004b). This difference was also reflected in automated home-cage learning tasks in which wood mice showed (moderately) better reversal of place-preference learning (Galsworthy et al. 2004). However,

Amrein et al. (2004a,b) also demonstrated a balance between proliferation activity and survival of newly generated cells. Yellow-necked wood mice showed an almost excessive daily proliferation rate of 1.1% of the resident granule cell population (Fig. 1a,b) that was correlated with an equally high proportion of apoptotic cells. In old wood mice, the reduced proliferation activity was compensated by reduced apoptosis, entailing an increased survival of the newly generated cells. In bank voles and European pine voles, the daily cell proliferation accounts for about 0.3% of the resident granule cell population, despite significant differences in total dividing cells, total granule cell numbers, and habitat type (Amrein et al. 2004a,b). Thus, the neurogenesis-dependent functionality of the hippocampus in relation to lifestyle and complexity of habitats in a sample of related species can only be inferred when taking into account the total number of granule cells in the DG and, when possible, seasonal variations.

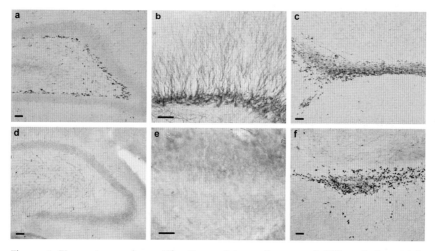

Figure 1. Two extremes in proliferation activity and neuronal differentiation. Adult yellow-necked wood mice (*Apodemus flavicollis; a,b,c*) show extreme proliferation activity in the dentate gyrus (DG) (*a*), where dividing cells are visualized with immunohistochemistry against Ki-67. Many of these cells in the subgranular layer (SGL) differentiate into neurons, as visualized with doublecortin immunohisto-chemistry (*b*). Proliferation and migration of cells can also be seen in the rostral migratory stream (RMS), where Ki-67-positive cells line up and travel to the olfactory bulb (OB) (*c*). In adult short-tailed bats (*Carollia perspicillata; d,e,f*), a few Ki-67-positive, proliferating cells can be seen in the hilus, but no dividing cells are found in the subgranular layer of the DG (*d*). Accordingly, no ongoing neuronal differentiation (doublecortin) can be found in the DG (*e*). However, proliferation activity in the RMS (*f*) is similar to that in yellow-necked wood mice. Bar, 50 μm.

Most bat species thus far do not fit the picture of adult hippocampal neurogenesis thought to be essential for spatial cognitive abilities. Most bats lack neurogenesis in the hippocampus (Fig. 1d,e), despite their obvious need for navigational ability in territories far greater than the size of those found in similar-sized rodents. Whether this reflects another functionality of the bat hippocampus or less-investigated general functions of the hippocampus such as cognitive flexibility (not prominent in bats) remains an open question. In any case, the sole emphasis on spatial navigation as neurogenesis-dependent behavior may not be warranted. In fact, studies in laboratory rats that used irradiation to ablate new neurons in adults show no deficit in spatial navigation, even though long-term spatial memory was impaired. Spatial, episodic, and contextual memory, rather than spatial navigation, may be more dependent on neurogenesis (see Chapter 21 by Abrous and Wojtowicz in this volume). This may be particularly important for predated rodents that must constantly relate danger to changing locations and stimuli, thus creating a particularly high demand for neurogenesis. The scarce adult neurogenesis of bats might also be linked to their astonishing longevity (Wilkinson and South 2002), as age is the most common down-regulator of mammalian adult neurogenesis.

WHAT IS REQUIRED TO STUDY ADULT NEUROGENESIS IN WILD-LIVING POPULATIONS?

Comparisons of neurogenesis across species should take age of the animals into account. Age may dictate not only rates of cell proliferation, which are clearly age-dependent, but also rates of differentiation and maturation.

Methods that minimize confounds due to handling stress and permit standardization across species should be used. The use of BrdU as a proliferation marker is impractical for the study of wild-living populations. Its popularity in rodent studies is due to the fact that its incorporation into the DNA permits one to follow the fate of the labeled cells. However, reliable labeling, quantification, and follow-up of a representative pool of dividing cells can only be achieved if the animals can be kept in captivity. Allowing for an approximate cell cycle time of 24 hours in rodents and perhaps longer in other species would require holding captured animals for days. This creates stress at the time of incorporation of BrdU that may influence the rate of neurogenesis. For example, stressed marmoset monkeys show a distinctly lower number of BrdU-labeled cells in the DG, even with a postinjection survival time of only 2 hours (Gould et al. 1998). Furthermore, dosage, uptake, toxicity, and penetration of BrdU or other thymidine analogs through the blood-brain barrier would have to be tested in each species in advance, substantially increasing the number of animals to be trapped and sacrificed and excessively inflating the costs in terms of manpower and infrastructure.

Recent advances in immunohistochemical labeling clearly show that the use of BrdU is no longer a methodological necessity. It has been shown that the number of cells labeled by Ki-67 in the rat DG correlates closely with the number of cells revealed by BrdU injections (Kee et al. 2002; Wojtowicz and Kee 2006). Other studies have proven the reliability of endogenous markers versus thymidine analogs for identifying proliferating cells and young migrating or differentiating neurons (Tye 1999; Rao and Shetty 2004). In particular, Ki-67 and MCMs (replication-initiating factors) are evolutionarily highly conserved proteins that can be visualized across a wide range of species from different phylogenetic families. Thus, a simple combination of available markers for different stages of replication, glial/neuronal differentiation, and developmental stages of neurons, including apoptosis, would be sufficient for a detailed comparative analysis without having to introduce unnecessary confounds of the BrdU technique. Finally, comparative data contributing to any functional conclusion or theory about the role of newly formed granule cells in the mammalian hippocampus need some form of standardization. Finding 1000 proliferating cells in a DG comprising 500,000 granule cells implies an other functionality than finding the same number in a larger-sized hippocampus containing 10 million granule cells. Up to now, there have been large methodological differences in referencing proliferating cells to the remaining DG—if this is done at all. Data have been presented as density measurements relative to area or volume, as ratio to the total number of resident granule cells, or as a fraction of mature granule cells. We suggest that the preferred requirement for species comparison of proliferation activity should be to the area of the subgranular zone, whereas neurogenesis and cell turnover should ideally be standardized to an estimate of the total granule cell number.

The value of these methods notwithstanding, an ideal study of adult neurogenesis in a natural population would include BrdU. Many of the species that could be studied live in restricted areas that can be surveyed by dedicated neuroscientists. Animals can be trapped, injected with BrdU, tagged, fitted with telemetry transmitters, and retrapped again. Seasonal variation can be taken into account and appropriate numbers of animals can be killed and processed for BrdU immunohistochemistry in combination with phenotypic markers. Even more challenging, but still feasible, are studies that would involve reduction of neurogenesis by irradiation, for example, and rereleasing the animals into their natural environment. This experiment would provide an ultimate test of the adaptive significance of neurogenesis.

CONCLUSIONS AND FUTURE DIRECTIONS

Despite the paucity of studies on adult hippocampal neurogenesis in wild-living populations, some conclusions can already be drawn:

1. Across mammalian orders, such as primates, rodents, and bats, there are enormous species differences in terms of basal proliferation and survival of newly generated granule cells that cannot yet be explained by cognitive ability and behavioral specializations.
2. Neurogenesis is age-dependent in all species examined thus far.
3. The species-specific levels of adult hippocampal neurogenesis may also depend on other species-specific cytoarchitectonic characteristics of the hippocampus. For example, a relative size of the DG to the rest of the hippocampal formation may indicate how important the classical trisynaptic pathway is in comparison to alternative afferent inputs into the hippocampus (Abrous and Wojtowicz, Chapter 21 in this volume). Size variations of the infrapyramidal mossy fiber (IIP-MF) projections in the hippocampus of rodents, i.e., of the axons of the dentate granule cells, should also be considered (for review, see Lipp 2007).

SUMMARY

There is an increasing awareness shown in the literature that if we are to make headway in our understanding of adult neurogenesis and of its significance, a broad range of species should be examined (see, e.g., Boonstra et al. 2001; Nottebohm 2002; Lindsey and Tropepe 2006). We suggest that at least three approaches would be profitable.

First, undertake detailed observational studies on each of a broad range of species from the natural world, particularly rodents, as this is the group that has been the major focus in laboratory studies. Such studies integrating detailed knowledge of animal's ecology, behavior, physiology, and evolutionary relatedness will help us to understand how season, sex, and environmental variables change rates of production and survival of new neurons.

Second, carry out experimental manipulations on these species to discriminate cause from effect in real-world scenarios (e.g., manipulating stressors such as conspecifics or predators) or semireal-world settings. Such studies should be intraspecific, comparing the same species but from areas where the environmental needs may select for different rates of production and survival.

Third, carry out interspecies studies, both of closely related species with markedly different life histories and of distantly related species. Research on laboratory animals provides useful conceptual and methodological guidelines that can now be employed in studies of natural populations.

ACKNOWLEDGMENTS

I.A. and H.-P.L. were supported by the Swiss National Science Foundation and the NCCR "Neural Plasticity and Repair." We appreciate the help of Dina Dechmann in trapping bats and the help of Inger Drescher-Lindh and Natascha Bologova in histological processing. R.B. and J.M.W. were supported by NSERC, Canada.

REFERENCES

Altman J. and Das G.D. 1967. Postnatal neurogenesis in the guinea-pig. *Nature* **214:** 1098–1101.

Amrein I. 2005. "Functional and neuroanatomical correlates of adult neurogenesis in the dentate gyrus of domesticated and wild rodents." Ph.D. thesis, University of Zurich, Switzerland.

Amrein I., Slomianka L., and Lipp H.P. 2004a. Granule cell number, cell death, and cell proliferation in the dentate gyrus of wild-living rodents. *Eur. J. Neurosci.* **20:** 3342–3350.

Amrein I., Dechmann D.K., Winter Y., and Lipp H.P. 2006. Absent or low rate of adult neurogenesis in the hippocampus of bats (*Microchiroptera*). In *Abstracts from the 36th Annual Meeting of the Society for Neurocience,* Atlanta, Georgia. (Abstr. 713.10).

Amrein I., Slomianka L., Poletaeva, II, Bologova N.V., and Lipp H.P. 2004b. Marked species and age-dependent differences in cell proliferation and neurogenesis in the hippocampus of wild-living rodents. *Hippocampus* **14:** 1000–1010.

Barker J.M., Wojtowicz J.M., and Boonstra R. 2005. Where's my dinner? Adult neurogenesis in free-living food-storing rodents. *Genes Brain Behav.* **4:** 89–98.

Barnea A. and Nottebohm F. 1994. Seasonal recruitment of hippocampal neurons in adult free-ranging black-capped chickadees. *Proc. Natl. Acad. Sci.* **91:** 11217–11221.

Berry R.J. 1981. Population dynamics of the house mouse. *Symp. Zool. Soc. Lond.* **47:** 395–425.

Boonstra R. 2005. Equipped for life: The adaptive role of the stress axis in male mammals. *J. Mammal.* **86:** 236–247.

Boonstra R., Galea L., Matthews S.G., and Wojtowicz J.M. 2001. Adult neurogenesis in natural populations. *Can. J. Physiol. Pharmacol.* **79:** 297–302.

Bronson F.H. and Heideman P.D. 1994 Seasonal regulation of reproduction in mammals. In *The physiology of reproduction* (ed. E. Knobil), pp. 541–583. Raven Press, New York.

Cameron H. A. and McKay R.D.G. 2001. Adult neurogenesis produces a large pool of new granule cells in the dentate gyrus. *J. Comp. Neurol.* **435:** 406–417.

Dawirs R.R., Teuchert-Noodt G., Hildebrandt K., and Fei F. 2000. Granule cell proliferation and axon terminal degradation in the dentate gyrus of gerbils (*Meriones unguiculatus*) during maturation, adulthood and aging. *J. Neural Transm.* **107:** 639–647.

Dell'Omo G. and Shore R.F. 1996. Behavioral effects of acute sublethal exposure to dimethoate on wood mice, *Apodemus sylvaticus*. II. Field studies on radio-tagged mice in a cereal ecosystem. *Arch. Environ. Contam. Toxicol.* **31:** 538–542.

Dobzhansky T. 1973. Nothing in biology makes sense except in the light of evolution. *Am. Biol. Teach.* **35:** 125–129.

Fahrner A., Kann G., Flubacher A, Heinrich C., Freiman T.M., Zentner J., Frotscher M., and Haas C.A. 2007. Granule cell dispersion is not accompanied by enhanced neurogenesis in temporal lobe epilepsy patients. *Exp. Neurol.* **203:** 320–332.

Fowler C.D., Liu Y., Ouimet C., and Wang Z. 2002. The effects of social environment on adult neurogenesis in the female prairie vole. *J. Neurobiol.* **51:** 115–128.

Galea L.A. and McEwen B.S. 1999. Sex and seasonal differences in the rate of cell proliferation in the dentate gyrus of adult wild meadow voles. *Neuroscience* **89:** 955–964.

Galea L.A., Spritzer M.D., Barker J.M., and Pawluski J.L. 2006. Gonadal hormone modulation of hippocampal neurogenesis in the adult. *Hippocampus* **16:** 225–232.

Galsworthy M.J., Amrein I., Kuptsov P.A., Poletaeva I.I., Zinn P., Rau A., Vyssotski A.L., and Lipp H.P. 2004. A comparison of wild-caught wood mice and bank voles in the Intellicage: Assessing exploration, daily activity patterns and place learning paradigms. *Behav. Brain Res.* **157:** 211–217.

Gould E., McEwen B.S., Tanapat P., Galea L.A., and Fuchs E. 1997. Neurogenesis in the dentate gyrus of the adult tree shrew is regulated by psychosocial stress and NMDA receptor activation. *J. Neurosci.* **17:** 2492–2498.

Gould E., Tanapat P., McEwen B., Flügge G., and Fuchs E. 1998. Proliferation of granule cell precursors in the dentate gyrus of adult monkeys is diminished by stress. *Proc. Natl. Acad. Sci.* **95:** 3168–3171.

Gould E., Reeves A.J., Fallah M., Tanapat P., Gross C.G., and Fuchs E. 1999. Hippocampal neurogenesis in adult Old World primates. *Proc. Natl. Acad. Sci.* **96:** 5263–5267.

Gueneau G., Privat A., Drouet J., and Court L. 1982. Subgranular zone of the dentate gyrus of young rabbits as a secondary matrix. A high-resolution autoradiographic study. *Dev. Neurosci.* **5:** 345–358.

Guidi S., Ciani E., Severi S., Contestabile A., and Bartesaghi R. 2005. Postnatal neurogenesis in the dentate gyrus of the guinea pig. *Hippocampus* **15:** 285–301.

Harman A., Meyer P., and Ahmat A. 2003. Neurogenesis in the hippocampus of an adult marsupial. *Brain Behav. Evol.* **62:** 1–12.

Huang L., DeVries G.J., and Bittman E.L. 1998. Photoperiod regulates neuronal bromodeoxyuridine labeling in the brain of seasonally breeding mammal. *J. Neurobiol.* **36:** 410–420.

Kennaway D.J., Voultsios A., Varcoe T.J., and Moyer R.W. 2002. Melatonin in mice: Rhythms, response to light, adrenergic stimuation, and metabolism. *Am. J. Physiol. Regul. Integr. Comp. Physiol.* **282:** R358–R365.

Kee N., Sivalingam S., Boonstra R., and Wojtowicz J.M. 2002. The utility of Ki-67 and BrdU as proliferative markers of adult neurogenesis. *J. Neurosci. Methods* **115:** 97–105.

Kempermann G., Kuhn H.G., and Gage F.H. 1998. Experience-induced neurogenesis in the senescent dentate gyrus. *J. Neurosci.* **18:** 3206–3212.

Kikkawa J. 1964. Movement, activity and distribution of the small rodents *Clethrionomys glareolus* and *Apodemus sylvaticus* in woodland. *J. Anim. Ecol.* **33:** 259–299.

Kornack D.R. and Rakic P. 1999. Continuation of neurogenesis in the hippocampus of the adult macaque monkey. *Proc. Natl. Acad. Sci.* **96:** 5768–5773.

Kuhn H.G., Dickinson-Anson H., and Gage F.H. 1996. Neurogenesis in the dentate gyrus of the adult rat: Age-related decrease of neuronal progenitor proliferation. *J. Neurosci.* **16:** 2027–2033.

Künzl C., Kaiser W., Meier E., and Sachser N. 2001. Is a wild mammal kept and reared in captivity still a wild animal? *Horm. Behav.* **43:** 187–196.

Lavenex P., Steele M.A., and Jacobs L.F. 2000. The seasonal pattern of cell proliferation and neuron number in the dentate gyrus of wild adult eastern grey squirrels. *Eur. J. Neurosci.* **12:** 643–648.

Lindsey B.W. and Tropepe V. 2006. A comparative framework for understanding the biological principles of adult neurogenesis. *Prog. Neurobiol.* **80:** 281–307.

Lipp H.-P., Amrein I., and Wolfer D.P. 2007. Natural genetic variation of hippocampal structures and behavior—An update. In *Neurobehavioral genetics: Methods and applications*, 2nd edition (ed. B.C. Jones and P. Mormède), pp. 389–410. CRC Press, Boca Raton, Florida.

Lledo P.-M., Alonso M., and Grubb M.S. 2006. Adult neurogenesis and functional plasticity in neuronal circuits. *Nat. Neurosci.* **7:** 179–193.

Miller R.A., Harper J.M., Galecki A., and Burke D.T. 2002. Big mice die young: Early life body weight predicts longevity in genetically heterogenous mice. *Aging Cell* **1:** 22–29.

McDonald H.Y. and Wojtowicz J.M. 2005. Dynamics of neurogenesis in the dentate gyrus of adult rats. *Neurosci. Lett.* **385:** 70–75.

Ngwenya L.B., Peters A., and Rosene D.L. 2006. Maturational sequence of newly generated neurons in the dentate gyrus of the young adult rhesus monkey. *J. Comp. Neurol.* **498:** 204–216.

Niethammer J. and Krapp F. 1978. *Handbuch der Säugetiere Europas,* vol. 1: Rodentia I, pp. 325–358. Akademische Verlagsgesellschaft, Wiesbaden, Germany.

———. 1982. *Handbuch der Säugetiere Europas,* vol. 2: Rodentia II, pp. 109–146. Akademische Verlagsgesellschaft, Wiesbaden, Germany.

Nottebohm F. 2002. Why are some neurons replaced in adult brain? *J. Neurosci.* **22:** 624–628.

Ormerod B.K. and Galea L.A. 2001. Reproductive status influences cell proliferation and cell survival in the dentate gyrus of adult female meadow voles: A possible regulatory role for estradiol. *Neuroscience* **102:** 369–379.

———. 2003. Reproductive status influences the survival of new cells in the dentate gyrus of adult male meadow voles. *Neurosci. Lett.* **346:** 25–28.

Ormerod B.K., Lee T.T., and Galea L.A. 2004. Estradiol enhances neurogenesis in the dentate gyri of adult male meadow voles by increasing the survival of young granule neurons. *Neuroscience* **128:** 645–654.

Piatti V.C., Esposito M.S., and Schinder A.F. 2006. The timing of neuronal development in adult hippocampal neurogenesis. *Neuroscientist* **12:** 463–468.

Rao M.S. and Shetty A.K. 2004. Efficacy of doublecortin as a marker to analyse the absolute number and dendritic growth of newly generated neurons in the adult dentate gyrus. *Eur. J. Neurosci.* **19:** 234–246.

Seki T. and Arai Y. 1995. Age-related production of new granule cells in the adult dentate gyrus. *NeuroReport* **6:** 2479–2482.

Smulders T.V. 2006. A multi-disciplinary approach to understanding hippocampal function in food-hoarding birds. *Rev. Neurosci.* **17:** 53–69.

Tonchev A.B. and Yamashima T. 2006. Differential neurogenic potential of progenitor cells in dentate gyrus and CA1 sector of the postischemic adult monkey hippocampus. *Exp. Neurol.* **198:** 101–113.

Tonchev A.B., Yamashima T., Zhao L., Okano H.J., and Okano H. 2003. Proliferation of neural and neuronal progenitors after global brain ischemia in young adult macaque monkeys. *Mol. Cell. Neurosci.* **23:** 292–301.

Tye B.K. 1999. MCM proteins in DNA replication. *Annu. Rev. Biochem.* **68:** 649–686.

Wilkinson G.S. and South J.M. 2002. Life history, ecology and longevity in bats. *Aging Cell* **1:** 124–131.

Wingfield J.C. and Sapolsky R.M. 2003. Reproduction and resistance to stress: When and how. *J. Neuroendocrinol.* **15:** 711–724.

Wojtowicz J.M. and Kee N. 2006. BrdU assay for neurogenesis in rodents. *Nat. Protoc.* **1:** 1399–1405.

Wolff J. 2003. Laboratory studies with rodents: Facts or artifacts? *Bioscience* **53:** 421–427.

Zhu H., Wang Z.Y., and Hansson H.A. 2003. Visualization of proliferating cells in the adult mammalian brain with the aid of ribonucleotide reductase. *Brain Res.* **977:** 180–189.

Index